Methods in Molecular Biology

Series Editor
John M. Walker
School of Life Sciences
University of Hertfordshire
Hatfield, Hertfordshire, AL10 9AB, UK

For further volumes:
http://www.springer.com/series/7651

Immunoinformatics

Second Edition

Edited by

Rajat K. De

Indian Statistical Institute, Kolkata, West Bengal, India

Namrata Tomar

Indian Statistical Institute, Kolkata, West Bengal, India

Humana Press

Editors
Rajat K. De
Indian Statistical Institute
Kolkata, West Bengal, India

Namrata Tomar
Indian Statistical Institute
Kolkata, West Bengal, India

ISSN 1064-3745 ISSN 1940-6029 (electronic)
ISBN 978-1-4939-1114-1 ISBN 978-1-4939-1115-8 (eBook)
DOI 10.1007/978-1-4939-1115-8
Springer New York Heidelberg Dordrecht London

Library of Congress Control Number: 2014942969

Printed on acid-free paper

Humana Press is a brand of Springer
Springer is part of Springer Science+Business Media (www.springer.com)

Dedication

Dedicated to my Mother
Rajat K. De

Dedicated to my Parents
Namrata Tomar

Preface

The immune system evolves as a defense mechanism against foreign particles and cancer cells in an organism. It interacts with self and foreign components and mounts adequate responses against pathogenic foreign and mutated self. At the same time, it should tolerate self and most of the other environmental particles so that an organism can maintain a healthy state. Immunology, a branch of biomedical science, deals with the structural and functional studies of all aspects of immune systems and their components. It includes physiological studies of the immune system in both healthy and diseased states as well as in immunological disorders. Immunology is a combinatorial science due to the diverse range of interactions involving immune system components and their targets. The combinatoriality also lies in the arrangements of immunoglobulins (Ig) in an individual, where the number of such arrangements is more than 10^9. The dynamic behavior of these interactions makes the systems even more complex.

In recent years, traditional approaches in science, due to the advent of high-throughput technology, have been complemented by computer-aided research. *In silico* analyses of the biological problems aid experimental research to reduce time and cost. Increasing amount of genomic sequence and functional annotation data are fuelling immunological research. There is also an abundance of large-scale projects for investigating host-pathogen and host-antigen interactions. Immunology, as in the case of molecular biology, has now moved from being a traditional qualitative science to more quantitative one. The requirement of storing, managing, and analyzing continuously growing experimental, clinical, and epidemiologic data has led to form a new research discipline known as "immunoinformatics." Due to the combinatorial nature of immunological data, efficient immunoinformatic databases and tools are required. The discipline "immunoinformatics," like bioinformatics, lies at the intersection of experimental and computational sciences. *In silico* models are increasingly being used to simulate immune system behavior as well as for analysis of host and pathogen genomes and their interactions. Simulating immune systemic models has certain applications, e.g., finding the course of infection and optimization of clinical protocols. Immunoinformatics is at the heart of the research areas of immunogenomics, immunoproteomics, and computational vaccinology. The most important task of immunoinformatics is to analyze immunological data using computational tools to generate biologically significant and rational interpretations.

Immunomics, in which we combine traditional immunology with computer science, mathematics, statistics, chemistry, biochemistry, genomics, and proteomics, offers large-scale analysis of immune system for further translation of basic immunology research into clinical practices. Although immunoinformatics is still in an evolving stage, it clearly has the potential to accelerate immunology research. Computational models also help in selecting appropriate laboratory experiments and formulating novel and testable hypotheses that could not be achieved using traditional approaches alone earlier. High complexity of immunological processes may lead to imprecise biochemical measurements and the inherent scientific biases and misconceptions. Therefore, care should be taken while developing computational tools for investigating and modeling underlying immunological processes. Otherwise, biases and misconceptions encoded in computational tools might result in the wrong biological interpretations.

Content and General Outline of the Book

We have tried to make this edition of the book self-contained. In principle, it aims at students and researchers from diverse background and levels interested in working with immunological problems. It provides biological insights into a certain extent as well as a simpler way to implement approaches and algorithms in the immunoinformatics research domain. There are 30 chapters distributed in five sections that cover various aspects of basic immunology to immunoinformatics.

Part I is dedicated to describing the transition from basic and traditional immunology to immunoinformatics. It includes three chapters that introduce a basic immune system, its interaction with metabolic machinery, and informatics related to immune system. Part II contains comprehensive detail on most of the existing databases related to an immune system and its components. Similarly, most of the possible approaches/tools/algorithms for the prediction of T/B-cell epitopes, allergenic proteins, and virulence factors are described in Part III. In Part IV, systems biology approaches in the immunoinformatics domain have been explained, particularly for inflammation and personalized medicine. Part V deals with some applications of immunoinformatics research. In this section, we have provided applications of immunoinformatics in cancer diagnosis and therapy, HIV pathogenesis, and methods to investigate the mechanisms of host-pathogen interactions. Part V also includes the description of the role of structure-based clustering of MHC molecules as well as small RNA in vaccine designing.

Chapter 1 introduces the basic immune system to the readers. It describes two distinct yet interrelated branches of an immune system, which gets activated at the time of antigen attack upon host system.

Chapter 2 depicts various investigations related to the behaviors of lymphocytes and other leukocytes regulated by metabolic activities of cells at different levels. Investigations on the molecular aspects of immunological-metabolic cross talk have become an interesting research topic. The role of glucose in an immune system and metabolic dependency in lymphocyte activation is explained in this chapter, along with the description of the role of nutrient sensors, adipose tissue, and toll-like receptors in maintaining immune-metabolic interactions.

Chapter 3 shows the need to handle the large accumulation of high-throughput data that has given rise to the field known as immunoinformatics. Thus this chapter reviews classical immunology, different databases, and prediction tools. Further, it briefly describes applications of immunoinformatics in reverse vaccinology, immune system modeling, cancer diagnosis, and therapy.

Chapter 4 provides details on the IMGT® system that was first developed in 1989. Since its development, it has been considered an interface between immunogenetics and immunoinformatics. This chapter reviews IMGT® definitive system for V, C, and G domains based on the IMGT-ONTOLOGY concepts. The web resource of IMGT provides data for nucleotide and protein sequences, genetic polymorphisms, as well as tools for analyzing immunoglobulins, T-cell receptors (TCR), major histocompatibility complex (MHC), and related components of an immune system.

Chapter 5 explains that Immuno Polymorphism Database (IPD) is based on the IMGT® model, which includes databases related to the study of polymorphic genes in an immune system. IPD currently consists of four databases: IPD-KIR, IPD-MHC, IPD-HPA, and IPD-ESTDAB.

Chapter 6 overviews publicly available databases of T-cell epitopes, including general databases, pathogen- and tumor-specific databases, and 3D structure databases. These databases include sequences, alleles, source organisms, structures, and diseases. Thus they are important data sources helping in the analysis of immune system components, functionalities, and development of prediction methods.

In Chap. 7, an overview of important databases for B-cell epitopes is provided, which also demonstrates the way to compile datasets for development of B-cell epitope prediction tools. Identification and characterization of B-cell epitopes in antigens are important in epitope-driven vaccine design, immunodiagnostic tests, and antibody production.

Chapter 8 describes a database, called AgAbDb, which includes an account of antigen-antibody interactions, a type of protein-protein interaction. These interactions are characterized by high affinity and specificity of antibodies towards their antigens. The chapter identifies and lists residues of binding sites of antigens and antibodies. It also compiles, curates, and analyzes determinants of interactions between the respective antigen-antibody molecules.

Chapter 9 deals with some allergen databases that can be classified into two types: biological and molecular databases. In this chapter, five popular allergen databases have been described. Among them, one is a biological database and the remaining four are molecular databases.

Chapter 10 introduces an ensemble learning-based method using antigenic sequences, which can predict the conformational B-cell epitopes. It also describes the properties of some existing data resources and computational methods for the same.

Chapter 11 provides a comprehensive set of 13 recent approaches for predicting linear B-cell epitopes and 4 methods for predicting conformational B-cell epitopes from the antigen sequences. It also provides some practical insights towards the use of these B-cell epitope predictors.

Chapter 12 narrates some fundamental of B-cell epitopes and use of SVM techniques for their prediction. It provides an example of linear B-cell prediction system based on physicochemical features and amino acid combinations.

Chapter 13 introduces mimotopes, the peptides that mimic epitopes on the corresponding antigen and can be obtained via panning the phage-display peptide library against the corresponding monoclonal antibody. This chapter describes mimotope-based prediction of B-cell epitopes under three conditions. It also provides details on protocols for retrieving and decoding the data obtained using phage-display technology.

Chapter 14 emphasizes on key physicochemical and biological considerations for B-cell epitope prediction that are relevant from an application perspective. It helps researchers in implementing computational tools for more practical purposes.

Chapter 15 shows a way to build a hybrid classifier for improved prediction of linear B-cell epitopes. It is further mentioned in the same chapter that this method can easily be applied for predicting conformational epitopes.

Chapter 16 contains the information regarding the B-cell epitope mapping and its wide usage to determine antibody-binding sites, diagnostic peptide development, and vaccine design. Three methods are described in this chapter, which are characterized by the simultaneous analysis of multiple peptides.

Chapter 17 deals with highly polymorphic human leukocyte antigen (HLA) genes, with diverse peptide-binding HLA specificities. Identification of new antigenic peptides that can bind to HLA class I and II molecules is important in vaccine development. Different HLA molecules are classified into "HLA supertypes" in order to reduce complexity. This chapter focuses on classification of HLA supertypes and their application in development of peptide-based vaccines.

Chapter 18 describes that peptide binding to MHC molecules is the most important selective step in T-cell recognition. This chapter explains how to derive peptide-MHC-binding motif profiles in EPIMHC and to use them in predicting peptide-MHC binding and T-cell epitopes.

Chapter 19 contains information related to the challenges involved in the task of T-cell epitope prediction due to MHC polymorphism and disparity encountered in the generation and presentation of T-cell epitopes. This chapter explains principles of some of the methods/algorithms for T-cell epitope prediction as well as procedural and practical aspects of their usage.

Chapter 20 describes a protocol to perform the calculation of electrostatic energy, followed by an illustration on the outer surface protein A of *Borrelia burgdorferi*, a pathogenic organism causing lyme disease.

Chapter 21 emphasizes on the importance of allergen prediction tools as there is an increase in the usage of genetically modified (GM) food and biopharmaceuticals in the population. Thus the allergen prediction tools are being used to assess the safety of GM crops, therapeutics, and biopharmaceuticals. This chapter describes the way to use four popular allergenic prediction servers, viz. Structural Database of Allergenic Proteins (SDAP), Allermatch, Evaller 2, and AlgPred.

Chapter 22 includes information on adhesins, the virulence factors secreted from the pathogen, which are of immunological interest. This chapter describes the bioinformatics approaches for adhesin prediction, which include specific adhesin prediction algorithms.

Chapter 23 deals with an application area of immunoinformatics. It describes a *Candida albicans*–zebrafish interactive infectious network, as an example, to demonstrate how a systems biology approach can be used to study systemic inflammation.

Chapter 24 explains the sampling of the mucosal tissues and analyses of immune responses as an integral step towards vaccine development strategies against HIV. This chapter describes commonly used practices of immunizations and of obtaining important mucosal tissue samples in nonhuman primates.

Chapter 25 provides a scenario of the major knowledgebases, as one can find continuous creation, usage, and, later, discontinuation of biological tools and databases. Thus, there should be a clear picture of the major knowledgebases that provide information about the functional existence of these databases and tools for the researchers from diverse backgrounds. This chapter provides an overview of information sources that also include a description of InnateDB. It helps researchers in selecting databases and tools related to immunoinformatics and systems biology, which can be further used in personalized medicine.

Chapter 26 provides details on small RNA molecules that play a vital role in defense systems. The detailed study of RNA gene silencing mechanisms has revealed that the small RNAs are the chief executioners for antiviral immunity in an organism. This chapter reviews the possibility of engineering small RNAs to enhance the immunity against specific viral pathogens.

Chapter 27 describes the use of structure-based clustering techniques in identifying superfamilies of major histocompatibility complex (MHC) proteins with similar binding specificities, which later help in vaccine development. This chapter provides a summary for grouping MHC proteins according to their structural interactions.

Chapter 28 includes information on cytotoxic T-cell (CTL) epitopes that are found to be important in the form of an immunotherapeutic product as they might help in tumor cell destruction. This chapter focuses on several different sequence-, structure-, and molecular modeling-based prediction tools to extract a list of peptide epitopes from tumor-specific or tumor-associated antigens (TSA or TAA).

Chapter 29 describes a protocol that delineates a process of genome-scale metabolic modeling, using flux balance analysis, for the analysis of host-pathogen behavior and interactions. The methods for biological interpretations of computed cell phenotypes, in the context of individual host and pathogen models and their integrations, are also discussed.

Chapter 30 provides details on mathematical models for in vivo dynamics of HIV infection and some recent concepts of disease progression. Initially, it discusses a basic mathematical model for investigating HIV dynamics, along with estimation of key parameters that characterize the infection. It also includes a review on some recent concepts related to disease progression that involves multiple infection of cells and the direct cell-to-cell transmission of virus through the formation of virological synapses.

Kolkata, West Bengal, India

Rajat K. De
Namrata Tomar

Contents

Part V Applications of Immunoinformatics

Contributors

RAQUEL ALMANSA • *Unidad de Investigacion Biomedica, Hospital Clinico Universitario de Valladolid, Valladolid, Spain*
SCOTT M. ANTHONY • *Department of Immunology, The University of Texas MD Anderson Cancer Centre, Houston, TX, USA*
JESÚS F. BERMEJO-MARTIN • *Unidad de Investigacion Biomedica, Hospital Clinico Universitario de Valladolid, Valladolid, Spain*
SWATI GUPTA BHATTACHARYA • *Division of Plant Biology, Bose Institute, Kolkata, India*
SALVADOR EUGENIO C. CAOILI • *Department of Biochemistry and Molecular Biology, College of Medicine, University of the Philippines, Manila, Philippines*
AJEET CHAUDHARY • *Department of Genetics, University of Delhi, New Delhi, India*
RUPANJALI CHAUDHURI • *CSIR-Institute of Genomics and Integrative Biology, Delhi, India*
BOR-SEN CHEN • *Lab of Control and Systems Biology, Department of Electrical Engineering, National Tsing Hua University, HsinChu, Taiwan*
MOGENS H. CLAESSON • *Laboratory of Experimental Immunology, Faculty of Health Sciences, Panum Institute, University of Copenhagen, Copenhagen, Denmark*
RAJAT K. DE • *Machine Intelligence Unit, Indian Statistical Institute, Kolkata, India*
MEREL DERKSEN • *Department of Biomolecular Chemistry, Radboud Institute for Molecular Life Sciences, Institute for Molecules and Materials and Netherlands Proteomics Centre, Radboud University Nijmegen, Nijmegen, The Netherlands*
DATTATRAYA V. DESAI • *Bioinformatics Centre, University of Pune, Pune, India*
YASSER EL-MANZALAWY • *Department of Systems and Computer Engineering, Al-Azhar University, Cairo, Egypt*
SÉBASTIEN FIORUCCI • *Faculté des Sciences, UMR-CNRS 7272, Institut de Chimie de Nice, Université de Nice-Sophia Antipolis, Nice, France*
JIANZHAO GAO • *School of Mathematical Sciences, Nankai University, Tianjin, China*
JASON A. HALLIWELL • *Anthony Nolan Research Institute, Royal Free Hospital, London, UK*
PATRICK HANLEY • *Department of Veterinary Sciences, The University of Texas MD Anderson Cancer Centre, Bastrop, TX, USA*
HONG HE • *Stem Cell Transplantation Research, The University of Texas MD Anderson Cancer Centre, Houston, TX, USA*
BIFANG HE • *Center of Bioinformatics (COBI), Key Laboratory for NeuroInformation of Ministry of Education, University of Electronic Science and Technology of China, Chengdu, China*
SANNE M.M. HENSEN • *Department of Biomolecular Chemistry, Radboud Institute for Molecular Life Sciences, Institute for Molecules and Materials and Netherlands Proteomics Centre, Radboud University Nijmegen, Nijmegen, The Netherlands*
VASANT HONAVAR • *College of Information Sciences and Technology, Huck Institutes of the Life Sciences, Pennsylvania State University, University Park, PA, USA*
JIAN HUANG • *Center of Bioinformatics (COBI), Key Laboratory for NeuroInformation of Ministry of Education, University of Electronic Science and Technology of China, Chengdu, China*

MENG KE • *School of Computer, Wuhan University, Wuhan, China*
URMILA KULKARNI-KALE • *Bioinformatics Centre, University of Pune, Pune, India*
LUKASZ KURGAN • *Department of Electrical and Computer Engineering, University of Alberta, Edmonton, AB, Canada*
ESTHER M. LAFUENTE • *Facultad de Medicina, Universidad Complutense de Madrid, Madrid, Spain*
MARIE-PAULE LEFRANC • *IMGT®, The International ImMunoGeneTics information system®, Laboratoire d'ImmunoGénétique Moléculaire, LIGM, Institut de Génétique Humaine IGH, Université Montpellier 2, Montpellier, Cedex, France*
JUAN LIU • *School of Computer, Wuhan University, Wuhan, China; State Key Laboratory of Hybrid Rice, College of Life Sciences, Wuhan University, Wuhan, China*
GUILLERMO LOPEZ-CAMPOS • *Health and Biomedical Informatics Centre, The University of Melbourne, Parkville, VIC, Australia*
STEVEN G.E. MARSH • *Anthony Nolan Research Institute, Royal Free Hospital, London, UK; UCL Cancer Institute, University College London, London, UK*
FERNANDO MARTIN-SANCHEZ • *Health and Biomedical Informatics Centre, The University of Melbourne, Parkville, VIC, Australia*
SEEMA MISHRA • *Department of Biochemistry, School of Life Sciences, University of Hyderabad, Hyderabad, Telangana, India*
MAGDALENA MOLERO-ABRAHAM • *Facultad de Medicina, Universidad Complutense de Madrid, Madrid, Spain*
SUNIL KUMAR MUKHERJEE • *Department of Genetics, University of Delhi, New Delhi, India*
GIRIJA NATEKAR-KALANTRE • *Department of Chemical Engineering, Indian Institute of Technology, Mumbai, India*
BHARTI NEHETE • *Department of Veterinary Sciences, The University of Texas MD Anderson Cancer Centre, Bastrop, TX, USA*
PRAMOD NEHETE • *Department of Veterinary Sciences, The University of Texas MD Anderson Cancer Centre, Bastrop, TX, USA*
YANQING NIU • *School of Mathematics and Statistics, South-Central University for Nationalities, Wuhan, China*
TUN-WEN PAI • *Department of Computer Science and Engineering, National Taiwan Ocean University, Keelung, Taiwan*
GER J.M. PRUIJN • *Department of Biomolecular Chemistry, Radboud Institute for Molecular Life Sciences, Institute for Molecules and Materials and Netherlands Proteomics Centre, Radboud University Nijmegen, Nijmegen, The Netherlands*
ANU RAGHUNATHAN • *Chemical Engineering Division, National Chemical Laboratory, Pune, India*
SRINIVASAN RAMACHANDRAN • *CSIR-Institute of Genomics and Integrative Biology, Delhi, India*
SHOBA RANGANATHAN • *Department of Biochemistry, Yong Loo Lin School of Medicine, National University of Singapore, Singapore, Singapore; Department of Chemistry and Biomolecular Sciences and ARC Center of Excellence in Bioinformatics, Macquarie University, Sydney, NSW, Australia*
SNEHAL RASKAR-RENUSE • *Agilent Technologies, Bangalore, India*
PEDRO RECHE • *Facultad de Medicina, Universidad Complutense de Madrid, Madrid, Spain*
JAMES ROBINSON • *Anthony Nolan Research Institute, Royal Free Hospital, London, UK; UCL Cancer Institute, University College London, London, UK*

Priyanka P. Sadhukhan • *Chemical Engineering Division, National Chemical Laboratory, Pune, India*
Bodhisattwa Saha • *Division of Plant Biology, Bose Institute, Kolkata, India*
Sudipto Saha • *Bioinformatics Centre, Bose Institute, Kolkata, India*
Debasree Sarkar • *Bioinformatics Centre, Bose Institute, Kolkata, India*
K. Jagannadha Sastry • *Department of Immunology, The University of Texas MD Anderson Cancer Centre, Houston, TX, USA; Department of Veterinary Sciences, The University of Texas MD Anderson Cancer Centre, Bastrop, TX, USA*
Smita A. Saxena • *Bioinformatics Centre, University of Pune, Pune, India*
Kimberly S. Schluns • *Department of Immunology, The University of Texas MD Anderson Cancer Centre, Houston, TX, USA*
Shailbala Singh • *Department of Immunology, The University of Texas MD Anderson Cancer Centre, Houston, TX, USA*
Subrata Sinha • *National Brain Research Centre, Gurgaon, Haryana, India*
Gaurab Sircar • *Division of Plant Biology, Bose Institute, Kolkata, India*
Tin Wee Tan • *Department of Biochemistry, Yong Loo Lin School of Medicine, National University of Singapore, Singapore, Singapore*
Namrata Tomar • *Machine Intelligence Unit, Indian Statistical Institute, Kolkata, India*
Joo Chuan Tong • *Institute of High Performance Computing, Singapore, Singapore; Department of Biochemistry, Yong Loo Lin School of Medicine, National University of Singapore, Singapore, Singapore*
Chun-Wei Tung • *School of Pharmacy & Ph.D. Program in Toxicology, Kaohsiung Medical University, Kaohsiung, Taiwan*
Hsin-Wei Wang • *Department of Computer Science and Engineering, National Taiwan Ocean University, Keelung, Taiwan*
Mingjun Wang • *Center for Inflammation and Epigenetics, Houston Methodist Research Institute, Houston, TX, USA*
Dominik Wodarz • *Department of Ecology and Evolutionary Biology, University of California, Irvine, CA, USA*
Chia-Chou Wu • *Lab of Control and Systems Biology, Department of Electrical Engineering, National Tsing Hua University, HsinChu, Taiwan*
Yi Xiong • *Department of Biological Sciences, Purdue University, West Lafayette, IN, USA*
Guojun Yang • *Department of Immunology, The University of Texas MD Anderson Cancer Centre, Houston, TX, USA*
Martin Zacharias • *Physics Department, Technische Universität München, Garching, Germany*
Wen Zhang • *School of Computer, Wuhan University, Wuhan, China*
Peng Zhou • *Center of Bioinformatics (COBI), Key Laboratory for NeuroInformation of Ministry of Education, University of Electronic Science and Technology of China, Chengdu, China*

Part I

Immunoinformatics: Transition from Basic Biology to Informatics

Chapter 1

A Brief Outline of the Immune System

Namrata Tomar and Rajat K. De

Abstract

The various cells and proteins responsible for immunity constitute the immune system, and their orchestrated response to defend foreign/non-self substances (antigen) is known as the immune response. When an antigen attacks the host system, two distinct, yet interrelated, branches of the immune system are active—the nonspecific/innate and specific/adaptive immune response. Both of these systems have certain physiological mechanisms, which enable the host to recognize foreign materials to itself and to neutralize, eliminate, or metabolize them. Innate immunity represents the earliest development of protection against antigens. Adaptive immunity has again two branches—humoral and cell mediated. It should be noted that both innate and adaptive immunities do not work independently. Moreover, most of the immune responses involve the activity and interplay of both the humoral and the cell-mediated immune branches of the immune system. We have described these branches in detail along with the mechanism of antigen recognition. This chapter also describes the disorders of immune system in brief.

Key words Immune response, Immune system, Adaptive immunity, Innate immunity, Antibody, T cells, B cells, Allergy, Antigen, Humoral immune system, Cell-mediated immune system

1 Introduction

The defense system consists of a wide variety of cells and molecules that have evolved to protect animals from invading pathogenic microorganisms and cancer. Recognition and response are two major activities of immune system. Immune recognition is quite specific. Moreover, it is able to discriminate between foreign molecules and the body's own cells and proteins. After the recognition of a foreign organism, it mounts an effector response through recruiting a variety of cells and molecules to eliminate the invader organism. Later exposure to the same foreign organism induces a memory response, characterized by a more rapid and heightened immune reaction that serves to eliminate the pathogen and prevent disease.

Historical perspective: The discipline of immunology developed through the observation when individuals who had recovered from certain infectious diseases were thereafter found to be protected

Rajat K. De and Namrata Tomar (eds.), *Immunoinformatics*, Methods in Molecular Biology, vol. 1184,
DOI 10.1007/978-1-4939-1115-8_1, © Springer Science+Business Media New York 2014

from the disease. The term "immunity" originated from the Latin term "immunis," meaning "exempt," that is, the state of protection from infectious disease. The earliest literary reference to immunology goes back to 430 bc in writings of Thucydides, where he wrote that only those who had recovered from the plague could nurse the sick because they would not contract the disease a second time [1]. In 1798, Edward Jenner found that some milkmaids were immune to smallpox as they had earlier contracted cowpox (a mild disease). The next major advancement in immunology came with the induction of immunity to cholera by Louis Pasteur. He demonstrated the possibility of administrating a weaken pathogen as a vaccine through a classic experiment. In 1881, he first vaccinated one group of sheep with heat-attenuated *Bacillus anthracis* and then challenged the vaccinated sheep and some unvaccinated sheep with a virulent culture of the bacillus. All the vaccinated sheep lived, and all the unvaccinated animals died. In 1885, after applying weakened pathogen to animals, he administered a dose of vaccine to a boy bitten by a rabid dog and later found that the boy survived. However, Pasteur could not explain its mechanism. In 1890, experiments of Emil Von Behring and Shibasaburo Kitasato led to the understanding of the mechanism of immunity. Their experiments described how antibodies present in the serum provided protection against pathogens. These experiments are described as milestone as the beginnings of the discipline of immunology.

2 Types of Immune System: A Layered Defense System

This line of defense against foreign invader microbes has been divided into two general types of immune responses: innate immunity and adaptive immunity. These two differ in time taken and duration of response, effector cell types, and its specificity for different classes of foreign microbes. Innate immune system represents a nonspecific response to a potentially harmful foreign particle; and the adaptive immune system displays a high degree of memory and specificity. Types of immune system have been shown through line diagram in Fig. 1. Table 1 provides the differences between the innate and adaptive immunity. Below is the brief description of innate immunity.

2.1 Innate Immunity (Nonspecific)

The innate immunity is an evolutionarily older defense system that is a dominant one in plants, fungi, insects, and primitive multicellular organisms [2, 3]. The innate system represents the first line of defense to an intruding pathogen. Innate immune systems are found in all plants and animals. The response evolved is therefore rapid and is unable to memorize. It comprises four types of defensive barriers, namely anatomic (e.g., skin and mucous membranes), physiological (e.g., temperature, low pH), phagocytic (e.g., blood

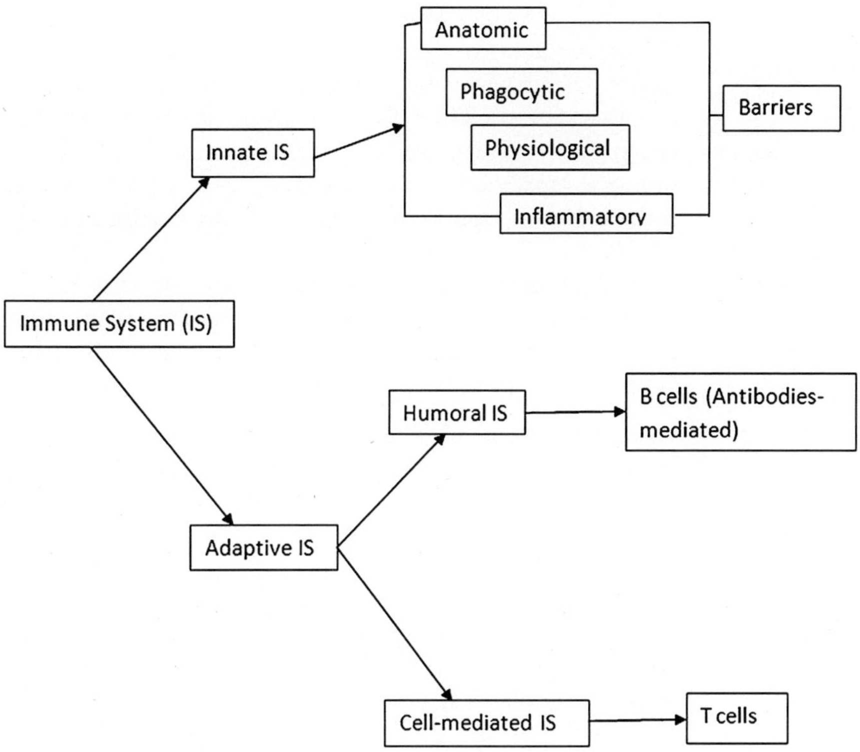

Fig. 1 Types of immune system (IS)

Table 1
Difference between innate and adaptive immune systems

Innate immune system	Adaptive immune system
Nonspecific response	Specific response
Immediate response	Lag time between antigen exposure and response
Retains no immunological memory	Retains immunological memory
Found in nearly all forms of life	Found in only jawed vertebrates

monocytes, neutrophils, tissue macrophages), and inflammatory (e.g., serum proteins).

Cells of the innate immune system: Phagocytes, neutrophils, macrophages, natural killer cells, mast cells, basophils, dendritic cells, eosinophils.

2.2 Adaptive Immunity (Acquired/Specific Immunity)

The adaptive immune system is activated by innate immunity. The components of the adaptive immune system possess slower temporal dynamics with high degree of specificity and a more potent secondary response. The adaptive immune system frequently incorporates cells and molecules of the innate system in its fight against harmful foreign bodies. For example, complement system (molecules of the innate system) may be activated by antibodies (molecules of the adaptive system). The cells of the acquired immune system are T and B lymphocytes that we will describe later. It is of two types: (1) humoral (antibody-mediated system) and (2) cell mediated. Below is the brief description of the types of adaptive immune system.

2.2.1 Humoral Immune System (Antibody-Mediated Immune System)

It involves substances found in the humors, or body fluids; therefore, the name is humoral immune system. This kind of immunity is mediated by macromolecules found in extracellular fluids such as secreted antibodies, complement proteins, and certain antimicrobial peptides.

Complement system: The complement system is involved in the responses of both innate immunity and acquired immunity. It is named so as it helps or "complements" the ability of antibodies and phagocytic cells to clear pathogens from an organism. It is a biochemical cascade of the innate immune system that helps clear pathogens from an organism. Activation of this system leads to cytolysis, chemotaxis, opsonization, immune clearance, and inflammation. Three biochemical pathways activate the complement system: the classical complement pathway, the alternate complement pathway, and the mannose-binding lectin pathway [3].

B cells: B cells belong to a group of white blood cells known as lymphocytes. The abbreviation "B," in B cell, comes from the bursa of Fabricius in birds, where they mature. In mammals, immature B cells are formed in the bone marrow, which is used as a backronym for the cells' name [5]. There is a random gene rearrangement during B cell maturation in the bone marrow that generates more than 10^{10} number of B cells with different antigenic specificities. Later, there is a selection process to eliminate any B cells with membrane-bound antibody that recognizes self-components. This ensures that self-reactive antibodies (autoantibodies) are not produced.

Somatic hypermutation: When a B cell recognizes an antigen, it starts proliferating. During proliferation, the B cell receptor (BCR) locus undergoes somatic mutation in the hypervariable regions, of 10^5- to 10^6-fold greater than the normal rate of mutation across the genome [6, 7]. Hypermutation enhances the ability of immunoglobulin receptors present on B cells to recognize and bind a specific antigen [3].

Antibodies: The production of antibodies is the main function of the humoral immune system [4]. Antibodies are secreted by plasma cell, a type of white blood cell. These are the large Y-shaped protein molecules secreted by B cells, also known as immunoglobulins (Ig). The antibody recognizes a unique part of the foreign target, called an antigen [2, 3]. Antibody has a "Y"-structured tip for a specific epitope, known as paratope. The structural diagram of antibody has been shown in Fig. 2. Isoforms of Igs have been described in Table 2.

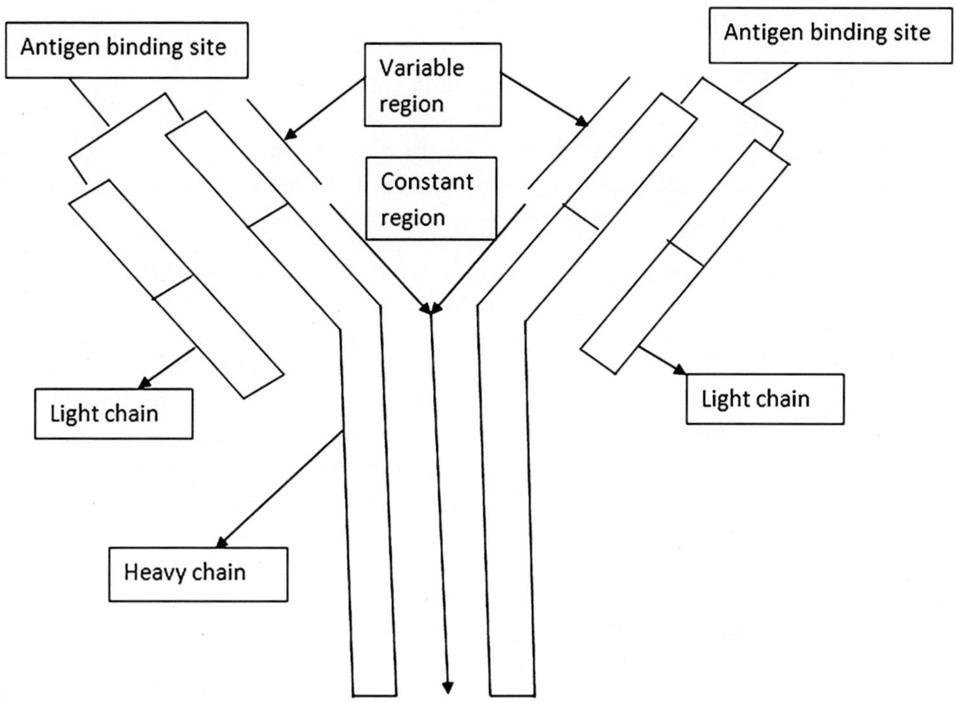

Fig. 2 Antibody structure

Table 2
Antibody isotypes

Type names	Description
IgA	Found in mucosal areas of gut, respiratory tract, and urogenital tract, including saliva, tears, and breast milk
IgD	Functions mainly as an antigen receptor on B cells that have not been exposed to antigens
IgE	Involves in allergy, binds to allergens, and triggers histamine release from mast cells and basophils
IgG	Only antibody that can cross the placenta to give passive immunity to the fetus
IgM	Secreted pentamer form, expressed on the surface of B cells (monomer). Eliminates pathogens in the early stages of B cell-mediated (humoral) immunity before there is sufficient IgG

Class switch recombination (CSR) (immunoglobulin class switching/isotype switching/isotypic commutation): B cell's production of antibody from one class to another can be changed through a biological mechanism called as CSR binding, for example, from an isotype called IgM to an isotype called IgG. During this process, the constant region portion of the antibody heavy chain is changed, but the variable region of the heavy chain stays the same; hence it does not affect the antigen specificity.

2.2.2 Cell-Mediated Immune System

It does not involve antibodies, rather activates phagocytes and antigen-specific cytotoxic T lymphocytes and releases various cytokines in response to an antigen attack.

T lymphocytes: Although T lymphocytes arise in the bone marrow, it migrates to the thymus gland to mature unlike B cells [1]. Within the thymus, it expresses a unique antigen-binding molecule on its membrane, called as T cell receptor (TCR). TCRs can recognize only antigen that is bound to cell-membrane proteins called major histocompatibility complex (MHC) molecules, unlike B cells. There are two well-defined subpopulations of T cells: T helper (Th) and T cytotoxic (Tc) cells. It becomes an effector cell (activated) that secretes various growth factors known collectively as cytokines, after a Th cell recognizes and interacts with an antigen–MHC class II molecule complex. The secreted cytokines play an important role in activating B cells, Tc cells, macrophages, and various other cells that participate in the immune response.

TCR–MHC molecule interaction to present antigen to T cell has been shown in Fig. 3.

Under the influence of TH-derived cytokines, a Tc cell recognizes an antigen and MHC class I and further proliferates and differentiates into an effector cell called as a cytotoxic T lymphocyte (CTL). It has cytotoxic activity and usually does not secrete cytokines. The CTL has a vital function in eliminating antigen-displaying cell, such as virus-infected cells, tumor cells, and cells of a foreign tissue graft.

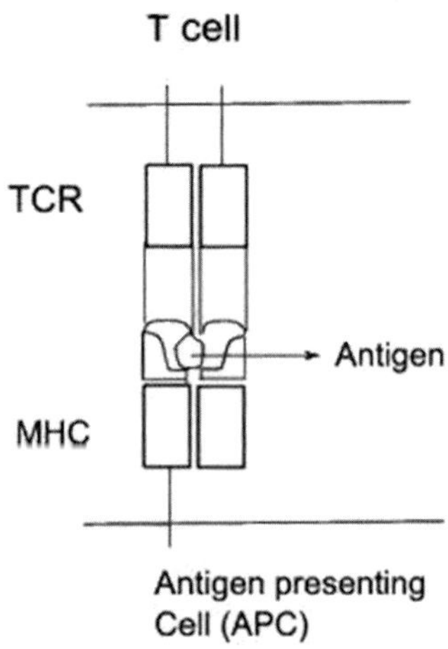

Fig. 3 TCR–MHC interaction for antigen presentation

T cell maturation also includes random rearrangements of a series of gene segments that encode the cell's antigen-binding receptor, like B cell maturation. The random rearrangement of the TCR genes is capable of generating on the order of 10^9 unique antigenic specificities. Each T lymphocyte cell expresses about 10^5 receptors, and all of the receptors on the cell and its clonal progeny have identical specificity for antigen. However, it is later diminished through a selection process to ensure that only T cells with receptors capable of recognizing antigen associated with MHC molecules will be able to mature [1].

The MHC: The MHC is a large genetic complex with multiple loci and encodes for three major classes of membrane-bound glycoproteins: class I, class II, and class III MHC molecules. These molecules do not have fine specificity for antigen characteristic; instead of this, it binds to a spectrum of antigenic peptides derived from the intracellular degradation of antigen molecules. In both class I and class II MHC molecules posses variable regions;, a cleft within which the antigenic peptide binds and is presented to T lymphocytes. As mentioned above, Th cells generally recognize antigen combined with class II molecules, whereas Tc cells generally recognize antigen combined with class I molecules.

Below are the major differences among these three classes: (1) Class I MHC genes encode glycoproteins expressed on the surface of nearly all nucleated cells; the major function of the class I gene products is presentation of peptide antigens to Tc cells. (2) Class II MHC genes encode glycoproteins expressed primarily on antigen-presenting cells (macrophages, dendritic cells, and B cells), where they present processed antigenic peptides to Th cells. (3) Class III MHC genes encode various secreted immune system-related proteins, including components of the complement system and molecules involved in inflammation.

Another important aspect is their structural features, where class I and class II MHC molecules have common structural features and both have roles in antigen processing. However, the class III MHC region encodes molecules that have little in common with class I or II molecules.

3 Disorders of Human Immunity

Although, the immune system is a remarkably specific and adaptive, however, it may lead to develop autoimmunity, hypersensitivities and immunodeficiencies, upon deregulation.

3.1 Autoimmunity

Autoimmunity arises when immune system fails to distinguish between self and non-self. Here, immune system attacks on self-antigens, instead of reacting against foreign antigens. The result is an inappropriate response of the immune system against self-components termed autoimmunity. Normal healthy individuals have been shown to possess self-reactive lymphocytes in periphery, where its presence does not inevitably result in autoimmune reactions [1]. However, their activity is regulated through clonal anergy or clonal suppression. Its deregulation can lead to the activation of humoral or cell-mediated responses against self-antigens. These reactions can damage cells and organs, sometimes with fatal consequences. Lymphocytes or antibodies bind to cell-membrane antigens and lead to cellular lysis and/or an inflammatory response in the affected organ. The damaged cellular structure is gradually replaced by connective tissue (scar tissue), and thereby the function of the organ declines.

Many autoimmune diseases are characterized by tissue destruction mediated directly by T cells. For example in rheumatoid arthritis, self-reactive T cells attack the tissue in joints, causing an inflammatory response that results in swelling and tissue destruction. In Hashimoto's thyroiditis, autoantibodies reactive with tissue-specific antigens such as thyroid peroxidase and thyroglobulin cause severe tissue destruction. Other examples include insulin-dependent diabetes mellitus and multiple sclerosis. The immune response is directed to a target antigen unique to a single organ or gland in an organ-specific autoimmune disease. This way, the effects are largely limited to that organ. In case of damage by humoral or cell-mediated effector mechanisms, the antibodies may overstimulate or block the normal function of the target organ.

3.2 Hypersensitivity

The ability of the immune system to respond inappropriately to antigenic challenge is known as hypersensitivity or allergy. It refers to undesirable reactions produced by the normal immune system, including allergies and autoimmunity. The four-group classification was given by Gell and Coombs in 1963 [8]. Table 3 gives brief description of this classification, along with an additional type.

Table 3
Allergy classification

Type	Names	Mediators
I	Allergy, IgE mediated	IgE and IgG4
II	Cytotoxic, antibody dependent	IgM and IgG
III	Immune complex disease	IgG
IV	Delayed-type hypersensitive (DTH)	T cells
V	Autoimmune disease, receptor mediated	IgM or IgG

3.3 Immunodeficiencies

Immunodeficiency is a state in which the immune system compromises or is unable to fight infectious disease. In this case, the system fails to protect the host from diseases or from malignant cells. A condition that occurs from a genetic or a developmental defect in the immune system is called a primary immunodeficiency. Secondary immunodeficiency, or acquired immunodeficiency, is the loss of immune function and results from exposure to various agents. Till date, the most common secondary immunodeficiency is acquired immunodeficiency syndrome, or AIDS, which results from infection with the human immunodeficiency virus 1 (HIV-1) [1].

Primary immunodeficiency: A primary immunodeficiency may affect either adaptive or innate immune functions. Most of the primary immunodeficiencies are inherited, and the genetic defects are determined. The consequences of primary immunodeficiency depend on the number and type of immune system components involved. Defects in components early in the hematopoietic developmental scheme affect the entire immune system. Deficiencies involving components of adaptive immunity, effector T or B cells, while phagocytes or complement, are impaired in innate immunity.

Secondary immunodeficiency: Agent-induced immunodeficiency results from the exposure to any of a number of chemical and biological agents that induce an immunodeficient state. These agents can be immunosuppressive medicines. The drugs that are used to combat autoimmune diseases such as rheumatoid arthritis or lupus erythematosis induce the abovementioned kind of immunodeficiency. Cytotoxic drugs or radiation treatments given to cancer patients damage the immune cells and thereby induce a state of immunodeficiency.

4 Conclusion

We have described immune system and its branches briefly in this chapter. We have described the difference between the two said branches of the immune system in a tabular way. We have also highlighted the immune system disorders.

Acknowledgment

Ms. Namrata Tomar, one of the authors, gratefully acknowledges CSIR, India, for providing her a Senior Research Fellowship (9/93(0145)/12, EMR-I).

References

1. Thomas K, Goldsby J, Osborne RA, Barbara A, Kuby J (2006) Kuby immunology, 6th edn. WH Freeman and Co., New York, NY
2. Litman GW, Cannon JP, Dishaw LJ (2005) Reconstructing immune phylogeny: new perspectives. Nat Rev Immunol 5(11):866–879
3. Janeway C, Travers P, Walport M, Shlomchik M (2001) Immunobiology, 5th edn. Garland Science, New York, NY
4. Pier GB, Lyczak JB, Wetzler LM (2004) Immunology, infection, and immunity. ASM Press, Washington, DC
5. Alberts B, Johnson A, Lewis J, Raff M, Roberts K, Walter P (2002) Molecular biology of the cell. Garland Science, New York, NY, p 1367
6. Li Z, Wool CJ, Iglesias-Ussel MD, Ronai D, Scharff MD (2004) The generation of antibody diversity through somatic hypermutation and class switch recombination. Genes Dev 18(1):1–11
7. Oprea M (1999) Antibody repertoires and pathogen recognition: the role of germline diversity and somatic hypermutation (Thesis) University of Leeds.
8. Gell PGH, Coombs RRA (eds) (1963) Clinical aspects of immunology, 1st edn. Blackwell, Oxford

Chapter 2

Cross Talk Between the Metabolic and Immune Systems

Namrata Tomar and Rajat K. De

Abstract

Understanding the interplay between metabolic and cellular signaling systems has emerged as a focus in the study of metabolic disorders, cancer, and immune responses. Immune system is active in the regulation of metabolism. Lymphocyte activation initiates a program of cell growth, proliferation, and differentiation that increase metabolic demand. Activated lymphocytes must alter their metabolism to support these increased synthetic activities. In this chapter, we describe how signaling via the immune system integrates with metabolic functions to control immune response and vice versa. It has been explained mainly in the context of T lymphocyte activation and, to a lesser detail, in other immune cell types.

Key words Immune system, Metabolic system, Lymphocyte, T cells, mTOR, TLR, Adipose tissue, Obesity, Inflammation

1 Introduction

Immune system is required to ward off tumors and infectious particles attacking the host. It is a very balanced homeostatic system and also guards against immune dysregulation, such as in allergy and autoimmunity. There is increase in observations to investigate how immune cells affect certain nonimmune functions, including neurodegeneration, cardiovascular function, and metabolism. Thus, immune metabolism is an emerging field of investigation, which is at the interface between the distinct disciplines of immunology and metabolism. Hepatocytes and myocytes are two cell types in which metabolic pathways have been well studied. Unlike these two, resting lymphocytes do not store glycogen in a larger amount. It makes them highly dependent on the import of extracellular glucose to meet increased metabolic needs [1–3]. The behaviors of lymphocytes and other leukocytes are controlled by metabolic activities of the cells at different levels.

Investigations on the molecular aspects of immunological-metabolic cross talk have become an important field of research. During the activation of a resting lymphocyte, large metabolic

Rajat K. De and Namrata Tomar (eds.), *Immunoinformatics*, Methods in Molecular Biology, vol. 1184,
DOI 10.1007/978-1-4939-1115-8_2,

demands are placed on the cell as it initiates proliferation and cytokine production [4]. The cell grows to approximately double its resting size and then enters into a program of rapid proliferation while also differentiating from a quiescent cell to a highly secretory one.

The role of glucose in immune system is explored initially in this chapter. Moreover, the metabolic dependency in lymphocyte activation is explained. Metabolic alterations and disturbances affect immunity of an individual. Thus obesity-associated inflammation, type 2 diabetes (T2D), and cardiovascular disease (CVD) are being explored as metabolic alterations, which result in the impairment of immune system. We have also described the role of nutrient sensors, adipose tissue, and toll-like receptors in maintaining immune–metabolic interactions.

2 Role of Glucose in the Immune System

In addition to acting as a defense mechanism for a human being, immune system also participates in the control of the resident colonizing microflora, which is essential for immunologic and metabolic health. These regulatory processes are energy demanding, and immune cells from both innate and adaptive immune systems use numerous extracellular molecules and signals as fuels [5, 6]. The exact nature of the energetic demands differs among immune cells and the nature of the required response. For example, energy demand is different from that of proliferative/secretory (B or T lymphocytes) than that of non-proliferative/secretory (macrophages or neutrophils). Observations using lymphocytes, stimulated with B- or T-specific mitogens (such as pokeweed mitogen for B cells, concanavalin-A for T cells), have revealed that the glucose uptake and catabolism are necessary to provide energy for their proliferation, biosynthesis, and secretory activities [1, 2]. It has been found that mitogen-induced lymphocyte activation leads to an increase in glucose consumption, which mostly metabolizes to lactate within 1 h of stimulation [7]. Moreover, other pathways of glucose utilization, such as the pentose phosphate pathway (PPP), have also been shown to be functional during lymphocyte stimulation and have peaked at 48 h after stimulation.

The metabolism of resting lymphocytes is limited by the availability of trophic signals and does not depend upon the availability of nutrients, such as glucose [8]. Once T cells approximately double their resting size and start proliferating, they start differentiating from a quiescent to a highly secretory state, after getting activation. These processes lead to increase in glucose consumption and hence activation of glycolysis [9].

Regulation of energy metabolism in immune cells requires coordination by signal transduction pathways as the functions of these pathways directly have an impact on the modulation of nutrient uptake and metabolism. Glucose transporter (GLUT) and insulin receptor (InsR) proteins are expressed in immune cells, like monocytes /macrophages, neutrophils, and B and T lymphocytes [10, 11]. It has been shown that physiological doses of insulin have led to increased expression of GLUT3 and GLUT4 in monocytes and B lymphocytes [12]. In contrast, insulin does not alter GLUT expression in resting T cells and in neutrophils. However, in vitro mitogen or LPS (the ligand for TLR4) stimulation of immune cells enhances the expression of membranes GLUT1, GLUT3, and GLUT4 [13, 14]. It has been observed that expression of InsR is essential for immune cell division, size, and survival [15].

3 Role of Immune Cells in Metabolism

There has been a fair amount of increase in the understanding of the immune system organization as well as its regulators. There is a close concordance between host nutritional status and immunity. Thus the investigation on the relationship among nutrition, health, and the immune system of an individual has now become a topic of study.

In the absence of B cells or IgA and in the presence of the microbiota, the intestinal epithelium upregulates interferon-inducible immune response pathways and represses Gata4-related metabolic functions [16]. It leads to lower absorption of lipid. Further, network analysis reveals the presence of two inversely expressed and interconnected epithelial cell gene networks—for lipid metabolism and regulating immunity. The authors have also observed similarities between the gene expression patterns in gut biopsies from individuals with common variable immunodeficiency (CVID)/HIV infection and intestinal malabsorption and from B cell-deficient mice. It possibly explains a relation between immunodeficiency and defective lipid absorption in humans.

Immune deficiency has been observed in leptin-deficient obese (ob/ob) mice. It has found to be associated with an impairment of dendritic cell (DC) function. The ob/ob mice have demonstrated reduced cellular and humoral response and an altered cytokine secretion profile following keyhole limpet hemocyanin (KLH) immunization. Variations have been observed in the cytokine profile secretion in both in vivo and in vitro experiments [17]. For example, more IL-10 and IFN-γ have been secreted by splenic cells from obese animals in an antigen-specific response. However, higher amounts of IL-10 and of IL-4 have been detected in control supernatant in a protocol of mixed lymphocyte reaction (MLR). Authors have also analyzed epidermal sheets of obese mice and found higher number of dendritic cells in obese mice compared with control one.

4 Metabolic Dependency in Lymphocyte Activation

Naive and memory T cells have metabolic activities for housekeeping functions, such as the transportation and turnover of biomaterials, maintenance of cytoskeleton, among others. Glucose oxidation through tricarboxylic acid (TCA) cycle and fatty acid β-oxidation provide most of the metabolic support for these basic cellular functions in naive and memory cells [18, 19]. Immune signaling from T cell receptor (TCR), co-stimulatory molecules, and cytokine receptors activate resting T cells upon antigen exposure. Upon activation, quiescent naive T cells undergo a growth phase followed by clonal expansion and differentiation. These changes are essential for accurate immune defense and regulation. Initial growth and rapid proliferation during the expansion phase increase bioenergetic and biosynthetic demands. It requires a metabolic rewiring during the transition between resting and activation stages. It also makes active T cells to use certain metabolic pathways in the ways that naive and memory T cells do not. In naive and memory T cells, the majority of pyruvate enters into the mitochondria, where it is converted to acetyl-CoA through oxidative decarboxylation, and later fluxes into TCA cycle to generate ATP. However, in active T cells, a major portion of pyruvate moves away from the TCA cycle to produce lactate. Thus it is clear that the production of lactate via glycolysis is significantly upregulated following T cell activation. It may be noted that this change is not restricted to low oxygen (anaerobic) in the environment and is actively regulated by signal transduction pathways when oxygen is plentiful (aerobic glycolysis) [20, 21]. Glutaminolysis, the glutamine catabolic pathway, is another major carbohydrate catabolism that is significantly elevated in T cells after their activation [22, 23].

5 Effects of Metabolic Alteration on Immune Reactivity

Metabolic disturbances, like obesity, have serious effects on immunity. Obesity and related disease and disease-like symptoms, such as insulin resistance in T2D and cardiovascular diseases, have become like an epidemic. Fatty acids and glucose enter into the blood after taking a meal. For an obese individual, the body has higher levels of fat and glucose, and it alters responsiveness of the immune system. This impairment of the immune system associated with human obesity has also been demonstrated in several animal models. Leptin is an adipocyte-derived cytokine. It is secreted proportionally to the amount of fat to finely regulate body weight [24]. Complete congenital absence of leptin leads to hyperphagia and morbid obesity in both humans and rodents [25]. A study has shown that obese animals have a delayed wound healing associated with increased polymorphonuclear cell infiltration [26]. In addition, both T and B cell-mediated immune responses are impaired in leptin-deficient obese mice (ob/ob) and diabetic db/db mice [27].

Imbalance in the cytokine network is another feature of obesity, which results in a low-grade systemic inflammatory status. It has been observed in both obese humans and animals [28]. The inflammatory cytokines interleukin 6 (IL6), IL1, and tumor necrosis factor-α (TNF-α) have found to be abnormally elevated in obesity, which mostly originate from the activated macrophages infiltrating the white adipose tissue [29, 30]. Investigations may be carried out to explore the reason behind the obesity-associated inflammation, the extent of obesity and inflammation being related, and the pathway(s) responsible for inflammation-induced T2D, cardiovascular disease, and other related pathologies. On the practical side, as inflammation mediates many pathological consequences of obesity, it may lead to exploration of anti-inflammatory drug discovery and drugs for the patients with obesity-associated metabolic and cardiovascular disorders.

6 Role of Nutrient Sensors, Adipose Tissue, and Toll-Like Receptors in Maintaining Immune–Metabolic Cross Talk

In most of the cases, immune cells use and respond to nutrients similarly as found in other cells. There are cell-intrinsic metabolic processes that influence the performance of immune cells [31]. The interesting aspect is to have a completely different perspective on the immunological metabolic interface to find out the extent and the precise mechanisms of typical cell-intrinsic metabolic processes that influence the functional performance of immune cells.

AKT1-3, AMPK-activated protein kinase (AMPK), mammalian target of rapamycin (mTOR), and *LKB1*: The serine/threonine kinases AKT1-3, AMPK, mTOR, and LKB1 are cellular nutrient sensors that help to maintain energy homeostasis.

Finlay and Cantrell [32] have suggested that AKT1-3, AMPK, and LKB1 control a fate switch, from cytotoxic effector to memory CD8+ T cells, in addition to providing nutrient responses. According to the authors, AKT proteins regulate repertoires of adhesion molecules and chemokine receptors in CD8+ T cells and control trafficking and migration. This, in turn, determines decision for the memory versus terminally differentiated effector CD8+ T cells. Considering LKB1, it is mentioned that an lkb1−/− bone marrow transplant was unable to reconstitute the hematopoietic system in irradiated mice. This observation suggests that the survival of hematopoietic stem cells (HSCs) depends on LKB1 [33]. An lkb1−/− bone marrow transplant was unable to reconstitute the hematopoietic system in irradiated mice, again suggesting that the survival of HSCs depends on LKB1. Moreover, a study shows that CD28 co-stimulation of human peripheral blood T cells enhances expression of glucose transporters, glucose uptake, and glycolysis. This increase depends on PI3K activity. Further, the majority of glucose processed by CD28-co-stimulated T cells is converted to lactate. It is not used for biosynthesis or oxidized for maximal energy extraction [34]. These observations have shown that under certain conditions, immune cells may use metabolic pathways to control fate and function in the ways that are different from other cells.

Adipose tissue and Toll-like receptors (TLRs) of the innate immune system, which are found on immune cells, intestinal cells, and adipocytes, are being studied as essential factors in the complex balance of immune and metabolic health.

6.1 Toll-Like Receptors

TLRs are broadly expressed in cells of the innate immune system, such as macrophages, epithelial and endothelial cells, and organ parenchyma cells. They have specific roles in local innate immune defense [35]. TLRs of the innate immune system, which are found in immune cells, intestinal cells, and adipocytes, are observed as essential for maintaining the complex balance of immune and metabolic systems [36]. Lipid is one of the components, which is recognized by TLRs. Some of the mammalian TLRs also regulate energy metabolism, mostly through acting on adipose tissue. This has opened a wide scope of research on the role of TLRs in pathologies related to metabolism, such as obesity, insulin resistance, and atherosclerosis. A study has reported that saturated fatty acids can induce the activation of TLR2 and TLR4, whereas unsaturated fatty acids have shown to inhibit TLR-mediated signaling pathways and gene expression [37].

6.2 Adipose Tissue

Adipose tissue is observed as an immunocompetent organ and adipocytes as components of the innate immune system. Adipocytes secrete classical cytokines (TNF- α, IL-6, IL-1 receptor antagonist, and TGF-β), levels of which are significantly increased in obesity, which contribute to the overall inflammatory status of obese persons [38]. In addition, leptin has also been shown to play an essential role in both innate and adaptive immune responses [39].

Adipocytes and macrophages have recently been described to originate from a common ancestral progenitor and to share several features as follows [40, 41]. Macrophages express some adipocyte-specific gene products, such as ap2, while adipocytes secrete macrophage-specific gene products, such as IL-6 or TNF-α. This common gene expression results in some analogous functional activities, such as lipid accumulation by macrophages in atherosclerotic lesions or phagocytic capacities exhibited by adipocytes towards certain pathogens, thereby revealing an apparent coordinated activity between these two cell types during the course of an innate immune response. Adipocytes, isolated from diet-induced obese mice or genetically obese animals, exhibited increased TLR expression [42–44], together with higher cytokine production upon stimulation. TGF-β is positively correlated with obesity and up-regulated both in human and in ob/ob mice white adipose tissue [45].

7 Conclusions

Fluctuations in blood glucose occur in inflammatory diseases, such as obesity, diabetes, and insulin resistance. It is now becoming clear that the emerging field of immune metabolism has theoretical and practical implications for future research. Generating an efficient and effective immune response involves large increase in cellular proliferative, biosynthetic, and secretory activities and processes, which require high energy consumption. As mentioned, adaptive as well as innate immune cells must be able to rapidly respond to the presence of pathogens, shifting from a quiescent phenotype to a highly active state within hours after stimulation. For this purpose, cells must dramatically alter their metabolism in order to support these increased synthetic activities based on extracellular signals as fuels, among which glucose is the most essential one. Since activated lymphocytes have high metabolic demands, manipulation of the lymphocyte-specific metabolic control pathways may be useful in treating diseases characterized by immune hyperactivation, autoimmune disorders, and graft rejection.

Acknowledgment

Ms. Namrata Tomar, one of the authors, gratefully acknowledges CSIR, India, for providing her a Senior Research Fellowship (9/93(0145)/12, EMR-I).

References

1. Culvenor JG, Weidemann MJ (1976) Phytohaemagglutinin stimulation of rat thymus lymphocyte glycolysis. Biochim Biophys Acta 437:354–363
2. Roos D, Loos JA (1970) Changes in the carbohydrate metabolism of mitogenically stimulated human peripheral lymphocytes. I. Stimulation by phytohaemagglutinin. Biochim Biophys Acta 222:565–582
3. Hedeskov CJ (1968) Early effects of phytohaemagglutinin on glucose metabolism of normal human lymphocytes. Biochem J 110:373–380
4. Krauss S, Brand MD, Buttgereit F (2001) Signaling takes breath- new quantitative perspectives on bioenergetics and signal transduction. Immunity 15:497–502
5. Calder PC (1995) Fuel utilization by cells of the immune system. Proc Nutr Soc 54:65–82
6. Frauwirth KA, Thompson CB (2004) Regulation of T lymphocyte metabolism. J Immunol 172:4661–4665
7. Sagone L Jr, BoBuglio AF, Balcerzak SP (1974) Alterations in hexose monophosphate shunt during lymphoblastic transformation. Cell Immunol 14:443–452
8. Buttgereit F, Burmester GR, Brand MD (2000) Bioenergetics of immune functions: Fundamental and therapeutic aspects. Immunol Today 21:194–199
9. Hume DA, Radhik JL, Ferber RE, Weidemann MJ (1978) Aerobic glycolysis and lymphocyte transformation. Biochem J 174:703–709

10. Chakrabarti R, Jung CY, Lee TP et al (1994) Changes in glucose transport and transporter isoforms during the activation of human peripheral blood lymphocytes by phytohemagglutinin. J Immunol 152:2660–2668
11. Fu Y, Maianu K, Melbert BR et al (2004) Facilitative glucose transporter gene expression in human lymphocytes, monocytes, and macrophages: a role for GLUT isoforms 1, 3, and 5 in the immune response and foam cell formation. Blood Cell Mol Dis 32:182–190
12. Leroux JP, Marchand JC, Ha RHT, Cartier P (1975) The influence of insulin on glucose permeability and metabolism of human granulocytes. Eur J Biochem 58:367–373
13. Ercolani L, Lin HL, Ginsberg BH (1985) Insulin-induced desensitization at the receptor and postreceptor level in mitogen-activated human T-lymphocytes. Diabetes 34:931–937
14. Maratou E, Dimitriadis G, Kollias A et al (2007) Glucose transporter expression on the plasma membrane of resting and activated white blood cells. Eur J Clin Invest 37: 282–290
15. Knutson VP (1991) Cellular trafficking and processing of the insulin receptor. FASEB J 5:2130–2138
16. Shulzhenko N, Morgun A, Hsiao W et al (2011) Crosstalk between B lymphocytes, microbiota and the intestinal epithelium governs immunity versus metabolism in the gut. Nat Med 17:1585–1593
17. Macia L, Melacre M, Abboud G et al (2006) Impairment of dendritic cell functionality and steady-state number in obese mice. J Immunol 177:5997–6006
18. van der Windt GJW, Everts B, Chang CH (2012) Mitochondrial respiratory capacity is a critical regulator of CD8+ T cell memory development. Immunity 36:68–78
19. Wang R and Green DR (2012) The immune diet: meeting the metabolic demands of lymphocyte activation. F1000 Biology Reports 4: 1–23
20. Wang R, Dillon CP, Shi LZ (2011) The Transcription Factor Myc Controls Metabolic Reprogramming upon T Lymphocyte Activation. Immunity 35:871–882
21. Jacobs SR, Herman CE, Maciver NJ et al (2008) Glucose uptake is limiting in T cell activation and requires CD28-mediated Akt dependent and independent pathways. J Immunol 180:4476–4486
22. Carr EL, Kelman A, Wu GS et al (2011) Glutamine uptake and metabolism are coordinately regulated by ERK/MAPK during T lymphocyte activation. J Immunol 185: 1037–1044
23. Newsholme EA, Crabtree B, Ardawi MS (1985) Glutamine metabolism in lymphocytes: its biochemical, physiological and clinical importance. Q J Exp Physiol 70:473–489
24. Goren I, Kampfer H, Podda M et al (2003) Leptin and wound inflammation in diabetic ob/ob mice: differential regulation of neutrophil and macrophage influx and a potential role for the scab as a sink for inflammatory cells and mediators. Diabetes 52:2821–2832
25. Mandel MA, Mahmoud AA (1978) Impairment of cell-mediated immunity in mutation diabetic mice (db/db). J Immunol 120:1375–1377
26. Friedman JM, Halaas JL (1998) Leptin and the regulation of body weight in mammals. Nature 395:763–770
27. Finlay D, Cantrell DA (2011) Metabolism, migration and memory in cytotoxic T cells. Nat Rev Immunol 11:109–117
28. Farooqi IS, Matarese G, Lord GM et al (2002) Beneficial effects of leptin on obesity, T cell hyporesponsiveness, and neuroendocrine/metabolic dysfunction of human congenital leptin deficiency. J Clin Invest 110: 1093–1103
29. Aronson D, Bartha P, Zinder O et al (2004) Obesity is the major determinant of elevated C-reactive protein in subjects with the metabolic syndrome. Int J Obes Relat Metab Disord 28:674–679
30. Weisberg SP, McCann D, Desai M et al (2003) Obesity is associated with macrophage accumulation in adipose tissue. J Clin Invest 112:1796–1808
31. Neels JG, Olefsky JM (2006) Inflamed fat: what starts the fire? J Clin Invest 116:33–35
32. Nakada D, Saunders TL, Morrison SJ (2010) Lkb1 regulates cell cycle and energy metabolism in haematopoietic stem cells. Nature 468:653–658
33. Andonegui G, Bonder CS, Green F et al (2003) Endothelium derived Toll-like receptor-4 is the key molecule in LPS induced neutrophil sequestration into lungs. J Clin Invest 111: 1011–1020
34. Wolowczuk I, Verwaerde C, Viltart O, Delanoye A, Delacre M, Pot B, Grangette C (2008) Feeding our immune system: Impact on metabolism. Clin Dev Immunol 2008: 1–19
35. Lee JY, Sohn KH, Rhee SH, Hwang D (2001) Saturated fatty acids, but not unsaturated fatty acids, induce the expression of cyclooxygenase-2 mediated through Toll-like receptor 4. J Biol Chem 276:16683–16689
36. Rondinone CM (2006) Adipocyte-derived hormones, cytokines, and mediators. Endocrine 29:81–90

37. Lam QL, Lu L (2007) Role of leptin in immunity. Cell Mol Biol 4:1–13
38. Cousin B, Munoz O, Andre A et al (1999) A role for preadipocytes as macrophage-like cells. FASEB J 13:305–312
39. Charriere G, Cousin B, Arnaud E et al (2003) Preadipocyte conversion to macrophage: evidence of plasticity. J Biol Chem 278: 9850–9855
40. Shi H, Kokoeva MV, Inouye K et al (2006) TLR4 links innate immunity and fatty acid-induced insulin resistance. J Clin Invest 116: 3015–3025
41. Batra A, Pietsch J, Fedke I et al (2007) Leptin-dependent Toll-like receptor expression and responsiveness in preadipocytes and adipocytes. Am J Pathol 170:1931–1941
42. Song MJ, Kim KH, Yoon JM et al (2006) Activation of Toll-like receptor 4 is associated with insulin resistance in adipocytes. Biochem Biophys Res Commun 346:739–745
43. Samad F, Yamamoto K, Pandey M et al (1997) Elevated expression of transforming growth factor-β in adipose tissue from obese mice. Mol Med 3:37–48
44. Frauwirth KA, Riley JL, Harris MH et al (2002) The CD28 signaling pathway regulates glucose metabolism. Immunity 16:769–777
45. Mathis D (2011) Immunometabolism: an emerging frontier. Nat Rev Immunol 11:81–83

Chapter 3

Immunoinformatics: A Brief Review

Namrata Tomar and Rajat K. De

Abstract

A large volume of data relevant to immunology research has accumulated due to sequencing of genomes of the human and other model organisms. At the same time, huge amounts of clinical and epidemiologic data are being deposited in various scientific literature and clinical records. This accumulation of the information is like a goldmine for researchers looking for mechanisms of immune function and disease pathogenesis. Thus the need to handle this rapidly growing immunological resource has given rise to the field known as immunoinformatics. Immunoinformatics, otherwise known as computational immunology, is the interface between computer science and experimental immunology. It represents the use of computational methods and resources for the understanding of immunological information. It not only helps in dealing with huge amount of data but also plays a great role in defining new hypotheses related to immune responses. This chapter reviews classical immunology, different databases, and prediction tool. Further, it briefly describes applications of immunoinformatics in reverse vaccinology, immune system modeling, and cancer diagnosis and therapy. It also explores the idea of integrating immunoinformatics with systems biology for the development of personalized medicine. All these efforts save time and cost to a great extent.

Key words Systems biology, Immunomics, In silico models, T cells, B cells, Allergy, Reverse vaccinology, Personalized medicine

1 Introduction

The human immune system is very complex and operates at multiple levels, viz., molecules, cells, organs, and organisms. Each individual has a unique immune system and will respond differently to immune challenges. It has a combination of biological structures and processes within an organism to protect it against disease. The earliest literary reference to immunology goes back to 430 b.c., courtesy Thucydides [1]. In 1798, Edward Jenner found some milkmaids immune to smallpox because earlier they contacted cowpox (a mild disease). The next major advancement in immunology came with the induction of immunity to cholera by Louis Pasteur. After applying weakened pathogen to animals, he administered a dose of vaccine to a rabid dog-bitten boy who later survived. But Pasteur could not explain its mechanism.

Rajat K. De and Namrata Tomar (eds.), *Immunoinformatics*, Methods in Molecular Biology, vol. 1184,
DOI 10.1007/978-1-4939-1115-8_3, © Springer Science+Business Media New York 2014

In 1890, experiments of Emil Von Behring and Shibasaburo Kitasato led to the understanding of the mechanism of immunity. Their experiments described that antibodies present in the serum provided protection against pathogens [1].

According to the traditional dogma of immunology, vertebrates have both innate and adaptive immunology. Innate immune system acts more rapidly and is older and more evolutionarily conserved in comparison with adaptive immune system. It provides the backbone on which adaptive immune system was able to evolve. Innate immune system is less specific and works as a first line of defense [2]. It comprises four types of defensive barriers, viz., anatomic (e.g., skin and mucous membranes), physiologic (e.g., temperature, low pH), phagocytic (e.g., blood monocytes, neutrophils, tissue macrophages), and inflammatory (e.g., serum proteins). Adaptive immune responses in vertebrates are generated within 5 or 6 days after the initial exposure to the pathogen. It is coordinated by a network of highly specialized cells that communicate through cell surface molecular interactions and a complex set of intercellular communication molecules known as cytokines and chemokines. Later exposure to the same pathogen induces a heightened and more specific response because it retains memory [1]. Adaptive immune system has two parts: the cellular immune response of T cells and humoral response of B cells [1, 3]. An antigen has a specific small part, known as epitope that is recognized by the corresponding receptor present on B or T cells. B cell epitopes can be linear and discontinuous amino acids. T cell epitopes are short linear peptides. Most of the T cells can be in either of the two subsets, distinguished by the presence of one or the other of the two glycoproteins on their surface, designated as CD8 or CD4. CD4 T cells function as T helper (Th) cells that recognize peptides displayed by MHC class II molecules. On the other hand, CD8 functions as Tc (cytotoxic T) cells which recognize peptides displayed by MHC class I molecules.

The complexity of the immune system arises from its hierarchical and combinatorial properties. Thus huge amount of data related to immune systems is being generated. Immunologic research needs to deal with this complexity. Immunologists have been using high-throughput experimental techniques for quite a long time, which have generated a vast amount of functional, clinical, and epidemiological data. Therefore the development of new computational approaches to store and analyze these data is needed. This gives rise to the field called immunoinformatics. Immunogenomics, immunoproteomics, epitope prediction, and in silico vaccination are different areas of computational immunological research. Recently, systems biology approaches are being applied to investigate the properties of dynamic behavior of an immune system network [4].

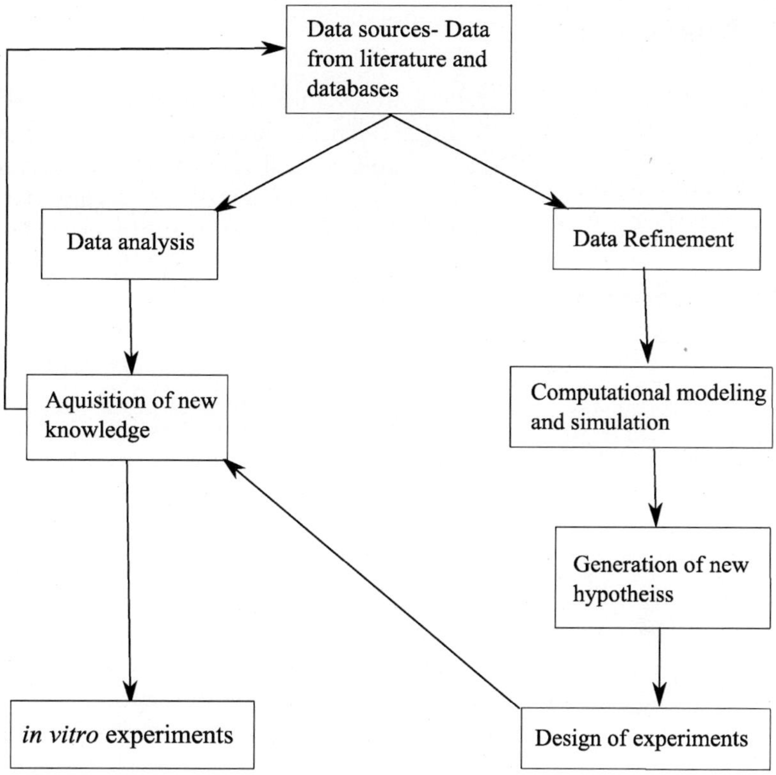

Fig. 1 A possible work flow in immunomics

It includes the study and design of algorithms for mapping potential B and T cell epitopes. These also can lead to exploring the potential binding sites for the development of new vaccines. This methodology is termed as "reverse vaccinology" [5]. It is quite advantageous because conventional methods need to cultivate pathogen and then to extract its antigenic proteins.

All the genes and proteins taking part in immune responses are referred to as "immunome," and it excludes genes and proteins that are expressed in cell types other than in immune cells [6]. All immune reactions due to interaction between host and antigenic peptides are referred to as "immunome reactions," and their study is called as "immunomics" [7]. Like genomics and proteomics, immunomics is a new discipline, which uses high-throughput techniques to understand immune system mechanism [8, 9]. Figure 1 shows work flow in immunomics. This chapter describes various available information regarding classical immunology, different immunomic databases, B and T cell epitope prediction tools and software, and applications of immunoinformatics.

2 Data Sources

Data sources include availability of data from lab experiments through scientific literature, molecular databases, tools and web servers, and clinical records. In this section, we focus on various immune system-related data types and databases. The section starts with some experimental techniques and results.

2.1 Data from Lab Experiments

Immunological experimental and high-throughput molecular biology techniques help in finding the structure and function of immune genes and their products and thereby accumulating a vast amount of experimental data. Experiments involve many immunological techniques to understand the mechanism of an immune system and its responses to various infections, diseases, and drugs, viz., affinity chromatography [10], flow cytometry [11], radioimmunoassay (RIA) [12], enzyme-linked immunosorbent assay (ELISA), [12, 13], competitive inhibition assay [14], and Coombs test [15]. Here, we present some experimental findings that help to identify B and T cell epitopes and to study immune responses.

The ability to identify epitopes in the immune response has important implications in diagnosis of diseases. Thus epitopes for B and T cells need to be identified and mapped. In this context, Wanga et al. [16] mapped B cell epitope present on nonstructural protein (NS1), viz., NS1-18 and NS1-19, in Japanese encephalitis virus. For epitope mapping, a series of 51 partially overlapping fragments covering the entire NS1 protein were expressed with a glutathione *S*-transferase (GST) tag and then screened by a monoclonal antibody (mAb). They found that the motif of (146) EHARW (150) was the minimal unit of the linear epitope recognized by that mAb. Purification techniques like affinity chromatography are used to purify MHC–peptide from membrane MHC molecules, which can be analyzed by capillary high-pressure liquid chromatography electrospray ionization-tandem mass spectrometry [17]. They can be further used to find new tumor-associated antigens (TAA). One such approach to find TAA is based on transfection of expression library made from cDNA into cells expressing the desired MHC haplotypes [18]. The clones are selected on the basis of their ability to provoke immune response in T cells of the individuals with the same MHC type.

2.2 Exploring the Microarray Technology for Immunomics

"Immunomic microarray" is a microarray technique based on the principle of binding and measurement of target biological specimens to complementary probes. It helps in selecting proteins that cause autoimmunity from genomic sequences [19]. It is being applied to autoimmune disease diagnosis and treatment [20], allergy prediction [21], T and B cell epitope mapping [22], and vaccination [23] to name a few. It includes dissociable antibody microarray [24], serum microarray [25], and serological analysis of cDNA expression library (SEREX) [26]. An antibody microarray is used to measure concentration of antigen for a specific antibody probe and thereby consists of antibody probes and antigen targets. On the contrast, peptide microarray uses antigen peptides as fixed probes and serum antibodies as targets. The recent technology is peptide–MHC microarray or artificial antigen-presenting chip. In this technique, recombinant peptide–MHC complexes and co-stimulatory molecules are immobilized on a surface, and population of T cells is incubated with the microarray. The T cell spots act as artificial antigen-presenting cells [27] containing a defined MHC-restricted peptides. The advantage of using peptide–MHC is that it can map MHC-restricted T cell epitope.

The immunomic and genomic microarray data have some similarities, yet both of them also differ in several ways; for example, both of them have different designs. One can measure two or more signals simultaneously determined by a single feature, i.e., epitope in immunomic microarray [28, 29]. DNA microarrays measure one response value for each gene per sample; that is, mRNA concentration produced by the gene but a single epitope can generate different response values corresponding to different epitopes in peptide–MHC chips. In case of B cell epitope, it can be recognized by different isotypes of immunoglobulins, so here, one can measure both intensity and quality of antibody response.

2.3 Immunomic Databases

The property of an antigen to bind specifically complementary antibodies is known as the antigen's antigenicity. Likewise, the ability of an antigen to induce an immune response is called its immunogenicity. Immunomic databases include epitope information-related databases, analysis tools, and prediction algorithms, which are crucial for basic immunological studies, diagnosis, and treatment of various diseases and in vaccine research [30]. InnateDB [31] (http://www.innatedb.ca) has been created to understand complete network of pathways and interactions of innate immune system responses. It has ~18,000 annotated molecular interactions of relevance to innate immunity and >1,200 genes, involved in innate immunity according to the recent update till February 16, 2012. It has a newer version, called Cerebral [32], which is a Java plug-in for the cytoscape biomolecular interaction viewer version 2.8.2 [33] for automatically generating layouts of biological pathways. Table 1 lists some of the databases that deal with information related to B cell epitopes, T cell epitopes, allergy prediction, and evolution of immune system genes and proteins.

2.4 B Cell Epitope Databases

A brief detail on B cell epitope databases is provided here. Readers can find a detailed description in later chapters. Mapping B cell epitopes plays an important role in vaccine design, immunodiagnostic tests, and antibody production. It has been found that 90 %

Table 1
Databases on B cell epitopes, T cell epitopes, allergen, and molecular evolution of immune system components

Databases	Names	URLs
B cell epitopes	CED	http://www.immunet.cn/ced/log.html
	Bcipep	http://www.imtech.res.in/raghava/bcipep
	Epiotme	http://www.rostlab.org/services/epitome/
	IEDB	http://www.immuneepitope.org/
	IMGT®	http://www.imgt.org
T cell epitopes	Syfpeithi	http://www.syfpeithi.de
	IEDB	http://www.immuneepitope.org/
	IMGT®	http://www.imgt.org
Allergen	Database of IUIS	http://www.allergen.org
	SDAP	http://www.fermi.utmb.edu/SDAP/
Information related to molecular evolution of immune system components	ImmTree	http://www.bioinf.uta.fi/ImmTree
	Immunome database	http://www.bioinf.uta.fi/Immunome/
	ImmunomeBase	http://www.bioinf.uta.fi/ImmunomeBase
	Immunome Knowledge Base	http://www.bioinf.uta.fi/IKB/

of B cell epitopes are conformational or discontinuous; however, they may comprise linear amino acid chain of peptides, which is brought closure in 3D space [34]. Bcipep [35] (http://www.imtech.res.in/raghava/bcipep) gives comprehensive information about experimentally verified B cell epitopes and tools for mapping these epitopes on an antigen sequence. Conformational epitope database (CED) [36] has a collection of B cell epitopes from the literature, conformational epitopes defined by methods, like X-ray diffraction, NMR, scanning mutagenesis, overlapping peptides, and phage display. Epitome [37] (http://www.rostlab.org/services/epitome/) contains all known antigen–antibody complex structures. A semiautomated tool has also been developed which identifies the antigenic interactions within the known antigen–antibody complex structures. They compiled these interactions into Epitome. None of the other databases till now explicitly can locate the complementary determining regions (CDRs) or identify the antigenic residues semiautomatically. Epitome update follows update of SCOP, that is, Epitome is updated twice a year as soon as SCOP gets updated.

The difference between Epitome and CED lies in the source of collection of B cell epitopes. Epitome collects B cell epitopes only from PDB structures and includes CDR information. In contrast, CED takes data from the literature and from abovementioned methods. As their sources are different, one can use the complementary information.

2.5 T Cell Epitope Databases

A brief detail on T cell epitope databases is provided here. A detailed description can be found in later chapters. A functional T cell response requires MHC–peptide binding and a proper interaction of the MHC–peptide ligand with a specific T cell receptor. We need well-characterized data to model the process of binding of peptides to TAP and MHCs which function as T cell epitopes. Some recent investigations include finding and mapping of potential epitopes. Epitope mapping leads to designing effective vaccines. Syfpeithi database [38] (http://www.syfpeithi.de) has information on MHC class I and II anchor motifs and binding specificity. It calculates a score based on the following rules—calculated score values differentiate among anchor, auxiliary anchor, or preferred residues. IEDB [39] has more than 88382 peptidic epitopes and can be found at http://www.immuneepitope.org/ and ontology-related information (http://ontology.iedb.org/) which has been specifically designed to capture intrinsic, chemical, and biochemical information on immune epitopes and their interactions with molecules of the host immune system. A beta version of IEDB (Immune Epitope Database and Analysis Resource Database) (http://www.immuneepitope.org/) [30], sponsored by the National Institute for Allergy and Infectious Diseases (http://www.niaid.nih.gov) (NIAID), has different tools to find B and T cell epitopes. It had 88382 peptidic epitopes till February 2012. FRED [40] deals with the methods for data processing and to compare the performance of the prediction methods considering experimental values. IMGT® [41] (the international ImMuno GeneTics information system®) (http://www.imgt.org) has a good collection of IG, TR, MHC, and related proteins of the immune system of human and other vertebrates. It has five databases and 15 interactive online tools for sequence, genome, and 3D structure analysis. The IMGT/HLA Database [42] (http://www.ebi.ac.uk/imgt/hla/) provides a specialist database that has 5,518 HLA class I alleles and 1,612 HLA class II alleles. It is a part of the international ImMunoGeneTics project (IMGT).

2.6 Allergy Prediction Databases

Allergy is a steadily increasing health problem for all age groups caused by allergens. Allergens are proteins or glycoproteins recognized by IgE that is produced by the immune system in allergic individuals. Online allergen databases and allergy prediction tools are being used to find cross-reactivity between known allergens. Localization of B and T cells in the allergen may not coincide [43]. The differences between both kinds of epitopes present in an antigen are as follows: T cell epitopes are only linear (as mentioned earlier) and distributed throughout the primary structure of the allergen, whereas B cell epitopes can be either linear or conformational, recognized by IgE antibodies, and are located on the surface of the molecule accessible to antibodies. Moreover, in the case of B cell epitopes, predicting allergenicity in a molecule based on known conformational epitopes is a difficult task.

Here, we describe allergen prediction databases in brief. One may get details on allergy prediction databases in a later chapter. Allergen Nomenclature database of the International Union of Immunological Societies (IUIS) has allergen database [44] (http://www.allergen.org), which has been last updated in October 2009. AllergenPro database [45] contains 2,434 allergen-related information, e.g., allergens in rice microbes (712 records), animals (617 records), and plants (1,105 records). The web server Allergome 4.0 [46] (www.allergome.org) provides an exhaustive repository of IgE-binding compound data. It has a total of 1,736 allergen sources (updated in March 2010). The real-time monitoring of IgE sensitization module (ReTiME), in Allergome 4.0, enables one to upload raw data from both in vivo and in vitro experiments. This is the first attempt where IT has been applied to allergy data mining. SDAP [47] (Structural database of Allergenic Proteins) (http://fermi.utmb.edu/SDAP/) is a web server that provides cross-referenced access to the sequence and structure of IgE epitope of allergenic proteins. Its algorithm is based on conserved properties of amino acid side chains. In its latest update, it has 1,478 allergens and isoallergens.

3 Immunomic Tools and Algorithms

The property of an antigen to bind specifically complementary antibodies is known as the antigen's antigenicity; likewise, the ability of an antigen to induce an immune response is called its immunogenicity. The main objective of epitope prediction is to design a molecule that can replace an antigen in the process of either antibody production or antibody detection. Such a molecule can be synthesized or, in case of a protein, its gene can be cloned into an expression vector. Designed molecules are inexpensive and noninfectious in contrast to viruses or bacteria. Epitopes are important for understanding the disease mechanism, host--pathogen interaction analyses, antimicrobial target discovery, and vaccine design. Traditionally, determination of binding affinity of MHC molecules and antigenic peptides predicts epitopes. The experimental techniques are found to be difficult and time consuming. Due to this reason, several in silico methodologies are being developed and used to identify epitopes. Here, we throw some light on available immunology-related tools and algorithms. These techniques include matrix-driven methods, finding structural binding motifs, quantitative structure–activity relationship (QSAR) analysis, homology modeling, protein threading, docking techniques, and design of several machine-learning algorithms and tools. Table 2 lists some of the tools that deal with B and T cell epitope prediction, allergy prediction, and in silico vaccination. However, detailed description and discussion over the usages of them will be provided in next chapters.

Table 2
Web servers and tools for prediction of B and T cell epitopes, allergens, and in silico vaccination

Web servers and tools	Names	URLs
B cell epitope prediction	ABCpred	http://www.imtech.res.in/raghava/abcpred
	COBEpro	http://www.scartch.proteomics.uci.edu
	Bepipred	http://www.cbs.dtu.dk/services/BepiPred
	IMGT®	http://www.imgt.org
	Bcepred	http://www.imtech.res.in/raghava/bcepred/
	DiscoTope	http://www.cbs.dtu.dk/services/DiscoTope/
	CEP	http://www.115.111.37.205/cgi-bin/cep.pl
	AgAbDb	http://www.115.111.37.206:8080/agabdb2/home.jsp
	MIMOP	Request from franck.molina@cpbs.univ-montp1.fr
	MIMOX	http://www.immunet.cn/mimox/
	Pepitope	http://www.pepitope.tau.ac.il/
	3DEX	http://www.schreiber-abc.com/3dex/
	IEDB	http://www.immuneepitope.org
T cell epitope prediction	MMBPred	http://www.imtech.res.in/raghava/mmbpred/
	NetCTL	http://www.cbs.dtu.dk/services/NetCTL/
	NetMHC 3.0	http://www.cbs.dtu.dk/services/NetMHC/
	TAPPred	http://www.imtech.res.in/raghava/tappred/
	Pcleavage	http://www.imtech.res.in/raghava/pcleavage/
	ElliPro	http://www.tools.immuneepitope.org/tools/ElliPro
	MHCPred	http://www.ddg-pharmfac.net/mhcpred/MHCPred/
	Propred	http://www.imtech.res.in/raghava/propred1/
	EpiToolKit	http://www.epitoolkit.org
	Syfpeithi	http://www.syfpeithi.de
	IMGT®	http://www.imgt.org
	IEDB	http://www.immuneepitope.org/
	EpiJen v 1.0	http://www.ddg-harmfac.net/epijen/EpiJen/EpiJen.htm
Allergy prediction	AlgPred	http://www.imtech.res.in/raghava/algpred
	Allermatch	http://www.allermatch.org
	APPEL	http://www.jing.cz3.nus.edu.sg/cgi-bin/APPEL
	EVALLER	http://www.slv.se/en-gb/Group1/ Food-Safety/e-Testing-of-protein-allergenicity/
In silico vaccination	VaxiJen	http://www.ddg-pharmfac.net/vaxijen/
	DyNAVacs	http://www.miracle.igib.res.in/dynavac/
	NERVE	http://www.bio.unipd.it/molbinfo
	VIOLIN	http://www.violinet.org
	Vaxign	http://www.violinet.org/vaxign/

3.1 B Cell Epitope Prediction

Experimental determination of B cell epitopes is time consuming and expensive; there is a need for computational methods for reliable identification of putative B cell epitopes from antigenic sequences. B cell epitopes are antigenic determinants on the surface of pathogens that interact with B cell receptors (BCRs). BCR-binding site is hydrophobic, having six hypervariable loops of variable length and amino acid composition. B cell epitopes are classified as continuous/linear/sequential and discontinuous/conformational [48]. Linear epitopes are short peptides that correspond to a contiguous amino acid sequence fragment of a protein. However, most epitopes are discontinuous, where distant residues are brought into spatial proximity by protein folding within the folded 3D protein structure. Experiments are mostly based on linear epitopes. There are both sequence-based and structure-based prediction tools, but prediction tools are limited for discontinuous B cell epitopes [35, 49].

3.1.1 Prediction of Methodology for Continuous B Cell Epitopes

Methodologies for prediction of continuous B cell epitopes involve sequence-based methods, amino acid propensity scale-based methods, and machine-learning methods.

Sequence-Based Methods

Sequence-based methods generally look for the epitope surface that must be accessible for antibody binding. These methods are limited to the prediction of continuous epitopes. Sequence-based methods have been tested on prediction of two protective epitopes known in influenza A virus hemagglutinin HA1 [50]. The first continuous epitope is the 91–108 epitope (SKAFSNCYPYDVPDYASL), which is a protective epitope in rabbit able to elicit antibodies neutralizing infectivity of influenza viruses [51]. The second continuous epitope is the 127–133 epitope (WTGVTQN) protective against the influenza strain A/Achi/2/68 (H3N2) in mouse [52].

Amino Acid Propensity Scale-Based Methods

Parameters such as hydrophilicity, flexibility, accessibility, turns, exposed surface, polarity, and antigenic propensity of polypeptide chains have been correlated with the location of continuous epitopes. Thus the classical methods of identifying potential linear B cell epitopes from antigenic sequences typically rely on the use of amino acid propensity scales. Amino acid scale-based methods apply amino acid scales to compute the scores of a residue i in a given protein sequence. The i–$(n-1)/2$ neighboring residues on each side of residue i are used to compute the score for residue i in a window of size n. The final score for residue i is the average of the scale values for n amino acids in the window. Pellequer [53] compared several propensity scale methods using a dataset of 14 epitope-annotated proteins. He found that the scales of Parker et al. [54], Chou and Fasman, [55], Levitt [56], and Emini [57] provide better results than the other scales tested [48]. El-Manzalawy et al. [58] compared propensity scale-based methods with a naive Bayes classifier and used two datasets: one is propensity dataset, and the other is from BciPep [35].

Bepitope tool [59] predicts continuous epitopes based on the prediction of protein turns. It is a newer version of PREDITOP [60] and uses more than 30 propensity scale values. Bcepred server [61] (http://www.imtech.res.in/raghava/bcepred/) predicts linear B cell epitopes with 58.7 % accuracy based on combined amino acid properties, like accessibility, hydrophilicity, flexibility, polarity, exposed surface, and turns. Analyses of antigen–antibody interaction are done on antibody-binding sites on proteins, which help in predicting the linear and conformational B cell epitopes. Taking this into consideration, a database, viz., AgAbDb [62] (http://202.41.70.51:8080/agabdb2/), has been developed which is based on molecular interactions of antigen–antibody cocrystal structures.

Machine-Learning Methods

Machine-learning algorithms and tools are being used to retrieve characteristics of an epitope. Here we describe some of these approaches in brief. Saha and Raghava [63] used feed-forward and recurrent neural networks to predict continuous B cell epitopes in ABCpred (http://www.imtech.res.in/raghava/abcpred). COBEpro [64] is a two-step system for prediction of continuous B cell epitopes. In the first step, COBEpro assigns a fragment epitopic propensity score to protein sequence fragment using SVM. In the second step, it calculates an epitopic propensity score for each residue based on the SVM scores of the peptide fragment in the antigenic sequence. For Bepipred [65], (http://www.cbs.dtu.dk/services/BepiPred), three datasets of linear B cell epitopes were constructed, viz., annotated proteins from literature, AntiJen database [66] (http://www.ddg-pharmfac.net/antijen/AntiJen/antijenhomepage.htm), and Los Alamos HIV database (http://www.hiv.lanl.gov). They tested a number of propensity scale methods on Pellequer dataset [53] and found the best scale by Levitt [56]. Then, they used HMM to predict the location of linear B cell epitopes and tested HMMs on Pellequer dataset to find optimal parameters. HMM was combined with one set of the two best propensity scale methods, i.e., Parker [54] and Levitt [56], to get the more accurate predictions. Currently, ~60–66 % of accuracy has been found for continuous epitope prediction, applying combinations of either amino acid scales or machine-learning techniques. The higher accuracy could possibly be achieved by improving the quality of existing B cell epitope datasets [48].

3.1.2 Prediction Methodology for Discontinuous B Cell Epitopes

The characterization and prediction of B cell epitopes are mainly conformational dependent based on the knowledge of the protein three-dimensional structure; thus the task of prediction is more difficult compared to that of T cell epitopes. Changes in protein folding may lead to changes in the number of epitopes [43]. The most accurate way to identify B cell epitope is through X-ray crystallography. Here we describe some of the prediction methods for conformational B cell epitopes in brief. Anderson et al. presented a method called DiscoTope [67] (http://www.cbs.dtu.dk/services/DiscoTope/), which is a combination of amino acid statistics, spatial information, and surface exposure. It detects 15.5 % of residues located in discontinuous epitopes with a specificity of 95 %. It is said to be the first method developed for prediction of discontinuous B cell epitope with better performance than methods based only on sequence data. PEPITO [68] uses a weighted linear combination of amino acid propensity scores and half-sphere exposure values [69] which encode side chain orientation and solvent accessibility of amino acid residues for the prediction of conformational epitopes. Authors have also reported its improvement in performance over DiscoTope method.

Bublil et al. developed Mapitope [70] for conformational B cell epitope mapping. The hypothesis behind Mapitope is that the simplest meaningful fragment of an epitope is an amino acid pair (AAP) of residues that lie within the epitope, which are the results of folding. A set of affinity-isolated peptides was obtained by screening the phage display peptide libraries with the antibody of interest. This set was given as algorithm input, and 1–3 epitope candidates on the surface of the atomic structure of the antigens were obtained as output.

A computational method has been presented by Sollner et al. [71] to automatically select and rank peptides for the stimulation of otherwise functionally altered antibodies. They investigated the integration of B cell epitope prediction with the variability of antigen and the conservation of patterns for posttranslational modification (PTM) prediction. By their observation, they found high antigenicity, low variability, and low likelihood of PTM for the identification of biorelevant sites. Ponomarenko [48] assembled non-redundant datasets of repetitive 3D structure of antigen and antigen–antibody complexes from the PDB. CEP web interface [72] (http://www.115.111.37.205/cgi-bin/cep.pl) predicts conformational and sequential epitopes and also antigenic determinants. Less availability of the 3D structure data of protein antigens limits the utility of this server. A recent approach has focused on the impact of interior residues, different contributions of adjacent residues, and imbalanced data which contain much more non-epitope residues than epitope and applied random forest (RF) algorithm for the prediction of conformational B cell epitope prediction [73]. This tool is available at http://www.code.google.com/p/my-project-bpredictor/downloads/list.

Mimotope-Based Methodology

Phage display library is widely used for finding protein–protein interactions (specially in antibody–antigen interactions), protein function identification, and development of new drugs and vaccines [74]. Pizzi et al. [75] have proposed an approach for mapping B cell epitopes, in which a phage display library of random peptides is scanned against a desired antibody to obtain mimotopes that bind to the antibody with high affinity. It is assumed that this panel of mimotopes mimics the physicochemical properties and spatial organization of the genuine epitopes [34, 74 and 76]. Mimotopes and antigens are both recognized by the same antibody paratope. Mimotopes are said to be the imitated part of the epitope. It is possible that mimotope may have some valuable information about epitope. However, homology may not exist between the mimotope and the epitope of the native antigen. This mimicry exists due to similarities in physiochemical properties and spatial organization [76]. Considering these properties, mimotope pools are being used to mine information to predict an epitope.

Using the above concept, MIMOP tool [76] has been developed. MIMOP predicts linear and conformational epitopes based on two algorithms, viz., MimAlign uses degenerated alignment analyses, and MimCons is based on consensus identification. MIMOX [77] (http://web.kuicr.kyoto-u.ac.jp/~hjian/mimox) comes in the same category, which maps a single mimotope or a consensus sequence of a set of mimotopes onto the corresponding antigen structure. Then, it searches for all of the clusters of residues that could be the native epitope. Pepitope [74] (http://pepitope.tau.ac.il/) (an advanced server for mimotope-based epitope prediction approaches) uses two algorithms, viz., Pepsurf [78] and Mapitope [70]. It maps each mimotope to map them onto the solved structure of antigen surface. Alignment of mimotope is done first in MIMOX, so this step is different in Pepitope. If we compare it with MIMOP, MIMOP aligns the peptides to the antigen at the sequence level rather than directly to the 3D structure. The 3D structure is considered only following the alignment stage. Given the 3D structure of an antigen and a set of mimotopes (or a motif sequence derived from the set of mimotopes), Pep-3D-Search [79] (http://kyc.nenu.edu.cn/Pep3DSearch/) can be used in two modes: mimotope or motif. It can be used for localizing the surface region mimicked by the mimotopes.

Sometimes linear peptides mimic conformational epitopes. The same phage display peptide libraries by screening with the respective antibodies are used to select these mimotopes. Schreiber et al. [80] presented software, 3DEX (3D-Epitope-Explorer) (http://www.schreiber-abc.com/3dex/), that allows localizing linear peptide sequences within 3D structures of proteins. Its algorithm takes into account the physiochemical neighborhood of C-α or C-β atoms of individual amino acids and surface exposure of the amino acids. Authors were able to localize mimotopes from HIV-positive patient plasma within 3D structure of gp120.

Hybrid (Ensemble) Prediction Method

Ensemble methods combine the predictions of several predictors and often outperform individual predictors in many biomolecular sequence and structure classification studies [81]. Several strategies for combining a set of predictors, *S*, into a single consensus or meta-predictor exist: (1) majority voting, (2) weighted linear combination, and (3) meta-learning [82]. A large number of nearest neighbor- and decision tree-based classifiers are trained using different sets of training data features for developing an ensemble of linear B cell epitope classifiers [83].

3.2 T Cell Epitope Prediction

The current challenge in immunological prediction software is to predict interacting molecules to a high degree of accuracy. The most popular methods currently available are based on binding affinity predictions for a range of MHC molecules. It is necessary to bind antigenic peptides with MHC so that cytotoxic T cells can recognize them. Thus, identification of MHC-binding peptides is a central part of any algorithm which predicts T cell epitopes. There exist several methodologies for prediction of MHC-binding peptides, which are based on the idea of quantitative matrices, hidden Markov model (HMM), artificial neural networks (ANNs), support vector machine (SVM), and structure of the peptides. Here we describe the abovementioned approaches in brief. One may find the details of these methodologies, among others, in later chapters.

3.2.1 Matrix-Driven Methods

Huang and Dai [84] first investigated a new encoding scheme of peptides based on BLOSUM matrix with the amino acid indicator vectors for direct prediction of T cell epitopes. It replaced each nonzero entry in the amino acid indicator vector by the corresponding value appeared in the diagonal entries in BLOSUM matrix. MMBPred [85] (http://www.imtech.res.in/raghava/mmbpred/) server predicts the mutated promiscuous and high-affinity MHC-binding peptide. It uses the matrix data in a linear prediction model and ignores peptide conformation. The prediction is based on the quantitative matrices of 47 MHC alleles.

3.2.2 Hidden Markov Model-Based Method

Transfer-associated protein (TAP) is an important component of the MHC I antigen processing and presentation pathway. A TAP transporter can translocate peptides of 8–40 amino acids into endoplasmic reticulum (ER). Zhang et al. developed $PRED^{TAP}$ [86] for the prediction of peptide binding to hTAP. They used a three-layer back propagation network with the sigmoid activation function. The inputs were the binary strings, representing nonamer peptide. Secondly, they used second-order HMM. The results are both sensitive and specific.

3.2.3 Artificial Neural Network-Based Method

ANNs can identify each amino acid residue and interactions between adjacent ones in a potential epitope. An ANN for a particular MHC molecule is trained to recognize associated input sequence and outputs, viz., the binding affinity for that sequence with the MHC molecule [87]. Trained ANN can predict the binding affinity of novel peptide sequences. Neilson et al. [88] described an improved neural network model to predict T cell class I epitopes.

They combined a sparse encoding, BLOSUM encoding, and input derived from HMM. The dataset consists of 528 nine-mer amino acid peptides for which the binding affinity to the HLA I molecule A*0204 has been measured in a method described by Buus et al. [89]. NetCTL server [90] (http://www.cbs.dtu.dk/services/NetCTL/) has method to integrate the prediction of peptide MHC class I binding, proteasomal C terminal cleavage, and TAP transport efficiency. NetMHC server 3.0 [91] (http://www.cbs.dtu.dk/services/NetMHC/) uses ANN and weight matrices. It has been trained on data from 55 MHC peptides (43 human and 12 nonhuman) and position-specific scoring matrices (PSSMs) for additional 67 HLA alleles.

Prediction of MHC class II binding peptides is found to be difficult due to the reasons including variable length of reported binding peptides, undetermined core region for each peptide, and number of amino acids as primary anchor. Brusic et al. developed PERUN [92], a hybrid method for the prediction of MHC class II binding peptide. It uses available experimental data and expert knowledge of binding motifs, evolutionary algorithms, and ANNs. They used PlaNet package version 5.6 [93] to design and train a three-layered fully connected feed-forward ANN. The whole process of MHC class I ligands' degradation and presentation has been modeled in EpiJen [94] (http://www.ddg-pharmfac.net/epijen/EpiJen/EpiJen.htm) in an integrative approach. It is a multi-step algorithm for T cell epitope prediction, based on quantitative matrices, which belongs to the next generation of in silico T cell epitope identification methods.

3.2.4 Other Machine-Learning Methods

Ant colony search systems (ACSs) have been found useful for solving combinatorial optimization problems and can be applied to the identification of a multiple alignment of a set of peptides. Basically, ACSs [95] attempt to find an optimal alignment for a given set of peptides based on the search strategy. For TAPPred [96] (http://www.imtech.res.in/raghava/tappred/), nine features of amino acids have been analyzed to find the correlation between binding affinity and physiochemical properties. An SVM-based method to predict TAP binding affinity of peptides has been developed and found cascade SVM to be more reliable. Cascade SVM has two layers of SVMs, and its performance is better than the other available algorithms. Nanni [97] demonstrated the use of SVM and support vector (SV) data description to predict T cell epitope. It is experimentally established that the immunoproteasome is involved in the generation of the MHC class I ligand. For this purpose, Pcleavage [98] (http://www.imtech.res.in/raghava/pcleavage/) has been developed to predict both kinds of cleavage sites in antigenic proteins. It uses SVM [99], Parallel Exemplar based Learning (PEBLS) [100], and Waikato Environment for Knowledge Analysis (Weka) [101].

3.2.5 Structure-Based Prediction

Accurate identification of peptides that bind to specific MHC molecules is important for understanding the underlying mechanism of immune recognition, for developing effective peptide-based vaccines, and for immunotherapies for allergy and autoimmunity. Current methods are mostly based on peptide binding affinity to MHC for predicting T cell epitope. 3D QSAR technology CoMSIA has been applied to the problem of peptide–MHC binding [102]. It uses the interaction potential around aligned sets of 3D peptide structures to describe binding. TEPITOPE [103] is used to predict promiscuous and allele-specific HLA II-restricted T cell epitope in silico. TEPITOPE's user interface has a display and comparison of pocket profiles, and it finds similar HLA II differing in their binding capacity for a given peptide sequence. It can be applied to only 51 out of over 700 known HLA-DR molecules. A new method called as TEPITOPEpan (http://www.biokdd.fudan.edu.cn/Service/TEPITOPEpan/) is developed by extrapolating from the binding specificities of HLA-DR molecules characterized by TEPITOPE to those uncharacterized [104].

T epitope designer [105], a web server, uses a definition of virtual binding pockets to position specific peptide residue anchors and estimation of peptide residue virtual binding pocket compatibility. Zhao et al. [106] described a novel predictive model using information from 29 human MHCp crystal structures. The overall binding between peptide and MHC provides a cumulative measure of the physical and chemical compatibility between each residue in the peptide and the residue forming the virtual pockets. ElliPro [107] (http://www.tools.immuneepitope.org/tools/ElliPro) is a web tool that implements a modified version of Thornton method, residue clustering algorithm, the Modeller program, and the Jmol viewer. It predicts and visualizes the antibody epitope in protein sequence and structure. It implements three algorithms for approximation of the protein shape as an ellipsoid, calculation of the residue protrusion index (PI), and clustering of neighboring residue based on their PI values.

It is generally accepted that only peptides that bind to MHC with an affinity above a threshold value (typically 500 nM) function as T cell epitopes. Guan et al. in Edward Jenner Institute for Vaccine Research, UK, introduced MHCPred version 2.0 [108] (http://www.ddg-pharmfac.net/mhcpred/MHCPred/). It is a perl implementation of 2D QSAR application to peptide–MHC prediction and covers both class I and class II MHC allele peptide specificity models. Peptide that can bind to MHC on the tumor cell surface has the potential to initiate a host immune response against the tumor. Schiewe et al. [109] developed an algorithm PeSSI (peptide–MHC prediction of structure through solvated interfaces) for flexible structure prediction of peptide binding to MHC molecule. They used cancer testis (CT) antigens, KU-CT-1, that are potential to bind HLA-A2.

Jojic et al. [110] developed an improved structure-based model which used known 3D structures of a small number of MHC–peptide complexes, MHC class I sequence, known binding energies for MHC–peptide complexes, and larger binary dataset having information about strong binders and non-binders. They used adaptive double threading, where the parameters of the threading model are learnable, and both MHC and peptide sequences can be threaded onto the structure of other alleles. Furman et al. [111] used an approach that can be applied to a wide range of MHC class I alleles. In this algorithm, peptide candidates are threaded, and their binding compatibility is evaluated by statistical pairwise potentials. They used the pairwise potential table of Miyazawa and Jernigan [112].

Immunodominant peptides are being used for rational design of peptide vaccines focusing on T cell immunity. Altuvia et al. [113] focused on antigenic peptides recognized by cytotoxic T cells. They applied the threading approach to screen a library of peptide sequence and identified the ones that optimally fit within the MHC groove. Propred [114] (http://www.imtech.res.in/raghava/proped) is a graphical web tool for predicting MHC class II binding regions in antigenic protein sequences. They extracted the matrices for 51 HLA-DR alleles from a pocket profile database developed by Sturniolo et al. [115]. EpiToolKit [116] (http://www.epitoolkit.org) web server includes several prediction methods for MHC class I and class II ligands and minor histocompatibility antigens. It can also investigate the effect of mutation on T cell epitopes.

3.2.6 Molecular Dynamics-Based Prediction

Molecular dynamics (MD) describes single and collective motion of atoms within a molecular system and provides a means by which one can measure theoretically that cannot be measured experimentally [117]. It is particularly suitable for the simulation and analysis of the otherwise inaccessible details of MHC–peptide interaction and of the immune synapse. Zhang et al. [118] were among the first who uses MD as a tool to explore peptide–MHC binding. They focused on docking using MD as well as on calculating free energies. Free energy calculations of the wild-type and the variant human T cell lymphotropic virus type 1 Tax peptide (LLFGYPVYV—wild Tax and LLFGYAVYV—mutant Tax) presented by the MHC to the TCR have been performed using large-scale massively parallel molecular dynamics simulations [119].

3.3 Allergy Informatics

Allergy is caused by adverse immunological reaction, and the causative agents are known as allergens that are otherwise not harmful in nature. An allergen cross-links immunoglobulin E (IgE) antibodies on mast cells or basophils and releases inflammatory mediators that cause allergy symptoms. Biotechnology- and genetic engineering-derived food contains some foreign proteins, which can be allergic to many human beings. Evaluation of the potential allergenicity of food derived from biotechnology and genetic engineering is a current food safety assessment. Allergen sequence databases are essential tools for safety assessments of bioengineered foods. They can analyze the structural and physiochemical properties of food allergen proteins. Current efforts in allergy informatics are primarily focused on prediction of T and B cell epitopes and assessment of allergenicity.

Allergy occurs by both extrinsic and intrinsic factors. Type I hypersensitive reaction is induced by certain allergens that elicit IgE antibodies [1]. Use of genetically modified food and therapeutics makes allergenic protein prediction necessary. According to the proposed guidelines of World Health Organization (WHO) and Food and Agriculture Organization (FAO) in 2001, a protein that has at least six same contiguous amino acids or a window of 80 amino acids when compared with known allergens is considered as allergen. It has already been established that allergens do not share common structural characteristics. Thus allergen databases are being used as reference for finding the sequence similarity in allergenicity evaluation [120]. It is said that a protein is considered as an allergen if it has a region or peptides identical to a known IgE epitope.

Allergen prediction method proposed by Kong et al. [121] is based on the determination of a combination of two allergen motifs in a given protein sequence. They took 575 proteins for allergen dataset and 700 sequences for non-allergen test set from the given reference [122]. They developed a database which has all possible combinations of two motifs from the set of allergenic motifs by using motif length of 35 amino acids and motif number of 500. Zorzet et al. [123] introduced a computational approach for classifying the amino acid sequences in allergens and non-allergens. They identified preprocessed 91 food allergens from various specialized public repositories of food allergy and SWALL database (SWISSPROT and TrEMBL).

AlgPred [124] (http://www.imtech.res.in/raghava/algpred) uses SVM and a similarity-based approach for analysis and scanned all 183 IgE epitopes against all proteins of the dataset. The server allows using a hybrid option to predict allergen using combined approach (SVMc, IgE epitope, ARPs BLAST, and MAST).

Stadler et al. [120] used MEME motif discovery tool to identify the most relevant motif present in allergen sequence. If the query finds an allergen motif or scores better than an E-value of 10^{-8} in the pairwise sequence alignment step, it is considered as the allergenic sequence. Then, these are compared with the FAO/WHO guidelines by performing allergenicity prediction for the sequence in SWISSPROT, and a synthetic test database ALLERMATCH (http://www.allermatch.org) is a web tool that uses sliding window approach to predict potential allergenicity of proteins [125].

It is done according to the current recommendations of the FAO/WHO Expert Consultation, [126] as outlined in Codex alimentarius [127]; however, this method generates false-positive and false-negative hits, so it is advised by the FAO/WHO that the outcomes should be combined with other allergenicity assessment methods.

APPEL [128] (Allergen Protein Prediction E-Lab) (http://www.jing.cz3.nus.edu.sg/cgi-bin/APPEL) tool uses SVM to identify novel allergen proteins. This tool correctly classified 93 % of 229 allergens and 99.9 % of 6717 non-allergens. It is based on statistical method, and it has the potential to discover novel allergen proteins. EVALLER [129] web server (http://www.slv.se/en-gb/Group1/Food-Safety/e-Testing-of-protein-allergenicity/) uses filtered length-adjusted allergen peptides (DFLAP) method [130] (via ulfh@slv.se) to identify the potential allergen proteins. DFLAP extracts variable length allergen sequence fragments and employs SVM.

EVALLER and APPEL servers assigned all calmodulins or calmodulin-like proteins as presumably non-allergens [128, 129]. But a conventional alignment approach (e.g., 35 % similarity over 80 amino acid segments) gives preference to find sequence similarity between input proteins and known allergens and puts abovementioned proteins in allergen category. These proteins are presumable non-allergenic homologues to the polcalcin family (members being potential allergens involved in pollen–pollen cross-sensitization). Tools, based on structural and physical characteristics, are useful to identify potential cross-reacting proteins that may escape detection through sequence similarity method alone. Details related with allergen prediction approaches may be found in later chapters.

4 Applications of Immunoinformatics

The use of immunological databases and prediction software has become an important part of the scientific research as they allow us to predict the interaction of molecules involved in an immune response, thereby significantly shortening experimental procedure. In this section, we focus on applications of immunoinformatics. It includes in silico vaccine design and immune system modeling and immunoinformatics for cancer diagnosis and therapy. It also explores the idea of integrating systems biology with immunoinformatics.

4.1 Reverse Engineering for Vaccine Design

Vaccines can be live attenuated whole pathogens, subunits, or epitope based. It is possible to design attenuated pathogens by removing virulence factors or reducing their metabolic capacity. These procedures can be done through computational design and discovery. Several in silico techniques have been developed to identify suitable vaccine candidates, principally proteins within pathogen genomes that have antigenic properties. Generally used vaccines are live attenuated or killed bacteria or viruses (examples include cholera, polio, measles). Thus there is a concern about the safety of these vaccines; if they are incompletely attenuated or killed, they may revert their pathogenicity or cause undesirable immune reactions. On the other hand, synthetic peptides are considered as candidates for safe vaccines. Methods predicting immunogenic peptides could lead to rational vaccine design. Genome sequencing, comparative proteomics, and immunoinformatics tools are well developed to design new vaccines. "Reverse vaccinology," a new concept, analyzes the entire genome to identify potentially antigenic extracellular proteins and thus helps in saving time and money. It was pioneered for *Neisseria meningitides* responsible for sepsis and meningococcal meningitides, and the vaccine type is conjugate based on capsular polysaccharide. These vaccines are available for pathogenic *N. meningitides A, C, Y, and W135* [131].

Microarray technique for vaccine design: Through microarray technology, it is easy to screen genes of various pathogens in different growth states and conditions for vaccine design [132]. It reduces the number of genes useful for vaccine in a given genome. Signal peptides derived from genomic sequences, structural motifs, and immunogenicity are important for vaccine development.

Epitope-driven approaches for vaccine design: These are comparatively more useful as they have no lethal effect of the whole-protein vaccines. It may induce immune response against immunodominant epitopes [133]. This kind of vaccine has a single start codon with an epitope which can be inserted consecutively in the construct [134]. The prediction of promiscuous binding ligands is considered to be a prerequisite for the most subunit vaccine design strategies [135]. It is originally named as "reverse immunogenetics" where T cell epitope mapping tools were employed to find new protein candidates for vaccines and diagnostic tests [136]. Epitope-driven vaccine design allows the discovery of previously unknown and undescribed antigens and epitopes as vaccine candidates. The major disadvantage of the epitope-based approach is that algorithms may fail to predict all the relevant epitopes [137]. A web server, PEPVAC (Promiscuous EPitope-based VACcine) (http://immunax.dfci.harvard.edu/PEPVAC/) [138], is optimized for the formulation of multi-epitope vaccines with broad population coverage. This optimization is accomplished through the prediction of peptides that bind to several HLA molecules with similar peptide-binding specificity.

Peptide-based vaccine design: Small peptides derived from epitopes are used as peptide-based vaccines. These peptides are recognized by MHC class I and thus boost the immune response. Three novel classes of methods have been described to predict MHC-binding peptides and a voting scheme to integrate them for improved results [139]. The first method is based on quadratic programming applied to quantitative and qualitative data. Second method uses linear programming, and the third one considers sequence profiles obtained by clustering known epitopes to score candidate peptides. This method is found to be better than other sequence-based methods for finding the MHC binders.

Alignment-free approach for vaccine design: Some proteins have similar structure and biological properties, but they may lack sequence similarity. For these kinds of proteins, a new alignment-free approach for antigen prediction has been proposed, which uses three datasets—each for bacteria, viruses, and tumors [140]. The models were validated using leave-one-out cross-validation (LOO-CV) on the whole sets and by external validation using test sets and were implemented in a server called VaxiJen version 2.0 (http://www.ddg-pharmfac.net/vaxijen/).

DNA vaccines: DNA vaccines produce cell-mediated and humoral immune response and are very useful in defending intracellular pathogens. It uses plasmid DNA, which contains a DNA sequence coding for an antigen and a promoter for gene expression in the mammalian cell. Plasmid DNA does not need a viral vector for delivery. Naked DNA is safe and can be used to sustain the expression of antigen in cells for longer periods of time than RNA or protein vaccines. The DNA delivers antigen as well as activates innate immunity and an adaptive immunity against cancer antigens. DyNAVacs [141] (http://www.miracle.igib.res.in/dynavac/) incorporates different modules like codon optimization for heterologous expression of genes in bacteria, yeast, and plant, mapping restriction enzyme sites, primer design, Kozak sequence insertion, custom sequence insertion, and design of genes for gene therapy.

The crucial question in deciding vaccine protocol is the vaccination schedule, i.e., is to decide whether the chronic protocol is able to give 100 % protection or shorter protocols could be applied. Thus a mathematical model/simulator (SimTriplex) which describes the immune response activated by the triplex vaccine has been developed [142]. Immunological prevention of cancer has been obtained in HER-2/neu transgenic mice using a vaccine that combines three different immune stimuli (triplex vaccine) that is repeatedly administered for the entire life-span of the host (chronic protocol).

The software NERVE [143] (http://www.bio.unipd.it/molbinfo) helps in designing subunit vaccines against bacterial pathogens. It combines automation with an exhaustive treatment of vaccine candidate selection task by implementing and integrating six different kinds of analyses. Xiang et al. developed a web-based database system, VIOLIN [144] (Vaccine Investigation and Online Information Network) (http://www.violinet.org), which curates, stores, and analyzes published vaccine data. It contains four integrated literature mining and search programs, viz.,

Litsearch, Vaxpresso, Vaxmesh, and Vaxlert. They have developed a web-based vaccine design system called Vaxign [145], which predicts possible vaccine targets. Major predicted features include subcellular location of a protein, transmembrane domain, adhesion probability, sequence conservation among genomes, sequence similarity to host (human or mouse) proteome, and epitope binding to MHC class I and class II. However, synthetic vaccine candidates must be tested experimentally to demonstrate their ability to generate neutralizing antibodies.

4.2 Immune System Modeling

The immune system can be seen as a parallel, information processing system that learns through examples, constantly adapts itself to new situations, and possesses a distributive memory for patterns. For theoretical immunology, immune system models and simulations can describe more insights into various interactions resulting in immunological phenomena. These models can test and find out the antigen–antibody interactions and immune responses for a particular antigen, in case of drug administration or testing of a vaccine candidate. Using visual modeling application described by Gong and Cai [146] one can understand the adaptive immune system effectively. The hierarchical immune system consists of inherent immune tier, adaptive immune tier, and immune cell tier. It is designed and visualized with Java Applet technique for simulation. For further simulation purpose, the learning of the antibody is implemented through the evolutionary mechanism of the immune algorithm. ImmunoGrid (http://www.immunogrid.eu) and Virolab (http://www.virolab.org/) projects are working to simulate immune systems. ImmunoGrid tries to simulate immune processes by combining experiments and computational studies, while Virolab attempts to develop a virtual lab for infectious diseases by examining the genetic causes of human illnesses [132]. SIMISYS 0.3 [147] is another example of a software that models and simulates the innate and adaptive components of the immune system based on computational framework of cellular automata. It simulates healthy and disease conditions by interpreting interactions among the cells, including macrophages, dendritic cells, B cells, T helper cells, and pathogenic bacteria.

Exclusive computational approaches like mathematical modeling generate enormous amount of data, but there should be a balance between virtual and real experimental data. Computationally generated data needs to be formally tested and translated into real knowledge. Post-genomic era needs to exchange data from wet lab to simulation and vice versa [148]. The model should be accurate, easy to use, and understandable to both model designers and biologists who can verify their hypothesis through in silico experiments.

4.3 Immunoinformatics for Cancer Diagnosis and Therapy

Antigen presentation plays a central role in the immune response and as a result also in immunotherapeutic methods like antitumor vaccination. There is a need to rapidly screen the antigens and to design specific types of expression constructs for immunotherapy of cancer. Competent immune responses to cancer are likely to be restricted to the immunome of a specific cancer, including the set of antigens that drive successful immune responses. However, it is still difficult to find the set of antigens that varies between different tumors. Antitumor vaccination takes advantage of in vivo processes, and it harnesses the full power of the immune system, unlike the more artificial ex vivo expansion of T cells [149].

Changes in the cancer diagnosis and prevention are being supported by informatics [150]. For example, the Cancer Biomedical Informatics Grid (caBIG) connects a network of 500 individuals and 50 institutions who share data and analyze tools to speed up the development of innovative approaches for the prevention and treatment of cancer [151]. The 2005 database issue of Nucleic Acids Research lists 14 cancer-related molecular databases, which mainly focus on cancer-related genes and gene expression [152]. Listings of tumor antigens are also available [153]. This list includes antigens that have defined T cell epitopes. Tumor-associated antigens (TAA) have played a vital role in both diagnosis and treatment of human carcinomas, such as prostate-specific antigen (PSA) in the diagnosis of prostate cancer. Despite this, the process of TAA identification has often been hampered by the complicated lab procedures. To fasten the process of tumor antigen discovery, and improve diagnosis and treatment of human carcinoma, a publicly available database Human Potential Tumor Associated Antigen (HPtaa) database (http://www.hptaa.org) has been established [154]. Systems biology approaches target identification of a small number of antigens expressed by cancer cells that are suitable targets of immune responses against cancer. A proteomic mapping of in vivo targets for antibodies in lungs, and solid tumors in experimental animals define aminopeptidase-P and annexin A1 as targets of anticancer immune responses [155]. Informatic methods have also been used for classification of tumors into subtypes, which supports decision making for the selection of therapeutic approaches; however, such applications in cancer immunology are yet to come [156].

Vaccine against tumors: Reliable predictions of immunogenic T cell epitope peptides are crucial for rational vaccine design and represent a key problem in immunoinformatics. Computational approaches have been developed to facilitate the process of epitope detection and show potential applications to the immunotherapeutic treatment of cancer. Epitope-driven vaccine design employs these bioinformatics algorithms to identify potential targets of vaccines against cancer [157]. The development of epitope-based

DNA vaccines and their antitumor effects in preclinical research against B cell lymphoma have been described [158].

Most immunotherapeutic approaches work on the induction of antitumor $CD8^+$ T cells, which exhibit cytolytic activity towards tumor cells expressing tumor-specific or tumor-associated Ags. But the immunization strategies that focus solely on $CD8^+$ T cell immunity might prove to be insufficient because they will be unable to provide long-term protective immunity [159]. It has been shown that the peptides predicted to bind MHC can elicit a tumor-killing cytotoxic T lymphocyte (CTL) response [160]. Although CTLs have been found to be the key player in the generation of antitumor therapeutic effects, sometimes they also remain as suboptimal. $CD4^+$ T cells are critical for the generation and maintenance of CTL response through providing cytokines or by major pathway, i.e., dendritic cell licensing [161, 162]. Class II MHC-bound epitopes activate $CD4^+$ T cells and maintain effective CTL response that plays an important role in the antitumor response [163, 164].

$CD4^+$ T cells determine the functional status of both innate and adaptive immune responses; thus, the inclusion of appropriate $CD4^+$ T cell epitopes may be essential for vaccine efficacy. Idiotypic immunoglobulin M (IgM) expressed by B cell lymphoma is a clonal marker and a tumor-specific antigen. Thus, it can be used as an immune target. Specific immunogenic epitopes identified from these tumor antigens can be used as vaccines to activate an immune response against tumor cells [165]. Concerning to lymphoproliferative malignancies, tetanus toxin fragment C (TTFrC)-fusion vaccine design was able to activate anti-Id antibody responses and to suppress tumor growth in murine models [166, 167] as well as was effective in inducing $CD8^+$ CTL in several tumor models [168].

4.4 Immunoinformatics and Systems Biology for Personalized Medicine

The idea to integrate immunoinformatics with systems biology approaches is for the better understanding of immune-related diseases at various systems levels. This integration can open the path of several translational studies for better clinical practices. The association between a disease and genetic variations is one of the most important aspects in pharmacogenomics and development of personalized medicine. Figure 2 shows the integration that leads to the development of personalized medicine (inspired by 169). The information about allele frequencies of immune molecules in a human population is important as different patient subgroups can be identified with different vaccine or drug responses [169]. For example, an SNP (S427T) in the innate immune gene interferon regulatory factor 3 (IRF3) has been associated with increased risk of human papillomavirus (HPV) persistence and cervical cancer [170]. Genomic variation databases such as HapMap (http://www.snp.cshl.org/) and dbSNP (http://www.ncbi.nlm.nih.gov/SNP/) provide information on individual genotype data. The Allele Frequencies Database can be used to search for polymorphic regions of various populations on histocompatibility and immunogenetics

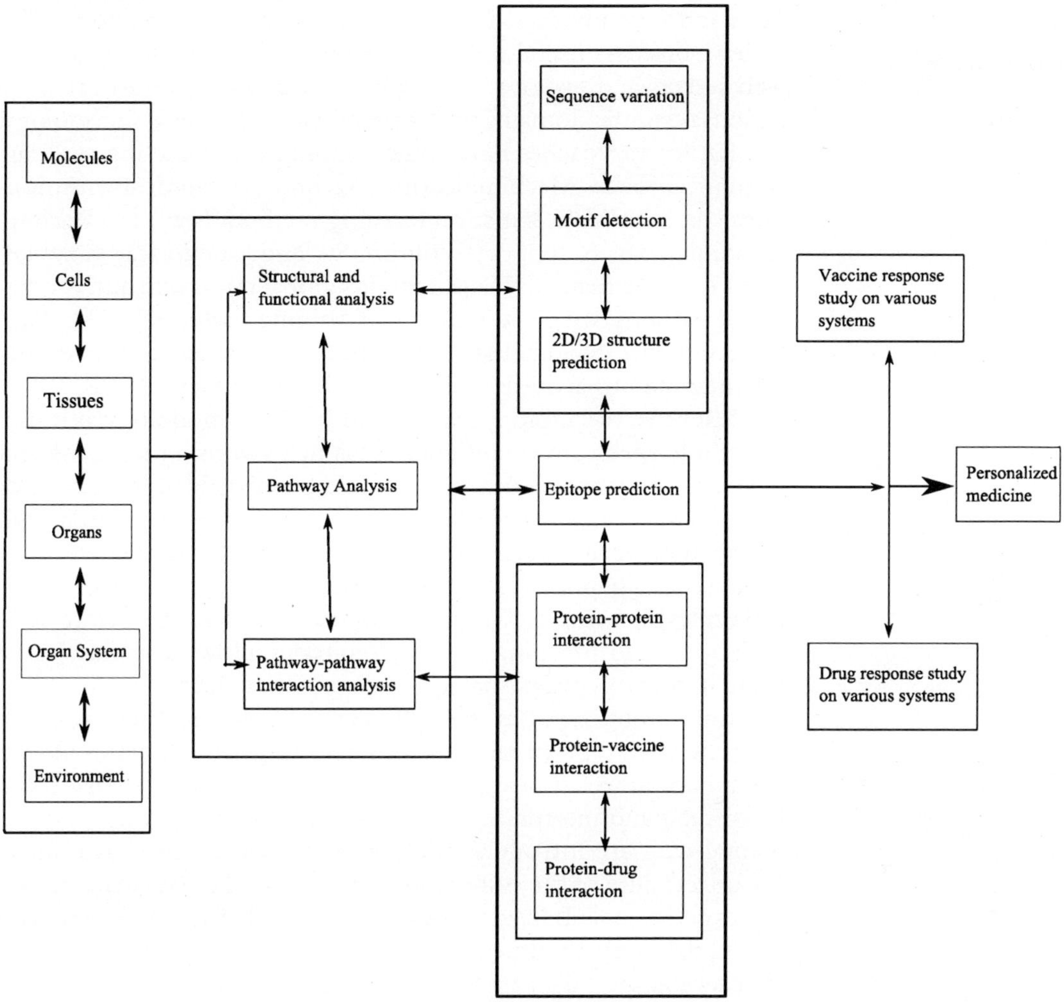

Fig. 2 An integration of immunoinformatics and systems biology, leading to the development of a personalized medicine, inspired by ref. [169]

(http://www.allelefrequencies.net/). This includes polymorphism information on HLA, cytokines, and killer-cell immunoglobulin-like receptors (KIR). Thus, there is a scope for the development of optimized vaccines and drugs tailored to personalized prevention and treatment through the integration of systems biology and immunoinformatics.

5 Conclusions and Discussions

High-throughput experimental techniques are combined with immunoinformatics, resulting in explosive growth of immunology. This is as similar as the event that has transformed genetics into genomics. Immunoinformatics may be placed at the junction point between experimental and computational approaches as it reduces time and cost involved in traditional study of immunology. This review considers useful online immunological databases, tools, and web servers and explores the application of immunoinformatics in various scientific domains with an emphasis on reverse vaccinology.

Earlier approaches have some limitations in handling real data (nonlinear data). Machine-learning techniques can deal with nonlinear data. SVM (a statistical learning methodology) is a learning technique which supports continuous and categorical variables. SVM is better than ANN, as it attains global minimum and is capable of working with less number of training patterns [171]. Thus both sequence characteristics and computational techniques should be integrated to acquire higher prediction accuracy.

"Reverse vaccinology" is a revolution in immunology as it uses the whole spectrum of antigens. This helps in using pools of vaccine candidates which otherwise would be missed (because of poor or no in vitro experimental information or facing problem in culturing the specific pathogen) [171]. Recently, the prediction of promiscuous peptides (capable of binding to a wide array of MHC molecules) is being given much emphasis. Screening of large-scale pathogens and mapping of T cell epitopes allow identification of prime target of epitope-based T cell vaccine design.

Immunoinformatics models simulate the real behavior of immune system processes and thus help to get the kinetics of cells during immune responses. It is engineered in such a way that it can be studied and interpreted easily and can be rebuilt if new experimental data are introduced. These mathematical models remove the uncertainty of the systems as they are found to be closed to wet lab experiments. It leads to design the path for refinement and model the new experiments. But they cannot be directly compared to real biological data as they rely on assumptions only. There is no data for extended time spans available to validate the model. This limits the accuracy of the results. Currently models are designed in such a way that they simulate the biological data only over a fixed time period [172]. It should have the ability to show the system's changes over an extended time period for immune response in case of antigen attack or drug administration. This will reduce the necessity of experimental research.

Drug response to a host's immune system can be better studied through computational models. Effect of drug administration can be added to model the immune system to find the drug efficacy [172]. Immune system/drug response study provides an idea about the dose composition, drug dosage duration, age of the patient, and other parameters. These modeling capabilities may lead to designing a drug, which can treat a disease without any side effects. Thus the idea of integrating systems biology with immunoinformatics can lead to better clinical trials.

Acknowledgment

Ms. Namrata Tomar, one of the authors, gratefully acknowledges CSIR, India, for providing her a Senior Research Fellowship (9/93(0145)/12, EMR-I).

References

1. Thomas K, Goldsby J, Osborne RA, Barbara A, Kuby J (2006) Kuby immunology, 6th edn. Freeman and Co., WH
2. Kimbrell DA, Beutler B (2001) The evolution and genetics of innate immunity. Nat Rev Genet 2:256–267
3. Korber B, LaBute M, Yusim K (2006) Immunoinformatics: comes of Age. PLoS Comput Biol 2:0484–0492
4. Gardy JL, Lynn DJ, Brinkman FSL, Rew H (2009) Enabling a systems biology approach to immunology: focus on innate immunity. Trends Immunol 30:249–262
5. Davies MN, Flower DR (2007) Harnessing bioinformatics to discover new vaccine. Drug Discov Today 12:389–395
6. Ortutay C, Vihinen M (2009) Immunome Knowledge base (IKB): An integrated service for immunome research. BMC Immunol 10
7. Sette A, Fleri W, Peters B, Sathiamurthy M, Bui HH (2005) A roadmap for the immunomics of category A-C pathogens. Immunity 22: 155–161
8. De Groot AS (2006) Immunomics: discovering new targets for vaccine and therapeutics. Drug Discov Today 11:203–209
9. Grainger DJ (2004) Immunomics: principles and practice. IRTL 2:1–6
10. No K, Everse J, Je D, Fe S, Cy L, Clt L, Ss T, Mosbach K (1974) Purification and separation of pyridine nucleotide-linked dehydrogenases by affinity chromatography techniques. Proc Natl Acad Sci U S A 71:3450–3454
11. Davey HM (2004) Flow cytometric techniques for the detection of microorganisms. Methods Cell Sci 24:91–97
12. Durkin MM, Connolly PA, Wheat LJ (1997) Comparison of radioimmunoassay and enzyme-linked immunoassay methods for detection of *histoplasma capsulatum* var. *capsulatum* antigen. J Clin Microbiol 35:2252–2255
13. Ma H, Shieh KJ, Lee SL (2006) Study of ELISA technique. Nature 4:36–37
14. Levine MA, Thornton P, Forman SJ, Hale PV, Holdorf D, Rouault CL, Powars D, Feinstein DI, Lukes RJ (1980) Positive Coombs test in Hodgkin's disease: significance and implications. Blood 55:607–611
15. Nishimaki T, Sagawa K, Motogi S, Saito K, Morito T, Yoshida H, Kasukawa R (1987) A competitive inhibition test of enzyme immunoassay for the anti-nRNP antibody. J Immunol Methods 100:157–160
16. Wanga B, Huaa RH, Tiana Z-J, Chena N-S, Zhaoa F-R, Liua T-Q, Wanga Y-F, Tong G-Z (2009) Identification of a virus-specific and conserved B-cell epitope on NS1 protein of Japanese encephalitis virus. Virus Res 141: 90–95
17. Admon A, Barnea E, Ziv T (2003) Tumor antigens and proteomics from the point of view of the major histocompatibility complex peptides. Mol Cell Proteomics 2:388–398
18. Boon T, Coulie PG, Eynde den BV (1997) Tumor antigens recognized by T cells. Immunol Today 18:267–268
19. De Groot AS, Sbai H, Aubin CS, Mcmurry J, Martin W (2002) Immuno-informatics: mining genomes for vaccine components. Immunol Cell Biol 80:225–269
20. Quintana FJ, Hagedorn PH, Gad E, Yifat M, Eutan D, Cohen IR (2004) Functional immunomics: microarray analysis of IgG autoantibody repertoires predicts the future response of mice to induced diabetes. Proc Natl Acad Sci 101:14615–14621
21. Sampson HA (2005) Food allergy-accurately identifying clinical reactivity. Allergy 60:19–24
22. de Vegvar HEN, Robinson WH (2004) Microarray profiling of antiviral antibodies for the development of diagnostics, vaccines, and therapeutics. J Clin Immunol 111: 196–201
23. Neuman de Vegvar HE, Amara RR, Steinman L, Utz PJ, Robinson HL, Robinson WH (2003) Microarray profiling of antibody responses against simian-human immunodeficiency virus: post challenge convergence of reactivities independent of host histocompatibility type and vaccine regimen. J Virol 77: 11125–11138

24. Wang Y (2004) Immunostaining with dissociable antibody microarrays. Proteomics 4:20–26
25. Magdalena J, Odling J, Qiang PH, Martenn S, Joakin L, Uhlen M, Hammarstrom L, Nilsson P (2005) Serum microarrays for large scale screening of protein levels. Mol Cell Proteomics 4:1942–1947
26. Sahin U, Tureci O, Pfreundschuh M (1997) Serological identification of human tumor antigens. Curr Opin Immunol 9:709–716
27. Oelke M, Maus MV, Didiano D, June CH, Mackensen A, Schneck JP (2003) *Ex vivo* induction and expansion of antigen-specific cytotoxic T cells by HLA-Ig coated artificial antigen-presenting cells. Nat Med 9:619–624
28. Braga-Neto UM, Marques ETA (2006) From functional genomics to functional immunomics: new challenges, Old problems, Big rewards. PLoS Comput Biol 2:651–662
29. Nahtman T, Jernberg A, Mahdavifar S, Zerweck J, Schutkowski M, Maeurer M, Reilly M (2007) Validation of peptide epitope microarray experiments and extraction of quality data. J Immunol Methods 328:1–13
30. Peters B, Sidney J, Bourne P et al (2005) The immune epitope database and analysis resource: from vision to blueprint. PLoS Biol 3: 1361–1370
31. Lynn DJ, Winsor GL, Chan C et al (2008) InnateDB: facilitating systems-level analyses of the mammalian innate immune response. Mol Syst Biol 4:1–11
32. Barsky S, Gardy JL, Hancock R, Munzer T (2007) Cerebral: a Cytoscape plugin for layout of and interaction with biological networks using subcellular localization annotation. Bioinformatics 23:1040–1042
33. Shanon P, Markiel A, Ozier O et al (2003) Cytoscape: a software environment for integrated models of biomolecular interaction networks. Genome Res 13:2498–2504
34. Evans MC (2008) Recent advances in immunoinformatics: application of *in silico* tools to drug development. Curr Opin Drug Discov Devel 11:233–241
35. Saha S, Bhasin M, Raghava GPS (2005) Bcipep: a database of B-cell epitopes. BMC Genomics 6
36. Huang J, Honda W (2006) CED: a conformational epitope. BMC Immunol 7:7
37. Schlessinger A, Ofran Y, Yachdav G, Rost B (2006) Epitome: database of structure-inferred antigenic epitopes. Nucleic Acids Res 34: D777–D780
38. Rammensee HG, Bachmann J, Emmerich NPN, Bachor OA, Stevanovic S (1999) SYFPEITHI: database for MHC ligands and peptide motifs. Immunogenetics 50:213–219
39. Sathiamurthy M, Peters B, Bui HH et al (2005) An ontology for immune epitopes: application to the design of a broad scope database of immune reactivities. Immunome Res 1
40. Feldhahn M, Donnes P, Thiel P, Kohlbacher O (2009) FRED-a framework for T-cell epitope detection. Bioinformatics 25:2758–2759
41. Lefranc M-P, Giudicelli V, Ginestoux C et al (2009) IMGT®, the international ImMuno GeneTics information system®. Nucleic Acids Res 37:D1006–D1012
42. Robinson J, Mistry K, McWilliam H, Lopez R, Parham P, Marsh SGE (2011) The IMGT/HLA database. Nucleic Acids Res 39(Suppl 1):D1171–D1176
43. Pomes A (2010) Relevant B cell epitopes in allergic disease. Int Arch Allergy Immunol 152:1–11
44. Hoffman D, Lowenstein H, Marsh DG, Platts-Mills TAE, Thomas W (1994) Allergen nomenclature. Bull World Health Organ 72:796–806
45. Kim C, Kwon S, Lee G, Lee H, Choi J, Kim Y, Hahn J (2009) A database for allergenic proteins and tools for allergenicity prediction. Bioinformation 3:344–345
46. Mari A, Scalab E, Palazzob P, Ridolfib S, Zennarob D, Carabella G (2006) Bioinformatics applied to allergy: Allergen databases, from collecting sequence information to data integration. The Allergome platform as a model. Cell Immunol 244:97–100
47. Ivanciuc O, Schein CH, Braun W (2003) SDAP: database and computational tools for allergenic proteins. Nucleic Acids Res 31: 359–362
48. Greenbaum JA, Andersen PH, Blythe M et al (2007) Towards a consensus on datasets and evaluation metrics for developing B-cell epitope prediction tools. J Mol Recognit 20: 75–82
49. Tong JC, Ren EC (2009) Immunoinformatics: current trends and future directions. Drug Discov Today 14:684–689
50. Bui HH, Peters B, Assarsson E, Mbawuike I, Sette A (2007) Ab and T cell epitopes of influenza A virus, knowledge and opportunities. Proc Natl Acad Sci U S A 104:246–251
51. Muller GM, Shapira M, Arnon R (1982) Anti-influenza response achieved by immunization with a synthetic conjugate. Proc Natl Acad Sci U S A 79:569–573
52. Naruse H, Ogasawara K, Kaneda R, Hatakeyama S, Itoh T, Kida H, Miyazaki T, Good RA, Onoe K (1994) A potential peptide vaccine against two different strains of influenza virus isolated at intervals of about 10 years. Proc Natl Acad Sci U S A 91:9588–9592

53. Pellequer J, Westhof E, Regenmortel MV (1991) Predicting the location of structure of continuous epitopes in proteins from their primary structure. Methods Enzymol 203: 176–201
54. Parker J, Guo D, Hodges R (1986) New hydrophilicity scale derived from high-performance liquid chromatography peptide retention data: correlation of predicted surface residues with antigenicity and X-ray-derived accessible sites. Biochemistry 25:5425–5432
55. Chou PY, Fasman GD (1978) Prediction of the secondary structure of proteins from their amino acid sequence. Adv Enzymol Relat Areas Mol Biol 47:45–148
56. Levitt M (1978) Conformational preferences of amino acids in globular proteins. Biochemistry 17:4277–4285
57. Emini E, Hughes J, Perlow D, Boger J (1985) Induction of hepatitis A virus-neutralizing antibody by a virus specific synthetic peptide. J Virol 55:836–839
58. EL-Manzalawy Y, Dobbs D, Honavar V (2008) Predicting protective linear b-cell epitopes using evolutionary information. In: IEEE International Conference on Bioinformatics and Biomedicine 289–292
59. Odorico M, Pellequer JL (2003) BEPITOPE: predicting the location of continuous epitopes and patterns in protein. J Mol Recognit 16: 20–22
60. Pellequer JL, Westhof E (1993) PREDITOP: a program for antigenicity predictions. J Mol Graph 11:204–210
61. Saha S, Raghava GPS (2004) BcePred: Prediction of Continuous B-Cell Epitopes in Antigenic Sequences Using Physico-chemical Properties. In: Nicosia G, Cutello V, Bentley PJ, Timis J (eds.) ICARIS Springer, LNCS 3239:197–204
62. Ghate AD, Bhagwat BU, Bhosle SG, Gadepalli SM, Kulkarni-Kale UD (2007) Characterization of antibody-binding sites on proteins: development of a knowledgebase and its applications in improving epitope prediction. Protein Pept Lett 14:531–535
63. Saha S, Raghava GPS (2006) Prediction of continuous B-cell epitopes in an antigen using recurrent neural network. Proteins 65:40–48
64. Sweredoski MJ, Baldi P (2009) COBEpro: a novel system for predicting continuous B-cell epitopes. Protein Eng Des Sel 22:113–120
65. Larsen JEP, Lund O, Nielsen M (2006) Improved method for predicting linear B cell epitopes. Immunome Res 2
66. Toseland CP, Clayton DJ, McSparron H et al (2005) AntiJen: a quantitative immunology database integrating functional, thermodynamic, kinetic, biophysical, and cellular data. Immunome Res 1
67. Anderson P, Nielsen M, Lund O (2006) Prediction of residues in discontinuous B cell epitopes using protein 3D structures. Protein Sci 15:2558–2567
68. Sweredoski M, Baldi P (2008) PEPITO: improved discontinuous B-cell epitope prediction using multiple distance thresholds and half sphere exposure. Bioinformatics 24: 1459–1460
69. Hamelryck T (2005) An amino acid has two sides: a new 2D measure provides a different view of solvent exposure. Proteins 59: 38–48
70. Bublil EM, Mayrose NTFI, Penn O, Berman AR (2007) Stepwise prediction of conformational discontinuous B-cell epitopes using the mapitope algorithm. Proteins 68:294–304
71. Sollner J, Grohmann R, Rapberger R, Perco P, Lukas A, Mayer B (2008) Analysis and prediction of protective continuous B cell epitopes on pathogen proteins. Immunome Res 4
72. Kale KU, Bhosle S, Kolaskar AS (2005) CEP: a conformational epitope prediction server. Nucleic Acids Res 33:W168–W171
73. Zhang W, Xiong Y, Zhao M, Zou H, Ye X, Liu J (2011) Prediction of conformational B-cell epitopes from 3D structures by random forests with a distance-based feature. BMC Bioinformatics 12:341
74. Mayrose I, Penn O, Erez E et al (2007) Pepitope: Epitope mapping from affinity-selected peptides. Bioinformatics 23:3244–3246
75. Pizzi E, Cortese R, Tramontano A (1995) Mapping epitopes on protein surfaces. Biopolymers 36:675–680
76. Moreau V, Granier C, Villard S, Laune D, Molina F (2006) Discontinuous epitope prediction based on mimotope analysis. Bioinformatics 22:1088–1095
77. Huang J, Gutteridge A, Honda W, Kanehisa M (2006) MIMOX: a web tool for phage display based epitope mapping. BMC Bioinformatics 7
78. Mayrose I, Shlomi T, Rubinstein ND, Gershoni JM, Ruppin E, Sharan R, Pupko T (2007) Epitope mapping using combinatorial phage-display libraries: a graph-based algorithm. Nucleic Acids Res 35:69–78
79. Huang YX, Bao YL, Guo SY, Wang Y, Zhou CG, Li YX (2008) Pep-3D-Search: a method for B-cell epitope prediction based on mimotope analysis. BMC Bioinformatics 9:538
80. Schreiber A, Humbert M, Benz A, Dietrich U (2005) 3D-Epitope-Explorer (3DEX): Localization of conformational epitopes within

three-dimensional structures of proteins. J Comput Chem 26:879–887

81. Caragea C, Sinapov J, Silvescu A, Dobbs D, Honavar V (2007) Glycosylation site prediction using ensembles of support vector machine classifiers. BMC Bioinformatics 8:438
82. EL-Manzalawy Y, Honavar V (2010) Recent advances in B-cell epitope prediction methods. Immunome Res 6(Suppl 2):2
83. Sollner J (2006) Selection and combination of machine learning classifiers for prediction of linear B-cell epitopes on proteins. J Mol Recognit 19:209–214
84. Huang L, Dai Y (2006) Direct prediction of T-cell epitopes using support vector machines with novel sequence encoding schemes. J Bioinform Comput Biol 4:93–107
85. Bhasin M, Raghava GPS (2003) Prediction of promiscuous and high-affinity mutated MHC binders. Hybrid Hybridomics 22:229–234
86. Zhang GL, Petrovsky N, Kwoh CK, August JT, Brusic V (2006) Pred^{TAP}: a system for prediction of peptide binding to the human transporter associated with antigen processing. Immunome Res 2
87. Buus S, Lauemoller SL, Worning P, Kesmir C, Frimurer T, Corbet S, Fomsgaard A, Hilden J, Holm A, Brunak S (2003) Sensitive quantitative predictions of peptide-MHC binding by a 'Query by Committee' artificial neural network approach. Tissue Antigens 62: 378–384
88. Neilsen M, Lundegaard C, Worning P, Lauemoller SL, Lamberth K, Buus S, Brunak S, Lund O (2003) Reliable prediction of T-cell epitopes using networks with novel sequence representations. Protein Sci 12: 1007–1017
89. Buus S, Stryhn A, Winther K, Kirkby N, Pedersen LO (1995) Receptor–ligand interactions measured by an improved spun column chromatography technique. A high efficiency and high throughput size separation method. Biochim Biophys Acta 1243:453–460
90. Larsen MV, Lundegaard C, Lamberth K, Buss S, Lund O, Nielsen M (2007) Large-scale validation of methods for cytotoxic T-lymphocyte epitope prediction. BMC Bioinformatics 8
91. Lundegaard C, Lamberth K, Harndahl M, Buus S, Lund O, Nielsen M (2008) NetMHC-3.0: accurate web accessible predictions of human mouse and monkey MHC class I affinities for peptides of length 8-11. Nucleic Acids Res 36:W509–W512
92. Brusic V, Rudy G, Honeyman M, Hammer J, Harrison L (1998) Prediction of MHC class II-binding peptides using an evolutionary and artificial neural network. Bioinformatics 14: 121–130
93. Miyata J (1991) A User's Guide to PlaNet Version 5.6.
94. Doytchinova IA, Guan P, Flower DR (2006) EpiJen: a server for multistep T cell epitope prediction. BMC Bioinformatics 7:131
95. Dorigo M, Maniezzo V, Colorni A (1996) Ant system: optimization by a colony of cooperating agents. IEEE Trans Syst Man Cybern B 26:29–41
96. Bhasin M, Raghava GPS (2004) Analysis and prediction of affinity of TAP binding peptides using cascade SVM. Protein Sci 13: 596–607
97. Nanni L (2006) Machine learning algorithms for T-cell epitopes prediction. Neurocomputing 69:866–868
98. Bhasin M, Raghava GPS (2005) Pcleavage: an SVM based method for prediction of constitutive proteasome and immunoproteasome cleavage sites in antigenic sequences. Nucleic Acids Res 33:W202–W207
99. Joachims T (1999) Marking large-scale support vector machine learning practical. In: Advances in Kernel methods: support vector learning. MIT Press, Cambridge, MA, pp 169–184
100. Cost S, Salzberg S (1993) A weighted nearest neighbor algorithm for learning with symbolic features. Mach Learn 10:57–78
101. Witten IH, Frank E (1999) Data mining: practical machine learning tools and techniques with java implementations, 2nd edn. Morgan Kaufman, San Francisco
102. Flower DR (2003) Towards *in silico* prediction of immunogenic epitopes. Trends Immunol 24:667–674
103. Bian H, Hammer H (2004) Discovery of promiscuous HLA restricted T cell epitope with TEPITOPE. Methods 34:468–475
104. Zhang L, Chen Y, Wong HS, Zhou S, Mamitsuka H, Zhu S (2012) TEPITOPEpan: extending TEPITOPE for peptide binding prediction covering over 700 HLA-DR molecules. PLoS One 7:e30483
105. Kangueane P, Sakharkar MK (2005) T epitope designer: HLA peptide binding prediction server. Bioinformation 1:21–24
106. Zhao B, Mathura VS, Ganapathy K, Moochhala S, Sakharkar MK, Kangneane P (2003) A novel MHCp binding prediction model. Hum Immunol 64:1123–1143
107. Ponomarenko JV, Bui HH, Li W, Fusseder N, Bourne PE, Sette A, Peters B (2008) ElliPro: a new structure-based tool for the prediction of antibody epitopes. BMC Bioinformatics 9

108. Guan P, Doytchinova IA, Zygouri C, Flower DR (2003) MHCPred: a server for quantitative prediction of peptide-MHC binding. Nucleic Acids Res 31:3621–3624
109. Schiewe AJ, Haworth IS (2007) Structure based prediction of MHC-peptide association: algorithm comparison and approach to cancer vaccine design. J Mol Graph Model 26:667–675
110. Jojic N, Gomez MR, Heckerman D, Kadle C, Furman OS (2006) Learning MHC-I peptide binding. Bioinformatics 22:e227–e235
111. Furman OS, Altuvia Y, Sette A, Margalit H (2000) Structure-based prediction of binding peptides to MHC class I molecules: Application to a broad range of MHC alleles. Protein Sci 9:1838–1846
112. Miyazawa S, Jernigan RL (1996) Residue-residue potentials with a favorable contact pair term and an unfavorable high packing density term, for simulation and threading. J Mol Biol 256:623–644
113. Altuvia Y, Margalit H (2004) A structure-based approach for prediction of MHC-binding peptides. Methods 34:454–459
114. Singh H, Raghava GPS (2001) Propred: prediction of HLA-DR binding sites. Trends Immunol 17:1236–1237
115. Sturniolo T, Bono E, Ding J et al (1999) Generation of tissue-specific and promiscuous HLA ligand database using DNA microarrays and virtual HLA class II matrices. Nat Biotechnol 17:555–561
116. Feldhahn M, Thiel P, Schuler MM, Hillen N, Stevanovic S, Rammensee HG, Ohlbacher O (2008) EpiToolKit-a web server for computational immunomics. Nucleic Acids Res 1:W519–W522
117. Flower DR, Phadwal K, Macdonald IK, Coveney PV, Davies MN, Wan S (2010) T-cell epitope prediction and immune complex simulation using molecular dynamics: state of the art and persisting challenges. Immunome Res 6(Suppl 2):S4
118. Zhang C, Anderson A (1998) DeLisi C: structural principles that govern the peptide-binding motifs of class I MHC molecules. J Mol Biol 281:929–947
119. Wan S, Coveney PV, Flower DR (2005) Molecular basis of peptide recognition by the TCR: affinity differences calculated using large scale computing. J Immunol 175: 1715–1723
120. Stadler MB, Stadler BM (2003) Allergenicity prediction by protein sequence. FASEB J 17: 1141–1143
121. Kong W, Tan TS, Tham L, Choo KW (2006) Improved prediction of allergenicity by combination of multiple sequence motifs. In Silico Biol 7:77–86
122. Bjorklund AK, Atmadja SD, Zorzet A, Hammerling U, Gustafsson MG (2005) Supervised identification of allergen-representative peptides for *in silico* detection of potentially allergenic proteins. Bioinformatics 21:39–50
123. Zorzet A, Gustafsson M, Hammerling U (2002) Prediction of food protein allergenicity: a bioinformatic learning systems approach. In Silico Biol 2:525–534
124. Saha S, Raghava GPS (2006) AlgPred: prediction of allergenic proteins and mapping of IgE epitopes. Nucleic Acids Res 34: W202–W209
125. Fiers MWEJ, Kleter GA, Nijland H, Peijnenburg AACM, Peter NJ, Ham RCHJV (2004) Allermatch^{TM}, a webtool for the prediction of potential allergenicity according to current FAO/WHO Codex alimentarius guidelines. BMC Bioinformatics 5
126. FAO/WHO: Allergenicity of Genetically Modified Foods. http://www.who.int/foodsafety/ publications/biotech/en/ec_jan 2001.pdf, 2001.
127. FAO/WHO: Codex Principles and Guidelines on Foods Derived from Biotechnology ftp://ftp.fao.org/codex/standard/en/Codex TextsBiotechFoods.pdf, 2003.
128. Cui J, Han LY, Li H, Ung CY, Tang ZQ, Zheng CJ, Cao ZW, Chen YZ (2007) Computer prediction of allergen proteins from sequence-derived protein structural and physicochemical properties. Mol Immunol 44:514–520
129. Barrio AM, Atmadja DS, Nistr A, Gustafsson MG, Hammerling U, Rudloff EB (2007) EVALLER: a web server for *in silico* assessment of potential protein allergenicity. Nucleic Acids Res 35:694–700
130. Soeria-Atmadja D, Lundell T, Gustafsson MG, Hammerling U (2006) Computational detection of allergenic proteins attains a new level of accuracy with *in silico* variable-length peptide extraction and machine learning. Nucleic Acids Res 34:3779–3793
131. Pizza M, Scarlato V, Masignani V et al (2000) Identification of vaccine candidates against serogroup B meningococcus by whole-genome sequencing. Science 287:1816–1820
132. De Groot AS, Rappuoli R (2003) Genome derived vaccines. Expert Rev Vaccines 3: 59–76

133. Gallimore A, Hengartner H, Zinkernagel R (1998) Hierarchies of antigen-specific cytotoxic T cell responses. Immunol Rev 164: 29–36
134. Morris S, Kelly C, Howard A, Li X, Collins F (2000) The immunogenicity of single and combination DNA vaccines against tuberculosis. Vaccine 18:2155–2163
135. Zhao B, Sakharkar KR, Lim CS, Kangueane P, Sakharkar MK (2007) MHC-peptide binding prediction for epitope based vaccine design. Int J Integr Biol 1:127–140
136. Davenport MP, Hill AV (1996) Reverse immunogenetics: from HLA disease associations to vaccine candidates. Mol Med Today 2:38–45
137. Iwai LK, Yoshida M, Sidney J et al (2003) *In silico* prediction of peptides binding to multiple HLA-DR molecules accurately identifies immunodominant epitopes from gp43 of *Paracoccidioides brasiliensis* frequently recognized in primary peripheral blood mononuclear cell responses from sensitized individuals. Mol Med 9:209–219
138. Reche PA, Reinherz EL (2005) PEPVAC: a web server for multi-epitope vaccine development based on the prediction of supertypic MHC ligands. Nucleic Acids Res 33: W138–W142
139. Florea L, Haldorsson B, Kohlbacher O, Schwarty R, Hoffman S, Istrail S (2003) Epitope prediction algorithm for peptide-based vaccine design. Proc IEEE Comput Soc Bioinform Conf 2:17–26
140. Doytchinova IA, Flower DR (2007) VaxiJen: a server for prediction of protective antigens, tumor antigens and subunit vaccines. BMC Bioinformatics 8
141. Nagarajan H, Gupta R, Agarwal P, Scaria V, Pillai B (2006) DyNAVacS: an integrative tool for optimized DNA vaccine design. Nucleic Acids Res 34:W264–W266
142. Lollini PL, Motta S, Pappalardo F (2006) Discovery of cancer vaccination protocols with a genetic algorithm driving an agent based simulator. BMC Bioinformatics 7:352
143. Vivona S, Bernante F, Filippini F (2006) NERVE: New enhanced reverse vaccinology environment. BMC Biotechnol 6
144. Xiang Z, Todd T, Ku KP et al (2008) VIOLIN: vaccine investigation and online information network. Nucleic Acids Res 36: 923–928
145. Xianga Z, He Y (2009) Vaxign: a web-based vaccine target design program for reverse vaccinology. Procedia in Vaccinol 1:23–29
146. Gong T, Cai Z (2005) Visual Modeling and Simulation of Adaptive Immune System. In: Proceedings of the 2005 IEEE Engineering in Medicine and Biology 27th Annual Conference, Shanghai, China 6:6116–6119
147. Kalita JK, Chandrashekar K, Hans R, Selvam P, Newell MK (2006) Computational modelling and simulation of the immune system. Int J Bioinform Res Appl 2:63–88
148. Castiglione F, Liso A (2005) The role of computational models of the immune system in designing vaccination strategies. Immunopharmacol Immunotoxicol 27:417–432
149. DeLuca DS, Blasczyk R (2007) The immunoinformatics of cancer immunotherapy. Tissue Antigens 70:265–271
150. Hu H, Brzeski H, Hutchins J et al (2004) Biomedical informatics: development of a comprehensive data warehouse for clinical and genomic breast cancer research. Pharmacogenomics 5:933–941
151. Sanchez W, Gilman B, Kher M, Lagou S, Covitz P (2004) caGRID White Paper (cancer biomedical informatics grid prototype project). National Cancer Institute Center for Bioinformatics (NCICB), USA
152. Galperin MY (2005) The molecular biology database collection: 2005 update. Nucleic Acids Res 33:D5–D24
153. Novellino L, Castelli C, Parmiani G (2005) A listing of human tumor antigens recognized by T-cells: March 2004 update. Cancer Immunol Immunother 54:187–207
154. Wang XS, Zhao HT, Xu QW et al (2006) HPtaa database-potential target genes for clinical diagnosis and immunotherapy of human carcinoma. Nucleic Acids Res 1:D607–D612
155. Oh P, Li Y, Yu J, Durr E, Krasinska KM, Carver LA, Testa JE, Schnitzer JE (2004) Subtractive proteomic mapping of the endothelial surface in lung and solid tumours for tissue-specific therapy. Nature 429:629–635
156. Camp RL, Dolled-Filhart M, Rimm DL (2004) X-tile: a new bio-informatics tool for biomarker assessment and outcome-based cut-point optimization. Clin Cancer Res 10: 7252–7259
157. Rosa DS, Ribeiro SP, Cunha-Neto E (2010) CD4+ T cell epitope discovery and rational vaccine design. Arch Immunol Ther Exp 58:121–130
158. Iurescia S, Fioretti D, Fazio VM, Rinaldi M (2012) Epitope-driven DNA vaccine design employing immunoinformatics against B-cell lymphoma: a biotech's challenge. Biotechnol Adv 30:372–383

159. Khanolkar A, Badovinac VP, Harty JT (2007) CD8 T cell memory development: CD4 T cell help is appreciated. Immunol Res 39:94–104

160. Lu J, Celis E (2000) Use of two predictive algorithms of the world wide web for the identification of tumor-reactive T-cell epitopes. Cancer Res 60:5223–5227

161. Smith CM, Wilson NS, Waithman J et al (2004) Cognate CD4(+) T cell licensing of dendritic cells in CD8(+) T cell immunity. Nat Immunol 5:1143–1148

162. Wan YY, Flavell RA (2009) How diverse–CD4 effector T cells and their functions. J Mol Cell Biol 1:20–36

163. Hung K, Hayashi R, Lafond-Walker A, Lowenstein C, Pardoll D, Levitsky H (1998) The central role of CD4+ T cells in the antitumor immune response. J Exp Med 188: 2357–2368

164. Kalams SA, Walker BD (1998) The critical need for CD4 help in maintaining effective cytotoxic T lymphocyte responses. J Exp Med 188:2199–2204

165. Houot R, Levy R (2009) Vaccines for lymphomas: idiotype vaccines and beyond. Blood Rev 23:137–142

166. King CA, Spellerberg MB, Zhu D et al (1998) DNA vaccines with single-chain Fv fused to fragment C of tetanus toxin induce protective immunity against lymphoma and myeloma. Nat Med 4:1281–1286

167. Thirdborough SM, Radcliffe JN, Friedmann PS, Stevenson FK (2002) Vaccination with DNA encoding a single-chain TCR fusion protein induces anticlonotypic immunity and protects against T-cell lymphoma. Cancer Res 62:1757–1760

168. Rice J, Elliott T, Buchan S, Stevenson FK (2001) DNA fusion vaccine designed to induce cytotoxic T cell responses against defined peptide motifs: implications for cancer vaccines. J Immunol 167:1558–1565

169. Yan Q (2010) Immunoinformatics and systems biology methods for personalized medicine. Methods Mol Biol 662:203–220

170. Wang SS, Bratti MC, Rodriguez AC et al (2009) Common variants in immune and DNA repair genes and risk for human papillomavirus persistence and progression to cervical cancer. J Infect Dis 199:20–30

171. Vivona S, Gardy JL, Ramachandran S, Brinkman FSL, Raghava GPS, Flower DR, Filippini F (2008) Computer-aided biotechnology: from immuno-informatics to reverse vaccinology. Trends Biotechnol 26:190–200

172. Daz P, Gillespie M, Krueger J, Prez J, Radebaugh A, Shearman T, Vo G, Wheatley C (2008) A mathematical model of the immune system's response in obesity-related chronic inflammation. McNair/MAOP Summer Research Symposium, Virginia Tech, Blacksburg VA 2:26–4.

Part II

Databases

Chapter 4

Immunoinformatics of the V, C, and G Domains: IMGT® Definitive System for IG, TR and IgSF, MH, and MhSF

Marie-Paule Lefranc

Abstract

By its creation in 1989, IMGT®, the international ImMunoGeneTics information system® (http://www.imgt.org, CNRS and Université Montpellier 2), marked the advent of immunoinformatics, which emerged at the interface between immunogenetics and bioinformatics. IMGT® is the global reference in immunogenetics and immunoinformatics. The accuracy and the consistency of the IMGT® data are based on the IMGT Scientific chart rules generated from the IMGT-ONTOLOGY axioms and concepts, which comprise IMGT standardized labels (DESCRIPTION), IMGT gene and allele nomenclature (CLASSIFICATION), IMGT unique numbering, and IMGT Collier de Perles (NUMEROTATION). The IMGT® standards have bridged the gap between genes, sequences, and three-dimensional (3D) structures for the receptors, chains, and domains. Started specifically for the immunoglobulins (IG) or antibodies and T cell receptors (TR), the IMGT-ONTOLOGY concepts have been extended to conventional genes of the immunoglobulin superfamily (IgSF) and major histocompatibility (MH) superfamily (MhSF), members of which are defined by the presence of at least one variable (V) or constant (C) domain, or two groove (G) domains, respectively. In this chapter, we review the IMGT® definitive system for the V, C, and G domains, based on the IMGT-ONTOLOGY concepts of IMGT unique numbering and IMGT Collier de Perles.

Key words IMGT, Immunoinformatics, Immunogenetics, IMGT-ONTOLOGY, IMGT Collier de Perles, IMGT unique numbering, Immunoglobulin, Antibody, T cell receptor, Major histocompatibility

1 Introduction

IMGT®, the international ImMunoGeneTics information system® (http://www.imgt.org) [1], was created in 1989 by Marie-Paule Lefranc at Montpellier, France (CNRS and Université Montpellier 2). The founding of IMGT® marked the advent of immunoinformatics, a new science, which emerged at the interface between immunogenetics and bioinformatics. For the first time, immunoglobulin (IG) or antibody and T cell receptor (TR) variable (V), diversity (D), joining (J) and constant (C) genes were officially recognized as "genes" as well as the conventional genes. This major breakthrough allowed genes and data of the complex

Rajat K. De and Namrata Tomar (eds.), *Immunoinformatics*, Methods in Molecular Biology, vol. 1184,
DOI 10.1007/978-1-4939-1115-8_4, © Springer Science+Business Media New York 2014

and highly diversified adaptive immune responses to be managed in genomic databases and tools.

The adaptive immune response was acquired by jawed vertebrates (or *gnathostomata*) more than 450 million years ago and is found in all extant jawed vertebrate species from fishes to humans. It is characterized by a remarkable immune specificity and memory, which are properties of the B and T cells owing to an extreme diversity of their antigen receptors. The specific antigen receptors comprise the immunoglobulins (IG) or antibodies of the B cells and plasmocytes, and the T cell receptors (TR) [2–5]. The IG recognize antigens in their native (unprocessed) form, whereas the TR recognize processed antigens which are presented as peptides by the highly polymorphic major histocompatibility (MH, in humans HLA for human leucocyte antigens) proteins.

The potential antigen receptor repertoire of each individual is estimated to comprise about 2×10^{12} different IG and TR, and the limiting factor is only the number of B and T cells that an organism is genetically programmed to produce [2, 3]. This huge diversity results from the complex molecular synthesis of the IG and TR chains (Fig. 1) and more particularly of their variable domains (V-DOMAIN) which, at their N-terminal end, recognize and bind the antigens [2, 3].

The IG and TR synthesis includes several unique mechanisms that occur at the DNA level: combinatorial rearrangements of the V, D, and J genes that code the V-DOMAIN (the V-(D)-J being spliced to the C gene that encodes the C-REGION in the transcript (Fig. 1)), exonuclease trimming at the ends of the V, D, and J genes, and random addition of nucleotides by the terminal deoxynucleotidyl transferase (TdT) that creates the junctional N-diversity regions, and later during B cell differentiation, for the IG, somatic hypermutations, and class or subclass switch [2, 3].

IMGT® manages the diversity and complexity of the IG and TR and the polymorphism of the MH of humans and other vertebrates. IMGT® is also specialized in the other proteins of the immunoglobulin superfamily (IgSF) and MH superfamily (MhSF) and related proteins of the immune system (RPI) of vertebrates and invertebrates. IMGT® provides a common access to standardized information from genes, sequences, genetics, two-dimensional (2D) and three-dimensional (3D) structures. It is a high-quality integrated knowledge resource in immunogenetics for exploring immune functional genomics. IMGT® (Fig. 2) comprises 7 databases (for sequences, genes and 3D structures) [6–11], 17 online tools [12–27], and more than 15,000 pages of Web resources (e.g., IMGT Scientific chart, IMGT Repertoire, IMGT Education > Aide-mémoire [28], The IMGT Immunoinformatics page) [1].

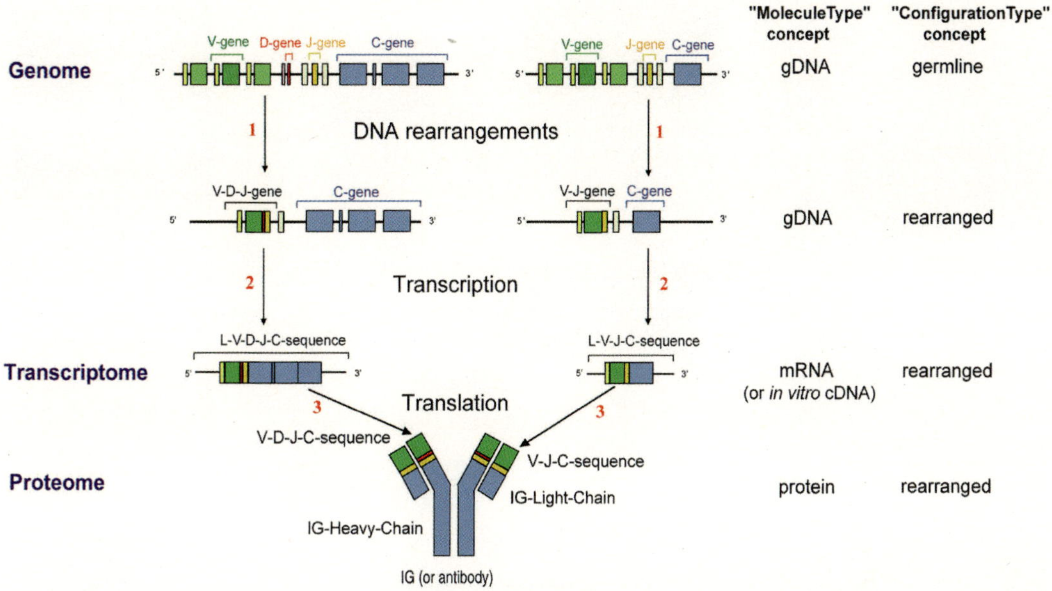

Fig. 1 Synthesis of an IG or antibody in humans. A human being has ~156 functional IG genes in his/her genome and potentially synthesizes 10^{12} different IG or antibody proteins [2] and ~185 functional TR genes and potentially synthesize 10^{12} different TR proteins [3]. The main steps of the IG synthesis, shown as example of antigen receptor synthesis, are indicated with numbers: *1*. DNA rearrangements. *2*. Transcription. *3*. Translation. The ten major molecular entities (V-gene, D-gene, J-gene, C-gene, V-D-J-gene, V-J-gene, L-V-D-J-C-sequence, L-V-J-C-sequence, V-D-J-C-sequence, V-J-C-sequence) are shown with standardized keywords and concepts of identification (*see* **Note 1**). Genomic DNA ("gDNA"), messenger RNA ("mRNA") (or in vitro complementary DNA (cDNA) in databases) are types of molecules ("MoleculeType") that are involved in the IG or TR synthesis, "germline" and "rearranged" are types of configuration ("ConfigurationType") (the configuration of C-gene is "undefined" (not shown)) (*see* **Note 1**) (IMGT® http://www.imgt.org, IMGT Education > Tutorials > Immunoglobulins and B cells; ibid > T cell receptors and T cells)

IMGT® is the global reference in immunogenetics and immunoinformatics [29–44]. Its standards have been endorsed by the World Health Organization–International Union of Immunological Societies (WHO-IUIS) Nomenclature Committee since 1995 (first IMGT® online access at the 9th International Congress of Immunology, San Francisco, USA) [45, 46] and the WHO International Nonproprietary Name (INN) Programme [47, 48]. The accuracy and the consistency of the IMGT® data are based on IMGT-ONTOLOGY [49–51], the first, and so far, unique ontology for immunogenetics and immunoinformatics [49–68]. IMGT-ONTOLOGY manages the immunogenetics knowledge through diverse facets that rely on seven axioms: IDENTIFICATION, DESCRIPTION, CLASSIFICATION, NUMEROTATION, LOCALIZATION, ORIENTATION, and OBTENTION [50, 51, 55]. The concepts generated from these axioms led to the elaboration of the IMGT® standards that constitute the IMGT Scientific chart,

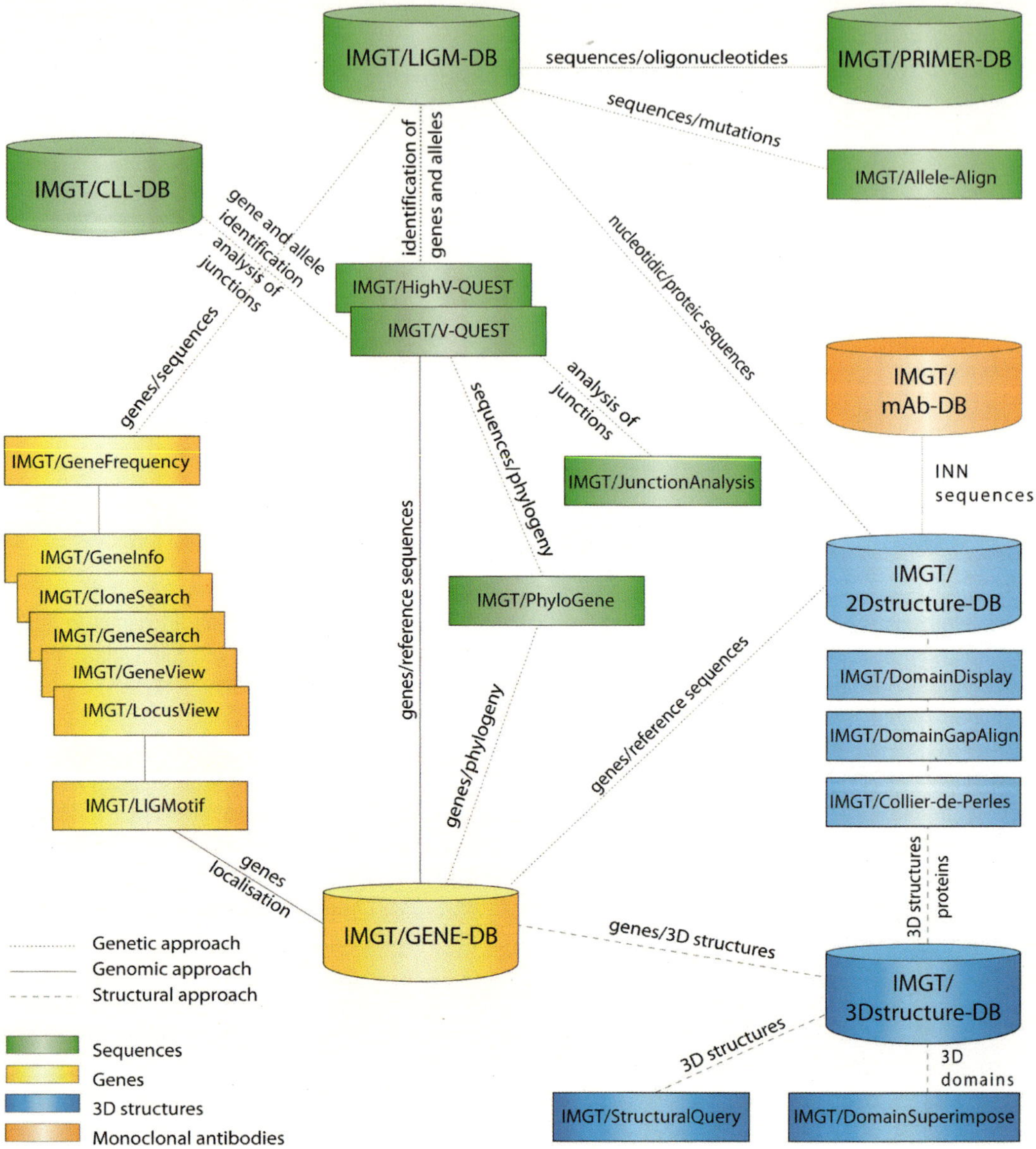

Fig. 2 IMGT®, the international ImMunoGenetics information system®, http://www.imgt.org [1]. Databases are shown as *cylinders* and tools as *rectangles*. The IMGT Repertoire and other Web resources (>15,000 pages) are not shown

e.g., IMGT® standardized keywords (IDENTIFICATION) [56] (*see* **Note 1**), IMGT® standardized labels (DESCRIPTION) [57] (*see* **Note 2**), standardized gene and allele nomenclature (CLASSIFICATION) [58] (*see* **Note 3**), and IMGT unique numbering [59–64] and standardized graphical 2D representation or IMGT Colliers de Perles [65–68] (NUMEROTATION).

The IMGT-ONTOLOGY CLASSIFICATION axiom was the trigger of immunoinformatics' birth. Indeed the IMGT® concepts of classification allowed, for the first time, to classify the antigen receptor genes (IG and TR) for any locus (e.g., immunoglobulin heavy (IGH), T cell receptor alpha (TRA) (*see* **Note 4**)), for any gene configuration (germline, undefined, or rearranged) (*see* **Note 1**) (Fig. 1) and for any species (from fishes to humans). Since the creation of IMGT® in 1989, at the 10th Human Genome Mapping Workshop (HGM10) (*see* **Note 5**), the standardized classification and nomenclature of the IG and TR of human and other vertebrate species have been under the responsibility of the IMGT Nomenclature Committee (IMGT-NC). The IMGT® IG and TR gene names [2–5] were approved by the Human Genome Organisation (HUGO) Nomenclature Committee (HGNC) in 1999 [69, 70] and were endorsed by the WHO-IUIS Nomenclature Subcommittee for IG and TR [45, 46].

The IMGT® IG and TR gene names are the official international reference and, as such, have been entered in IMGT/GENE-DB [7], the IMGT® gene database, in the Genome Database (GDB) [71], in LocusLink at the National Center for Biotechnology Information (NCBI) USA [72], in Entrez Gene (NCBI) when this database (now designated as "Gene") superseded LocusLink [73], in NCBI MapViewer, in Ensembl at the European Bioinformatics Institute (EBI) [74], and in the Vertebrate Genome Annotation (Vega) Browser [75] at the Wellcome Trust Sanger Institute (UK). HGNC, Gene (NCBI), Ensembl, and Vega have direct links to IMGT/GENE-DB [7]. IMGT® human IG and TR genes were also integrated in IMGT-ONTOLOGY on the National Center for Biomedical Ontology (NCBO) BioPortal and, on the same site, in the HUGO ontology and in the National Cancer Institute (NCI) Metathesaurus. Amino acid sequences of human IG and TR constant genes (e.g., *Homo sapiens* IGHM, IGHG1, IGHG2) were provided to UniProt in 2008. In June 2013, IMGT/GENE-DB [7] contains 3,107 IMGT® genes and 4,722 IMGT® alleles from 17 species (694 genes and 1,420 alleles for *Homo sapiens* and 868 genes and 1,318 alleles for *Mus musculus*). Since 2007, IMGT® gene and allele names have been used for the description of the therapeutic monoclonal antibodies (mAb, INN suffix -mab) and of the fusion proteins for immunological applications (FPIA, INN suffix -cept) of the WHO-INN programme [47, 48], with access from IMGT/mAb-DB [11] (*see* **Note 6**).

The IMGT-ONTOLOGY NUMEROTATION axiom is acknowledged as the "IMGT® Rosetta stone" that has bridged the biological and computational spheres in bioinformatics [37]. The IMGT® concepts of numerotation comprise the IMGT unique numbering [59–64] and the IMGT Collier de Perles [65–68]. Developed for and by the "domain," these concepts integrate sequences, structures, and interactions into a standardized knowledge for a modular and highly diverse functional genomics. The IMGT unique numbering has been defined for the variable V domain (V-DOMAIN of the IG and TR, and V-LIKE-DOMAIN of IgSF other than IG and TR) [59–61], the constant C domain (C-DOMAIN of the IG and TR, and C-LIKE-DOMAIN of IgSF other than IG and TR) [62], and the groove G domain (G-DOMAIN of the MH, and G-LIKE-DOMAIN of MhSF other than MH) [63]. Thus, the IMGT unique numbering and IMGT Collier de Perles provide a definitive and universal system for the V, C, and G domain of IG, TR, MH, IgSF, and MhSF [64, 68].

This chapter reviews the V, C, and G domain IMGT® definitive system and the IMGT® tools and databases which are widely used for standardized domain analysis and study: IMGT/Collier-de-Perles tool [26] for their 2D representation, IMGT/DomainGapAlign [9, 24, 25] for their amino acid sequence analysis, IMGT/V-QUEST [12–17] for the IG and TR V-DOMAIN nucleotide sequence analysis with results of the integrated IMGT/JunctionAnalysis [18, 19] and IMGT/Automat [20, 21], and its high-throughput version IMGT/HighV-QUEST for Next-Generation Sequencing (NGS) [22, 23], IMGT/3Dstructure-DB for their 3D structures [8–10] and its extension, IMGT/2Dstructure-DB (for antibodies and other proteins for which the 3D structure is not available). IMGT® tools and databases run against IMGT reference directories built from sequences annotated in IMGT/LIGM-DB, the IMGT® nucleotide database [6] (170,685 sequences from 335 species in June 2013) and from IMGT/GENE-DB [7]. The V, C, and G domain IMGT® definitive system allows standardized domain sequence, structure, and contact analysis. This is of major interest in: antibody engineering and humanization [32, 39–41, 43, 76–78], IG repertoire in normal and pathological situations [79–82], IG allotypes and immunogenicity [83–85], TR clonal diversity and expression [23, 86], NGS repertoire [22, 23], TR/peptide-MH (TR/pMH) interactions [87, 88], computational analysis of MH helices [89, 90], and evolution studies of the IgSF [91–95] and MhSF [96, 97].

2 V Domain IMGT® Definitive System

In the IMGT® definitive system, the V domain includes the V-DOMAIN of the IG (Fig. 3) [2] (*see* **Note 7**) and of the TR (Fig. 4) [3] (*see* **Note 8**), which correspond to the V-J-REGION or V-D-J-REGION encoded by V-(D)-J rearrangements [2, 3] (Fig. 1), and the V-LIKE-DOMAIN of the IgSF other than IG and TR [91–95].

The V domain description of any receptor, any chain and any species is based on the IMGT unique numbering for V domain (V-DOMAIN and V-LIKE-DOMAIN) [59–61, 64].

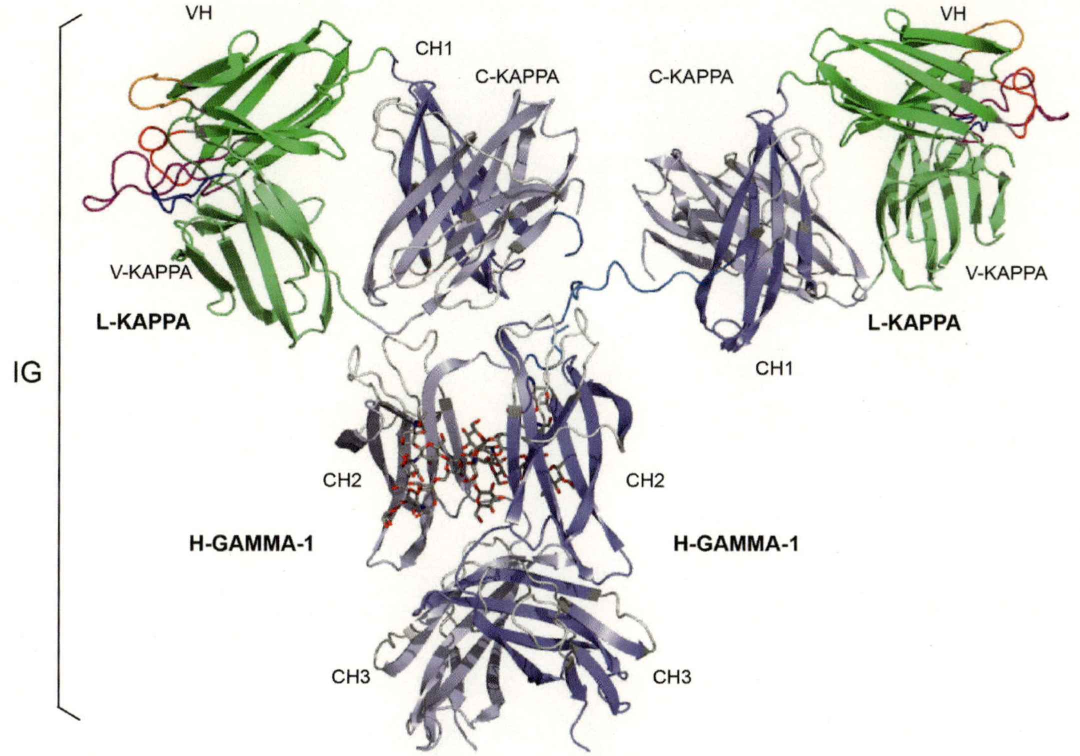

Fig. 3 An immunoglobulin (IG) or antibody. In vivo, an IG or antibody is anchored in the membrane of a B cell as part of a signaling B cell receptor (BcR = membrane IG + CD79) or, as shown here, is secreted [2]. An IG is made of two identical heavy (H, for IG-HEAVY) chains and two identical light (L, for IG-LIGHT) chains [2]. An IG comprises 12 domains (for example, IgG1, shown here) or 14 domains (IgM or IgE). The V-DOMAIN of each chain (*green* online) and the C-DOMAIN, one for each L chain and three for each H chain (*blue* online) are highlighted. The light chain (here, L-KAPPA) is made of a variable domain (V-DOMAIN, here, V-KAPPA) at the N-terminal end and a constant domain (C-DOMAIN, here, C-KAPPA) at the C-terminal end. The heavy chain (here, H-GAMMA-1) is made of a VH (at the N-terminal end) and of three CH (four for H-MU or H-EPSILON, *see* **Note 7**) [2]. The structure is that of the antibody b12, an IgG1-kappa, and so far the only complete human IG crystallized (1hzh from IMGT/3Dstructure-DB (http://www.imgt.org))

A V domain (Fig. 5) comprises about 100 amino acids and is made of nine antiparallel beta strands (A, B, C, C′, C″, D, E, F, and G) linked by beta turns (AB, CC′, C″D, DE, and EF) and three loops (BC, C′C″, and FG), forming a sandwich of two sheets [ABED] [GFCC′C″] [59–61, 64].

The sheets are closely packed against each other through hydrophobic interactions giving a hydrophobic core, and joined together by a disulfide bridge between a first highly conserved cysteine (1st-CYS) (*see* **Note 9**) in the B strand (in the first sheet) and a second equally conserved cysteine (2nd-CYS) in the F strand (in the second sheet) [59–61, 64].

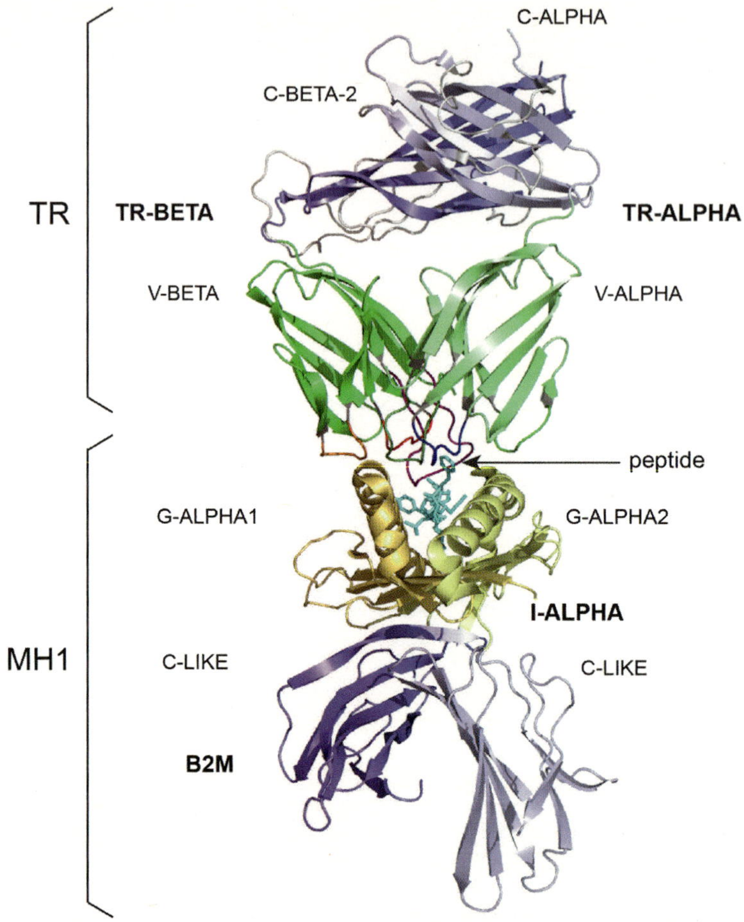

Fig. 4 A T cell receptor (TR)/peptide-major histocompatibility 1 (pMH1) complex. A TR (here, TR alpha_beta) is shown (on *top*, upside down) in complex with a MH (here, MH1) presenting a peptide in its groove. *In vivo*, a TR is anchored in the membrane of a T cell as part of the signaling T cell receptor (TcR = TR + CD3) A TR is made of two chains, each comprising a variable domain (V-DOMAIN) at the N-terminal end and a constant domain (C-DOMAIN) at the C-terminal end [3]. The V-DOMAIN (*green* online) and the C-DOMAIN (*blue* online) of each chain are highlighted. The domains are V-ALPHA and C-ALPHA for the TR-ALPHA chain, V-BETA and C-BETA for the TR-BETA chain (*see* **Note 8**) [3]. A MH1 is made of the I-ALPHA chain with two G-DOMAIN (G-ALPHA1 and G-ALPHA2) and a C-LIKE-DOMAIN (C-LIKE), non-covalently associated with the B2M (a C-LIKE-DOMAIN) [63]. The two G-DOMAIN (*yellow* online) and the C-LIKE (*blue* online) are highlighted. The TR/pMH1 complex structure is 3qfj from IMGT/3Dstructure-DB (http://www.imgt.org)

The V domain strands and loops and their delimitations and lengths, based on the IMGT unique numbering for V domain [59–61, 64], are shown in Table 1.

In the IG and TR V-DOMAIN, the three hypervariable loops BC, C′C″, and FG involved in the ligand recognition (antigen for IG and pMH for TR) are designated complementarity determining regions (CDR-IMGT) (*see* **Note 10**), whereas the strands form the framework region (FR-IMGT), which includes FR1-

IMGT, FR2-IMGT, FR3-IMGT, and FR4-IMGT (Table 1). For a V domain, the BC loop (or CDR1-IMGT in a V-DOMAIN) encompasses positions 27–38, the C′C″ loop (or CDR2-IMGT in a V-DOMAIN) positions 56–65, and the FG loop (or CDR3-IMGT) positions 105–117. In a V-DOMAIN, the CDR3-IMGT encompasses the V-(D)-J junction that results from a V-J or V-D-J rearrangement [2, 3] and is more variable in sequence and length than the CDR1-IMGT and CDR2-IMGT that are encoded by the V-REGION only. For CDR3-IMGT of length >13 AA, additional IMGT positions are added at the top of the loop between 111 and 112 (*see* **Note 11**).

The loop and strand lengths are visualized in the IMGT Colliers de Perles [65–68] which can be displayed on one layer (closer to the amino acid sequence) or on two layers (closer to the 3D structure) (Fig. 5). The lengths of the three loops, BC, C′C″, and FG (or CDR1-IMGT, CDR2-IMGT, and CDR3-IMGT for a V-DOMAIN) are delimited by the IMGT anchors, which are shown in square in the IMGT Colliers de Perles (*see* **Note 12**). In biological data, the lengths of the loops and strands are given by the number of occupied positions (unoccupied positions or "IMGT gaps" are represented with hatches in the IMGT Collier de Perles (Fig. 5) or by dots in alignments). The CDR-IMGT lengths are given in number of amino acids (or codons), into brackets and separated by dots: for example [9.6.9] means that the BC, C′C″, and FG loops (or CDR1-IMGT, CDR2-IMGT, and CDR3-IMGT for a V-DOMAIN) have a length of 9, 6, and 9 amino acids (or codons), respectively. Similarly [25.17.38.11] means that the FR1-IMGT, FR2-IMGT, FR3-IMGT, and FR4-IMGT have a length of 25, 17, 38, and 11 amino acids (or codons), respectively.

A V domain has five characteristic amino acids at given positions (positions with bold (online red) letters in the IMGT Colliers de Perles). Four of them are highly conserved and hydrophobic [28] (*see* **Note 13**) and are common to the C domain: 23 (1st-CYS), 41 (CONSERVED-TRP), 89 (hydrophobic), and 104 (2nd-CYS) (*see* **Note 9**). These amino acids contribute to the two major features shared by the V and C domain: the disulfide bridge (between the two cysteines 23 and 104) and the internal hydrophobic core of the domain (with the side chains of tryptophan W41 and amino acid 89). The fifth position, 118, is an anchor of the FG loop (*see* **Note 12**). It is occupied, in the V domains of IgSF other than IG or TR, by amino acids with different physicochemical properties [28]. In contrast, in IG and TR V-DOMAIN, that position 118 is occupied by remarkably conserved amino acids which consist in a phenylalanine or a tryptophan encoded by the J-REGION and therefore designated J-TRP or J-PHE 118 (*see* **Note 9**). The bulky aromatic side chains of J-TRP and J-PHE are internally orientated and structurally contribute to the V-DOMAIN hydrophobic core [61].

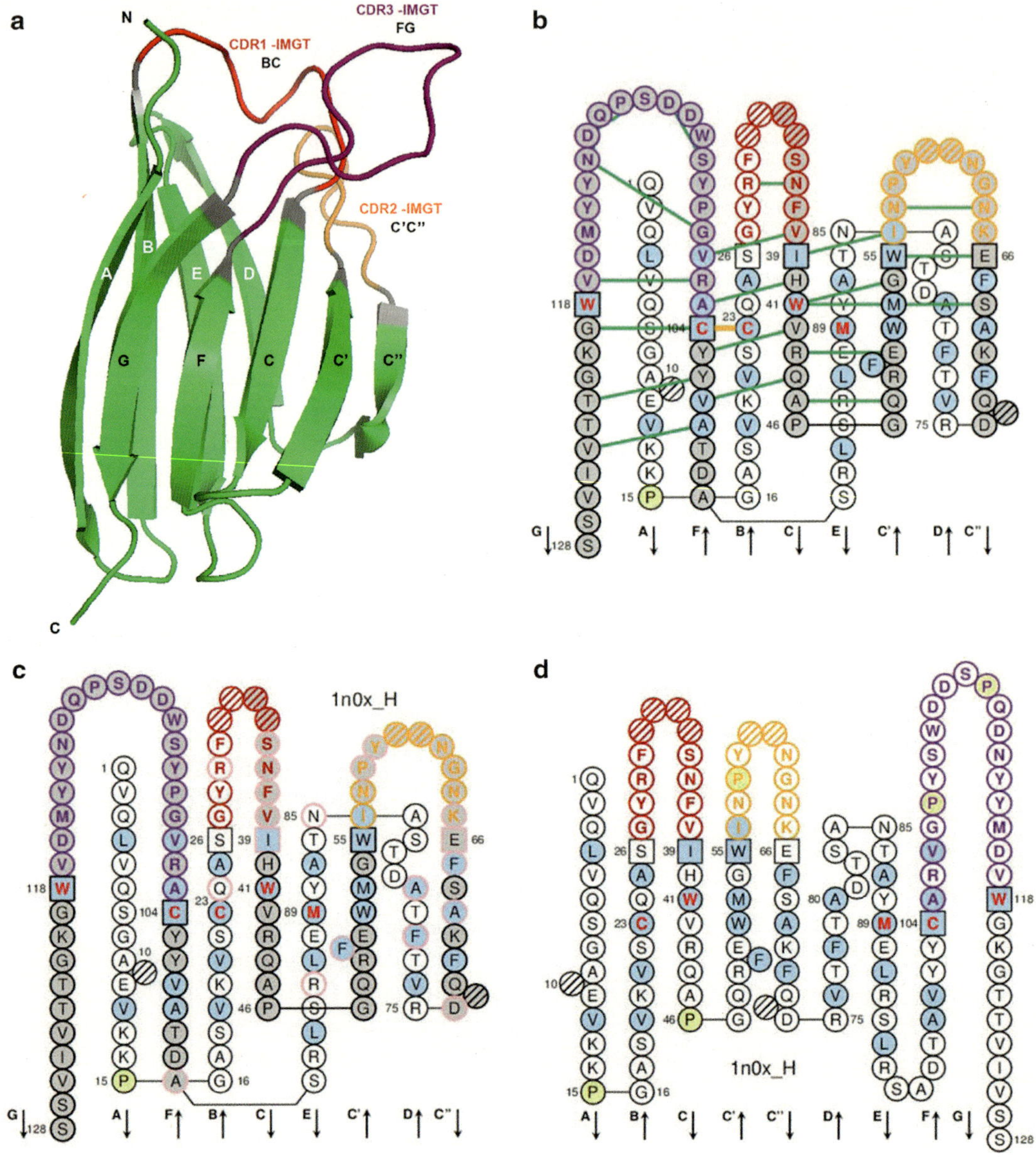

Fig. 5 Variable (V) domain. An IG VH (V-DOMAIN) is shown as example. (**a**) 3D structure ribbon representation with the IMGT strand and loop delimitations [61]. (**b**) IMGT Collier de Perles on two layers with hydrogen bonds. The IMGT Collier de Perles on two layers show, in the forefront, the GFCC′C″ strands (forming the sheet located at the interface VH/VL of the IG) and, in the back, the ABED strands. The BC, C′C″, and FG loops (for a V-DOMAIN, CDR1-IMGT, CDR2-IMGT, and CDR3-IMGT, respectively) are represented in color online (for a VH, *red*, *orange*, and *purple*). The IMGT Collier de Perles with hydrogen bonds (*green lines* online, here only shown for the GFCC′C″ sheet) is generated by the IMGT/Collier-de-Perles tool integrated in IMGT/3Dstructure-DB, from the experimental 3D structure data [8–10]. (**c**) IMGT Collier de Perles on two layers generated from IMGT/DomainGapAlign [9, 24, 25]. *Pink circles* (online) indicate amino acid changes compared to the closest genes and alleles from the IMGT reference directory. (**d**) IMGT Collier de Perles on one layer. The CDR-IMGT lengths are [8.8.20] and the FR-IMGT are [25.17.38.11]. Amino acids are shown in the one-letter abbreviation (*see* **Note 9**). All proline (P) are shown online in *yellow*. IMGT anchors are in square (*see* **Note 12**). *Hatched circles* are IMGT

Table 1
V domain strands and loops, IMGT positions and lengths, based on the IMGT unique numbering for V domain (V-DOMAIN and V-LIKE-DOMAIN) [59–61, 64]

V domain strands and loops[a]	IMGT positions	Lengths[b]	Characteristic IMGT Residue@ Position[c]	V-DOMAIN FR-IMGT and CDR-IMGT
A-STRAND	1–15	15 (14 if gap at 10)		FR1-IMGT
B-STRAND	16–26	11	1st-CYS 23	
BC-LOOP	27–38	12 (or less)		CDR1-IMGT
C-STRAND	39–46	8	CONSERVED-TRP 41	FR2-IMGT
C′-STRAND	47–55	9		
C′C″-LOOP	56–65	10 (or less)		CDR2-IMGT
C″-STRAND	66–74	9 (or 8 if gap at 73)		FR3-IMGT
D-STRAND	75–84	10 (or 8 if gaps at 81, 82)		
E-STRAND	85–96	12	hydrophobic 89	
F-STRAND	97–104	8	2nd-CYS 104	
FG-LOOP	105–117	13 (or less, or more)		CDR3-IMGT
G-STRAND	118–128	11 (or 10)	V-DOMAIN J-PHE 118 or J-TRP 118[d]	FR4-IMGT

[a]IMGT® labels (concepts of description) are written in capital letters (no plural) [57] (*see* **Note 2**). Beta turns (AB, CC′, C″D, DE, or EF) are individualized only if they have additional AA compared to the standard description. If not, they are included in the strands

[b]In number of amino acids (or codons)

[c]IMGT Residue@Position is a given residue (usually an amino acid) or a given conserved property amino acid class, at a given position in a domain, based on the IMGT unique numbering [64]

[d]In the IG and TR V-DOMAIN, the G-STRAND (or FR4-IMGT) is the C-terminal part of the J-REGION, with J-PHE or J-TRP 118 and the canonical motif F/W-G-X-G at positions 118–121 [2, 3]. The JUNCTION refers to the CDR3-IMGT plus the two anchors 2nd-CYS 104 and J-PHE or J-TRP 118 [60, 61]. The JUNCTION (positions 104–118) is therefore two amino acids longer than the corresponding CDR3-IMGT (positions 105–117) [2, 3]

Fig. 5 (continued) gaps according to the IMGT unique numbering for V domain [61, 64]. Positions with *bold* (online *red*) *letters* indicate the four conserved positions that are common to a V domain and to a C domain: 23 (1st-CYS), 41 (CONSERVED-TRP), 89 (hydrophobic), 104 (2nd-CYS) [59–62, 64], and the fifth conserved position that is specific to the IG and TR V-DOMAIN: 118 (J-TRP or J-PHE) [61, 64] (Table 1). In an IG or TR V-DOMAIN, the hydrophobic amino acids (hydropathy index with positive value: I, V, L, F, C, M, A) and tryptophan (W) [28] found at a given position in more than 50 % of sequences are shown (online with a *blue background color*). The FR4-IMGT is at least composed of nine or ten amino acids beyond the phenylalanine F (J-PHE 118) or tryptophan W (J-TRP 118) of the motif F/W-G-X-G that characterizes the J-REGION. Arrows indicate the direction of the beta strands and their designations in 3D structures. The identifier of the chain to which the VH domain belongs is 1n0x_H (from the *Homo sapiens* b12 Fab) in IMGT/3Dstructure-DB (http://www.imgt.org). The 3D ribbon representation was obtained using PyMOL (http://www.pymol.org) and "IMGT numbering comparison" of 1n0x_H (VH) from IMGT/3Dstructure-DB (http://www.imgt.org)

A last criterion used in the IMGT® definitive system for the characterization of a V domain is its delimitation taking into account the exon delimitations, if appropriate (*see* **Note 14**). This genomic approach integrates the strands A and G, in contrast to structural alignments that usually lack these strands due to their poor structural conservation, and bridges the gap between genomic data (exon) and 3D structure (domain).

3 C Domain IMGT® Definitive System

In the IMGT® definitive system, the C domain includes the C-DOMAIN of the IG (Fig. 3) [2] (*see* **Note 7**) and of the TR (Fig. 4) [3] (*see* **Note 8**) and the C-LIKE-DOMAIN of the IgSF other than IG and TR [91–95]. The C domain description of any receptor, any chain and any species is based on the IMGT unique numbering for C domain (C-DOMAIN and C-LIKE-DOMAIN) [62, 64].

A C domain (Fig. 6) comprises about 90–100 amino acids and is made of seven antiparallel beta strands (A, B, C, D, E, F, and G) linked by beta turns (AB, DE, and EF), a transversal strand (CD) and loops (BC and FG), and forming a sandwich of two sheets [ABED] [GFC] [62, 64].

A C domain has a topology and a three-dimensional structure similar to that of a V domain but without the C′ and C″ strands and the C′C″ loop [62].

The C domain strands, turns, and loops and their delimitations and lengths, based on the IMGT unique numbering for C domain [62, 64], are shown in Table 2.

The lengths of the strands and loops are visualized in the IMGT Colliers de Perles [66–68], on one layer and two layers (Fig. 6). The loops BC and FG and the transversal strand CD are delimited by the IMGT anchors (*see* **Note 12**).

In the IMGT® definitive system, the C domains (C-DOMAIN and C-LIKE-DOMAIN) are delimited taking into account the exon delimitation, if appropriate (*see* **Note 14**). As for the V domain, this genomic approach integrates the strands A and G which are absent of structural alignments.

Fig. 6 (continued) positions that are common to a V domain and to a C domain: 23 (1st-CYS), 41 (CONSERVED-TRP), 89 (hydrophobic), 104 (2nd-CYS) [59–62, 64] (Table 2) and position 118 which, as the V domain in general but in contrast to the V-DOMAIN, is not conserved in the C domain. The identifier of the chain to which the CH domain belongs is 1n0x_H (of the *Homo sapiens* b12 Fab) from IMGT/3Dstructure-DB (http://www.imgt.org). The 3D ribbon representation was obtained using PyMOL and "IMGT numbering comparison" of 1n0x_H (CH1) from IMGT/3Dstructure-DB (http://www.imgt.org)

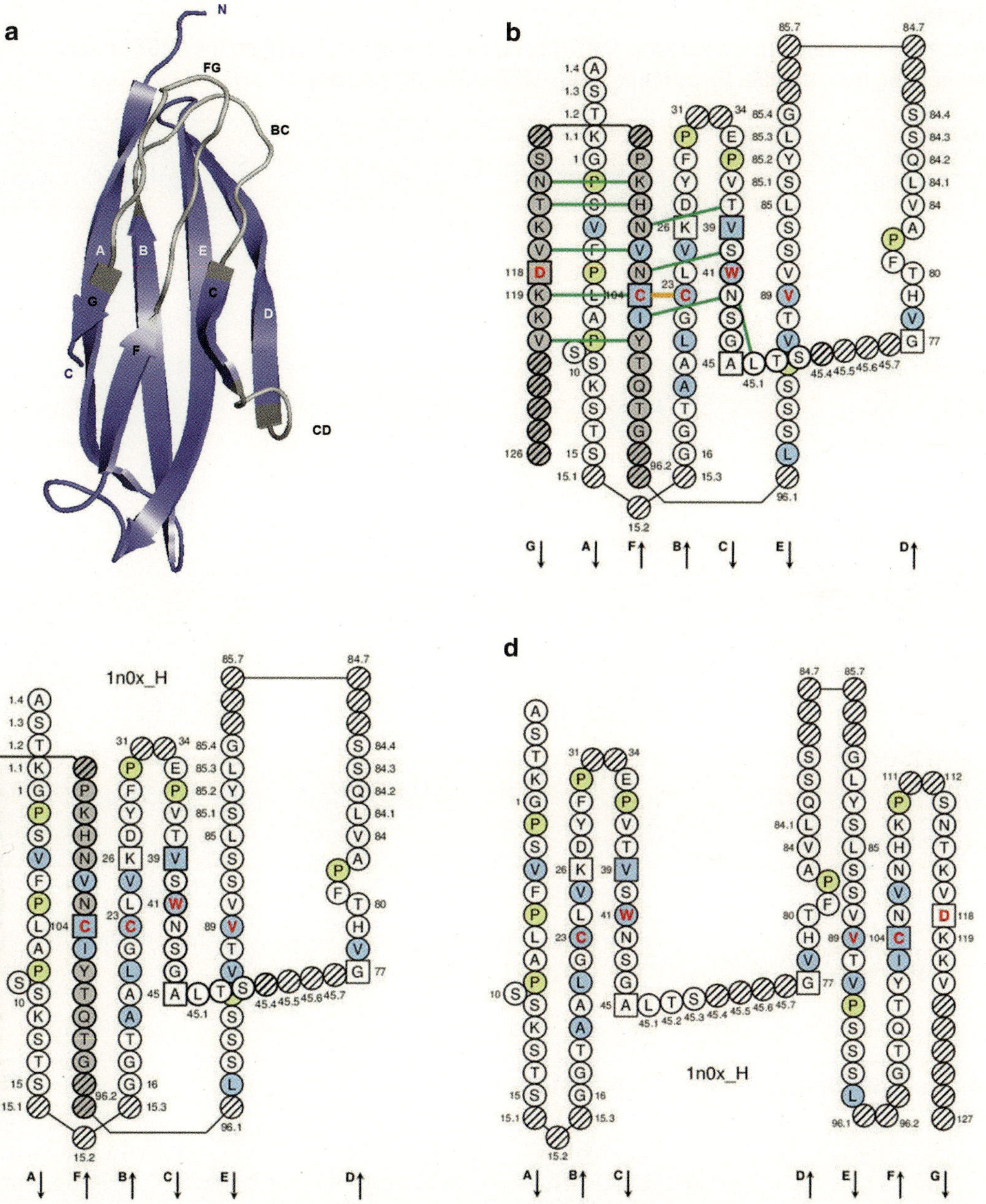

Fig. 6 Constant (C) domain. An IG CH (C-DOMAIN) is shown as example. (**a**) 3D structure ribbon representation with the IMGT strand and loop delimitations [62]. (**b**) IMGT Collier de Perles on two layers with hydrogen bonds. The IMGT Colliers de Perles on two layers show, in the forefront, the GFC strands and, in the back, the ABED strands (located at the interface CH1/CL of the IG), linked by the CD transversal strand. The IMGT Collier de Perles with hydrogen bonds (*green lines* online, here only shown for the GFC sheet) was generated by the IMGT/Collier-de-Perles tool integrated in IMGT/3Dstructure-DB, from the experimental 3D structure data [8–10]. (**c**) IMGT Collier de Perles on two layers from IMGT/DomainGapAlign [9, 24, 25]. (**d**) IMGT Colliers de Perles on one layer. Amino acids are shown in the one-letter abbreviation (*see* **Note 9**). All proline (P) are shown online in *yellow*. IMGT anchors are in *square* (*see* **Note 12**). *Hatched circles* are IMGT gaps according to the IMGT unique numbering for C domain [62, 64]. Positions with *bold* (online *red*) *letters* indicate the four conserved

Table 2
C domain strands, turns, and loops, IMGT positions and lengths, based on the IMGT unique numbering for C domain (C-DOMAIN and C-LIKE-DOMAIN) [62, 64]

C domain strands, turns, and loops[a]	IMGT positions	Lengths[b]	Characteristic IMGT Residue@Position[c]
A-STRAND	1–15	15 (14 if gap at 10)	
AB-TURN	15.1–15.3	0–3	
B-STRAND	16–26	11	1st-CYS 23
BC-LOOP	27–31 34–38	10 (or less)	
C-STRAND	39–45	7	CONSERVED-TRP 41
CD-STRAND	45.1–45.9	0-9	
D-STRAND	77–84	8 (or 7 if gap at 82)	
DE-TURN	84.1–84.7 85.1–85.7	0–14	
E-STRAND	85–96	12	Hydrophobic 89
EF-TURN	96.1–96.2	0-2	
F-STRAND	97–104	8	2nd-CYS 104
FG-LOOP	105–117	13 (or less, or more)	
G-STRAND	118–128	11 (or less)	

[a]IMGT® labels (concepts of description) are written in capital letters (no plural) [57] (*see* **Note 2**)
[b]In number of amino acids (or codons)
[c]IMGT Residue@Position is a given residue (usually an amino acid) or a given conserved property amino acid class, at a given position in a domain, based on the IMGT unique numbering [64]

4 G Domain IMGT® Definitive System

In the IMGT® definitive system, the G domain includes the G-DOMAIN of the MH (Fig. 4) (*see* **Note 15**) [63, 64] and the G-LIKE-DOMAIN of the MhSF other than MH (or RPI-MH1Like) (*see* **Note 16**) [96, 97]. The G domain description of any receptor, any chain and any species is based on the IMGT unique numbering for G domain (G-DOMAIN and G-LIKE-DOMAIN) [63, 64].

A G domain (Fig. 7) comprises about 90 AA and is made of four antiparallel beta strands (A, B, C, and D) linked by turns (AB, BC, and CD), and of a helix; the helix sits on the beta strands, its axis forming an angle of about 40° with the strands [87, 88].

Two G domains are needed to form the MhSF groove made of a “floor” and two “walls” [63, 64]. Each G domain contributes by

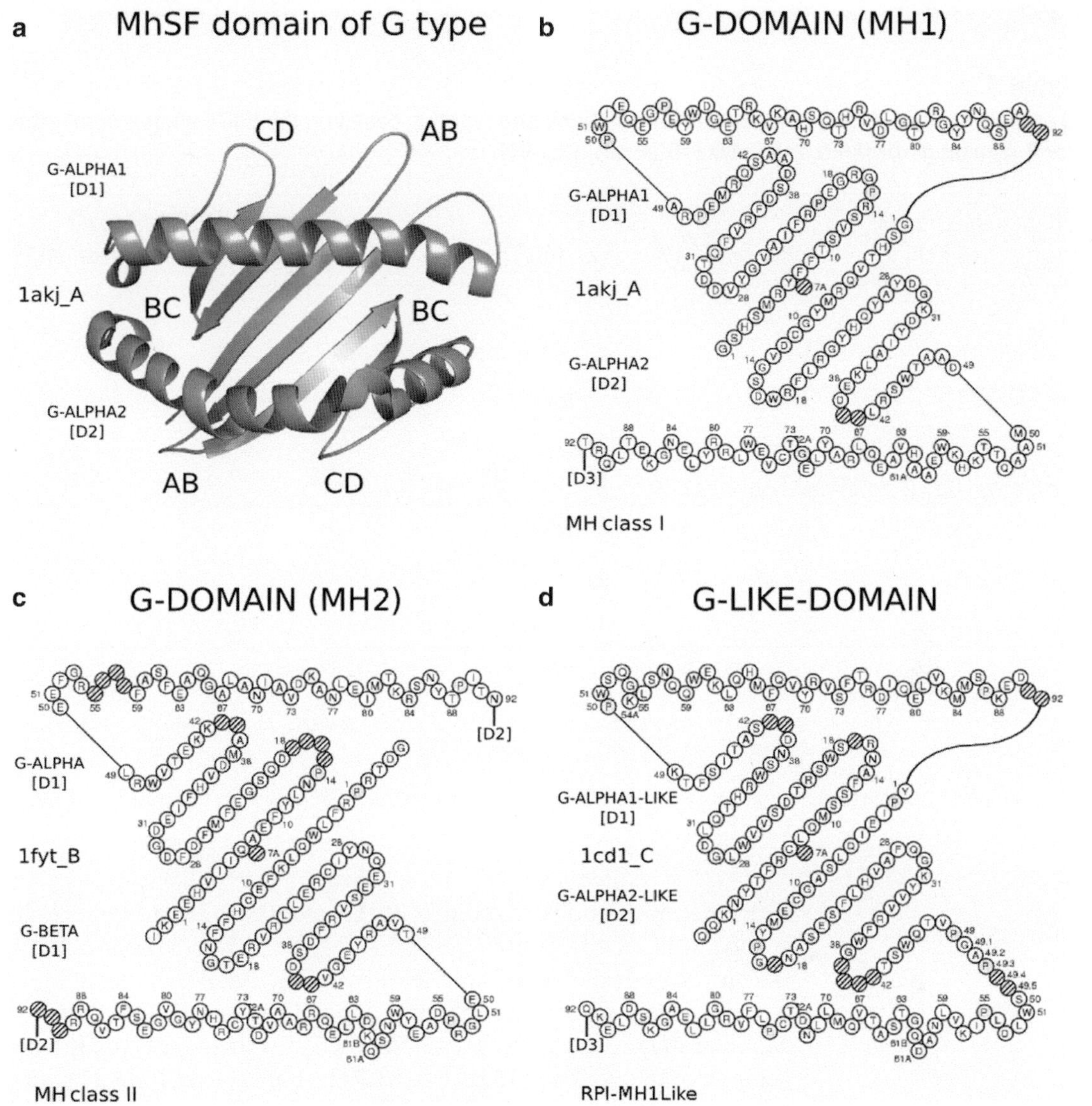

Fig. 7 Groove (G) domain. (**a**) 3D structure ribbon representation of the two G domains. The two domains form a groove with a "floor" (four strands from each domain) and two "walls" (one helix from each domain) [63]. The G domains characterize the proteins of the MhSF which comprises the MH (MH1 and MH2) and the RPI-MH1Like (MhSF other than MH) [63]. The two G-DOMAIN of a MH1 are shown as an example. The view is from above the cleft with the G-ALPHA1 (on top) and G-ALPHA2 (on bottom). (**b**) IMGT Colliers de Perles of the two G-DOMAIN of a MH1. G-ALPHA1 (on *top*) and G-ALPHA2 (on *bottom*) belong to the I-ALPHA chain [63]. (**b**) IMGT Colliers de Perles of the two G-DOMAIN of a MH2. G-ALPHA (on *top*) and G-BETA (on *bottom*) to the II-ALPHA chain and to the II-BETA chain, respectively [63]. (**c**) IMGT Colliers de Perles of the two G-LIKE-DOMAIN of a RPI-MH1Like. G-ALPHA1-LIKE (on *top*) and G-ALPHA2-LIKE (on *bottom*) belong to the I-ALPHA-LIKE chain. Helices are moved outside of the floor to make it visible. Amino acids are shown in the one-letter abbreviation (*see* **Note 9**). All proline (P) are shown online in *yellow. Hatched circles* are IMGT gaps according to the IMGT unique numbering for G domain [63, 64]. Domain numbers are shown between *brackets*. The 3D ribbon representation was obtained using PyMOL and "IMGT numbering comparison" of 1akj_A (G-ALPHA1 and G-ALPHA2) in IMGT/3Dstructure-DB (http://www.imgt.org). IMGT Colliers de Perles amino acid sequences are from 1akj_A for MH1 (*Homo sapiens* HLA-A*0201), 1fyt_A and 1fyt_B for MH2 (*Homo sapiens* HLA-DRA*0101 and HLA-DRB1*0101, respectively), and 1cd1_C for RPI-MH1Like (*Mus musculus* CD1D1). The IMGT Colliers de Perles were generated using the IMGT/Collier-de-Perles tool integrated in IMGT/3Dstructure-DB (http://www.imgt.org) [8–10]

Table 3
G domain strands, turns, and helix, IMGT positions and lengths, based on the IMGT unique numbering for G domain (G-DOMAIN and G-LIKE-DOMAIN) [63, 64]

G domain strands, turns, and helix[a]	IMGT positions	Lengths[b]	Characteristic IMGT Residue@Position[c] and additional positions[d]
A-STRAND	1–14	14	7A, CYS-11
AB-TURN	15–17	3 (or 2 or 0)	
B-STRAND	18–28	11 (or 10[e])	
BC-TURN	29–30	2	
C-STRAND	31–38	8	
CD-TURN	39–41	3 (or 1[f])	
D-STRAND	42–49	8	49.1–49.5
HELIX	50–92	43 (or less or more)	54A, 61A, 61B, 72A, CYS-74, 92A

[a]IMGT® labels (concepts of description) are written in capital letters (no plural) [57] (*see* **Note 2**)
[b]In number of AA (or codons)
[c]IMGT Residue@Position is a given residue (usually an amino acid) or a given conserved property amino acid class, at a given position in a domain, based on the IMGT unique numbering [64]
[d]For details on additional positions, *see* ref. 63
[e]Or 9 in some G-BETA [63]
[f]Or 0 in some G-ALPHA2-LIKE [63]

its four strands and turns to half of the groove floor and by its helix to one wall of the groove [63, 64, 87, 88]. The MH groove in which the peptide binds is made of two G-DOMAIN belonging to a single chain or to two chains, depending on the MH group, MH1 or MH2, respectively (*see* **Note 15**). In the MH1, the groove is made of two G-DOMAIN (G-ALPHA1 and G-ALPHA2) which belong to the same chain I-ALPHA (Fig. 7b), whereas in the MH2, the groove is made of two G-DOMAIN (G-ALPHA and G-BETA) which belong to two different chains, II-ALPHA and II-BETA, respectively (Fig. 7c). For the RPI-MH1Like (*see* **Note 16**), the two G-LIKE-DOMAIN also belong, as for the MH1, to the same chain (I-ALPHA-LIKE) [96, 97] (Fig. 7d).

The G domain strands, turns, and helix and their delimitations and lengths, based on the IMGT unique numbering for G domain [63, 64] are shown in Table 3.

The strands and helix of each domain are visualized in the IMGT Collier de Perles [66–68, 87, 88] (Fig. 7). The views are from above the cleft, (with the helices displaced to show the floor) and with on top and on bottom, respectively, G-ALPHA1 and G-ALPHA2 (MH1), G-ALPHA and G-BETA (MH2), and G-ALPHA1-LIKE and G-ALPHA2-LIKE (RPI-MH1Like). There is no link between G-ALPHA and G-BETA because they belong to different chains (II-ALPHA and II-BETA) (*see* **Note 15**). Two conserved cysteines,

CYS-11 (in the A strand) and CYS-74 (in the helix) (Table 3) are found in the G-ALPHA2, G-BETA, and G-ALPHA2-LIKE (Fig. 7), where they form a disulfide bridge fixing the helix to the floor.

In the IMGT® definitive system, the G domains (G-DOMAIN and G-LIKE-DOMAIN) are delimited taking into account the exon delimitations, if appropriate (alignment sequence comparison with previously identified genes are used when genomic data are not yet available as this was recently done for the rainbow trout (*Oncorhynchus mykiss*) MH1 and MH2 (IMGT® http://www.imgt.org, IMGT Repertoire (MH) > IMGT Proteins and alleles > Protein displays)).

5 IMGT® Tools for V, C, or G Domain Analysis

5.1 IMGT/Collier-de-Perles Tool

The IMGT/Collier-de-Perles tool [26], on the IMGT® Web site at http://www.imgt.org, allows the users to draw IMGT Colliers de Perles [65–68] starting from their own domain amino acid sequences (sequences already with IMGT gaps, using for example IMGT/DomainGapAlign (Table 4)).

IMGT Collier de Perles can be obtained for V and C domains (on one or two layers) and for G domains (with one or the two domains of the groove).

IMGT/Collier-de-Perles tool online can be customized to display the IG and TR CDR-IMGT according to the IMGT color menu and the amino acids according to their hydropathy or volume, or to the 11 IMGT physicochemical classes [28] (*see* **Note 13**). IMGT color menu for the CDR-IMGT of a V-DOMAIN indicates the type of rearrangement V-J or V-D-J [2, 3]. Thus, the IMGT color menu for CDR1-IMGT, CDR2-IMGT, and CDR3-IMGT is red, orange, and purple for the IG VH (*see* **Note 7**) and for the TR V-BETA or V-DELTA (*see* **Note 8**) (encoded by a V-D-J-REGION resulting from a V-D-J rearrangement), and blue, green, and green-blue for the IG V-KAPPA or V-LAMBDA (*see* **Note 7**) and for the TR V-ALPHA or V-GAMMA (see **Note 8**) (encoded by a V-J-REGION resulting from a V-J rearrangement). Arbitrarily the red, orange, and purple are used for the BC, C′C″, and FG loops of the V domain of IgSF other than IG or TR.

The IMGT/Collier-de-Perles tool is integrated in IMGT/DomainGapAlign [9, 24, 25] (users start from V, C, or G amino acid sequences) and in IMGT/V-QUEST [12–17] (users start from IG and TR V-DOMAIN nucleotide sequences) (Table 4). IMGT Colliers de Perles for V, C, and G domains are provided in IMGT/2Dstructure-DB (for amino acid sequences in the database) and in IMGT/3Dstructure-DB (on two layers with hydrogen bonds for the V or C domains or with the pMH contact sites for the G domains, for 3D structures in the database) [8–10] (Table 4).

Table 4
IMGT® tools and databases with functionalities for V, C, and G domain analysis

IMGT® tools and databases	Results of domain analysis	Domain type analysis[a]	Submitted entries	Detailed protocols
IMGT/Collier-de-Perles [26]	Graphical 2D representation of IMGT Colliers de Perles [65–68]	V, C, and G	User "IMGT gapped" amino acid sequences (one sequence)	[26]
IMGT/DomainGapAlign [9, 24, 25]	1. Introduction of IMGT gaps 2. Identification of the closest genes and alleles 3. Delimitation of the domains 4. Description of amino acid changes 5. IMGT Colliers de Perles [65–68] with highlighted AA changes (pink circles online)	V, C, and G	User amino acid sequences (one to several sequences of same domain type)	[24, 25]
IMGT/V-QUEST [12–17]	1. Introduction of IMGT gaps 2. Identification of the closest V, D, and J genes and alleles 3. IMGT/JunctionAnalysis results [18, 19] 4. Description of mutations and amino acid changes 5. Annotation by IMGT/Automat [20, 21] 6. IMGT Colliers de Perles [65–68]	V-DOMAIN (IG and TR)	User nucleotide sequences [1–50]	[16]
IMGT/HighV-QUEST [21–23]	1. Introduction of IMGT gaps 2. Identification of the closest V, D, and J genes and alleles 3. IMGT/JunctionAnalysis results [18, 19] 4. Description of mutations and amino acid changes 5. Annotation by IMGT/Automat [20, 21] 6. NGS statistical analysis [22] 7. Characterization of the IMGT clonotypes (AA) [23]	V-DOMAIN (IG and TR)	User nucleotide sequences (up to 50,000 and statistics on results	[22, 23]

IMGT/3Dstructure-DB [8–10]	1. Identification of the closest genes and alleles 2. IMGT/DomainGapAlign results [9, 24, 25] 3. IMGT Collier de Perles [65–68] on two layers with hydrogen bonds (for V and C) 4. IMGT Collier de Perles [65–68] with pMH contact sites (for G in pMH or in TR/pMH complexes) 5. Contact analysis between domain and ligand and between domains 6. Renumbered IMGT files 7. IMGT numbering comparison	V, C, and G	Nb of 3D structures entries in June 2014: 2,987	[10]
IMGT/2Dstructure-DB [10]*	1. Identification of the closest genes and alleles 2. IMGT/DomainGapAlign results [9, 24, 25] 3. IMGT Collier de Perles [65–68] 4. Renumbered IMGT files	V, C, and G	Nb of amino acid sequence entries in June 2014: 543	[10]*

An asterisk (*) indicates that parts of the protocol dealing with 3D structures (hydrogen bonds in IMGT Colliers de Perles on two layers, Contact analysis) are not relevant, otherwise all other queries and results are similar to IMGT/3Dstructure-DB

[a] *V* V domain (includes V-DOMAIN of IG and TR and V-LIKE-DOMAIN of IgSF other than IG and TR) [61]. *C* C domain (includes C-DOMAIN of IG and TR and C-LIKE-DOMAIN of IgSF other than IG and TR) [62]. *G* G domain (includes G-DOMAIN of MH and G-LIKE-DOMAIN of MhSF other than MH) [63]

5.2 IMGT/DomainGapAlign

IMGT/DomainGapAlign [9, 24, 25] is the IMGT® online tool for the analysis of amino acid sequences of V, C, and G domains (Table 4). IMGT/DomainGapAlign analyzes V, C, or G domain amino acid sequences (*see* **Note 17**) by comparison with the IMGT domain reference directory sets (*see* **Note 18**). IMGT/DomainGapAlign results include: introduction of "IMGT gaps" in the user amino acid sequences; alignments and identification of the genes and alleles by comparison with the closest domain(s); delimitation of the V, C, or G domain(s) in the user sequence (Fig. 8).

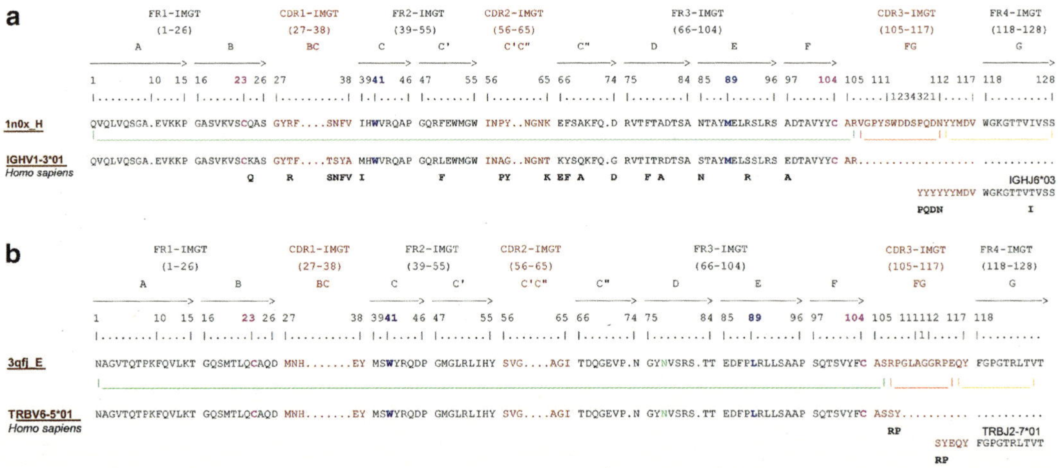

Fig. 8 IMGT/DomainGapAlign alignment results. The alignment results are shown for two V domains, VH (**a**) and V-BETA (**b**), as examples of V-DOMAIN which belong to different receptors and chains (IG-HEAVY and TR-BETA, respectively) (*see* **Notes 7** and **8**). The sequences submitted to IMGT/DomainGapAlign were ungapped amino acid sequences (*see* **Note 17**). The nine strands (A, B, C, C′, C″, D, E, F, and G) (*horizontal arrows*), and the three loops (BC, C′C″, and FG) are shown according to the IMGT unique numbering for V domain [59–61, 64], with the *upper line*, the V-DOMAIN FR-IMGT and CDR-IMGT delimitations (with start and end positions). The closest genes and alleles are identified automatically by IMGT/DomainGapAlign by comparison with the IMGT domain reference directory (V and J genes and alleles for a V-DOMAIN) (*see* **Note 18**). The VH sequence (from b12 Fab, 1n0x_H) is identified as having 79.6 % and 93.8 % identity (results online, above the alignment) with *Homo sapiens* IGHV1-3*01 and IGHJ6*03, respectively (*see* **Note 3**). The VH CDR-IMGT lengths are [8.8.20] and the FR-IMGT lengths [25.17.35.11] = 91 AA (results online, below the alignment). The V-BETA sequence (from A6 TR, 3qfj_E) is identified as having 100 % identity with *Homo sapiens* TRBV6-5*01 and TRBJ2-7*01 (*see* **Note 3**). The V-BETA CDR-IMGT lengths are [5.6.14] and the FR-IMGT lengths [26.17.37.10] = 90 AA (results online, below the alignment). The V-REGION of the b12 VH sequence is heavily mutated [2] as shown by the high number of amino acid changes (20, shown in *bold* below the alignment, and detailed per strand and per loop online in the IMGT/DomainGapAlign results) [9, 24, 25]. One AA change is also observed in the FR4-IMGT (T125 > I). In contrast the V-REGION of the V-BETA is unmutated, as expected for a TR [3]. The region localized in the CDR3-IMGT which results from the V-(D)-J rearrangement (Fig. 1) and which cannot be identified as being V or J is the (N-D)-REGION. Conserved AA are in *bold* and in color online: C23 (*pink*), W41 (*blue*), hydrophobic 89 (*blue*, here M, L), and C104 (*pink*). An N (*see* **Note 9**) in *green* online in the V-BETA (N77) indicates an N-glycosylation site (motif N-X-S/T). *Horizontal lines* below the user sequence indicate the domain and here, for a V-DOMAIN, its regions (in color online): *green* for V-REGION, *red* for (N-D)-REGION, and *yellow* for J-REGION (IMGT®, http://www.imgt.org, IMGT Scientific chart > IMGT color menu)

If several closest genes and/or alleles are identified, the user can select the display of each corresponding alignment. Clicking on the user sequence name in the alignment gives access to the IMGT/Collier-de-Perles tool [26] which automatically provides the IMGT Collier de Perles of the analyzed domain [65–68] with highlighted amino acid differences (in pink circles online) with the closest reference sequence (Fig. 5c).

The user amino acid sequence is also displayed, according to the IMGT color menu, with the delimitations of the domains (and for the V-DOMAIN, the V-REGION and J-REGION, and if present, the (N-D)-REGION) identified by the tool. The characteristics of the AA changes are shown in strands and loops (and for the V-DOMAIN, in FR-IMGT and CDR-IMGT).

IMGT/DomainGapAlign is very popular for antibody humanization as it allows the comparison of the user V-DOMAIN against reference sequences of the V and J regions of other species (e.g., mouse, rat, human) and the delimitation and characterization of the FR-IMGT and of the CDR-IMGT to be grafted [32, 39–41, 43].

5.3 IMGT/V-QUEST

IMGT/V-QUEST [12–17] is the IMGT® online tool for the analysis of nucleotide sequences of the IG and TR V-DOMAIN (Table 4). IMGT/V-QUEST identifies the variable (V), diversity (D) and junction (J) genes in rearranged IG and TR sequences and, for the IG, the nucleotide (nt) mutations and amino acid (AA) changes resulting from somatic hypermutations by comparison with the IMGT/V-QUEST reference directories (*see* **Note 19**). The tool integrates IMGT/JunctionAnalysis [18, 19] for the detailed characterization of the V-D-J or V-J junctions (*see* **Note 20**), IMGT/Automat [20, 21] for a complete sequence annotation, and IMGT/Collier-de-Perles [26].

IMGT/V-QUEST functionalities include: introduction of "IMGT gaps" in the user nucleotide sequences (and in its translation); alignments and identification of the genes and alleles with the closest germline V, D, and J genes (*see* **Note 3**), analysis of somatic hypermutations (*see* **Note 21**) and amino acid changes (*see* **Note 13**), analysis of the junctions (*see* **Note 22**), and identification of insertions and deletions (indels) and their correction (*see* **Note 23**).

Customized parameters and results provided by IMGT/V--QUEST and IMGT/JunctionAnalysis have been described elsewhere [12–17].

5.4 IMGT/HighV-QUEST

IMGT/HighV-QUEST [22] is the high-throughput version of IMGT/V-QUEST. It is so far the only online tool available for the direct analysis of complete IG and TR domain sequences from Next Generation Sequencing (NGS). It analyzes sequences, preferentially long sequences obtained e.g., from Roche 454, without the need of computational read assembly [21–23] (Table 4). IMGT/HighV-QUEST analyzes up to 50,000 sequences per run and performs statistical analysis on the results [22, 23], with the same degree of resolution and high quality results as IMGT/V-QUEST [12–17]. The option "Search for insertion/deletion" (*see* **Note 23**), added by default, allows an accurate V-DOMAIN analysis, despite the high frequency of indels due to homopolymer hybridization sequencing errors in NGS 454 sequences. IMGT/HighV-QUEST represents a major breakthrough for the analysis and the comparison of the huge repertoires of antigen receptor V-DOMAIN (potentially 2×10^{12} per individual), by the recent standardized characterization of clonotypes or "IMGT clonotypes (AA)," with for the first time for NGS data, a clear distinction between clonal diversity and expression [23]. Since its launch in October 2010, 846 users from 40 countries have been registered (2.5 billions of analyzed sequences in June 2014 with 62 % from the USA, 25 % from EU, 13 % from the remaining world).

6 V, C, and G Domain Analysis in IMGT® Databases

6.1 IMGT/3Dstructure-DB

IMGT/3Dstructure-DB [8–10], the IMGT® structure database, provides IMGT annotation and contact analysis on receptors and chains which contain V, C, and/or G domains and for which 3D structures are available (Table 4). The "PDB code" (four letters and/or numbers, e.g., 1hzh) is used as "IMGT entry ID" for the 3D structures obtained from the Research Collaboratory for Structural Bioinformatics (RCSB) Protein Data Bank (PDB) [98]. The IMGT/3Dstructure-DB card provides eight search/display options: "Chain details," "Contact analysis," "Paratope and epitope," "3D visualization Jmol or QuickPDB," "Renumbered IMGT files," "IMGT numbering comparison," "References and links," "Printable card."

The "Chain details" section comprises information first on the chain itself, then per domain (*see* **Notes 7**, **8** and **15**). Chain and domain annotation includes the IMGT gene and allele names (*see* **Note 3**), region and domain delimitations (*see* **Note 2**) and domain amino acid (AA) positions according to the IMGT unique numbering [59–64] (Subheadings 2–4). The closest IMGT genes and alleles (found expressed in each domain of a chain) are identified with the integrated IMGT/DomainGapAlign [9, 24, 25], which aligns the AA sequences of the 3D structures with the IMGT domain reference directory (*see* **Note 18**). "Contact analysis" (*see* **Note 24**) gives access, by Clicking on "Domain contact (overview)," to a table with the different "Domain pair contacts" of the 3D structure. "Domain pair contacts" are contacts between a pair of domains (V, C, and/or G) or between a domain and a ligand. Clicking on "DomPair" gives access to a given "Domain pair contacts," in which the atom pair contacts are described at the level of amino acids at a given position in a domain (or IMGT Residue@Position). Clicking on "R@P" gives access to an individual "IMGT Residue@Position" card (*see* **Note 25**). The IMGT Residue@

Position cards can also be accessed directly from the amino acid sequences of the IMGT/3Dstructure-DB card or from the IMGT Colliers de Perles, by clicking on one AA.

For IG/antigen [43] and TR/pMH [87, 88] complexes, a detailed and standardized description of paratope (amino acids of the V-DOMAIN in contact with the antigen) and epitope (residues of the antigen or of the pMH in contact with the paratope) is provided, on the basis of the contact analysis.

"Renumbered IMGT file" allows to view (or download) an IMGT coordinate file renumbered according to the IMGT unique numbering, and in which the chain and domain IMGT specific information (identical to that provided in "Chain details") is added in the "REMARK 410" lines (blue online). Tools associated to IMGT/3Dstructure-DB include IMGT/StructuralQuery [8] and IMGT/DomainSuperimpose, available online. IMGT/StructuralQuery allows to retrieve the IMGT/3Dstructure-DB entries, based on specific structural characteristics of the intramolecular interactions: phi and psi angles, accessible surface area, type of atom contacts, distance in angstrom between amino acids, IMGT Residue@Position contacts and, for V-DOMAIN, CDR-IMGT length or pattern [8]. IMGT/DomainSuperimpose allows to superimpose the 3D structures of two domains from IMGT/3Dstructure-DB.

6.2 IMGT/2Dstructure-DB

IMGT/2Dstructure-DB was created as an extension of IMGT/3Dstructure-DB [8–10] to describe and analyze amino acid sequences of chains and domains for which no 3D structures were available (Table 4). IMGT/2Dstructure-DB uses the IMGT/3Dstructure-DB informatics frame and interface which allow one to analyze, manage, and query IG or antibodies, TR and MH, as well as other IgSF and MhSF and engineered proteins (FPIA, CPCA), as polymeric receptors made of several chains, in contrast to the IMGT/LIGM-DB sequence database that analyzes and manages sequences individually [6]. The amino acid sequences are analyzed and managed with the IMGT® criteria of standardized nomenclature (*see* **Note 3**), description (*see* **Note 2**), and numerotation [59–64] (Subheadings 2–4). The current IMGT/2Dstructure-DB entries include amino acid sequences of antibodies from Kabat [99] (those for which there were no available nucleotide sequences), and amino acid sequences of mAb and FPIA from the WHO-INN programme [11, 47, 48] (*see* **Note 6**). Queries can be made on an individual entry, using the Entry ID or the Molecule name. The same query interface is used for IMGT/2Dstructure-DB and IMGT/3Dstructure-DB. Thus a "trastuzumab" query in "Molecule name" allows to retrieve three results: two INN ("trastuzumab" and "trastuzumab emtansine") from IMGT/2Dstructure-DB, and one 3D structure ("1nz8") from IMGT/3Dstructure-DB. The IMGT/2Dstructure-DB cards provide standardized IMGT information on chains and domains and IMGT Colliers de Perles on one or two layers, identical to that provided for the sequence analysis in IMGT/3Dstructure-DB, however the information on experimental structural data (hydrogen bonds in IMGT Collier de Perles on two layers, Contact analysis) is only available in the corresponding IMGT/3Dstructure-DB cards, if the antibodies have been crystallized.

7 V, C, and G Domain Annotation and Contact Analysis Using the IMGT® Definitive System

A TR/pMH complex (Fig. 4) provides an example of a structure that contains the three domain types: two V domains (V-ALPHA and V-BETA of the TR (*see* **Note 8**)), four C domains (C-ALPHA and C-BETA of the TR (*see* **Note 8**) and C-LIKE of the MH1 I-ALPHA and of the B2M (*see* **Note 15**)) and two G domains (G-ALPHA1 and G-ALPHA2 of the MH1 I-ALPHA (*see* **Note 15**)). The IMGT/3Dstructure card of the TR/pMH shown in Fig. 4 (*see* **Note 26**) can be accessed by typing its PDB code (3qfj) in the "Entry code" window of the IMGT/3Dstructure-DB and IMGT/2Dstructure-DB Query page (http://www.imgt.org). Snapshots of the IMGT domain annotation and IMGT contact analysis for the V, C, and G domains of this complex are described below.

7.1 IMGT Domain Annotation

In the IMGT/3Dstructure card for 3qfj, the TR is described as a "TR-ALPHA_BETA-2," with a TR-ALPHA chain (3qfj_D) and a TR-BETA chain (3qfj_E). The TR-ALPHA chain comprises the V-ALPHA (1–110) [D1]+C-ALPHA (112–200) [D2]. The V-REGION and J-REGION of the V-ALPHA have 100 % identity with the human TRAV12-2*01 and TRAJ24*02, respectively, and the V-ALPHA CDR-IMGT lengths are [6.6.11]. The C-ALPHA has 100 % identity with the human TRAC*01. The TR-BETA-2 chain comprises the V-BETA (1–115) [D1]+C-BETA-2 (116–244) [D2]. The V-REGION and J-REGION of the V-BETA have 100 % identity with the human TRBV6-5*01 and TRBJ2-7*01, respectively, and the V-BETA CDR-IMGT lengths are [5.6.14]. The C-BETA-2 has 98.40 % identity with TRBC2*01 owing to two in vitro AA changes, C85.1>A and N97>D.

The MH1 is described as MH1-ALPHA_B2M, with an I-ALPHA chain (3qfj_A) and the B2M (3qfj_B). The I-ALPHA chain comprises the G-ALPHA1 [1–90] [D1]+G-ALPHA2 (91–182) [D2]+C-LIKE (183–274) [D3], and each of the three domains has 100 % identity with HLA-A*0201. The B2M chain only comprises a C-LIKE [2–100] [D1] and has 100 % with the human B2M*01.

For each domain, the IMGT Colliers de Perles can be obtained by clicking on "IMGT Collier de Perles" (results in panel d in Figs. 9, 10 and 11) and, for the visualization of the hydrogen bonds (for V and C), on "IMGT Collier de Perles on 2 layers" (results in panel b in Figs. 9, 10 and 11).

IMGT Colliers de Perles can be also obtained via the *IMGT/DomainGapAlign results* (results in panel c in Figs. 9, 10 and 11). The IMGT Colliers de Perles of the TR V-BETA (Fig. 9) and V-ALPHA (Fig. 10) can be compared with those of the IG VH (Fig. 5) and the IMGT Colliers de Perles of the TR C-BETA (Fig. 11) with those of the IG CH1 (Fig. 6).

7.2 IMGT Contact Analysis

For G domains of pMH and TR/pMH complexes, a link to "IMGT pMH contact sites" [87, 88] is available that gives access to the IMGT Colliers de Perles of G domains with pMH contact sites (Fig. 12).

"IMGT pMH contact sites" are calculated from the experimental structural data and allow one to easily identify the peptide amino acids of the peptide which are effectively located in the groove. This display is of great interest for pMH2 and TR/pMH2 complexes in which the peptides are longer than the groove [63, 87, 88].

"Domain pair contacts (overview)" gives access, by clicking on a "DomPair," to the contacts between a pair of domains or between a domain and the ligand. Contacts between pairs of domains include, for examples, contacts between V-BETA and G-ALPHA1, V-BETA and G-ALPHA2 (Fig. 13), between V-ALPHA and G-ALPHA1, V-ALPHA and G-ALPHA2 (Fig. 14). Contacts between the TR domains (V-ALPHA, V-BETA) and the Ligand (peptide) are shown in Fig. 15.

The paratope includes the amino acids of the V-ALPHA and V-BETA which have contacts with the G-ALPHA1, G-ALPHA2 and peptide (displayed in Figs. 13, 14, and 15, and listed in the legends). Reciprocally, the epitope includes the amino acids of the G-ALPHA1, G-ALPHA2 and peptide that have contacts with the V-ALPHA and V-BETA (displayed in Figs. 13, 14 and 15, and listed in the legends).

8 Availability and Citation

Authors who use IMGT® databases and tools are encouraged to cite this article and to quote the IMGT® Home page, http://www.imgt.org. Online access to IMGT® databases and tools are freely available for academics and under licences and contracts for companies.

9 Notes

1. More than 325 IMGT® standardized keywords (189 for sequences and 137 for 3D structures) were precisely defined [56]. They represent the controlled vocabulary assigned dur-

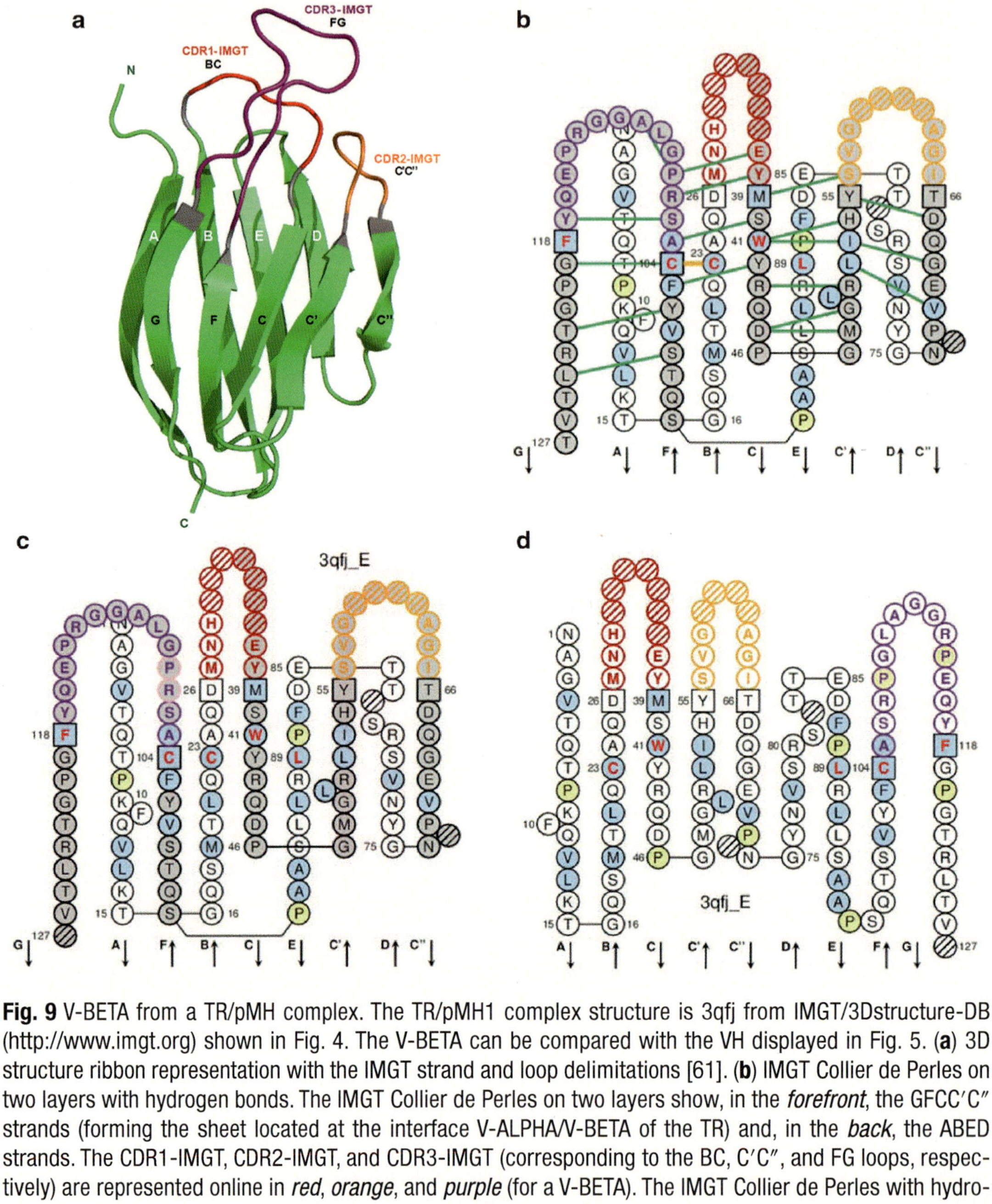

Fig. 9 V-BETA from a TR/pMH complex. The TR/pMH1 complex structure is 3qfj from IMGT/3Dstructure-DB (http://www.imgt.org) shown in Fig. 4. The V-BETA can be compared with the VH displayed in Fig. 5. (**a**) 3D structure ribbon representation with the IMGT strand and loop delimitations [61]. (**b**) IMGT Collier de Perles on two layers with hydrogen bonds. The IMGT Collier de Perles on two layers show, in the *forefront*, the GFCC′C″ strands (forming the sheet located at the interface V-ALPHA/V-BETA of the TR) and, in the *back*, the ABED strands. The CDR1-IMGT, CDR2-IMGT, and CDR3-IMGT (corresponding to the BC, C′C″, and FG loops, respectively) are represented online in *red*, *orange*, and *purple* (for a V-BETA). The IMGT Collier de Perles with hydrogen bonds (*green lines* online, here only shown for the GFCC′C″ sheet) was generated by the IMGT/Collier-de-Perles tool integrated in IMGT/3Dstructure-DB [8–10]. (**c**) IMGT Collier de Perles on two layers generated from IMGT/DomainGapAlign [9, 24, 25]. (**d**) IMGT Collier de Perles on one layer. The CDR-IMGT lengths are [5.6.14] and the FR-IMGT are [26.17.37.10]. Amino acids are shown in the one-letter abbreviation (*see* **Note 9**). All proline (P) are shown online in *yellow*. IMGT anchors are in *square* (*see* **Note 12**). *Hatched circles* are IMGT gaps according to the IMGT unique numbering for V domain [61, 64]. Positions with *bold* (online *red*) *letters* indicate the four conserved positions that are common to a V domain and to a C domain: 23 (1st-CYS), 41 (CONSERVED-TRP), 89 (hydrophobic), 104 (2nd-CYS) [59–62, 64], and the fifth conserved position that is specific to the IG and TR V-DOMAIN: 118 (here, J-PHE) which belongs to the motif F/W-G-X-G that

ing the annotation process and allow standardized search criteria for querying the IMGT® databases and for the extraction of sequences and 3D structures. Standardized keywords assigned to nucleotide sequences are found in the "DE" (definition) and "KW" (keyword) lines of the flat files of IMGT/LIGM-DB, the IMGT® nucleotide sequences database [6] (Fig. 2). They characterize for instance the gene type, the configuration type and the functionality type. There are six gene types: variable (V), diversity (D), joining (J), constant (C), conventional-with-leader, and conventional-without-leader. Four of them (V, D, J and C) identify the IG and TR genes and are specific to immunogenetics. There are four configuration types: germline (for the V, D, and J genes before DNA rearrangement), rearranged (for the V, D, and J genes after DNA rearrangement (Fig. 1)), partially-rearranged (for D gene after only one DNA rearrangement), and undefined (for the C gene and for the conventional genes which do not rearrange). The functionality type depends on the gene configuration. The functionality type of genes in germline or undefined configuration is functional (F), ORF (for "open reading frame"), or pseudogene (P). The functionality type of genes in rearranged or partially-rearranged configuration is either productive (no stop codon in the V-(D)-J region and in-frame junction) or unproductive (stop codon(s) in the V-(D)-J region, and/or out-of-frame junction). IMGT-ONTOLOGY concepts of identification have been entered in BioPortal at the National Center for Biomedical Ontology (NCBO) in 2010 (http://bioportal.bioontology.org/ontologies/1491).

2. More than 560 IMGT® standardized labels (277 for sequences and 285 for 3D structures) were precisely defined [57]. They are written in capital letters (no plural) to be recognizable without creating new terms. Standardized labels assigned to the description of sequences are found in the "FT" lines of the flat files of IMGT/LIGM-DB [6] (Fig. 2). Querying these labels represent a big plus compared to the generalist databases (GenBank/European Nucleotide Archive (ENA)/DNA Data Bank of Japan (DDBJ)). Thus it is possible to query for the "CDR3-IMGT" of the human rearranged productive sequences of IG-Heavy-Gamma (e.g., 1,733 CDR3-IMGT

Fig. 9 (continued) characterizes the J-REGION [61, 64] (Table 1). The hydrophobic amino acids (hydropathy index with positive value: I, V, L, F, C, M, A) and tryptophan (W) [28] found at a given position in more than 50 % of sequences are shown (online with a *blue background color*). *Arrows* indicate the direction of the beta strands and their designations in 3D structures. The identifier of the chain to which the V-BETA domain belongs is 3qfj_E (of the *Homo sapiens* A6 TR) in 3qfj from IMGT/3Dstructure-DB (http://www.imgt.org). The 3D ribbon representation was obtained using PyMOL and "IMGT numbering comparison" of 3qfj_E (V-BETA) from IMGT/3Dstructure-DB (http://www.imgt.org)

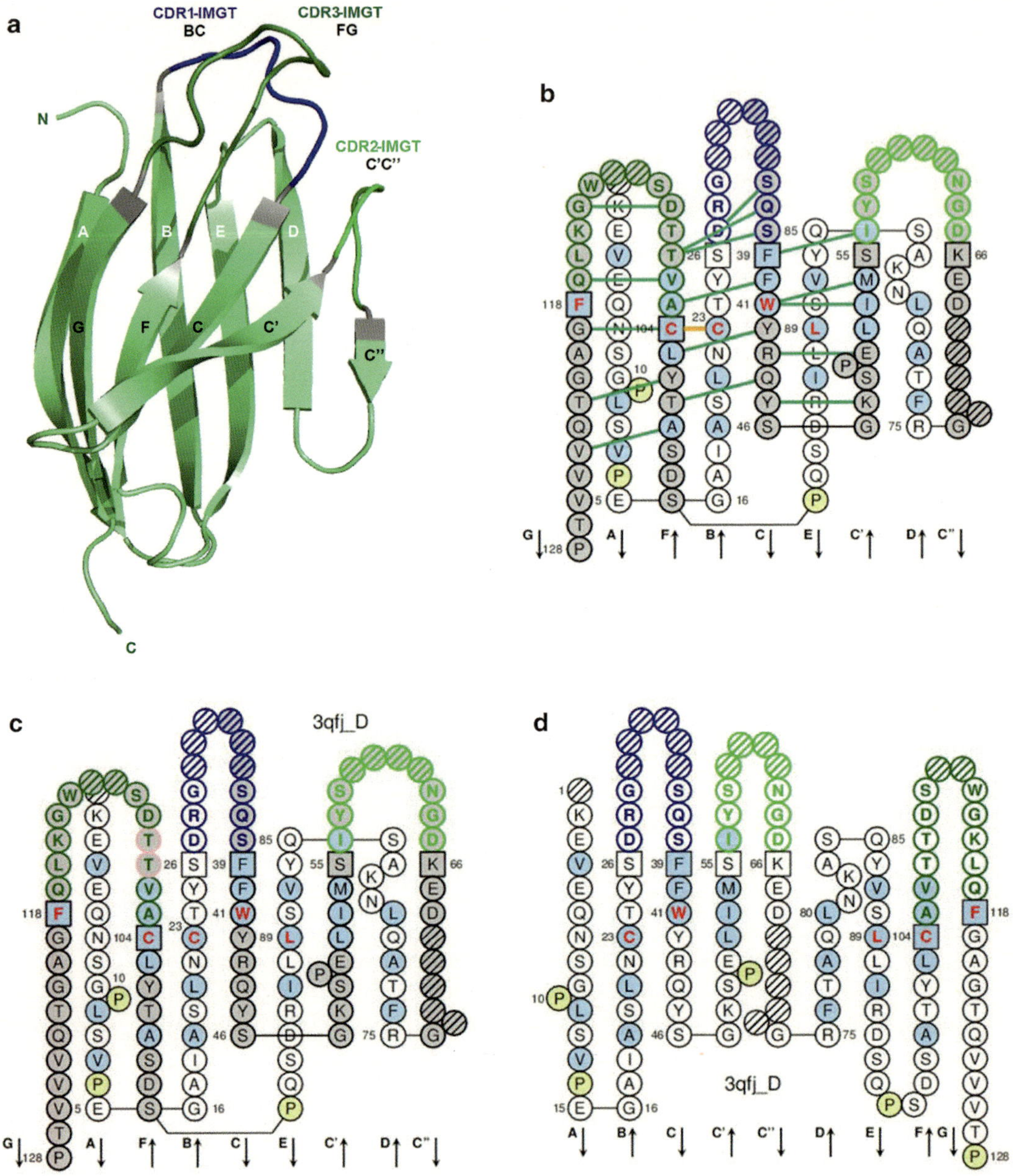

Fig. 10 V-ALPHA from a TR/pMH complex. The TR/pMH1 complex structure is 3qfj from IMGT/3Dstructure-DB (http://www.imgt.org) shown in Fig. 4. The V-ALPHA can be compared with the VH (Fig. 5) and with the V-BETA (Fig. 9). (**a**) 3D structure ribbon representation with the IMGT strand and loop delimitations [61]. (**b**) IMGT Collier de Perles on two layers with hydrogen bonds. The IMGT Collier de Perles on two layers show, in the *forefront*, the GFCC′C″ strands (forming the sheet located at the interface V-ALPHA/V-BETA of the TR) and, in the *back*, the ABED strands. The CDR1-IMGT, CDR2-IMGT, and CDR3-IMGT (corresponding to the BC, C′C″, and FG loops, respectively) are represented online in *blue*, *green*, and *green-blue* (for a V-ALPHA). The IMGT Collier de Perles with hydrogen bonds (*green lines* online, here only shown for the GFCC′C″ sheet) was generated by the IMGT/Collier-de-Perles tool integrated in IMGT/3Dstructure-DB [8–10]. (**c**) IMGT Collier de Perles on two layers generated from IMGT/DomainGapAlign [9, 24, 25]. (**d**) IMGT Collier de Perles on one layer. The CDR-IMGT lengths are [6.6.11] and the FR-IMGT are [25.17.34.11] (FR1-IMGT is 25 instead of 26, as Q1 is missing in 3qfj_D).

obtained, with their sequences at the nucleotide or amino acid level). The core labels include V-REGION, D-REGION, J-REGION and C-REGION which correspond to the coding region of the V, D, J and C genes.

3. IMGT® gene and allele names are based on the concepts of classification of "Group," "Subgroup," "Gene" and "Allele" [58]. "Group" allows to classify a set of genes which belongs to the same multigene family, within the same species or between different species. For example, there are ten groups for the IG of higher vertebrates: IGHV, IGHD, IGHJ, IGHC, IGKV, IGKJ, IGKC, IGLV, IGLJ, IGLC. "Subgroup" allows to identify a subset of genes which belong to the same group, and which, in a given species, share at least 75 % identity at the nucleotide level, e.g., *Homo sapiens* IGHV1 subgroup. Subgroups, genes and alleles are always associated to a species name. An allele is a polymorphic variant of a gene, which is characterized by the mutations of its sequence at the nucleotide level, identified in its core sequence (*see* **Note 2**) and compared to the gene allele reference sequence, designated as allele *01. For example, *Homo sapiens* IGHV1-2*01 is the allele *01 of the *Homo sapiens* IGHV1-2 gene that belongs to the *Homo sapiens* IGHV1 subgroup which itself belongs to the IGHV group. For the IGH locus, the constant genes are designated by the letter (and eventually number) corresponding to the encoded isotypes (IGHM, IGHD, IGHG3, etc.), instead of using the letter C. IMGT-ONTOLOGY concepts of classification have been entered in BioPortal at the National Center for Biomedical Ontology (NCBO) in 2013 (http://bioportal.bioontology.org/ontologies/1491). IG and TR gene names are managed in IMGT/GENE-DB, the IMGT® gene database [7]. IG and TR genes and alleles are not italicized in publications.
4. In higher vertebrates, there are seven IG and TR major loci (other loci correspond to chromosomal orphon sets, genes of

Fig. 10 (continued) The absence of four amino acids at positions 69–72 (strand C″) is a characteristic of the TRAV genes. Amino acids are shown in the one-letter abbreviation (*see* **Note 9**). All proline (P) are shown online in *yellow*. IMGT anchors are in *square* (*see* **Note 12**). *Hatched circles* are IMGT gaps according to the IMGT unique numbering for V domain [61, 64]. Positions with *bold* (online *red*) *letters* indicate the four conserved positions that are common to a V domain and to a C domain: 23 (1st-CYS), 41 (CONSERVED-TRP), 89 (hydrophobic), 104 (2nd-CYS) [59–62, 64], and the fifth conserved position that is specific to the IG and TR V-DOMAIN: 118 (here, J-PHE) which belongs to the motif F/W-G-X-G that characterizes the J-REGION [61, 64] (Table 1). The hydrophobic amino acids (hydropathy index with positive value: I, V, L, F, C, M, A) and tryptophan (W) [28] found at a given position in more than 50 % of sequences are shown (online with a *blue background color*). *Arrows* indicate the direction of the beta strands and their designations in 3D structures. The identifier of the chain to which the V-ALPHA domain belongs is 3qfj_D (of the *Homo sapiens* A6 TR) in 3qfj from IMGT/3Dstructure-DB (http://www.imgt.org). The 3D ribbon representation was obtained using PyMOL and "IMGT numbering comparison" of 3qfj_D (V-ALPHA) from IMGT/3Dstructure-DB (http://www.imgt.org)

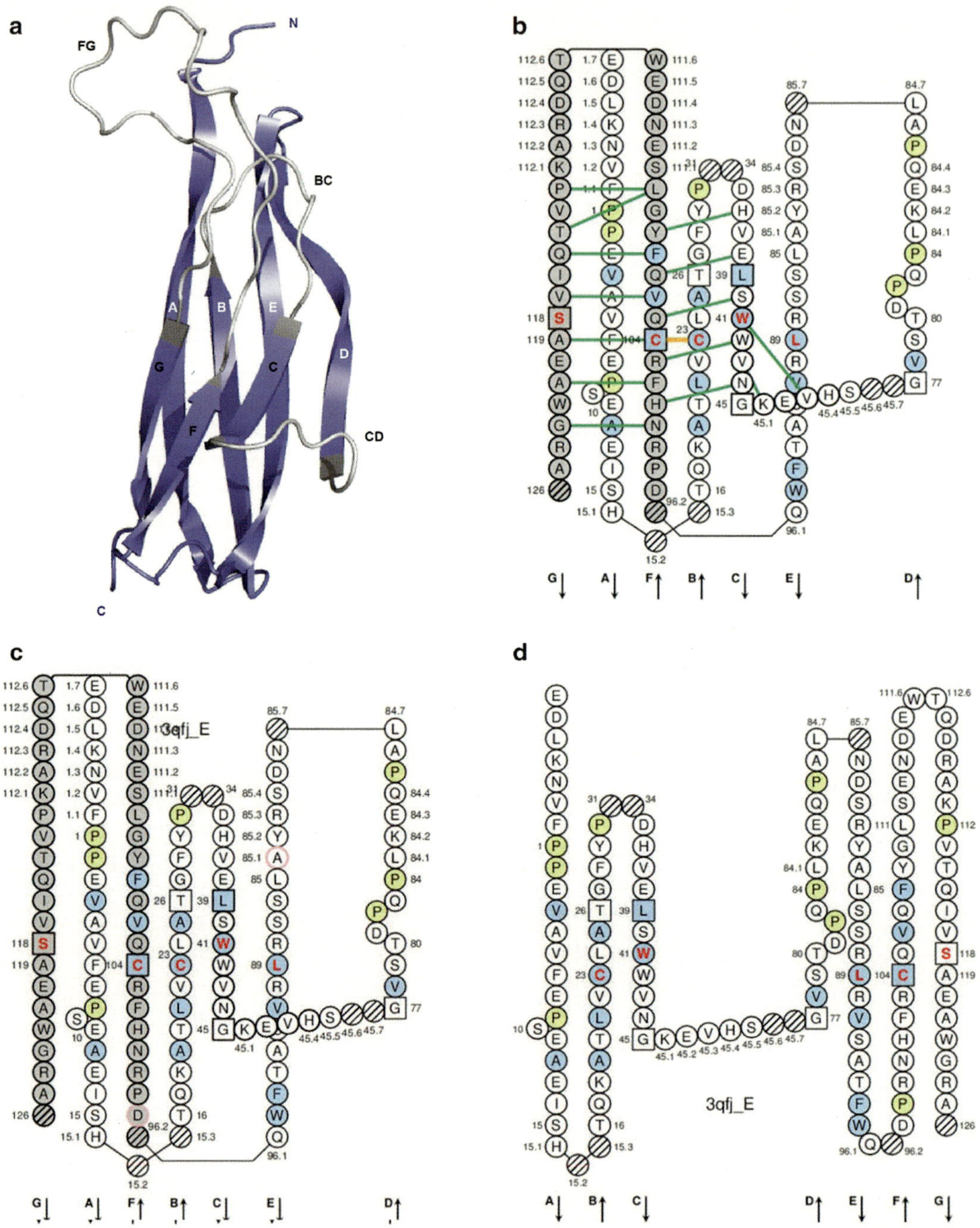

Fig. 11 C-BETA from a TR/pMH complex. The TR/pMH1 complex structure is 3qfj from IMGT/3Dstructure-DB (http://www.imgt.org) shown in Fig. 4. The C-BETA can be compared with the CH (Fig. 6). (**a**) 3D structure ribbon representation with the IMGT strand and loop delimitations [62]. (**b**) IMGT Collier de Perles on two layers with hydrogen bonds. The IMGT Colliers de Perles on two layers show, in the *forefront*, the GFC strands and, in the *back*, the ABED strands (located at the interface C-BETA/C-ALPHA of the TR), linked by the CD transversal strand. The C-BETA has extended F and G strands (six amino acids longer than other C domains). The IMGT Collier de Perles with hydrogen bonds (*green* lines online, here only shown for the GFC sheet) was generated by the IMGT/Collier-de-Perles tool integrated in IMGT/3Dstructure-DB [8–10]. (**c**) IMGT Collier de Perles on two

which are orphons, not used in the IG or TR chain synthesis). The IG major loci include the immunoglobulin heavy (IGH), and for the light chains, the immunoglobulin kappa (IGK) and the immunoglobulin lambda (IGL). The TR major loci include the T cell receptor alpha (TRA), the T cell receptor beta (TRB), the T cell receptor gamma (TRG), and the T cell receptor delta (TRD).

5. The Tenth Human Genome Mapping Workshop (HGM10) took place at Silliman College, Yale, New Haven, Connecticut, the USA, on June 11–17, 1989. The IG and TR data of the Laboratoire d'ImmunoGénétique Moléculaire (CNRS, Montpellier University, Montpellier) were entered in the HGM10 database (Cytogenetics and Cell Genetics 1989. Vol 51, A2336–A2344), with for the first time, the genes of a complete antigen receptor locus, the T cell receptor gamma locus ("The human T-cell receptor γ (TRG) genes" by M.-P. Lefranc and T.H. Rabbitts (TIBS vol 14, June 1989)). The official acceptance of these genes at HGM10 marked the birth of IMGT, which was decided in agreement with the HGM10 nomenclature and organizing committees, for bringing the special expertise required for the management of the diversity and complexity of the IG and TR genes and alleles.
6. IMGT/mAb-DB [11] has been developed to provide an easy access to amino acid sequences (links to IMGT/2Dstructure-DB) and structures (links to IMGT/3Dstructure-DB, if 3D structures are available) of therapeutic antibodies and FPIA from INN [47, 48] (Fig. 2). IMGT/mAb-DB data include mAb (an INN -mab is defined by the presence of at least an IG variable domain) and FPIA (an INN -cept is defined by a receptor fused to a Fc) [47, 48]. IMGT/mAb-DB also includes a few composite proteins for clinical applications (CPCA) (e.g., protein or peptide fused to an Fc for only increasing their half-life; the INN prefix ef- was recently adopted for these CPCA) and some related proteins of the immune system (RPI) (used, unmodified) for clinical applications.
7. An IG ("Receptor") (Fig. 3) [2] is made of two identical heavy (H, for IG-HEAVY) chains and two identical light (L, for

Fig. 11 (continued) layers from IMGT/DomainGapAlign [9, 24, 25]. (**d**) IMGT Colliers de Perles on one layer. Amino acids are shown in the one-letter abbreviation (*see* **Note 9**). All proline (P) are shown online in *yellow*. IMGT anchors are in square (*see* **Note 12**). Hatched circles are IMGT gaps according to the IMGT unique numbering for C domain [62, 64]. Positions with *bold* (online *red*) *letters* indicate the four conserved positions that are common to V and C domains: 23 (1st-CYS), 41 (CONSERVED-TRP), 89 (hydrophobic), 104 (2nd-CYS) [59–62, 64] (Table 2) and position 118 which, as the V domain in general but in contrast to the V-DOMAIN, is not conserved in the C domain. The identifier of the chain to which the C-BETA domain belongs is 3qfj_E (of the *Homo sapiens* A6 TR) from IMGT/3Dstructure-DB (http://www.imgt.org). The 3D ribbon representation was obtained using PyMOL and "IMGT numbering comparison" of 3qfj_E (C-BETA) from IMGT/3Dstructure-DB (http://www.imgt.org)

IMGT pMH contact sites for

Peptide chain ID: 3qfj_C and sequence: LLFGFPVYV

MH1 chain ID: 3qfj_A and domains: G-ALPHA1, G-ALPHA2

Click here for standards IMGT contact sites.

AA numbering in the groove	1	2	3	4	5	6	7	8	9
Peptide sequence	L	L	F	G	F	P	V	Y	V
Contact sites	C1	C3	C4	C5	C6	C8	C9	C10	C11
G-ALPHA1	59	7 9 45 63 66 67				69 70		72 73 76	77 80 81 84
G-ALPHA2	73 77 81		9 66 67 70			7	59 61A 63		26 33 55 58

IMGT Collier de Perles with pMH contact sites

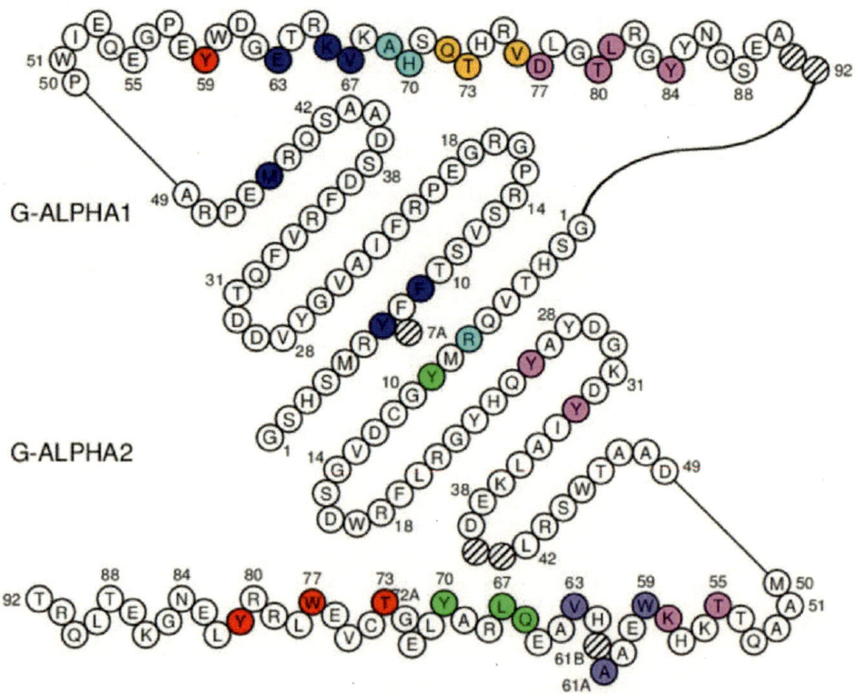

Fig. 12 pMH contact analysis from a TR/pMH complex. The TR/pMH1 complex structure is 3qfj from IMGT/3Dstructure-DB (http://www.imgt.org) shown in Fig. 4. The IMGT Colliers de Perles of the MH1 G-ALPHA1 and G-ALPHA2 domains are shown with pMH contact sites. Each domain is shown with its four strands and helix using the IMGT unique numbering for G domain [63, 64]. The view is from above the cleft, with G-ALPHA1 on top and G-ALPHA2 on bottom. The IMGT Colliers de Perles were generated using the IMGT/Collier-de-Perles tool integrated in IMGT/3Dstructure-DB [8–10]. The G-ALPHA1 and G-ALPHA2 amino acid positions were assigned

IG-LIGHT) chains ("Chain") and usually comprises 12 (e.g., IgG1) or 14 (e.g., IgM) domains. An IgG1 contains 12 domains whereas an IgM contains 14 domains. Each chain has an N-terminal V-DOMAIN (or V-(D)-J-REGION, encoded by the rearranged V-(D)-J genes (Fig. 1)), whereas the remaining of the chain is the C-REGION (encoded by a C gene). The IG C-REGION comprises one C-DOMAIN (C-KAPPA or C-LAMBDA) for the L chain, or several C-DOMAIN (CH) for the H chain [2]. IG receptor, chain and domain structure labels, and correspondence with sequence labels, are shown for two examples of IG (*Homo sapiens* IgG1-kappa (Fig. 3) and IgM-lambda).

IG structure labels (IMGT/3Dstructure-DB)				**Sequence labels (IMGT/LIGM-DB)**
Receptor	**Chain**	**Domain description type**	**Domain**[a]	**Region**
IG-GAMMA-1_ KAPPA	L-KAPPA[b]	V-DOMAIN	V-KAPPA	V-J-REGION
		C-DOMAIN	C-KAPPA	C-REGION
	H-GAMMA-1	V-DOMAIN	VH	V-D-J-REGION
		C-DOMAIN	CH1	C-REGION[c]
		C-DOMAIN	CH2	
		C-DOMAIN	CH3	
IG-MU_ LAMBDA	L-LAMBDA[b]	V-DOMAIN	V-LAMBDA	V-J-REGION
		C-DOMAIN	C-LAMBDA-1	C-REGION
	H-MU	V-DOMAIN	VH	V-D-J-REGION
		C-DOMAIN	CH1	C-REGION[c]
		C-DOMAIN	CH2	
		C-DOMAIN	CH3	
		C-DOMAIN	CH4[d]	

[a]The IG V-DOMAIN includes VH (for the IG heavy chain) and VL (for the IG light chain). In higher vertebrates, the VL is V-KAPPA or V-LAMBDA, whereas in fishes, the VL is V-IOTA. The C-DOMAIN includes CH (for the IG heavy chain, the number of CH per chain depending on the isotype [2]) and CL (for the IG light chain). In higher vertebrates, the CL is C-KAPPA or C-LAMBDA, whereas in fishes, the CL is C-IOTA. In humans, there are nine isotypes, H-MU, H-DELTA, H-GAMMA-3, H-GAMMA-1, H-ALPHA1, H-GAMMA2, H-GAMMA-4, H-EPSILON, H-ALPHA2 (listed in the order 5′–3′ in the IGH locus of the IGHC genes which encode the constant region of the heavy chains (IMGT® http://www.imgt.org, IMGT Repertoire))

[b]The kappa (L-KAPPA) or lambda (L-LAMBDA) light chains may associate to any heavy chain isotype (e.g., H-GAMMA-1, H-MU)

[c]The heavy chain C-REGION also includes the HINGE-REGION for the H-ALPHA, H-DELTA, and H-GAMMA chains and, for membrane IG (mIG), the CONNECTING-REGION (CO), the TRANSMEMBRANE-REGION (TM), and the CYTOPLASMIC-REGION (CY); for secreted IG (sIG), the C-REGION includes CHS instead of CO, TM, and CY

[d]For H-MU and H-EPSILON

Fig. 12 (continued) automatically to the "IMGT pMH contact sites" [87, 88] from the experimental structural data. They are shown (in *colors* online) in the IMGT Colliers de Perles (IMGT®, http://www.imgt.org, IMGT Scientific chart > IMGT color menu). In the table *above*, the numbers 1–9 refers to the peptide AA numbering in the groove which is determined automatically (here a 9-AA peptide LLFGFPVYV, 3qjj_C). The contact sites C1 to C11 refer to the 11 standard "IMGT pMH contact sites" defined for IMGT standardized analysis and comparison of pMH interactions [87, 88]. Here, there are no C2 and C7 in agreement with a MH1 binding a 9-AA peptide [87, 88]. In that 3D structure, there are no C5 and C6 contacts because the glycine G4 and phenylalanine F5 scores are too low

a

IMGT/3Dstructure-DB Domain pair contacts

Contacts of | Domain Chain **[D1] V-BETA 3qfj_E** | *with* | Domain Chain **[D1] G-ALPHA1 3qfj_A**

Summary:

Residue pair contacts	Number of residues			Atom pair contact types			
	Total	From 1	From 2	**Total**	**Polar**	**Hydrogen**	**Nonpolar**
3	4	1	3	**26**	**0**	**0**	**26**

List of the Residue@Position pair contacts:
Click 'R@P' for IMGT Residue@Position cards

	Order IMGT Num	Residue		Domain	Chain		Order IMGT Num	Residue		Domain	Chain	Atom pair contact types Total	Polar	Hydrogen	Nonpolar
R@P	110	LEU	L	V-BETA [D1]	3qfj_E	R@P	69	ALA	A	G-ALPHA1 [D1]	3qfj_A	11	0	0	11
R@P	110	LEU	L	V-BETA [D1]	3qfj_E	R@P	72	GLN	Q	G-ALPHA1 [D1]	3qfj_A	8	0	0	8
R@P	110	LEU	L	V-BETA [D1]	3qfj_E	R@P	73	THR	T	G-ALPHA1 [D1]	3qfj_A	7	0	0	7

b

IMGT/3Dstructure-DB Domain pair contacts

Contacts of | Domain Chain **[D1] V-BETA 3qfj_E** | *with* | Domain Chain **[D2] G-ALPHA2 3qfj_A**

Summary:

Residue pair contacts	Number of residues			Atom pair contact types			
	Total	From 1	From 2	**Total**	**Polar**	**Hydrogen**	**Nonpolar**
12	12	6	6	**87**	**16**	**3**	**71**

List of the Residue@Position pair contacts:
Click 'R@P' for IMGT Residue@Position cards

	Order IMGT Num	Residue		Domain	Chain		Order IMGT Num	Residue		Domain	Chain	Atom pair contact types Total	Polar	Hydrogen	Nonpolar
R@P	110	LEU	L	V-BETA [D1]	3qfj_E	R@P	58	LYS	K	G-ALPHA2 [D2]	3qfj_A	1	1	0	0
R@P	111	ALA	A	V-BETA [D1]	3qfj_E	R@P	58	LYS	K	G-ALPHA2 [D2]	3qfj_A	5	1	1	4
R@P	112.1	GLY	G	V-BETA [D1]	3qfj_E	R@P	61A	ALA	A	G-ALPHA2 [D2]	3qfj_A	9	1	0	8
R@P	112.1	GLY	G	V-BETA [D1]	3qfj_E	R@P	63	VAL	V	G-ALPHA2 [D2]	3qfj_A	1	0	0	1
R@P	112	GLY	G	V-BETA [D1]	3qfj_E	R@P	61A	ALA	A	G-ALPHA2 [D2]	3qfj_A	14	2	0	12
R@P	112	GLY	G	V-BETA [D1]	3qfj_E	R@P	62	HIS	H	G-ALPHA2 [D2]	3qfj_A	1	0	0	1
R@P	112	GLY	G	V-BETA [D1]	3qfj_E	R@P	66	GLN	Q	G-ALPHA2 [D2]	3qfj_A	1	0	0	1
R@P	113	ARG	R	V-BETA [D1]	3qfj_E	R@P	61A	ALA	A	G-ALPHA2 [D2]	3qfj_A	3	1	0	2
R@P	113	ARG	R	V-BETA [D1]	3qfj_E	R@P	62	HIS	H	G-ALPHA2 [D2]	3qfj_A	30	6	0	24
R@P	113	ARG	R	V-BETA [D1]	3qfj_E	R@P	65	GLU	E	G-ALPHA2 [D2]	3qfj_A	10	4	2	6
R@P	113	ARG	R	V-BETA [D1]	3qfj_E	R@P	66	GLN	Q	G-ALPHA2 [D2]	3qfj_A	5	0	0	5
R@P	114	PRO	P	V-BETA [D1]	3qfj_E	R@P	66	GLN	Q	G-ALPHA2 [D2]	3qfj_A	7	0	0	7

Fig. 13 IMGT/3Dstructure-DB Domain pair contacts between V-BETA and MH1 from a TR/pMH complex. The TR/pMH1 complex structure is 3qfj from IMGT/3Dstructure-DB (http://www.imgt.org) shown in Fig. 4. The V-BETA has contacts with the G-ALPHA1 (**a**) and G-ALPHA2 (**b**). There are a total of 113 atom pair contacts (16 polar including 3 hydrogen bonds and 97 nonpolar) for 15 pair contacts (sums of the two Summary tables). The results show that only amino acids of the CDR3-IMGT (*purple* color online) interact with the MH1 G-ALPHA1 and G-ALPHA2 and, as expected, only with helix positions. The "Domain pair contacts" shows that in (**a**) the V-BETA binds A69, Q72, and T73 of the G-ALPHA1 helix (Fig. 12) by a single amino acid of the CDR3-IMGT, L110 (Fig. 9) and in (**b**) the V-BETA binds K58, A61A, H62, V63, E65, Q66 of the G-ALPHA2 helix (Fig. 12) by six amino acids, L110, A111, G112.1, G112, R113, and P114, all located at the top of the CDR3-IMGT (Fig. 9)

8. A TR ("Receptor") (Fig. 4) [3] is made of two chains (alpha and beta, or gamma and delta) ("Chain") and comprises four domains. Each chain has an N-terminal V-DOMAIN (or V-(D)-J-REGION, encoded by the rearranged V-(D)-J genes [3]) whereas the remaining of the chain is the C-REGION (encoded by a C gene). The TR C-REGION comprises one C-DOMAIN [3]. TR receptor, chain and domain structure labels, and correspondence with sequence labels, are shown for two examples of TR (*Homo sapiens* TR alpha_beta (Fig. 4) and TR gamma_delta).

TR structure labels (IMGT/3Dstructure-DB)				**Sequence labels (IMGT/LIGM-DB)**
Receptor	**Chain**	**Domain description type**	**Domain**[a]	**Region**
TR-ALPHA_BETA	TR-ALPHA	V-DOMAIN	V-ALPHA	V-J-REGION
		C-DOMAIN	C-ALPHA	Part of C-REGION[b]
	TR-BETA	V-DOMAIN	V-BETA	V-D-J-REGION
		C-DOMAIN	C-BETA	Part of C-REGION[b]
TR-GAMMA_DELTA	TR-GAMMA	V-DOMAIN	V-GAMMA	V-J-REGION
		C-DOMAIN	C-GAMMA	Part of C-REGION[b]
	TR-DELTA	V-DOMAIN	V-DELTA	V-D-J-REGION
		C-DOMAIN	C-DELTA	Part of C-REGION[b]

[a]The TR V-DOMAIN includes V-ALPHA, V-BETA, V-GAMMA, and V-DELTA. The TR C-DOMAIN includes C-ALPHA, C-BETA, C-GAMMA, and C-DELTA (there are two isotypes for the TR-BETA and TR-GAMMA chains in humans, TR-BETA-1 and TR-BETA-2, and TR-GAMMA-1 and TR-GAMMA-2, the C-REGION of these chains being encoded by the TRBC1 and TRBC2 genes, and TRGC1 and TRGC2 genes, respectively) (IMGT® http://www.imgt.org, IMGT Repertoire) [3]

[b]The TR chain C-REGION also includes the CONNECTING-REGION (CO), the TRANSMEMBRANE-REGION (TM), and the CYTOPLASMIC-REGION (CY), which are not present in 3D structures

9. The 20 usual amino acids (AA) are designated by one-letter or three-letter abbreviations, or in full: A (Ala), alanine; C (Cys), cysteine; D (Asp), aspartic acid; E (Glu), glutamic acid; F (Phe), phenylalanine; G (Gly), glycine; H (His), histidine; I (Ileu), isoleucine; K (Lys), lysine; L (Leu), leucine; M (Met), methionine; N (Asn), asparagine; P (Pro), proline; Q (Gln), glutamine; R (Arg), arginine; S (Ser), serine; T (Thr), threonine; V (Val), valine; W (Trp), tryptophan; Y (Tyr), tyrosine. Highly conserved amino acids at a given position in a V, C, or G domain have IMGT labels [57] (*see* **Note 2**). They include 1st-CYS (position 23), CONSERVED-TRP (position 41) and 2nd-CYS (position 104) for the V and C domains [59–62, 64], J-PHE and J-TRP (position 118) for the V-DOMAIN [59–61, 64], CYS-11 and CYS-74 for the G domain (G-ALPHA2, G-BETA, and G-ALPHA2-LIKE) [63, 64].

a IMGT/3Dstructure-DB Domain pair contacts

Contacts of [D1] V-ALPHA 3qfj_D (Domain, Chain) with [D1] G-ALPHA1 3qfj_A (Domain, Chain)

Summary:

Residue pair contacts	Number of residues			Atom pair contact types			
	Total	From 1	From 2	Total	Polar	Hydrogen	Nonpolar
13	14	7	7	119	17	3	102

List of the Residue@Position pair contacts:

Click 'R@P' for IMGT Residue@Position cards

	Order						Order					Atom pair contact types			
	IMGT Num	Residue		Domain	Chain		IMGT Num	Residue		Domain	Chain	Total	Polar	Hydrogen	Nonpolar
R@P	3	GLU	E	V-ALPHA [D1]	3qfj_D	R@P	58	GLU	E	G-ALPHA1 [D1]	3qfj_A	2	2	0	0
R@P	27	ASP	D	V-ALPHA [D1]	3qfj_D	R@P	58	GLU	E	G-ALPHA1 [D1]	3qfj_A	2	1	0	1
R@P	37	GLN	Q	V-ALPHA [D1]	3qfj_D	R@P	66	LYS	K	G-ALPHA1 [D1]	3qfj_A	4	1	0	3
R@P	108	THR	T	V-ALPHA [D1]	3qfj_D	R@P	65	ARG	R	G-ALPHA1 [D1]	3qfj_A	7	3	1	4
R@P	108	THR	T	V-ALPHA [D1]	3qfj_D	R@P	66	LYS	K	G-ALPHA1 [D1]	3qfj_A	3	0	0	3
R@P	109	ASP	D	V-ALPHA [D1]	3qfj_D	R@P	62	GLY	G	G-ALPHA1 [D1]	3qfj_A	1	1	0	0
R@P	109	ASP	D	V-ALPHA [D1]	3qfj_D	R@P	65	ARG	R	G-ALPHA1 [D1]	3qfj_A	20	6	2	14
R@P	109	ASP	D	V-ALPHA [D1]	3qfj_D	R@P	66	LYS	K	G-ALPHA1 [D1]	3qfj_A	14	1	0	13
R@P	113	TRP	W	V-ALPHA [D1]	3qfj_D	R@P	65	ARG	R	G-ALPHA1 [D1]	3qfj_A	20	1	0	19
R@P	113	TRP	W	V-ALPHA [D1]	3qfj_D	R@P	68	LYS	K	G-ALPHA1 [D1]	3qfj_A	9	0	0	9
R@P	113	TRP	W	V-ALPHA [D1]	3qfj_D	R@P	69	ALA	A	G-ALPHA1 [D1]	3qfj_A	18	0	0	18
R@P	113	TRP	W	V-ALPHA [D1]	3qfj_D	R@P	72	GLN	Q	G-ALPHA1 [D1]	3qfj_A	10	0	0	10
R@P	114	GLY	G	V-ALPHA [D1]	3qfj_D	R@P	65	ARG	R	G-ALPHA1 [D1]	3qfj_A	9	1	0	8

b IMGT/3Dstructure-DB Domain pair contacts

Contacts of [D1] V-ALPHA 3qfj_D (Domain, Chain) with [D2] G-ALPHA2 3qfj_A (Domain, Chain)

Summary:

Residue pair contacts	Number of residues			Atom pair contact types			
	Total	From 1	From 2	Total	Polar	Hydrogen	Nonpolar
13	16	8	8	108	15	1	93

List of the Residue@Position pair contacts:

Click 'R@P' for IMGT Residue@Position cards

	Order						Order					Atom pair contact types			
	IMGT Num	Residue		Domain	Chain		IMGT Num	Residue		Domain	Chain	Total	Polar	Hydrogen	Nonpolar
R@P	28	ARG	R	V-ALPHA [D1]	3qfj_D	R@P	77	TRP	W	G-ALPHA2 [D2]	3qfj_A	13	1	0	12
R@P	28	ARG	R	V-ALPHA [D1]	3qfj_D	R@P	80	ARG	R	G-ALPHA2 [D2]	3qfj_A	5	1	0	4
R@P	29	GLY	G	V-ALPHA [D1]	3qfj_D	R@P	77	TRP	W	G-ALPHA2 [D2]	3qfj_A	7	0	0	7
R@P	37	GLN	Q	V-ALPHA [D1]	3qfj_D	R@P	70	TYR	Y	G-ALPHA2 [D2]	3qfj_A	6	0	0	6
R@P	37	GLN	Q	V-ALPHA [D1]	3qfj_D	R@P	73	THR	T	G-ALPHA2 [D2]	3qfj_A	11	1	0	10
R@P	38	SER	S	V-ALPHA [D1]	3qfj_D	R@P	66	GLN	Q	G-ALPHA2 [D2]	3qfj_A	1	1	0	0
R@P	57	TYR	Y	V-ALPHA [D1]	3qfj_D	R@P	66	GLN	Q	G-ALPHA2 [D2]	3qfj_A	22	2	0	20
R@P	57	TYR	Y	V-ALPHA [D1]	3qfj_D	R@P	69	ALA	A	G-ALPHA2 [D2]	3qfj_A	8	1	0	7
R@P	58	SER	S	V-ALPHA [D1]	3qfj_D	R@P	69	ALA	A	G-ALPHA2 [D2]	3qfj_A	8	2	0	6
R@P	63	ASN	N	V-ALPHA [D1]	3qfj_D	R@P	76	GLU	E	G-ALPHA2 [D2]	3qfj_A	5	2	0	3
R@P	82	LYS	K	V-ALPHA [D1]	3qfj_D	R@P	72A	GLY	G	G-ALPHA2 [D2]	3qfj_A	1	0	0	1
R@P	82	LYS	K	V-ALPHA [D1]	3qfj_D	R@P	73	THR	T	G-ALPHA2 [D2]	3qfj_A	7	2	0	5
R@P	82	LYS	K	V-ALPHA [D1]	3qfj_D	R@P	76	GLU	E	G-ALPHA2 [D2]	3qfj_A	14	2	1	12

Fig. 14 IMGT/3Dstructure-DB Domain pair contacts between V-ALPHA and MH1 from a TR/pMH complex. The TR/pMH1 complex structure is 3qfj from IMGT/3Dstructure-DB (http://www.imgt.org) shown in Fig. 4. The V-ALPHA has contacts with the G-ALPHA1 (**a**) and G-ALPHA2 (**b**). There is a total of 227 atom pair contacts (32 polar including 4 hydrogen bonds and 195 nonpolar) for 26 pair contacts (sums of the two Summary tables). The results show that the three CDR-IMGT of V-ALPHA interact with the MH1 G-ALPHA1 and G-ALPHA2 and,

10. In the IMGT® definitive system, the CDR-IMGT have accurate and unambiguous delimitations in contrast for the CDR described in the literature. Correspondences between the IMGT unique numbering with other numberings are available in the IMGT Scientific chart. These correspondences with other numberings are useful for the interpretation of previously published data but nowadays the usage of these numberings has become obsolete in regard of the development of immunoinformatics based on the IMGT® standards [59–68] (IMGT® http://www.imgt.org, IMGT Scientific chart > Numbering).

11. For CDR3-IMGT length > 13 AA, IMGT additional positions are created between positions 111 and 112 (in bold in the table below) at the top of the CDR3-IMGT loop in the following order 112.1,111.1, 112.2, 111.2, 112.3, 111.3, etc. (with two digits after the dot, if necessary).

CDR3-IMGT lengths	IMGT additional positions for CDR3-IMGT length > 13 AA									
21	**111**	111.1	111.2	111.3	111.4	112.4	112.3	112.2	112.1	**112**
20	**111**	111.1	111.2	111.3	–	112.4	112.3	112.2	112.1	**112**
19	**111**	111.1	111.2	111.3	–	–	112.3	112.2	112.1	**112**
18	**111**	111.1	111.2	–	–	–	112.3	112.2	112.1	**112**
17	**111**	111.1	111.2	–	–	–	–	112.2	112.1	**112**
16	**111**	111.1	–	–	–	–	–	112.2	112.1	**112**
15	**111**	111.1	–	–	–	–	–	–	112.1	**112**
14	**111**	–	–	–	–	–	–	–	112.1	**112**

For CDR3-IMGT length < 13 AA, IMGT gaps are created classically from the top of the loop, in the following order 111, 112, 110, 113, 109, 114, etc. (with two digits after the dot, if necessary).

12. IMGT anchors are positions that belong to strands and represent anchors for the loops of the V and C domains (and by extension to the CD strand of the C domains that do not have the C′–C″ loop) [62]. Anchor positions are shown in square in IMGT Colliers de Perles. Positions 26 and 39 are anchors of the BC

Fig. 14 (continued) as expected, only with helix positions. The "Domain pair contacts" shows that in (**a**) the V-ALPHA binds seven amino acids (E58, G62, R65, K66, K68, A69, and Q72) of the G-ALPHA1 helix (Fig. 12) by its CDR1-IMGT (D27 and Q37) and at a greater extent by its CDR3-IMGT (T108, D109, W113, and G114) (Fig. 10) and in (**b**) the V-ALPHA binds seven amino acids (Q66, A69, Y70, T73, E76, W77, R80) of the G-ALPHA2 helix (Fig. 12) by its CDR1-IMGT (R28, G29, Q37, S38) and by its CDR2-IMGT (Y57, S38, N63). One amino acid of the FR3-IMGT, the lysine K82 (in the V-ALPHA D strand) has contacts with G72A, T73, and E76 (22 atom pair contacts: 4 polar including 1 hydrogen bond and 18 nonpolar)

a IMGT/3Dstructure-DB Domain pair contacts

Contacts of [D1] V-ALPHA 3qfj_D (Domain, Chain) *with* (Ligand) 3qfj_C (Domain, Chain)

Summary:

Residue pair contacts	Number of residues			Atom pair contact types			
	Total	From 1	From 2	Total	Polar	Hydrogen	Nonpolar
15	13	7	6	123	16	3	107

List of the Residue@Position pair contacts:

Click 'R@P' for IMGT Residue@Position cards

	Order						Order					Atom pair contact types			
	IMGT Num	Residue		Domain	Chain		IMGT Num	Residue		Domain	Chain	Total	Polar	Hydrogen	Nonpolar
R@P	29	GLY	G	V-ALPHA [D1]	3qfj_D	R@P	1	LEU	L	(Ligand)	3qfj_C	5	0	0	5
R@P	37	GLN	Q	V-ALPHA [D1]	3qfj_D	R@P	1	LEU	L	(Ligand)	3qfj_C	4	0	0	4
R@P	37	GLN	Q	V-ALPHA [D1]	3qfj_D	R@P	2	LEU	L	(Ligand)	3qfj_C	6	2	1	4
R@P	37	GLN	Q	V-ALPHA [D1]	3qfj_D	R@P	3	PHE	F	(Ligand)	3qfj_C	12	2	0	10
R@P	37	GLN	Q	V-ALPHA [D1]	3qfj_D	R@P	4	GLY	G	(Ligand)	3qfj_C	7	2	0	5
R@P	37	GLN	Q	V-ALPHA [D1]	3qfj_D	R@P	5	PHE	F	(Ligand)	3qfj_C	13	0	0	13
R@P	38	SER	S	V-ALPHA [D1]	3qfj_D	R@P	5	PHE	F	(Ligand)	3qfj_C	10	0	0	10
R@P	107	THR	T	V-ALPHA [D1]	3qfj_D	R@P	5	PHE	F	(Ligand)	3qfj_C	2	0	0	2
R@P	108	THR	T	V-ALPHA [D1]	3qfj_D	R@P	4	GLY	G	(Ligand)	3qfj_C	5	2	0	3
R@P	108	THR	T	V-ALPHA [D1]	3qfj_D	R@P	5	PHE	F	(Ligand)	3qfj_C	5	0	0	5
R@P	109	ASP	D	V-ALPHA [D1]	3qfj_D	R@P	4	GLY	G	(Ligand)	3qfj_C	15	3	0	12
R@P	109	ASP	D	V-ALPHA [D1]	3qfj_D	R@P	5	PHE	F	(Ligand)	3qfj_C	13	0	0	13
R@P	110	SER	S	V-ALPHA [D1]	3qfj_D	R@P	4	GLY	G	(Ligand)	3qfj_C	8	2	2	6
R@P	110	SER	S	V-ALPHA [D1]	3qfj_D	R@P	5	PHE	F	(Ligand)	3qfj_C	15	1	0	14
R@P	110	SER	S	V-ALPHA [D1]	3qfj_D	R@P	6	PRO	P	(Ligand)	3qfj_C	3	2	0	1

b IMGT/3Dstructure-DB Domain pair contacts

Contacts of [D1] V-BETA 3qfj_E (Domain, Chain) *with* (Ligand) 3qfj_C (Domain, Chain)

Summary:

Residue pair contacts	Number of residues			Atom pair contact types			
	Total	From 1	From 2	Total	Polar	Hydrogen	Nonpolar
9	10	6	4	101	7	2	94

List of the Residue@Position pair contacts:

Click 'R@P' for IMGT Residue@Position cards

	Order						Order					Atom pair contact types			
	IMGT Num	Residue		Domain	Chain		IMGT Num	Residue		Domain	Chain	Total	Polar	Hydrogen	Nonpolar
R@P	37	GLU	E	V-BETA [D1]	3qfj_E	R@P	8	TYR	Y	(Ligand)	3qfj_C	12	2	1	10
R@P	109	GLY	G	V-BETA [D1]	3qfj_E	R@P	6	PRO	P	(Ligand)	3qfj_C	1	1	0	0
R@P	110	LEU	L	V-BETA [D1]	3qfj_E	R@P	6	PRO	P	(Ligand)	3qfj_C	11	2	0	9
R@P	110	LEU	L	V-BETA [D1]	3qfj_E	R@P	7	VAL	V	(Ligand)	3qfj_C	9	1	0	8
R@P	110	LEU	L	V-BETA [D1]	3qfj_E	R@P	8	TYR	Y	(Ligand)	3qfj_C	35	1	1	34
R@P	111	ALA	A	V-BETA [D1]	3qfj_E	R@P	7	VAL	V	(Ligand)	3qfj_C	2	0	0	2
R@P	111	ALA	A	V-BETA [D1]	3qfj_E	R@P	8	TYR	Y	(Ligand)	3qfj_C	12	0	0	12
R@P	112.1	GLY	G	V-BETA [D1]	3qfj_E	R@P	7	VAL	V	(Ligand)	3qfj_C	9	0	0	9
R@P	114	PRO	P	V-BETA [D1]	3qfj_E	R@P	5	PHE	F	(Ligand)	3qfj_C	10	0	0	10

Fig. 15 IMGT/3Dstructure-DB Domain pair contacts between the TR V-ALPHA and V-BETA and the Ligand (a 9-mer peptide) from a TR/pMH1 complex. The TR/pMH1 complex structure is 3qfj from IMGT/3Dstructure-DB (http://www.imgt.org) shown in Fig. 4. The V-ALPHA and B-BETA interact with the 9-mer peptide by only their CDR1-IMGT and to a greater extent by their CDR3-IMGT. No other amino acid is involved. In (**a**), the "Domain pair contacts" shows that the V-ALPHA binds AA 1–5 (LLFGF) of the peptide (AA positions in the groove)

loop of the V domain (CDR1-IMGT in V-DOMAIN) and C domain. Positions 55 and 66 are anchors of the C′–C″ loop of the V domain (CDR2-IMGT in V-DOMAIN), whereas positions 45 and 77 are anchors of the CD strand of the C domain. Positions 104 in F strand (2nd-CYS) and 118 in G strand (J-PHE or J-TRP in V-DOMAIN) are anchors of the FG loop of the V domain (CDR3-IMGT in V-DOMAIN) and C domain. The JUNCTION of an IG or TR V-DOMAIN includes the anchors 104 and 118 and is therefore two amino acids longer than the corresponding CDR3-IMGT (positions 105–117).

13. The 20 usual amino acids (*see* **Note 9**) have been classified in 11 IMGT physicochemical classes which are based on "Hydrophathy," "Volume," and "Chemical" characteristics (IMGT® http://www.imgt.org, IMGT Education > Aide-mémoire > Amino acids). The amino acid (AA) changes are described according to the hydropathy, volume, and IMGT physicochemical classes [28]. For example Q1 > E (++−) means that in the AA change (Q > E), the two amino acids belong to the same hydropathy (+) and volume (+) classes but to different IMGT physicochemical properties (−) classes. Four types of AA changes are identified in IMGT®: very similar (+++), similar (++−, +−+), dissimilar (−−+, −+−, +−−), and very dissimilar (−−−).
14. The exon rule is not used for the delimitation of the 5′ end of the first N-terminal domain of proteins with a leader (this includes the V-DOMAIN of the IG and TR chains). In those cases, the 5′ end of the first N-terminal domain corresponds to the proteolytic site between the leader (L-REGION) and the coding region of the mature protein. The IG and TR V-DOMAIN is therefore delimited in 5′ by a proteolytic site and in 3′ by the splicing site of the J-REGION. The exon rule takes into account the fact that a domain may be encoded by two exons as found in IgSF other than IG and TR.
15. A MH ("Receptor") [63] depending on the MH group is made of one chain (I-ALPHA) noncovalently associated to the beta2-microglobulin (B2M) (MH1 group, in the literature MHC class I) (Fig. 4) or of two chains (II-ALPHA and II-BETA) (MH2 group, in the literature MHC class II). The I-ALPHA chain has two G-DOMAIN, whereas each II-ALPHA and II-BETA has one G-DOMAIN. MH receptor, chain and domain structure

Fig. 15 (continued) (Fig. 12) by its CDR1-IMGT (G29, Q37, S38) and binds AA 4–6 (GFP) of the peptide by its CDR3-IMGT (T107, T108, D109, S110) (Fig. 10). On the 123 atom pair contacts (16 polar including 3 hydrogen bonds and 107 nonpolar) ("Summary"), 93 atom pair contacts (10 polar including 2 hydrogen bonds and 83 nonpolar) are engaged between V-ALPHA and two amino acids (G4 and F5) of the peptide. In (**b**) the "Domain pair contacts" shows that the V-BETA binds AA 5–8 by its CDR1-IMGT (E37) and at a greater extent by its CDR3-IMGT (G109, L110, A111, G112.1, P114). On the 101 atom pair contacts (7 polar including 2 hydrogen bonds and 94 nonpolar) ("Summary"), 59 atom pair contacts (3 polar including 2 hydrogen bonds and 56 nonpolar) are engaged between the V-BETA and one amino acid (Y8) of the peptide

labels, and correspondence with sequence labels, are shown for examples of members of the MH1 and MH2 groups.

	MH structure labels (IMGT/3Dstructure-DB)					Sequence labels (IMGT/LIGM-DB)
MH group	**Receptor**	**Chain**	**Domain description type[a]**	**Domain**	**Domain number**	**Region**
MH1	MH1-ALPHA_ B2M	I-ALPHA	G-DOMAIN	G-ALPHA1	[D1]	Part of REGION[b]
			G-DOMAIN	G-ALPHA2	[D2]	
			C-LIKE-DOMAIN	C-LIKE	[D3]	
		B2M	C-LIKE-DOMAIN	C-LIKE	[D]	REGION
MH2	MH2-ALPHA_ BETA	II-ALPHA	G-DOMAIN	G-ALPHA	[D1]	Part of REGION[b]
			C-LIKE-DOMAIN	C-LIKE	[D2]	
		II-BETA	G-DOMAIN	G-BETA	[D1]	Part of REGION[b]
			C-LIKE-DOMAIN	C-LIKE	[D2]	

[a]The domain description type shows that the MH proteins belong to the MhSF by their G-DOMAIN and to the IgSF by their C-LIKE-DOMAIN. The B2M associated to the I-ALPHA chain in MH1 has only a single C-LIKE-DOMAIN and only belongs to the IgSF

[b]The REGION of the I-ALPHA, II-ALPHA, and II-BETA chains also includes the CONNECTING-REGION (CO), the TRANSMEMBRANE-REGION (TM), and the CYTOPLASMIC-REGION (CY) which are not present in the 3D structures

16. MhSF proteins other than MH only include RPI-MH1Like proteins (there is no "RPI-MH2Like" identified so far) [96, 97]. The RPI-MH1Like in humans comprise: AZGP1 (that regulates fat degradation in adipocytes), CD1A to CD1E proteins (that display phospholipid antigens to T cells and participate in immune defense against microbian pathogens), FCGRT (that transports maternal immunoglobulins through placenta and governs neonatal immunity), HFE (that interacts with transferring receptor and takes part in iron homeostasis by regulating iron transport through cellular membranes), MICA and MICB (that are induced by stress and involved in tumor cell detection), MR1 (that may regulate mucosal immunity), PROCR, previously EPCR (that interacts with activated C protein and is involved in the blood coagulation pathway), RAET1E, RAETG, and RAET1L (that are inducible by retinoic acid and stimulate cytokine/chemokine production and cytotoxic activity of NK cells), and ULBP1, ULBP2, and ULBP3 (that are ligands for NKG2D receptor).
17. In the IMGT/DomainGapAlign Welcome page, amino acid sequences are submitted in FASTA format (pasted in a text

area or uploaded in a file). A precise delimitation of the domain sequences is not required, however if the sequence contains several domains, the sequence should be split between the different domains. Several domain amino acid sequences can be analyzed simultaneously (up to 50) provided that each sequence has a distinct name and that they all belong to the same domain type (V, C, or G). If not, the query needs to be launched for each domain type, successively. If the limits and the numbers of domains of an amino acid sequence are unknown, the protein can be analyzed progressively, shortening the sequence once a domain has been identified by the tool (it should be reminded that the first domain identified by the tool is not necessarily the first one in the protein).

18. The IMGT domain reference directory is the IMGT reference directory for V, C, and G domains. It is manually curated and contains the amino acid sequences of the domains delimited according to the IMGT rules (based on the exon delimitations). Sequences are from the IMGT Repertoire [1] and from IMGT/GENE-DB [7]. Owing to the particularities of the V-DOMAIN synthesis [2, 3] there is no V-DOMAIN in the IMGT reference directory. Instead, the directory comprises the translation of the IG and TR germline V and J genes (V-REGION and J-REGION, respectively). The IMGT domain reference directory provides the IMGT "gene" and "allele" names ("CLASSIFICATION" axiom) (*see* **Note 3**). Data are comprehensive for human and mouse IG and TR and human MH whereas for other species and IgSF and MhSF they are added progressively. The IMGT domain reference directory comprises domain sequences of functional (F), ORF (open reading frame) and in frame pseudogene (P) genes (*see* **Note 1**). As IMGT alleles are characterized at the nucleotide level (*see* **Note 3**), identical sequences at the amino acid level may therefore correspond to different alleles, in the IMGT domain reference directory. These reference amino acid sequences can be displayed by querying IMGT/DomainDisplay (http://www.imgt.org).

19. The IMGT/V-QUEST reference directory sets include IMGT reference sequences from all functional (F) genes and alleles, all open reading frame (ORF) and all in-frame pseudogenes (P) alleles. By definition, the IMGT reference directory sets contain one sequence for each allele (*see* **Note 3**). By default, the user sequences are compared with all genes and alleles. However, the option "With allele *01 only" is useful for: (1) "Detailed view," if the user sequences need to be compared with different genes, and (2) "Synthesis view," if the user sequences which use the same gene need to be aligned together (independently of the allelic polymorphism).

    IMGT/QUEST reference directories have been set up for species which have been extensively studied, such as human and mouse. This also holds for the other species or taxons with incomplete IMGT reference directory sets. In those cases, results should be interpreted considering the status of the IMGT reference directory (information on the updates on the IMGT® Web site). Links to the IMGT/V-QUEST reference directory sets are available from the IMGT/V-QUEST Welcome page.

20. The way to identify the closest germline D is different between IMGT/V-QUEST and IMGT/JunctionAnalysis since the evaluation of the alignment score is different. In case of discrepancy, the results of IMGT/JunctionAnalysis are the most accurate. If the option "with full list of eligible D-GENE" was selected in "Display view," its results allow comparing the IMGT/JunctionAnalysis D gene identification with all D genes which match the junction with their corresponding score. The alignment provided by IMGT/V-QUEST is still provided, although less accurate, as it is less stringent and displays several D genes and alleles, and therefore may help solving some ambiguous cases.

21. The number of silent and nonsilent mutations is evaluated, as well as each type of transition (a>g, g>a, c>t, t>c) and transversion (a>c, c>a, a>t, t>a, g>c, c>g, g>t, t>g). The number of identical AA and of AA changes is evaluated, as well as each type of AA changes (*see* **Note 13**). Mutation hot spots are identified in the germline V-REGION with their positions. They include (a/t)a, t(a/t), (a/g)g(c/t)(a/t), (a/t)(a/g)c(c/t) (or wa, tw, rgyw, wrcy). IMGT/V-QUEST is frequently used by clinicians for the analysis of somatic hypermutations in leukemia, lymphoma, and myeloma, and more particularly in chronic lymphocytic leukemia (CLL) [80–82] in which the percentage of mutations of the rearranged IGHV gene in the VH of the leukemic clone has a patient prognostic value. IMGT/V-QUEST is the recommended standard recommended by ERIC for comparative analysis between laboratories [80].

22. The sequences of the V-(D)-J junctions determined by IMGT/JunctionAnalysis [18, 19] are also used in the characterization of stereotypic patterns in CLL [81, 82] and for the junction synthesis of specific probes for the follow-up of residual diseases in leukemias and lymphomas.

23. Potential insertions or deletions are suspected by IMGT/V-QUEST when the V-REGION score is very low (less than 200), and/or the percentage of identity is less than 85 %, and/or when the input sequence has different CDR1-IMGT and/or CDR2-IMGT lengths, compared to those of the closest germline V. In those cases, the user can go back to the IMGT/V-QUEST Search page and select the option "Search for insertions and deletions" in "Advanced parameters." If indeed insertions and/or deletions are detected, they

will be described in the "Result summary" row with their localization in FR-IMGT or CDR-IMGT, the nb of inserted or deleted nt and, for insertions, the inserted nt, the presence or absence of frameshift, the V-REGION codon from which the insertion or deletion starts and the nt position in the user submitted sequence. The insertions are highlighted in capital letters in the user sequence and the tool runs a classical IMGT/V-QUEST search after having removed the insertion(s) from the user sequence. In case of deletions, the tool adds gaps to replace the identified deletions before running a classical IMGT/V-QUEST search. Users should be aware that an insertion or a deletion at the beginning of FR1-IMGT or at the end of the FR3-IMGT may not be detected.

24. In IMGT/3Dstructure-DB, contacts are described as atom pair contacts. Atom pair contacts are obtained by a local program in which atoms are considered to be in contact when no water molecule can take place between them [8, 9]. Atom pair contacts are provided by atom contact types (Non-covalent, Polar, Hydrogen bond, etc.) and/or atom contact categories ((BB) Backbone/backbone, (SS) Side chain/side chain, etc.) [8, 9, 87, 88].

25. In an IMGT Residue@Position card (or "R@P"), the "IMGT Residue@Position" is defined by the IMGT position numbering in a domain, or if not characterized, in the chain, the AA name (three-letter and between parentheses one-letter abbreviation), the IMGT domain description and the IMGT chain ID, e.g., "110–LEU(L)–V-BETA–3qfj_E." The characteristics reported in an "R@P" includes (1) general information (PDB file numbering, IMGT file numbering, residue full name and formula), (2) structural information "IMGT LocalStructure@Position" (secondary structure, Phi and Psi angles (in degrees), and accessible surface area (ASA) (in square angstrom)), and (3) detailed contact analysis with amino acids of other domains.

26. The first *Homo sapiens* TR/pMH complex crystallized is that of the TR alpha_beta A6 [100] (1ao7 in IMGT/3Dstructure-DB). The TR alpha_beta A6 recognizes a 9-mer peptide LLFGYPVYV of the Tax protein of the human T cell lymphotropic virus-1 (HTLV-1) presented by the human MH1, HLA-A*0201. Several TR/pMH complexes containing the same TR A6 with the same MH1 (HLA-A*0201) but different peptide variants or ligands were then crystallized and these represent interesting data to compare specificity, cross-reactivity and binding mechanisms of these complexes. The IMGT/3Dstructure-DB entry 3qfj (Fig. 4), is one of these variants in which the Tax peptide has one amino acid change Y5>F [101]. In the IMGT/3Dstructure-DB card the peptide is described as "Tax peptide 11-19 (Q82235), Y5>F [HTLV1]."

Acknowledgements

We are grateful to Gérard Lefranc for helpful discussion, Souphatta Sasorith for help in the figures and the IMGT® team for its expertise and constant motivation. We thank Cold Spring Harbor Protocol Press for the pdf of the IMGT Booklet available in IMGT references. IMGT® is a registered trademark of CNRS. IMGT® is member of the International Medical Informatics Association (IMIA). IMGT® was funded in part by the BIOMED1 (BIOCT930038), Biotechnology BIOTECH2 (BIO4CT960037), 5th PCRDT Quality of Life and Management of Living Resources (QLG2-2000-01287), and 6th PCRDT Information Science and Technology (ImmunoGrid, FP6 IST-028069) programmes of the European Union (EU). IMGT® is currently supported by the Centre National de la Recherche Scientifique (CNRS), the Ministère de l'Enseignement Supérieur et de la Recherche (MESR), the University Montpellier 2, the Agence Nationale de la Recherche (ANR) Labex MabImprove (ANR-10-LABX-53-01), the Région Languedoc-Roussillon (Grand Plateau Technique pour la Recherche (GPTR)). This work was granted access to the HPC resources of CINES under the allocation 036029 (2010–2014) made by GENCI (Grand Equipement National de Calcul Intensif).

References

1. Lefranc M-P, Giudicelli V, Ginestoux C, Jabado-Michaloud J, Folch G, Bellahcene F, Wu Y, Gemrot E, Brochet X, Lane J, Regnier L, Ehrenmann F, Lefranc G, Duroux P (2009) IMGT®, the international ImMunoGeneTics information system®. Nucleic Acids Res 37:D1006–D1012
2. Lefranc M-P, Lefranc G (2001) The Immunoglobulin FactsBook. Academic, London, pp 1–458
3. Lefranc M-P, Lefranc G (2001) The T cell receptor FactsBook. Academic, London, pp 1–398
4. Lefranc M-P (2000) Nomenclature of the human immunoglobulin genes. In: Coligan JE, Bierer BE, Margulies DE, Shevach EM, Strober W (eds) Current protocols in immunology. Wiley, Hoboken, NJ, pp A.1P.1–A.1P.37
5. Lefranc M-P (2000) Nomenclature of the human T cell Receptor genes. In: Coligan JE, Bierer BE, Margulies DE, Shevach EM, Strober W (eds) Current protocols in immunology. Wiley, Hoboken, NJ, pp A.1O.1–A.1O.23
6. Giudicelli V, Duroux P, Ginestoux C, Folch G, Jabado-Michaloud J, Chaume D, Lefranc M-P (2006) IMGT/LIGM-DB, the IMGT® comprehensive database of immunoglobulin and T cell receptor nucleotide sequences. Nucleic Acids Res 34:D781–D784
7. Giudicelli V, Chaume D, Lefranc M-P (2005) IMGT/GENE-DB: a comprehensive database for human and mouse immunoglobulin and T cell receptor genes. Nucleic Acids Res 33:D256–D261
8. Kaas Q, Ruiz M, Lefranc M-P (2004) IMGT/3Dstructure-DB and IMGT/StructuralQuery, a database and a tool for immunoglobulin, T cell receptor and MHC structural data. Nucleic Acids Res 32:D208–D210
9. Ehrenmann F, Kaas Q, Lefranc M-P (2010) IMGT/3Dstructure-DB and IMGT/DomainGapAlign: a database and a tool for immunoglobulins or antibodies, T cell receptors, MHC, IgSF and MhcSF. Nucleic Acids Res 38:D301–D307
10. Ehrenmann F, Lefranc M-P (2011) IMGT/3Dstructure-DB: Querying the

IMGT Database for 3D Structures in Immunology and Immunoinformatics (IG or Antibodies, TR, MH, RPI, and FPIA). Cold Spring Harb Protoc 6:750–761. doi:10.1101/pdb.prot5637, pii: pdb.prot5637

11. Poiron C, Wu Y, Ginestoux C, Ehrenmann F, Duroux P, Lefranc M-P (2010) IMGT/mAb-DB: the IMGT® database for therapeutic monoclonal antibodies. Poster n°101, 11èmes Journées Ouvertes de Biologie, Informatique et Mathématiques (JOBIM), Montpellier, 7–9 Sept 2010
12. Giudicelli V, Chaume D, Lefranc M-P (2004) IMGT/V-QUEST, an integrated software for immunoglobulin and T cell receptor V-J and V-D-J rearrangement analysis. Nucleic Acids Res 32:W435–W440
13. Giudicelli V, Lefranc M-P (2005) Interactive IMGT on-line tools for the analysis of immunoglobulin and T cell receptor repertoires. In: Veskler BA (ed) New research on immunology. Nova Science Publishers Inc, New York, pp 77–105
14. Brochet X, Lefranc M-P, Giudicelli V (2008) IMGT/V-QUEST: the highly customized and integrated system for IG and TR standardized V-J and V-D-J sequence analysis. Nucleic Acids Res 36:W503–W508
15. Giudicelli V, Lefranc M-P (2008) IMGT® standardized analysis of immunoglobulin rearranged sequences. In: Ghia P, Rosenquist R, Davi F (eds) Immunoglobulin gene analysis in chronic lymphocytic leukemia, chap 2. Wolters Kluwer Health Italy, Italy, pp 33–52
16. Giudicelli V, Brochet X, Lefranc M-P (2011) IMGT/V-QUEST: IMGT standardized analysis of the immunoglobulin (IG) and T cell receptor (TR) nucleotide sequences. Cold Spring Harb Protoc 6:695–715. doi:10.1101/pdb.prot5633, pii: pdb.prot5633
17. Alamyar E, Duroux P, Lefranc M-P, Giudicelli V (2012) IMGT® tools for the nucleotide analysis of immunoglobulin (IG) and T cell receptor (TR) V-(D)-J repertoires, polymorphisms, and IG mutations: IMGT/V-QUEST and IMGT/HighV-QUEST for NGS. In: Christiansen F, Tait B (eds) Immunogenetics. Humana, New York. Meth Mol Biol 882: 569–604
18. Yousfi Monod M, Giudicelli V, Chaume D, Lefranc M-P (2004) IMGT/JunctionAnalysis: the first tool for the analysis of the immunoglobulin and T cell receptor complex V-J and V-D-J JUNCTIONs. Bioinformatics 20: i379–i385
19. Giudicelli V, Lefranc M-P (2011) IMGT/JunctionAnalysis: IMGT standardized analysis of the V-J and V-D-J junctions of the rearranged immunoglobulins (IG) and T cell receptors (TR). Cold Spring Harb Protoc 6:716–725. doi:10.1101/pdb.prot5634, pii: pdb.prot5634
20. Giudicelli V, Protat C, Lefranc M-P (2003) The IMGT strategy for the automatic annotation of IG and TR cDNA sequences: IMGT/Automat. In: INRIA (DISC/Spid), Paris, DKB-31. Proceedings of the European Conference on Computational Biology (ECCB 2003), pp 103–104
21. Giudicelli V, Chaume D, Jabado-Michaloud J, Lefranc M-P (2005) Immunogenetics sequence annotation: the strategy of IMGT based on IMGT-ONTOLOGY. Stud Health Technol Inform 116:3–8
22. Alamyar E, Giudicelli V, Shuo L, Duroux P, Lefranc M-P (2012) IMGT/HighV-QUEST: the IMGT® web portal for immunoglobulin (IG) or antibody and T cell receptor (TR) analysis from NGS high throughput and deep sequencing. Immunome Res 8(1):26
23. Li S, Lefranc M-P, Miles JJ, Alamyar E, Giudicelli V, Duroux P, Freeman JD, Corbin V, Scheerlinck J-P, Frohman MA, Cameron PU, Plebanski M, Loveland B, Burrows SR, Papenfuss AT, Gowans EJ (2013) IMGT/HighV-QUEST paradigm for T cell receptor IMGT clonotype diversity and next generation repertoire immunoprofiling. Nat Commun 4:2333
24. Ehrenmann F, Lefranc M-P (2011) IMGT/DomainGapAlign: IMGT standardized analysis of amino acid sequences of variable, constant, and groove domains (IG, TR, MH, IgSF, MhSF). Cold Spring Harb Protoc 6:737–749. doi:10.1101/pdb.prot5636, pii: pdb.prot5636
25. Ehrenmann F, Lefranc M-P (2012) IMGT/DomainGapAlign: the IMGT® tool for the analysis of IG, TR, MHC, IgSF and MhcSF domain amino acid polymorphism. In: Christiansen F, Tait B (eds) Immunogenetics. Humana, New York, chap 33. Methods Mol Biol 882:605–633
26. Ehrenmann F, Giudicelli V, Duroux P, Lefranc M-P (2011) IMGT/Collier de Perles: IMGT standardized representation of domains (IG, TR, and IgSF variable and constant domains, MH and MhSF groove domains). Cold Spring Harb Protoc 6:726–736. doi:10.1101/pdb.prot5635, pii: pdb.prot5635
27. Lane J, Duroux P, Lefranc M-P (2010) From IMGT-ONTOLOGY to IMGT/LIGMotif: the IMGT® standardized approach for immunoglobulin and T cell receptor gene identification and description in large genomic sequences. BMC Bioinformatics 11:223

28. Pommié C, Levadoux S, Sabatier R, Lefranc G, Lefranc M-P (2004) IMGT standardized criteria for statistical analysis of immunoglobulin V-REGION amino acid properties. J Mol Recognit 17:17–32
29. Lefranc M-P (2003) IMGT, the international ImMunoGeneTics information system. In: Bock G, Goode J (eds) Immunoinformatics: bioinformatic strategies for better understanding of immune function. Novartis Foundation Symposium, Wiley, Chichester, UK, 254:126–126, discussion 136–142, 216–222, 250–252
30. Lefranc M-P, Giudicelli V, Ginestoux C, Chaume D (2003) IMGT, the international ImMunoGeneTics information system: the reference in immunoinformatics. Stud Health Technol Inform 95:74–79
31. Lefranc M-P (2003) IMGT databases, web resources and tools for immunoglobulin and T cell receptor sequence analysis. Leukemia 17(1):260–266
32. Lefranc M-P (2004) IMGT, the international ImMunoGenetics information system®. In: Lo BKC (ed) Antibody engineering methods and protocols, 2nd edn. Humana, Totowa NJ. Methods Mol Biol 248:27–49
33. Lefranc M-P (2004) IMGT-ONTOLOGY and IMGT databases, tools and Web resources for immunogenetics and immunoinformatics. Mol Immunol 40(10):647–660
34. Lefranc M-P (2005) IMGT, the international ImMunoGeneTics information system: a standardized approach for immunogenetics and immunoinformatics. Immunome Res 1:3
35. Lefranc M-P (2007) IMGT®, the international ImMunoGeneTics information system® for immunoinformatics. Methods for querying IMGT® databases, tools and Web resources in the context of immunoinformatics. In: Flower DR (ed) Immunoinformatics: predicting immunogenicity in silico, chap 2. Humana, Totowa NJ. Methods Mol Biol 409:19–42
36. Lefranc M-P (2008) IMGT-ONTOLOGY, IMGT® databases, tools and Web resources for Immunoinformatics. In: Schoenbach C, Ranganathan S, Brusic V (eds) Immunoinformatics, vol 1, chap 1. Immunomics reviews, Series of Springer Science and Business Media LLC. Springer, New York. pp 1–18
37. Lefranc M-P, Giudicelli V, Regnier L, Duroux P (2008) IMGT®, a system and an ontology that bridge biological and computational spheres in bioinformatics. Brief Bioinform 9:263–275
38. Lefranc M-P (2008) IMGT®, the international ImMunoGeneTics information system® for immunoinformatics. Methods for querying IMGT® databases, tools and Web resources in the context of immunoinformatics. Mol Biotechnol 40:101–111
39. Lefranc M-P (2009) Antibody databases and tools: The IMGT® experience. In: Zhiqiang A (ed) Therapeutic monoclonal antibodies: from Bench to Clinic, chap 4. Wiley, Hoboken, NJ. pp 91–114
40. Lefranc M-P (2009) Antibody databases: IMGT®, a French platform of world-wide interest [in French]. Bases de données anticorps: IMGT®, une plate-forme française d'intérêt mondial. Médecine/Sciences 25:1020–1023
41. Ehrenmann F, Duroux P, Giudicelli V, Lefranc M-P (2010) Standardized sequence and structure analysis of antibody using IMGT®. In: Kontermann R, Dübel S (eds) Antibody engineering, vol 2, chap 2. Springer, Berlin. pp 11–31
42. Lefranc M-P (2011) IMGT, the International ImMunoGeneTics Information System. Cold Spring Harb Protoc 6:595–603. doi:10.1101/pdb.top115, pii: pdb.top115
43. Lefranc M-P, Ehrenmann F, Ginestoux C, Duroux P, Giudicelli V (2012) Use of IMGT® databases and tools for antibody engineering and humanization. In: Chames P (ed) Antibody engineering, chap 1. Humana, New York. Methods Mol Biol 907:3–37
44. Lefranc M-P (2013) IMGT® Information System. In: Dubitzky W, Wolkenhauer O, Cho K-H, Yokota H (eds) Encyclopedia of systems biology. doi:10.1007/978-1-4419-9863-7. Springer Science + Business Media, LLC012, pp. 959–964
45. Lefranc M-P (2007) WHO-IUIS Nomenclature Subcommittee for immunoglobulins and T cell receptors report. Immunogenetics 59: 899–902
46. Lefranc M-P (2008) WHO-IUIS Nomenclature Subcommittee for immunoglobulins and T cell receptors report August 2007, 13th International Congress of Immunology, Rio de Janeiro, Brazil. Dev Comp Immunol 32:461–463
47. World Health Organization (2012) International Nonproprietary Names (INN) for biological and biotechnological substances (a review). INN Working Document 05.179. Update 2012. http://www.who.int/medicines/services/inn/BioRev2012.pdf
48. Lefranc M-P (2011) Antibody nomenclature: from IMGT-ONTOLOGY to INN definition. MAbs 3(1):1–2

49. Giudicelli V, Lefranc M-P (1999) Ontology for immunogenetics: IMGT-ONTOLOGY. Bioinformatics 15:1047–1054
50. Giudicelli V, Lefranc M-P (2012) IMGT-ONTOLOGY (2012). Frontiers in bioinformatics and computational biology. Front Genet 3:79
51. Giudicelli V, Lefranc M-P (2013) IMGT-ONTOLOGY. In: Dubitzky W, Wolkenhauer O, Cho K-H, Yokota H (eds) Encyclopedia of systems biology. doi:10.1007/978-1-4419-9863-7. Springer Science + Business Media, LLC012, pp. 964–972
52. Giudicelli V, Lefranc M-P (2003) IMGT-ONTOLOGY: gestion et découverte de connaissances au sein d'IMGT. In: Hacid M-S, Kodratoff Y, Boulanger D (Eds.) Extraction et gestion des connaissances (EGC'2003), Actes des troisièmes journées Extraction et Gestion des Connaissances, Lyon, France, 22–24 janvier 2003. Revue des Sciences et Technologies de l'Information, RSTI, série Revue d'Intelligence Artificielle- Extraction des Connaissances et Apprentissage (RIA-ECA), ISBN 2-7462-0631-5. Hermès Science Publications 17(1-2-3). pp 13–23
53. Lefranc M-P, Giudicelli V, Ginestoux C, Bosc N, Folch G, Guiraudou D, Jabado-Michaloud J, Magris S, Scaviner D, Thouvenin V, Combres K, Girod D, Jeanjean S, Protat C, Yousfi Monod M, Duprat E, Kaas Q, Pommié C, Chaume D, Lefranc G (2004) IMGT-ONTOLOGY for immunogenetics and immunoinformatics. In Silico Biol 4:17–29
54. Lefranc M-P, Clément O, Kaas Q, Duprat E, Chastellan P, Coelho I, Combres K, Ginestoux C, Giudicelli V, Chaume D, Lefranc G (2005) IMGT-Choreography for immunogenetics and immunoinformatics. In Silico Biol 5:45–60
55. Duroux P, Kaas Q, Brochet X, Lane J, Ginestoux C, Lefranc M-P, Giudicelli V (2008) IMGT-Kaleidoscope, the formal IMGT-ONTOLOGY paradigm. Biochimie 90:570–583
56. Lefranc M-P (2011) From IMGT-ONTOLOGY IDENTIFICATION axiom to IMGT standardized keywords: for immunoglobulins (IG), T cell receptors (TR), and conventional genes. Cold Spring Harb Protoc 6:604–613. doi:10.1101/pdb.ip82, pii: pdb.ip82
57. Lefranc M-P (2011) From IMGT-ONTOLOGY DESCRIPTION axiom to IMGT standardized labels: for immunoglobulin (IG) and t cell receptor (TR) sequences and structures. Cold Spring Harb Protoc 6:614–626. doi:10.1101/pdb.ip83, pii: pdb.ip83
58. Lefranc M-P (2011) From IMGT-ONTOLOGY CLASSIFICATION axiom to IMGT standardized gene and allele nomenclature: for immunoglobulins (IG) and T cell receptors (TR). Cold Spring Harb Protoc 6:627–632. doi:10.1101/pdb.ip84, pii: pdb.ip84
59. Lefranc M-P (1997) Unique database numbering system for immunogenetic analysis. Immunol Today 18:509
60. Lefranc M-P (1999) The IMGT unique numbering for immunoglobulins, T cell receptors and Ig-like domains. Immunologist 7:132–136
61. Lefranc M-P, Pommié C, Ruiz M, Giudicelli V, Foulquier E, Truong L, Thouvenin-Contet V, Lefranc G (2003) IMGT unique numbering for immunoglobulin and T cell receptor variable domains and Ig superfamily V-like domains. Dev Comp Immunol 27:55–77
62. Lefranc M-P, Pommié C, Kaas Q, Duprat E, Bosc N, Guiraudou D, Jean C, Ruiz M, Da Piedade I, Rouard M, Foulquier E, Thouvenin V, Lefranc G (2005) IMGT unique numbering for immunoglobulin and T cell receptor constant domains and Ig superfamily C-like domains. Dev Comp Immunol 29:185–203
63. Lefranc M-P, Duprat E, Kaas Q, Tranne M, Thiriot A, Lefranc G (2005) IMGT unique numbering for MHC groove G-DOMAIN and MHC superfamily (MhcSF) G-LIKE-DOMAIN. Dev Comp Immunol 29: 917–938
64. Lefranc M-P (2011) IMGT Unique Numbering for the Variable (V), Constant (C), and Groove (G) Domains of IG, TR, MH, IgSF, and MhSF. Cold Spring Harb Protoc 6:633–642. doi:10.1101/pdb.ip85, pii: pdb.ip85
65. Ruiz M, Lefranc M-P (2002) IMGT gene identification and Colliers de Perles of human immunoglobulins with known 3D structures. Immunogenetics 53:857–883
66. Kaas Q, Lefranc M-P (2007) IMGT Colliers de Perles: standardized sequence-structure representations of the IgSF and MhcSF superfamily domains. Curr Bioinform 2:21–30
67. Kaas Q, Ehrenmann F, Lefranc M-P (2007) IG, TR and IgSf, MHC and MhcSF: what do we learn from the IMGT Colliers de Perles? Brief Funct Genomic Proteomic 6:253–264
68. Lefranc M-P (2011) IMGT Collier de Perles for the variable (V), constant (C), and groove (G) domains of IG, TR, MH, IgSF, and MhSF. Cold Spring Harb Protoc 6:643–651. doi:10.1101/pdb.ip86, pii: pdb.ip86
69. Wain HM, Bruford EA, Lovering RC, Lush MJ, Wright MW, Povey S (2002) Guidelines

for human gene nomenclature. Genomics 79:464–470

70. Bruford EA, Lush MJ, Wright MW, Sneddon TP, Povey S, Birney E (2008) The HGNC Database in 2008: a resource for the human genome. Nucleic Acids Res 36:D445–D448
71. Letovsky SI, Cottingham RW, Porter CJ, Li PW (1998) GDB: the Human Genome Database. Nucleic Acids Res 26(1):94–99
72. Maglott DR, Katz KS, Sicotte H, Pruitt KD (2000) NCBI's LocusLink and RefSeq. Nucleic Acids Res 28(1):126–128
73. Maglott D, Ostell J, Pruitt KD, Tatusova T (2007) Entrez gene: gene-centered information at NCBI. Nucleic Acids Res 35:D26–D31
74. Stabenau A, McVicker G, Melsopp C, Proctor G, Clamp M, Birney E (2004) The Ensembl core software libraries. Genome Res 14: 929–933
75. Wilming LG, Gilbert JG, Howe K, Trevanion S, Hubbard T, Harrow JL (2008) The vertebrate genome annotation (Vega) database. Nucleic Acids Res 36:D753–D760
76. Magdelaine-Beuzelin C, Kaas Q, Wehbi V, Ohresser M, Jefferis R, Lefranc M-P, Watier H (2007) Structure-function relationships of the variable domains of monoclonal antibodies approved for cancer treatment. Crit Rev Oncol Hematol 64:210–225
77. Pelat T, Bedouelle H, Rees AR, Crennell SJ, Lefranc M-P, Thullier P (2008) Germline humanization of a non-human Primate antibody that neutralizes the anthrax toxin, by in vitro and in silico engineering. J Mol Biol 384:1400–1407
78. Pelat T, Hust M, Hale M, Lefranc M-P, Dübel S, Thullier P (2009) Isolation of a human-like antibody fragment (scFv) that neutralizes ricin biological activity. BMC Biotechnol 9:60
79. Robert R, Lefranc M-P, Ghochikyan A, Agadjanyan MG, Cribbs DH, Van Nostrand WE, Wark KL, Dolezal O (2010) Restricted V gene usage and VH/VL pairing of mouse humoral response against the N-terminal immunodominant epitope of the amyloid β peptide. Mol Immunol 48(1–3):59–72
80. Ghia P, Stamatopoulos K, Belessi C, Moreno C, Stilgenbauer S, Stevenson FI, Davi F, Rosenquist R (2007) ERIC recommendations on IGHV gene mutational status analysis in chronic lymphocytic leukemia. Leukemia 21:1–3
81. Agathangelidis A, Darzentas N, Hadzidimitriou A, Brochet X, Murray F, Yan XJ, Davis Z, van Gastel-Mol EJ, Tresoldi C, Chu CC, Cahill N, Giudicelli V, Tichy B, Pedersen LB, Foroni L, Bonello L, Janus A, Smedby K, Anagnostopoulos A, Merle-Beral H et al (2012) Stereotyped B-cell receptors in one third of chronic lymphocytic leukemia: towards a molecular classification with implications for targeted therapeutic interventions. Blood 119(19):4467–4475
82. Kostareli E, Gounari M, Janus A, Murray F, Brochet X, Giudicelli V, Pospisilova S, Oscier D, Foroni L, di Celle PF, Tichy B, Pedersen LB, Jurlander J, Ponzoni M, Kouvatsi A, Anagnostopoulos A, Thompson K, Darzentas N, Lefranc M-P, Belessi C et al (2012) Antigen receptor stereotypy across B-cell lymphoproliferations: the case of IGHV4-59/IGKV3-20 receptors with rheumatoid factor activity. Leukemia 26(5):1127–1131
83. Jefferis R, Lefranc M-P (2009) Human immunoglobulin allotypes: Possible implications for immunogenicity. MAbs 1(4): 332–338
84. Lefranc M-P, Lefranc G (2012) Human Gm, Km and Am allotypes and their molecular characterization: a remarkable demonstration of polymorphism. In: Christiansen F, Tait B (eds) Immunogenetics. Humana, New York. Meth Mol Biol 882:635–680
85. Dechavanne C, Guillonneau F, Chiappetta G, Sago L, Lévy P, Salnot V, Guitard E, Ehrenmann F, Broussard C, Chafey P, Le Port A, Vinh J, Mayeux P, Dugoujon J-M, Lefranc M-P, Migot-Nabias F (2012) Mass spectrometry detection of G3m and IGHG3 alleles and follow-up of differential mother and neonate IgG3. PLoS One 7(9):e46097
86. Giest S, McWinnie A, Lefranc M-P, Little AM, Grace S, Mackinnon S, Madrigal JA, Travers PJ (2012) CMV specific CD8+ T cells targeting different peptide/HLA combinations demonstrate varying T cell receptor diversity. Immunology 135(1):27–39
87. Kaas Q, Lefranc M-P (2005) T cell receptor/peptide/MHC molecular characterization and standardized pMHC contact sites in IMGT/3Dstructure-DB. In Silico Biol 5: 505–528
88. Kaas Q, Duprat E, Tourneur G, Lefranc M-P (2008) IMGT standardization for molecular characterization of the T cell receptor/peptide/MHC complexes. In: Schoenbach C, Ranganathan S, Brusic V (eds) Immunoinformatics, chap 2. Immunomics reviews, Series of Springer Science and Business Media LLC. Springer, New York, pp 19–49
89. Hischenhuber B, Frommlet F, Schreiner W, Knapp B (2012) MH2c: Characterization of major histocompatibility α-helices—an information criterion approach. Comput Phys Commun 183(7):1481–1490
90. Hischenhuber B, Havlicek H, Todoric J, Höllrigl-Binder S, Schreiner W, Knapp B (2013) Differential geometric analysis of

alterations in MH α-helices. J Comput Chem 34(21):1862–1879. doi:10.1002/jcc.23328, Epub 2013 May 24

91. Duprat E, Kaas Q, Garelle V, Lefranc G, Lefranc M-P (2004) IMGT standardization for alleles and mutations of the V-LIKE-DOMAINs and C-LIKE-DOMAINs of the immunoglobulin superfamily. In: Pandalai SG (ed) Recent research developments in human genetics, vol 2. Research Signpost, Trivandrum, Kerala, India, pp 111–136
92. Bertrand G, Duprat E, Lefranc M-P, Marti J, Coste J (2004) Characterization of human FCGR3B*02 (HNA-1b, NA2) cDNAs and IMGT standardized description of FCGR3B alleles. Tissue Antigens 64:119–131
93. Bernard D, Hansen JD, du Pasquier L, Lefranc M-P, Benmansour A, Boudinot P (2005) Costimulatory receptors in jawed vertebrates: conserved CD28, odd CTLA4 and multiple BTLAs. Dev Comp Immunol 31: 255–271
94. Garapati VP, Lefranc M-P (2007) IMGT Colliers de Perles and IgSF domain standardization for T cell costimulatory activatory (CD28, ICOS) and inhibitory (CTLA4, PDCD1 and BTLA) receptors. Dev Comp Immunol 31:1050–1072
95. Hansen JD, Pasquier LD, Lefranc M-P, Lopez V, Benmansour A, Boudinot P (2009) The B7 family of immunoregulatory receptors: a comparative and evolutionary perspective. Mol Immunol 46:457–472
96. Frigoul A, Lefranc M-P (2005) MICA: standardized IMGT allele nomenclature, polymorphisms and diseases. In: Pandalai SG (ed) Recent research developments in human genetics, vol 3. Research Signpost, Trivandrum, Kerala, India, pp 95–145
97. Duprat E, Lefranc M-P, Gascuel O (2006) A simple method to predict protein binding from aligned sequences—application to MHC superfamily and beta2-microglobulin. Bioinformatics 22:453–459
98. Rose PW, Beran B, Bi C, Bluhm WF, Dimitropoulos D, Goodsell DS, Prlic A, Quesada M, Quinn GB, Westbrook JD, Young J, Yukich B, Zardecki C, Berman HM, Bourne PE (2011) The RCSB Protein Data Bank: redesigned web site and web services. Nucleic Acids Res 39:D392–D401
99. Kabat EA, Wu TT, Perry HM, Gottesman KS, Foeller C (1991) Sequences of proteins of immunological interest. U.S. Department of Health and Human Services (USDHHS), Washington, DC. National Institute of Health NIH Publication, 91-3242
100. Garboczi DN, Ghosh P, Utz U, Fan QR, Biddison WE, Wiley DC (1996) Structure of the complex between human T-cell receptor, viral peptide and HLA-A2. Nature 384:134–141
101. Scott DR, Borbulevych OY, Piepenbrink KH, Corcelli SA, Baker BM (2011) Disparate degrees of hypervariable loop flexibility control T-cell receptor cross-reactivity, specificity, and binding mechanism. Mol Biol 414(3):385–400

Chapter 5

IMGT/HLA and the Immuno Polymorphism Database

James Robinson, Jason A. Halliwell, and Steven G.E. Marsh

Abstract

The IMGT/HLA Database (http://www.ebi.ac.uk/ipd/imgt/hla/) was first released over 15 years ago, providing the HLA community with a searchable repository of highly curated HLA sequences. The HLA complex is located within the 6p21.3 region of human chromosome 6 and contains more than 220 genes of diverse function. Many of the genes encode proteins of the immune system and are highly polymorphic, with some genes currently having over 3,000 known allelic variants. The Immuno Polymorphism Database (IPD) (http://www.ebi.ac.uk/ipd/) expands on this model, with a further set of specialist databases related to the study of polymorphic genes in the immune system. The IPD project works with specialist groups or nomenclature committees who provide and curate individual sections before they are submitted to IPD for online publication. IPD currently consists of four databases: IPD-KIR contains the allelic sequences of killer-cell immunoglobulin-like receptors; IPD-MHC is a database of sequences of the major histocompatibility complex of different species; IPD-HPA, alloantigens expressed only on platelets; and IPD-ESTDAB, which provides access to the European Searchable Tumour Cell-Line Database, a cell bank of immunologically characterized melanoma cell lines. Through the work of the HLA Informatics Group and in collaboration with the European Bioinformatics Institute we are able to provide public access to this data through the website http://www.ebi.ac.uk/ipd/.

Key words Immunogenetics, Database, Polymorphism, Variation, Sequence, Allele, MHC, HLA, KIR

1 Introduction

The Immuno Polymorphism Database (IPD) is a set of specialist databases related to the study of polymorphic genes in the immune system. The IPD project [1] works with specialist groups or nomenclature committees who provide and curate individual sections before they are submitted to IPD for online publication. The IPD project stores all the data in a set of related databases. IPD currently consists of five databases: IMGT/HLA contains sequences of the human major histocompatibility complex; IPD-KIR contains the allelic sequences of killer-cell immunoglobulin-like receptors; IPD-MHC is a database of sequences of the MHC of different species; IPD-HPA, alloantigens expressed only on

Rajat K. De and Namrata Tomar (eds.), *Immunoinformatics*, Methods in Molecular Biology, vol. 1184,
DOI 10.1007/978-1-4939-1115-8_5, © Springer Science+Business Media New York 2014

platelets; and IPD-ESTDAB, which provides access to the European Searchable Tumour Cell-Line Database, a cell bank of immunologically characterized melanoma cell lines.

The IMGT/HLA Database [2] was established to provide a locus-specific database (LSDB) for the allelic sequences of the genes in the HLA system, also known as the human major histocompatibility complex (MHC). The core genes of interest in the HLA system are 21 highly polymorphic HLA genes, found within the 6p21.3 region of the short arm of human chromosome 6, whose protein products mediate human responses to infectious disease and influence the outcome of cell and organ transplants. The MHC is one of the most complex and polymorphic regions of the human genome, with in excess of 220 genes [3]. Three distinct regions have been identified within the MHC. The class I region is located at the telomeric end of the MHC and encodes the genes for the HLA class I molecules, HLA-A, -B, and -C. These are codominantly expressed on the cell surface and responsible for presenting intracellularly derived peptides to CD8-positive T cells. The class II region lies at the centromeric end of the MHC and encodes HLA class genes HLA-DRA, -DRB1, -DRB3, -DRB4, -DRB5, -DQA1, -DQB1, -DPA1, and -DPB1. HLA class II expression is limited to cells involved in immune responses, where these molecules present extracellular derived peptides to CD4-positive T cells. Located between the class I and class II regions lies the class III region where a number of non-HLA genes with immune function are located. The HLA molecules play a key role in transplantation, with the success of kidney and bone marrow transplantation correlated with the degree to which donors and recipient are HLA matched. It has been shown that HLA matching is recognized as a critical determinant of outcome for patients receiving unrelated donor hematopoietic stem cell for hematological disorders [4]. This has led to progressive improvements in the level of resolution achieved by HLA class I and II typing methods. The typing of HLA now focuses on distinguishing differences at both synonymous and the non-synonymous level, for the nucleotide sequences encoding the protein domains of HLA class I and II, which bind peptides and interact with variable lymphocyte receptors. The consequence of these improvements has required the development, for each polymorphic HLA class I and II gene, of a nucleotide sequence database that is both accurate and comprehensive. The first public release of the IMGT/HLA Database was made on 16 December 1998 [5]. This centralized and curated LSDB manages these highly polymorphic variants and with a nomenclature now covering more than 50 genes and almost 10,000 alleles. Since its inception the database has been updated every 3 months, with over 60 releases, to include all the publicly available sequences officially named by the WHO Nomenclature Committee.

2 IMGT/HLA Nomenclature

The naming of new HLA genes and allele sequences and their quality control is the responsibility of the WHO Nomenclature Committee for Factors of the HLA System, which first met in 1968. This committee meets regularly to discuss issues of nomenclature and has published 19 major reports [6–24] initially documenting the serologically defined HLA antigens and more recently the genes and alleles defined by nucleotide sequences. The IMGT/HLA Database provides the nomenclature committee with the online tools necessary for its task. The dissemination of new allele names and sequences is of paramount importance in the clinical transplant setting, because the variation that distinguishes HLA alleles can have a critical impact on the outcome of a hematopoietic stem cell transplant [25, 26]. The identification, verification, and publication of the sequences of these variants through a centralized resource are necessary for accurate identification of HLA alleles in a clinical setting. Sequencing of HLA alleles began in the late 1970s predominantly using protein-based techniques to determine the sequences of HLA class I allotypes. The first complete HLA class I allotype sequence, B7.2, now known as *B*07:02:01*, was published in 1979 [27]. The first HLA class II allele, *DRA*01:01*, was defined by protein sequencing and later in 1982 by DNA sequencing [28–30]. The first HLA DNA sequences or alleles were named by the WHO Nomenclature Committee for Factors of the HLA System (10) in 1987. At that time 12 class I alleles and 9 class II alleles were named: in the first 9 months of 2013 the WHO Nomenclature Committee was able to assign names to 1,029 alleles; *see* Fig. 1.

3 IMGT/HLA as a Model for Other Highly Polymorphic Gene Systems

The HLA Nomenclature and its publication through the IMGT/HLA Database have been taken as a model by other groups working in the field. The MHC sequences of many different species have been reported [31–42], along with different nomenclature systems used in the naming and identification of new genes and alleles in each species [43]. The nomenclature for MHC genes and alleles in species other than humans [24, 44] and mice [45, 46] has historically been overseen either informally by groups generating sequences or by formal nomenclature committees set up by the International Society for Animal Genetics (ISAG) [47]. This work is now overseen by the Comparative MHC Nomenclature Committee and is supported by ISAG and the Veterinary Immunology Committee (VIC) of the International Union of Immunological Societies (IUIS) [48]. The sequences of the MHC

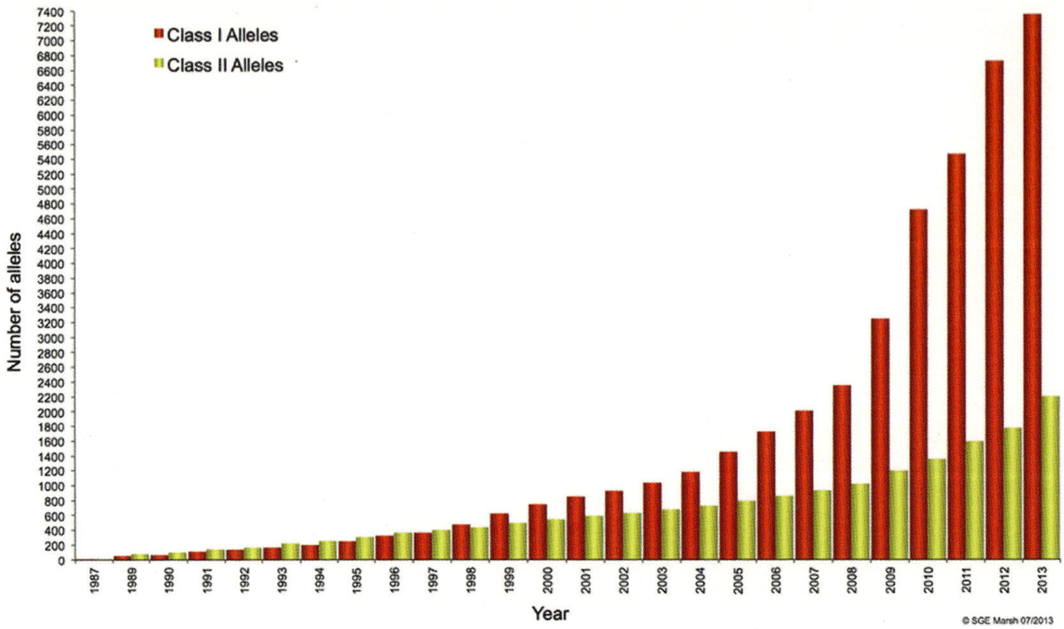

Fig. 1 Graph of HLA allele numbers. Graph showing the numbers of antigens and alleles named by year from 1987 to the end of March 2013. The numbers of HLA class I alleles are shown in *red* and the HLA class II alleles in *yellow*

from a number of different species are highly conserved between species [49], and by bringing the work of different nomenclature committees and the sequences of different species together it is hoped to provide a central resource that will facilitate further research on the MHC of each species and on their comparison [50]. The first version of the IPD-MHC database involved the work of groups specializing in non-human primates (NHP) [41], canines (DLA) [37], and felines (FLA) [51] and incorporated all data previously available in the IMGT/MHC Database [50]. Since the first version we have been able to add sequences from cattle (BoLA) [42], teleost fish [52], rats (RT1) [53], sheep (OLA) [40], and swine (SLA) [39]. In 2012 the nomenclature used to describe the alleles of NHP was extensively revised and updated [41]. This was accompanied by updating the IPD-MHC NHP section to complement the publication; IPD-MHC NHP currently contains over 4,000 alleles covering 47 species of apes and Old World and New World Monkeys. The management of the sequences within IPD-MHC and the provision of an online submission tool have enabled these databases to grow, the number of sequences increasing by at least 10 % each year and the nomenclature to expand since the inclusion of a species within IPD. This has resulted in regular publications reporting updates or changes to the nomenclature [40–42, 54].

The principles behind the IMGT/HLA model can also be applied outside the MHC; this is seen in the IPD-KIR database. The Killer-cell Immunoglobulin-like Receptors (KIR) are members of the immunoglobulin super family (IgSF) formerly called killer-cell inhibitory receptors. KIRs have been shown to be highly polymorphic both at the allelic and haplotypic levels [55]. They are composed of two or three Ig domains, a transmembrane region, and cytoplasmic tail, which can in turn be short (activatory) or long (inhibitory). The leukocyte receptor complex (LRC), which encodes KIR genes, has been shown to be polymorphic, polygenic, and complex in a manner similar to the MHC. Because of the complexity in the KIR region and KIR sequences a KIR Nomenclature Committee was established in 2002 to undertake the naming of human KIR allele sequences. The first KIR Nomenclature report was published in 2003 [56], which coincided with the first release of the IPD-KIR database. The number of officially named human KIR alleles has increased since the initial release which contained 89 alleles. As of September 2013, there are over 600 alleles, which code for over 320 unique protein sequences.

4 IPD Data Sources

IPD receives submissions from laboratories across the world. These submissions are curated and analyzed, and if they meet the strict requirements, an official allele designation is assigned. The IMGT/HLA Database is the official repository for the WHO Nomenclature Committee for Factors of the HLA System and is the only way of receiving an official allele designation for a sequence. The other IPD sections work in the same way with official nomenclature committees for KIR and different nonhuman MHC committees. The sequences are then incorporated into the periodic releases of the database. Since its release in December 1998 the IMGT/HLA Database has received over 17,700 submissions. These submissions have come from a variety of sources; the majority are from laboratories involved in clinical HLA typing for hospitals or donor registries or commercial organizations performing contract HLA typing for large hematopoietic stem cell donor registries. Further data has been submitted following large-scale genome sequencing projects [3, 57]. For all projects the submissions must meet strict acceptance criteria before the sequence receives an official designation. These minimum standards cover the methodologies used to define the sequence, the length of sequence submitted, and the source of the sequence; the full list of the minimum criteria can be seen online. Within IMGT/HLA, around 3 % of the submissions received fail to meet these criteria and are rejected. In addition all the submissions received by the IPD are also available from the International Nucleotide Sequence Database Collaboration (INSDC) [58].

The INSDC consists of DNA DataBank of Japan (DDBJ) (Japan), GenBank (USA), and the EMBL-European Nucleotide Archive (ENA) (UK) [59–61]. The ENA entries also contain database cross-references to the IPD entries. Cross-references to the IMGT/HLA Database are also included in ENSEMBL [62] and VEGA entries [63].

5 Tools Available at IPD

IPD provides a large number of tools for the analysis of HLA, KIR, and nonhuman MHC sequences. These tools are either custom written for the database or are incorporated into existing tools on the European Bioinformatics Institute (EBI) website [64, 65].

These tools include the following:

- Sequence alignments—Access to alignment tool, which filters pre-generated alignments to the users' specification; provides alignments at the protein, cDNA, and gDNA level.
- Allele queries—Access to detailed information on any allele, including information on database cross-references and seminal publications.
- Sequence similarity search tools—Integration into EBI's suite of search tools including FASTA [66] and BLAST [67].
- Downloads—Access to an FTP directory containing all the data from the current and previous releases in a variety of commonly used formats like FASTA, MSF, and PIR.

There are core tools, which are common to all projects, and other tools specific for individual sections. For example tools have also been developed to support the laboratories that sequence HLA. The use of sequence-based typing (SBT) as a method for defining the HLA type is well documented [68, 69]; most SBT typing strategies currently employed use the exon 2 and exon 3 sequences for HLA class I analysis and exon 2 alone for HLA class II analysis. Due to the heterozygous nature of the SBT analysis the combinations of many pairs of alleles may give an ambiguous typing result; currently there are nearly 80,000 recognized ambiguous combinations. The IMGT/HLA maintains and regularly updates a listing of these ambiguous allele combinations. The document also includes a list of all alleles which are identical over exons 2 and 3 for HLA class I and exon 2 for HLA class II.

6 Clinical Algorithms

The IPD project also collaborates with clinicians to provided Web-based version of published algorithms which have a clinical impact on transplant outcome. Two examples of this are the IMGT/HLA

Database—DPB1 T-Cell Epitope Algorithm and the IPD-KIR—Donor B Content Algorithm.

Recent developments on the IPD-KIR website include online tools to assist in the prediction of transplant outcome in an unrelated hematopoietic stem cell transplant based on the KIR content of the individuals involved. In 2008 a tool was added to the website to help predict NK cell alloreactivity based on the KIR ligands present in the patient and donor, as transplant strategies based on KIR-ligand mismatches had been shown to influence relapse, graft vs. host disease (GvHD), and survival in patients with acute myeloid leukaemia (AML) [70]. In 2010, with the goal of developing a donor selection strategy to improve transplant outcome, Cooley et al. [71] compared the contribution of KIR gene motifs to the clinical benefit conferred by donors with a particular haplotype. Donor KIR genotype influenced transplantation outcome for some forms of leukaemia after a T-cell replete unrelated donor transplant. KIR genotyping several HLA-matched potential donors could substantially increase the frequency of transplants using unrelated donor grafts with favorable KIR gene content. In order to implement this strategy the IPD-KIR database was asked to provide an online version of the algorithm described in the paper. The B-Content calculator (http://www.ebi.ac.uk/ipd/kir/donor_b_content.html) allows the user to enter the KIR genotypes for up to five prospective donors and receive their B-Content assignments and a prediction result of the effect of the KIR genotype on transplant outcome. To ensure that only valid KIR genotypes are submitted, all genotypes submitted are compared to a list of predicted genotypes based on known KIR haplotypes. In addition this list has been supplemented with a number of additional KIR genotypes that have been defined in routine KIR typing. If a prospective donor's KIR typing does not match any of the genotypes on this list a warning is issued.

Recent data has suggested that certain HLA mismatches may be permissive (i.e., do not result in a poor clinical outcome), while others are non-permissive (do result in a poor clinical outcome) [72]. The classification of HLA-DPB1 mismatches based on T-cell epitope (TCE) groups has been shown to identify permissive mismatches and non-permissive mismatches for HLA-DPB1 after unrelated donor hematopoietic stem cell transplantation (HSCT). With the strong clinical data showing a survival disadvantage in patients who receive a transplant from a non-permissive HLA-DPB1 TCE mismatched donor, defined on the basis of functional data, matching of DPB1 TCE groups can be routinely included in the donor selection process [73–76]. The IMGT/HLA Database provides an online, freely available tool, which was developed to help those selecting donors to predict the immunogenicity of any given patient–donor HLA-DPB1 types [77]. The aim of the tool is to provide a web interface to predict HLA-DPB1 immunogenicity based on the published algorithms. Tables in the original publi-

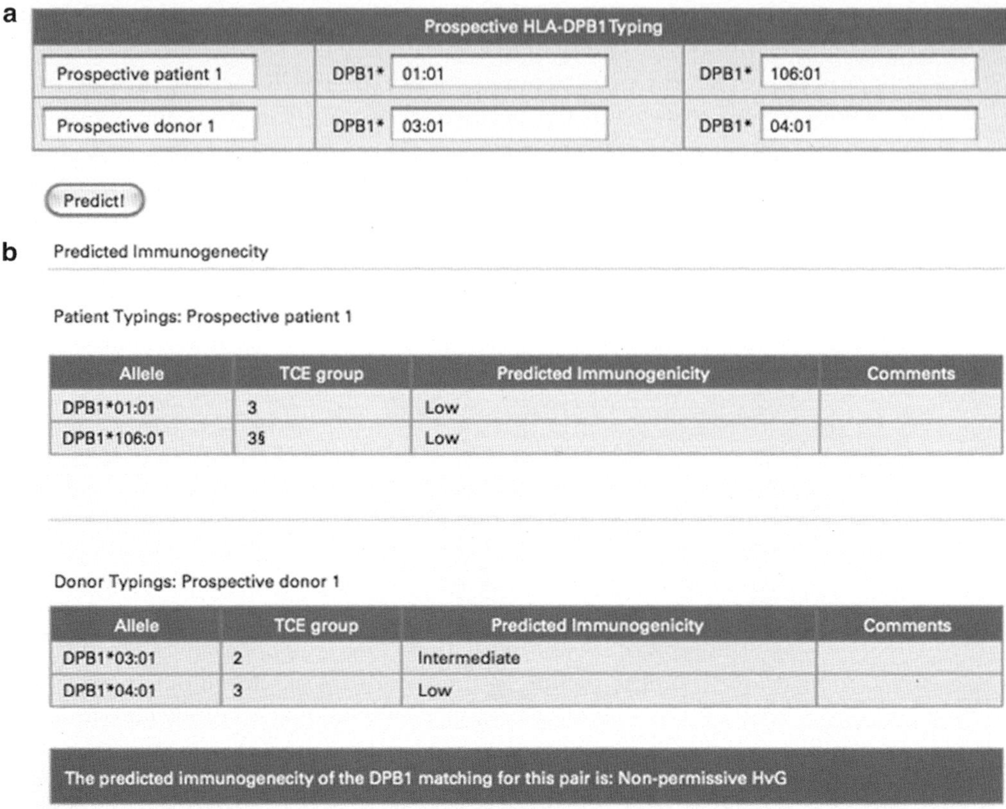

Fig. 2 Example of the DPB1-T-Cell Epitope Algorithm web page. A graphic example of the prediction of immunogenicity and permissivity by the tool. (**a**) The input screen with the HLA-DPB1 typing of the prospective patient (HLA-DPB1*01:01, 106:01) and prospective donor 1 (HLA-DPB1*03:01, 04:01). (**b**) The output screen showing that the two alleles of the prospective patients are both predicted to have "low" immunogenicity, while the HLA-DPB1*03:01 allele of the prospective donor 1 is predicted to have "intermediate" immunogenicity, indicating that the HLA-DPB1 matching status for this patient and donor is "non-permissive host vs. graft disease"

cations provide details of the TCE groups, functionally defined on the basis of alloreactive T-cell cross-reactivity patterns and predicted immunogenicity hierarchies for a number of HLA-DPB1 proteins. These tables are then queried for the TCE groups, and these results generate the predicted immunogenicity. The search tool allows users to enter the HLA-DPB1 data for a single prospective patient and up to five prospective donors (Fig. 2). The predicted immunogenicity of the HLA-DPB1 matching for each patient–donor pair is provided. If the input includes nonexistent alleles, null alleles, or the unstudied TCE groups, a warning detailing the problem is given. The tool also allows for labeling the patient and donors with user-defined identification numbers. The results can therefore be printed and stored. The web tool is hosted on the IMGT/HLA Database website and can be accessed at http://www.ebi.ac.uk/ipd/imgt/hla/dpb.html.

7 Conclusion

The IPD project provides a resource for those interested in the study of polymorphic sequences in the immune system. By accommodating related systems in a single database, data can be made available in common formats aiding the use and interpretation. As the projects grow and more sections are added, the benefit of having expertly curated sequences from related areas stored in a single location is becoming more apparent. This is particularly true of the IPD-MHC project, where cross-species studies are able to utilize the high-quality sequences provided by the different nomenclature committees in a common standardized format, ready for use. The initial release of the IPD Database contained only four sections and a small number of tools; however as the database has grown and more sections and species have been added, more tools have been added to the website. We plan to use the existing database structures to house data for new sections of the IPD project as they become available. The files will also be made available in different formats to download from the website, FTP server, and included different web services at the EBI [65].

The IMGT/HLA Database provides a centralized resource for the study of the HLA system, whether this is clinically or scientifically focussed. The database and accompanying tools allow the study of HLA alleles from a single site on the World Wide Web. It aids in the management and development of HLA nomenclature, providing a continuing and updated resource for the WHO Nomenclature Committee. The challenges for the database are to keep up with this increase in submitted sequences, keep pace with the increasing difficulties in performing analyses on the larger datasets, and develop new tools for the visualization of the sequences while maintaining the high standards set in the presentation and quality of the HLA sequences and nomenclature to the research community.

8 Licensing

The IPD is covered by the Creative Commons Attribution-NoDerivs Licence, which is applicable to all copyrightable parts of the database, which includes the sequence alignments. This means that users are free to copy, distribute, display, and make commercial use of the databases in all legislations, provided that they give the appropriate credit [78, 79]. If users intend to distribute a modified version of the data in any form, then they must ask us for permission; this can be done by contacting hla@alleles.org for further details of how modified data can be reproduced.

Acknowledgements

The authors would like to thank Angie Dahl of the Be The Match Foundation for her continuing work in securing ongoing funding for the IMGT/HLA Database. We would like to thank all of the individuals and organizations that support our work financially.

The authors would also like to thank Shirley Ellis, John Hammond, Chak-Sum Ho, Lorna Kennedy, Hans Dijkstra, Natasja de Groot, Nel Otting and Ronald Bontrop, Lutz Walter, Keith Ballingall, Donald Miller, Paul Metcalfe, Nick Watkins, Graham Pawelec, Libby Guethlein, Peter Parham, Jeff Miller, and Sarah Cooley for their involvement in the IPD project.

Finally the authors would like to thank Rodigo Lopez and Hamish McWilliam and the European Bioinformatics Institute for technical and infrastructure support.

Funding: This work was supported by Histogenetics; One Lambda Inc.; Conexio; Abbott Molecular Laboratories Inc.; Life Technologies; the American Society for Histocompatibility and Immunogenetics; DKMS; Olersup SSP; 454 Sequencing; Lab Corp; Lifecodes + Immunocor Gamma; the European Federation for Immunogenetics; Zentrum Knochenmarkspender-Register Deutschland; Anthony Nolan; the Asia-Pacific Histocompatibility and Immunogenetics Association; BAG Healthcare; Be the Match Foundation; the National Marrow Donor Program; Rose; Innotrain Diagnostik GMBH; and GenDX. Initial support for the IMGT/HLA Database project was from the Imperial Cancer Research Fund (now Cancer Research UK) and a EU Biotech grant (BIO4CT960037).

Appendix. Access and Contact

IMGT/HLA homepage: http://www.ebi.ac.uk/ipd/
IMGT/HLA FTP site: ftp://ftp.ebi.ac.uk/pub/databases/ipd/
Contact: hla@alleles.org

References

1. Robinson J, Halliwell JA, McWilliam H et al (2013) IPD—the Immuno-Polymorphism Database. Nucleic Acids Res 41:D1234–D1240
2. Robinson J, Halliwell JA, McWilliam H et al (2013) The IMGT/HLA Database. Nucleic Acids Res 41:D1222–D1227
3. Horton R, Wilming L, Rand V et al (2004) Gene map of the extended human MHC. Nat Rev Genet 5:889–899
4. Shaw BE, Arguello R, Garcia-Sepulveda CA et al (2010) The impact of HLA genotyping on survival following unrelated donor haematopoietic stem cell transplantation. Br J Haematol 150: 251–258
5. Robinson J, Bodmer JG, Malik A et al (1998) Development of the international immunogenetics HLA database. Hum Immunol 59:17
6. WHO Nomenclature Committee (1968) Nomenclature for factors of the HL—a system. Bull World Health Organ 39:483–486
7. WHO Nomenclature Committee (1970) WHO Terminology Report. In: Terasaki PI (ed) Histocompatibility testing. Munksgaard, Copenhagen. p 49

8. WHO Nomenclature Committee (1972) Nomenclature for factors of the HL-A system. Bull World Health Organ 47:659–662
9. WHO IUIS Terminology-Committee (1975) Nomenclature for factors of the HLA system. Bull World Health Organ 52:261–265
10. WHO Nomenclature Committee (1978) Nomenclature for factors of the HLA system, 1977. Tissue Antigens 11:81–86
11. WHO Nomenclature Committee (1980) Nomenclature for Factors of the HLA System. In: Terasaki PI, (ed). *Histocompatibility Testing, 1980.* UCLA Tissue Typing Laboratory, Los Angeles: pp 18–20
12. WHO Nomenclature Committee (1984) Nomenclature for factors of the HLA system 1984. Tissue Antigens 24:73–80
13. WHO Nomenclature Committee (1988) Nomenclature for factors of the HLA system, 1987. Tissue Antigens 32:177–187
14. Bodmer JG, Marsh SGE, Parham P et al (1990) Nomenclature for factors of the HLA system, 1989. Tissue Antigens 35:1–8
15. Bodmer JG, Marsh SGE, Albert ED et al (1991) Nomenclature for factors of the HLA system, 1990. Tissue Antigens 37:97–104
16. Bodmer JG, Marsh SGE, Albert ED et al (1992) Nomenclature for factors of the HLA system, 1991. Hum Immunol 34:4–18
17. Bodmer JG, Marsh SGE, Albert ED et al (1994) Nomenclature for factors of the HLA system, 1994. Tissue Antigens 44:1–18
18. Bodmer JG, Marsh SGE, Albert ED et al (1995) Nomenclature for factors of the HLA system, 1995. Tissue Antigens 46:1–18
19. Bodmer JG, Marsh SGE, Albert ED et al (1997) Nomenclature for factors of the HLA system, 1996. Tissue Antigens 49:297–321
20. Bodmer JG, Marsh SGE, Albert ED et al (1999) Nomenclature for factors of the HLA system, 1998. Tissue Antigens 53:407–446
21. Marsh SGE, Bodmer JG, Albert ED et al (2001) Nomenclature for factors of the HLA system, 2000. Tissue Antigens 57:236–283
22. Marsh SGE, Albert ED, Bodmer WF et al (2002) Nomenclature for factors of the HLA system, 2002. Tissue Antigens 60:407–464
23. Marsh SGE, Albert ED, Bodmer WF et al (2005) Nomenclature for factors of the HLA system, 2004. Tissue Antigens 65:301–369
24. Marsh SGE, Albert ED, Bodmer WF et al (2010) Nomenclature for factors of the HLA system, 2010. Tissue Antigens 75:291–455
25. Lee SJ, Klein J, Haagenson M et al (2007) High-resolution donor-recipient HLA matching contributes to the success of unrelated donor marrow transplantation. Blood 110:4576–4583
26. Shaw BE, Mayor NP, Russell NH et al (2010) Diverging effects of HLA-DPB1 matching status on outcome following unrelated donor transplantation depending on disease stage and the degree of matching for other HLA alleles. Leukemia 24:58–65
27. Orr HT, Lopez de Castro JA, Lancet D et al (1979) Complete amino acid sequence of a papain-solubilized human histocompatibility antigen, HLA-B7. 2. Sequence determination and search for homologies. Biochemistry 18: 5711–5720
28. Lee JS, Trowsdale J, Travers PJ et al (1982) Sequence of an HLA-DR alpha-chain cDNA clone and intron-exon organization of the corresponding gene. Nature 299:750–752
29. Wake CT, Long EO, Strubin M et al (1982) Isolation of cDNA clones encoding HLA-DR alpha chains. Proc Natl Acad Sci U S A 79: 6979–6983
30. Yang C, Kratzin H, Gotz H et al (1982) Primary structure of class II human histocompatibility antigens. 2nd Communication. Amino acid sequence of the N-terminal 179 residues of the alpha-chain of an HLA-Dw2/DR2 alloantigen (author's transl). Hoppe Seylers Z Physiol Chem 363:671–676
31. Longenecker BM, Mosmann TR (1981) Nomenclature for chicken MHC (B) antigens defined by monoclonal antibodies. Immunogenetics 13:25–28
32. Briles WE, Bumstead N, Ewert DL et al (1982) Nomenclature for chicken major histocompatibility (B) complex. Immunogenetics 15: 441–447
33. 1991 Leukocyte antigens in cattle, sheep and goats. Nomenclature. Vet Immunol Immunopathol. 27:15–16
34. Davies CJ, Andersson L, Joosten I et al (1992) Characterization of bovine MHC class II polymorphism using three typing methods: serology, RFLP and IEF. Eur J Immunogenet 19:253–262
35. Naessens J (1993) Leukocyte antigens of cattle and sheep. Nomenclature. Vet Immunol Immunopathol 39:11–12
36. Kennedy LJ, Altet L, Angles JM et al (2000) Nomenclature for factors of the dog major histocompatibility system (DLA), 1998: first report of the ISAG DLA Nomenclature Committee. Anim Genet 31:52–61
37. Kennedy LJ, Angles JM, Barnes A et al (2001) Nomenclature for factors of the dog major histocompatibility system (DLA), 2000: second report of the ISAG DLA Nomenclature Committee. Anim Genet 32:193–199
38. Miller MM, Bacon LD, Hala K et al (2004) 2004 Nomenclature for the chicken major histocompatibility (B and Y) complex. Immunogenetics 56:261–279
39. Smith DM, Lunney JK, Ho CS et al (2005) Nomenclature for factors of the swine leukocyte antigen class II system, 2005. Tissue Antigens 66:623–639

40. Ballingall KT, Herrmann-Hoesing L, Robinson J et al (2011) A single nomenclature and associated database for alleles at the major histocompatibility complex class II DRB1 locus of sheep. Tissue Antigens 77:546–553
41. de Groot NG, Otting N, Robinson J et al (2012) Nomenclature report on the major histocompatibility complex genes and alleles of Great Ape, Old and New World monkey species. Immunogenetics 64:615–631
42. Hammond JA, Marsh SGE, Robinson J et al (2012) Cattle MHC nomenclature: is it possible to assign sequences to discrete class I genes? Immunogenetics 64:475–480
43. Klein J, Bontrop RE, Dawkins RL et al (1990) Nomenclature for the major histocompatibility complexes of different species: a proposal. Immunogenetics 31:217–219
44. Robinson J, Halliwell JA, McWilliam H et al (2013) The IMGT/HLA Database. Nucleic Acids Res 41:D1222–D1227
45. Rodgers JR, Levitt JM, Cresswell P et al (1999) A nomenclature solution to mouse MHC confusion. J Immunol 162:6294
46. Eppig JT, Blake JA, Bult CJ et al (2012) The Mouse Genome Database (MGD): comprehensive resource for genetics and genomics of the laboratory mouse. Nucleic Acids Res 40: D881–D886
47. Ellis SA, Bontrop RE, Antczak DF et al (2006) ISAG/IUIS-VIC Comparative MHC Nomenclature Committee report, 2005. Immunogenetics 57:953–958
48. Ballingall KT (2012) Progress of the Comparative MHC Committee and a summary of the Comparative MHC Workshops held at the 32nd ISAG, Edinburgh and the 9th IVIS, Tokyo, 2010. Vet Immunol Immunopathol 148:202–208
49. Parham P (1999) Virtual reality in the MHC. Immunol Rev 167:5–15
50. Robinson J, Waller MJ, Parham P et al (2003) IMGT/HLA and IMGT/MHC: sequence databases for the study of the major histocompatibility complex. Nucleic Acids Res 31: 311–314
51. Drake GJ, Kennedy LJ, Auty HK et al (2004) The use of reference strand-mediated conformational analysis for the study of cheetah (Acinonyx jubatus) feline leucocyte antigen class II DRB polymorphisms. Mol Ecol 13: 221–229
52. Lukacs MF, Harstad H, Bakke HG et al (2010) Comprehensive analysis of MHC class I genes from the U-, S-, and Z-lineages in Atlantic salmon. BMC Genomics 11:154
53. Fujii H, Kakinuma M, Yoshiki T et al (1991) Polymorphism of the class II gene of rat major histocompatibility complex, RT1: partial sequence comparison of the first domain of the RT1.B beta 1 alleles. Immunogenetics 33: 399–403
54. Ho CS, Lunney JK, Ando A et al (2009) Nomenclature for factors of the SLA system, update 2008. Tissue Antigens 73:307–315
55. Garcia CA, Robinson J, Guethlein LA et al (2003) Human KIR sequences 2003. Immunogenetics 55:227–239
56. Marsh SGE, Parham P, Dupont B et al (2003) Killer-cell immunoglobulin-like receptor (KIR) nomenclature report, 2002. Tissue Antigens 62:79–86
57. Mungall AJ, Palmer SA, Sims SK et al (2003) The DNA sequence and analysis of human chromosome 6. Nature 425:805–811
58. Karsch-Mizrachi I, Nakamura Y, Cochrane G (2012) The international nucleotide sequence database collaboration. Nucleic Acids Res 40:D33–D37
59. Kodama Y, Mashima J, Kaminuma E et al (2012) The DNA Data Bank of Japan launches a new resource, the DDBJ Omics Archive of functional genomics experiments. Nucleic Acids Res 40:D38–D42
60. Amid C, Birney E, Bower L et al (2012) Major submissions tool developments at the European Nucleotide Archive. Nucleic Acids Res 40: D43–D47
61. Benson DA, Karsch-Mizrachi I, Clark K et al (2012) GenBank. Nucleic Acids Res 40: D48–D53
62. Flicek P, Amode MR, Barrell D et al (2012) Ensembl 2012. Nucleic Acids Res 40:D84–D90
63. Wilming LG, Gilbert JG, Howe K et al (2008) The vertebrate genome annotation (Vega) database. Nucleic Acids Res 36:D753–D760
64. Goujon M, McWilliam H, Li W et al (2010) A new bioinformatics analysis tools framework at EMBL-EBI. Nucleic Acids Res 38: W695–W699
65. McWilliam H, Valentin F, Goujon M et al (2009) Web services at the European Bioinformatics Institute-2009. Nucleic Acids Res 37:W6–W10
66. Pearson WR, Lipman DJ (1988) Improved tools for biological sequence comparison. Proc Natl Acad Sci U S A 85:2444–2448
67. Altschul SF, Madden TL, Schaffer AA et al (1997) Gapped BLAST and PSI-BLAST: a new generation of protein database search programs. Nucleic Acids Res 25:3389–3402
68. Santamaria P, Lindstrom AL, Boyce-Jacino MT et al (1993) HLA class I sequence-based typing. Hum Immunol 37:39–50
69. Rozemuller EH, Bouwens AG, van Oort E et al (1995) Sequencing-based typing reveals new insight in HLA-DPA1 polymorphism. Tissue Antigens 45:57–62

70. Ruggeri L, Capanni M, Casucci M et al (1999) Role of natural killer cell alloreactivity in HLA-mismatched hematopoietic stem cell transplantation. Blood 94:333–339
71. Cooley S, Weisdorf DJ, Guethlein LA et al (2010) Donor selection for natural killer cell receptor genes leads to superior survival after unrelated transplantation for acute myelogenous leukemia. Blood 116: 2411–2419
72. Kawase T, Morishima Y, Matsuo K et al (2007) High-risk HLA allele mismatch combinations responsible for severe acute graft-versus-host disease and implication for its molecular mechanism. Blood 110:2235–2241
73. Fleischhauer K, Shaw BE, Gooley T et al (2012) Effect of T-cell-epitope matching at HLA-DPB1 in recipients of unrelated-donor haemopoietic-cell transplantation: a retrospective study. Lancet Oncol 13:366–374
74. Crocchiolo R, Zino E, Vago L et al (2009) Nonpermissive HLA-DPB1 disparity is a significant independent risk factor for mortality after unrelated hematopoietic stem cell transplantation. Blood 114:1437–1444
75. Zino E, Frumento G, Marktel S et al (2004) A T-cell epitope encoded by a subset of HLA-DPB1 alleles determines nonpermissive mismatches for hematologic stem cell transplantation. Blood 103:1417–1424
76. Zino E, Vago L, Di Terlizzi S et al (2007) Frequency and targeted detection of HLA-DPB1 T cell epitope disparities relevant in unrelated hematopoietic stem cell transplantation. Biol Blood Marrow Transplant 13:1031–1040
77. Shaw BE, Robinson J, Fleischhauer K et al (2013) Translating the HLA-DPB1 T-cell epitope matching algorithm into clinical practice. Bone Marrow Transplant 48(12): 1510–1512
78. Robinson J, Malik A, Parham P et al (2000) IMGT/HLA database—a sequence database for the human major histocompatibility complex. Tissue Antigens 55:280–287
79. Robinson J, Waller MJ, Fail SC et al (2009) The IMGT/HLA database. Nucleic Acids Res 37:D1013–D1017

Chapter 6

Databases for T-Cell Epitopes

Chun-Wei Tung

Abstract

Modem immunology and vaccinology incorporate immunoinformatics techniques to give insights into immune systems and accelerate vaccine design. Databases managing epitope data in a structured form with immune-related annotations including sequences, alleles, source organisms, structures, and diseases could be the most crucial part of immunoinformatics offering data sources for the analysis of immune systems and development of prediction methods. This chapter provides an overview of publicly available databases of T-cell epitopes including general databases, pathogen- and tumor-specific databases, and 3D structure databases.

Key words Database, Immunogenicity, Immunoinformatics, Major histocompatibility complex, Pathogen, T-cell epitope, Transporter associated with antigen processing, Tumor, Vaccine

1 Introduction

T-cell epitopes are processed antigens presented in the surface of antigen-presenting cells (APCs) that can be recognized by T cells leading to T-cell activation. Generally, there are two major antigen processing and presentation pathways responsible for endogenous and exogenous antigens. Major histocompatibility complex (MHC) molecules play major roles in both recognition of antigens and presentation of antigens to T cells for both endogenous and exogenous antigens. To be recognized by T cells, endogenous antigens should be cleaved by proteasome, transported into endoplasmic reticulum by transporter associated with antigen processing (TAP), and presented to the cell surface by MHC class I molecules. For exogenous antigens, they should be processed by lysosome and presented to the cell surface by MHC class II molecules to be immunogenic.

Two major T cells of cytotoxic T (Tc) and T helper (Th) cells are responsible for recognizing endogenous and exogenous antigens presented by MHC class I and II molecules, respectively. The Tc cells play a critical role in protective immunity by recognizing and

Rajat K. De and Namrata Tomar (eds.), *Immunoinformatics*, Methods in Molecular Biology, vol. 1184,
DOI 10.1007/978-1-4939-1115-8_6, © Springer Science+Business Media New York 2014

eliminating self-altered cells, which recognize processed antigens derived from intracellular degradation of foreign antigens and bound to MHC class I molecules. In contrast, the activation of Th cells causes the proliferation and differentiation of the Th cells into different Th subtypes secreting various cytokines that assist B-cell maturation, Tc-cell activation, and macrophage activation.

The identification and analysis of T-cell epitopes are important for vaccine development [1, 2]. Various assays were developed to detect features of T-cell activation induced by T-cell epitopes. The cytotoxic activity of activated Tc cells can be directly evaluated by measuring the specific lysis by Tc cells. Tc cells cause apoptosis of target cells via the release of lytic granules containing perforin and granzymes or Fas/Fas ligand interactions. There are three commonly used assays for Tc-cell activation including the chromium-release assay, just another method (JAM) test, and in vivo T-cell cytotoxicity assay [3]. The chromium-release assay measures radioactivity released from the lysis of target cells labeled with ^{51}Cr [4]. The JAM test measures the amount of DNA retained in target cells labeled with [^{3}H]thymidine that are not killed by Tc cells [5]. For in vivo Tc-cell cytotoxicity assay, target cells are firstly labeled with carboxyfluorescein succinimidyl ester (CFSE). Subsequently, flow cytometry is utilized to evaluate the Tc-cell activity according to the loss of CFSE-bright cells [3].

During T-cell activation, cytokines and cytokine receptors are produced and lead to the proliferation of activated T cells. The T-cell activation is commonly measured by the clonal size. Proliferation assays are reliable and simple methods that have been widely used to assess overall T-cell responses [6]. The incorporation of [^{3}H]thymidine or BrdU can be utilized to analyze T-cell proliferation. The CFSE-based methods are also applicable for T-cell proliferation [7]. The cytokine signatures produced during T-cell activation are also practical indicators of T-cell activation. For example, IL-2 is required for conventional T-cell proliferation that can be used as an indicator. The elicited cytokines can be profiled using enzyme-linked immunosorbent spot (ELISPOT) assay to monitor immune responses [8, 9]. Both cytokine release and T-cell proliferation could be indicators of Th-cell activation.

The mapping of T-cell epitopes could provide useful information for the design of peptide-based vaccines. In order to provide better understanding of immune responses associated with T-cell epitopes, several high-throughput methods have been developed for the large-scale identification of T-cell epitopes [10]. For example, the construction of peptide libraries comprised synthetic overlapping peptides for screening T-cell epitopes [11] and 15-mer peptides with 11 amino acid overlap are concluded to be good compromise for stimulating both Tc and Th cells [12]. With the high-throughput methods, data of T-cell epitopes grows fast. It is desirable to develop informatics techniques for organization and utilization of the produced epitope data.

Immunoinformatics aims to analyze and model immunological problems using information techniques of database, data mining, and machine learning. Databases providing centralized, structured, and searchable information of T-cell epitopes could help the mapping of T-cell epitopes on new pathogens and serve as data sources for analyzing T-cell epitopes and constructing computational prediction models. With benefit from the low costs, high efficiency, and high accuracy, computational prediction models are becoming essential tools for T-cell epitope mapping in modern immunology [13]. The utilization of a larger dataset and more relevant data for constructing computational prediction models could improve their prediction performances [14, 15]. The T-cell epitope databases hence play a vital role in providing accurate and detailed data for constructing prediction models (*see* **Note 1**).

Several important databases of T-cell epitopes have been developed to meet the urgent need of data storage and sharing. This chapter summarized databases focused on T-cell epitopes with brief descriptions of their content and functionality. According to their main contents, databases are classified into three categories of general databases, pathogen- and tumor-specific databases, and 3D structure databases.

2 Databases

General protein sequence and structure databases such as UniProt and PDB databases could be valuable resources of T-cell epitopes. Keywords can be utilized to search for T-cell epitope-related information. However, this chapter focuses on only specialized databases of T-cell epitopes. General protein databases will not be included.

2.1 Databases of T-Cell Epitopes

Databases of T-cell epitopes collecting information of MHC-binding peptides, T-cell epitopes, and complexes of T-cell receptor (TCR)–peptide–MHC are listed in Table 1. Several pioneer databases were developed more than 10 years ago. Some of their services are no longer available. For example, FIMM database [16] containing data relevant to functional molecular immunology is no longer accessible. MHCPEP database [17] is one of the earliest T-cell epitope databases whose maintenance and update are discontinued. Fortunately, most of their contents were collected and integrated into newly developed databases. This section describes the databases that are still accessible.

MHCPEP, probably the first specialized database for MHC-binding peptides, is a curated database comprising over 13,000 peptide sequences known to bind MHC molecules [17]. Its contents were collected from published literatures and experimental data with information of peptide sequences, associated MHC alleles, anchor positions, peptide sources, and references.

Table 1
General databases of T-cell epitopes

Database	Description	Availability
MHCPEP	Database of MHC-binding peptides	ftp://ftp.wehi.edu.au/pub/biology/mhcpep/
SYFPEITHI	Database of MHC ligand and peptide motifs	http://www.syfpeithi.de/
AntiJen (JenPep)	Quantitative immunology database	http://www.ddg-pharmfac.net/antijen
FIMM	Functional immunology database	Not available
MHCBN	Database of MHC/TAP-binding peptides and T-cell epitopes	http://www.imtech.res.in/raghava/mhcbn/
EPIMHC	Database for customized computational vaccinology	http://bio.dfci.harvard.edu/epimhc/
IEDB	Immune epitope database	http://www.iedb.org/

One of the unique features of MHCPEP is that T-cell responses were collected and classified into six categories of high, medium, little, none, immunogenic-not-quantified, and unknown. For MHC class I binding peptides, the classifications are according to the concentration of peptides giving 50 % of maximum specific lysis by Tc cells of target cells displaying the peptide. For MHC class II binding peptides, the concentration of peptides giving 50 % of maximum proliferation is utilized to classify T-cell responses induced by the epitopes. MHC-binding affinity is also classified into five categories of high, medium, low, none, and unknown. The predefined categories provide useful information for analyzing the correlation between T-cell responses and MHC-binding affinities and constructing classifiers [15]. Several newly developed databases integrated MHCPEP contents into their databases such as MHCBN [18] and EPIMHC [19] (*see* **Note 2**).

SYFPEITHI provides a publicly accessible resource for curated MHC ligands and peptide motifs [20]. In addition to basic information of MHC alleles, MHC-binding peptides, T-cell epitopes, sources, and references, the most significant features are the information of peptide motifs and their prediction function. In contrast to MHCPEP database containing both published and preliminary data, SYFPEITHI only collects epitopes with published functional evidences making it a popular and reliable resource for T-cell epitope research. Currently, there are more than 8,000 MHC-binding peptides with qualitative data of MHC-binding peptides and T-cell epitopes in SYFPEITHI. The usage of SYFPEITHI for searching and mapping of T-cell epitopes has been demonstrated in a recent article [21].

JenPep is a family of relational databases containing quantitative data on peptides binding to MHC and TAP and T-cell epitopes aiming to support the development of computational vaccinology [22, 23]. Instead of classifying peptides into several categories of MHC-binding and T-cell responses, the quantitative data provided in JenPep could be useful for developing quantitative prediction methods. AntiJen, the successor to JenPep, contains a wider spectrum of immunological data and advanced search functions [24]. More than 31,000 entries have been collected in AntiJen database with thermodynamic and kinetic measures of peptides binding to MHC and TAP, MHC–peptide–TCR complexes, and general immunological protein–protein interactions. With hyperlinks to major databases including Swiss-Prot, NCBI protein database, and PUBMED reference database, users can easily retrieve related information. There are more than 4,000 T-cell epitopes with experimental information available in AntiJen. However, it seems that AntiJen database has not been updated for a long time. The hyperlinks to Swiss-Prot and IMGT/HLA are broken.

FIMM is a functional immunology database consisting of protein antigens, MHC molecules, MHC-binding peptides, and relevant disease associations [16]. The major sources of MHC-binding peptides include MHCPEP, SYFPEITHI, HIV Molecular Immunology Database [25], and literatures. FIMM focuses on human MHCs (human leukocyte antigens, HLAs) and associated diseases as the most distinctive feature (*see* **Note 3**).

MHCBN was developed to serve as a comprehensive database of MHC-binding peptides integrating information from MHCPEP, FIMM, SYFPEITHI, and HIV Molecular Immunology database with hyperlinks to major databases of GenBank, Swiss-Prot, PDB, IMGT/HLA, and PUBMED. The latest version 4.0 of MHCBN contains more than 25,000 peptide entries of binders and nonbinders for MHC and TAP molecules and T-cell epitopes [18]. Search function is available for each data field. Advanced tools for the mapping of T-cell epitopes and dataset creation are also available at the website of MHCBN. Please refer to the article describing the detailed tutorial for epitope mapping using MHCBN [26].

EPIMHC focuses on T-cell epitopes of naturally occurring proteins [19]. EPIMHC was compiled from MHCPEP, SYFPEITHI, JenPep, MHCBN, FIMM, and literatures using the same data scheme of MHCPEP. There are more than 2,000 T-cell epitopes out of 4,875 distinct MHC-binding peptides whose source organisms are known. More than 80 T-cell epitopes are derived from tumor-associated antigens. A useful function for generating position-specific scoring matrices from query results enables the development of motif predictors of interests (*see* **Note 4**).

IEDB, the immune epitope database, is a versatile and comprehensive database with the largest collection of immune epitopes [27]. Epitope information is curated from literatures into the

Table 2
Pathogen- and tumor-specific databases of T-cell epitopes

Database	Description	Availability
AntigenDB	Database of pathogen antigens	http://www.imtech.res.in/raghava/antigendb/
Protegen	Database of protective antigens	http://www.violinet.org/protegen/
HIV Molecular Immunology Database	HIV database	http://www.hiv.lanl.gov/content/immunology/
HCV Immunology Database	HCV database	http://hcv.lanl.gov/content/immuno/immuno-main.html
Cancer Immunity Peptide Database	Database of tumor T-cell antigens	http://www.cancerimmunity.org/peptide/
TANTIGEN	Database of tumor T-cell antigens	http://cvc.dfci.harvard.edu/tadb/

structured database with detailed experimental information including T-cell assays and MHC-binding assays. Hyperlinks to major databases are available for each entry. Different T-cell assays could conclude divergent results; hence the detailed information of T-cell assays should be carefully curated instead of pulling assays altogether. IEDB database compiling more than 200,000 T-cell assays and 230,000 MHC-binding assays from literatures with detailed experimental information is an essential resource for developing computational prediction methods for both MHC binding and T-cell activation. Numerous functions have been developed and integrated into IEDB database including the IEDB-3D structure database (*see* Subheading 2.3) [28] and immune epitope database analysis resource (IEDB-AR) [29]. IEDB-AR provides several T-cell epitope prediction tools for proteasome cleavage, TAP binding, and MHC binding that could help the identification and design of T-cell epitopes. IEDB is recently expanded to include non-peptidic epitopes and hyperlinks to ChEBI, a database and ontology of chemical entities of biological interest, enabling the analysis of non-peptidic epitopes.

2.2 Pathogen- and Tumor-Specific Databases of T-Cell Epitopes

The aforementioned general databases tried to collect T-cell epitopes as many as possible without focusing on specific applications (*see* **Notes 5** and **6**). For developing treatments against pathogens and diseases, it is desirable to collect and analyze pathogen- or tumor-specific T-cell epitopes. Several databases have been created to fulfill the need of storage and analysis of pathogen- and tumor-specific T-cell epitopes as listed in Table 2.

AntigenDB puts a special emphasis on pathogen antigens [30] with or without epitope information. In addition to basic information of T-cell epitopes and MHC-binding peptides, several useful features have been integrated including gene-expression and posttranslational modifications (PTMs) to facilitate vaccine development. Highly expressed antigens are suitable vaccine candidate. PTM information could give insights into the recognition of TCR–peptide–MHC. For each antigen containing T-cell epitopes, its induced immunogenicity of Tc or Th cells, T-cell epitopes, associated PTMs, MHC-binding affinity, TAP binders, and cleavage sites are collected from literatures and available at AntigenDB. AntigenDB contains more than 500 antigens from 44 important pathogenic species. In addition to protein antigens, glycoproteins and lipoproteins are also collected in AntigenDB. It provides numerous hyperlinks to major databases that could be a useful database for computational vaccinology.

Protegen is a database for protective antigens capable of inducing immune responses in the host against infectious and non-infectious diseases [31]. Currently, 708 protective antigens are available against over 100 infectious diseases, cancers, and allergies that could be a useful resource for developing vaccines, biomarkers for disease diagnosis, and analysis of protective antigens. In contrast to AntigenDB that includes epitopes of both protective and non-protective antigens, Protegen contains only protective antigens.

HIV Molecular Immunology Database [25] and HCV Immunology Database [32] are specifically designed for HIV- and HCV-related information including T-cell epitopes and their interactions with the host immune system. The built-in search functions allow users to efficiently extract related information. Subheading 3 of both databases provides summarized experimental information extracted from literature that enables in-depth exploration of epitopes. The numbers of Tc and Th epitopes are 7,537 and 1,315 for the HIV Database and 383 and 222 epitopes for the HCV Database, respectively.

For the development of T-cell epitope-based cancer vaccines, the Cancer Immunity Peptide Database was constructed with a collection of 129 tumor antigens with T-cell epitopes [33]. The tumor antigens are categorized into four types of unique, tumor-specific, differentiation, and overexpressed antigens with hyperlinks to GeneCard and PubMed. In spite of the small size of the database, it collected only epitopes with required experimental evidences for inducing T-cell responses serving as a useful resource for validated epitopes. TANTIGEN (tumor T-cell antigen database) comprises 4,006 antigen entries from 251 unique antigens with information of T-cell epitopes and MHC-binding peptides [34]. The integration of prediction tools for MHC-binding peptides could help the identification of potential T-cell epitopes.

Table 3
3D structure databases of T-cell epitopes

Database	Description	Availability
MPID-T2	Database of crystal structures of peptide–MHC and TCR–peptide–MHC	http://biolinfo.org/mpid-t2/
IMGT/3Dstructure-DB	3D structure database of IMGT	http://www.imgt.org/3Dstructure-DB/
IEDB-3D	3D structure database of IEDB	http://www.iedb.org/
CrossTope	Database of experimental and modeled pMHC-I structures	http://www.crosstope.com.br

2.3 Three-Dimensional (3D) Structure Databases of T-Cell Epitopes

The collection and analysis of 3D structures of TCR–peptide–MHC complexes could provide a better understanding of TCR–peptide–MHC interactions. Three-dimensional structure databases of T-cell epitopes are listed in Table 3. MPID is firstly created by collecting structures of peptide–MHC complexes from PDB database [35]. The updated version MPID-T extended its contents to include TCR–peptide–MHC complexes [36]. Currently, the latest version MPID-T2 (Nov, 2010) comprises 353 peptide–MHC and 62 TCR–peptide–MHC structures. Intermolecular parameters were pre-calculated and available for the analysis of structures in MPID-T2 including hydrogen bonds, gap index, gap volume, binding energy, molecular surface electrostatic potential, TCR docking angle, and contact area. WebLogo tool [37] is utilized to represent peptide motifs obtained from MPID-T2. The pre-calculated structural alignments of peptide–MHC and TCR–peptide–MHC complexes could provide insights into the interactions of TCR–peptide–MHC.

IMGT/3Dstructure-DB [38] is the 3D structure database of IMGT, the international ImMunoGenetics information system. Both its 3D structures of TCR–peptide–MHC and peptide–MHC complexes were collected from the PDB database and stored in IMGT/3Dstructure-DB. Its most distinctive feature is that all structures were annotated with the IMGT-ONTOLOGY concepts of classification and domain information obtained by applying IMGT/DomainGapAlign [38]. Non-peptidic epitopes are also included. Pre-calculated contact residues are available for investigating the structural features of peptide–MHC complexes with or without TCR. Currently, IMGT/3Dstructure-DB contains 84 and 486 entries of

TCR–peptide–MHC and peptide–MHC structures (Apr 6, 2013), respectively. IMGT database integrating various tools and databases of its own is a comprehensive system for analyzing T-cell epitopes. The detailed protocol for querying the IMGT/3Dstructure-DB is available in a published book chapter [39].

IEDB-3D, the 3D structure database of IEDB, collects published 3D structures of TCR or MHC complexes with epitopes curated in IEDB [28]. All the 3D structures were curated from PDB database including complexes of TCR–peptide–MHC and peptide–MHC. IEDB-3D is fully integrated with IEDB enabling the cross-reference of detailed information of structures, epitopes, references, T-cell assays, and MHC binding. The integrated EpitopeViewer provides intuitive user interface for the analysis of contacting atoms in 3D structures [40]. IEDB-3D can be easily accessed from the link of "Browse by 3D structure." Notably, both peptidic and non-peptidic T-cell epitope are curated in IEDB-3D. On the date of access (Jun 11, 2013), there were 62 and 337 non-redundant structures for TCR–peptide–MHC and peptide–MHC complexes, respectively.

Due to the lack of available 3D structures of peptide–MHC complexes (*see* **Note 7**), CrossTope was developed as a curated database collecting 3D structures of immunogenic peptide–MHC class I complexes (pMHC-I) from the public PDB database and computational modeling [41]. All pMHC-I complexes are curated from literatures with experimentally verified T-cell responses. Except for the pMHC-I complexes with available 3D structures in PDB that can be directly curated into CrossTope, a three-step modeling method is applied to reconstruct pMHC-I complexes [42]. For each entry, the structure types of "crystal" and "model" indicate the sources of pMHC-I complexes from PDB crystal structures and computational modeling, respectively. The pMHC-I structures could serve as useful resources for structure-based virtual screening of cross-reactive targets as demonstrated by the authors [41–43]. Currently, CrossTope contains 182 non-redundant pMHC-I complexes from two human and two murine alleles.

3 Notes

This chapter summarized three kinds of databases related to T-cell epitopes including general databases, pathogen- and tumor-specific databases, and 3D structure databases. The efforts of large-scale identification of T-cell epitopes will continue producing a vast amount of T-cell epitope data. The databases of T-cell epitopes will be more important than ever as data sources for the analysis of immune responses, development of computational prediction methods, and vaccine design. Several notes are provided as follows:

1. For the development of computational prediction methods for T-cell epitopes, one of the most crucial parts is the dataset construction that is usually integrated from different databases [14]. However, different databases could use distinct criteria for data curation and annotation. Also, there are various assays for determining T-cell responses induced by epitopes as described in Subheading 1 and results from different assays may not be directly comparable. The integration of heterogeneous data should be carefully processed.
2. The web service of MHCPEP is discontinued. However, its data is still available at an FTP site as shown in Table 1.
3. FIMM is no longer accessible, and its data has been integrated into MHCBN and EPIMHC.
4. EPIMHC web server is currently not working, while its main web page is still accessible.
5. Most general databases of T-cell epitopes also contain information of protein sources and host organisms. By querying the databases with keywords related to pathogens, pathogen-specific information can be retrieved.
6. The IEDB database, being the largest general database of immune epitopes, provides also disease information related to epitopes. For extracting disease-specific information, the function of "disease finder" can be utilized to filter epitope data related to the disease of interests such as cancers from IEDB.
7. The 3D structures of T-cell epitopes and related MHCs and TCRs could provide important clues to the molecular mechanism of antigen presentation and T-cell activation. However, the available structures from existing databases are scarce. Computational modeling methods could be alternative ways to accomplish structure databases. More experimental and computational efforts are desirable to improve knowledge in this field.

Acknowledgments

The author would like to acknowledge the financial support from National Science Council of Taiwan (NSC 101-2311-B-037-001-MY2) and Kaohsiung Medical University Research Foundation (KMU-Q102012).

References

1. Sette A, Peters B (2007) Immune epitope mapping in the post-genomic era: lessons for vaccine development. Curr Opin Immunol 19(1):106–110
2. Salit RB, Kast WM, Velders MP (2002) Ins and outs of clinical trials with peptide-based vaccines. Front Biosci 7:e204–e213
3. Wonderlich J, Shearer G, Livingstone A, Brooks A (2006) Induction and measurement of cytotoxic T lymphocyte activity. Curr Protoc Immunol. Chapter 3:Unit 3.11
4. Brunner KT, Mauel J, Cerottini JC, Chapuis B (1968) Quantitative assay of the lytic action of immune lymphoid cells on 51-Cr-labelled allogeneic target cells in vitro; inhibition by isoantibody and by drugs. Immunology 14(2): 181–196
5. Matzinger P (1991) The JAM test. A simple assay for DNA fragmentation and cell death. J Immunol Methods 145(1–2):185–192
6. Kruisbeek AM, Shevach E, Thornton AM (2004) Proliferative assays for T cell function. Curr Protoc Immunol. Chapter 3:Unit 3.12
7. Anthony DD, Milkovich KA, Zhang W, Rodriguez B, Yonkers NL, Tary-Lehmann M, Lehmann PV (2012) Dissecting the T cell response: proliferation assays vs cytokine signatures by ELISPOT. Cells 1(2):127–140
8. Czerkinsky CC, Nilsson LA, Nygren H, Ouchterlony O, Tarkowski A (1983) A solid-phase enzyme-linked immunospot (ELISPOT) assay for enumeration of specific antibody-secreting cells. J Immunol Methods 65(1–2): 109–121
9. Kalyuzhny AE (2005) Handbook of ELISPOT: methods and protocols. Humana, Totowa, NJ
10. Li Pira G, Ivaldi F, Moretti P, Manca F (2010) High throughput T epitope mapping and vaccine development. J Biomed Biotechnol 2010: 325720
11. Rodda SJ (2002) Peptide libraries for T cell epitope screening and characterization. J Immunol Methods 267(1):71–77
12. Kiecker F, Streitz M, Ay B, Cherepnev G, Volk HD, Volkmer-Engert R, Kern F (2004) Analysis of antigen-specific T-cell responses with synthetic peptides—what kind of peptide for which purpose? Hum Immunol 65(5): 523–536
13. Lundegaard C, Lund O, Nielsen M (2012) Predictions versus high-throughput experiments in T-cell epitope discovery: competition or synergy? Expert Rev Vaccines 11(1):43–54
14. Tung CW, Ziehm M, Kamper A, Kohlbacher O, Ho SY (2011) POPISK: T-cell reactivity prediction using support vector machines and string kernels. BMC Bioinformatics 12:446
15. Tung CW, Ho SY (2007) POPI: predicting immunogenicity of MHC class I binding peptides by mining informative physicochemical properties. Bioinformatics 23(8):942–949
16. Schonbach C, Koh JL, Flower DR, Wong L, Brusic V (2002) FIMM, a database of functional molecular immunology: update 2002. Nucleic Acids Res 30(1):226–229
17. Brusic V, Rudy G, Harrison LC (1998) MHCPEP, a database of MHC-binding peptides: update 1997. Nucleic Acids Res 26(1): 368–371
18. Lata S, Bhasin M, Raghava GP (2009) MHCBN 4.0: a database of MHC/TAP binding peptides and T-cell epitopes. BMC Res Notes 2:61
19. Reche PA, Zhang H, Glutting JP, Reinherz EL (2005) EPIMHC: a curated database of MHC-binding peptides for customized computational vaccinology. Bioinformatics 21(9):2140–2141
20. Rammensee H, Bachmann J, Emmerich NP, Bachor OA, Stevanovic S (1999) SYFPEITHI: database for MHC ligands and peptide motifs. Immunogenetics 50(3–4):213–219
21. Schuler MM, Nastke MD, Stevanovikc S (2007) SYFPEITHI: database for searching and T-cell epitope prediction. Methods Mol Biol 409:75–93
22. Blythe MJ, Doytchinova IA, Flower DR (2002) JenPep: a database of quantitative functional peptide data for immunology. Bioinformatics 18(3):434–439
23. McSparron H, Blythe MJ, Zygouri C, Doytchinova IA, Flower DR (2003) JenPep: a novel computational information resource for immunobiology and vaccinology. J Chem Inf Comput Sci 43(4):1276–1287
24. Toseland CP, Clayton DJ, McSparron H, Hemsley SL, Blythe MJ, Paine K, Doytchinova IA, Guan P, Hattotuwagama CK, Flower DR (2005) AntiJen: a quantitative immunology database integrating functional, thermodynamic, kinetic, biophysical, and cellular data. Immunome Res 1(1):4
25. Korber B, Moore J, Brander C, Koup R, Haynes B, Walker BD (1998) HIV Molecular Immunology Database. Los Alamos National Laboratory, Theoretical Biology and Biophysics, Los Alamos, New Mexico. LA-UR 99-586
26. Bhasin M, Lata S, Raghava GP (2007) Searching and mapping of T-cell epitopes, MHC binders, and TAP binders. Methods Mol Biol 409:95–112

27. Vita R, Zarebski L, Greenbaum JA, Emami H, Hoof I, Salimi N, Damle R, Sette A, Peters B (2010) The immune epitope database 2.0. Nucleic Acids Res 38(Database issue): D854–D862
28. Ponomarenko J, Papangelopoulos N, Zajonc DM, Peters B, Sette A, Bourne PE (2011) IEDB-3D: structural data within the immune epitope database. Nucleic Acids Res 39(Database issue):D1164–D1170
29. Kim Y, Ponomarenko J, Zhu Z, Tamang D, Wang P, Greenbaum J, Lundegaard C, Sette A, Lund O, Bourne PE, Nielsen M, Peters B (2012) Immune epitope database analysis resource. Nucleic Acids Res 40(Web Server issue):W525–W530
30. Ansari HR, Flower DR, Raghava GP (2010) AntigenDB: an immunoinformatics database of pathogen antigens. Nucleic Acids Res 38(Database issue):D847–D853
31. Yang B, Sayers S, Xiang Z, He Y (2011) Protegen: a web-based protective antigen database and analysis system. Nucleic Acids Res 39(Database issue):D1073–D1078
32. Hraber PT, Leach RW, Reilly LP, Thurmond J, Yusim K, Kuiken C, Los Alamos HIV database team (2007) Los Alamos hepatitis C virus sequence and human immunology databases: an expanding resource for antiviral research. Antivir Chem Chemother 18(3):113–123
33. van der Bruggen P, Stroobant V, Vigneron N, Van den Eynde B (2013) Peptide database: T cell-defined tumor antigens. Cancer Immun 13:15
34. Bioinformatics Core at Cancer Vaccine Center D-FCI (2009) TANTIGEN: Tumor T cell Antigen Database
35. Govindarajan KR, Kangueane P, Tan TW, Ranganathan S (2003) MPID: MHC-Peptide Interaction Database for sequence-structure-function information on peptides binding to MHC molecules. Bioinformatics 19(2): 309–310
36. Tong JC, Kong L, Tan TW, Ranganathan S (2006) MPID-T: database for sequence-structure-function information on T-cell receptor/peptide/MHC interactions. Appl Bioinformatics 5(2):111–114
37. Crooks GE, Hon G, Chandonia JM, Brenner SE (2004) WebLogo: a sequence logo generator. Genome Res 14(6):1188–1190
38. Ehrenmann F, Kaas Q, Lefranc MP (2010) IMGT/3Dstructure-DB and IMGT/Domain GapAlign: a database and a tool for immunoglobulins or antibodies, T cell receptors, MHC, IgSF and MhcSF. Nucleic Acids Res 38 (Database issue):D301–D307
39. Ehrenmann F, Lefranc MP (2011) IMGT/3Dstructure-DB: querying the IMGT database for 3D structures in immunology and immunoinformatics (IG or antibodies, TR, MH, RPI, and FPIA). Cold Spring Harb Protoc 2011(6):750–761
40. Beaver JE, Bourne PE, Ponomarenko JV (2007) EpitopeViewer: a Java application for the visualization and analysis of immune epitopes in the Immune Epitope Database and Analysis Resource (IEDB). Immunome Res 3:3
41. Sinigaglia M, Antunes DA, Rigo MM, Chies JA, Vieira GF (2013) CrossTope: a curate repository of 3D structures of immunogenic peptide: MHC complexes. Database, Oxford 2013:bat002
42. Antunes DA, Vieira GF, Rigo MM, Cibulski SP, Sinigaglia M, Chies JA (2010) Structural allele-specific patterns adopted by epitopes in the MHC-I cleft and reconstruction of MHC:peptide complexes to cross-reactivity assessment. PLoS One 5(4):e10353
43. Antunes DA, Rigo MM, Silva JP, Cibulski SP, Sinigaglia M, Chies JA, Vieira GF (2011) Structural in silico analysis of cross-genotype-reactivity among naturally occurring HCV NS3-1073-variants in the context of HLA-A*02:01 allele. Mol Immunol 48(12–13): 1461–1467

Chapter 7

Databases for B-Cell Epitopes

Juan Liu and Wen Zhang

Abstract

Identification and characterization of B-cell epitopes in target antigens is one of the key steps in epitope-driven vaccine design, immunodiagnostic tests, and antibody production. For localizing epitopes by experimental methods is time consuming and cost expensive, researchers have been developing in silico or computational models for the prediction of B-cell epitopes, enabling immunologists and clinicians to identify the most promising epitopes for characterization in the laboratory. A sufficient number of available B-cell epitopes are indispensable for establishing the prediction models. To our knowledge, some popular databases associated with the B-cell epitopes are proposed and widely used in the immunoinformatics. In this chapter, we present an overview of the important databases and introduce how to compile datasets for the development of B-cell epitope prediction tools.

Key words B-cell epitope, Mimotope, Databases, Immune response

1 Introduction

The interactions between antibodies and antigens play important roles in the immunological reaction, and the interaction sites can reveal the mechanism of the immune system and help to design the vaccines [1–4]. A B-cell epitope is the region or the segment of an antigen which is recognized by B cells and thus activates the B-cell immune response. With growing need of vaccine design, the recognition of B cell epitope has become more and more important. In general, B-cell epitopes can be categorized into two classes, linear (continuous) and conformational (discontinuous). A continuous epitope is a segment of sequential residues in an antigen sequence, while a discontinuous epitope is a segment of antigen residues that are far away from each other in the primary sequence but are brought to spatial proximity by polypeptide folding. According to crystallographic studies, the discontinuous epitopes take the majority of all epitopes (~90 %).

The B-cell epitope is rather important to immunodetection and immunotherapeutic applications since an epitope as the minimal immune unit is strong enough to elicit a potent humoral immune

Rajat K. De and Namrata Tomar (eds.), *Immunoinformatics*, Methods in Molecular Biology, vol. 1184,
DOI 10.1007/978-1-4939-1115-8_7, © Springer Science+Business Media New York 2014

response with no harmful side effects to human body [5]. The ultimate goal of epitope prediction is to aid the design of molecules that can mimic the structure and function of a genuine epitope and replace it in medical diagnostics and therapeutics and also in vaccine design [6]. The most reliable methods for identification of an epitope are X-ray crystallography and NMR techniques [7, 8], but they are costly and time consuming. Hence, computational methods and tools, with the virtues of low cost and high speed, are employed to predict B-cell epitopes in silico.

Since two classes of B-cell epitopes are quietly different, the computational methods can be divided into the linear-epitope prediction methods and conformational-epitope prediction methods. The linear-epitope prediction methods are usually constructed on the linear-epitope sequences, and typical models utilize the amino acid propensities (hydrophilicity, flexibility, beta turns, surface accessibility, etc.) to make the prediction [9–13]. The recent machine learning-based models utilize sequence-derived features (amino acid composition, amino acid cooperativeness, etc.) [14–20]. Since the lengths of linear epitopes are not fixed and can vary from 5 to 20 amino acids, all the epitope sequences have been set to the specified epitope length via trimming and extending operations, respectively, to build the prediction models. The conformational-epitope prediction methods are usually built on the crystal structures of antigen–antibody complexes. The binding sites (conformational-epitope residues) are first annotated by analyzing the antigen–antibody complexes, and then prediction models are constructed based on the structures with annotated conformational epitopes. Typical methods use the structural features (secondary structure, RSA, neighbor count, half-sphere neighbor count, protrusion index, etc.) to make the prediction [21–32].

There is a vast and increasing number of biological data in the last decades, which provide abundant data resources for the development of immunoinformatics. As the critical component of the epitope-based vaccine design, the B-cell epitopes are of the most important. Data resource is critical for the development of the B-cell epitope prediction models. This chapter briefly introduces popular databases for the B-cell epitopes and helps the researchers get access to the data resources for the development of useful computational tools.

2 The Popular B-Cell Epitope Databases

The availability of experimental data plays a pivotal role in B-cell epitope prediction. With the development of biological technique, a great number of B-cell epitope-related data are being released and available on the Internet or in the publications. The popular databases are listed in Table 1.

Table 1
The popular B-cell epitope-related databases

Databases	URLs	Comments
PDB	http://www.rcsb.org/pdb	Protein data bank
IEDB	http://www.iedb.org/	Immune Epitope Database
Bcipep	http://www.imtech.res.in/raghava/bcipep	B-cell epitope database
CED	http://immunet.cn/ced/	Conformational epitope database
EPITOME	http://www.rostlab.org/services/epitome/	Database of structurally inferred antigenic epitopes in proteins
AntiJen	http://www.ddg-pharmfac.net/antijen/AntiJen/aj_bcell.htm	Kinetic, thermodynamic, and cellular database
HIV Molecular Immunology	http://www.hiv.lanl.gov/content/immunology/index	HIV molecular immunology database

2.1 Protein Data Bank

The 3D structure of antigen or the complex of antigen–antibody is stored in the Protein Data Bank (PDB) database [33]; PDB was established by Brookhaven National Laboratories in 1971, subsequently managed and maintained by the RCSB. PDB database compiles the compounds derived from the X-ray crystallography and NMR experiments (*see* **Note 1**). The main server and all the mirrors around the world provide database search and download service as well as the description of the PDB data file formats. The construction and development of PDB database meet the need of researchers in every field of bioinformatics including epitope prediction.

2.2 The Immune Epitope Database

The Immune Epitope Database (IEDB) [34] is established in 2004 by the La Jolla Institute for Allergy and Immunology as a part of the National Institutes of Health. IEDB is the most commonly used and most authoritative database in epitope prediction [35]. As a big project, this database provides a catalog of experimentally characterized B-cell epitopes (both linear epitope and conformational epitope), T-cell epitopes, as well as major histocompatibility complex (MHC) binding, which are collected from peer-viewed publications or directly submitted by research groups. At present, IEDB includes 159,339 B-cell assays. Each epitope is linked to its reference source, and the epitope structure, source antigen, and organism from which the epitope is derived are all described. For published manuscripts, some information such as the authors, article title, journal name, and abstract are provided. The interface for the IEDB database is shown in Fig. 1.

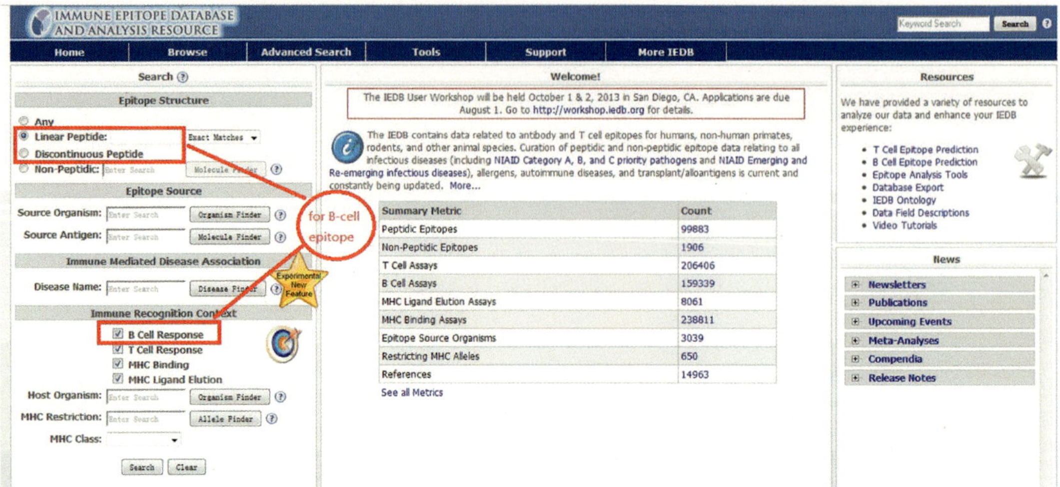

Fig. 1 The interface for the IEDB database

In addition to databases, IDEB provides some tools to predict linear B-cell epitopes by using amino acid scales as well as some tools to predict conformational epitopes by using crystal structures (*see* **Note 2**).

2.3 Bcipep

Bcipep [36] is a database established by the Institute of Microbial Technology, Chandigarh, in India (shown in Fig. 2). This database contains the experimentally determined linear B-cell epitopes, which are collected from literature and other publicly available databases. At present, there are nearly 555 epitopes in Bcipep, covering a wide range of pathogenic organisms like viruses, bacteria, protozoa, and fungi. For each entry, some details such as peptide sequence, source protein, and pathogen group are described. For data obtained from other databases, hyperlinks to the original resources are also provided.

The database also supports the use of keyword search, peptide mapping, and BLAST search for the analysis and extraction of data.

2.4 Conformational Epitope Database

CED [37] is a conformational epitope database (as shown in Fig. 3). At present, CED contains 293 entries, and all entries are manually curated from publications in PubMed and ScienceDirect. Specifically, more than 3,000 references are analyzed manually, and the conformational epitopes with high resolution and completeness are extracted into the database. CED provides related information on epitopes including location of the epitope, the immunological property of the epitope, the source antigen, and corresponding antibody of the epitope.

In addition, the database can be browsed or searched through a user-friendly web interface. Most epitopes in CED can also be viewed interactively in the context of their 3D structures.

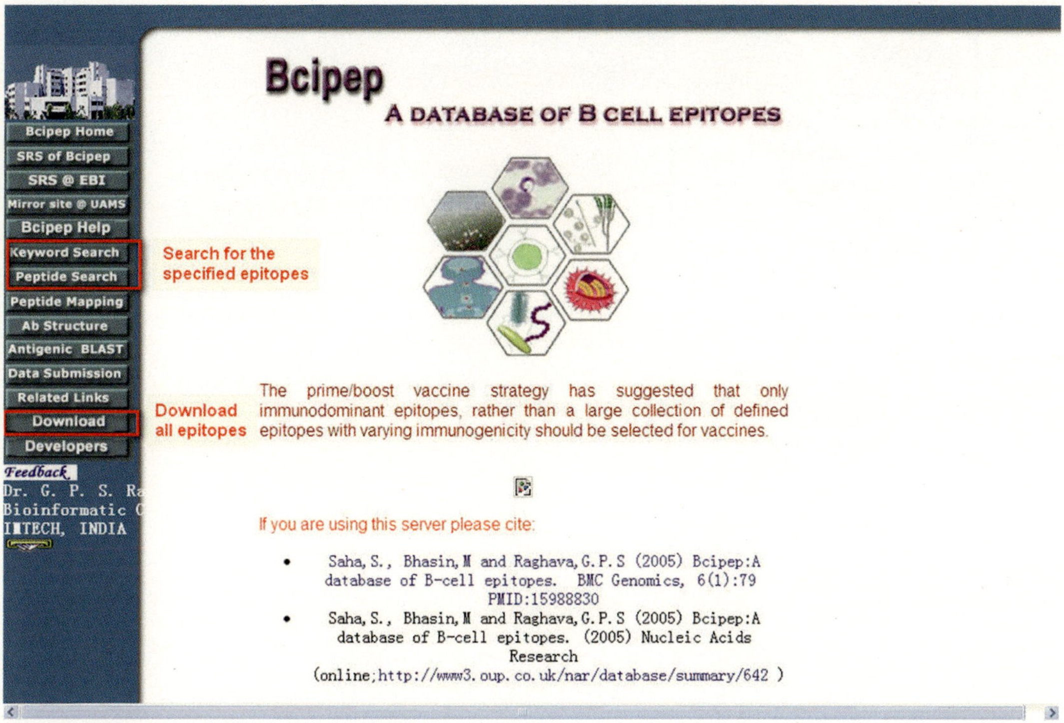

Fig. 2 The interface for the Bcipep database

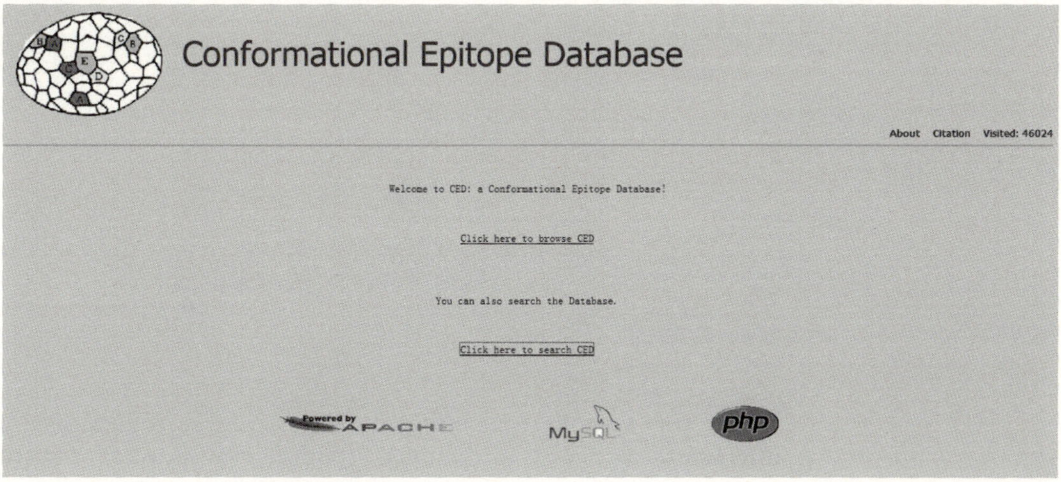

Fig. 3 The interface for the CED database

CED also provides hyperlinks to several external databases, such as PDB database and PubMed, providing wide background information for each entry.

2.5 Epitome

Epitome [38] is a database of all known antigenic epitopes inferring from the antigen–antibody complexes as well as the antibodies that interact with them (as shown in Fig. 4).

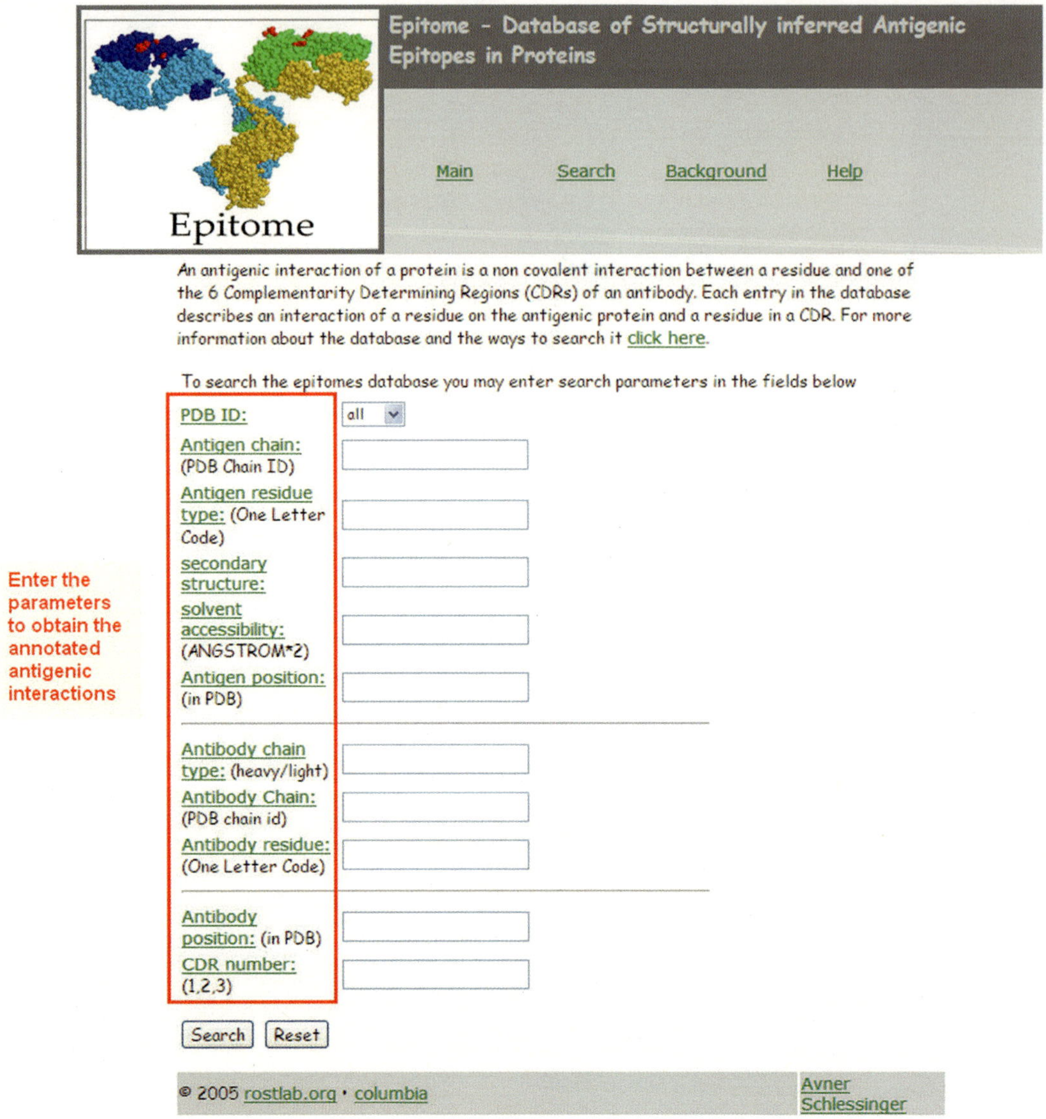

Fig. 4 The interface for the Epitome database

In this database, an antigenic interaction is defined as the interaction between an antigen residue and one of the six complementarity-determining regions (CDRs) of an antibody, and antigenic epitopes consist of the antigenic interaction sites. Thus, antigenic epitopes infer from the antigen–antibody complexes. Specifically, all available structures of antibodies are first aligned and analyzed so as to identify CDRs, and then all antigen residue proteins in PDB that bind to CDRs are identified. By doing this, the structures of all known antigen–antibody complexes in the PDB are analyzed, and antigenic interactions are annotated and extracted into the database.

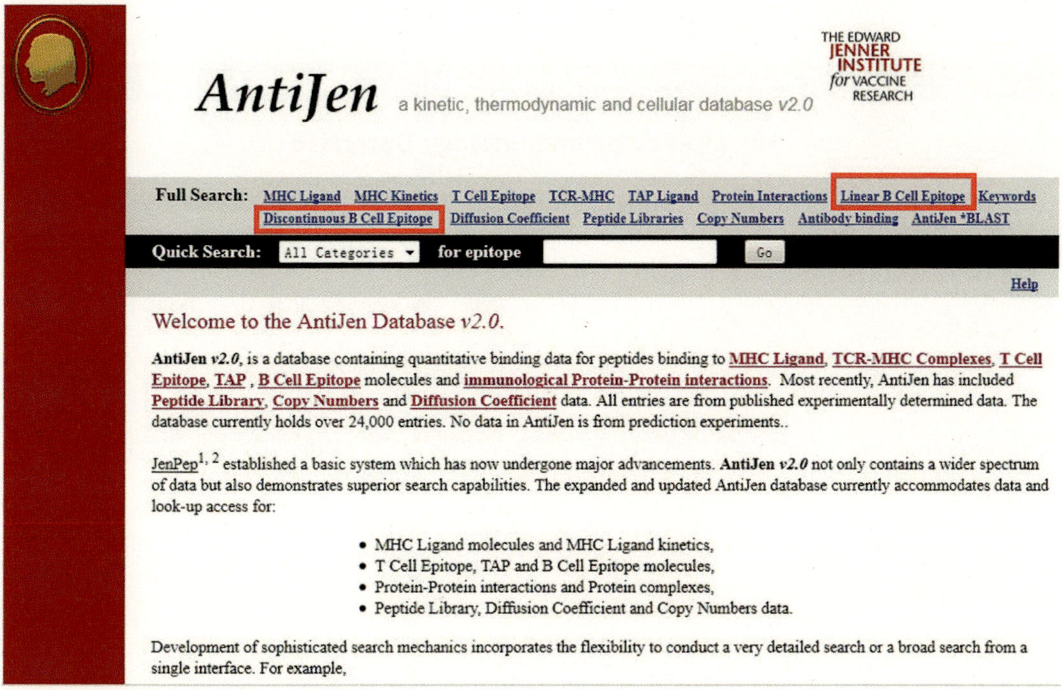

Fig. 5 The interface for the AntiJen database

Epitome contains 142 antigens from protein–antibody complex structures with 10,180 annotated antigenic interactions (*see* **Note 3**). The related information such as PDB ID, PDB chain ID, and PDB position is provided for the entries. Additionally, Epitome provides the interface-friendly tool to visualize interactions in Jmol.

2.6 AntiJen

AntiJen [39] is a comprehensive database focused on the integration of kinetic, thermodynamic, functional, and cellular data within the context of immunology and vaccinology (as shown in Fig. 5). The database currently contains totally 24,000 entries that were collected from the experimentally determined data reported in PubMed publications, including quantitative binding data for peptides binding to MHC ligand, TCR–MHC complexes, T-cell epitope, TAP, B-cell epitope molecules, and immunological protein–protein interactions. The present version (AntiJen v2.0) contains 3,541 B-cell epitopes (linear and conformational epitopes) and provides user-friendly retrieval interface. Each epitope is described by its peptide source, Ab source, antibody, comment, and external hyperlink.

2.7 HIV Molecular Immunology Database

HIV Molecular Immunology Database [40] contains HIV virus epitopes which are extracted from the HIV immunology literature (as shown in Fig. 6). At present, there are nearly 11,361 HIV-specific B-cell and T-cell responses summarized and annotated in this database (*see* **Note 4**). The annotation includes information

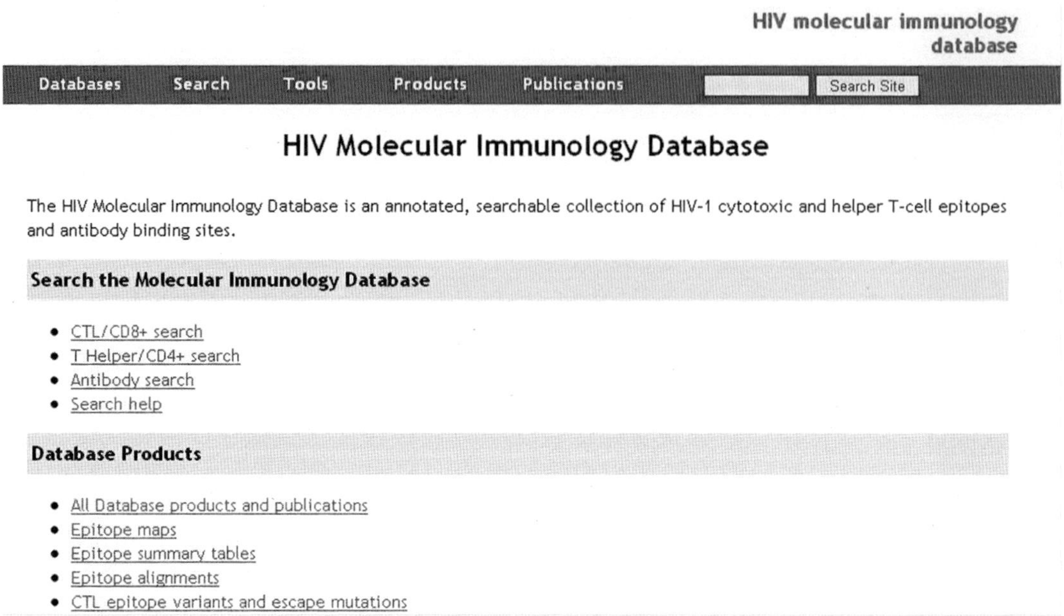

Fig. 6 The interface for the HIV Molecular Immunology database

such as cross-reactivity, escape mutations, antibody sequence, TCR usage, functional domains that overlap with an epitope, immune response associations with rates of progression and therapy, and how specific epitopes were experimentally defined.

3 The Mimotope Databases

The aforementioned databases are important resources for linear/conformational B-cell epitope prediction. The data from these databases provide the resources for computational biologists to derive benchmark and customized datasets for new algorithm development and tool evaluation. In recent years, mimotopes are also widely used in immunoinformatics. A mimotope is a macromolecule, often a peptide, which mimics the structure of a genuine epitope. It causes an antibody response similar to the one elicited by the genuine epitope. That means, an antibody for a given epitope antigen will recognize a mimotope which mimics that epitope. Moreover, the selected mimotopes commonly share high sequential similarity which implies that certain key binding motifs and physicochemical preferences exist during the interaction with antibody. Therefore mapping these mimotopes back to the source antigen can help finding the genuine epitopes more accurately. Mimotopes are commonly obtained from phage display libraries through bio-panning. There have been several databases containing the information of released mimotopes which are summarized in Table 2 [41].

Table 2
Mimotope databases

Databases	URLs	Comments
ASPD	http://wwwmgs.bionet.nsc.ru/mgs/gnw/aspd	Artificially selected protein/ peptide database, first mimotope database
RELIC Peptides	http://www.northeastern.edu/xray/ downloads/	Small molecule-oriented peptide database
PepBank	http://pepbank.mgh.harvard.edu	Includes but not limited to peptide sequences
MimoDB	http://immunet.cn/mimodb/	Largest mimotope database currently
Sun's Benchmark datasets	http://cs.nenu.edu.cn/bioinfo/benchmark%20 datasets/index.html	Datasets for mimotope-based B-cell epitope prediction

3.1 Artificial Selected Peptides/Proteins Database

Artificial Selected Peptides/Proteins Database (ASPD) (as shown in Fig. 7) is a curated database that incorporates data on full-length proteins, protein domains, and peptides that were obtained through in vitro-directed evolution processes (mainly by means of phage display) [42]. ASDP is the first database for mimotopes, currently containing 195 entries which were described in 112 original papers. For each entry, the following information is provided: target, template, links to external databases (SWISS-PROT, PDB), aligned sequences of peptides which retrieved from in vitro evolution and relevant native or constructed sequences, rounds of selection, and occurrences of clones with each sequence. ASPD has a user-friendly interface and can be searched by means of the SRS system. In addition, ASDP provides a BLAST search tool for looking directly for homologies. ASPD database has not been updated for years.

3.2 PepBank

PepBank (as shown in Fig. 8) is a database of peptides based on sequential text mining and public peptide data sources [43]. Only peptides with available sequences and with 20 amino acids or shorter are stored. At present, it contains 21,691 individual peptide entries originated from PubMed, ASPD, UniProt, and PDF. The database has a Web-based user interface with a simple, Google-like search function, advanced text search, BLAST and Smith–Waterman search capabilities.

3.3 MimoDB

MimoDB (as shown in Fig. 9) is a database of peptides that have been selected from random peptide libraries based on their abilities to bind with small compounds, nucleic acids, proteins, cells, tissues, etc. through phage display [44, 45]. The core data of the MimoDB database are mimotope sets and related information such

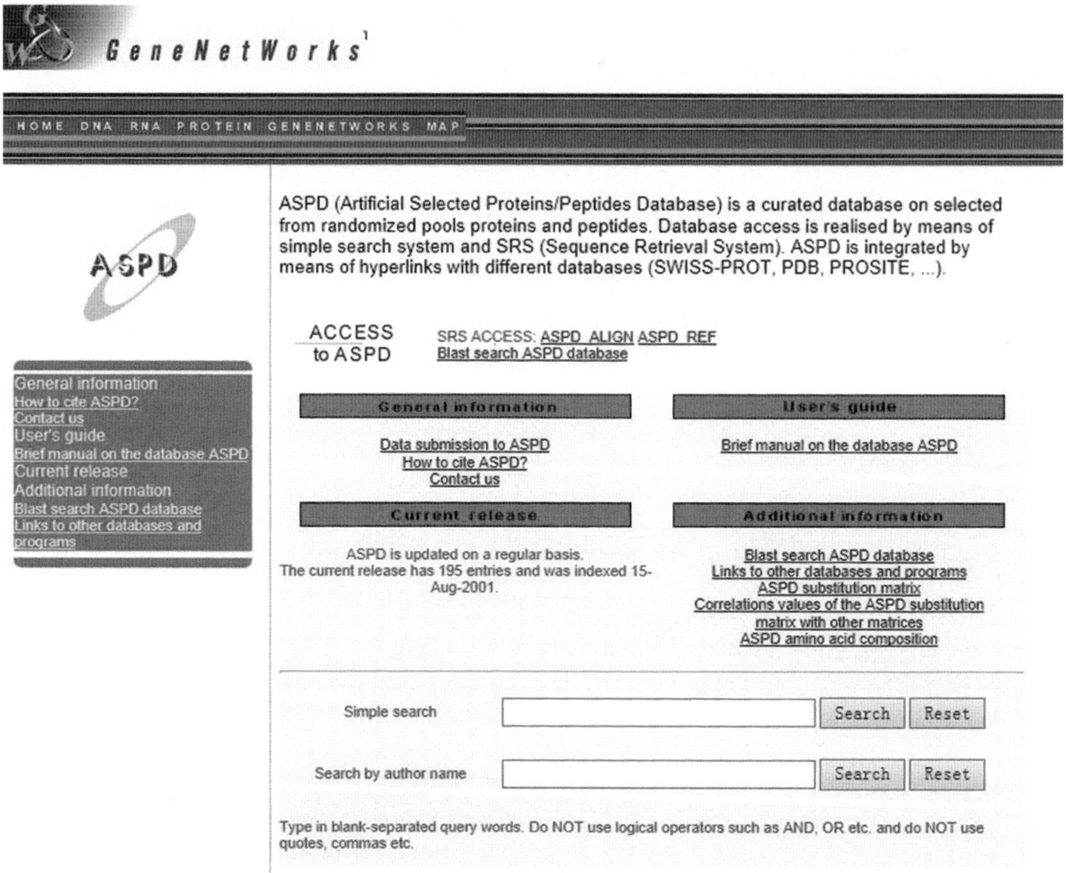

Fig. 7 The interface for the ASPD database

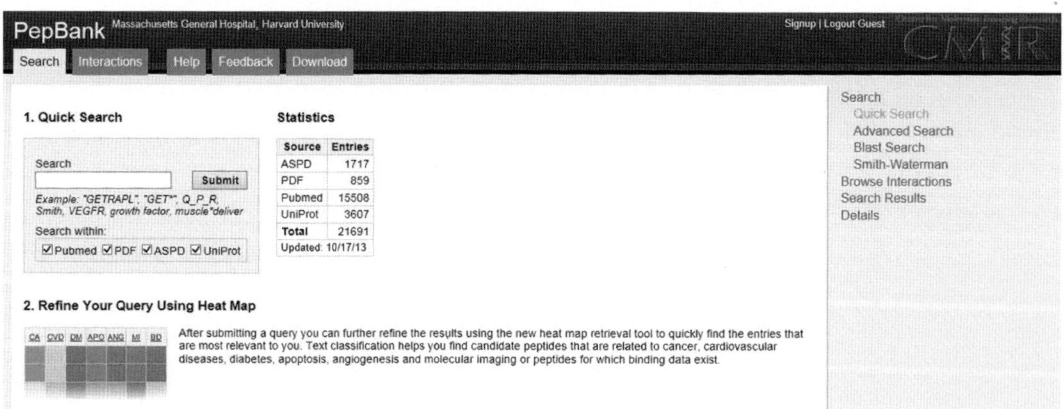

Fig. 8 The interface for the PepBank database

as sequences, structures, targets, templates, and libraries. Peptides are grouped into a mimotope set if they are from the same independent experiment. In this database, (1) only peptides with available sequences are stored; (2) only peptides that are 40 amino acids or shorter are stored; (3) only peptides selected from phage display

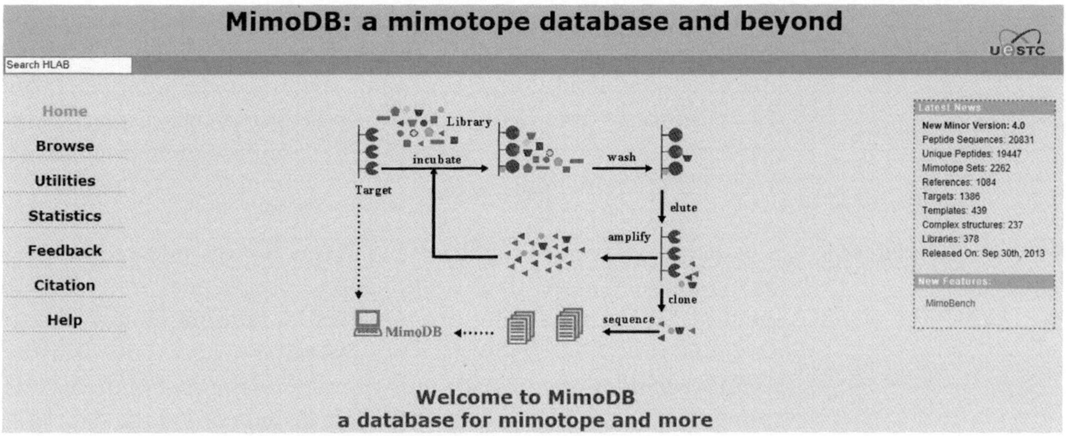

Fig. 9 The interface for the MimoDB database

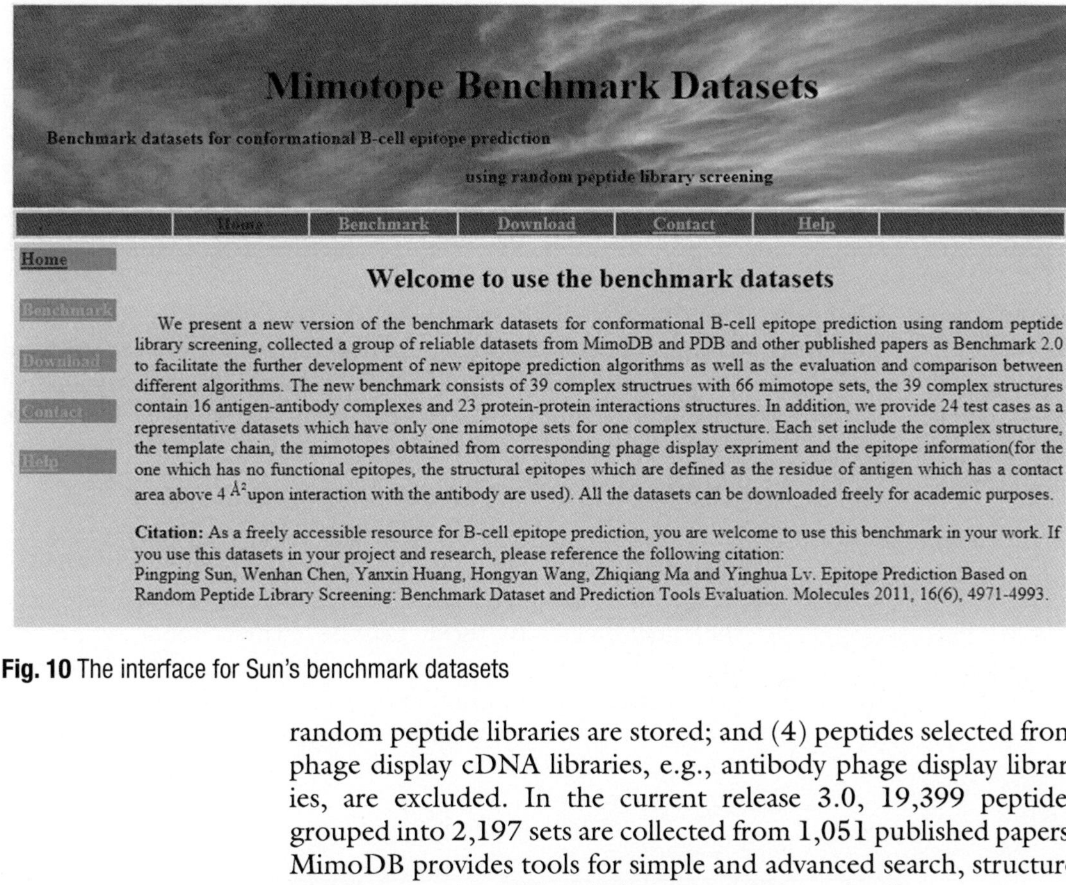

Fig. 10 The interface for Sun's benchmark datasets

random peptide libraries are stored; and (4) peptides selected from phage display cDNA libraries, e.g., antibody phage display libraries, are excluded. In the current release 3.0, 19,399 peptides grouped into 2,197 sets are collected from 1,051 published papers. MimoDB provides tools for simple and advanced search, structure visualization, BLAST, and alignment view on the fly.

3.4 Sun's Benchmark Datasets

Sun's benchmark datasets (as shown in Fig. 10), constructed by Sun et al. [46], are special for conformational B-cell epitope prediction based on mimotope analysis. The current version 2.0 consists of 39 complex structures (16 antigen–antibody complexes and 23 protein–protein interaction structures) with 66 mimotope sets.

In addition, 24 cases each with only one mimotope set for one complex structure are also provided as the test data. Each set includes information on the complex structure, the template chain, the mimotopes obtained from corresponding phage display experiment, and the epitope (*see* **Note 5**). All datasets can be downloaded freely for academic purposes.

3.5 RELIC Peptides Database

The RELIC peptides database (*see* **Note 5**) was released in 2004 and contained more than 5,000 peptide sequences selected with small-molecule metabolite drugs as well as random clones from its parent libraries. A web interface was provided to access the database. RELIC peptides were usually indispensable as the part of the RELIC suite for many tools in the database heavily depend on the data [47].

4 Notes

1. In the PDB database, searchable structures are updated over time as some structures become out of date and are removed from the database.
2. IEDB database provides some state-of-the-art tools to analyze the B-cell epitopes. Specifically, the tool "Antibody Epitope Prediction" can be used to predict the linear epitopes; Discotope and ElliPro can be used for the conformational-epitope prediction.
3. If the interested protein has not a known complex with an antibody in the database, user can blast its sequence against all the sequences in the database. All known complexes between antibodies and proteins that are similar to the interested sequence will be returned.
4. In the HIV molecular immunology database, only B-cell responses to HIV-1 and HIV-2 are summarized and annotated.
5. RELIC web server was shut down in October 2010. To replace the functionality of those peptide analysis tools, Makowski et al. have written a set of stand-alone programs for Windows platforms. All the executable versions of the programs, instructions for use, and sample input and output files for the programs can be downloaded via http://www.northeastern.edu/xray/downloads/.

Acknowledgments

This work was supported by the National Science Foundation of China (61272274, 61103126), Program for New Century Excellent Talents in Universities (NCET-10-0644), the Open

Research Fund of State Key Laboratory of Hybrid Rice (Wuhan University) (KF201301), the Ph.D. Programs Foundation of Ministry of Education of China (20100141120049), and Natural Science Foundation of Hubei Province (No. 2011CDB454).

References

1. Van Regenmortel MH (1989) The concept and operational definition of protein epitopes. Philos Trans R Soc Lond B Biol Sci 323: 451–466
2. Van Regenmortel MH (2004) Pitfalls of reductionism in the design of peptide-cased vaccines. Vaccine 19:2369–2374
3. Walter G (1986) Production and use of antibodies against synthetic peptides. J Immunol Methods 88:149–161
4. Wiesmuller KH, Fleckenstein B, Jung G (2001) Peptide vaccines and peptide libraries. Biol Chem 382:571–579
5. Irving MB, Pan O, Scott JK (2001) Random-peptide libraries and antigen-fragment libraries for epitope mapping and the development of vaccines and diagnostics. Curr Opin Chem Biol 5:314–324
6. Gomara MJ, Haro I (2007) Synthetic peptides for the immunodiagnosis of human diseases. Curr Med Chem 14:531–546
7. Rus JJ, Burnett RM (2000) Type-specific epitope locations revealed by X-ray crystallographic study of adenovirus type 5 hexon. Mol Ther 1:3–4
8. Mayer M, Meyer B (2001) Group epitope mapping by saturation transfer difference NMR to identify segments of a ligand in direct contact with a protein receptor. J Am Chem Soc 123:6108–6117
9. Chen J, Liu H, Yang J, Chou K (2007) Prediction of linear B-cell epitopes using amino acid pair antigenicity scale. Amino Acids 33:423–428
10. Karplus PA, Schulz GE (1985) Prediction of chain flexibility in proteins—a tool for the selection of peptide antigens. Naturwissenschaften 72:212–213
11. Kolaskar AS, Tongaonkar PC (1990) A semi-empirical method for prediction of antigenic determinants on protein antigens. FEBS Lett 276:172–174
12. Pellequer J, Westhof E, Van Regenmortel M (1993) Correlation between the location of antigenic sites and the prediction of turns in proteins. Immunol Lett 36:83–99
13. Saha S, Raghava GP (2004) BcePred: prediction of continuous B-cell epitopes in antigenic sequences using physico-chemical properties. Lect Notes Comput Sci 3239:197–204
14. Saha S, Raghava GP (2006) Prediction of continuous B-cell epitopes in an antigen using recurrent neural network. Proteins 65: 40–48
15. Sollner J, Mayer B (2006) Machine learning approaches for prediction of linear B-cell epitopes on proteins. J Mol Recognit 19:200–208
16. Sweredoski MJ, Baldi P (2009) COBEpro: a novel system for predicting continuous B-cell epitopes. Protein Eng Des Sel 22:113–120
17. Wee LJ, Simarmata D, Kam YW, Ng LF, Tong JC (2010) SVM-based prediction of linear B-cell epitopes using Bayes Feature Extraction. BMC Genomics 11(Suppl 4):S21
18. Zhang W, Liu J, Zhao M, Li Q (2012) Predicting linear B-cell epitopes by using sequence-derived structural and physicochemical features. Int J Data Min Bioinform 6: 557–569
19. El-Manzalawy Y, Dobbs D, Honavar V (2008) Predicting linear B-cell epitopes using string kernels. J Mol Recognit 21:243–255
20. Larsen JE, Lund O, Nielsen M (2006) Improved method for predicting linear B-cell epitopes. Immunome Res 2:2
21. Kulkarni-Kale U, Bhosle S, Kolaskar AS (2005) CEP: a conformational epitope prediction server. Nucleic Acids Res 33(Web Server issue):W168–W171
22. Andersen PH, Nielsen M, Lund O (2006) Prediction of residues in discontinuous B cell epitopes using protein 3D structures. Protein Sci 15:2558–2567
23. Sweredoski MJ, Baldi P (2008) PEPITO: improved discontinuous B-cell epitope prediction using multiple distance thresholds and half sphere exposure. Bioinformatics 24: 1459–1460
24. Sun J, Wu D, Xu T, Wang X, Xu X, Tao L, Li Y, Cao Z (2009) SEPPA: a computational server for spatial epitope prediction of protein antigens. Nucleic Acids Res 37(Suppl 2): W612–W616
25. Rubinstein ND, Mayrose I, Pupko T (2009) A machine learning approach for predicting B-cell epitopes. Mol Immunol 46:840–847

26. Rubinstein ND, Mayrose I, Martz E, Pupko T (2009) Epitopia: a web-server for predicting B-cell epitopes. BMC Bioinformatics 10:287
27. Liang S, Zheng D, Zhang C, Zacharias M (2009) Prediction of antigenic epitopes on protein surfaces by consensus scoring. BMC Bioinformatics 10:302
28. Liang S, Zheng D, Standley DM, Yao B, Zacharias M, Zhang C (2010) EPSVR and EPMeta: prediction of antigenic epitopes using support vector regression and multiple server results. BMC Bioinformatics 11:381
29. Zhao L, Li J (2010) Mining for the antibody-antigen interacting associations that predict the B cell epitopes. BMC Struct Biol 10(Suppl 1):S6
30. Zhang W, Xiong Y, Zhao M, Zou H, Liu J (2011) Prediction of B-cell epitopes on 3D structure by random forests with combined features. BMC Bioinformatics 12:341
31. Zhang W, Niu Y, Xiong Y, Zhao M, Liu J (2012) Computational prediction of conformational B-cell epitopes from antigen primary structures by ensemble learning. PLoS One 7:e43575
32. Gao J, Faraggi E, Zhou Y, Ruan J, Kurgan L (2012) BEST: improved prediction of B-cell epitopes from antigen sequences. PLoS One 7:e40104
33. Berman HM, Westbrook J, Feng Z et al (2000) The Protein Data Bank. Nucleic Acids Res 28(1):235–242
34. Vita R, Zarebski L, Greenbaum JA, Emami H, Hoof I, Salimi N, Damle R, Sette A, Peters B (2009) The immune epitope database 2.0. Nucleic Acids Res 38(Database issue): D854–D862
35. Ponomarenko J, Papangelopoulos N, Dirk M et al (2011) IEDB-3D: structural data within the immune epitope database. Nucleic Acids Res 39(Database issue):D1164–D1170
36. Saha S, Bhasin M, Raghava GPS (2005) Bcipep: a database of B-cell epitopes. BMC Genomics 6:79
37. Huang J, Honda W (2006) CED: a conformational epitope database. BMC Immunol 7:7
38. Schlessinger A, Ofran Y, Yachdav G, Rost B (2006) Epitome: database of structure-inferred antigenic epitopes. Nucleic Acids Res 34 (Database issue):D777–D780
39. Toseland CP, Clayton DJ, McSparron H, Hemsley SL, Blythe MJ, Paine K, Doytchinova IA, Guan P, Hattotuwagama CK, Flower DR (2005) AntiJen: a quantitative immunology database integrating functional, thermodynamic, kinetic, biophysical, and cellular data. Immunome Res 1:4
40. Yusim K, Bette TM, Korber CB, Haynes BF, Koup R, Moore JP, Walker BD, Watkins DI (eds) (2009) HIV molecular immunology 2009. Los Alamos National Laboratory, Theoretical Biology and Biophysics, Los Alamos, New Mexico, LA-UR 09-05941
41. Huang J, Ru B, Dai P (2011) Bioinformatics resources and tools for phage display. Molecules 16:694–709
42. Valuev VP, Afonnikov DA, Ponomarenko MP, Milanesi L, Kolchanov NA (2002) ASPD (Artificially Selected Proteins/Peptides Database): a database of proteins and peptides evolved in vitro. Nucleic Acids Res 30:200–202
43. Shtatland T, Guettler D, Kossodo M, Pivovarov M, Weissleder R (2007) PepBank—a database of peptides based on sequence text mining and public peptide data sources. BMC Bioinformatics 8:280
44. Ru B, Huang J, Dai P, Li S, Xia Z, Ding H, Lin H, Guo F, Wang X (2010) MimoDB: a new repository for mimotope data derived from phage display technology. Molecules 15: 8279–8288
45. Huang J, Ru B, Zhu P, Nie F, Yang J, Wang X, Dai P, Lin H, Guo FB, Rao N (2012) MimoDB 2.0: a mimotope database and beyond. Nucleic Acids Res 40(Database issue): D271–D277
46. Sun P, Chen W, Huang Y, Wang H, Ma Z, Lv Y (2011) Epitope prediction based on random peptide library screening: benchmark dataset and prediction tools evaluation. Molecules 16(6):4971–4993
47. Mandava S, Makowski L, Devarapalli S, Uzubell J, Rodi DJ (2004) RELIC—a bioinformatics server for combinatorial peptide analysis and identification of protein-ligand interaction sites. Proteomics 4:1439–1460

Chapter 8

Antigen–Antibody Interaction Database (AgAbDb): A Compendium of Antigen–Antibody Interactions

Urmila Kulkarni-Kale, Snehal Raskar-Renuse, Girija Natekar-Kalantre, and Smita A. Saxena

Abstract

Antigen–Antibody Interaction Database (AgAbDb) is an immunoinformatics resource developed at the Bioinformatics Centre, University of Pune, and is available online at http://bioinfo.net.in/AgAbDb.htm. Antigen–antibody interactions are a special class of protein–protein interactions that are characterized by high affinity and strict specificity of antibodies towards their antigens. Several co-crystal structures of antigen–antibody complexes have been solved and are available in the Protein Data Bank (PDB). AgAbDb is a derived knowledgebase developed with an objective to compile, curate, and analyze determinants of interactions between the respective antigen–antibody molecules. AgAbDb lists not only the residues of binding sites of antigens and antibodies, but also interacting residue pairs. It also helps in the identification of interacting residues and buried residues that constitute antibody-binding sites of protein and peptide antigens. The Antigen–Antibody Interaction Finder (AAIF), a program developed in-house, is used to compile the molecular interactions, viz. van der Waals interactions, salt bridges, and hydrogen bonds. A module for curating water-mediated interactions has also been developed. In addition, various residue-level features, viz. accessible surface area, data on epitope segment, and secondary structural state of binding site residues, are also compiled. Apart from the PDB numbering, Wu–Kabat numbering and explicit definitions of complementarity-determining regions are provided for residues of antibodies. The molecular interactions can be visualized using the program Jmol. AgAbDb can be used as a benchmark dataset to validate algorithms for prediction of B-cell epitopes. It can as well be used to improve accuracy of existing algorithms and to design new algorithms. AgAbDb can also be used to design mimotopes representing antigens as well as aid in designing processes leading to humanization of antibodies.

Key words Antigen, Antibody, Antigen–antibody complex, Antigen–antibody interactions, B-cell epitope, Paratope, Antibody-binding site, Conformational or discontinuous epitope, Immunoinformatics, Bioinformatics, Derived database

Abbreviations

AAIF	Antigen–Antibody Interaction Finder
Ab	Antibody
Ag	Antigen
Ag–Ab	Antigen–antibody

Rajat K. De and Namrata Tomar (eds.), *Immunoinformatics*, Methods in Molecular Biology, vol. 1184,
DOI 10.1007/978-1-4939-1115-8_8, © Springer Science+Business Media New York 2014

AgAbDb	Antigen–Antibody Interaction Database
ASA	Accessible surface area
BR	Buried residue
BS	Binding site
CDR	Complementarity-determining region on heavy chain
CE	Conformational epitope
CEP	Conformational epitope prediction
Ig	Immunoglobulin
IR	Interacting residues
LDR	Complementarity-determining region on light chain
PDB	Protein Data Bank

1 Introduction

Antibodies are produced by vertebrates in response to antigens. Antigens are usually foreign molecules of invading pathogens. Antibodies are produced in billions of forms by B cells and are collectively referred to as immunoglobulins (abbreviated as Ig). The clonal selection theory states that all the antibodies produced by an individual B cell have the same antigen-binding site. Furthermore, every B cell produces a single species of antibody having a unique antigen-binding site.

1.1 Antibody Structure

An antibody molecule is a polymer of two light and two heavy chains. The two light chains are identical and are of a length of ~220 amino acids each. Similarly, the two heavy chains are identical with a typical length of ~440 amino acids each. The four chains are held together by various noncovalent and covalent (disulfide) bonds. Every light chain has one variable and one constant region, whereas heavy chains have one variable and two to three constant regions. As a result, two identical antigen-binding sites are formed by the N-terminal variable regions of a pair of light and heavy chains. The tail (Fc) and hinge regions are however formed by the constant regions of two heavy chains. The antigen-binding site of an antibody is referred to as a "paratope" [1, 2].

1.2 Types of Antibodies

There are five classes of antibodies such as IgA, IgD, IgE, IgG, and IgM, which are based on five types of heavy chains such as α, δ, ε, γ, and μ. Each of these heavy chains is known to invoke a specific cascade of reactions upon binding to an antigen. However, there are only two types of light chains (κ or λ) that pair with one of the heavy chains. Therefore, the type of light chain does not seem to affect the properties of the antibody, other than its specificity for the antigen [1, 2].

1.3 Paratope

Paratope, the antigen-binding site of an antibody, is typically a region on the surface of the antibody that interacts with a region on the surface of the antigen (epitope) through non-covalent interaction. The paratope region is observed to be unique to every antibody and is said to be complementary to the "epitope" of the antigen. A paratope is made of six discontinuous regions, which are referred to as complementarity-determining regions (CDRs). There are three CDRs each on light and heavy chains. These regions are highly variable and are the loops connecting beta strands of the immunoglobulin fold.

1.4 Properties of Antigen–Antibody Interactions

The binding of an antigen to an antibody is reversible, and both the molecules can exist independently. The antigen–antibody interactions are thus mediated by many relatively weak, non-covalent forces such as hydrogen bonds, hydrophobic interactions, van der Waals forces, and ionic interactions. Of all the forces, van der Waals forces are the weakest and can attract all kinds of molecules. Hydrogen or ion–dipole bonds are formed between oppositely charged atoms, whereas "hydrophobic" interactions are formed between atoms of nonpolar amino acids which do not form electric dipole [3, 4]. These weak forces are effective only when the antigen molecule is close enough to allow some of its atoms to fit into complementary niches on the surface of the antibodies. The attractive forces exerted by ionic and hydrophobic bonds help the molecules to overcome hydration energies. This leads to the expulsion of water molecules and results in bringing the epitope and paratope closer. This spatial proximity facilitates van der Waals interactions. The overall strength of binding depends on goodness of fit between the epitope and paratope and the total area of contact between them [3, 4].

1.5 Characterization of Binding Sites

Antigen–antibody interactions are highly specific, and understanding the molecular basis of the specificity has been one of the goals of immunology. A large number of high-resolution X-ray structures of several antigens have been solved in the native (uncomplexed) form as well as in complex with antibody, and the data are archived in Protein Data Bank (PDB) [5]. Analyses of these structures have helped in understanding characteristics of both epitopes (antibody-binding site on antigen) and paratopes (antigen-binding site of antibody), which are complementary to each other and are relational entities [6–8].

Epitopes are of two types, viz. sequential or contiguous and conformational or discontinuous. The sequential epitopes are a stretch of amino acid residues linked by the peptide bonds and are recognized by an antibody. The other type is called conformational or discontinuous epitope where the antibody recognizes multiple sequential regions that come together due to folding of the polypeptide chain and a few independent residues [9–11]. Availability of crystal structures enabled the study of various features of binding sites such as size, shape, and complementarity of interacting surfaces of the antigens and antibodies [12–14]. These features and data, in an implicit and explicit manner, also served as a knowledgebase to develop and benchmark algorithms for prediction of sequential (continuous) and conformational (discontinuous) B-cell epitopes. These algorithms have been extensively reviewed elsewhere [15, 16]. Several attempts have been made to compile and curate the immunological data at various levels of complexity, which has resulted in the development of several useful databases that could themselves be grouped into categories based on the data archived, viz. antibody sequences and crystal structures: IMGT/LIGM-DB [17] and IMGT/3D structure DB [18]; experimentally characterized linear and conformational epitopes: IEDB [19], Epitome [20], CED [21], and BCIpep [22]; and Antigen–Antibody Interaction Database: AgAbDb [23], BEID [24], and IEDB3D [25]. Since humoral or antibody-based immune response is the first line of defense against most of the bacterial and viral pathogens, development of well-designed immunoinformatics databases in this area has been considered as one of the most important activities in the realm of reverse vaccinology and vaccine informatics [26, 27]. Importance of these databases is further substantiated since computational modeling of B-cell epitopes is complex due to posttranslational modifications of B-cell epitopes and the role of carbohydrates in antigen–antibody interactions.

The first version of AgAbDb was published only with curated data of Ag–Ab complexes where antigens are proteins. The first version included limited data on interactions [23]. It is the first database that compiled various non-covalent atomic interactions, which facilitates the binding of antibodies to antigens. AgAbDb also documented the interacting residues (IR) and buried residues (BR) specifically [23]. It is known that many residues of an antigen get buried under an antibody and may not necessarily be a part of any intermolecular interactions. However, these residues are important in maintaining the scaffold of binding sites. The residues of binding site (BS) however are obtained by summation of IR and BR, which help to determine the area of an antigen buried under the footprint of an antibody [28–30]. AgAbDb was also instrumental in providing the interacting residue pairs of antigens and antibodies. Most of the immunoinformatics databases and servers mentioned earlier [19–22, 24, 25] list interacting residues of the epitopes and paratopes independently and lack data on equivalence. It was further noticed that most of the databases listing the interaction data are not specially designed for Ag–Ab interaction studies and provide data on interactions of other immunological molecules as well. As a result of this, there is a lag in updation and several Ag–Ab complexes are not included in their versions posted online.

This chapter documents features of the current version of AgAbDb, which has significant additions in terms of not only curated data of peptide antigen–antibody complexes but also water-mediated interactions, epitope segment data with secondary structural states of participating residues, etc. The content, format, browsing, and retrieval of data from AgAbDb are explained using suitable examples.

2 Materials

2.1 Data Collection

PDB archives high-resolution structures of several antigens in both native (free unbound state) and complex (bound to antibody) states [5]. There are a few structures in the PDB where residues of either antigen or antibody molecules are mutated to study the effect of mutations on antigen–antibody interactions. The PDB is searched using text-based queries to retrieve entries of antigen–antibody complexes. It is a multi-step process, and scripts are written to compile all the structures. These structures are broadly grouped into two types based on the length of antigen sequence. Antigens having length ≤35 amino acids were referred to as peptide antigens. Antigen sequences with length >35 were grouped as protein antigens. The atomic coordinate file for each antigen–antibody co-crystal structure is downloaded from the PDB (www.rcsb.pdb.org). The data are parsed through a series of Perl scripts to curate the derived data of an antigen, antibody, and various intermolecular interactions.

The most recent release of AgAbDb (Aug 3, 2013) includes data of 427 antigen–antibody co-crystal structure complexes. There are 289 and 138 entries, respectively, for protein and peptide antigens in complex with respective antibodies. Of the protein antigens, majority are monomers (266) whereas 21 are dimers and only 2 are multimers. AgAbDb is updated regularly based on the release of antigen–antibody co-crystal structure complexes in the PDB.

2.2 Data Curation

AgAbDb compiles and curates derived data of antigen–antibody interactions, and the PDB is the primary source of experimental data. Various tables are populated with the data of antigen, antibody, and antigen–antibody interactions. The derived data includes the residues of epitope and paratope, interacting residue pairs, and types of interactions between them. Derived data of molecular interactions is generated using Antigen–Antibody Interaction Finder (AAIF), a Perl program developed in-house [23]. Various geometrical and stereochemical criteria used to curate interactions are described earlier [23]. Perl scripts are also written to curate water-mediated interactions, sequential epitope segments, secondary structural state of the residues of epitope, etc.

3 Methods

3.1 Database Design and Organization

The AgAbDb is implemented as a relational database using MySQL Server 5. The database comprises of 12 tables to compile, curate, and archive data on antigens, antibodies, and interactions and is normalized up to Third Normal Form (3NF). The lists of interacting and buried residues of the respective antibody and antigen are stored in four tables. The various types of atomic interactions between antigens and antibodies are stored in two tables. An additional table is used to store "water-mediated interactions," a feature which is added recently. Two tables are used to store the epitope segments and secondary structures of amino acid residues, while the remaining three tables are used to store the annotation data of every antigen, antibody, and entire complex. The query system has been developed using JSP, JSTL, HTML, and JavaScripts. Perl scripts are written to retrieve the data from the PDB and to populate database tables.

3.2 Characterization of Binding Sites

The residues of the binding site of both the antigen and antibody are compiled. The residues of BS of an antigen (epitope) are classified as IR and BR based on their role in the complex formation. The binding site of an antibody (paratope) comprises three CDRs each on the variable domain of heavy (CDR 1–3) and light (LDR 1–3) chains.

3.3 Interacting Residues (IR)

The residues of an antigen that form non-covalent interaction(s) with residue(s) of an antibody molecule are defined as IR. AAIF calculates non-covalent interactions, viz. van der Waals interactions, hydrogen bonds, salt bridges, short contacts using distance, and geometry-based criteria described earlier [23, 31–34]. The positions of hydrogen atoms are predicted using the fourth atom fixation algorithm. The hydrogen bond donors and acceptors are defined as per HBplus program [35].

3.4 Buried Residues (BR)

In addition to the IR, a few more residues of an antigen are buried under the footprint of an antibody. These residues do not directly participate in the antigen–antibody interactions but are part of the scaffold to maintain the binding site. These residues are identified based on loss of solvent-accessible surface area (ASA) upon antibody binding. Solvent ASA of antigen and antibody molecules was computed using the Voronoi polyhedron algorithm [36] in both unbound and bound states. The difference in the solvent ASA of every residue of the antigen (and antibody) in uncomplexed as well as complexed states needs to be computed to determine the area of interaction and the list of BR. The residues that lose ASA greater than or equal to 0.1 $Å^2$ upon formation of the complex are defined as BR.

3.5 Water-Mediated Interactions

Water-mediated hydrogen bonds between antigen and antibody molecules are computed. Only the crystallographic water molecules present in the PDB files are included in the computation. Potential hydrogen bond donors and acceptors were defined as per HBplus program [35]. Both bond distance and angle criteria are used to curate ionic interactions between charged residues and trapped water molecules.

3.6 Epitope Segments

The antibody-binding sites of antigens are typically made of a few sequential epitopes that come together due to folding of the polypeptide chain along with a few individual residues referred as singleton residues [28–30]. The continuous sequential segments of conformational epitope (antibody-binding site) are listed along with the individual residues, if any.

3.7 Secondary Structural State of Binding Site Residues

Single-letter codes of the secondary structural state of every residue of the sequential epitope are compiled. The secondary structural states defined by the DSSP program [37] are used for this purpose.

3.8 Antibody Re-numbering

The residues of light and heavy chains of antibodies are renumbered based on the CDR definitions put forward by Wu and Kabat using the AbCheck server [38]. The AgAbDb tables are populated such that the correspondence between both the PDB and Kabat numbering is maintained.

4 AgAbDb: Need and Scope

4.1 Antigen–Antibody Complexes

The structures of more than 1,000 complexes of antigens with respective antibodies have been solved to date, and the data is deposited in the PDB [5]. Several antibodies have been co-crystallized with various antigens such as proteins, peptides, small molecules, nucleotides, and DNA. These co-crystal structures have been solved with different objectives like mapping the antibody-binding sites, studying the mode of interactions between the two molecules, identifying critically important residues, examining cross-reactivity of antibodies towards antigens, and assessing conformational changes in the antigen, antibody, or both upon formation of complexes.

The vast amount of co-crystal structure complex data have also been collectively used for studying the properties of interacting interfaces such as epitopes and paratopes in particular and protein–protein interactions in general [14]. The analysis of data was also instrumental in the development of B-cell epitope prediction algorithms for both sequential and conformational epitopes. Apart from serving as a knowledgebase for epitope predictions, the data have also been used to validate and benchmark the performance of epitope prediction algorithms. The data of antibody structures in free and bound forms have been used to develop dedicated homology-modeling programs for the prediction of three-dimensional structures of antibodies [40, 41]. The importance of prediction of 3D structures of antibodies is ever increasing as the antibodies are increasingly being used for diagnostic and therapeutic purposes in diseases such as cancer. Humanization of antibodies is another important area where high-resolution curated data of the antigen–antibody structures and interactions are desirable.

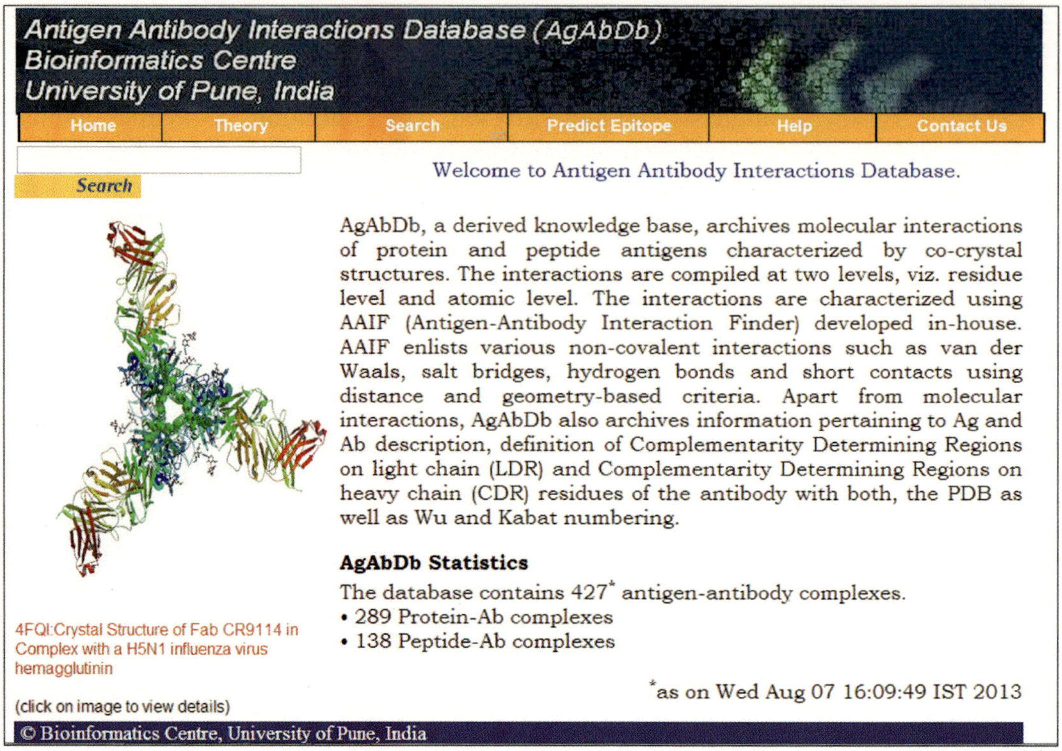

Fig. 1 A snapshot of the home page of AgAbDb

Note: A list of all the available antigen–antibody complexes in PDB could be made by searching PDB using keyword-based searches. However, there is always a possibility of missing out on a few entries since all the three keywords, antigen, antibody, and complex, may not be explicitly present in every PDB file. Most often, the type of antigen and its description such as the lysozyme or a peptide sequence are mentioned rather than the word antigen. Therefore, compilation of PDB files of Ag–Ab complexes becomes a multi-step curation exercise. In AgAbDb, the scripts have been written to automate PDB searches and curation. These searches are performed every week to corroborate with the weekly schedule of updation of PDB.

4.2 AgAbDb: Design and Contents

Every antigen–antibody co-crystal structure helps in understanding how an antibody interacts with an antigen at an atomic level and illustrates specificity of interaction. AgAbDb catalogs the antigen–antibody interactome data individually and collectively. The home page of AgAbDb is shown in Fig. 1. The current version of AgAbDb archives data on protein and peptide antigens only. Furthermore, AgAbDb curates data of only those complexes where both the antibody chains (heavy and light) are part of the complex. Figure 2 shows the growth of antigen–antibody co-crystal structure

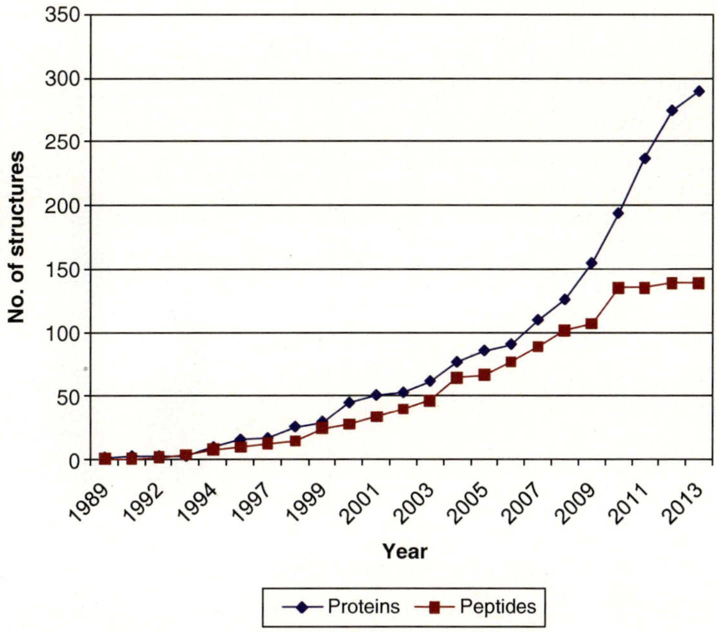

Fig. 2 The growth of co-crystal structures of protein- and peptide–antibody complexes in AgAbDb

complexes of protein and peptide antigens over a period of time. Figure 3 shows a bar chart of the number of structures against resolution at which they were solved. It can be seen that a majority of the structures have been solved with a resolution better than 3 Å.

Note: For data on complexes of other antigens such as carbohydrates, RNA, and DNA, the users are suggested to use the PDB. It is planned to populate AgAbDb with data on all the antigen–antibody complexes, regardless of the antigen type in future.

4.3 AgAbDb: Search and Browse

A user-friendly web-enabled interface for AgAbDb (http://115.111.37.206:8080/agabdb2) has been designed and tested for all the web browsers. A "quick search box" is provided on all the web pages of the interface. The "quick search" supports the database search using the PDB ID or the keywords. This in turn opens a page listing the search results. AgAbDb can be browsed by clicking on the PDB ID. The search results page also provides links to view the antigen–antibody interactions archived in AgAbDb using Jmol [V], to view the corresponding complex at the RCSB PDB site [P], and to download the file from the RCSB PDB [D].

The "Search" option on the menu bar pops down three options (*see* Fig. 4). Text-based queries can be performed in the "Advanced search" as shown in panel "A." This page also provides an option to search Ag–Ab complexes for a pair of interacting residues.

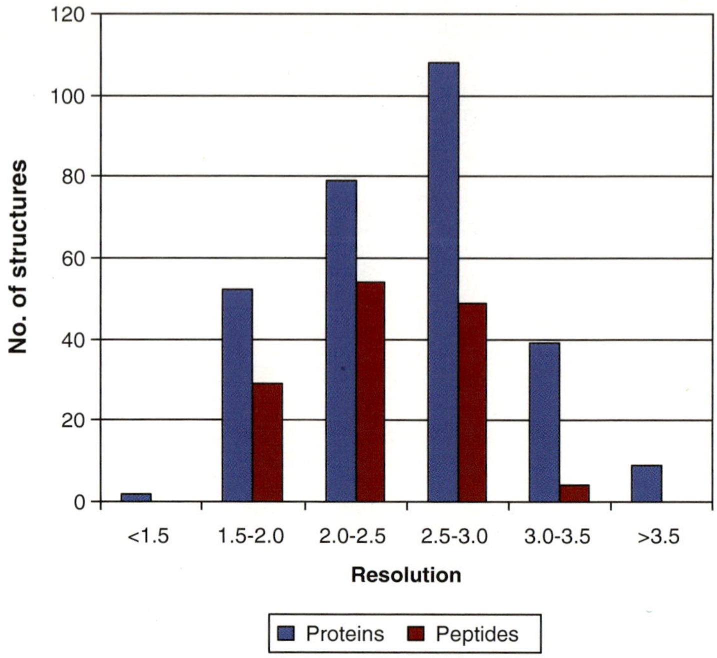

Resolution graph for 427 complexes

Fig. 3 Distribution of co-crystal structures of antigen–antibody complexes based on resolution

The "Residue wise search: Ag and Ab" option permits retrieval of entries for any interacting residues (panel "B"). The "Residue-wise interactions: CDR" search provides quick summary of interacting residues for selected CDRs of antibody (panel "C"). Detailed help for searching and browsing is also provided on the AgAbDb website.

Note: It is advisable to use keyword-based searches when either the antigen or the antibody is known. One can quickly view Ag–Ab interactions, if the PDB ID is known. AgAbDb, however, is the only database which facilitates querying of AgAb interactions using residues of epitope, paratope, or both.

4.4 AgAbDb: Data Formats and Displays

AgAbDb archives data of antigens, antibodies, and molecular interactions under eight categories, viz. Summary, IR: Epitope-Paratope, IR: Epitope Segments, Binding Site: IR + BR, Atomic Level Interactions, Water-Mediated Interactions, View Interactions, and Statistics. The tables displaying interaction data under each of these eight categories can be exported as Excel files. The complex, binding site residues of antigen and antibody along with subsets of various interactions can be visualized using Jmol (http://www.jmol.org/). The snapshots of screens based on eight categories are shown in Fig. 5. AgAbDb records for a complex of the antibody

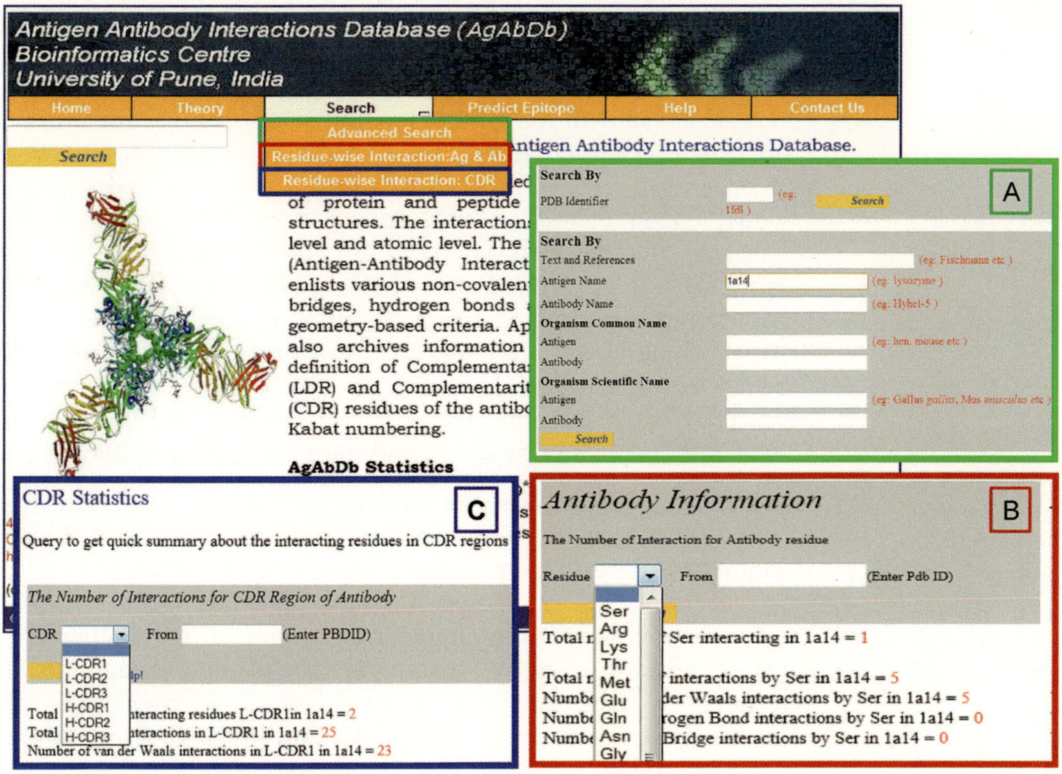

Fig. 4 Snapshots of various search strategies in AgAbDb

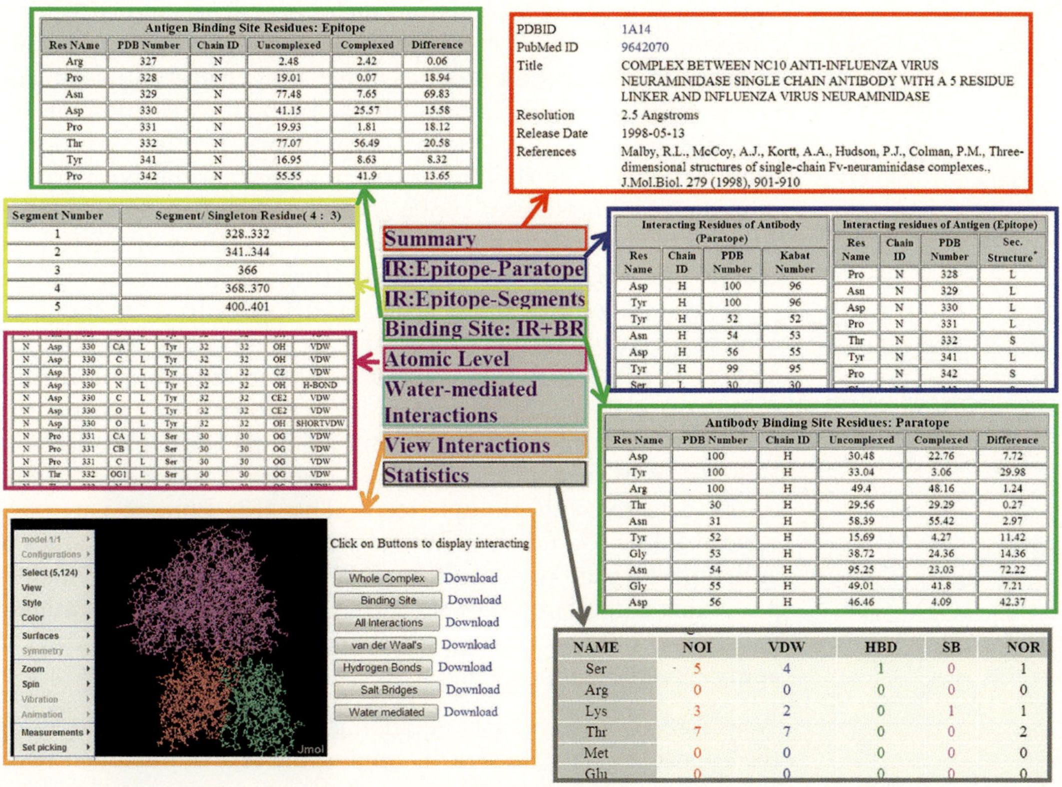

Fig. 5 Snapshots of various data archived in AgAbDb. The PDB ID: 1A14 (complex of neuraminidase and antibody NC10) is used as a case study

NC10 Fv and neuraminidase from influenza virus ([39], PDB ID: 1A14) are shown. AgAbDb uses PDB ID as a unique identifier to archive interaction data.

4.4.1 Summary

This section provides overall information of the complex, the antigen, and the antibody. Data are curated from the PDB and typically lists PDB ID, PubMed ID, resolution, release date, and citation information. The data on antibody includes name, class/type, scientific and common names of the source, and the PDB chain identifiers for light and heavy chains. The data on antigen includes name, scientific and common names of the source, antigen type (protein or peptide), and the PDB chain identifiers.

Note: The data on class/type of the antibody, if available in the PDB, is curated. It is observed that class/type of antibody is mentioned only occasionally in the PDB.

4.4.2 IR: Epitope–Paratope

This section lists all the interacting residues of the binding sites. The residues of antibody (paratope) that are interacting with the residues of antigen (epitope) are provided. For example, the numbers of interacting residues of paratope (NC10 Fv) and epitope (neuraminidase) are 12 and 17, respectively (PDB ID: 1A14). The paratope residues are listed with chain type (heavy or light chain), PDB numbering, and Kabat numbering. It is preferred to have both the numbering systems and their equivalence known as far as antibody numbering is concerned. The table also lists equivalence between the interacting residues of the antigen and antibody. This is one of the unique features of AgAbDb. It is very useful and facilitates interesting analyses as a residue may interact with one or more residues. The residues of both antigen or antibody having minimum and maximum contacts can be identified. For example, Asn400 of the antigen interacts with two residues of CDR2 and one residue of CDR1 of heavy chain. Identification of such important residues or hot spots may have applications in mutation analysis, which is a prerequisite for designing antigen scaffolds and/or peptide/subunit vaccines. Other immunoinformatics resources, viz. IEDB-3D and IMGT/3Dstructure-DB, do not provide the list of pairs of interacting residues in an explicit fashion. Generation of such a list using these resources calls for processing of the data through multiple steps. The "IR: Epitope-Paratope" table also lists secondary structural states of interacting residues of antigen, which are obtained from DSSP assignments [37]. Analysis pertaining to preference of secondary structural states of antigens has always been the area of interest and has been used effectively in epitope prediction programs.

Note: AgAbDb curates data of binding sites and interactions parsing the coordinate data and not by mining the text of published references. It was observed that there are a lot of variations in the way in which the authors have listed epitope and paratope residues and interactions in the published papers. Some publications listed only the interacting residues, while others listed both interacting and buried residues. Furthermore, it was noted that different programs and varied criteria are being used by the authors to enlist residues of binding sites. Hence, for the purpose of objectivity and uniformity in defining binding site residues, all the complexes were parsed through the program, AAIF, which is developed in-house. Other resources, viz. IEDB-3D, provide both "curated" and "calculated" contacts.

4.4.3 IR: Epitope Segments

This section enlists the segments of the binding site. Antibody-binding site of antibody NC10 characterized in PDB ID: 1A14 consists of four segments and three individual residues. Most often antibody-binding sites on antigens are conformational epitopes, where multiple sequential epitopes and a few individual residues are brought together due to the folding of polypeptide chain. The segments or the sequential epitopes are defined where consecutive amino acids (at least two) are a part of the binding site. The conformational epitope prediction (CEP) server, the first program to predict conformational or discontinuous epitopes, developed by our group (http://bioinfo.net.in/cep.htm or http://117.239.43.116/index.html), successfully used the distance-based criteria to predict conformational epitopes using sequential epitopes and individual residues [29, 30].

Note: Various resources may use different criteria and cutoffs for listing segments, sequential epitopes, and hence conformational epitopes.

4.4.4 Binding Site: IR+BR

This section lists all the residues of the respective binding sites of the antigens and antibodies. Separate tables for antigen and antibody molecules are generated. In addition to the interacting residues, several residues of epitope are buried under the footprint of an antibody. Such residues are a part of the binding site scaffold and may not directly interact with residues of CDRs and LDRs of an antibody. Similarly, CDR and LDR also have only a few interacting residues while the other residues forming the scaffold, though not interacting explicitly, are used to calculate the area of interface of antibody with antigen.

Note: It is advisable to know the composition of both the epitope and paratope in terms of interacting and buried residues for a variety of purposes and applications.

4.4.5 Atomic Level Interactions

This section displays various non-covalent interactions between residues of the epitopes and paratopes. For example, NC10 antibody (PDB ID: 1A14) has about 107 non-covalent interactions of the types such as salt bridges (1), hydrogen bonds (7), short van der Waals interactions (2), and van der Waals interactions (97). These interactions are curated using the program AAIF.

Note: Though several programs are available for characterizing various interactions between residues of antigen and antibodies, it calls for analysis that requires pre- and post-processing. Curation and summary of residues involved in non-covalent interactions is a value added feature of AgAbDb.

4.4.6 Water-Mediated Interactions

This table lists the interactions mediated through crystallographic water molecules. It is known that the water molecules are observed in cavities of binding sites of antigens and antibodies. Such trapped water molecules often form bridging hydrogen bonds between the antibody and antigen. AgAbDb now provides a utility to enlist water-mediated interactions.

Note: Since there are no trapped water molecules in the complex of NC10 and neuraminidase (PDB ID: 1A14), a snapshot of this table is not included in Fig. 5.

4.4.7 Statistics

This section provides a residue-wise summary of various interactions. Separate tables are provided for the antigen (epitope) and antibody (paratope), which list the residues that contribute maximally to the antigen–antibody interactions. This section provides a summary of interactions for every residue and includes data on the total number of interactions, which is a sum of the total number of hydrogen bonds, van der Waals interactions, and salt bridges. The table also lists the total number of residues (from the partner molecule) with which a given residue is interacting. This section also helps to quickly enlist which of the 20 amino acids are parts of the paratope and epitope. For example, NC10 antibody CDRs have only 7 (S, T, N, F, L, D, Y) amino acids whereas the neuraminidase epitope has 11 (S, K, T, N, G, A, D, I Y, P, W) amino acids as characterized in the complex 1A14 [39].

AgAbDb also helps in analyzing how every CDR participates in binding to the epitope. This utility is provided under the "Search" option on the main menu bar. Three CDRs on light chain are termed as LDR 1–3. There are three LDRs (light chain) and three CDRs (heavy chain). Since the PDB numbering may or may not be in accordance with the position of a given residue in sequence and/or Kabat scheme of numbering, AgAbDb provides equivalence between PDB and Kabat numbering. "CDR statistics" for NC10 antibody (PDB ID: 1A14) reveals that two of the six CDRs such as LDR2 and CDR1 do not participate in the antigen binding at all. The LDR1, LDR3, CDR2, and CDR3, respectively, have 2, 4, 3, and 3 residues interacting with various residues of the antigen. Of the 107 total interactions, 25, 34, 27, and 21 interactions are contributed by LDR1, LDR3, CDR2, and CDR3, respectively. Thus, AgAbDb can be used to perform various queries and to study the multiple aspects of antigen–antibody interactions.

4.4.8 Update

AgAbDb is updated every week. Curation of antigen–antibody interaction data of the antigens other than proteins and peptides is under process. The interaction data of the antigens such as small molecules, carbohydrates, RNA, and DNA will be curated and made available in future.

Acknowledgements

Dr. Urmila Kulkarni-Kale gratefully acknowledges financial support from the Department of Biotechnology (DBT), Government of India, and the Department of Science and Technology (DST), Government of India. Ms. Snehal Raskar-Renuse and Ms. Smita A. Saxena acknowledge the Department of Information Technology (DeitY), Ministry of Communications and Information Technology (MCIT), Government of India, for fellowship. Ms. Girija Natekar-Kalantre acknowledges DBT for fellowship.

Funding: This work was supported by the Center of Excellence (CoE) grant by the DBT, Govt. of India. Some modules of the AgAbDb are developed under the PURSE program of the DST, Government of India.

References

1. Goldsby RA, Kindt TJ, Osborne BA (2000) Kuby immunology, 4th edn. W. H. Freeman and Company, New York
2. Janeway CA Jr, Travers P, Walport M, Shlomchik MJ (2001) Immunobiology, 5th edn. Garland Science, New York
3. Van Oss CJ (1995) Hydrophobic, hydrophilic and other interactions in epitope-paratope binding. Mol Immunol 32:199–211
4. Reverberi R, Reverberi L (2007) Factors affecting the antigen-antibody reaction. Blood Transfus 5:227–240. doi:10.2450/2007.0047-07
5. Berman HM, Westbrook J, Feng Z, Gilliland G, Bhat TN, Weissig H, Shindyalov IN, Bourne PE (2000) The Protein Data Bank. Nucleic Acids Res 28:235–242
6. Davies DR, Cohen GH (1996) Interactions of protein antigens with antibodies. Proc Natl Acad Sci U S A 93:7–12
7. Wilson IA, Stanfield RL (1994) Antibody-antigen interactions: new structures and new conformational changes. Curr Opin Struct Biol 4:857–867
8. MacCallum RM, Martin AC, Thornton JM (1996) Antibody-antigen interactions: contact analysis and binding site topography. J Mol Biol 262:732–745
9. Van Regenmortel MH (2009) What is a B-cell epitope? Methods Mol Biol 524:3–20
10. Al-Lazikani B, Lesk AM, Chothia C (1997) Standard conformations for the canonical structures of immunoglobulins. J Mol Biol 273:927–948
11. Chothia C, Gelfand I, Kister A (1998) Structural determinants in the sequences of immunoglobulin variable domain. J Mol Biol 278:457–479
12. Janin J, Chothia C (1990) The structure of protein-protein recognition sites. J Biol Chem 265:16027–16030
13. Janin J, Miller S, Chothia C (1988) Surface, subunit interfaces and interior of oligomeric proteins. J Mol Biol 204:155–164
14. Jones S, Thornton JM (1996) Principles of protein-protein interactions. Proc Natl Acad Sci U S A 93:13–20
15. Greenbaum JA, Andersen PH, Blythe M, Bui HH, Cachau RE, Crowe J, Davies M, Kolaskar AS, Lund O, Morrison S et al (2007) Towards a consensus on datasets and evaluation metrics

for developing B-cell epitope prediction tools. J Mol Recognit 20:75–82

16. El-Manzalawy Y, Honavar V (2010) Recent advances in B-cell epitope prediction methods. Immunome Res 6(Suppl 2):S2
17. Giudicelli V, Duroux P, Ginestoux C, Folch G, Jabado-Michaloud J, Chaume D, Lefranc MP (2006) IMGT/LIGM-DB, the IMGT® comprehensive database of immunoglobulin and T cell receptor nucleotide sequences. Nucleic Acids Res 34:D781–D784
18. Kaas Q, Ruiz M, Lefranc MP (2004) IMGT/3Dstructure-DB and IMGT/Structural Query, a database and a tool for immunoglobulin, T cell receptor and MHC structural data. Nucleic Acids Res 32:D208–D210
19. Kim Y, Ponomarenko J, Zhu Z, Tamang D, Wang P, Greenbaum J, Lundegaard C, Sette A, Lund O, Bourne PE et al (2012) Immune epitope database analysis resource. Nucleic Acids Res 40:W525–W530
20. Schlessinger A, Ofran Y, Yachdav G, Rost B (2006) Epitome: database of structure-inferred antigenic epitopes. Nucleic Acids Res 34: D777–D780
21. Huang J, Honda W (2006) CED: a conformational epitope database. BMC Immunol 7:7
22. Saha S, Bhasin M, Raghava GP (2005) Bcipep: a database of B-cell epitopes. BMC Genomics 6:79
23. Ghate AD, Bhagwat BU, Bhosle SG, Gadepalli SM, Kulkarni-Kale UD (2007) Characterization of antibody-binding sites on proteins: development of a knowledgebase and its applications in improving epitope prediction. Protein Pept Lett 14:531–535
24. Tong JC, Song CM, Tan PT, Ren EC, Sinha AA (2008) BEID: database for sequence-structure-function information on antigen-antibody interactions. Bioinformation 3:58–60
25. Ponomarenko J, Papangelopoulos N, Zajonc DM, Peters B, Sette A, Bourne PE (2011) IEDB-3D: structural data within the immune epitope database. Nucleic Acids Res 39: D1164–D1170
26. He Y, Rappuoli R, De Groot AS, Chen RT (2010) Emerging vaccine informatics. J Biomed Biotechnol 2010:218590. doi:10.1155/2010/218590
27. Kulkarni-Kale U, Waman V, Raskar S, Mehta S, Saxena S (2012) Genome to vaccinome: role of bioinformatics, immunoinformatics & comparative genomics. Curr Bioinformatics (CBIO) 7:454–466
28. Davies DR, Padlan EA, Sheriff S (1990) Antibody-antigen complexes. Annu Rev Biochem 59:439–473
29. Kolaskar AS, Kulkarni-Kale U (1999) Prediction of three-dimensional structure and mapping of conformational epitopes of envelope glycoprotein of Japanese encephalitis virus. Virology 261:31–42
30. Kulkarni-Kale U, Bhosle S, Kolaskar AS (2005) CEP: a conformational epitope predic-tion server. Nucleic Acids Res 33: W168–W171
31. Ramachandran GN, Sasisekharan V (1968) Conformation of polypeptides and proteins. Adv Protein Chem 23:283–438
32. Barlow DJ, Thornton JM (1983) Ion-pairs in proteins. J Mol Biol 168:867–885
33. Sheriff S (1993) Some methods for examining the interactions between two molecules. Immunomethods 3:191–196
34. Tsumoto K, Ogasahara K, Ueda Y, Watanabe K, Yutani K, Kumagai I (1996) Role of salt bridge formation in antigen-antibody interaction. Entropic contribution to the complex between hen egg white lysozyme and its monoclonal antibody HyHEL10. J Biol Chem 271:32612–32616
35. McDonald IK, Thornton JM (1994) Satisfying hydrogen bonding potential in proteins. J Mol Biol 238:777–793
36. McConkey BJ, Sobolev V, Edelman M (2002) Quantification of protein surfaces, volumes and atom-atom contacts using a constrained Voronoi procedure. Bioinformatics 18: 1365–1373
37. Kabsch W, Sander C (1983) Dictionary of protein secondary structure: pattern recognition of hydrogen-bonded and geometrical features. Biopolymers 22:2577–2637
38. Martin AC (1996) Accessing the Kabat antibody sequence database by computer. Proteins 25:130–133
39. Malby RL, McCoy AJ, Kortt AA, Hudson PJ, Colman PM (1998) Three-dimensional structures of single-chain Fv-neuraminidase complexes. J Mol Biol 279:901–910
40. Kuroda D, Shirai H, Jacobson MP, Nakamura H (2012) Computer-aided antibody design. Protein Eng Des Sel 25:507–521
41. Sircar A, Kim ET, Gray JJ (2009) RosettaAntibody: antibody variable region homology modeling server. Nucleic Acids Res 37:W474–W479. doi:10.1093/nar/gkp387

Chapter 9

Allergen Databases

Gaurab Sircar, Debasree Sarkar, Swati Gupta Bhattacharya, and Sudipto Saha

Abstract

In this chapter, five popular allergen databases have been described: (1) Allergome is based on basic and clinical information on allergens causing an IgE-mediated disease; (2) AllergenOnline allows online search of peer-reviewed allergen list; (3) International Union of Immunological Societies Allergen nomenclature subcommittee database contains systematic nomenclature and molecular details of well-characterized allergens; (4) AllFam allows classifying allergens into protein families based on domain information; and (5) SDAP provides in detail structural information of the allergens.

Key words Allergen nomenclature, IgE-mediated disease, Domain, 3D structure

1 Introduction

Allergens are basically nonparasitic antigens capable of triggering a type-I hypersensitivity reaction in individuals with genetic predisposition to allergy. This immune response is mediated by inappropriate production of immunoglobulin E (IgE). The hereditary tendency of an individual to make IgE derived from plasma cells in response to stimulation of Th2 cells by common environmental allergens. There are many different types of allergens that could trigger an allergic reaction and may require clinical care. The common sources of allergen are dust mite excretion, pollen, latex, mould, insect stings, and some foods including peanuts, seafood, and shellfish. There are certain important features that make an antigen to be allergen: (1) can induce Th2 type response, (2) activation of IL-4-producing CD4+ T cells, and (3) contains peptides that bind host MHC class II molecule to prime T cells [1, 2]. Clinico-immunological and molecular data related to allergy and allergens are increasing with advancement of genomic and proteomic technologies. Sequences and three-dimensional structures of several potential food and aeroallergens have been determined in recent years. Information related to allergens is stored in specialized databases

Rajat K. De and Namrata Tomar (eds.), *Immunoinformatics*, Methods in Molecular Biology, vol. 1184, DOI 10.1007/978-1-4939-1115-8_9, © Springer Science+Business Media New York 2014

and repositories to support allergy research [3]. Basically, allergen databases can be classified into two types [4]: (A) Biological database, which provides only clinical or physiological information about allergens: It may not contain molecular information. Some of the examples of this class are Allergome (http://www.allergome.org/), Informall (http://farrp.unl.edu/resources/gi-fas/informall), and AllAllergy (http://allallergy.net/). (B) Molecular databases, which focused on sequences and structures of allergens: International Union of Immunological Societies (IUIS) allergen nomenclature subcommittee, Allergen Database for Food Safety (ADFS) (http://allergen.nihs.go.jp/ADFS/), Allergen Online (http://allergenonline.com), AllerMatch (http://www.allermatch.org), and Structural Database of Allergen Proteins (SDAP) (http://fermi.utmb.edu/SDAP/sdap_ver.html) are some of the molecular databases. In this chapter, we discuss about five commonly used allergen databases of which the first one is a biological database (Allergome) and the remaining four are molecular databases (AllergenOnline, IUIS allergen nomenclature database, AllFam, and SDAP) (*see* **Note 1**).

2 Materials and Methods

2.1 Allergome Database

2.1.1 Description of Allergome Database

The Allergome Database is available at http://www.allergome.org/. The web version 4.0 is free, but a registration is required and users need to choose a username (max 16 spaces) and a password (max 16 characters). The menus are in the top side of the page and are interlinked [5, 6]. Followings are the brief description of menus.

Allergome Home

It links to the home page of Allergome and allows access to the following menus: allergens, real-time monitoring of IgE sensitization (ReTiME), RefArray, Tools, History, and statistics. Access to historical copies of the database needs registration, and new users need to log in for further access.

Allergens

This menu allows to access search engine of Allergome. In this search page, users can input query in the allergen database. There are two types of search: (1) quick search and (2) advanced search. Details of the two searches are described in Subheading 2.1.2.

ReTiME

The "ReTiME" links to a module created to acquire and store real-time data related to IgE sensitization.

RefArray

RefArray is a module created to access the references in the Allergome reference archive that contains all the processed papers available from the literature.

Tools

Allergome aligner allows comparing query sequence with the Allergome sequence dataset.

History

History page allows access to historical copies of the allergome database, starting from the year 2005.

Statistics

The number of allergen sources and its composition available in Allergome database starting from 2005 are available in this link. The updated version on March 20, 2013, contains 2,275 entries.

Links

It contains updated links to events, scientific associations, and journals.

Help

This page contains general information about the database menus and how to use it. There are other information related to specification and requirements for accessing the data from Allergome.

2.1.2 Usage of Allergome Database

There are two ways to search the database: (1) quick search and (2) advanced search. The query forms have been shown in Fig. 1a, b.

(A) In the "Quick Search" page, users can search the scientific or common name of the allergen source or the common name or IUIS defined name or Allergome code of the allergen molecule (*see* **Notes 2** and **3**).

(B) The Allergome search engine retrieves monographs, which contain all the entries showing matches with words being searched for (i.e., "pollen birch" will list all the monographs containing both the words).

(C) In the "Advanced Search," fields are considered as a single string of character (i.e., "*Dermatophagoides farinae*" does not retrieve allergens of the "*Dermatophagoides pteronyssinus*" species).

(D) Users can perform advanced search on "All" the archives of the Allergome database (default choice in the Select-a-field from pop-down menu). This search may be slower, but searches for the queried text in any part of the Allergome database. Alternatively, a specific archive may be selected if the term being searched for is known to be in that archive (e.g., term "Pollen" in the "Tissues" archive).

(E) Users can also perform a refined advanced search by using Allergenicity Scoring parameters like Species of Interest, Data Generation, Sequence, and Epidemiology from Literature.

2.1.3 Query Result of Allergome

In the output, each allergen molecule is described in a monograph that includes general features of the allergen and data on allergenicity of the native and cloned molecule. An example of an allergen monograph is shown in Fig. 1c. The monograph is divided into three parts: (1) The "General Information" page contains data for the identification of the allergen and its relationship with other allergens within the Allergome. (2) The "Native Form" page contains

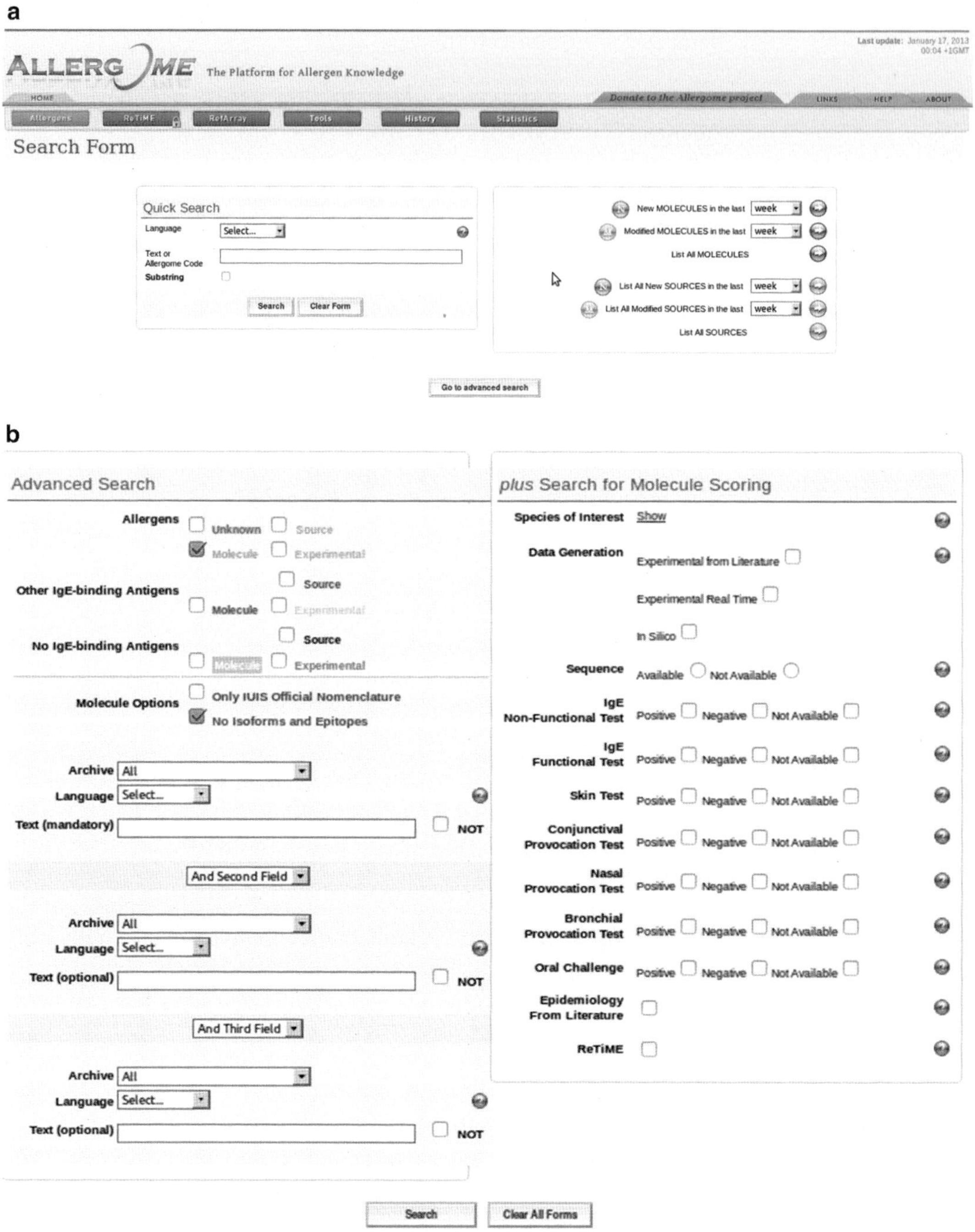

Fig. 1 Screenshots of Allergome (**a**) query submission form for "Quick Search"; (**b**) form for "Advanced Search"; (**c**) example of a molecule monograph

c

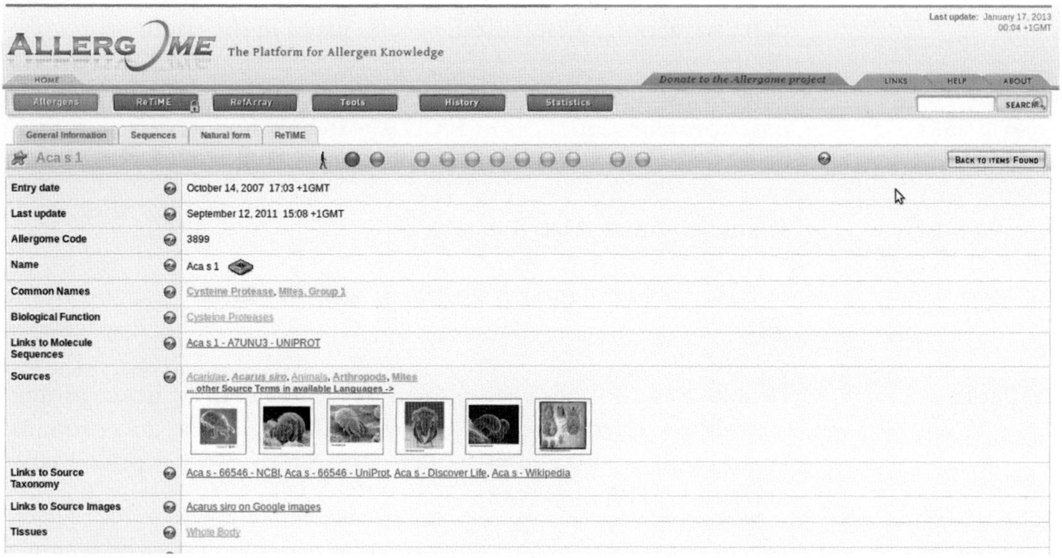

Fig. 1 (continued)

data on allergenicity of the allergen in its natural conformation. (3) The "Recombinant Form" page(s) contains (contain) data on allergenicity of the allergen obtained by means of molecular biology techniques. "Recombinant Form" pages are named by the expression vector used to produce the recombinant molecule.

2.2 AllergenOnline Database

2.2.1 Description of the AllergenOnline Database

The AllergenOnline database is accessible at http://www.allergenonline.org/. The version 13 as on February 13, 2012, contains 1,630 peer-reviewed sequences and 612 taxonomic protein groups [7]. The menus are on the left side of the page and are interlinked. Following are the brief description of menus.

AllergenOnline Home

It links to the home page of AllergenOnline, and it describes briefly about the features and tools available in the database. There are other information related to tools and contact of peer reviewers.

About AllergenOnline

This page shows the brief overview about the AllergenOnline database, including processing data entries and references.

Contact Page

It shows the e-mail address of the database developers.

Browse the Database

It allows the user to access all the entries in one page. The data is presented in eight columns: species, common name, IUIS Allergen, type, group, length, GI number, and version release number. There are filters in each column for quick search. More details about this query page are described in Subheading 2.2.2.

Version History

This page shows statistics of previous and current version, release date, sequences, groups, and species listed in the database.

Sequence Search Allergen Database

This menu allows the user to perform query search using one or more protein sequences in FASTA format. More details about this query page are described in Subheading 2.2.2.

Database and GMO

It links to other related databases.

FARRP Home

It links to Food Allergy Research and Resource Program home page.

Celiac Disease

It links to a tool of celiac disease risk assessment of novel protein and allows users to browse by peptides, references, and proteins. In addition, it also allows sequence search by exact peptide match and full FASTA sequence.

2.2.2 Usage of AllergenOnline Database

There are two ways to search the database: (1) browse entries and (2) sequence search.

(A) For browsing all the AllergenOnline database entries, click on the "Browse the Database" hyperlink under the "Navigation" options along the left-hand side of the home page.

(B) A summary page containing an outline of the whole database is displayed in a table with the following columns: Species, Common, IUIS Allergen, Type, Group, Length, GI number, and Version release number. Under each column, a blank field allows filtering of the table by that column using a particular keyword (e.g., filtering with the keyword "*Actinidia chinensis*" in the "Species" column lists all the entries in the database for that particular species).

(C) Clicking on each of the entries in the "Group" column opens new window containing information about the published references used to classify the protein as an allergen as well as the individual sequences clustered into the group.

(D) Clicking on the "gi" number for each entry opens the page containing the complete NCBI entry of that particular protein.

(E) For sequence search option, click on the "Sequence Search allergen Database" hyperlink under the "Navigation" options along the left-hand side of the home page.

(F) Users can enter one or more protein sequences in FASTA format and use any one of the search method options: (1) Full FASTA, (2) Sliding 8mer window, and (3) 8mer exact search.

2.2.3 AllergenOnline Database Query Result

The page displaying the entries by browsing options in the AllergenOnline database is shown in Fig. 2. It contains information about the Allergen Source (columns "Species" and "Common"

Fig. 2 Screenshot of AllergenOnline database

enlisting the scientific and common names of the source organism, respectively), IUIS Allergen Nomenclature, Type of Allergen (e.g., aero, plant, food animal, venom, or salivary), Allergen Groups and References describing the evidence of allergenicity for the group, and the length and NCBI gi number of the allergen molecule and the database version in which the specific allergen was entered. Sequences of allergens are compiled in a table under "species," shown in the left-hand column (scientific name: genus and species). The common name of the source is also listed. Each sequence is listed separately, and there can be multiple different isoforms or partial sequences for a single type of allergen (e.g., *Actinidia deliciosa* Act d 1). IUIS designation or name is shown if known. The allergen "Group" is linked to more information including the published references describing the information used to classify the protein as an allergen as well as the individual sequences clustered into the group. The gi number in the table of allergens is hyperlinked to the NCBI page to display the complete NCBI entry. For groups with multiple sequence entries, all entries and gi numbers are listed along with the publication references. The references may apply to a single sequence or to one or more sequences in the group. In some cases, they provide information of the allergenicity of the source. Additional columns supply information on the number of amino acids in the allergen protein sequence and the database version in which the specific allergen was entered. In case of sequence query search, the expected result is the best hit protein name based on high *Z* score, percentage of identity, and similarity values.

2.3 IUIS Allergen Nomenclature Subcommittee Database

The IUIS subcommittee has proposed a unique, unambiguous, and systematic nomenclature of well-characterized allergenic proteins published in peer-reviewed journals and maintains a database, which is available at http://www.allergen.org/. This database contains all the allergens officially approved by World Health Organization (WHO) and IUIS [8–10]. The menus are on the top side of the page and are interlinked. Following is the brief description of menus.

2.3.1 Description of the IUIS Allergen Nomenclature Subcommittee Database

Home

It links to the home page of IUIS Allergen Nomenclature Subcommittee database. It contains a brief description of the database and also allows users to search the database by allergen name and source.

Search

It links to query page and allows users to search by (1) IUIS name of the allergen, (2) allergen source (scientific name or common name), and (3) major taxonomic group in the form of drop-down box for example "Plantae Liliopsida," which can further be filtered by orders from respective drop-down menu. Figure 3a shows the search form for the database.

Tree View

This menu links to "Tree view" page that has an updatable list of allergens with their official nomenclature arranged according to Linnaean system, viz. kingdoms—Plantae, Fungi, and Animalia; each is further subdivided into relevant orders containing link to the list of allergenic source organisms.

Publications

This page contains the allergen nomenclature publication list.

Standardization

This page contains WHO/IUIS allergen standardization committee member list.

Executive Committee

This page contains IUIS executive committee member address list including chairman, secretary, and committee members.

Submission Form

This page allows users to submit a new allergen to the IUIS allergen nomenclature database.

Log-In

It allows members to log in to the IUIS database.

2.3.2 Usage of the IUIS Allergen Nomenclature Subcommittee Database

(A) Users can search by allergen name and by allergen source (common or scientific name).

(B) Alternatively, the major taxonomic group of the source organism may be selected from the drop-down box to retrieve a list of allergenic molecules from organisms belonging to that group. An example of such a list is shown in Fig. 3b. This list may be filtered by choosing the taxonomic order of the organism from the next drop-down menu.

(C) Users can get detailed information of each allergen molecule by clicking on the allergen name.

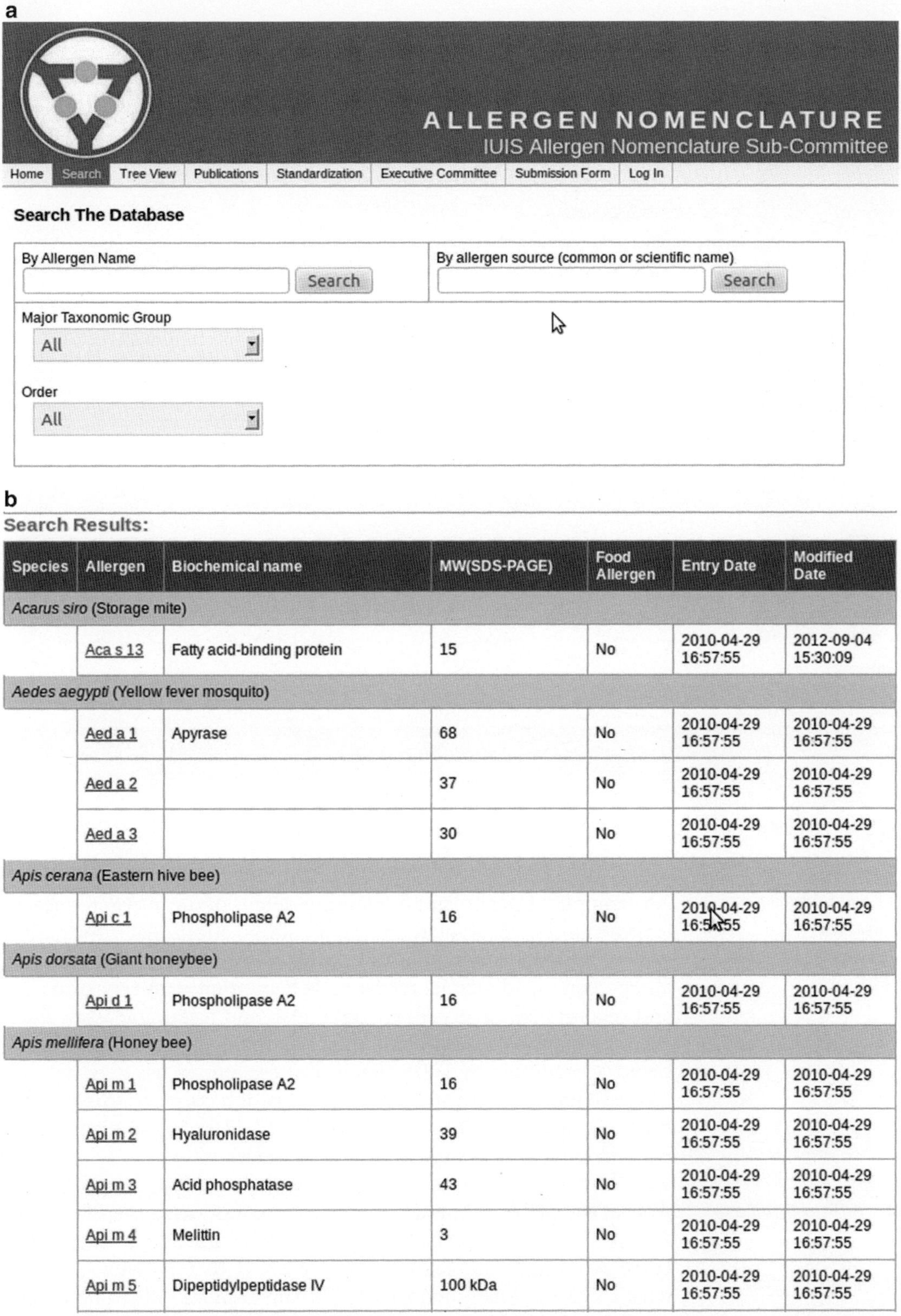

Species	Allergen	Biochemical name	MW(SDS-PAGE)	Food Allergen	Entry Date	Modified Date
Acarus siro (Storage mite)						
	Aca s 13	Fatty acid-binding protein	15	No	2010-04-29 16:57:55	2012-09-04 15:30:09
Aedes aegypti (Yellow fever mosquito)						
	Aed a 1	Apyrase	68	No	2010-04-29 16:57:55	2010-04-29 16:57:55
	Aed a 2		37	No	2010-04-29 16:57:55	2010-04-29 16:57:55
	Aed a 3		30	No	2010-04-29 16:57:55	2010-04-29 16:57:55
Apis cerana (Eastern hive bee)						
	Api c 1	Phospholipase A2	16	No	2010-04-29 16:5[illegible]55	2010-04-29 16:57:55
Apis dorsata (Giant honeybee)						
	Api d 1	Phospholipase A2	16	No	2010-04-29 16:57:55	2010-04-29 16:57:55
Apis mellifera (Honey bee)						
	Api m 1	Phospholipase A2	16	No	2010-04-29 16:57:55	2010-04-29 16:57:55
	Api m 2	Hyaluronidase	39	No	2010-04-29 16:57:55	2010-04-29 16:57:55
	Api m 3	Acid phosphatase	43	No	2010-04-29 16:57:55	2010-04-29 16:57:55
	Api m 4	Melittin	3	No	2010-04-29 16:57:55	2010-04-29 16:57:55
	Api m 5	Dipeptidylpeptidase IV	100 kDa	No	2010-04-29 16:57:55	2010-04-29 16:57:55

Fig. 3 Screenshots of IUIS Allergen nomenclature subcommittee database (**a**) query submission form; (**b**) results of a broad search with only the major taxonomic group of the source organism selected from the drop-down menu; (**c**) the page containing detailed information about a particular allergenic molecule

c

Animalia Arthropoda > Astigmata > *Acarus siro* > Aca s 13

Allergen Details:

Allergen name:	Aca s 13
Lineage:	Source: Animalia Arthropoda Order: Astigmata Species: Acarus siro (Storage mite)
Biochemical name:	Fatty acid-binding protein
MW(SDS-PAGE):	15
Allergenicity:	3 out of 13 (23%) A. siro RAST-positive patients showed strong IgE binding to rAca s 13 on immunoblot
Allergenicity ref.:	10474032
Food allergen:	No
Date Created:	2010-04-29 16:57:55
Last Updated:	2012-09-04 15:30:09
Submitter Info:	
Name:	
Institution:	
City:	
Date:	

Comments

Table of IsoAllergens:

+/-	Isoallergen and variants	GenBank Nucleotide	UniProt	PDB
▷	Aca s 13.0101	AJ006774	O76821	

Fig. 3 (continued)

2.3.3 IUIS Allergen Nomenclature Subcommittee Database Query Result

An example of the output containing details of the allergen queried is shown in Fig. 3c. The search result shows additional information such as biochemical name, molecular weight of the allergen, and allergenicity evidence in terms of IgE binding property of native and recombinant allergen, basophil test, and histamine assay. Each entry may also contain a list of isoallergens approved and numbered accordingly by IUIS. Each of these entries has external link to GenBank, UniProt sequence data through corresponding accession number, and, if available, the PDB IDs. These entries also contain external links to PubMed references.

2.4 AllFam Database

2.4.1 Description of the AllFam Database

AllFam is available online at http://www.meduniwien.ac.at/allergens/allfam/. All the allergens in AllFam were assigned and classified to corresponding Pfam families. There are 1,091 allergens, out of which 995 were assigned to 186 AllFam families [11]. The menus are on the left side of the page and are interlinked. Following is the brief description of menus.

AllFam Home

It links to home page of AllFam, a database of allergen families. It describes AllFam statistics and how to use AllFam database and AllFam news.

Browse/Search AllFam

This page links to query page of the database. It allows users to search by two options: (1) Get AllFam family chart and (2) search by allergen families. Detailed information about query search of AllFam database is available in Subheading 2.4.2.

About AllFam

This page links to information about AllFam database including background, how AllFam was created, how to cite AllFam, and the AllFam team.

FAQ

This page links to information about AllFam construction and algorithms, AllFam user interface, problems, and errors.

Papers Citing AllFam

This page links to papers citing AllFam database.

2.4.2 Usage of the AllFam Database

(A) The menu "Browse/search AllFam" allows users to access the query submission form as shown in Fig. 4a. There are two search options: (1) Get AllFam family chart and (2) search by allergen families.

(B) In the "Get AllFam family chart" option, users can browse all the AllFam data by clicking in the "Browse AllFam" button. The output result is a list of all the protein family names along with the number of allergens in each family. The search can be restricted by allergen source or route of exposure to be selected from the respective drop-down menu.

(C) Users can get detailed information for each allergen by clicking on the allergen name in the listed allergen name.

(D) In "Search for allergen families" option, users can search AllFam database by Pfam ID, AllFam ID, and keywords. For example, the keywords "inhalant fungal allergens" gave output with a list of 64 AllFam families containing total 132 allergens.

2.4.3 AllFam Database Query Result

The AllFam Allergen Family Chart output page is shown in Fig. 4b. Each allergen family has two links. The one with the "Fact sheet" links to the page containing information about corresponding Pfam ID, biochemical properties of the allergenic protein, and their allergological significance along with references as shown in Fig. 4c. "List allergens" links to a new page displaying the list of allergens reported under that specific allergen family as shown in Fig. 4d. In that list, the allergens are arranged with their corresponding IUIS name, source organism, and routes of exposure. In addition, the output page links to Allergome and IUIS databases.

2.5 Structural Database of Allergenic Proteins Database

2.5.1 Description of the SDAP Database

The SDAP database is available from https://fermi.utmb.edu/SDAP/ and contains information of 1,526 allergens, out of which 92 allergens have PDB structures [12, 13]. It is free for academic and nonprofit use; however licenses for commercial use can be obtained by contacting W. Braun (webraun@utmb.edu). This database also provides prediction tools for allergens including FAO/WHO allergenicity test and IgE epitopes. The menus are on the left side of the page. Following is the brief description of menus.

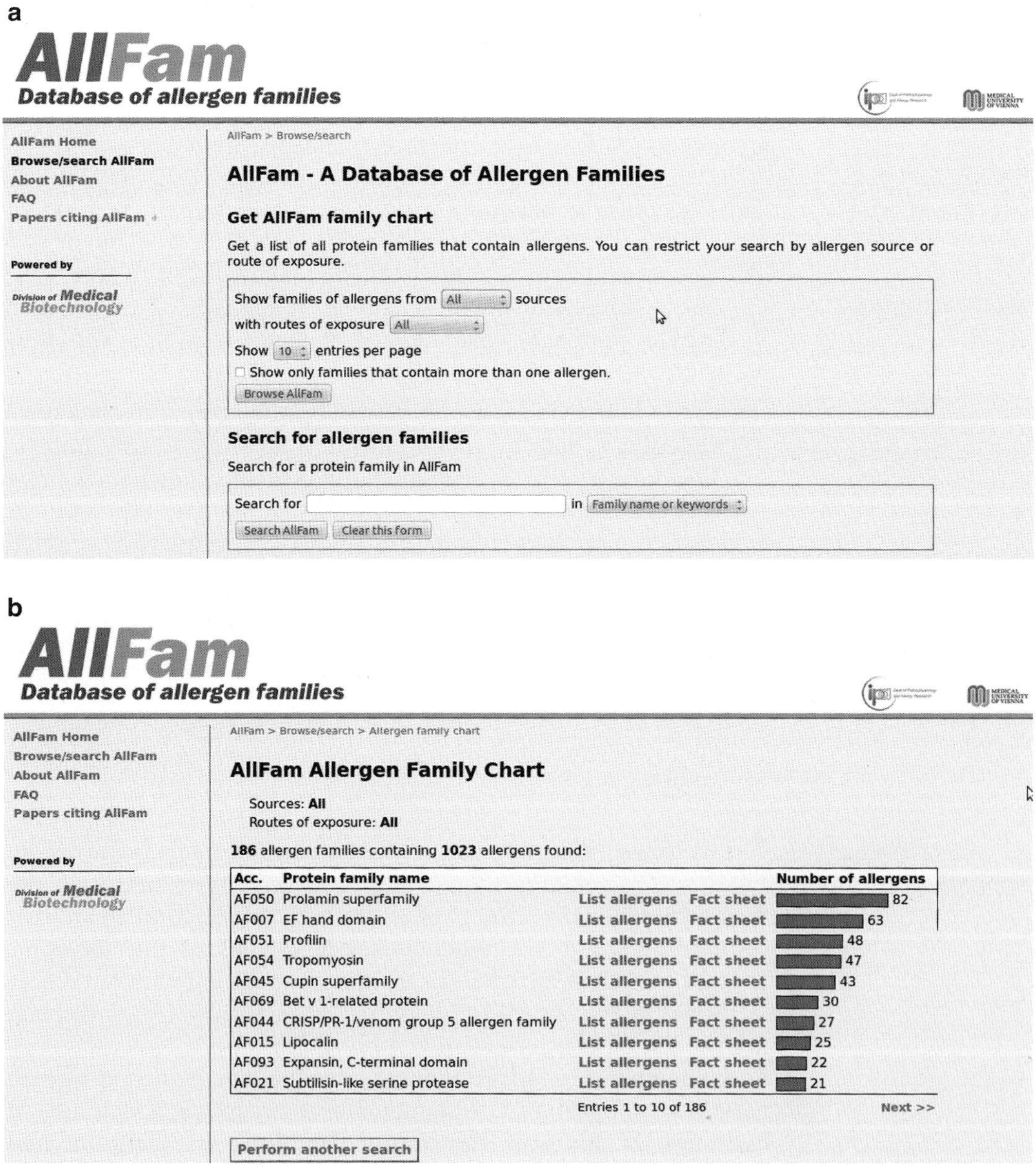

Fig. 4 Screenshots of AllFam (**a**) query submission form; (**b**) output page showing AllFam Allergen Family Chart; (**c**) "Fact sheet" of a protein family; (**d**) page containing list of all allergens from a particular protein family

c

AF050: Prolamin superfamily

Pfam domains

PF00234: Protease inhibitor/seed storage/LTP family

Biochemical properties

The prolamin superfamily derives its name from the alcohol-soluble proline and glutamine rich storage proteins of cereals. Members of the this family are characterized by the presence of an α-helical globular domain that contains a conserved pattern of six or eight cysteine residues that form three or four intra-molecular disulfide bonds [1]. Apart from the conserved cysteine pattern, there exist litte sequence similarities between members of different subfamilies. Members of the prolamin superfamily include the cereal prolamin seed storage proteins and several families of disulfide-rich small proteins. The prolamin seed storage proteins (gliadins and glutenins) contain a repetitive coiled-coil domain rich in proline and glutamine residues and a globular disulfide-rich domain. Families of low molecular weight sulfur-rich proteins are the grain softness proteins, indolines, non-specific lipid transfer proteins, soybean hydrophobic protein, bifunctional α-amylase/protease inhibitors, and 2S albumin seed storage proteins.

Allergological significance

Several families that belong to the prolamin superfamily were described as allergens.

Cereal prolamins: These proteins rarely account for allergic reactions. IgE reactivity to these proteins was observed in patients with wheat-induced atopic dermatitis or exercise-induced anaphylaxis [2].

2S albumins: The 2S albumins are a major group of seed storage proteins from a botanically diverse range of dicotyledonous plants. Many of the seed and tree nut allergens belong to the 2S albumins such as Sin a 1 from yellow mustard, Ber e 1 from Brazil nut, Jug r 1 from English walnut, and Ara h 2 and Ara h 6 from peanut [1].

Non-specific lipid transfer proteins: nsLTPs have been suggested to mediate the transfer of phospholipids between vesicles and membranes. However, plants have used the three-dimensional scaffold of the nsLTPs in a promiscuous fashion and many nsLTPs are not able to transfer lipids. Instead, they may play a role in plant defense against fungi and bacteria. nsLTPs are found in high concentrations in epidermal tissues of fruits. Hence, they are major allergens of fruits from the Rosaceae family. In addition, allergenic nsLTPs were found in nuts, seeds, vegetables, pollen and *Hevea brasiliensis* latex [3, 4].

d

AllFam

Database of allergen families

AllFam Home
Browse/search AllFam
About AllFam
FAQ
Papers citing AllFam

Powered by
Division of Medical Biotechnology

AllFam > Browse/search > Allergen list

AllFam Allergen List

AF050: Prolamin superfamily [view allergen family fact sheet]

Sources: **All**
Routes of exposure: **All**

82 allergens found (sorted by allergen name)

Name			Source	Kingdom	Routes of exposure
Amb a 6	ALLERGOME	IUIS	*Ambrosia artemisiifolia* (short ragweed)	Plants	Inhalation
Ana o 3	ALLERGOME	IUIS	*Anacardium occidentale* (cashew)	Plants	Ingestion
Api g 2	ALLERGOME	IUIS	*Apium graveolens* (celery)	Plants	Ingestion
Ara h 2	ALLERGOME	IUIS	*Arachis hypogaea* (peanut)	Plants	Ingestion
Ara h 6	ALLERGOME	IUIS	*Arachis hypogaea* (peanut)	Plants	Ingestion
Ara h 7	ALLERGOME	IUIS	*Arachis hypogaea* (peanut)	Plants	Ingestion
Ara h 9	ALLERGOME	IUIS	*Arachis hypogaea* (peanut)	Plants	Ingestion
Ara t 3	ALLERGOME		*Arabidopsis thaliana* (mouse-ear cress)	Plants	Inhalation
Art v 3	ALLERGOME	IUIS	*Artemisia vulgaris* (mugwort)	Plants	Inhalation
Ber e 1	ALLERGOME	IUIS	*Bertholletia excelsa* (Brazil nut)	Plants	Ingestion

Fig. 4 (continued)

SDAP Home Page It links to the home page of SDAP, which allows browsing of allergens alphabetically and provides links for citation and recent developments. At the top of the page, it also provides option to go and search "SDAP all proteins" and "SDAP food allergens."

Fig. 5 Screenshots of SDAP (**a**) query submission page; (**b**) output page showing alphabetical listing of allergens; (**c**) sample output page containing detailed information on the allergen molecule

c

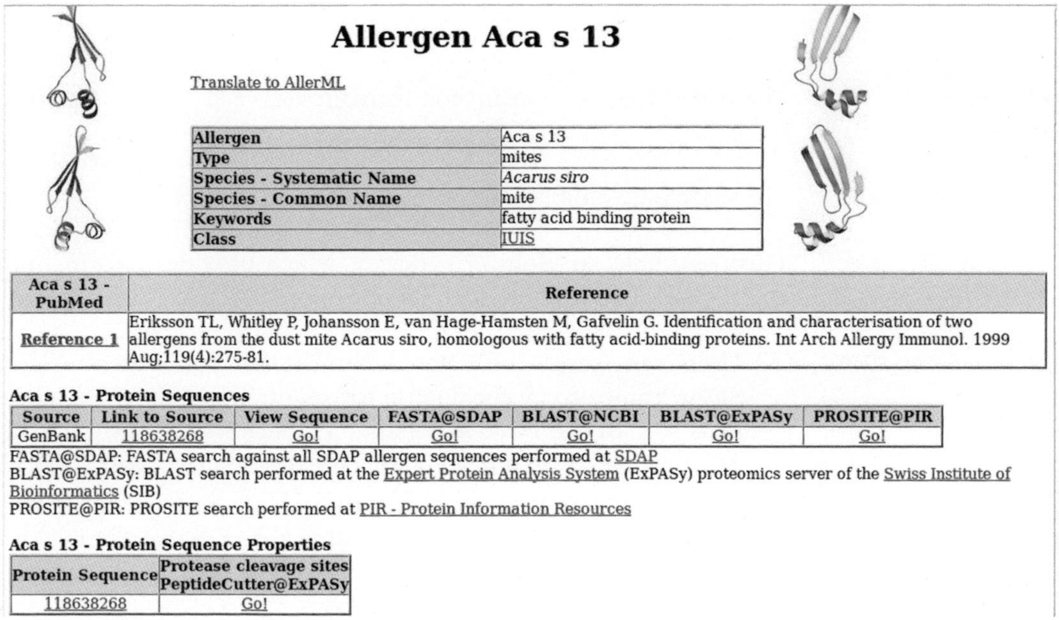

Allergen Aca s 13

Translate to AllerML

Allergen	Aca s 13
Type	mites
Species - Systematic Name	*Acarus siro*
Species - Common Name	mite
Keywords	fatty acid binding protein
Class	IUIS

Aca s 13 - PubMed	Reference
Reference 1	Eriksson TL, Whitley P, Johansson E, van Hage-Hamsten M, Gafvelin G. Identification and characterisation of two allergens from the dust mite Acarus siro, homologous with fatty acid-binding proteins. Int Arch Allergy Immunol. 1999 Aug;119(4):275-81.

Aca s 13 - Protein Sequences

Source	Link to Source	View Sequence	FASTA@SDAP	BLAST@NCBI	BLAST@ExPASy	PROSITE@PIR
GenBank	118638268	Go!	Go!	Go!	Go!	Go!

FASTA@SDAP: FASTA search against all SDAP allergen sequences performed at SDAP
BLAST@ExPASy: BLAST search performed at the Expert Protein Analysis System (ExPASy) proteomics server of the Swiss Institute of Bioinformatics (SIB)
PROSITE@PIR: PROSITE search performed at PIR - Protein Information Resources

Aca s 13 - Protein Sequence Properties

Protein Sequence	Protease cleavage sites PeptideCutter@ExPASy
118638268	Go!

Fig. 5 (continued)

SDAP Overview

This page provides information about the content of SDAP database including the lists of allergens, list of allergens with protein sequences, list of allergens with PDB structures, list of allergens with 3D models, list of allergens with IgE epitopes, and list of allergens with Pfam classes. This page also allows browsing of allergens alphabetically.

Use SDAP (SDAP All and SDAP Food)

These menus allow users to search SDAP all allergens and SDAP food allergens. Detailed information about its usage is available in Subheading 2.5.2.

SDAP Tools

There are many important web tools including FAO/WHO allergenicity test, FASTA search in SDAP, peptide match, peptide similarity, peptide-protein PD index, AllerML (markup languages for allergens), and SDAP list available in this links.

About SDAP

It links to pages about general information, manual, FAQ, publications list, team, and advisory board members of SDAP database.

Allergy Links

This page links to other important allergy-related databases.

Other Software Tools

There are many other software tools available including homology modelling, energy minimizations, calculation of solvent-accessible areas, and mapping of conformational epitopes.

Protein Databases Link

These are links to important protein databases including PDB, NCBI-Entrez, SWISS-PROT, and PIR.

Protein Classification Link

These are links to important protein classifications including CATH, ProtoMap, TOPS, and VAST.

Link to Bioinformatics Servers

These are links to popular bioinformatics servers.

Link to Bioinformatics Tools

These are links to macromolecular structural views tools.

Other Bioinformatics Links

This page links to bioinformatics.ca, which provides information about Canadian bioinformatics workshops.

2.5.2 Usage of the SDAP Database

(A) Users can search the database by the left panel menu "Use SDAP" and also by clicking on top menu links "SDAP All allergens" or "SDAP Food allergens." The snapshot of the "SDAP-All allergens" search page is shown in Fig. 5a (*see* **Note 2**).

(B) The query search allows users to search a term or a phrase. It provides a filters search by choosing any of the selected fields: Allergen—scientific name; Source—scientific name; Source—common name; Allergen description.

(C) It also allows users to browse the data according to the first letter of the allergen name arranged alphabetically from the home page.

(D) Users can get full detailed information about sequence and structure of each allergen by clicking in the allergen name from the search results.

2.5.3 SDAP Database Query Result

Each search result will appear as a tabular list of allergens along with their homologues in a new page. The list contains preliminary information on allergens including its IUIS status and the biochemical nature of the protein under the heading "Keywords." All information about the allergens starting with the alphabet "A" is displayed in Fig. 5b. Users can browse more detailed information of each allergen; for example information on "allergen Aca s 13" is shown in Fig. 5c.

3 Notes

1. The users can download all the entries from the search result, registration may be required, and users need to choose a username and a password.
2. The default values set for query search were the optimized values and can be changed by the users.
3. The database searching is not case sensitive.

References

1. Galli JS (2000) Allergy. Curr Biol 10(3): R93–R95
2. Kay AB (2000) Overview of allergy and allergic diseases: with a view to the future. Br Med Bull 56(4):843–864
3. Ghosh D, Gupta Bhattacharya S (2011) Allergen bioinformatics: recent trends and developments. In: Xia X (ed) Selected Works in Bioinformatics. InTechOpen, Croatia
4. Mari A, Rasi C, Palazzo P et al (2009) Allergen databases: current status and perspectives. Curr Allergy Asthma Rep 9:376–383
5. Mari A, Scala E, Palazzo P et al (2007) Bioinformatics applied to allergy: allergen databases, from collecting sequence information to data integration. The allergome platform as a model. Cell Immunol 244: 97–100
6. Mari A, Scala E (2006) Allergome: a unifying platform. Arb Paul Ehrlich Inst Bundesamt Sera Impfstoffe Frankf A M 95:29–39, discussion 39–40
7. Goodman RE (2006) Practical and predictive bioinformatics methods for the identification of potentially cross-reactive protein matches. Mol Nutr Food Res 50:655–660
8. Marsh DG, Goodfriend L, King TP et al (1986) Allergen nomenclature. Bull World Health Organ 64:767–774
9. Larsen JN, Lowenstein H (1996) Allergen nomenclature. J Allergy Clin Immunol 97: 577–578
10. Chapman MD, Pomés A, Breiteneder H et al (2007) Nomenclature and structural biology of allergens. J Allergy Clin Immunol 119:414–420
11. Radauer C, Bublin M, Wagner S et al (2008) Allergens are distributed into few protein families and possess a restricted number of biochemical functions. J Allergy Clin Immunol 121:847–852
12. Ivanciuc O, Schein CH, Braun W (2003) SDAP: database and computational tools for allergenic proteins. Nucleic Acids Res 31(1): 359–362
13. Ivanciuc O, Schein CH, Braun W (2002) Data mining of sequences and 3D structures of allergenic proteins. Bioinformatics 18(10): 1358–1364

Part III

Tools for Prediction

Chapter 10

Prediction of Conformational B-Cell Epitopes

Wen Zhang, Yanqing Niu, Yi Xiong, and Meng Ke

Abstract

Conformational B-cell epitopes play an important role in the epitope-based vaccine design. The increase of available data promotes the development of computational methods. Compared with the wet experiments, the computational methods are faster and more economic. In the past few years, a number of computational methods (especially the machine learning-based methods) have been developed to predict the conformational B-cell epitopes. In this chapter, we introduce important data resources and computational methods, which are publicly available. Moreover, we introduce our ensemble learning-based method that can predict the conformational epitopes from sequences. These promising methods may assist immunologists in identifying potential vaccine candidates.

Key words Conformational B-cell epitopes, Machine learning, Epitope-based vaccine design

1 Introduction

Antigen–antibody interaction is a critical event in the immune process, which may elucidate the underlying mechanism of immune recognition [1–4]. The sites on antigens recognized and bound by B cell-produced antibodies are well known as B-cell epitopes. The location of B-cell epitopes is useful for synthesizing peptides that can elicit the immune response with specific cross-reacting antibodies. For this reason, the identification of B-cell epitopes facilitates the design of the potentially safer peptide-based vaccines. B-cell epitopes can be classified into two categories: linear (continuous) epitopes and conformational (discontinuous) epitopes. Linear epitopes are formed by continuous amino acid sequences, while conformational epitopes consist of residues that are distantly separated in the sequences but spatially proximal.

In the last decade, the increase of available data promotes the development of computational methods, which may be fast and economic [5]. Although the majority of all epitopes (about 90 %) are conformational, the study began fairly late. In the prediction work, there are several definitions ever used for the conformational

Rajat K. De and Namrata Tomar (eds.), *Immunoinformatics*, Methods in Molecular Biology, vol. 1184,
DOI 10.1007/978-1-4939-1115-8_10, © Springer Science+Business Media New York 2014

epitopes inferred from the X-ray structures of antigen–antibody complexes. By definitions, the epitope residue is an antigen residue with area loss upon antibody binding more than a given threshold or an antigen residue separated from any antibody residue by a Euclidean distance less than 4 Å. The study revealed that these definitions do not make significant difference. Here, we must emphasize that the epitopes in the computational work are not functional but structural, and structural epitopes cannot definitely lead to the immune response. Currently, the prediction of functional epitopes is a tough task. Thereafter, the epitopes mean the structural epitopes.

Although some protein docking methods (such as Patch Dock [6] and ClusPro [7]) can be used to predict conformational epitopes, these methods are different from those which are specially designed for the conformational epitope prediction. The docking methods require the structures of both antigens and antibodies to make prediction, while the specially designed methods attempt to predict the epitopes from antigens in the absence of antibodies.

CEP is the pioneering method proposed for the prediction of conformational epitopes [8], which uses the residue solvent accessibility. DiscoTope [9] exploits the surface accessibility, spatial information, and amino acid statistics information to identify epitopes. PEPITO [10] combines amino acid propensities and half-sphere exposure values at multiple distances to make prediction. ElliPro [11] uses Thornton's propensities and residue clustering to make prediction. In SEPPA [12], two concepts, "unit patch of residue triangle" and "clustering coefficient," are introduced to describe the local spatial context and spatial compactness. EPITOPIA [13] combines structural and physicochemical features and then adopts naive Bayes classifier to make prediction. EPCES [14] uses the consensus score of several structural and physicochemical terms. EPSVR [15] uses support vector machine (SVM) and combines various features for prediction. EPMeta [16] is a meta method combining the predictions from several existing servers. Liu et al. [17] adopted the logistic regression to predict the conformational epitopes. We [18] proposed a random forest-based method by dealing with the imbalanced dataset and combining various features. Above methods construct the prediction models based on antigen structures.

Although a great number of structure-based methods have been developed, their application is undermined by the limited number of available structures, and the experimental techniques that determine structures are costly and time consuming. Instead of making predictions from structures, Ansari et al. made the first attempt on sequence-based conformational epitope prediction [19]. Gao et al. developed a method based on averaging selected scores generated from sliding 20-mers by SVMs [20]. Recently, we proposed an ensemble learning model using the antigen sequences [21].

2 Materials

2.1 Database

(a) Immune Epitope Database (IEDB) (http://www.iedb.org/) [22] can provide a highly annotated set of B-cell epitopes curated from crystal structures of antigen–antibody complexes.

(b) Conformational Epitope Database (CED) (http://immunet.cn/ced/) [23] collected the conformational epitopes thoroughly sourced from articles published in the peer-reviewed journals. Initially, references were obtained by exhaustive querying on PubMed and ScienceDirect. The references were further manually filtered to annotate conformational epitopes.

(c) AntiJen [24] is a database with the published experimentally determined conformational B-cell epitopes (http://www.ddg-pharmfac.net/antijen/).

2.2 Dataset

In the conformational epitope prediction, the antigen–antibody complexes are analyzed to annotate the binding sites (epitope residues) on the antigens, and then only the antigens (structures or sequences) are used to develop the prediction models.

Several datasets are widely used in the conformational epitope prediction. The structure datasets can be classified into two kinds: bound dataset and unbound dataset. A bound dataset consists of the antigen–antibody complex structures, and the epitopes on antigen are annotated according to the definition of the conformational epitope. Then, the structures of the antigens are directly extracted from the complexes for modeling. An unbound dataset consists of complex structures and unbound structures of antigens. Annotated epitope residues on complexes (calculated according to the definition) are aligned to the residues on unbound structures of antigens. Then, the unbound antigen structures are used for modeling. One popular bound dataset is published by Rubinstein, which consists of 66 non-redundant complex structures, available at http://epitopia.tau.ac.il/trainData/. Liang's unbound dataset including 48 complexes and the unbound structures of antigens are available at http://sysbio.unl.edu/services/. The antigen sequences can be extracted from the antigen–antibody complexes for the sequence-based prediction. Ansari et al. published benchmark sequence datasets available at http://www.imtech.res.in/raghava/cbtope/supple.php.

3 Method

In this section, we introduce widely used conformational epitope prediction methods and their public servers (*see* **Note 1**).

3.1 DiscoTope

DiscoTope [9] is a structure-based method for conformational epitope prediction. The method uses the amino acid propensity (Parker hydrophilicity scale), spatial information (contact numbers), and surface accessibility to make prediction.

Parker hydrophilicity scale is an amino acid propensity, which can be obtained from AAIndex database. The residue contact number is the number of Cα atoms in the antigen within a distance of 10 Å of the residue Cα atom. The relative solvent-accessible surface area per antigen residue is calculated using the NACCESS program with a probe radius of 1.4 Å.

Given an antigen–antibody complex structure, the contact number score and surface accessibility score of each antigen residue are calculated. Here, the Parker hydrophilicity score of each residue is calculated over a smoothing window of seven residues. For a candidate residue, the weighted sum of the Parker hydrophilicity score, contact number score, and surface accessibility score is used for prediction. According to a preset threshold, the residue is predicted as epitope or non-epitope.

The web server of DiscoTope is available at http://www.cbs.dtu.dk/services/DiscoTope/. The users can use the PDB IDs of antigen–antibody complexes or the PDB files as input, and the server will return the prediction results. Users can specify the threshold for epitope identification.

3.2 EPITOPIA

For a given structure, a patch of 20 amino acids is constructed around each solvent-accessible antigen residue. Rubinstein et al. statistically evaluated a wide range of amino acid physicochemical and structural-geometrical properties [13]. These properties are (1) the ratio between the frequencies of some amino acid types in the patch and the remaining antigen surface, (2) the ratio between the frequency of helix secondary structures in the patch and the remaining antigen surface, (3) the average relative accessibility of the patch to the solvent, (4) the average accessibility of the patch, (5) the average curvature of the patch atoms, and (6) several amino acid propensities.

Then, Rubinstein et al. use the feature selection technique to obtain the optimal property subset. Starting with all properties, one property for which the deletion had the least effect on prediction accuracy is removed at each iteration. Finally, the subset of properties with the highest number of successful predictions was selected as the optimal set. The optimal property subset is used to represent patches as feature vectors. Then, naive Bayes is used as the classification engine to build prediction model. Thus, a server named "Epitopia" is constructed to predict conformational epitopes.

Epitopia is available at http://epitopia.tau.ac.il. Users can enter the PDB ID or upload the PDB file for prediction.

3.3 EPCES

In EPCES [14], a patch (with 20 residues) is formed around each candidate antigen residue. EPCES uses consensus score from six different scoring terms to make prediction. These scoring terms are residue epitope propensity, conservation score, side-chain energy score, contact number, surface planarity score, and secondary structure composition.

The residue epitope propensity was calculated as the product of the normalized solvent-accessible surface of the residue and the logarithm ratio of the epitopic area to the rest area. The conservation score was calculated by the position-specific substitution matrix generated from PSIBLAST and the diagonal element of BLOSUM62. The side-chain energy score was calculated from the side-chain energies of all possible rotamers. The contact number is as same as the introduction in Subheading 3.1. The planarity of each patch was calculated as the root mean squared deviation of all the Cα atoms in the patch from the least squares plane through the atoms. The secondary structure composition was the fraction of patch residues forming turns or loops in all 20 patch residues.

For each candidate residue, the residue epitope propensity, conservation score, and side-chain energy score were calculated at the residue level and distance-based averaged over all residues in the patch by following distance-based equation

$$E_{\text{patch}}(i) = \sum_{k=1}^{20} E_{\text{residue}}(K) \cdot e^{\frac{-d}{T}}$$

where $E_{\text{residue}}(K)$ is the score of residue K in the patch, d is the distance between K and the central residue of the patch, and T is the parameter needed to be optimized.

Each scoring term can predict a candidate residue as epitope or non-epitope according to its score and a given threshold. For a residue, if more than five scoring terms yield the scores greater than a given threshold, it is finally predicted as the epitope residue.

A web-based EPCES application is available at http://sysbio.unl.edu/services/EPCES/. The PDB ID of an unbound structure or the PDB file is used as the input. The output will be displayed on this web page when the prediction is completed. The output includes the predicted antigen residue and its possibility of being an epitope residue.

3.4 EPSVR

EPSVR [15] uses a support vector regression (SVR) method to integrate six scoring terms ever used in the EPCES.

For each surface patch, the number of epitopic residues could be any integer value between 0 and the patch size (i.e., 20). Therefore, each patch is assigned a real value associated with the number of epitopic residues, and the prediction of conformational epitopes is transformed as a problem of regression. Each surface patch had six SVR attributes, whose values were calculated with the six scoring terms: residue epitope propensity, conservation score, side-chain energy score, contact number, surface planarity score, and secondary structure composition. The six scores and the number of observed epitope residues in the patch were scaled to 0–1. Then, the SVR-based model is built to make prediction.

The web server of EPSVR is available at http://sysbio.unl.edu/EPSVR/. The input and output of EPSVR are same as those of EPCES.

3.5 CBTOPE

CBTOPE [19] is the first method of predicting conformational B-cell epitopes from antigen sequences. The fixed-length window is shifted over the antigen sequences to generate residue segments (peptides). According to the central residues (epitope or non-epitope), the peptides are labeled as positive or negative. Then, each peptide can be represented as a feature vector by several encoding schemes, including binary profile, physicochemical profile, and composition profile.

Binary profile represents each amino acid as a 21-dimensional vector. Physicochemical profile uses Grantham polarity, Karplus–Schulz flexibility, Kolaskar antigenicity, Parker hydrophobicity, and Ponnuswami polarity index to represent amino acids. Amino acid composition is the percentage of each amino acid type in a peptide. Three encoding schemes are used for peptide representation, and the prediction models are constructed by using SVM. Among all encoding schemes, the composition profile can produce the best results.

A web server CBTOPE has been developed to predict conformational epitopes, available at http://www.imtech.res.in/raghava/cbtope/. Users can enter antigen sequences for prediction.

4 The Sequence-Based Ensemble Learning Method

We follow the work pioneered by CBTOPE and focus on two aspects concerning the sequence-based prediction [21]. One is to explore more potential sequence-derived features relevant to conformational epitopes. The other is to effectively use various features which may share redundant information. In order to address these issues, we evaluate several sequence-derived features, which are ever used in the epitope prediction or similar tasks. Second, we consider the ensemble learning technique that can incorporate useful features, and the weighted scoring approach is adopted to build the prediction model.

4.1 The Basic Idea of Ensemble Learning Method

The overlapping residue segments (peptides) are generated from the antigen sequences by using a sliding window of the length L. For simplifying, let L be an odd integer. For a sequence with N residues, a total of $N-L+1$ peptides are extracted, and each peptide is labeled as positive or negative according to the label of its central residue (epitope residue or non-epitope residue). The prediction of conformational epitopes from sequences is formulated as the problem of binary classification. We consider several sequence-derived features, which are described as follows.

Physicochemical propensities: These physicochemical propensities are flexibility scale, hydrophilicity scale, surface-exposed residue scale, polarity scale, beta-turn scale, and accessibility scale.

Sparse profile: Sparse profile is a widely used representation of amino acids. Each amino acid type (20 common types in all) can be represented by a 20-bit binary string, in which the value at one bit is 1 and others are 0.

Amino acid composition: According to the previous study, some amino acid types are significantly overrepresented in epitopes, and others are underrepresented; thus the amino acid composition can be used to differentiate epitope regions from non-epitope regions. Here, we use the amino acid composition of the residue segments (also called as sliding windows or samples) extracted from the whole sequences.

Amino acid function group: Since contacts between antibodies and the antigens are mostly determined through functional moieties of the R-groups, functional moieties can influence the location of antibody–antigen-binding sites. According to different R-groups, 20 amino acid types are classified into 13 classes. In order to take antigen–antibody interaction into consideration, we present a novel feature named "amino acid function group" and use 13-bit binary strings to represent 13 functional classes.

Amino acid functional composition: By incorporating both amino acid function group and amino acid composition, we present a novel feature "amino acid functional composition," which represents the percentage of each amino acid functional type in a sequence.

Evolutionary profile: The evolutionary conservation is represented by the position-specific scoring matrix (PSSM), which is obtained by aligning the target sequence against NCBI non-redundant reference sequences with PSI-BLAST tool. For an amino acid sequence with L residues, the PSSM has L rows and 20 columns. PSSM values in each row are rescaled by the standard logistic function $f(x) = 1/(1 + e^{-x})$. When using the evolutionary profile, a residue is represented by its corresponding 20-dimensional row vector in the matrix.

Amino acid pair profile: The amino acid pair profile is usually observed to be associated with the protein functions. Amino acid pair profile of a sequence represents the percentage of each amino acid pair type.

Although structural information cannot be directly obtained from antigen sequences, some state-of-the-art tools can help to predict it. Here, the SABLE program [25] is adopted, for the

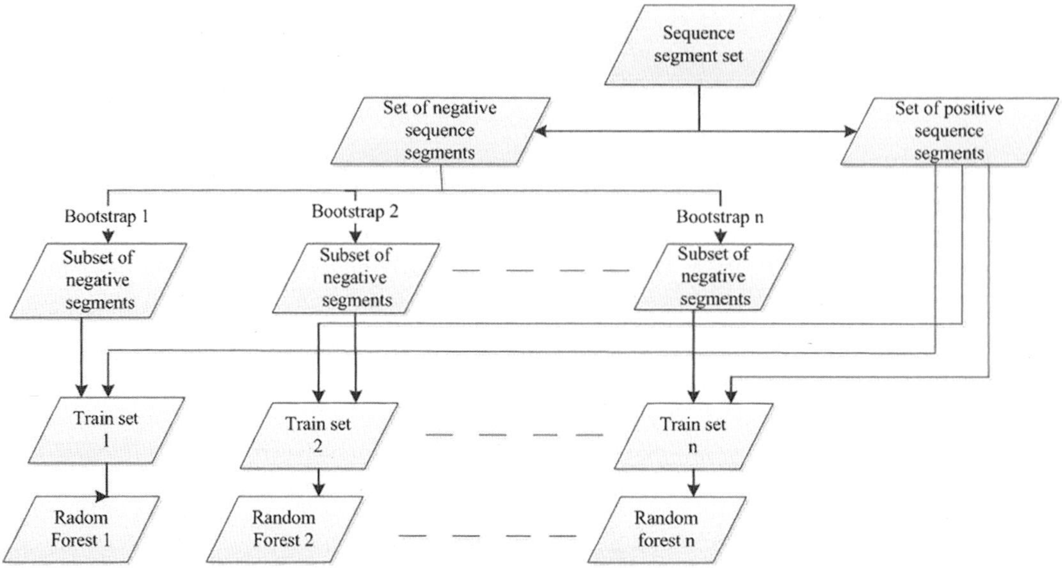

Fig. 1 The classification model based on the random forest and data bootstrap

stand-alone tool is publicly available. With the given sequences as input, the software can predict the secondary structures (SS) and relative accessible surface areas (RASA) of residues. The predicted SS of a residue is denoted as H, E, or C (helix, sheet, coil), and (1, 0, 0), (0, 1, 0), and (0, 0, 1) are, respectively, used to represent three types. The predicted RASA of a residue is a real value between 0 and 100, representing the percentage of exposed area of the residue over its full area.

The statistical study indicates that all features have the ability of differentiating epitope regions from non-epitope regions [21]. Since the amino acid functional composition incorporates both amino acid composition and amino acid group, seven groups of features including physicochemical propensities, evolutionary profile, amino acid functional composition, sparse profile, amino acid pair, sequence-predicted secondary structure, and sequence-predicted relative solvent accessibility are finally used for the development of prediction models.

Obviously, there are much more non-epitopes than epitopes, and the instances are seriously imbalanced. A strategy based on the data bootstrap is used to deal with the imbalanced data, and random forest [26] is used as the classification engine. Thus, a classification model which consists of multiple random forests is constructed (described in Fig. 1) and used as the base module for ensemble learning.

Since a peptide can be represented as different feature vectors by different descriptors (features), multiple base modules can be constructed. We adopt a simple ensemble strategy named weighted scoring [27] to integrate modules and develop the ensemble model

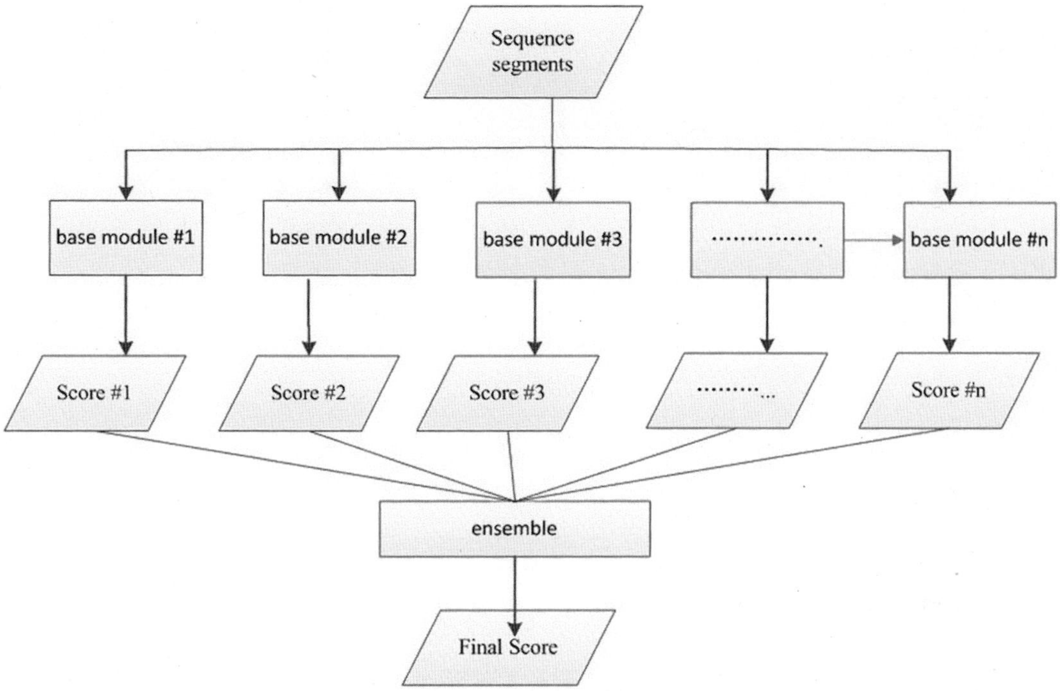

Fig. 2 The schematic diagram of ensemble model by integrating base modules

(described in Fig. 2). Given an instance, each base module will produce a score, and then these scores are normalized. Further, a weight is assigned to the normalized score yielded by a base module, and the sum of weighted scores is adopted as the final prediction (*see* **Note 2**).

4.2 The Construction of Web Server

The web server is constructed by JavaScript and Tomcat. In order to calculate the conservation score, secondary structures, and relative accessible surface areas, we have to use some external tools (i.e., PSI-BLAST and SABLE program). PSI-BLAST is a Windows version executive program; SABLE [25] is written in Perl. The outputs from external tools are parsed to obtain feature values used for sequence representation.

We adopt the Weka package [28] to implement the machine learning methods. Weka is a collection of java code implementing machine learning algorithms, including data preprocessing, classification, regression, clustering, association rules, and visualization. Here, we use the random forest class in Weka to develop our ensemble learning model. The inputs of the model are the feature vectors representing sequences, and probability of being an epitope residue is returned for each residue. The server is available at http://bcell.whu.edu.cn.

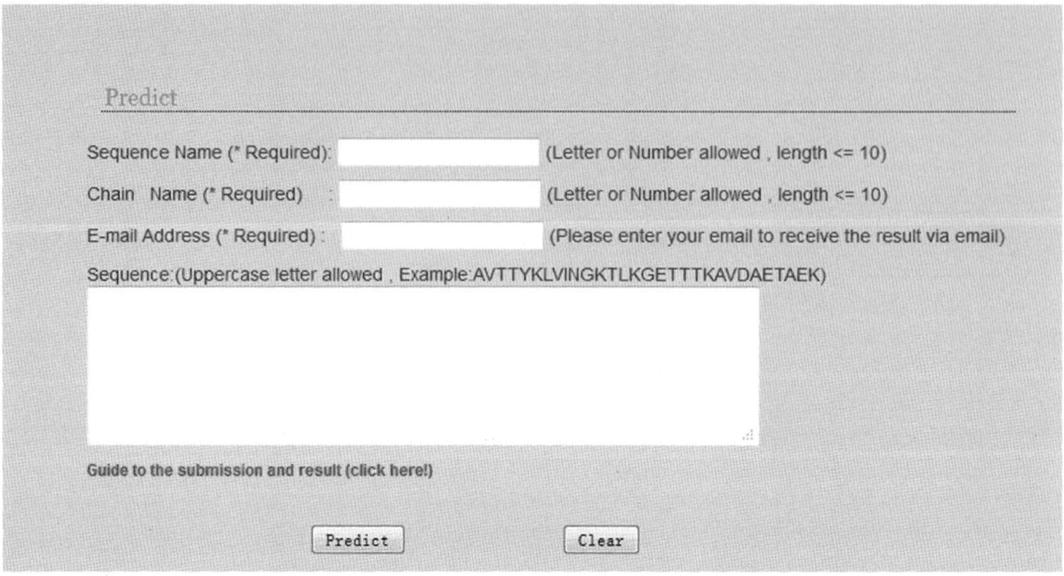

Fig. 3 The web page of the server

Result:

0	A	0.4933333333333334
1	V	0.6533333333333332
2	T	0.2866666666666667
3	T	0.24666666666666667
4	Y	0.263525434170393
5	K	0.31678109640819324
6	L	0.08668649400409438

Fig. 4 An example of the returned result

In the web page of prediction (shown in Fig. 3), users can enter an antigen sequence and its information (sequence name and chain name). In addition, the e-mail address should be specified to receive the prediction result. A typical task (a sequences of 30 residues) takes about 15–20 min. The running time depends on the length of the submitted sequence. In the returned result (shown in Fig. 4), the first column is the residue id; the second column is the residue name; and the third column is the probability for the residue to be the epitope residue.

5 Conclusion

This chapter introduces the data resources and computational methods related with the conformational B-cell epitope prediction, especially our sequence-based conformational epitope prediction method and the public server. The above-discussed methods have large potential for the practical use. The publicly available servers will assist immunologists in identifying potential vaccine candidates.

6 Notes

1. As far as we know, some structure-based methods are trained and evaluated on the bound dataset (DiscoTope, SEPPA, Epitopia), and others are constructed and tested on the unbound dataset (EPSVR, EPCES). CBTOPE and our ensemble method are developed by using antigen sequences.
2. The sequence-based ensemble learning method has some advantages. First, the ensemble model provides a flexible frame that incorporates individual feature-based classifiers. Second, the ensemble model can select the features by itself and integrate them based on the discriminative power. According to the optimal weights, we can approximately know the components of the ensemble model. Therefore, this ensemble model is easy to not only implement but also explain.

Acknowledgments

This work is supported by the National Science Foundation of China (61103126), the Ph.D. Programs Foundation of Ministry of Education of China (20100141120049), and Natural Science Foundation of Hubei Province (No. 2011CDB454).

References

1. Van Regenmortel MH (1989) The concept and operational definition of protein epitopes. Philos Trans R Soc Lond B Biol Sci 323(1217): 451–466
2. Walter G (1986) Production and use of antibodies against synthetic peptides. J Immunol Methods 88(2):149–161
3. Van Regenmortel MH (2004) Pitfalls of reductionism in the design of peptide-cased vaccines. Vaccine 19:2369–2374
4. Flower DR (2007) Immunoinformatics: predicting Immunogenicity in silico, 1st edn. Humana, Totowa, NJ
5. Blythe MJ, Flower DR (2005) Benchmarking B cell epitope prediction: underperformance of existing methods. Protein Sci 14(1): 246–248
6. http://bioinfo3d.cs.tau.ac.il/PatchDock/
7. http://cluspro.bu.edu/login.php
8. Kulkarni-Kale U, Bhosle S, Kolaskar AS (2005) CEP: a conformational epitope prediction server. Nucleic Acids Res 33(Web Server issue): W168–W171
9. Andersen PH, Nielsen M, Lund O (2006) Prediction of residues in discontinuous B cell epitopes using protein 3D structures. Protein Sci 15(11):2558–2567
10. Sweredoski MJ, Baldi P (2008) PEPITO: improved discontinuous B-cell epitope prediction using multiple distance thresholds and half sphere exposure. Bioinformatics 24(12): 1459–1460
11. Ponomarenko J, Bui HH, Li W et al (2008) ElliPro: a new structure-based tool for the prediction of antibody epitopes. BMC Bioinformatics 9:514
12. Sun J, Wu D, Xu T et al (2009) SEPPA: a computational server for spatial epitope

prediction of protein antigens. Nucleic Acids Res 37 (Suppl 2):W612–W616

13. Rubinstein ND, Mayrose I, Pupko T (2009) A machine learning approach for predicting B-cell epitopes. Mol Immunol 46(5):840–847
14. Rubinstein ND, Mayrose I, Martz E et al (2009) Epitopia: a web-server for predicting B-cell epitopes. BMC Bioinformatics 10:287
15. Liang S, Zheng D, Zhang C et al (2009) Prediction of antigenic epitopes on protein surfaces by consensus scoring. BMC Bioinformatics 10:302
16. Liang S, Zheng D, Standley DM et al (2010) EPSVR and EPMeta: prediction of antigenic epitopes using support vector regression and multiple server results. BMC Bioinformatics 11:381
17. Liu R, Hu J (2011) Prediction of discontinuous B-cell epitopes using logistic regression and structural information. J Proteomics Bioinformatics 4:10–15
18. Zhang W, Xiong Y, Zhao M et al (2011) Prediction of conformational B-cell epitopes from 3D structures by random forests with a distance-based feature. BMC Bioinformatics 12:341
19. Ansari HR, Raghava GP (2010) Identification of conformational B-cell Epitopes in an antigen from its primary sequence. Immunome Res 6:6
20. Gao J, Faraggi E, Zhou Y et al (2012) BEST: improved prediction of B-cell epitopes from antigen sequences. PLoS One 7(6):e40104
21. Zhang W, Niu Y, Xiong Y et al (2012) Computational prediction of conformational B-cell epitopes from antigen primary structures by ensemble learning. PLoS One 7(8):e43575
22. Vita R, Zarebski L, Greenbaum JA et al (2010) The immune epitope database 2.0. Nucleic Acids Res 38(1):D854–D862
23. Huang J, Honda W (2006) CED: a conformational epitope database. BMC Immunol 7:7
24. Toseland CP, Clayton DJ, McSparron H et al (2005) AntiJen: a quantitative immunology database integrating functional, thermodynamic, kinetic, biophysical, and cellular data. Immunome Res 1(1):4
25. Sable server available at: http://sable.cchmc.org/
26. Breiman L (2001) Random forests. Mach Learn 45:5–32
27. Polikar R (2006) Ensemble based systems in decision making. IEEE Circuits Syst Mag 6(3):21
28. Hall M, Frank E, Holmes G et al (2009) The WEKA Data Mining Software: an update. SIGKDD Explorations 11(1)

Chapter 11

Computational Prediction of B Cell Epitopes from Antigen Sequences

Jianzhao Gao and Lukasz Kurgan

Abstract

Computational identification of B-cell epitopes from antigen chains is a difficult and actively pursued research topic. Efforts towards the development of method for the prediction of linear epitopes span over the last three decades, while only recently several predictors of conformational epitopes were released. We review a comprehensive set of 13 recent approaches that predict linear and 4 methods that predict conformational B-cell epitopes from the antigen sequences. We introduce several databases of B-cell epitopes, since the availability of the corresponding data is at the heart of the development and validation of computational predictors. We also offer practical insights concerning the use and availability of these B-cell epitope predictors, and motivate and discuss feature research in this area.

Key words B-cell epitope, Linear epitope, Conformational epitope, Antigen, Immunotherapeutic, Vaccine, Prediction, Database

1 Introduction

One of the key aspects of an immune system is the antibody-mediated ability to identify foreign, infectious objects, such as bacteria and viruses. This is implemented through binding of the antibodies and antigens (e.g., proteins from the pathogenic entity) at sites known as B-cell epitopes. Ability to identify these binding areas in the antigen sequence or on its surface is important for the development of vaccines and immunotherapeutics [1]. The B-cell epitopes are categorized into two classes: linear/continuous and conformational/discontinuous. The former B-cell epitope is a short segment in the corresponding amino acid sequence (Fig. 1a). Majority of the B-cell epitopes are conformational, which means that they are distributed over multiple segments in the protein chain that are located in close proximity in the folded three-dimensional structure (Fig. 1b) [2].

Although several experimental techniques can be used to identify the B-cell epitopes [3], they are relatively time consuming and

Rajat K. De and Namrata Tomar (eds.), *Immunoinformatics*, Methods in Molecular Biology, vol. 1184,
DOI 10.1007/978-1-4939-1115-8_11, © Springer Science+Business Media New York 2014

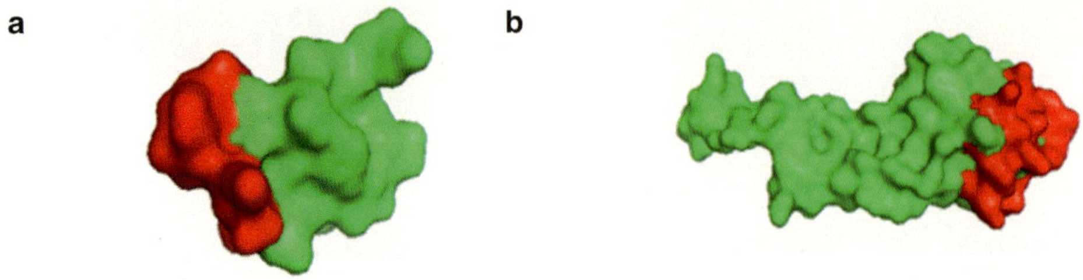

NIYNCEP**ANPSEKNSP**STQYCYSIQ

MAPMLSGLLARLVKLLLGRHGSALHWRAAGAATVLLVIVLLAGS
YLAV**LA**E**RGAPGAQ**L**ITY**P**R**ALWWSVETATTVGYGDLYPVTLWG
RCVAVVVMVAGITSFGLVTAALATWFVGREQERRGH

Fig. 1 Example linear and conformational epitopes. Panel (**a**) shows linear epitope for the B-lymphocyte antigen CD20 from *H. sapiens* (IEDB ID: 161083l; PDB ID: 3PP4:P). Panel (**b**) gives conformational epitope for the voltage-gated potassium channel from *S. lividans* (IEDB ID: 142362; PDB id: 1K4D:C). Annotations of epitopes were extracted from the Immune Epitope DataBase (IEDB) [8] and the protein structures were collected from the Protein Data Bank PDB [24]. *Red* color denotes localization of the B-cell epitope on the surface of the antigen protein and *red* and *bold font* shows the epitope in the corresponding sequence

expensive, particularly when considering to do that on large, genomic scale. Computational methods are a viable alternative to provide a fast and cost-effective way to predict the B-cell epitopes [4]. A fairly large number of computational B-cell epitope predictors, which are characterized by varying degrees of success and scope, have been developed over the last three decades [4–7]. Although progress has been accomplished in the context of the development and applications of these computational methods, much remains to be done, particularly considering modest predictive performance of these approaches (*see* **Note 1**). In parallel, a few efforts to collect, annotate, and deposit B-cell epitopes into publicly accessible databases are currently under way [8–10] and integrated resources that provide access to multiple tools for prediction and analysis of epitopes are available [11, 12]. Such efforts should make these technologies more accurate (more data allows for building more accurate predictive models) and more convenient (freely available and integrated) for the end users.

The algorithms that predict the B-cell epitopes are classified into sequence based and structure based. The structure-based methods use the three-dimensional structure of the antigen to perform the prediction, while the sequence-based methods utilize only the sequence of the antigen. While the structure-based predictors usually provide higher predictive performance when directly compared with the sequence-based methods [13–15], they are constrained to a relatively small set of targets for which the structure is available. They also suffer from a limited availability of the annotated data. Recent years have witnessed a revival of the development of the sequence-based methods, which currently are capable of finding both linear and conformational epitopes. To this end, we overview major relevant databases and summarize a comprehensive set of 17 sequence-based predictors of the B-cell epitopes, which expands over the coverage of recent predictors offered by the prior reviews [4, 6].

2 Databases of B-Cell Epitopes

Several databases that store experimentally annotated B-cell epitopes were developed over the last decade. They differ in scope and sources of data. These databases provide data that are used to develop and evaluate new and improved predictors of B-cell epitopes (*see* **Note 2**). We briefly summarize, in chronological order, six publicly available databases.

2.1 AntiJen

This repository was developed in 2001 at the Edward Jenner Institute for Vaccine Research in the UK [16]. It was later updated to version 2.0 [10, 17]. It stores experimental thermodynamic binding data concerning the interaction of peptides including B-cell receptors, T-cell receptors, major histocompatibility complexes (MHCs), TAP transporters, and immunological protein–protein interactions. The B-cell and T-cell epitopes are also included. As of January 2013, there were total of 24,000 entries in this database, and according to [17] 816 entries were related to B-cell epitopes. Users can search for the relevant data utilizing BLAST [18] and a variety of specialized search options that allow defining specific experimental conditions and molecules. Based on the Web of Knowledge as of June 2013, this resource accumulated 211 citations across the three publications.

Availability: http://www.ddg-pharmfac.net/antijen/.

2.2 IEDB

IEDB (Immune Epitope DataBase) was established in 2004 at the La Jolla Institute of Allergy and Immunology in San Diego [19, 20] and it was recently upgraded to version 2.0 [8]. This comprehensive resource provides integrated access to experimentally characterized B-cell epitopes, T-cell epitopes, and data on the MHC binding. The data are extracted from epitope-related articles available in PubMed and from direct submissions from scientists. The database includes epitope sequence and structure, source antigen and organism from which the epitope is derived, and details concerning experiments describing recognition of an epitope and related assays including MHC ligand elution assays and MHC binding assays. Users can conveniently query the database through a web interface utilizing a variety of criteria, such as the source antigen, source organism, epitope structure, immune recognition context, and host organism. Based on the Web of Knowledge as of June 2013, this database is highly cited with the combined number of citations for the three articles totaling to 332.

Availability: http://www.iedb.org.

2.3 Bcipep

This resource was developed in 2004 at the Institute of Microbial Technology, Chandigarh, in India [21]. It provides access to experimentally determined linear B-cell epitopes, which were extracted from literature in PubMed and collected from other publicly available databases. As of January 2013, it contained 3,031 entries including 539 entries from bacteria, 2,046 from viruses, 236 from protozoa, 53 from fungi, and 157 from other organisms. Users can search the database through a variety of options including keywords related to the relevant publications, sequence, entry number, and source organism, by utilizing sequence similarity with BLAST, and by scanning through the associated protein structures.

Availability: http://www.imtech.res.in/raghava/bcipep/.

2.4 CED

CED (Conformational Epitope Database) was built in 2005 by Huang and Honda at the University of Electronic Science and Technology in China [22]. This database focuses on the conformational epitopes. The entries were extracted from peer-reviewed journal articles collected from PubMed and ScienceDirect. CED provides the location of the epitope in the sequence and structure, immunological properties of the epitope, source antigen, and corresponding antibody. The database can be browsed or searched using keywords through a website interface. As of January 2013, CED included 293 entries.

Availability: http://immunet.cn/ced/.

2.5 Epitome

This database was established in 2005 by Rost Group at the Columbia University [23]. Epitome provides access to a collection of antigen–antibody complex structures, including annotation and visualization of residues that are involved in the interactions and information concerning certain structural characteristics of the binding regions. The entries were collected from Protein Data Bank (PDB) [24]. User can search the database utilizing keywords with options to specify chain and certain structural properties of antigen and antibody, and also by finding similar sequence with BLAST. This resource contains 142 antigens from protein–antibody complexes [23].

Availability: http://www.rostlab.org/services/epitome/.

2.6 SEDB

Structural Epitope Database (SEDB) was developed in 2011 at the Pondicherry University in India [9]. It provides access to a comprehensive set of structures of B-cell, T-cell, and MHC binding proteins. The data was collected from PDB, PDBsum [25], MHCBN [26], IMGT/3D [27], Bcipep, and IEDB databases. SEDB includes information concerning epitope sequence and position, antigen–antibody interacting residues, and corresponding taxonomic identifiers, and is cross-linked to relevant databases such as IEDB, UniprotKB [28], PDB, and NCBI [29]. The database can be either browsed or searched by finding, using BLAST, similar chains. It currently includes 614 entries with 273 B-cell epitopes.

Availability: http://sedb.bicpu.edu.in/.

3 Sequence-Based Predictors of Linear B-Cell Epitopes

Prediction of linear B-cell epitopes from the antigen sequences dates back to 1980s. The trailblazing methods were fairly simple and utilized a single propensity (flexibility, solvent accessibility, etc.) of the underlying chain or chain fragment [2, 30–35]. A new generation of methods that combined multiple physicochemical propensities to predict B-cell epitopes has surfaced in 1990s. They include PREDITOP [36], PEOPLE [37], BEPITOPE [38], BcePred [39], and LEP-LP [40] predictors. Predictive quality of these approaches was questioned in 2005 in a study by Blythe and Flower [41]. They analyzed predictive performance of close to 500 amino acid propensity scales on 50 antigens and determined that these propensities performed only slightly better than random. Since then this field has observed a revival that resulted in the development of more sophisticated knowledge-based methods, particularly in the context of the predictive models that they utilize. The considered models included a neural network in ABCpred [42], hidden Markov model in BepiPred [43], and naïve Bayes that was used in Epitopia [13, 14]. The dominant model used in recent years is the support vector machine (SVM), which was applied in a wide range of methods, such as AAP [44], BCPred [45], FBCPred [46], COBEpro [47], BayesB method [48], BROracle [49], LEPS [50], SVMTriP [51], and LBtope [52]. These approaches differ in the formulation and scope of information extracted from the input antigen sequence, in the size of data that were used to compute the SVM model, and in the type of SVM kernel function used. Table 1 summarizes methods that were developed since 2005 and includes one representative older method, BEPITOPE (*see* **Note 3**). COBEpro can also predict conformational epitopes and thus it is discussed later in this chapter. Several predictors of linear B-cell epitopes are widely cited in the literature, relative to when they were published. Based on the Web of Knowledge as of June 2013, ABCpred and BepiPred that were published in 2006 were already cited 139 and 145, respectively. The AAP method that was published in 2007 was cited 106 times, and the newer articles for BCPred and Epitopia that were released in 2008 and 2008 already accumulated 54 and 47 (for the two

Table 1
Summary of sequence-based predictors of linear B-cell epitopes

Method	Year	Model	Type[a]	Input[b]	Availability
BEPITOPE	2003	Scoring function	SP	SC	By contacting the authors
ABCpred	2006	Neural network	WS	SC	http://www.imtech.res.in/raghava/abcpred/
BepiPred	2006	Hidden Markov model	WS + SP	MC	http://www.cbs.dtu.dk/services/BepiPred/
AAP	2007	Support vector machine	WS + SP	SC	http://ailab.cs.iastate.edu/bcpreds/
LEP-LP	2008	Scoring function	WS	Unknown	http://biotools.cs.ntou.edu.tw/lepd_antigenicity.php[c]
BCPred	2008	Support vector machine	WS + SP	SC	http://ailab.cs.iastate.edu/bcpreds/
FBCPred	2008	Support vector machine	WS + SP	SC	http://ailab.cs.iastate.edu/bcpreds/
Epitopia	2009	Naïve Bayes	WS + SP	SC	http://epitopia.tau.ac.il
BayesB	2010	Support vector machine	WS	SC	http://www.immunopred.org/bayesb/
BROracle	2011	Support vector machine	SP	Unknown	https://sites.google.com/site/oracleclassifiers/[c]
LEPS	2011	Support vector machine	WS	SC	http://leps.cs.ntou.edu.tw
SVMTriP	2012	Support vector machine	WS	SC	http://sysbio.unl.edu/SVMTriP
LBtope	2013	Support vector machine	WS	MC	http://crdd.osdd.net/raghava/lbtope/

The methods are sorted chronologically
[a]*SP* stand-alone program, *WS* web server
[b]*SC* method predicts a single chain, i.e., prediction has to be restarted for each chain, *MC* multiple chains can be predicted at the same time
[c]A given predictor is currently unavailable

publications combined) citations, respectively. Most of the above-mentioned recent sequence-based linear B-cell epitope predictors, except BROracle, are available as convenient web servers that require the end user only to provide an input antigen sequence. Five methods, BepiPred, AAP, BCPred, FBCPred, and Epitopia, can also be downloaded as stand-alone applications, which would appeal to the users who would like to incorporate such tools into their computational pipelines. Following, we summarize the 13 predictors from Table 1 in the chronological order.

3.1 BEPITOPE

BEPITOPE was published in 2003 by Pellequer's group at the Centre de Marcoule at CEA in France [38]. BEPITOPE utilizes a scoring function that combines information from over 30 selected physicochemical propensities including hydrophilicity, flexibility, propensity to form beta turns, and surface accessibility. User can define sequence motifs to filter the predictions.

Inputs: Protein sequence in FASTA format or accession number.

Outputs: Numerical profile over the input chain where putative epitopes are indicated by peaks.

Architecture: Scoring function.

Availability: This program is available for free for academic use and has to be requested from the authors. User is required to sign a license agreement before receiving a copy of the software. Web server is not available.

3.2 ABCpred

ABCpred was developed in 2006 by Raghava's group at the Institute of Microbial Technology, Chandigarh, in India [42]. This method was one of the first to use a more sophisticated, machine learning-based prediction model. This model is a recurrent neural network that has a single hidden layer with 35 neurons. It utilizes a segment of 16 consecutive residues to perform prediction.

Inputs: Amino acid sequence using since-letter encoding. User can also set values of several parameters including threshold to identify epitopes and segment length. Default values are used in case if user does not want to set parameter values.

Outputs: Starting position and numeric score for predicted epitope(s).

Architecture: Recurrent neural network.

Availability: Web server at http://www.imtech.res.in/raghava/abcpred/.

3.3 BepiPred

BepiPred was created in 2006 by Lund's group at the Technical University of Denmark [43]. This is the first and so far the only method that utilizes hidden Markov model. This model combines multiple physicochemical propensities including antigenicity, hydrophilicity, hydrophobicity, solvent accessibility, and secondary structure, which are preprocessed using a running mean window.

Inputs: Protein sequence or a set of sequences (up to 2000) in FASTA format. Each sequence has to have at least 10 and no more than 6,000 amino acids. User can also set value of threshold to identify epitopes; default value (0.35) is used otherwise.

Outputs: Numeric score for each residue in the query protein sequence. The predicted epitope is composed of residues with scores higher than the threshold.

Architecture: Hidden Markov model.

Availability: Web server at http://www.cbs.dtu.dk/services/BepiPred/. Stand-alone version for UNIX platform is also available at this website.

3.4 AAP

AAP (amino acid pair antigenicity) predictor was developed in 2007 at the Shanghai Jiaotong University in China [44]. This is the first method that utilizes the SVM-based prediction model. The authors introduced antigenicity propensity scale, which was empirically shown to improve over previously used physicochemical propensities, that was utilized to convert the query sequence into numerical inputs for the SVM.

*Input*s: Amino acid sequence using since-letter encoding. User can also select the length of the epitope to be predicted, with default value set to 20 and allowed values of 12, 14, 16, 18, 20, and 22.

Outputs: Predicted epitope segments with the predefined length.

Architecture: Support vector machine with RBF kernel.

Availability: The authors do not provide the software. However, a web server that is a part of BCPREDS platform can be found at http://ailab.cs.iastate.edu/bcpreds/. Stand-alone version is also available at this website.

3.5 LEP-LP

LEP-LP was released in 2008 by Tun-Wen Pai's group at the National Taiwan Ocean University [40]. The authors utilized mathematical morphology to extract local peaks from a numerical profile that implements combination of several weighted physicochemical propensity scales, such as hydrophilicity, solvent accessibility, polarity, flexibility, antigenicity, and secondary structure.

*Input*s: Amino acid sequence using since-letter encoding.

Outputs: Ranked putative epitope segments with the associated numeric scores.

Architecture: Scoring function based on mathematical morphology.

Availability: Web server at http://biotools.cs.ntou.edu.tw/lepd_antigenicity.php (currently unavailable).

3.6 BCPred

BCPred was published in 2008 at the Iowa State University [45]. This is the second method that applied SVM-based prediction model; however this model is customized to use string kernel. The authors utilized a specific type of the string kernel, subsequence kernel, which considers a feature (input) space generated by a set of k-mer subsequences of the input chain.

Inputs: Amino acid sequence using since-letter encoding. User can also select the length of the epitope to be predicted, with default value set to 20 and allowed values of 12, 14, 16, 18, 20, and 22.

Outputs: Predicted epitope segments with the predefined length and with the associated numeric scores.

Architecture: Support vector machine with string kernel.

Availability: Web server at http://ailab.cs.iastate.edu/bcpreds/. Stand-alone version is also available at this website.

3.7 FBCPred

FBCPred was developed in 2008 at the Iowa State University [46]. Similar to BCPred, this method also uses SVM model with the subsequence kernel. FBCPred targets prediction of linear B-cell epitopes of variable length, in contrast to BCPred that assumes fixed (user-defined) length.

Inputs: Amino acid sequence using since-letter encoding. User can also select the length of the epitope to be predicted, with default value set to 14.

Outputs: Predicted epitope segments with the predefined length and with the associated numeric scores.

Architecture: Support vector machine with string kernel.

Availability: Web server at http://ailab.cs.iastate.edu/bcpreds/. Stand-alone version is also available at this website.

3.8 Epitopia

This predictor was published in 2009 by Tal Pupko group at the Tel Aviv University in Israel [13, 14]. Epitopia predicts linear B-cell epitopes from either a protein structure or sequence; here we focus on the sequence-based version. This method uses naïve Bayes classifier by considering a small sliding window of seven residues. The inputs for the classifier are generated from this window by using 14 physicochemical propensities including polarity, flexibility, antigenicity, hydrophilicity, solvent accessibility, secondary structure, and ratio between the frequency of selected amino acid in the window and the remaining part of the sequence.

Inputs: Protein sequence in FASTA format and an e-mail address of the user.

Outputs: Numeric immunogenicity score and corresponding probability for each amino acid in the query protein sequence. The immunogenicity scores are used to derive a ranked list of epitope segments.

Architecture: Naïve Bayes classifier.

Availability: Web server at http://epitopia.tau.ac.il. Stand-alone version for LINUX platform is also available at this website.

3.9 BayesB

This method was created in 2010 at the National University of Singapore [48]. BayesB utilizes the SVM model and employs Bayes feature extraction that is based on differences in the frequency of occurrence of amino acid types at each position in a predefined (training) set of epitopes and non-epitope segments.

Inputs: Protein sequence in FASTA format or using since-letter encoding. User can also select the length of the epitope to be predicted, with default value set to 20.

Outputs: Predicted epitope segments with the predefined length.

Architecture: Support vector machine with RBF kernel.

Availability: Web server at http://www.immunopred.org/bayesb/.

3.10 BROracle

B-Cell Epitope Oracle (BROracle) method was developed in 2011 at the Dana-Farber Cancer Institute [49]. This predictor is implemented using SVM model. The input to the model were generated from the sequence and a variety of sequence-derived characteristics including evolutionary information calculated from PSI-BLAST output [53], secondary structure predicted with PSI-PRED [54], solvent accessibility predicted with ACCpro [55], disorder predicted with VSL2 algorithm [56], and sequence complexity computed with SEG algorithm [57].

Inputs: Protein sequence.

Outputs: Unknown.

Architecture: SVM classifier with polynomial kernel.

Availability: Stand-alone program at https://sites.google.com/site/oracleclassifiers/ (currently unavailable). Web server is not available.

3.11 LEPS

LEPS (Linear Epitope prediction based on Propensities scale and SVM) was created in 2011 by Tun-Wen Pai's group at the National Taiwan Ocean University [50]. This method extends the LEP-LP predictor by the same group. First, candidate epitopes are predicted with LEP-LP. Next, SVM model is used to remove less probable candidates utilizing their amino acid sequences.

Inputs: Protein sequence in FASTA format or using since-letter encoding. The user has an option to adjust 32 parameters related to the setup of the propensities considered in LEP-LP. Default parameter values are used in case if user does not want to set parameter values.

Outputs: Ranked list of predicted epitope segments.

Architecture: Support vector machine with RBF kernel.

Availability: Web server at http://leps.cs.ntou.edu.tw.

3.12 SVMTriP

SVMTriP was created in 2012 by Chi Zhang's group at the University of Nebraska, Lincoln [51]. This predictor is based on SVM model that utilizes similarity, calculated with Blosum62 matrix, and frequency of tripeptides (3-mers) from the input antigen chain.

Inputs: Protein sequence in FASTA format or using since-letter encoding. User can select the length of the epitope to be predicted, with default value set to 20.

Outputs: Predicted epitope segments with the predefined length and with the associated numeric scores.

Architecture: Support vector machine with string kernel.

Availability: Web server at http://sysbio.unl.edu/SVMTriP.

3.13 LBtope

LBtope was published in 2013 by Raghava group at the Institute of Microbial Technology, Chandigarh, in India [52]. This method converts the antigen chain into numerical features (descriptors) that are based on dipeptide (2-mer) profiles. These features are fed into the SVM model that predicts epitopes.

Inputs: Protein sequence or a set of sequences, in FASTA format. User can also select model type, using fixed size epitope fragments (20 residues long) or variable length epitopes (user-defined between 5 and 30); default value (variable length with 15 residues segment) is used otherwise.

Outputs: Predicted epitope segments with the predefined length and with the associated numeric scores.

Architecture: Support vector machine with undisclosed type of kernel.

Availability: Web server at http://crdd.osdd.net/raghava/lbtope/.

4 Sequence-Based Predictors of Conformational B-Cell Epitopes

A few methods were recently developed to predict the conformational B-cell epitopes from protein chains. This is a challenging problem given the fact that the corresponding epitopic residues are potentially distributed over an entire protein chain, without necessarily being clustered into longer segments. The prediction methods score each amino acid in an input protein chain (using a numeric or a binary value) to indicate whether it is part of an epitope. A drawback of this prediction is that these programs do not group the predicted epitopic residues into the corresponding epitopes, which could be an issue if a given chain contains more than one epitope. The sequence-based predictors of conformational epitopes, which are summarized in Table 2, include COBEpro that was designed to predict linear epitopes and extended to predict conformational epitopes [47], CBTOPE [58], BEST [15], and Bprediction [59] (*see* **Note 4**). The first three methods apply the SVM model, while the most recent Bprediction is based on the random forest model, which utilizes a set of decision trees. Based on the Web of Knowledge as of June 2013, the oldest sequence-based predictor of conformational B-cell epitopes, COBEpro, which was published in 2009, was already cited 30 times. The other methods are too recent to accumulate citations. COBEpro, CBTOPE, and

Table 2
Summary of sequence-based predictors of conformational B-cell epitopes

Method	Year	Model	Type[a]	Input[b]	Availability
COBEpro	2009	Support vector machine	WS	SC	http://scratch.proteomics.ics.uci.edu
CBTOPE	2010	Support vector machine	WS + SP	MC	http://www.imtech.res.in/raghava/cbtope/
BEST	2012	Support vector machine	SP	MC	http://biomine.ece.ualberta.ca/BEST/
Bprediction	2012	Random forest	WS	SC	http://bcell.whu.edu.cn

The methods are sorted chronologically
[a]*SP* stand-alone program, *WS* web server
[b]*SC* method predicts a single chain, i.e., prediction has to be restarted for each chain, *MC* multiple chains can be predicted at the same time

Bprediction are available to the end users via web servers. Two of the methods, CBTOPE and BEST, are provided as stand-alone software that the end users would install and use on their computers. Next, we summarize these four predictors in the chronological order.

4.1 COBEpro

COBEpro was published in 2009 by Baldi's group at the University of California [47]. COBEpro has a two-tier architecture where the first layer applies SVM to predict short segments (5–18 residues long) in the input chain utilizing information based on their similarity to epitopic segments in a training database, and secondary structure and solvent accessibility predicted with SSpro [60, 61] and ACCpro [55], respectively. The second layer is used to combine the above predictions to calculate epitopic propensity score for each amino acid. This allows COBEpro to be used for the prediction of discontinuous B-cell epitopes.

Inputs: Protein sequence or a set of sequences, using since-letter encoding, and an e-mail address of the user.

Outputs: Ranked (according to propensity) list of most likely predicted epitopes, including their predicted secondary structure and solvent accessibility, and numeric propensity scores for each amino acid in the query protein sequence.

Architecture: Support vector machine with Gaussian kernel.

Availability: COBEpro is incorporated into the SCRATCH web server suite at http://scratch.proteomics.ics.uci.edu/.

4.2 CBTOPE

CBTOPE was released in 2010 by Raghava's group at the Institute of Microbial Technology, Chandigarh, in India [58]. This method applies a sliding window (a segment of 19 residues that is moved along the input antigen sequence) to predict the epitopic score for the residues in the middle of a given window. CBTOPE computes amino acid composition, which is represented using a binary vector, of the residues in the window and these values are inputted into the SVM model that predicts epitopic propensity.

Inputs: Protein sequence in FASTA format or using since-letter encoding. User can select a threshold for the output scores from the predictor, with a default value set to −0.3. Residues with scores above the threshold are assumed to be epitopic.

Outputs: Numeric propensity scores for each amino acid in the query antigen chain. The scores are integers between 0 and 9, where higher value denotes a higher likelihood of a given residue to be in an epitope.

Architecture: Support vector machine with Gaussian kernel.

Availability: Web server at http://www.imtech.res.in/raghava/cbtope/. Stand-alone version for Windows operating system is also available at this website.

4.3 BEST

BEST (B-cell Epitope prediction using Support vector machine Tool) was published in 2012 by Kurgan's group at the University of Alberta in Canada [15]. This method utilizes SVM model and a comprehensive set of sequence-derived characteristics of the antigen chain. BEST is implemented using a two-layer architecture; *see* Fig. 2. In the first layer, the input antigen sequence is processed using sliding widows of 20 amino acids. Each 20-mer segment is encoded by a numerical feature vector that utilizes sequence conservation computed based on Weighted Observation Percentage

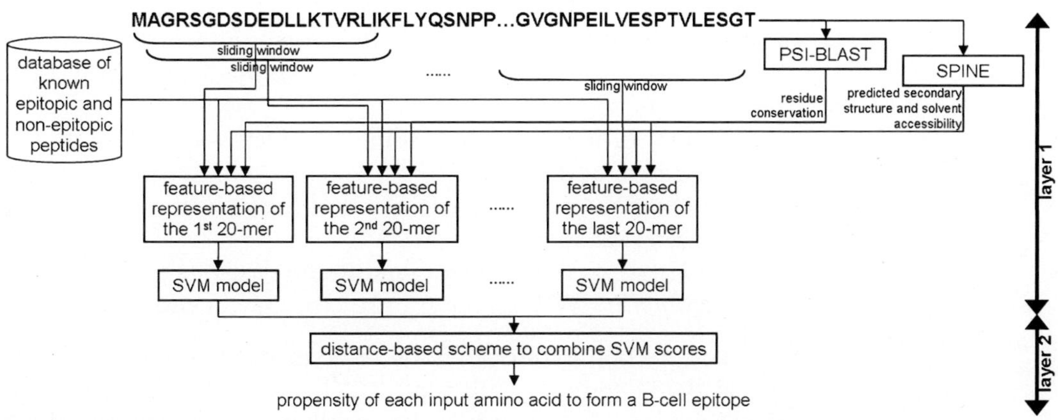

Fig. 2 Architecture of the BEST predictor of conformational B-cell epitopes. SVM stands for support vector machine

(WOP) matrix generated with PSI-BLAST [53], similarity to training epitopes based on measure proposed in [47], and secondary structure and relative solvent accessibility predicted with SPINE [62, 63]. This vector is inputted into SVM model and the predictions from SVM are combined to generate the epitopic propensities in the second layer.

Inputs: Protein sequence or a set of sequences, in FASTA format.

Outputs: Numeric propensity scores for each amino acid in the query protein sequence.

Architecture: Support vector machine with RBF kernel

Availability: Stand-alone software for Linux platform is available at http://biomine.ece.ualberta.ca/BEST/. Web server is not available.

4.4 Bprediction

Bprediction was made available in 2012 by Zhang's group at the Wuhan University in China [59]. This predictor has a two-level design and applies an ensemble of random forest models that take a set of numerical features computed from sliding windows of size 9 (9-mers) generated over the antigen chain as their inputs. The inputs are divided into nine sets, where each set is utilized by a different random forest model, which include (1) physicochemical propensities including flexibility, hydrophilicity, solvent accessibility, polarity, and propensity for formation of beta turns; (2) amino acid composition of the residues in the window represented using binary vectors and (3) real-valued vectors; (4) composition of amino acid sets defined based on their R-groups; (5) values from the position-specific scoring matrix (PSSM) generated by PSI-BLAST [53]; (6) composition of dipeptides (2-mers) in the window; and (7) secondary structure and (8) relative solvent accessibility predicted with SABLE [64]. The second level generates the output propensity scores by computing weighted average of normalized, based on *z* scores, values of predictions from these nine models; *see* Fig. 3.

Inputs: Protein sequence using since-letter encoding and an e-mail address of the user.

Outputs: Numeric propensity scores for each amino acid in the query protein sequence.

Architecture: Ensemble of random forests.

Availability: Web server at http://bcell.whu.edu.cn.

The overall architectures of the two most recent conformational B-cell epitope predictors, BEST and Bprediction, are relatively similar (Figs. 2 and 3). Both utilize the two-layered design and use multiple sequence alignments computed with PSI-BLAST and predictions of secondary structure and solvent accessibility. The main differences are in the fact that they use different prediction

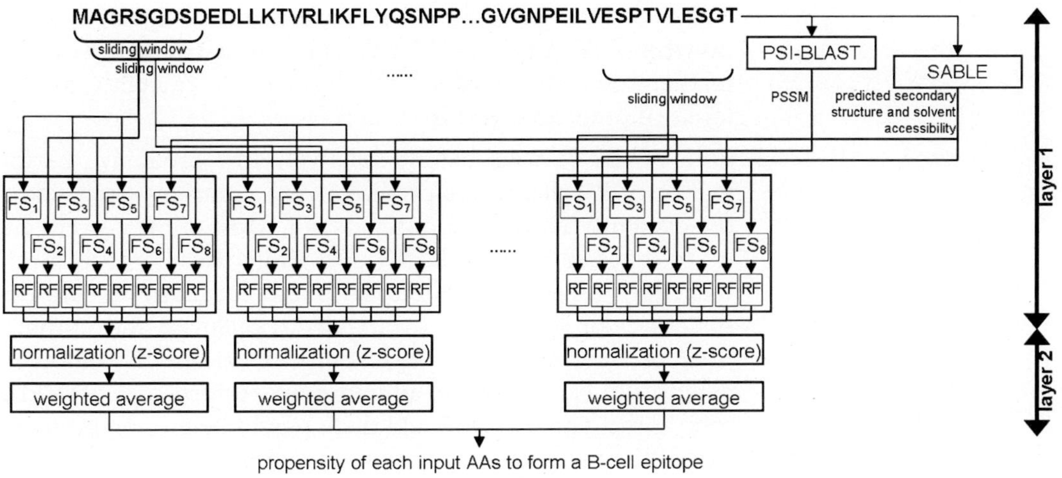

Fig. 3 Architecture of the Bprediction method for the prediction of conformational B-cell epitopes. FS_i refers to *i*th feature set, where $i = 1$ (physicochemical propensities), 2 (binary amino acid composition), 3 (real-valued amino acid composition), 4 (composition of amino acid sets), 5 (composition of dipeptides), 6 (PSSM values), 7 (predicted secondary structure), 8 (predicted relative solvent accessibility). RF stands for random forest

models (SVM vs. ensemble of decision forests) and several different inputs (similarity scores vs. physicochemical propensities and various amino acid compositions). In spite of utilizing these relatively sophisticated architectures, the predictive performance of these and other predictors of conformational epitopes is at modest levels (*see* **Note 1**). This calls for more research towards the development of more accurate methods (*see* **Note 5**).

5 Notes

1. We sampled recent publications that evaluated predictive performance of the current B-cell epitope predictors. For simplicity we concentrate on the area under the ROC curve (AUC) measure [4]. AUC values range between 0.5 and 1, with 0.5 denoting a random prediction and higher values corresponding to better predictive performance. Five methods that predict epitopes from antigen sequences were compared side by side in [15] and were shown to achieve AUC between 0.52 and 0.57 on a benchmark dataset consisting of 149 antigens. In another study, six and two methods that predict epitopes from antigen structures and sequences, respectively, were evaluated on a small dataset with 19 antigens; their AUC values were in the 0.57–0.63 range [59]. A recent review of predictors that utilize antigen structure demonstrates that AUC

values for the prediction of conformational epitopes range between 0.57 and 0.64 [5]. Overall, these results reveal that further research is needed to improve the currently modest levels of predictive performance.

2. One of the reasons behind relatively low predictive performance of B-cell epitope predictors is a relatively small size of the currently available annotated data. Most of the current and more successful methods are knowledge based, which means that they utilize annotated, with the location of epitopes, structures or sequences of antigens to calculate and optimize their predictive models. Availability of additional annotated data would likely result in an improved performance of predictors, as the data used to build them would be more representative of the complete population of epitopes.
3. When testing sequence-based predictors of linear B-cell epitopes we found that two of them, LEP-LP and BROracle, were no longer available. The web server implementations of the remaining methods allow predictions for a single chain. In case a user wants to predict a set of chains, he or she has to supply and predict them one at a time. The two exceptions are BepiPred and LBtope that simultaneously process prediction of multiple chains, with a limit of up to 2,000 sequences for a single run of BepiPred. Moreover, the BayesB predictor cannot predict peptides shorter than 25 residues.
4. Three sequence-based predictors of conformational B-cell epitopes are available to the end users as web servers and two as stand-alone applications. Two of them, COBEpro and Bprediction, are limited in the sense that they can predict only one sequence at the time. The other two, BEST and CBTOPE, are capable of predicting multiple chains in a single run. A further limitation of COBEpro is that it can be used to predict chains shorter than 1,500 residues.
5. There are potentially many ways to pursue the development of more accurate predictors of the B-cell epitopes. One possibility is to utilize a consensus of different predictors. Although Bprediction already implements a consensus approach, it is limited to the same predictive models and the same prediction flow. Instead, the consensus should consider combining outputs of multiple methods that use different models and flows, say BEST, Bprediction, BCTOPE, and COBEpro. Similar attempts were shown to be successful for related prediction tasks, such as prediction of MHC class II peptide binding [65] and T-cell epitopes [66]. Another potential direction is to find new and useful sources of information that are helpful in identifying epitopic regions. Examples include predicted disordered regions and flexible residues, predicted regions involved in protein–protein interactions, and results generated through homology modeling.

Acknowledgements

This work was supported by National Science Foundation of China (NSFC) grants 31050110432 and 31150110577 to L.K. J.G. was supported by the Fundamental Research Funds for the Central Universities grant 65011491.

References

1. Kringelum JV, Lundegaard C, Lund O, Nielsen M (2012) Reliable B cell epitope predictions: impacts of method development and improved benchmarking. PLoS Comput Biol 8(12): e1002829
2. Pellequer JL, Westhof E, van Regenmortel MH (1991) Predicting location of continuous epitopes in proteins from their primary structures. Methods Enzymol 203:176–201
3. Reineke U, Schutkowski M (2009) Epitope mapping protocols. Methods Mol Biol, vol 524
4. El-Manzalawy Y, Honavar V (2010) Recent advances in B-cell epitope prediction methods. Immunome Res 6(Suppl 2):S2
5. Yao B, Zheng D, Liang S, Zhang C (2013) Conformational B-cell epitope prediction on antigen protein structures: a review of current algorithms and comparison with common binding site prediction methods. PLoS One 8(4):e62249
6. Ansari HR, Raghava GP (2013) In silico models for B-cell epitope recognition and signaling. Methods Mol Biol 993:129–138
7. Yang X, Yu X (2009) An introduction to epitope prediction methods and software. Rev Med Virol 19(2):77–96
8. Vita R, Zarebski L, Greenbaum JA, Emami H, Hoof I, Salimi N, Damle R, Sette A, Peters B (2010) The immune epitope database 2.0. Nucleic Acids Res 38:D854–D862
9. Sharma OP, Das AA, Krishna R, Kumar SM, Mathur PP (2012) Structural Epitope Database (SEDB): a Web-based database for the epitope, and its intermolecular interaction along with the tertiary structure information. J Proteomics Bioinform 5:84–89
10. Toseland CP, Clayton DJ, McSparron H, Hemsley SL, Blythe MJ, Paine K, Doytchinova AI, Guan P, Hattotuwagama CK, Flower DR (2005) AntiJen: a quantitative immunology database integrating functional, thermodynamic, kinetic, biophysical, and cellular data. Immunome Res 1:4
11. Kim Y, Ponomarenko J, Zhu Z, Tamang D, Wang P, Greenbaum J, Lundegaard C, Sette A, Lund O, Bourne PE, Nielsen M, Peters B (2012) Immune epitope database analysis resource. Nucleic Acids Res 40:W525–W530
12. Zhang Q, Wang P, Kim Y, Haste-Andersen P, Beaver J, Bourne PE, Bui HH, Buus S, Frankild S, Greenbaum J, Lund O, Lundegaard C, Nielsen M, Ponomarenko J, Sette A, Zhu Z, Peters B (2008) Immune epitope database analysis resource (IEDB-AR). Nucleic Acids Res 36:W513–W518
13. Rubinstein ND, Mayrose I, Martz E, Pupko T (2009) Epitopia: a web-server for predicting B-cell epitopes. BMC Bioinformatics 10:287
14. Rubinstein ND, Mayrose I, Pupko T (2009) A machine-learning approach for predicting B-cell epitopes. Mol Immunol 46:840–847
15. Gao J, Faraggi E, Zhou Y, Ruan J, Kurgan L (2012) BEST: improved prediction of B-cell epitopes from antigen sequences. PLoS One 7:e40104
16. Blythe MJ, Doytchinova IA, Flower DR (2002) JenPep: a database of quantitative functional peptide data for immunology. Bioinformatics 18:434–439
17. McSparron H, Blythe MJ, Zygouri C, Doytchinova IA, Flower DR (2003) JenPep: a novel computational information resource for immunobiology and vaccinology. J Chem Inf Comput Sci 43:1276–1287
18. Altschul SF, Gish W, Miller W, Myers EW, Lipman DJ (1990) Basic local alignment search tool. J Mol Biol 215(3):403–410
19. Peters B, Sidney J, Bourne P, Bui HH, Buus S, Doh G, Fleri W, Kronenberg M, Kubo R, Lund O, Nemazee D, Ponomarenko JV, Sathiamurthy M, Schoenberger S, Stewart S, Surko P, Way S, Wilson S, Sette A (2005) The immune epitope database and analysis resource: from vision to blueprint. PLoS Biol 3(3):e91
20. Peters B, Sette A (2007) Integrating epitope data into the emerging web of biomedical knowledge resources. Nat Rev Immunol 7(6): 485–490
21. Saha S, Bhasin M, Raghava GP (2005) Bcipep: a database of B-cell epitopes. BMC Genomics 6(1):79

22. Huang J, Honda W (2006) CED: a conformational epitope database. BMC Immunol 7:7
23. Schlessinger A, Ofran Y, Yachdav G, Rost B (2006) Epitome: database of structure-inferred antigenic epitopes. Nucleic Acids Res 34: D777–D780
24. Berman HM, Westbrook J, Feng Z, Gilliland G, Bhat TN, Weissig H, Shindyalov IN, Bourne PE (2000) The protein data bank. Nucleic Acids Res 28(1):235–242
25. Laskowski RA (2001) PDBsum: summaries and analyses of PDB structures. Nucleic Acids Res 29:221–222
26. Bhasin M, Singh H, Raghava GP (2003) MHCBN: a comprehensive database of MHC binding and non-binding peptides. Bioinformatics 19:665–666
27. Kaas Q, Ruiz M, Lefranc MP (2004) IMGT/3Dstructure-DB and IMGT/StructuralQuery, a database and a tool for immunoglobulin, T cell receptor and MHC structural data. Nucleic Acids Res 32: D208–D210
28. Magrane M, UniProt Consortium (2011) UniProt Knowledgebase: a hub of integrated protein data. Database:bar009
29. Pruitt KD, Tatusova T, Brown GR, Maglott DR (2012) NCBI Reference Sequences (RefSeq): current status, new features and genome annotation policy. Nucleic Acids Res 40:D130–D135
30. Hopp TP, Woods KR (1981) Prediction of protein antigenic determinants from amino acid sequences. Proc Natl Acad Sci U S A 78: 3824–3828
31. Welling GW, Weijer WJ, van der Zee R, Welling-Wester S (1985) Prediction of sequential antigenic regions in proteins. FEBS Lett 188:215–218
32. Karplus PA, Schulz GE (1985) Prediction of chain flexibility in proteins: a tool for the selection of peptide antigen. Naturwissenschaften 72:212–213
33. Parker JM, Guo D, Hodges RS (1986) New hydrophilicity scale derived from high-performance liquid chromatography peptide retention data: correlation of predicted surface residues with antigenicity and X-ray derived accessible sites. Biochemistry 25:5425–5432
34. Kolaskar AS, Tongaonkar PC (1990) A semi empirical method for prediction of antigenic determinants on protein antigens. FEBS Lett 276:172–174
35. Pellequer JL, Westhof E, van Regenmortel MH (1993) Correlation between the location of antigenic sites and the prediction of turns in proteins. Immunol Lett 36(1):83–99
36. Pellequer JL, Westhof E (1993) PREDITOP: a program for antigenicity prediction. J Mol Graph 11:191–202
37. Alix AJ (1999) Predictive estimation of protein linear epitopes by using the program PEOPLE. Vaccine 18:311–314
38. Odorico M, Pellequer JL (2003) BEPITOPE: predicting the location of continuous epitopes and patterns in proteins. J Mol Recognit 16(1): 20–22
39. Saha S, Raghava GP (2004) BcePred: prediction of continuous b-cell epitopes in antigenic sequences using physico-chemical properties. Third Intern Conf on Artificial Immune Systems. pp 197–204
40. Chang HT, Liu CH, Pai TW (2008) Estimation and extraction of B-cell linear epitopes predicted by mathematical morphology approaches. J Mol Recognit 21(6):431–441
41. Blythe MJ, Flower D (2005) Benchmarking B cell epitope prediction: underperformance of existing methods. Protein Sci 14:246–248
42. Saha S, Raghava GP (2006) Prediction of continuous B-cell epitopes in an antigen using recurrent neural network. Proteins 65(1):40–48
43. Larsen JE, Lund O, Nielsen M (2006) Improved method for predicting linear B-cell epitopes. Immunome Res 2:2
44. Chen J, Liu H, Yang J, Chou KC (2007) Prediction of linear B-cell epitopes using amino acid pair antigenicity scale. Amino Acids 33(3): 423–428
45. El-Manzalawy Y, Dobbs D, Honavar V (2008) Predicting linear B-cell epitopes using string kernels. J Mol Recognit 21(4):243–255
46. El-Manzalawy Y, Dobbs D, Honavar V (2008) Predicting flexible length linear B-cell epitopes. Comput Syst Bioinformatics Conf 7:121–132
47. Sweredoski MJ, Baldi P (2009) COBEpro: a novel system for predicting continuous B-cell epitopes. Protein Eng Des Sel 22(3):113–120
48. Wee LJ, Simarmata D, Kam YW, Ng LF, Tong JC (2010) SVM-based prediction of linear B-cell epitopes using Bayes feature extraction. BMC Genomics 11(Suppl 4):S21
49. Wang Y, Wu W, Negre NN, White KP, Li C, Shah PK (2011) Determinants of antigenicity and specificity in immune response for protein sequences. BMC Bioinformatics 12:251
50. Wang HW, Lin YC, Pai TW, Chang HT (2011) Prediction of B-cell linear epitopes with a combination of support vector machine classification and amino acid propensity identification. J Biomed Biotechnol 2011:432830
51. Yao B, Zhang L, Liang S, Zhang C (2012) SVMTriP: a method to predict antigenic epitopes using support vector machine to integrate

tri-peptide similarity and propensity. PLoS One 7(9):e45152
52. Singh H, Ansari HR, Raghava GP (2013) Improved method for linear B-cell epitope prediction using antigen's primary sequence. PLoS One 8(5):e62216
53. Altschul SF, Madden TL, Schäffer AA, Zhang J, Zhang Z, Miller W, Lipman DJ (1997) Gapped BLAST and PSI-BLAST: a new generation of protein database search programs. Nucleic Acids Res 25(17):3389–3402
54. Jones DT (1999) Protein secondary structure prediction based on position-specific scoring matrices. J Mol Biol 292(2):195–202
55. Cheng J, Randall AZ, Sweredoski MJ, Baldi P (2005) SCRATCH: a protein structure and structural feature prediction server. Nucleic Acids Res 33:W72–W76
56. Vucetic S, Brown CJ, Dunker AK, Obradovic Z (2003) Flavors of protein disorder. Proteins 52:573–584
57. Wootton JC, Federhen S (1996) Analysis of compositionally biased regions in sequence databases. Methods Enzymol 266:554–571
58. Ansari HR, Raghava GP (2010) Identification of conformational B-cell epitopes in an antigen from its primary sequence. Immunome Res 6:6
59. Zhang W, Niu Y, Xiong Y, Zhao M, Yu R, Liu J (2012) Computational prediction of conformational B-cell epitopes from antigen primary structures by ensemble learning. PLoS One 7(8):e43575
60. Pollastri G, Baldi P, Fariselli P, Casadio R (2002) Prediction of coordination number and relative solvent accessibility in proteins. Proteins 47(2):142–153
61. Pollastri G, Przybylski D, Rost B, Baldi P (2002) Improving the prediction of protein secondary structure in three and eight classes using recurrent neural networks and profiles. Proteins 47(2):228–235
62. Faraggi E, Xue B, Zhou Y (2009) Improving the prediction accuracy of residue solvent accessibility and real-value backbone torsion angles of proteins by guided-learning through a two-layer neural network. Proteins 74:847–856
63. Dor O, Zhou Y (2007) Achieving 80 % ten-fold cross-validated accuracy for secondary structure prediction by large-scale training. Proteins 66:838–845
64. Adamczak R, Porollo A, Meller J (2005) Combining prediction of secondary structure and solvent accessibility in proteins. Proteins 59:467–475
65. Wang P, Sidney J, Dow C, Mothé B, Sette A, Peters B (2008) A systematic assessment of MHC class II peptide binding predictions and evaluation of a consensus approach. PLoS Comput Biol 4(4):e1000048
66. Moutaftsi M, Peters B, Pasquetto V, Tscharke DC, Sidney J, Bui HH, Grey H, Sette A (2006) A consensus epitope prediction approach identifies the breadth of murine T(CD8+)-cell responses to vaccinia virus. Nat Biotechnol 24(7):817–819

Chapter 12

Machine Learning-Based Methods for Prediction of Linear B-Cell Epitopes

Hsin-Wei Wang and Tun-Wen Pai

Abstract

B-cell epitope prediction facilitates immunologists in designing peptide-based vaccine, diagnostic test, disease prevention, treatment, and antibody production. In comparison with T-cell epitope prediction, the performance of variable length B-cell epitope prediction is still yet to be satisfied. Fortunately, due to increasingly available verified epitope databases, bioinformaticians could adopt machine learning-based algorithms on all curated data to design an improved prediction tool for biomedical researchers. Here, we have reviewed related epitope prediction papers, especially those for linear B-cell epitope prediction. It should be noticed that a combination of selected propensity scales and statistics of epitope residues with machine learning-based tools formulated a general way for constructing linear B-cell epitope prediction systems. It is also observed from most of the comparison results that the kernel method of support vector machine (SVM) classifier outperformed other machine learning-based approaches. Hence, in this chapter, except reviewing recently published papers, we have introduced the fundamentals of B-cell epitope and SVM techniques. In addition, an example of linear B-cell prediction system based on physicochemical features and amino acid combinations is illustrated in details.

Key words B-cell epitope, Machine learning, Support vector machine, Propensity scale, Kernel function

1 Introduction of B-Cell Epitopes

The immune system is a collection of organs, tissues, cells, and molecules that work together to protect the body from various foreign pathogens such as bacteria, viruses, parasites, and fungi. This defense system against pathogens has been divided into two main strategies in vertebrates: innate immunity and adaptive immunity mechanisms. The innate immune system is considered as the first defending process against invading pathogens, while the adaptive immune system of the second defending layer creates immunological memories after an initial response to a specific pathogen and induces an enhanced response to subsequent encounters regarding the same pathogen. The latter adaptive immunity is classified into two branches of immune responses including cellular

Rajat K. De and Namrata Tomar (eds.), *Immunoinformatics*, Methods in Molecular Biology, vol. 1184,
DOI 10.1007/978-1-4939-1115-8_12, © Springer Science+Business Media New York 2014

immunity mediated by T-cell lymphocytes that eliminate infected cells and humoral immunity mediated by B-cell lymphocytes secreting antibodies which neutralize pathogens in the body fluid. Epitopes or antigenic determinants are defined as clusters of amino acid segments located on the surface of an antigen that bind to antigen-specific membrane receptors on lymphocytes or to secreted antibodies, and which elicit either cellular or humoral immune response and are recognized by specific antibodies [1]. Due to expensive and time-consuming factors of biomedical and immunological experiments, in silico epitope prediction and analysis prior to biological experiments become practical and standard strategies for both biomedical researchers and immunologists regarding various immunology-related applications such as epitope-based vaccine design and disease prevention, diagnosis, and treatment. There are several good review articles for both T-cell and B-cell epitope prediction analysis based on computational approaches as well as several useful epitope databases [2–8]. Among all published papers, epitope prediction methods can be simply categorized into four major types: sequence-based, structure-based, hybrid of sequence-based and structure-based, and consensus methods. It is in general expected that the prediction accuracy could be improved if an antigen structure has been determined. This is mainly due to easy validation of the surface characteristics of candidate epitopes on an antigen from the resolved structure. Hence, combination of sequence and structure features simultaneously should provide better prediction results than using sequence-based or structure-based along methods. Furthermore, combining several prediction methods and summarizing all individual prediction result through a voting mechanism could be anticipated to achieve an even better prediction accuracy since each prediction method held its own strength. Nevertheless, due to limited numbers of determined antibody–antigen complex structures and integrating difficulties for various computational limitations, there is yet no such a successfully integrated system for both B-cell and T-cell epitope prediction. Most of the prediction systems still focus on identifying one specific type of epitope according to its own characteristics.

T-cell epitopes are defined as peptide sequences presented on the surface of an antigen-presenting cell, and they are bound to major histocompatibility complex (MHC) class I and II molecules. Known as a structural basis for peptide binding to MHC molecules, T-cell epitopes are typically composed by continuous amino acids ranging from 9 to 11 in length for MHC class I binding and a length ranging from 13 to 25 amino acids for MHC class II binding [9, 10, 4]. For B-cell epitopes, it is generally categorized into two types: linear epitope (LE), a segment composed of a continuous stretch of amino acid residues, and conformational epitope (CE) constituted by several sequentially discontinuous segments that are dispersed among discontinuous regions, but become

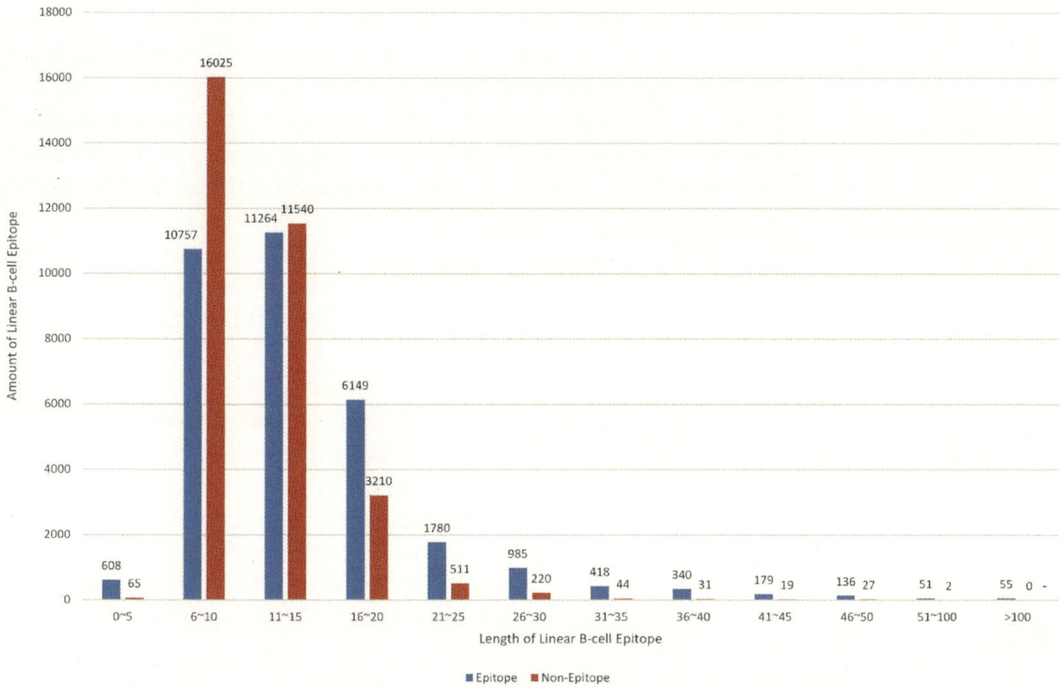

Fig. 1 Length distribution of linear B-cell epitopes and non-epitopes collected from IEDB database (version 2.4)

aggregated on the protein surface [11, 12]. Compared to continuous T-cell epitopes, linear B-cell epitopes possess significantly various peptide lengths from 2 to 829 residues from verified LE data statistics (IEDB: http://www.eiedb.org/) [13]. Length distribution of verified linear B-cell epitopes from IEDB database is shown in Fig. 1. Near 95 % of verified linear B-cell epitopes possess flexible lengths ranging from 6 to 30 residues. Even several annotated epitopes are with lengths larger than 100 residues. It was also reported that the proportion of LEs is considered with only 10 % of all B-cell epitopes [11], while the majority of B-cell epitope belongs to the discontinuous CE type with epitope size ranging from 6 to 29 residues [14]. However, in contrast to less complex features of T-cell epitope prediction systems and superior achievement for T-cell epitope prediction, the performance of predicting B-cell epitopes is yet to be satisfied and all proposed approaches still face a lot of challenges in computational immunology. Besides, only a small set of verified CEs are curated, a small set of resolved antibody–antigen complex structures, and not many convincible CE prediction systems are available. Therefore, in this chapter, we mainly discuss most of the published linear B-cell epitope prediction methods, and demonstrate how to adopt machine learning-based approaches for linear B-cell epitope prediction. It is also noticed that the support vector machine (SVM)-based learning method is one of the most popular approaches in recent reports.

In addition, the SVM-based system provided better performance compared to other machine learning methods. To demonstrate the usage of in silico prediction on linear B-cell epitopes through machine learning approaches, we choose to introduce the SVM classifier and hope that readers can fully understand the complete procedures and fundamental knowledge of linear B-cell epitope prediction.

Currently, various computational approaches and software for linear B-cell epitope prediction have been boomingly proposed in the last decade. Table 1 shows available methods, applicable websites, and kernel methods applied for LE prediction in a chronological order.

Most of the LE prediction focused on sequence contents and their corresponding propensity scales including surface accessibility [35], hydrophilicity [36], flexibility [37], and secondary structure [38] have been heavily considered in epitope predictive algorithms. The distinguishing characteristics among currently available programs such as BEPITOPE [17], PEOPLE [16], and BcePred [18] are mainly dealing with computation of different weighting scales over a sliding window along a query protein sequence. However, Blythe and Flower hypothesized that "single-scale amino acid propensity profiles cannot be used to predict epitope locations reliably" [39], a conclusion based on the observation that in the field of epitope prediction, even the best combinations of physicochemical propensity scales were not accurate enough to estimate and predict qualified B-cell epitopes. Therefore, several methods integrating the concept of amino acid propensity scales with machine learning technologies were proposed. For example, Saha and Raghava used recurrent artificial neural networks based on amino acid sequence information in ABCPred [19]; Larsen employed hidden Markov model (HMM) in BepiPred [20]; Chen et al. adopted SVM classifier on amino acid pairs [22]; Söllner and Mayer utilized a molecular operating environment with the decision tree and nearest neighbor approaches [21]; El-Manzalawy et al. developed BCPred [23] and FBCPred [24] employing SVM with a subsequence kernel for both fixed and flexible length epitopes; Sweredoski and Baldi developed COBEpro [26]; Wang et al. designed LEPS [30]; and Gao et al. presented BEST [31]; the last three approaches applied an SVM classifier in a two-step system to predict LEs based on an improved propensity scale approach; similarly, the BEEPro system designed by Lin et al. [33] and the LBtope system provided by Singh et al. [34] also adopted SVM classifiers by combining different propensity scales to enhance the prediction accuracies.

In the ABCPred system, two artificial neural network methods were developed, feed-forward (FNN) and recurrent neural network (RNN), for the prediction of continuous B-cell epitopes. Both FNN and RNN networks were used to achieve B-cell epitope prediction

Table 1
Linear B-cell epitope prediction methods

Name	URL	Method	Year	Reference
Antigenic	http://www.emboss.bioinformatics.nl/cgi-bin/emboss/antigenic	Physicochemical properties, occurrence of amino acid residues	1990	[15]
PEOPLE	n/a	Physicochemical properties	1999	[16]
BEPITOPE	Stand-alone program can be obtained freely to academics jlpellequer@cea.fr	Physicochemical properties	2003	[17]
BcePred	http://www.imtech.res.in/raghava/bcepred/	Physico-chemical properties	2004	[18]
ABCpred	http://www.imtech.res.in/raghava/abcpred/	ANN	2006	[19]
BepiPred	http://www.cbs.dtu.dk/services/BepiPred/	HMM	2006	[20]
Söllner	n/a	MOE, KNN, Decision tree	2006	[21]
Chen	n/a	SVM, AAP	2007	[22]
BCPred	http://www.ailab.cs.iastate.edu/bcpreds/	SVM, String kernel	2008	[23]
FBCPred	http://www.ailab.cs.iastate.edu/bcpreds/	SVM, String kernel	2008	[24]
LEPD	http://www.lepd.cs.ntou.edu.tw/	Physicochemical properties, mathematical morphology	2008	[25]
COBEpro	http://www.ics.uci.edu/~baldig/scratch/index.html	SVM	2009	[26]
Epitopia	http://epitopia.tau.ac.il	Naïve Bayes classifier	2009	[27, 28]
BayesB	http://www.immunopred.org/bayesb/index.html	SVM, Bayes feature extraction	2010	[29]
LEPS	http://leps.cs.ntou.edu.tw/	Physicochemical properties, mathematical morphology, SVM	2011	[30]
BEST	http://biomine.ece.ualberta.ca/BEST/	SVM	2012	[31]
SVMTriP	http://sysbio.unl.edu/SVMTriP/	SVM, tripeptide similarity and propensity	2012	[32]
BEEPro	n/a	Physicochemical properties, SVM, PSSM	2013	[33]
LBtope	http://crdd.osdd.net/raghava/lbtope/	SVM, binary profile, dipeptide composition, AAP	2013	[34]

ANN artificial neural network, *HMM* hidden Markov model, *MOE* molecular operating environment, *KNN* k-nearest neighbor, *PSSM* position-specific scoring matrix, *n/a* not applicable

using different window lengths from 10 to 20 amino acids, and the best performance of 66 % accuracy evaluated on a dataset of 700 B-cell epitopes and 700 non-epitopes was obtained by adopting an RNN trained on peptides of 16 amino acids in length. The BepiPred combined two amino acid propensity scales and an HMM trained on LEs to gain a slightly improved prediction accuracy rate over the propensity scale only-based methods by Parker et al. and Levitt et al. on the Pellequer dataset of 14 proteins and 83 epitopes. In Chen's approach, the observed certain amino acid pairs (AAPs) tend to appear more frequently in known B-cell epitopes than in non-epitope peptides. They utilized an AAP propensity scale based on such observation and trained with an SVM classier to increase an improved prediction accuracy rate of 71 % from the datasets of 872 B-cell epitopes and 872 non-epitopes. In the method of Söllner and Mayer, each epitope is represented using a set of propensity scales, neighborhood matrices, and respective probability and likelihood values. This approach combined several parameters previously associated with antigenicity, and included novel parameters based on frequencies of amino acids and amino acid neighborhood propensities. In their report, the best performance of 72 % was achieved utilizing a nearest-neighbor classifier with feature selection from datasets of 1,211 B-cell epitopes and 1,211 non-epitopes. For the BCPred developed by El-Manzalawy et al., they applied five different kernel methods to evaluate SVM classifiers on a homology-reduced dataset of 701 linear B-cell epitopes and 701 non-epitopes, and they demonstrated that the BCPred outperformed the ABCPred and Chen's methods. In addition to BCPred, El-Manzalawy et al. also developed another FBCPred for predicting flexible length linear B-cell epitopes using the subsequence kernels. Two machine learning approaches were adopted in their study: one approach utilized four sequence kernels for determining a similarity score between any arbitrary pair of variable length sequences, and the other approach applied four different methods of mapping a variable length sequence into a fixed length feature vector. The FBCPred was demonstrated with an improved performance of 73 % accuracy rate on the homology-reduced dataset of flexible length linear B-cell epitopes. In the COBEpro system, Sweredoski applied SVM to make predictions on short peptide fragments within the query antigen sequence and calculated an epitopic propensity score for each residue based on the fragment predictions. The accuracy rates and AUC values of COBEpro possessed better performance than Chen, BCpred, and BepiPred regarding different benchmark datasets. The LEPS system designed by Wang et al. combined improved propensity scale method, local high antigenicity profile, occurring frequencies of amino acid segments (AASs), and SVM classifier to predict LEs with flexible length. Using several benchmark datasets, LEPS has shown its competitive performance comparing to BepiPred, ABCPred, BCPred, and FBCPred. For the BEEPro developed by Lin et al., authors have claimed that both linear and conformational epitopes could be predicted by the SVM-based system which employed the features mainly based on evolutionary information, amino acid ratio propensity scale, and 14 specifically selected physicochemical propensity scales. The results have shown a superior performance compared to BepiPred, ABCPred, BCPred, FBCPred, and LEPS. For the BEST system presented by Gao et al., authors constructed an SVM training architecture based on features of averaging selected propensity scores by a 20-mer sliding window, sequence similarity score, predicted secondary structure, and solvent accessibility. The prediction performance was compared to Chen, BCPred, COBEpro, BayseB, and CBTOPE with an accuracy rate around 74 % for fragment-based LE prediction. For the latest LBtope system, authors provided five various training datasets, and they emphasized on experimentally verified non-epitope datasets compared to previously random peptides used in other studies. In this study, they applied SVM and K-nearest-neighbor learning models using various physicochemical propensity scales and amino acid composition-transition-distribution properties, and the LBtope prediction system obtained accuracy rates ranged from 54 to 86 % on the created datasets. Since most of machine learning-based approaches applied SVM classifiers to improve the performance of B-cell epitope prediction and the results showed that SVM-based methods possessed a better performance than other approaches, here we will briefly introduce basic theories of SVM in the next section for readers interested in related fields. An example of prediction system will also be applied to illustrate the combination of propensity scales and machine learning kernel method for LE prediction.

2 A Supervised Learning Method: SVM Classifier

Machine learning is a subfield of applied statistics, which trains on a collected sample dataset and generalizes rules from previous experiences. The training data with unknown probability distribution is usually applied to extract some general principles and perhaps the distribution for future predictions on new testing data. There are several types of machine learning algorithm based on trained inputs or desired outcomes, such as supervised, unsupervised, semi-supervised, and reinforcement learning mechanisms. Recently, one of the most popular computer algorithms for a variety of biological applications including epitope prediction is the SVM kernel method, a supervised learning model and learned by known epitope contents to predict novel epitopes within a query protein sequence [40]. To build an epitope prediction model, users have to provide a set of training examples including two classes, named as true epitopes and non-epitopes. The constructed SVM model is a representation of the trained examples as points in the selected feature space, and these sample points are divided by a hyperplane with a separable margin as wide as possible.

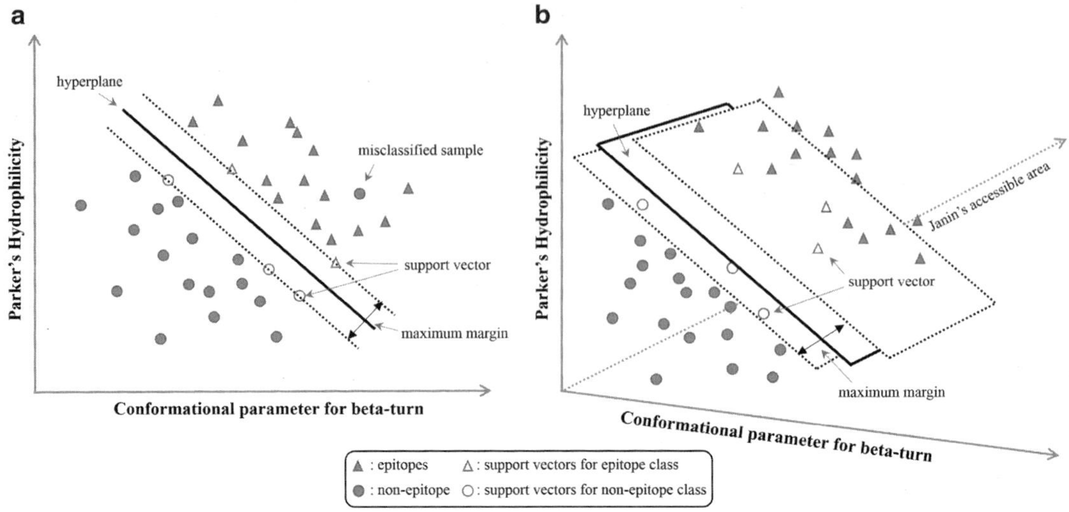

Fig. 2 Two examples of two-class SVM classifiers. (**a**) The first example of two-dimensional feature space and two classes were separated by a straight line with the maximum margin. (**b**) The second example of adding one more feature to a three-dimensional feature space and the two class samples were separated by a hyperplane with maximum margin. Each *circle* and *triangle* element represents samples from two different classes, and *empty circle* and *triangle* objects represent the support vectors for each class. The hyperplane was defined with a maximum margin between two planes constructed by support vectors

Query protein sequence segments are then mapped into the same feature space and assigned to one of the two defined categories based on the locations of the testing segment.

2.1 The Hyperplane of an SVM Model

Figure 2a shows a simple example of mapped points in a two-dimensional feature space. In this example, it is assumed that each peptide was calculated and mapped into a corresponding feature point by two selected feature values: secondary structure (conformation parameter for beta turn) and hydrophilicity (Parker's parameters [36]). The feature profile of each known epitope or non-epitope peptide is calculated according to the residue contents and the feature values are mapped into the two-dimensional space and represented by triangle and circle objects, respectively. In this case, it is quite easy to draw a line between two clusters geometrically, and an unknown data point could be predicted easily according to the query feature point falling on the epitope or the non-epitope sides of this separating line. If we add one more different feature such as Janin's accessible area to classify a peptide into two clusters, the feature space becomes a three-dimensional space, and we need a plane to divide the space into two parts as shown in Fig. 2b. Definitely, similar procedures could be extended to higher dimensions by adding more features. Hence, the original straight line in two-feature space can be extended to a hyperplane in a higher dimensional space which represents the border line to separate two clusters.

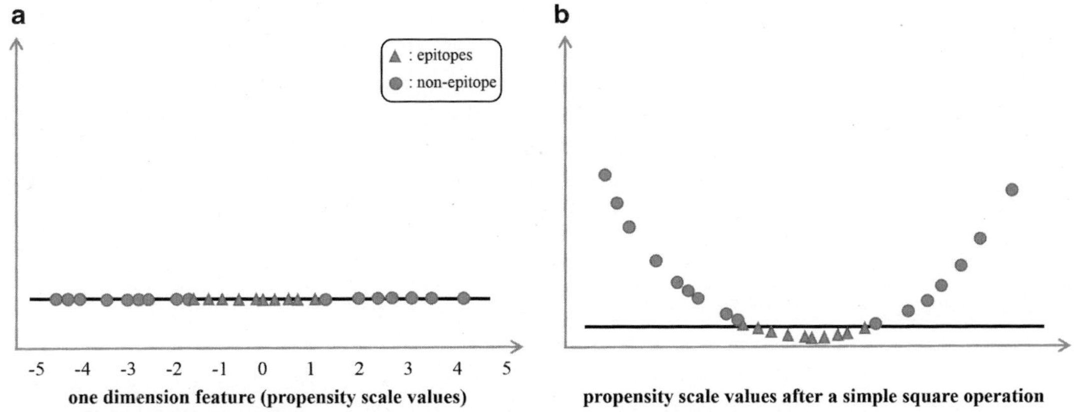

Fig. 3 An example of applying degree-2 polynomial kernel on all data points. (**a**) One-dimensional feature space and hard to find a single line to separate all data into two classes. (**b**) Applying square operation on all data points, and a clear hyperplane could separate all data points into two classes

2.2 Maximum Margins of a Hyperplane

It is obvious that the hyperplanes are not unique in an SVM model. How to select an optimal hyperplane between two clusters is the main goal of adopting SVM predictor and it serves as the key factor of a successful SVM classifier. Based on general statistical assumptions and the definition of a margin as the distance between the hyperplane to the nearest points (support vectors) within one cluster, the SVM model could find an optimal hyperplane possessing the maximal margin from any one of the training data points within two clusters. Hence, the selected hyperplane could maximize the performance of the SVM classifier to predict query samples. Nevertheless, several outlier data points might reside in wrong clusters from real applications and are called misclassified samples, and it might be solved by introducing an ε-insensitive loss function [41] which balances the number of hyperplane violations and the size of the margin.

2.3 Selection of Kernel Functions

Sometimes a tolerant margin could not support to find an optimal hyperplane to separate two clusters since the data points are crossly distributed in a feature space. In that case, there might exist a kernel function which provides a solution by adding an additional dimension for the data points. The original points could be transferred by a kernel function in order to find a better hyperplane to separate two clusters in a higher dimensional feature space. For an example shown in Fig. 3, the one-dimensional feature points could apply a simple square operation to transfer all data points into a two-dimensional space, and therefore an optimal hyperplane could be observed clearly. There are several frequently applied standard kernel functions, such as linear, polynomial, radial basis function (RBF), and sigmoid which can help to transfer the data points into a higher dimension to find a better hyperplane [42, 43]. However, it should be noticed that a very high-dimensional kernel function may cause overfitting problems and generally lead poor predictive performance. To avoid too many irrelevant dimensions, the selection of types and degrees of kernel functions should be carefully considered. Nevertheless, the traditional way to find a better kernel function is usually achieved by a trial-and-error approach and verified by cross-validation processes. But the selected so-called best kernel function still does not guarantee the optimal performance.

3 A Practical Example of Predicting Liner Epitope Based on SVM Classifier

To demonstrate how to apply SVM in predicting LEs, we selected the features used by Wang et al. [30], including physicochemical and AAS propensity scales. The first step is to discover all segments with global high or local high antigenicities according to the corresponding physicochemical properties. Once the potential segments were identified, the frequently appeared AASs were evaluated according to previously identified LE candidates. Based on the SVM method and the constructed models, the potential candidate segments were classified into epitopes or non-epitopes. Here we go through more details and learn how to apply machine learning technology intuitively in the application of LE prediction.

3.1 Antigenicity Analysis

An antigenic peptide possesses physicochemical properties of hydrophilicity, polarity, charge, flexibility, accessibility, secondary structure, and some other miscellaneous factors. For each category of specific physicochemical property, the individual score was given by sliding a window of a specified length along the query protein sequence from the *N* to *C* terminal direction and applying respective assigned weighting coefficients to each residue. The mean value of the assigned physicochemical feature within a sliding window was then calculated, and the average value was considered as a representative score at the midpoint of the window [44]. The boundary problems will be faced at both the *N*- and *C*-termini since the length of the neighboring residues was not sufficient to be considered within a fixed sliding window size. Hence, only the covered neighboring amino acids were applied to calculate the antigenicity value. Once an individual scale for each physicochemical feature was determined, a combination of different weighted coefficients on various scales at each position was calculated to achieve a final antigenicity profile. Different weighting coefficient assignment definitely affects the final antigenicity profile at a certain level. Users could assign the weightings according to his/her special concerns or simply apply equally distributed weightings. Here we applied different weighting coefficients to enhance and distinguish the importance of antigenic features with respect to LE prediction. In this example, we applied the beta turn [45], hydrophilicity [46], flexibility [47], and surface accessibility [48] with weighting coefficients of 0.4, 0.3, 0.15, and 0.15, respectively [16] (*see* **Note 1**).

3.2 Mathematical Morphology and Local Peak Determination

Most of the LE prediction systems focused on identifying the global high antigenicity segments. However, Wang et al. found that some of the experimental verified epitopes are located within the local high antigenicity profile. They applied some filtering processes to identify segments with global or local high antigenicity as epitope candidates. Their proposed filtering processes were completed employing mathematical morphology algorithm which is a nonlinear filter for signal analysis built on lattice theory and topology with applications to one-, two-, or n-dimensional signals. An antigenicity profile was interacted with a predetermined structuring element under three basic operations: erosion, dilation, and opening. Details of operating descriptions could be referred to refs. [49–52]. Nevertheless, the segments with local high or global high antigenicity were detected and extracted for next processes. It should be noticed that the default settings of window size for calculation of antigenicity scale, extraction of local peaks, and filtering of minimal size of epitope candidates played an important role at the initial stage. These default window sizes were selected according to the optimal performance in terms of accuracy analysis from known datasets [53]. The global antigenicity was defined as the average of the whole protein sequence antigenicity, and the low-to-moderate antigenicity meant the antigenicity of a predicted peptide lower than that of global antigenicity. Once the antigenic scale of each amino acid was calculated applying a running mean window by default settings, all epitope candidates were extracted when the average antigenicity of residues was continuously higher than that of the entire sequence or when the residues were located within peptides with locally high antigenicity compared to their neighboring segments. For fragments with globally or locally high antigenic residues, a merging function was performed to identify the candidates of LEs. All extracted segments with either globally or locally high antigenicity scales would be further filtered by the next classifier according to the SVM learning model based on previously statistical features. One example for extracting all possible epitope candidates is shown in Fig. 4. The figure shows a set of identified epitope candidates by mathematical morphology approaches on the *P30* protein. The original antigenicity profile according to the default weighting coefficient settings was shown in Fig. 4a. Eroded antigenicity profile by an erosion operator was shown in Fig. 4b, and a following dilation filter was applied on the previously eroded profile and an opened antigenicity profile was obtained and shown in Fig. 4c. All local peaks in Fig. 4d could be detected by taking the difference between the original and opened antigenicity profiles at the corresponding positions. These local peak segments were further filtered by a scanning window and the filtered segments were regarded as initially predicted candidates as shown in Fig. 4e. Finally, the predicted candidate LEs were obtained according to the locally high antigenic characteristics as shown in Fig. 4f.

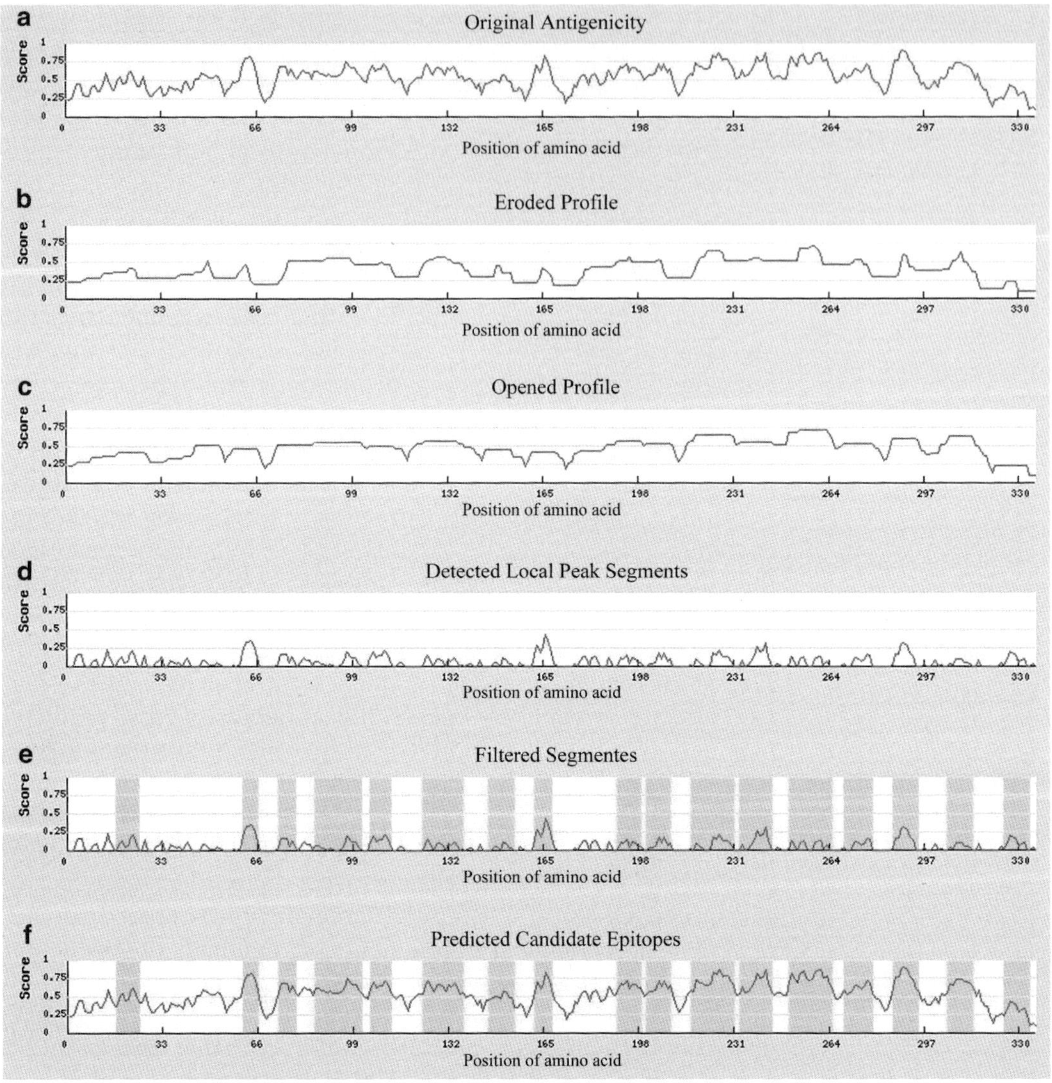

Fig. 4 An example of applying combination of morphological filters to extract segments with globally or locally high antigenicity characteristics on the *P30* protein. (**a**) The original antigenicity profile for *P30* protein. (**b**) Eroded antigenicity profile by an erosion operator. (**c**) Followed by a dilation filter and an opened antigenicity profile was obtained. (**d**) All local peaks detected by taking the difference between (**a**) and (**c**) at the corresponding positions. (**e**) Filtering local peaks with a default scanning window and the highlighted segments were considered as the candidate epitope locations. (**f**) All predicted candidate LEs for *P30* protein based on selected physicochemical propensity scales

3.3 SVM Classifier on Candidate Epitopes

The processes of adopting an SVM machine learning-based approach for epitope prediction usually comprise two major stages. The first stage requires a collection of training datasets and selected features. The training dataset includes samples in two categories: positive samples (true epitope segments) and negative samples (non-epitope segments). These samples will be trained to construct an SVM model according to the selected feature set. It should be noticed that a successful collection of training samples leads to good performance for all machine learning-based classifiers. In other words, lacking verified knowledge for collecting either positive or negative class samples affects the performance of the classifier dramatically and yields a biased estimation on evaluating system performance. As we know, most of the machine learning-based classification applications in biological fields fall short in negative samples. To balance the collection of both positive and negative class training samples, sometimes the generation of artificial negative samples is required [54]. Here, we adopted the Chen's dataset [22] containing 872 epitopes and 872 non-epitopes, for training the SVM classifier. All epitopes and non-epitopes within this dataset were restricted to a length of 20 residues. For the feature selection problem, since the physicochemical properties were already considered in the previous epitope candidate selection, here we only choose the amino acid combination propensity scales as the training features. We evaluated the statistical characteristics that determined the frequencies of occurrence of AASs with various lengths from another B-cell LE dataset, Bcipep [55], and the Chen's non-epitope dataset. Next, an SVM model was built based on the statistical features of the epitopes and non-epitopes. It should be noticed that the requirement of fixed window size for training and prediction sometimes considered as a deficiency in the machine learning-based approaches. Here, all collected epitopes and non-epitopes within the training dataset were restricted to a length of 20 residues. These verified epitopes were retrieved employing a "truncation-extension treatment." That is, when the length of an LE was longer than 20 residues, an equal number of superfluous residues were truncated from both the *N*- and *C*-termini to preserve the central 20 residues. Conversely, when the length of an LE was shorter than 20 residues, an equal number of neighboring residues were added to both the *N*- and *C*-termini according to its original sequences until the epitope comprised 20 residues. Both epitopes and non-epitopes with fixed length were then used to analyze their corresponding features and trained to produce an SVM model for future prediction.

3.4 Statistical Analysis of Amino Acid Segments and Corresponding Epitope Indexes

For constructing an SVM model in this example, we simply considered three statistical features by calculating the occurrence frequencies of combined residues in different lengths for both epitopes and non-epitopes. For the first feature of amino acid segment with two residues (AAS^2), 400 possible combinations of residue pairs should be analyzed for their corresponding occurrence frequencies in both the collected epitope and non-epitope segments. The epitope index $Epidex_i^2$ of the *i*th pattern (AAS_i^2) is defined by taking logarithmic value of the ratio of the number of AAS_i^2 among all epitopes $AASs^2$ compared to the same ratio in the non-epitope AASs^2 group. It can be formulated as the following equation:

$$\text{Epidex}_i^2 = \log\left(\frac{f_i^{2^+} / \sum_i f_i^{2^+}}{f_i^{2^-} / \sum_i f_i^{2^-}}\right), \qquad i = 1, 2, \ldots, 400$$

where f_i^{2+} and f_i^{2-} are the numbers of AAS_i^2 in the epitope and non-epitope datasets, respectively, and $\sum_i f_i^{2+}$ and $\sum_i f_i^{2-}$ denote the total number of AAS_i^2 in the corresponding dataset. Finally, the values of Epidex_i^2 are normalized to the range of [0, 1] to avoid dominance of any individual Epidex_i^2 in the classifier learning processes. For the next two features of amino acid segments with three and four residues (AAS^3 and AAS^4), there are a total of 8,000 and 160,000 possible combinations, respectively. In this case, a large portion of AAS^3 or AAS^4 do not appear in the non-epitope dataset and it would cause a problem of dividing by zero. Hence, the definitions of Epidex_i^3 and Epidex_i^4 are modified from the definition of Epidex_i^2, and the corresponding epitope indices for AAS^3 and AAS^4 are defined as the following formula. Both obtained Epidex_i^3 and Epidex_i^4 will be normalized to the range of [0, 1] as well:

$$\text{Epidex}_i^l = \frac{f_i^{l^+}}{\sum_i f_i^{l^+}}, \qquad l = 3 \text{ or } 4.$$

3.5 SVM Features and Kernel Selection

There are a variety of choices of open-source SVM software for feature training, model selection, and cross validation [56]. Users are able to select a suitable SVM tool based on their own requirements. Here one of the most popular open-source toolboxes, LIBSVM (Library for Support Vector Machines) developed by Chang and Lin [42], is adopted to demonstrate the application on LE prediction. In LIBSVM, each instance in the training set possessed one target value (class label) and several features (attributes). In the testing set, only the features are required for each instance. The objective of SVM is to generate a model from the training set that facilitated the prediction of the target value of each instance in the testing set. A peptide corresponded to an instance and the target value (1 or −1) represents whether that peptide is an epitope. Each peptide contains three feature values including Epidex_i^2, Epidex_i^3, and Epidex_i^4. For example, a 20-mer peptide is decomposed into 19 AAS_i^2 subsegments, and the corresponding epitope index of this peptide is obtained by taking the average of 19 Epidex_i^2 from the corresponding AAS_i^2. Similarly, the feature values of Epidex_i^3 and Epidex_i^4 can be obtained by calculating the averages of 18 Epidex_i^3 and 17 Epidex_i^4 subsegments, respectively.

As previously described, sometimes the sample data points are crossly distributed in a feature space and cannot be separated by a linear hyperplane. In that case, a kernel function transformation might be able to provide a solution by adding an additional dimension for sample points. However, there is no straightforward decision or theoretical methods to decide what kind of kernel functions provides the best results for a given dataset; trial and error on experimenting with different kernel functions is the only way to find the best function. In this example, the experimental dataset was used to construct an SVM model based on three feature values and the target values of each epitope and non-epitope. Four common kernel functions including linear, polynomial, RBF, and sigmoid were provided by LIBSVM. We examined all these four kernel functions with a fivefold cross-validation (*see* **Note 2**). The training dataset was equally divided into five different subsets; four of the subsets were used for training the model and the last one was used for testing the model. These processes were repeated five times with each individual subset used as the testing subset. Based on the cross-validation results in this case, the RBF kernel function provided the best performance regarding the collected samples and it was selected as the default kernel function. Subsequently, the RBF kernel function was applied to train the whole collected positive and negative datasets again and construct the final SVM classifier for future LE prediction.

3.6 Performance Measurement

To evaluate the performance of an epitope prediction system, either peptide- or residue-level evaluation could be applied according to the characteristics of prediction system and testing databases. For example, several epitope/non-epitope datasets provided by LBtope only contain fragments of antigen proteins and are required to be verified as LEs or not. Definitely, the peptide-level evaluation will be an appropriate selection in this application. However, if a whole-antigen protein sequence was considered as the query data and the prediction system could provide flexible length LE candidates, then a residue-level evaluation method is more suitable. Therefore, residue-level evaluation method was applied to the LEPS prediction system. There are five commonly used indicators for measuring effectiveness of a prediction system, which include (1) *sensitivity* (*SEN*), defined as the percentage of epitopes that are correctly predicted as epitopes; (2) *specificity* (*SPE*), defined as the percentage of non-epitopes that are correctly predicted as non-epitopes; (3) *positive predictive value* (*PPV*), defined as the probability that a predicted epitope is an epitope; (4) *accuracy* (*ACC*), defined as the proportion of correctly predicted peptides; and (5) *Matthews correlation coefficient* (*MCC*), which is a measure of the predictive performance incorporating both SEN and SPE into a single value between −1 and +1. A merged and non-redundant testing dataset called AHP dataset was created by Wang et al. from AntiJen, HIV, and PC datasets, which contained 193 proteins with 843 non-overlapping epitopes [30]. These three datasets were selected to balance the variations in each dataset including variations in epitope length and the physicochemical properties of antigens. It should be noticed that all antigen proteins selected in testing dataset must be different from the training dataset and all repeated proteins should be removed in advance. In this example the SVM-based learning system could achieve a performance of SEN of 27.0 %, SPE of 84.2 %, ACC of 72.5 %, PPV of 32.1 %, and MCC of 10.4 %. One point should be mentioned here: The PPV indicated the rate of identifying real epitopes among all positive predicted candidates, and it is one of the most important factors for immunologists in conducting vaccine development. Reduction of the false-positive candidates can significantly improve the effectiveness and efficiency of identifying the real epitopes. Compared to other systems, the LEPS also showed its excellent performance for all different testing datasets. All the comparison details can be referred to Wang et al. [30].

4 Conclusion

In silico linear B-cell epitope prediction is definitely an important procedure for designing peptide-based vaccine, diagnostic test, disease prevention, treatment, antibody production, and other related applications. Successful prediction facilitates biomedical researchers and immunologists in reduction of experimental time and overall costs. In this chapter, current linear B-cell epitope prediction methods, collected databases, and available online systems based on machine learning techniques are comprehensively reviewed. Especially, one of the most applied machine learning methods, the support vector machine classifier, is also briefly introduced for non-computer-background readers. To demonstrate the usage of combining popularly used propensity scales and machine learning techniques, an LE prediction system proposed by Wang et al. was also introduced through step-by-step description. We hope that the details can help beginners to find some important materials, to clarify some fundamental questions, and to gain a better understanding of applying machine learning approaches on epitope prediction. Though research on epitope analysis and related prediction systems were booming in the last two decades, the performance on B-cell epitope prediction is yet to be satisfied comparing to T-cell epitope prediction. This is mainly due to the complexity of B-cell epitope binding mechanisms, variable lengths of B-cell epitopes, and limited availability of resolved antigen and antibody–antigen complex structures. In addition, still several other existing problems are waiting for solutions to improve recent prediction tools. The first problem is the deficiency of a comprehensive learning dataset containing both verified epitope and verified non-epitope peptides. Especially the verified non-epitope dataset plays an important role to improve prediction accuracy. In general, many trained non-epitope samples were generated by artificially random approaches which might lead to a wrong and biased learning model. Except the verified epitopes on both positive and negative classes, the total number of non-redundant epitopes should be large enough for reliable training and similar sequences should be avoided for appearing in both training and testing datasets. It must be very careful to remove redundant and high similarity sequences from a testing dataset comparing to the training dataset. Overly optimistic performance may be obtained if similar protein sequences appeared in both datasets. Most importantly, if more three-dimensional antigen structures or antigen–antibody complexes could be crystallized and determined, the binding mechanisms between antibodies and antigens could be understood in more details. Hence, it might be possible to categorize B-cell epitopes into several different interaction mechanisms and various levels of immunogenic potency. The machine learning approaches could also be advanced from a two-class to a multiple-class classifier. With all these considerations, the integrated prediction system based on sequence and structural features could become more reliable and practical.

5 Notes

1. Since the original propensity scores are in different scales, a normalization procedure needs to be performed before combining each antigenicity score. The final antigenicity scores for each reside therefore appear within a range of [0, 1].
2. In addition to the selection of kernel functions, several parameters for each kernel function are required to be identified. LIBSVM provides a simple parameter selection tool based on grid-search approach to try different combinations of parameters heuristically.

Acknowledgements

This work is supported by the Center of Excellence for the Oceans, National Taiwan Ocean University, and National Science Council, Taiwan, R.O.C. (NSC 102-2321-B-019-001 and NSC 102-2221-E-019-059 to T.-W. Pai).

References

1. Davies DR, Cohen GH (1996) Interactions of protein antigens with antibodies. Proc Natl Acad Sci U S A 93(1):7–12
2. Korber B, LaBute M, Yusim K (2006) Immunoinformatics comes of age. PLoS Comput Biol 2(6):e71. doi:10.1371/journal.pcbi.0020071
3. Greenbaum JA, Andersen PH, Blythe M, Bui HH, Cachau RE, Crowe J, Davies M, Kolaskar AS, Lund O, Morrison S, Mumey B, Ofran Y, Pellequer JL, Pinilla C, Ponomarenko JV, Raghava GP, van Regenmortel MH, Roggen EL, Sette A, Schlessinger A, Sollner J, Zand M, Peters B (2007) Towards a consensus on datasets and evaluation metrics for developing B-cell epitope prediction tools. J Mol Recognit 20(2):75–82. doi:10.1002/jmr.815
4. Yang X, Yu X (2009) An introduction to epitope prediction methods and software. Rev Med Virol 19(2):77–96. doi:10.1002/rmv.602
5. Salimi N, Fleri W, Peters B, Sette A (2010) Design and utilization of epitope-based databases and predictive tools. Immunogenetics 62(4):185–196. doi:10.1007/s00251-010-0435-2
6. El-Manzalawy Y, Honavar V (2010) Recent advances in B-cell epitope prediction methods. Immunome Res 6(Suppl 2):S2. doi:10.1186/1745-7580-6-S2-S2
7. Caoili SE (2010) Benchmarking B-cell epitope prediction for the design of peptide-based vaccines problems and prospects. J Biomed Biotechnol, vol. 2010, Article ID 910524:1–14, doi:10.1155/2010/910524
8. Yao B, Zheng D, Liang S, Zhang C (2013) Conformational B-cell epitope prediction on antigen protein structures: a review of current algorithms and comparison with common binding site prediction methods. PLoS One 8(4):e62249. doi:10.1371/journal.pone.0062249
9. Jardetzky TS, Brown JH, Gorga JC, Stern LJ, Urban RG, Strominger JL, Wiley DC (1996) Crystallographic analysis of endogenous peptides associated with HLA-DR1 suggests a common, polyproline II-like conformation for bound peptides. Proc Natl Acad Sci U S A 93(2):734–738
10. Patronov A, Doytchinova I (2013) T-cell epitope vaccine design by immunoinformatics. Open Biol 3(1):120139. doi:10.1098/rsob.120139
11. Barlow DJ, Edwards MS, Thornton JM (1986) Continuous and discontinuous protein antigenic determinants. Nature 322(6081):747–748
12. Van Regenmortel MH (2006) Immunoinformatics may lead to a reappraisal of the nature of B cell epitopes and of the feasibility of synthetic peptide vaccines. J Mol Recognit 19(3):183–187
13. Salimi N, Fleri W, Peters B, Sette A (2012) The immune epitope database: a historical retrospective of the first decade. Immunology 137(2):117–123. doi:10.1111/j.1365-2567.2012.03611.x
14. Kringelum JV, Nielsen M, Padkjaer SB, Lund O (2013) Structural analysis of B-cell epitopes in antibody:protein complexes. Mol Immunol 53(1–2):24–34. doi:10.1016/j.molimm.2012.06.001
15. Kolaskar AS, Tongaonkar PC (1990) A semi-empirical method for prediction of antigenic determinants on protein antigens. FEBS Lett 276(1–2):172–174
16. Alix AJ (1999) Predictive estimation of protein linear epitopes by using the program PEOPLE. Vaccine 18(3–4):311–314
17. Odorico M, Pellequer JL (2003) BEPITOPE: predicting the location of continuous epitopes and patterns in proteins. J Mol Recognit 16(1):20–22
18. Saha S, Raghhava GPS (2004) BcePred: prediction of continuous B-cell epitopes in antigenic sequences using physico-chemical properties. LNCS 3239:197–204
19. Saha S, Raghava GP (2006) Prediction of continuous B-cell epitopes in an antigen using recurrent neural network. Proteins 65(1):40–48
20. Larsen JE, Lund O, Nielsen M (2006) Improved method for predicting linear B-cell epitopes. Immunome Res 2:2
21. Sollner J, Mayer B (2006) Machine learning approaches for prediction of linear B-cell epitopes on proteins. J Mol Recognit 19(3):200–208
22. Chen J, Liu H, Yang J, Chou KC (2007) Prediction of linear B-cell epitopes using amino acid pair antigenicity scale. Amino Acids 33(3):423–428. doi:10.1007/s00726-006-0485-9
23. El-Manzalawy Y, Dobbs D, Honavar V (2008) Predicting linear B-cell epitopes using string kernels. J Mol Recognit 21(4):243–255. doi:10.1002/jmr.893
24. El-Manzalawy Y, Dobbs D, Honavar V (2008) Predicting flexible length linear B-cell epitopes. Comput Syst Bioinformatics Conf 7:121–132
25. Chang HT, Liu CH, Pai TW (2008) Estimation and extraction of B-cell linear epitopes predicted by mathematical morphology approaches. J Mol Recognit 21(6):431–441. doi:10.1002/jmr.910
26. Sweredoski MJ, Baldi P (2009) COBEpro: a novel system for predicting continuous B-cell

epitopes. Protein Eng Des Sel 22(3):113–120. doi:10.1093/protein/gzn075
27. Rubinstein ND, Mayrose I, Martz E, Pupko T (2009) Epitopia: a web-server for predicting B-cell epitopes. BMC Bioinformatics 10:287. doi:10.1186/1471-2105-10-287
28. Rubinstein ND, Mayrose I, Pupko T (2009) A machine-learning approach for predicting B-cell epitopes. Mol Immunol 46(5):840–847. doi:10.1016/j.molimm.2008.09.009
29. Wee LJ, Simarmata D, Kam YW, Ng LF, Tong JC (2010) SVM-based prediction of linear B-cell epitopes using Bayes feature extraction. BMC Genomics 11(Suppl 4):S21. doi:10.1186/1471-2164-11-S4-S21
30. Wang HW, Lin YC, Pai TW, Chang HT (2011) Prediction of B-cell linear epitopes with a combination of support vector machine classification and amino acid propensity identification. J Biomed Biotechnol 2011:432830. doi:10.1155/2011/432830
31. Gao J, Faraggi E, Zhou Y, Ruan J, Kurgan L (2012) BEST: improved prediction of B-cell epitopes from antigen sequences. PLoS One 7(6):e40104. doi:10.1371/journal.pone.0040104
32. Yao B, Zhang L, Liang S, Zhang C (2012) SVMTriP: a method to predict antigenic epitopes using support vector machine to integrate tri-peptide similarity and propensity. PLoS One 7(9):e45152. doi:10.1371/journal.pone.0045152
33. Lin SY, Cheng CW, Su EC (2013) Prediction of B-cell epitopes using evolutionary information and propensity scales. BMC Bioinformatics 14(Suppl 2):S10
34. Singh H, Ansari HR, Raghava GP (2013) Improved method for linear B-cell epitope prediction using antigen's primary sequence. PLoS One 8(5):e62216. doi:10.1371/journal.pone.0062216
35. Emini EA, Hughes JV, Perlow DS, Boger J (1985) Induction of hepatitis A virus-neutralizing antibody by a virus-specific synthetic peptide. J Virol 55(3):836–839
36. Parker JM, Guo D, Hodges RS (1986) New hydrophilicity scale derived from high-performance liquid chromatography peptide retention data: correlation of predicted surface residues with antigenicity and X-ray-derived accessible sites. Biochemistry 25(19):5425–5432
37. Vihinen M, Torkkila E, Riikonen P (1994) Accuracy of protein flexibility predictions. Proteins 19(2):141–149
38. Debelle L, Wei SM, Jacob MP, Hornebeck W, Alix AJ (1992) Predictions of the secondary structure and antigenicity of human and bovine tropoelastins. Eur Biophys J 21(5):321–329
39. Blythe MJ, Flower DR (2005) Benchmarking B cell epitope prediction: underperformance of existing methods. Protein Sci 14(1):246–248
40. Noble WS (2006) What is a support vector machine? Nat Biotechnol 24(12):1565–1567. doi:10.1038/nbt1206-1565
41. Vapnik VN (1995) The nature of statistical learning theory. Springer, New York
42. Chang C-C, Lin C-J (2011) LIBSVM: a library for support vector machines. ACM Trans Intell Syst Technol 2(3):1–27. doi:10.1145/1961189.1961199
43. Joachims T (1999) Making large-scale support vector machine learning practical. Advances in kernel methods. MIT Press, Cambridge, MA, pp 169–184
44. Gasteiger E, Hoogland C, Gattiker A, Duvaud S, Wilkins MR, Appel RD, Bairoch A (2005) Protein identification and analysis tools on the ExPASy server. In: Walker JM (ed) The proteomics protocols handbook. Humana, Totowa, NJ, pp 571–607
45. Deleage G, Roux B (1987) An algorithm for protein secondary structure prediction based on class prediction. Protein Eng 1(4):289–294
46. Kyte J, Doolittle RF (1982) A simple method for displaying the hydropathic character of a protein. J Mol Biol 157(1):105–132
47. Karplus PA, Schulz GE (1987) Refined structure of glutathione reductase at 1.54 A resolution. J Mol Biol 195(3):701–729
48. Alix AP (1997) Molecular modeling of globular proteins: strategy 1D ⇒ 3D: secondary structures and epitopes. In: Vergoten G, Theophanides T (eds) Biomolecular structure and dynamics, vol. 342. NATO ASI series. Springer, Netherlands, pp 121–150. doi:10.1007/978-94-011-5484-0_6
49. Giardina CR, Dougherty ER (1988) Morphological methods in image and signal processing. Prentice-Hall, Inc., Upper Saddle River, NJ
50. Maragos P, Schafer RW (1987) "Morphological Filters" part I and II. IEEE Trans Signal Process 35(8):1153–1184
51. Serra J (1982) Image analysis and mathematical morphology, vol 1. Academic, New York
52. Serra J (1988) Image analysis and mathematical morphology, vol 2. Academic, New York

53. Liu C-H (2007) Mathematical morphology based biochemical property filters for linear epitope prediction. National Taiwan Ocean University, Keelung, Taiwan
54. Yousef M, Jung S, Showe LC, Showe MK (2008) Learning from positive examples when the negative class is undetermined–microRNA gene identification. Algorithms Mol Biol 3:2. doi:10.1186/1748-7188-3-2
55. Saha S, Bhasin M, Raghava GP (2005) Bcipep: a database of B-cell epitopes. BMC Genomics 6:79. doi:10.1186/1471-2164-6-79
56. Ivanciuc O (2007) Applications of support vector machines in chemistry. Reviews in computational chemistry. Wiley, Hoboken, NJ, pp 291–400. doi:10.1002/9780470116449.ch6

Chapter 13

Mimotope-Based Prediction of B-Cell Epitopes

Jian Huang, Bifang He, and Peng Zhou

Abstract

Mimotopes are peptides mimicking epitopes on the corresponding antigen. They can be obtained via panning the phage-displayed random peptide library against the corresponding monoclonal antibody or specific sera. Besides mimotopes however, the experimental results also include all kinds of unwanted sequences called "target-unrelated peptides," which often interfere with the subsequent experimental and computational analyses. Nevertheless, the prediction of B-cell epitopes based on the experimental result of phage display has shown to be a promising and reliable strategy with acceptable precision. In this chapter, we summarize mimotope-based prediction of B-cell epitopes under three conditions and focus on protocols and tips for retrieving, cleaning, and decoding the data from phage display technology.

Key words B-cell epitope, Epitope prediction, Mimotope, Target-unrelated peptide, Phage display, Peptide library

1 Introduction

1.1 B-Cell Epitopes

B-cell epitopes are special regions on an antigen that can bind to the corresponding B-cell receptors or antibodies [1]. Traditionally, they are grouped into two categories: continuous and discontinuous epitope. A continuous epitope is also known as a sequential or a linear epitope since it is just a continuous segment of antigen sequence. On the contrary, a discontinuous epitope includes a few separate residues and several segments that are not continuous at the sequence level but adjacent at the 3D-structure level due to protein folding. Accordingly, this type of B-cell epitope is also called conformational epitope. As the basis of the interaction between antigen and antibody, mapping B-cell epitopes on the antigen is a basic task in immunology study.

Though there are several experimental methods for B-cell epitope mapping, they are labor intensive, time consuming, and costly. Even worse, the experimental methods sometimes are technically difficult or even impossible. For example, some antigens such as

Rajat K. De and Namrata Tomar (eds.), *Immunoinformatics*, Methods in Molecular Biology, vol. 1184, DOI 10.1007/978-1-4939-1115-8_13,

membrane proteins are hard to get crystallized and to obtain a crystal of any antigen-antibody complex partly depends on luck. In such conditions, X-ray diffraction method is not always applicable. The 3D structure of protein in solution can be resolved using nuclear magnetic resonance (NMR) spectroscopy; however, it is only available for small proteins rather than big antigen-antibody complexes. In addition, experimental methods sometimes are not suitable for an immense scale. Therefore, B-cell epitope prediction has been widely used to lower the experimental workload [1].

1.2 B-Cell Epitope Prediction

More than 30 years ago, the study of B-cell epitope prediction started from locating continuous B-cell epitope in silico using propensity scales of amino acids to profile an antigen sequence. Such methods were thought to be helpful for finding peptide candidates capable of eliciting antibodies that were also cross-reacting with the whole antigen, therefore benefiting the development of novel epitope-based vaccines or diagnostics [2]. However, the systemic evaluation work by Blythe and Flower showed that even the best scales had only marginally better performances than random, implying that better scales and methods were needed for predicting continuous B-cell epitopes [3]. Indeed, new scales [4, 5] and machine-learning methods demonstrate improved performances for the prediction of continuous B-cell epitopes [2].

In recent years, more attention is paid to the prediction of discontinuous B-cell epitopes given the fact that nearly 90 % native B-cell epitopes are conformational [6]. At present, dozens of algorithms and programs for the prediction of discontinuous B-cell epitopes are available [2]. These programs, based on either antigen structure or sequence, have succeeded in some case studies.

The prediction of B-cell epitopes mentioned above, either for continuous or for discontinuous epitopes, is based only on antigen sequence or structure and tries to map major regions on the antigen that may induce humoral immune response. However, it is context dependent for a given B-cell epitope. Therefore, a new paradigm for B-cell epitope prediction needs not only antigen information but also relevant data such as sequences of corresponding antibody and mimotopes.

1.3 Mimotope-Based Prediction of B-Cell Epitopes

Mimotopes are peptides mimicking an epitope on the corresponding antigen [7]. They are usually obtained through screening the random peptide library using special sera or a monoclonal antibody. Since an antigen and its mimotopes competitively bind to the same monoclonal antibody or sera, their physicochemical characteristics and spatial arrangement are believed to be similar [8]. Thus, the native epitope can reasonably be located when its mimotopes are compared to the antigen sequence or structure.

The random peptide library used most widely nowadays is constructed by displaying inserted peptides on the coat proteins of phages. The technology, now called phage display, was first introduced by George Smith [9]. Screening a phage-displayed random peptide library is termed as biopanning or panning in short, which usually includes the following steps. First of all, special sera or a monoclonal antibody is fixed on the surface of disks or beads and then incubated with a random peptide library. The antibody used to screen the library is termed as target, and the corresponding antigen is called template. Then phages with no affinity to the target are washed away with buffer. Later on, bound phages are eluted with the target, the template, or stronger buffer only. At last, the bound phages are amplified by infecting bacteria to build a secondary library, which is then used for the next round of panning. After several rounds, eluted phage clones are randomly picked and sequenced [10].

However, there are not only mimotopes but also "target-unrelated peptides" in the sequencing results [11]. Target-unrelated peptides (TUPs) creep into the results due to growth advantage or binding to other components of the screening system rather than binding to the target. If mimotopes are signal that can be used to predict the corresponding epitope, TUPs are noise that will interfere with the prediction.

Nevertheless, mimotope-based prediction of B-cell epitopes represents a new trend in B-cell epitope prediction, which utilizes not only antigen information but also context data relevant to the corresponding antibody [12]. Therefore, it has shown to be a promising and reliable approach with acceptable precision among various types of methods for B-cell epitope prediction so far. In this chapter, we focus on protocols and tips for retrieving, cleaning, and decoding the data from phage display technology to interpret and predict B-cell epitopes more reasonably and accurately.

2 Data and Methods

2.1 Data Retrieval

1. Retrieve and manually check all sequencing data of the phage display experiment provided by your sequencer or contracted company (*see* **Note 1**).
2. Retrieve the antigen sequence file from the UniProt Knowledgebase (http://www.uniprot.org/) if it is known [13].
3. Retrieve the antigen structure file from the PDB database (http://www.rcsb.org) if it is resolved.
4. For computational biologists who want to develop and evaluate tools for mimotope-based prediction of B-cell epitopes, get data from the MimoDB database (http://immunet.cn/mimodb) [14].

2.2 Data Cleaning (See Note 2)

1. Use TUPScan (http://immunet.cn/sarotup/cgi-bin/TUPScan.pl) to clean peptides with any known TUP motif [11].
2. Use MimoSearch (http://immunet.cn/sarotup/cgi-bin/MimoSearch.pl) to clean peptides identical to those in the MimoDB database with various targets [14].
3. Use MimoBlast (http://immunet.cn/sarotup/cgi-bin/MimoBlast.pl) to clean peptides highly similar to those in the MimoDB database with various targets [14].
4. Use PhD7Faster (http://immunet.cn/sarotup/cgi-bin/PhD7Faster.pl) to clean peptides possibly with growth advantage if they are from the Ph.D.-7 library (New England Biolabs) [15].
5. Use SABinder (http://immunet.cn/sarotup/cgi-bin/SABinder.pl) to clean peptides that possibly bind to strept avidin if this protein is a component of the screening system rather than the target.

2.3 Data Decoding

2.3.1 When the Antigen Structure Is Known

1. Use EpiSearch (http://curie.utmb.edu/episearch.html) [16], Pepitope (http://pepitope.tau.ac.il/) [17], and PepMapper (http://informatics.nenu.edu.cn/PepMapper) [18] to decode all mimotopes and predict the conformational epitopes on the antigen structure (*see* **Note 3**).
2. Find common results from the above predictions if possible.

2.3.2 When Only the Antigen Sequence Is Known

1. Use JalView (http://www.jalview.org) to align all mimotopes and identify the consensus sequence [19].
2. Use the SAROTUP suite to check if the consensus sequence is specific enough [11].
3. Format all mimotopes and the consensus sequence for BLAST (http://blast.ncbi.nlm.nih.gov).
4. Perform a local BLAST analysis to see if any segment of the antigen is homologous to the consensus sequence or any mimotope which indicates linear B-cell epitopes (*see* **Note 4**).
5. Use PRATT (http://www.ebi.ac.uk/Tools/pfa/pratt) to discover patterns that are conserved in mimotopes [20].
6. Use MimoScan (http://immunet.cn/sarotup/cgi-bin/MimoScan.pl) to check if the patterns mentioned above are specific enough.
7. Use the ScanProsite tool (http://prosite.expasy.org/scanprosite) to scan if the antigen contains patterns that are conserved in mimotopes. The matching segment also indicates linear B-cell epitopes [21].
8. Perform a BLAST against the PDB database using the antigen sequence.

9. Use Swiss-Model (http://swissmodel.expasy.org) to build homology models for the antigen if the highest similarity of **step 8** is above 20 % [22].
10. Repeat Subheading 2.3.1 to predict possible conformational epitopes on the antigen structure model.

2.3.3 When the Antigen Is Unknown

1. Use JalView (http://www.jalview.org) to align all mimotopes and identify the consensus sequence [19].
2. Use the SAROTUP suite to check if the consensus sequence is specific enough [11].
3. Perform a BLAST analysis for the consensus sequence and for each peptide, respectively, as well to see if it is homologous to any known protein (http://blast.ncbi.nlm.nih.gov). The consensus sequence or some mimotopes could have high similarities with the antigen that induces the sera or the monoclonal antibody (*see* **Note 4**).
4. Use PRATT (http://www.ebi.ac.uk/Tools/pfa/pratt) to discover patterns that are conserved in mimotopes [20].
5. Use MimoScan (http://immunet.cn/sarotup/cgi-bin/MimoScan.pl) to check if the patterns mentioned above are specific enough.
6. Use the ScanProsite tool (http://prosite.expasy.org/scanprosite) to scan if any known protein matches patterns that are conserved in mimotopes [21]. The antigen that induces the sera or the monoclonal antibody may contain patterns that are conserved in mimotopes.
7. If the antigen is determined after the analyses above, repeat **steps 8–10** of Subheading 2.3.2.

3 Notes

1. All sequencing data should be manually checked, especially when there are ambiguous amino acids. For example, sequences interrupted by a stop codon are often reported by sequencing companies when they are panning from the Ph.D. series of libraries (New England Biolabs). However, the stop codon TAG should be translated into a glutamine when the library was amplified in ER2738 strain or any supE strain.
2. TUP is not an absolute concept. It depends on the context. For example, the phage clones binding to plastic are usually taken as TUPs. However, they are signals rather than noise when plastic is just the intended target. Besides, both experimental and computational methods cannot eliminate TUPs completely.

In addition, there might be phage clones that have special affinity to the target but also have growth advantage at the same time. Thus, some signals might be excluded too.

3. Both Pepitope and PepMapper have integrated two programs, respectively, i.e., PepSurf and Mapitope, MimoPro and Pep-3D-Search. It is natural to select the common results from these tools as the reliable prediction if their results intersect. According to our tests, all these programs have their own pros and cons. For example, Mapitope is not suitable for a single mimotope or a very small set of mimotopes due to its statistical method; PepSurf cannot decode mimotope longer than 14 amino acids. However, these tools complement each other, succeeding in overlapping but different cases. Thus, it is very difficult to determine beforehand which is better when they give quite different prediction results.
4. As mimotopes are usually 6–20 amino acids long, it should be considered to optimize parameters for short nearly exact matches when performing BLAST. According to the BLAST documentations at NCBI, the following parameters are recommended: expect value cutoff 20,000, word size 2, scoring matrix PAM30, composition-based statistics off, and the filters of low-complexity regions off.

Acknowledgments

This chapter was supported in part by the National Natural Science Foundation of China under Grant 61071177 and the Program for New Century Excellent Talents in University (NCET-12-0088).

References

1. Tomar N, De RK (2010) Immunoinformatics: an integrated scenario. Immunology 131(2): 153–168. doi:10.1111/j.1365-2567.2010.03330.x
2. Castelli M, Cappelletti F, Diotti RA, Sautto G, Criscuolo E, Dal Peraro M, Clementi N (2013) Peptide-based vaccinology: experimental and computational approaches to target hypervariable viruses through the fine characterization of protective epitopes recognized by monoclonal antibodies and the identification of T-cell-activating peptides. Clin Dev Immunol 2013:12. doi:10.1155/2013/521231
3. Blythe MJ, Flower DR (2005) Benchmarking B cell epitope prediction: underperformance of existing methods. Protein Sci 14(1):246–248. doi:10.1110/ps.041059505
4. Huang J, Honda W, Kanehisa M (2007) Predicting B cell epitope residues with network topology based amino acid indices. Genome Inform Int Conf Genome Inform 19:40–49
5. Huang J, Kawashima S, Kanehisa M (2007) New amino acid indices based on residue network topology. Genome Inform Int Conf Genome Inform 18:152–161
6. Huang J, Honda W (2006) CED: a conformational epitope database. BMC Immunol 7:7. doi:10.1186/1471-2172-7-7
7. Geysen HM, Rodda SJ, Mason TJ (1986) A priori delineation of a peptide which mimics a discontinuous antigenic determinant. Mol Immunol 23(7):709–715
8. Huang J, Gutteridge A, Honda W, Kanehisa M (2006) MIMOX: a web tool for phage display

based epitope mapping. BMC Bioinformatics 7:451
9. Smith GP (1985) Filamentous fusion phage: novel expression vectors that display cloned antigens on the virion surface. Science 228(4705): 1315–1317
10. Smothers JF, Henikoff S, Carter P (2002) Tech. sight. Phage display. Affinity selection from biological libraries. Science 298(5593):621–622. doi:10.1126/science.298.5593.621
11. Huang J, Ru B, Li S, Lin H, Guo FB (2010) SAROTUP: scanner and reporter of target-unrelated peptides. J Biomed Biotechnol 2010:101932. doi:10.1155/2010/101932
12. He B, Mao C, Ru B, Han H, Zhou P, Huang J (2013) Epitope mapping of Metuximab on CD147 using phage display and molecular docking. Comput Math Methods Med 2013:6. doi:10.1155/2013/983829
13. UniProt C (2013) Update on activities at the Universal Protein Resource (UniProt) in 2013. Nucleic Acids Res 41(Database issue):D43–D47. doi:10.1093/nar/gks1068
14. Huang J, Ru B, Zhu P, Nie F, Yang J, Wang X, Dai P, Lin H, Guo FB, Rao N (2012) MimoDB 2.0: a mimotope database and beyond. Nucleic Acids Res 40(Database issue):D271–D277. doi:10.1093/nar/gkr922
15. Ru B, 't Hoen PAC, Nie F, Lin H, Guo FB, Huang J (2014) PhD7Faster: predicting clones propagating faster from the Ph.D.-7 phage display peptide library. J Bioinform Comput Biol 12(1): 1450004. doi:10.1142/S021972001450005X
16. Negi SS, Braun W (2009) Automated detection of conformational epitopes using phage display peptide sequences. Bioinform Biol Insights 3:71–81
17. Mayrose I, Penn O, Erez E, Rubinstein ND, Shlomi T, Freund NT, Bublil EM, Ruppin E, Sharan R, Gershoni JM, Martz E, Pupko T (2007) Pepitope: epitope mapping from affinity-selected peptides. Bioinformatics 23(23):3244–3246
18. Chen W, Guo WW, Huang Y, Ma Z (2012) PepMapper: a collaborative web tool for mapping epitopes from affinity-selected peptides. PLoS One 7(5):e37869. doi:10.1371/journal.pone.0037869
19. Waterhouse AM, Procter JB, Martin DM, Clamp M, Barton GJ (2009) Jalview Version 2—a multiple sequence alignment editor and analysis workbench. Bioinformatics 25(9):1189–1191. doi:10.1093/bioinformatics/btp033
20. Jonassen I (1997) Efficient discovery of conserved patterns using a pattern graph. Comput Appl Biosci CABIOS 13(5):509–522
21. de Castro E, Sigrist CJ, Gattiker A, Bulliard V, Langendijk-Genevaux PS, Gasteiger E, Bairoch A, Hulo N (2006) ScanProsite: detection of PROSITE signature matches and ProRule-associated functional and structural residues in proteins. Nucleic Acids Res 34(1):W362–W365. doi:10.1093/nar/gkl124
22. Bordoli L, Kiefer F, Arnold K, Benkert P, Battey J, Schwede T (2009) Protein structure homology modeling using SWISS-MODEL workspace. Nat Protoc 4(1):1–13. doi:10.1038/nprot.2008.197

Chapter 14

Hybrid Methods for B-Cell Epitope Prediction

Approaches to the Development and Utilization of Computational Tools for Practical Applications

Salvador Eugenio C. Caoili

Abstract

Many computational approaches to B-cell epitope prediction have been published, including combinations of previously proposed methods, which complicates the tasks of further developing such computational approaches and of selecting those most appropriate for practical applications (e.g., the design of novel immunodiagnostics and vaccines). These tasks are considered together herein to clarify their close but often overlooked interrelationship, thereby providing a guide to their performance in mutual support of one another, with emphasis on key physicochemical and biological considerations that are relevant from an applications perspective. This aims to assist investigators in performing either or both tasks, with the overall goals of successfully applying computational tools towards practical ends and of generating informative new data towards iterative improvement of the tools, particularly as regards the design of peptide-based immunogens for eliciting the production of antipeptide antibodies that modulate biological activity of protein targets via functionally relevant cross-reactivity in relation to the phenomena of protein folding and protein disorder.

Key words Epitope prediction, B-cell epitopes, Proteins, Peptides, Antibodies, Protein folding, Antibody–antigen binding, Immunogenicity, Cross-reactivity, Biological activity

1 Introduction

1.1 B-Cell Epitope Prediction

B-cell epitope prediction is the computational identification of molecular or supramolecular (e.g., protein quaternary) structural features that are potential targets for immune recognition via binding by immunoglobulins (e.g., antibodies). The said features (i.e., B-cell epitopes) each comprise atoms that come into direct contact with the paratope (i.e., epitope-binding site) upon binding by immunoglobulin, although they may also comprise other atoms that are nonetheless important for the binding process (e.g., to maintain or assume a particular conformational state that is recognized by the paratope) [1]. Whereas virtually any sufficiently large

Rajat K. De and Namrata Tomar (eds.), *Immunoinformatics*, Methods in Molecular Biology, vol. 1184,
DOI 10.1007/978-1-4939-1115-8_14, © Springer Science+Business Media New York 2014

structure (e.g., biomolecule or synthetic molecule) may contain one or more B-cell epitopes, the published literature has focused mainly on peptidic B-cell epitopes (i.e., of proteins and peptides), for which their constituent amino-acid residues (rather than individual atoms) are often the structural units of practical interest in computational analyses (e.g., of molecular sequences) and experimental procedures (e.g., for B-cell epitope construction by either peptide synthesis or protein engineering). Likewise, the discussion herein is primarily concerned with peptidic B-cell epitopes described in terms of amino-acid residue sequences, albeit noting that three-dimensional structure and atomic-level details are crucial considerations for B-cell epitope prediction (e.g., where solvent-accessible surface area of atoms is used to estimate the affinity of paratope–epitope binding [2, 3]).

B-cell epitope prediction may be conceptualized in categorical and deterministic terms, as if structural features could be dichotomized into mutually exclusive epitope and non-epitope categories whose members consistently behaved as such empirically. The simplicity of this approach is appealing in that it readily lends itself to very straightforward computational analyses (e.g., for binary classification), but it overlooks the highly context-dependent and stochastic nature of B-cell epitope recognition by immunoglobulins, which is subject to biological variability that manifests even under tightly controlled experimental conditions (e.g., where immunization experiments are performed on essentially homogeneous animal populations maintained in the same uniform laboratory environment). A more realistic alternative approach is to identify physicochemically plausible candidate structural features (e.g., accessible surface patches having dimensions conceivably compatible with binding by a paratope) as putative B-cell epitopes, which may be further characterized quantitatively as to particular properties of interest (e.g., potential to actually elicit the production of antibodies having a particular biological activity, such as the ability to neutralize a pathogen or virulence factor thereof) on the basis of pertinent physicochemical and biological information (e.g., juxtaposition of various putative B-cell epitopes vis-á-vis functional correlates of their structure). This can be understood in relation to the concepts of antigenicity, immunogenicity, and cross-reactivity, as discussed below.

1.2 Antigenicity, Immunogenicity, and Cross-Reactivity

As typically understood in the context of B-cell epitope prediction, antigenicity is the potential for immune recognition via binding by immunoglobulin, whereas immunogenicity is the potential to actually elicit the production of such immunoglobulin as antibodies. Hence, immunogenicity implies antigenicity insofar as binding by surface immunoglobulins of B-cell receptors is a prerequisite for B-cell activation and consequent antibody production. However, antigenicity may be poorly correlated with immunogenicity (e.g., in the setting of immune tolerance to B-cell epitopes among self proteins). Nevertheless, antigenicity may be manifest via cross-reaction whereby antibodies produced in response to one structurally unique B-cell epitope also bind another, which itself may be of low immunogenicity. This cross-reaction, the potential for which is hereafter referred to as cross-reactivity, may occur as a consequence of functional similarity between the two B-cell epitopes that enables the same antibodies to bind both, although the structural basis for such similarity may be inapparent at the sequence level (e.g., where highly divergent sequences adopt dissimilar conformations, but in doing so place B-cell epitope atoms in spatial configurations that enable binding by the same paratope). More generally, sequence similarity may be poorly correlated with cross-reactivity, especially considering the possibility of low cross-reactivity between B-cell epitopes differing from each other in only a single chemical group [4] and even between those of identical sequence where the overall structural contexts (e.g., as regards conformation and surface accessibility) are sufficiently dissimilar (e.g., where the same sequence forms part of either a short unfolded peptide or a large folded protein, such that the sequence becomes conformationally constrained and at least partially buried in the protein).

Antigenicity and immunogenicity are often regarded as all-or-none binary functions of molecular sequence and possibly other variables, which is implicit in the selective labeling of sequences as predicted antigenic or immunogenic B-cell epitopes and is tantamount to the problematic conceptualization of B-cell epitope prediction in categorical and deterministic terms that has been alluded to in Subheading 1.1. To avoid this, both antigenicity and immunogenicity may be cast alternatively as continuous variables that in turn are functions of key variables defined by relevant experimental contexts. Minimally, such a context could be described as a biphasic process comprising two consecutive temporal phases, namely an initial immunization phase and a subsequent immunoassay phase, with each phase associated with a particular antigen (i.e., structure potentially recognizable by the immune system and possibly containing one or more B-cell epitopes) [2]. The immunization phase would be associated with an antigen (hereafter referred to as the immunogen) administered to elicit an immune response, particularly the production of antibodies that are capable of binding the said antigen (i.e., the immunogen) while the immunoassay phase would be associated with an antigen (hereafter referred to as the immunoassay antigen) serving as an immunologic probe to detect antibodies via an immunoassay.

According to the biphasic scheme outlined thus far (which assumes the placement of appropriate experimental controls, e.g., by obtaining antibodies from unimmunized animals for use as negative-control antibodies in the immunoassay phase), antigenicity may be equated with the affinity of antibodies (generated during the immunization phase) for the immunoassay antigen while immunogenicity may be expressed as the quantity (e.g., expressed as either the concentration or fraction) of such antibodies. Insofar as the affinity would depend on the particular immunoassay antigen used, immunogenicity would be evaluated using an immunoassay antigen that is either the immunogen itself (used in the immunization phase) or an appropriate surrogate thereof (e.g., a fragment of the immunogen for which the antibodies have sufficiently high affinity that enables their detection via the immunoassay) to avoid possible failure to detect the antibodies. Where the immunogen and immunoassay antigen are nonidentical, cross-reactivity may be expressed as the affinity of the antibodies for the immunoassay antigen relative to their affinity for the immunogen or some antigenically similar surrogate thereof (again with the proviso that the surrogate is bound by the antibodies with sufficiently high affinity). Affinity is thus fundamental to B-cell epitope prediction and therefore warrants further elaboration below.

1.3 Affinity of Paratope–Epitope Interactions

For a bimolecular reversible-association reaction between a receptor R (e.g., antibody) and a ligand L (e.g., antigen) for the reversible formation of a receptor–ligand complex RL (e.g., antibody–antigen complex), affinity (i.e., strength of binding) may be expressed as the equilibrium association constant K_A, or, equivalently, as the equilibrium dissociation constant K_D, noting that:

$$K_A = 1 / K_D \tag{1}$$

and also:

$$K_A = k_{on} / k_{off} \tag{2}$$

where k_{on} and k_{off} are the on- and off-rate constants for the association and dissociation reactions, respectively, which under equilibrium binding conditions implies:

$$K_D = [R][L] / [RL] \tag{3}$$

with the square brackets denoting the molar concentrations of the indicated molecular species (in which case K_D is equivalent to the value of [R] at which [RL] = [L], i.e., at which [RL] is half-maximal). Furthermore, K_A is related to the free-energy change ΔG of association, under equilibrium binding conditions as:

$$K_A = \exp(-\Delta G / RT) \tag{4}$$

where R is the gas constant and T the temperature. Hence, antigenicity may be expressed as K_A or K_D for some combination of immunogen, immunoassay antigen, and possibly other experimental parameters, notably in relation to the constraint of B-cell affinity maturation [5], which in vivo serves as a means of increasing antibody affinity for antigen yet also tends to limit the said affinity by imposing both an upper bound on k_{on} (corresponding to the diffusion-control limit) and a lower bound on k_{off} (related to the kinetics of receptor-mediated antigen endocytosis by B cells via surface immunoglobulin) [6] in accordance with Eq. 2, as detailed below.

Again denoting receptor and ligand by R and L, respectively (cf. Eq. 3), the on-rate constant for diffusion-limited collisional encounters between R and L is given by:

$$k_{\text{on}}^{\text{max}} = 4\pi a(D_{\text{R}} + D_{\text{L}})(N / 1{,}000) \quad (5)$$

where a is the encounter distance; D_{R} and D_{L} are the diffusion constants; and N is Avogadro's number (i.e., 6.02×10^{23} mol^{-1}); and $k_{\text{on}}^{\text{max}}$ is thus obtained in M^{-1} s^{-1} for a in cm and both D_{R} and D_{L} in cm^{2} s^{-1} [7]. For binding of antibodies to small protein antigens in solution, $k_{\text{on}}^{\text{max}}$ is typically in the range of 10^{5}–10^{6} M^{-1} s^{-1} [8, 9], and antibodies in general are thus unlikely to have much higher values of $k_{\text{on}}^{\text{max}}$ [6]. For capture of IgG-class antibodies from solution by sufficiently large or immobilized antigens where the antigen diffusion constant is practically zero, $k_{\text{on}}^{\text{max}}$ may be estimated from Eq. 5 using an encounter distance of 1.57×10^{-8} cm and an antibody diffusion constant of 4×10^{-7} cm^{2} s^{-1}, yielding a value of 4.75×10^{7} M^{-1} s^{-1} [7].

To estimate the lower bound for k_{off} during affinity maturation, endocytic antigen uptake may be modeled to a first approximation with classical Michaelis–Menten kinetics applied to transmembrane transport [5], in which case the Michaelis–Menten constant is given by:

$$K_{\text{M}} = (k_{\text{off}} + k_{\text{in}}) / k_{\text{on}} \quad (6)$$

where k_{in} is the rate constant for endocytic internalization of surface immunoglobulin-bound antigen. As K_{M} is numerically equivalent to the antigen concentration at which the steady-state rate of antigen internalization is half-maximal, a decrease in K_{M} confers a competitive advantage upon B cells thus enabled to internalize antigen more rapidly, such that k_{on} may approach $k_{\text{on}}^{\text{max}}$ (from Eq. 5) during affinity maturation; but k_{off} is unlikely to decrease much further below k_{in} as the gain in competitive advantage would become negligible [6], such that K_{M} approaches the lower limit of $k_{\text{in}}/k_{\text{on}}$ according to Eq. 6. Considering the reported half-life of 8.5 min for surface immunoglobulins endocytosed on Epstein–Barr virus-transformed B-lymphoblastoid cells [10], the lower bound for k_{off} during affinity maturation is estimated to be in the range of 10^{-4} to 10^{-3} s^{-1} under the assumption that two to three surface-immunoglobulin half-lives is the upper limit beyond which increased immune-complex stability confers no competitive advantage [6].

Affinity maturation thus tends to limit antibody affinity for antigen with a ceiling value on K_A, although higher values may be realized where affinity maturation is bypassed (e.g., through protein engineering or artificial affinity selection via yeast display [11]).

2 Theoretical Framework

2.1 General Considerations

From the standpoint of translational research, B-cell epitope prediction is useful insofar as it facilitates efficient utilization of available empirical data (e.g., on biomolecular sequences and structures) as bases for developing practical applications, notably immunodiagnostic and immunization (e.g., vaccination) strategies for health care. B-cell epitope prediction provides support for the design of antigens either as immunologic probes for antibody detection or as immunogens for eliciting the production of antibodies that in turn serve as immunologic probes for antigen detection. This aids in the development of immunodiagnostics where the goal is either detection of antibodies as indicators of immune status (e.g., in relation to past vaccination) and as markers of disease (e.g., due to infection) or detection of antigens (e.g., of pathogens) as markers of disease. The role of B-cell epitope prediction in the design of antigens as immunogens also provides support for the development of strategies for prophylactic or therapeutic induction of antibody-mediated immunity, either using the immunogens as vaccine components for active immunization or to elicit the production of antibodies for passive immunization. These all necessitate application-specific and highly context-dependent analyses of potential paratope–epitope interactions, with attention to antigenicity, immunogenicity, and cross-reactivity as defined in Subheading 1.2, as these relate to safety and efficacy for biomedical applications in particular.

Given a set of candidate epitopes (e.g., defined as oligopeptidic subsequences of a polypeptide chain corresponding to a protein of interest), B-cell epitope prediction may thus be regarded as primarily concerned with estimating antigenicity as the affinity (cf. Eqs. 1–4) for binding of each candidate epitope by a complementary paratope, considering that sufficiently high affinity is necessary for practically significant binding (e.g., detectable antibody–antigen binding in a diagnostic immunoassay, or pathogen neutralization by antibodies elicited via immunization with a vaccine). Additionally, estimation of immunogenicity among the candidate epitopes is also often practically relevant, particularly where immune responses might be strongly biased towards antibody production against highly immunogenic epitopes, possibly at the expense of antibody production against other epitopes (e.g., overlapping or neighboring the said immunogenic epitopes [12]) that may thus be less readily exploited as targets for both immunodiagnosis and immunization.

Furthermore, estimation of cross-reactivity between candidate epitopes is crucial where envisioned applications entail cross-reaction of antibodies with epitopes different from those against which the antibodies were produced (e.g., where antipeptide antibodies are to cross-react with protein antigens); and this may be approached by estimating cross-reactivity as affinity for binding of a particular target epitope (e.g., on a protein) by a paratope complementary to a given candidate epitope (e.g., of a peptide), insofar as cross-reactivity is a conceptual generalization of antigenicity as paratope affinity for an epitope.

In view of the preceding considerations, B-cell epitope prediction is ultimately useful mainly as a means to avoid undesirable antibody–antigen interactions, particularly nonspecific ones, in immunodiagnosis and immunization. In immunodiagnosis, nonspecific antibody–antigen interactions (involving reagents that are either antibodies for antigen detection or antigens for antibody detection) tend to produce potentially misleading false-positive results (e.g., the incorrect diagnosis of disease where none is present). In immunization, where the desired outcome is the production or administration of antibodies that prevent or otherwise control disease, the antibodies may fail to confer protective immunity and even produce unintended deleterious effects (e.g., autoimmune disease or paradoxic antibody-mediated enhancement of infectious disease), especially where the antibodies bind nonspecifically (e.g., cross-reacting with self antigens, thereby producing autoimmune disease). These undesirable interactions can be avoided by restricting the repertoire of epitopes presented for paratope binding, either in an immunoassay or in the course of an immunization process. Where the goal is to design antigens as immunologic probes for antibody detection, B-cell epitope prediction aids in identifying epitopes that are sufficiently antigenic yet of restricted cross-reactivity in the sense of having acceptably low potential for binding by nonspecifically cross-reactive antibodies (which are irrelevant as immunodiagnostic markers but may nonetheless produce false-positive results). Likewise, where the goal is to design immunogens for the production of antibodies (either as immunologic probes for antigen detection or as mediators of immunity), B-cell epitope prediction aids in identifying epitopes that are sufficiently antigenic (and, by extension, immunogenic in some biologically realistic context) yet elicit the production of antibodies with adequately restricted cross-reactivity (i.e., reacting or cross-reacting strongly with their intended targets but only minimally with other antigens). In all cases, the particular quantitative criteria for sufficient antigenicity and restricted cross-reactivity must be operationally defined in an application-specific manner, in relation to physicochemical and biological constraints (e.g., on antibody concentration in the system under consideration, possibly distinguishing between physically feasible and medically acceptable upper limits [5]).

2.2 Physicochemical and Biological Correlates

In the overall scheme of B-cell epitope prediction, paratope–epitope binding affinity (Eqs. 1–4) links underlying biomolecular structure (e.g., described in terms of protein sequences and atomic coordinates) to biological function (i.e., biological outcomes of paratope–epitope binding). B-cell epitope prediction may thus be resolved into two sequential steps, namely estimation of antigenicity as paratope–epitope binding affinity and estimation of biological impact as regards potential for paratope–epitope binding. These steps are discussed in turn below, noting they may be performed only implicitly in certain (e.g., machine learning) approaches to B-cell prediction. Throughout the discussion, the binding affinity is equated with the free-energy change ΔG of paratope–epitope binding (as introduced in Eq. 4).

To estimate ΔG, one or more candidate epitopes must first be defined, possibly by partitioning one or more antigen structures of interest into physicochemically plausible candidate B-cell epitopes. Considering a hypothetical monomeric protein antigen that exists in either completely unfolded (e.g., denatured) or completely folded (e.g., native) forms, the unfolded form (i.e., a single polypeptide chain) may be partitioned into a set of overlapping oligopeptide sequences while the folded form may be partitioned into a set of overlapping surface patches, such that each oligopeptide sequence and surface patch is regarded as a candidate epitope. The candidate epitopes may be more precisely defined in relation to geometric constraints on antibody–antigen binding. For example, assuming a typical circular antibody footprint diameter of 20 Å [13, 14], each hexapeptide sequence may be regarded as a candidate epitope for the unfolded form given a peptide contour length of 3.5 Å per residue [3, 15]; and each patch may be defined as being centered on the C^{α} atom of a solvent-accessible (i.e., surface-exposed) residue and encompassing all other solvent-accessible residues whose C^{α} atoms are within a 10-Å radius of the central residue [2], possibly applying additional constraints to avoid defining physically implausible fragmented or ring-like patches [16] as well as patches comprising residues located within paratope-inaccessible concavities (e.g., deep and narrow crevices that may be solvent-accessible yet paratope-inaccessible) [13]. Such residue-oriented analyses yield arbitrarily defined candidate epitopes, especially considering that B-cell epitopes are actually defined at an atomic level of detail (albeit with unavoidably imprecise delineation of their boundaries) [1]; nonetheless, the resulting candidate epitopes are reasonably representative structures amenable to production via conventional residue-oriented experimental procedures (e.g., of peptide synthesis and protein engineering). This approach can be generalized to proteins with quaternary structure and those forming parts of supramolecular structures such as viral capsids and biological membranes, such that surface patches may comprise residues of more than one polypeptide chain. Furthermore, intrinsic disorder (i.e., conformational flexibility as dynamic random coils) of proteins at the levels of intradomain sequence segments (e.g., surface-exposed loops), domains and entire polypeptide chains may be a prominent feature of B-cell epitopes even in the native state, such that surface patches may comprise disordered residues. Where the antigen of primary interest is a single protein (e.g., a virulence factor of a pathogen), additional antigens (e.g., proteins of a host organism infected by the pathogen) may also be considered for the purpose of defining a set of application-relevant candidate epitopes insofar as antibodies targeting the protein might also cross-react with any of the other antigens (e.g., thereby producing false-positive immunodiagnostic results or mediating autoimmune reactions).

Once a set of candidate epitopes has thus been defined, ΔG may be estimated on the basis of their structure. In principle, a set of plausible paratopes could be defined (e.g., by stochastic simulation of adaptive immunoglobulin-diversity generation via germline heavy- and light-chain variable-region gene rearrangement, with paratope structure prediction); subsequently, docking between the paratope and candidate-epitope structures could be performed and evaluated to estimate affinity, for example, using structural energetics [17, 18], which quantitatively relates ΔG to the associated changes in apolar and polar solvent-accessible surface area that occur in the binding process. However, this would require detailed organism-specific immunobiological knowledge on the species selected for antibody production and would also be very computationally expensive. As an alternative, much less computationally demanding approximations may be employed that treat the paratope implicitly on the basis of candidate-epitope structure. For instance, structural energetics may still be employed by assuming that all paratope and epitope solvent-accessible surface area is completely lost (i.e., buried at the paratope–epitope interface) upon binding and that the paratope thus loses the same amounts of apolar and polar solvent-accessible surface area as the epitope [2, 3]. Where affinity maturation is anticipated to limit the maximum binding affinity (e.g., during immunization in vivo) as discussed in Subheading 1.3, an affinity ceiling can be applied such that any initial affinity estimates (e.g., obtained using structural energetics) exceeding the ceiling value are revised downward to the said value (i.e., such that ΔG is set to the value corresponding to the affinity ceiling). Additionally, cross-reactivity may be estimated as ΔG for binding alternative candidate epitopes that differ from one another in terms of conformational state (e.g., in the case of an antipeptide paratope cross-reacting with a folded protein comprising the sequence of the unfolded peptide recognized by the antipeptide paratope) or sequence (e.g., in the case of an antipathogen paratope cross-reacting with a host self antigen comprising a sequence similar to that recognized by the antipathogen paratope).

Having estimated ΔG for the candidate epitopes, the biological impact of paratope–epitope binding may in turn be estimated. In particular, immunogenicity may be at least partially inferred from antigenicity estimated in terms of ΔG, insofar as primary antibody responses to B-cell epitopes are driven by affinity of paratope–epitope binding [19]. However, potential for high-affinity binding (e.g., as suggested by estimated ΔG values) may fail to manifest as immunogenicity where immune tolerance is established towards particular epitopes (e.g., of self antigens or other normally tolerated antigens), yet such tolerance might also be broken by immunization with the epitopes (e.g., if these are covalently coupled to highly immunogenic carrier molecules); in either case, B-cell epitope prediction can facilitate identification of possible problems related to immune tolerance (e.g., by identifying candidate epitopes that are likely to be tolerated and for which breaking of tolerance might result in autoimmune or allergic reactions). Moreover, differences in ΔG among physically overlapping B-cell epitopes may manifest as immunodominance (i.e., bias of immune responses towards a subset of so-called immunodominant epitopes) whereby antibody responses are mounted against immunodominant epitopes, thus suppressing antibody responses to nonimmunodominant epitopes such that the immunogenicity of epitopes may be masked.

The mechanistic basis of immunodominance among B-cell epitopes may be understood in terms of the affinity-driven competition among B-cell clones for antigen to recruit T-cell help (as expressed in Eq. 6), with higher paratope–epitope affinity favoring B-cell clonal expansion and antibody production; this leads to earlier and more extensive binding of immunodominant epitopes by antibodies that interferes with antigen capture by B-cells whose surface immunoglobulins bind nonimmunodominant epitopes, particularly where the nonimmunodominant epitopes physically overlap with the immunodominant epitopes such that binding of the immunodominant epitopes by antibodies sterically blocks subsequent binding of nonimmunodominant epitopes by paratopes [3, 12]. Hence, immunodominance among candidate epitopes of an antigen may be modeled to a first approximation as a thermodynamically determined hierarchical steric-exclusion phenomenon, by ranking the candidate epitopes in order of decreasing estimated paratope affinity and subsequently defining the subset of predicted immunodominant epitopes to include each candidate epitope whose paratope affinity exceeds that of every candidate epitope with which it physically overlaps, assuming that the affinity ranking is maintained in the course of immunization (including affinity maturation where applicable) [3]. However, the actual situation may be complicated by tolerance to particular epitopes (in which case tolerated epitopes might be better excluded from the affinity ranking, assuming that tolerance towards them would likely be maintained rather than broken in the course of immunization), original antigenic sin (i.e., immune imprinting with consequent bias towards production of antibodies against epitopes that are identical or antigenically similar to epitopes recognized by memory B cells generated during previous immune responses) [20, 21] and quantitatively similar paratope affinities among epitopes (such that the deterministic affinity ranking may fail to capture the stochastic emergence of immunodominance, e.g., where potential for bistability arises due to comparable paratope affinities between epitopes); and steric exclusion conceivably can occur where immunodominant and nonimmunodominant epitopes are non-overlapping yet placed sufficiently close to (e.g., abutting) one another.

Given the immediately preceding considerations, immunodominance thus poses a major challenge for B-cell epitope prediction. Nevertheless, the problem of predicting immunodominance may be at least partially circumvented for certain applications, particularly where immunogen structure is investigator-determined (e.g., where the immunogen is a vaccine component). Towards this end, the most straightforward approach is to physically isolate an epitope such that it becomes essentially the only epitope presented to the immune system during immunization; this approach is applicable where peptide-based immunogens are administered to elicit the production of antipeptide antibodies (e.g., that cross-react with protein antigens), with oligopeptide sequences serving as immunogenic epitopes, although the typically low intrinsic immunogenicity of these sequences often necessitates their incorporation into larger immunogenic structures such as multiple antigenic peptides (each comprising multiple copies of the same epitope) [22–25] or carrier-containing constructs (wherein the epitope of interest is covalently linked to a macromolecular or particulate immunogenic carrier, in which case carrier- and linker-associated epitopes may elicit extraneous antibody production) [3, 26]. Alternatively, epitopes may be selectively modified (e.g., via site-directed mutagenesis) in order to decrease their immunogenicity (e.g., by substituting alanine residues for residues with larger sidechains), thereby favoring immunodominance of a particular epitope of interest (e.g., that physically overlaps with the modified epitopes); such modification of epitopes may be employed for immune refocusing [27], whereby immune imprinting (e.g., original antigenic sin) is overcome by selectively deleting epitopes recognized by memory B cells, thereby unmasking the immunogenicity of other epitopes for targeting by antibody responses.

Although issues of immunogenicity and immunodominance are fundamental to B-cell epitope prediction, it is more generally concerned with biological function of antibodies produced via immunization (often without much regard to actual degrees of immunogenicity and immunodominance, provided that an adequate supply of antibodies can be generated, e.g., as monoclonal antibodies for passive immunization [28]). Very broadly construed, such biological function encompasses binding of antigen and any downstream events (e.g., modulation of antigen biological activity, or activation of various immune effector mechanisms), either in vitro or in vivo. In vitro, antigen binding itself may thus be regarded as the minimal criterion for biological function; and such binding in itself (i.e., exclusive of downstream events, except perhaps the binding of appropriately labeled detector constructs) may be sufficient for the purpose of detecting either antigens or antibodies (e.g., via an immunoassay for some immunodiagnostic application).

Resolving antigen binding by antibodies into epitope binding by paratopes, each paratope–epitope binding event may be regarded either as a reaction if the epitope is identical to that which elicited production of the antibodies during immunization or as a cross-reaction in all other cases. Where cross-reactions are due to an antibody capable of binding both a peptide and a cognate protein (i.e., comprising the sequence of the peptide), the antibody may be either an antipeptide antibody (e.g., produced by immunization with the peptide) cross-reactive with the protein or an antiprotein antibody (e.g., produced by immunization with the protein) cross-reactive with peptide. With regard to such peptide–protein cross-reactions, a distinction has been proposed between so-called genuine and apparent cross-reactions, which purportedly involve native and denatured proteins, respectively [29–31]. As originally proposed, the distinction is rooted in the classical paradigm of perceived dichotomy between completely folded native and completely unfolded denatured proteins (as evident in the reference to denatured-protein epitopes as unfoldons [29]), which has been more recently supplanted by a much more nuanced view of protein disorder observed at various levels of both native and denatured protein structure, with protein structural and functional versatility manifest as intrinsic protein disorder [32, 33] as well as coupled protein folding and binding [34]. Hence, genuine and apparent cross-reactions might be more meaningfully distinguished on the basis of antibody-mediated modulation of native-protein function [35] rather than the presence or degree of protein disorder; but such a distinction would nonetheless be problematic operationally insofar as protein denaturation occurs to varying degrees in real biological samples, possibly without complete loss of protein function where only partial denaturation occurs. At any rate, some denaturation is likely to occur among protein immunogens administered in vivo (e.g., during vaccination) prior to paratope–epitope binding, such that polyclonal antiprotein antibodies may inevitably bind denatured protein to some extent [36]. As regards disordered regions of proteins, these may serve as crucial targets (i.e., epitopes or epitope-containing sites) for binding by (i.e., cross-reaction of) antipeptide antibodies elicited by oligopeptide subsequences of the proteins; for although antipeptide antibodies may fail to cross-react with protein epitopes that are conformationally constrained due to folding (e.g., because the paratopes recognize the epitope sequences in conformations different from those in the protein [37]), such antibodies may nonetheless cross-react with disordered protein epitopes (which can readily adopt conformations recognized by the paratopes [5]).

Beyond antigen binding per se, antibody biological function may manifest as modulation of antigen biological function, as already alluded to above for cross-reaction of antipeptide antibodies with native cognate proteins. Thus, where the antigen is a protein, binding of the antigen by antibody may modulate protein function (e.g., enzyme catalysis, in which case the antibody might function as a competitive inhibitor binding an active site or as a noncompetitive inhibitor binding an allosteric site, notwithstanding the possibility that the antibody might function instead as an allosteric activator), possibly via relatively nonspecific mechanisms (e.g., where antibodies target neither active sites nor allosteric-inhibitor binding sites on an enzyme, yet nonetheless interfere with its catalytic activity by binding it and thus hindering its diffusion especially through gel-like biological matrices, possibly also sequestering it within immune complexes and thereby restricting substrate access to its active sites). Such functional modulation may be regarded as protective if it favors a desirable prophylactic or therapeutic outcome (e.g., neutralization of a pathogen). Other protective effects may be less directly realized via downstream immune effector mechanisms such as complement pathways and immune-complex clearance by professional phagocytes (e.g., macrophages). For example, initial activation of the classical complement pathway by immune complexes may be augmented by consequent activation of the alternative complement pathway, leading to opsonization of the immune complexes (thus favoring their immune clearance via internalization by professional phagocytes) and possibly even the formation of membrane attack complexes (thus perforating target biological membranes such as the outer membranes of Gram-negative bacterial pathogens); and incorporation of IgG into the immune complexes may itself lead to their opsonization. Such downstream immune mechanisms may be activated regardless of the target antigen conformational state (i.e., native or denatured), provided that antibodies bind the target antigen to form suitable immune complexes. Accordingly, antibody-mediated protective effects may be classified as instances of either class-I or -II protectivities [38], which, respectively, correspond to direct effects of antibody binding and to indirect effects mediated by immune mechanisms activated by immune complexes. Although this suggests that antigen binding by antibodies protects against disease, such binding may actually promote or enhance disease, as exemplified by the phenomena of autoantibody-mediated hypersensitivity and of antibody-dependent enhancement of infection. Antibody-dependent enhancement of infection, which occurs in infections due to a wide variety of pathogens ranging from viruses to protozoa [39], tends to exhibit nonlinear antibody-concentration dependence, such that enhancement occurs when antibody concentrations fall below some critical threshold for protection [40], with the threshold itself possibly dependent on factors such as complement-component concentrations [41] and pathogen state (e.g., developmental stage) [42, 43].

2.3 Quantitative Biological Effects of Paratope–Epitope Binding

To quantitatively describe the biological impact of paratope–epitope binding and thus complete the task of B-cell epitope prediction for a particular practical application, affinity (as expressed in Eqs. 1–4) can serve as the fundamental basis for the entire theoretical framework [2, 3]. An affinity value in the form of an association-constant value K_A is in itself a quantitative description of a molecular binding process, yet it is practically meaningful only in relation to an application-specific cutoff value K_A^{cut}, at or above which the binding process would be deemed useful for its intended purpose.

Typically, K_A^{cut} must be sufficiently high to elicit antibody production in the first place (which for in-vivo immunization likely requires K_A^{cut} in the submillimolar range [19]), unless immunization is somehow bypassed (e.g., via genetic engineering of antibody-secreting cells). At any rate, all applications dependent on antibody–antigen binding interactions entail the formation of immune complexes, such that K_A^{cut} must correspond to some required minimum extent of immune-complex formation, which typically can be expressed as a fraction f of antigen bound by antibody. For example, the equilibrium value of f under conditions of antibody excess relative to antigen may be obtained from Eq. 3 as:

$$f = \frac{1}{1 + (K_D / [Ab])} \tag{7}$$

where K_D is the equilibrium dissociation constant (as in Eqs. 1 and 3) and [Ab] is the antibody concentration, such that K_D is the value of [Ab] at which half of the binding sites for antibody are occupied. For extension of applicability to more complex cases where cooperative binding interactions occur, Eq. 7 may be generalized to a form of the Hill equation [44, 45]:

$$f = \frac{1}{1 + (K_D / [Ab])^n} \tag{8}$$

where n is an interaction coefficient whose value is a lower bound on the number of binding sites for antibody on the antigen (such that

Eq. 8 reduces to Eq. 7 where n is unity). Combining Eqs. 1 and 8, the equilibrium association constant K_A may be obtained as:

$$K_A = [\text{Ab}]^{-1} \left(\frac{f}{1-f} \right)^{\left(\frac{1}{n}\right)} \quad (9)$$

such that K_A^{cut} is thus obtained if the minimum anticipated values are assumed for [Ab] and f. The minimum anticipated value for [Ab] may, for instance, correspond to a threshold antibody titer for ascertaining immunity (e.g., following vaccination) or for diagnosing disease (e.g., following infection). As to the minimum anticipated value for f, this may be selected on the basis of technical considerations (e.g., for immunodiagnostics to detect antibodies, in relation to the amount of antigen and sensitivity limit of equipment used, so as to ensure antibody detection at the threshold antibody titer).

For immunodiagnostics wherein formed immune complexes may irreversibly dissociate (e.g., in immunosorbent assays wherein immobilized immune complexes are subject to extensive washing), an important additional consideration is that the off-rate k_{off} for immune-complex dissociation must be sufficiently low for the immune complexes to persist long enough to enable their detection, in which case K_A^{cut} might be computed according to Eq. 2 using the upper (i.e., diffusion-control) limit for the on-rate k_{on} as given by Eq. 5 (which is plausible where antibodies are obtained via immunization that entails affinity maturation) and some technically appropriate upper bound on k_{off} (noting that k_{off} is the reciprocal of the mean lifetime of an individual immune complex); hence, decreasing K_A^{cut} (e.g., by increasing the upper bound on k_{off}) would tend to increase the analytical detection limit, thus rendering the diagnostic test less sensitive and therefore more prone to yielding false-negative results.

For applications that aim to neutralize the biological activity of a particular target antigen (e.g., toxin or other pathogen), K_A^{cut} may be computed on the basis of a practically acceptable upper bound on the probability p of some biological outcome (e.g., lethality or infection). In the absence of antibodies to the target antigen, p may be estimated phenomenologically as:

$$p = \frac{1}{1 + (C_{\text{m}} / C)^b} \quad (10)$$

where b is an empirical coefficient, C is the concentration of causative agent, and C_{m} is the value of C at which p is half-maximal, such that C_{m} is the median effective concentration (in the sense of tending to produce the biological effect in half the members of a given test population, e.g., of whole organisms or of cells in vitro) and may, for example, represent the median lethal concentration

(LC_{50}) or the median infectious concentration (IC_{50}, distinct from the median inhibitory concentration in pharmacological studies) where the biological effect is lethality or infection, respectively; more generally [46], C may represent the dose of causative agent, with C_m thus representing the median effective dose (e.g., the median lethal dose LD_{50} or the median infective dose ID_{50}), typically normalized per unit body mass.

For a simple bimolecular association (i.e., as described by Eqs. 1–4) between a causative agent and a neutralizing antibody thereto, with binding equilibrium rapidly attained upon introduction of the causative agent into the system or subsystem of interest (e.g., a cell culture, or the total extracellular body fluid or circulating blood plasma of a living host organism), the concentration C of free causative agent (i.e., not bound by antibody) may be estimated from the total concentration C_0 of free and antibody-bound causative agent combined assuming an excess of antibody, as:

$$C = \frac{C_0}{1 + ([\mathrm{Ab}] / K_\mathrm{D})} \tag{11}$$

where [Ab] is the antibody concentration and K_D is the equilibrium dissociation constant (as in Eqs. 1 and 3). Combining Eqs. 1, 10, and 11, the equilibrium association constant K_A may be obtained as:

$$K_\mathrm{A} = [\mathrm{Ab}]^{-1} \left(\frac{C_0}{C_\mathrm{m}} \left(\frac{1-p}{p} \right)^{\left(\frac{1}{b}\right)} - 1 \right) \tag{12}$$

such that $K_\mathrm{A}^{\mathrm{cut}}$ is thus obtained if the maximum anticipated values are assumed for C_0, [Ab] and p. The maximum anticipated values for C_0 and p may be selected on the basis of clinical or experimental exposure or infection data in conjunction with Eq. 10, whereas the corresponding values for [Ab] may be selected in line with physiological constraints (e.g., upper bounds on endogenous antibody production, where active immunization is employed) subject to technical and safety considerations (e.g., as regards upper bounds on administration of exogenous antibodies, where passive immunization is employed) [5]. For communicable infectious diseases, the value for p may be selected on the basis of the critical immunization threshold q_c (i.e., the minimum proportion of immune individuals in a host population required for herd immunity, that is, population-level resistance to epidemic spread of disease), which for well-mixed host populations (i.e., wherein individual hosts are homogeneously interacting with one another) may be estimated as:

$$q_c = 1 - \frac{1}{R_0} \tag{13}$$

where R_0 is the basic reproduction number (i.e., the number of secondary cases that one case is expected to produce in a completely susceptible host population); hence, the complement of q_c (i.e., $1 - q_c$) may be regarded as an upper bound on acceptable values of p, considering that p is the expected failure rate for attempts to achieve protective antibody-mediated immunity among individual hosts in the context of Eq. 12 (such that $p \leq 1 - q_c$ would be acceptable assuming that the maximum anticipated antibody concentration were to be realized in every member of the entire host population).

With regard to all scenarios considered thus far, the central task of quantitatively estimating paratope–epitope affinity is further complicated where the problem of potential cross-reaction arises. For example, whereas the application of structural energetics to estimate affinity of an antibody reacting with an epitope (i.e., where the epitope itself elicited production of the antibody) is relatively straightforward, it is much less so for the same antibody cross-reacting with another structurally distinct epitope of a different conformation or sequence, as the structural difference between the epitopes introduces additional contributions of uncertain magnitude (e.g., to account for structural adjustments upon binding) into the calculation of affinity for the cross-reaction [2, 3]. Typically, the affinity of a cross-reaction is lower than that of the corresponding reaction due to suboptimal paratope–epitope complementarity, but the affinity of cross-reaction may exceed that of reaction (e.g., where cross-reaction involves a more conformationally constrained epitope, such that cross-reaction is more thermodynamically favorable due to a decrease in the loss of conformational entropy upon binding by antibody). Hence, predicted affinities are likely to be less accurate for cross-reactions than for reactions, although this may be a relatively minor problem for cross-reaction of antipeptide antibodies with disordered cognate protein regions, particularly where the protein epitopes can readily adopt conformations recognized by the antibodies [5]. At the same time, the possibility of unintended cross-reaction (e.g., of antipathogen antibodies cross-reacting with host self antigens) must be carefully considered in an application-specific manner (e.g., with reference to self antigens of the relevant host species). Taking the example of antibodies to be produced against a peptide fragment of a pathogen protein, computational evaluation of potential for deleterious cross-reactions in a particular host might initially proceed on the basis of sequence comparisons to search for sequences matching that of the peptide among proteins of the relevant host proteome and possibly also proteomes of organisms contributing to the host diet or forming part of the host environment, with the host and other proteomes relevant as regards potential autoimmune and allergic reactions, respectively. Any exact or partial sequence matches thus found could be further investigated as to their possible biological significance, for example, with attention to relative abundance as well as histologic and subcellular localization, considering that low-abundance intracellular host self antigens conceivably are unlikely targets for autoimmune antibodies (but also noting that the contents of at least some host intracellular compartments are accessible to antibodies [47]).

The theoretical framework developed thus far can be incrementally extended on the basis of greater mechanistic detail by means of modeling and simulating immune and infectious processes at multiple levels of structural and functional organization, with paratope–epitope binding affinity still as the basic foundation. This could provide increasingly more appropriate methods for particular specialized applications (e.g., immunodiagnostics and vaccines). Although the present work emphasizes the role of B-cell epitope prediction methods in supporting biomedical and other practical applications, the quantitative approaches described herein for this purpose apply as well to basic experimental research (e.g., studies on antibody-mediated toxin- and pathogen-neutralization in vitro) that nonetheless can yield results useful as benchmark data for evaluation and further refinement of B-cell epitope prediction methods [35], as discussed further below.

2.4 Benchmarking B-cell Epitope Prediction Using Quantitative Data

In the context of B-cell epitope prediction, benchmarking is the process whereby computational predictions of paratope–epitope binding are evaluated against empirical data (i.e., benchmark data). Benchmarking thus facilitates comparison between alternative prediction methods, thereby providing an objective basis for the iterative refinement of such methods. The judicious selection and utilization of benchmark data is a crucial prerequisite to avoid benchmarking errors (i.e., overrating or underrating of prediction methods, which can mislead investigators in their decisions on selecting between methods and modifying these towards superior computational performance) [35, 48].

Numerous and diverse forms of empirical data on paratope–epitope binding are available for use as benchmark data. Among these data, those of primary interest are on outcomes of antibody–antigen binding assays. Such data may be broadly categorized as either qualitative or quantitative. The qualitative data are typically dichotomous, such that they are designated either as positive data if they are associated with empirical evidence of paratope–epitope binding or as negative data otherwise. The quantitative data are much more heterogeneous and are often associated with various units of measurement reflecting the diverse experimental approaches by which they were obtained, although these data are amenable to normalization that enables their coherent aggregation for benchmarking; furthermore, quantitative data may be readily transformed into dichotomous qualitative data by applying a threshold or cutoff value, but this results in loss of information and statistical power [49].

Because of the ease with which qualitative data can be generated experimentally and quantitative data can be converted into qualitative data, qualitative data have historically predominated as the basis for benchmarking B-cell prediction; however, this approach is fundamentally problematic and error-prone [35, 48], for which reason the present work focuses on the use of quantitative benchmark data instead.

The benchmark data are customarily organized into records, each of which comprises three minimal data components, namely structural data on an immunogen, structural data on an antigen used in an immunoassay, and data on the outcome of the immunoassay, with the concept of an immunoassay broadly defined to include structural determination (e.g., X-ray crystallography and NMR spectroscopy for elucidating immune-complex structure); often, the only structural data provided are in the form of sequences. Arguably, other data on immunization conditions (e.g., species of immunized animal) and on immunoassay conditions (e.g., antigen and antibody concentrations as well as temperature) are equally important as input for B-cell epitope prediction methods (e.g., in view of Eqs. 3, 4, and 7–12), but benchmarking has historically focused almost exclusively on the minimal data. Where the immunogens and antigens are peptidic (i.e., peptides or proteins), each putative epitope is a set of amino-acid residues that may be either continuous (i.e., consisting of a single unbroken sequence of contiguous residues, which is typical where at least the immunogen or the antigen is an oligopeptide) or discontinuous (i.e., comprising two or more noncontiguous sequence segments, which is typical where both the immunogen and the antigen are proteins).

Peptidic epitopes may be delineated for monoclonal antibodies either via structural determination (e.g., on inspection of paratope–epitope atomic contacts in NMR structures) or via binding studies (e.g., mapping epitope residues by observing decreases in binding affinity following removal or substitution of residues in antigens), but such epitopes may be much more difficult to delineate unambiguously for polyclonal antibodies due to coexistence of multiple paratopes that bind overlapping epitope sequences; yet, an individual monoclonal antibody or even a panel of monoclonal antibodies from a common source may be unrepresentative of the polyclonal repertoire from which it was originally derived. This dilemma between monoclonal and polyclonal antibodies can be deliberately minimized for polyclonal antipeptide antibodies by using immunogens containing only short oligopeptides, such that the oligopeptides are each likely to contain only one epitope. Hence, benchmarking with quantitative data is considered herein for antipeptide antibodies reacting with peptide antigens and cross-reacting with protein antigens, for clarity of illustration.

Quantitative benchmark data are values of continuous variables, and B-cell epitope prediction methods themselves render predictions (i.e., computational approximations of the benchmark data) as values of continuous variables, although this is often obscured by subsequent dichotomization of predictions for compatibility with available qualitative data. The predictions thus can be used directly as values of continuous variables to evaluate a performance measure such as the Pearson correlation coefficient (PCC) [35, 48]. For two continuous variables X and Y of which paired values (X_i, Y_i) define n data points, the PCC (denoted by r) can be generalized as a weighted PCC (wPCC), such that:

$$r = \frac{\sum_{i=1}^{n} w_i (X_i - \bar{X})(Y_i - \bar{Y})}{\sqrt{\sum_{i=1}^{n} w_i (X_i - \bar{X})^2 \sum_{i=1}^{n} w_i (Y_i - \bar{Y})^2}} \tag{14}$$

where w is a nonnegative weight while $\bar{X}$ and $\bar{Y}$ are weighted arithmetic means both of the form:

$$\bar{Z} = \frac{\sum_{i=1}^{n} w_i Z_i}{\sum_{i=1}^{n} w_i} \tag{15}$$

where Z is a generic continuous variable. If the values of X are empirically obtained while those of Y are corresponding computational predictions, each data point (X_i, Y_i) may be assigned a weight w_i representing the appraised worth of X_i relative to other values of X, such that zero weight is assigned to data points deemed completely worthless while progressively more positive weights are assigned to other data points of increasing appraised worth. The weight thus could be defined as a function of both measurement quality (e.g., as regards accuracy and precision) and informativeness (i.e., the potential usefulness of a particular empirically obtained numeric value in the benchmarking of predictions). For simplicity, the present work focuses mainly on informativeness to define an upper limit on the weight assuming maximum measurement quality (e.g., perfect accuracy and precision).

To benchmark B-cell epitope prediction for antipeptide antibodies reacting with peptide antigens, empirical values for a standard measure of affinity such as K_A or K_D (cf. Eqs. 1–4) may serve as benchmark data, such that the predictions themselves should be approximations of the said values computed on the basis of other pertinent data (e.g., antigen sequence and immunoassay temperature, for structural-energetic calculations), in which case a correlation-coefficient value may be obtained using Eqs. 14 and 15 with all weights set to unity (i.e., using an unweighted PCC). However, this approach conceivably would be problematic if applied to benchmark B-cell epitope prediction for antipeptide antibodies cross-reacting with protein antigens; because the antibodies might bind either or both native and denatured cognate protein antigens, the appropriate selection of protein-antigen structure to be used as the basis for rendering affinity predictions (e.g., using structural energetics) would be unclear if antigen binding per se (as affinity) were measured. Nevertheless, this problem can be circumvented by benchmarking with continuous antibody dose-response data that reflect antibody-mediated modulation (e.g., inhibition) of native-protein biological function (e.g., enzyme catalysis), which is especially appropriate where the envisioned practical application would be to identify target epitopes on native proteins (e.g., to support the design and development of vaccines that elicit the production of antipeptide antibodies binding native proteins and thereby neutralizing protein biological activity).

To facilitate the utilization of benchmark datasets comprising continuous dose-response data on antibody-mediated modulation of biological activity, such data typically can be normalized to yield quotients in the range of zero to unity that represent the magnitude of an observed antibody-mediated biological effect relative to its theoretical or empirically determined maximum magnitude [35, 48]. Each quotient may thus be obtained as:

$$q = B / B_0 \tag{16}$$

where B and B_0 are the observed and maximum magnitudes of the antibody-mediated biological effect, respectively. For antibody-mediated inhibition of biological activity (e.g., enzyme catalytic activity or pathogen infectivity), B may be equated with the observed fractional activity loss due to binding by antibody, such that B_0 is unity (corresponding to complete loss of activity). Likewise, for antibody-mediated host protection against lethal challenge (e.g., with a toxin or pathogen), B may be equated with the observed fractional host survival (i.e., proportion of surviving hosts) due to binding by antibody (e.g., antitoxin or pathogen-neutralizing antibody), such that B_0 is again unity (corresponding to complete protection against lethality).

More generally, B and B_0 are readily defined where q can be interpreted as the probability of a particular functional state (e.g., catalytically active versus inactive, or viable versus nonviable). In the mechanistically simplest cases, this functional state directly corresponds to the binding state (i.e., either free or antibody-bound) of the antigen of interest (e.g., an enzyme with a single catalytic site that is active in the free state but completely inactivated in the antibody-bound state). In such cases, the probability of the functional state is equivalent to the fraction of antigen that is either free or antibody-bound, with the equilibrium value of the antibody-bound fraction f approximated under conditions of antibody excess relative to the antigen according to Eq. 7 or 8. Alternatively, Eqs. 10 and 11 can be used jointly to evaluate the probability p of some biological outcome (e.g., lethality or infection) due to some causative agent (e.g., toxin or pathogen), by first estimating the free concentration C of the said agent (using Eq. 11) and subsequently estimating p (using Eq. 10). Thus, predictive estimates of f and p can be used as values for the empirical quotient q in Eq. 16.

Granted that Eqs. 7, 8, 10, and 11 may be applicable only to relatively simple cases, they nonetheless illustrate the importance of antibody concentration [Ab] in the rendering of predictions that are to be benchmarked against continuous dose-response data normalized as the empirical quotient q according to Eq. 16. In particular, values of q approaching either zero or unity correspond to extremes of [Ab] (i.e., low or high values of [Ab] with negligible or near-maximal biological effects, respectively) and are thus relatively uninformative insofar as estimation of q (e.g., using Eqs. 7, 8, 10, and 11) becomes insensitive to variation in [Ab] in the limit of low or high [Ab]; conversely, the most informative value of q is half unity, which corresponds to the point of maximal sensitivity to variation in [Ab] (e.g., at which the second derivative of f in Eqs. 7 and 8 is zero) in the estimation of q.

Returning to the problem of assigning the weight w per data point for Eq. 14 in light of the immediately preceding considerations, if X is equated with the empirical quotient q in Eq. 16 while Y is obtained as a predictive estimate of q (e.g., by means of Eqs. 7, 8, 10, and 11), w should be maximal where q is half unity and zero where q is either zero or unity; these constraints are satisfied by the Shannon information entropy [50] calculated in bits as:

$$H = -(q\log_2 q + (1-q)\log_2(1-q)) \tag{17}$$

assuming two possible alternative states of the mathematically modeled system (e.g., an enzyme that is either active when free or inactivated when antibody-bound, or a cell that has either survived or died following challenge with a toxin). If the values of q (i.e., benchmark data) under consideration are all of maximum measurement quality, w may be equated with H; otherwise, w may be assigned a value less than H according to limitations of measurement quality (e.g., of accuracy and precision). In other words, H may be regarded as an upper bound on w in the limit of perfect measurement quality.

Further clarification is warranted regarding the choice of H as a measure of informativeness in the present work considering that H has long been recognized instead as a measure of uncertainty, particularly in line with the view of statistical mechanics as an application of information theory [51, 52]. This view holds that uncertainty may be quantitatively expressed as H in terms of a probability distribution for the occupancy of microscopic states available to a thermodynamic system, following the form of Eq. 17 for a two-state system; accordingly, the uncertainty is least if occupancy of exactly one state is completely certain (i.e., with probability equal to unity, corresponding to zero entropy), whereas the uncertainty is greatest for a uniform probability distribution over all the available states (i.e., with all states being equiprobable, e.g., having a probability of half unity for each state in a two-state system). The notion of entropy as uncertainty may be extended to systems for which the states under consideration are mutually exclusive outcomes (e.g., death or survival), such that completely certain outcomes are associated with zero entropy while maximally uncertain (i.e., equiprobable) outcomes are associated with maximum entropy (e.g., one bit for two equiprobable outcomes). From the standpoint of predictively estimating the empirical quotient q in Eq. 16, zero and maximum entropy, respectively, correspond to the most and least trivial predictive tasks in that tolerance for error (e.g., in the estimation of the dissociation constant K_D for use in Eqs. 7, 8, 10, and 11) increases without bound as q approaches either zero or unity, at which values q thus becomes completely uninformative for benchmarking (consistent with the use of H as the weight w in Eqs. 14 and 15 (*see* **Note 1**).

3 Computational Resources

3.1 Databases

As discussed in Subheading 2.4, empirical data on paratope–epitope binding serve as the basis for benchmarking B-cell epitope prediction. Furthermore, these data may be utilized directly to develop B-cell epitope prediction methods (e.g., via machine learning approaches [53, 54]). Databases containing such data are therefore crucial (albeit underutilized) computational resources supporting the advancement of B-cell epitope prediction.

Adopting a broadly inclusive view of databases as organized collections of data in any physical form, the earliest published databases of B-cell epitope data were small collections of qualitative binding data compiled for the express purpose of benchmarking particular prediction methods [55–58]. These databases, which were published as tables in printed manuscripts, comprised data on peptide sequences as objects of peptide–protein cross-reactivity, including data on both antipeptide antibodies cross-reacting with protein antigens and antiprotein antibodies cross-reacting with peptide antigens. The said data were positive data (i.e., based on empirical evidence of cross-reaction). For benchmarking purposes, these positive data (i.e., peptide sequences associated with empirical evidence of cross-reaction) were typically combined with other data assumed to be negative (i.e., peptide sequences for which empirical evidence of cross-reaction was unavailable, either because they yielded negative results in experiments to detect cross-reactions or because such experiments had not been performed in the first place). This approach to the definition of negative data poses a problem where data are thus mislabeled as negative when in fact the associated peptide sequences may actually comprise B-cell epitopes, which might be detectable only under particular experimental conditions (e.g., different from those used for published work reporting negative results) [35]. The problem is further compounded by conflation of data on cross-reactions of antipeptide antibodies with protein antigens and of antiprotein antibodies with peptide antigens, notably where negative results on the latter type of cross-reaction are interpreted as implying that the peptide antigens in question are devoid of B-cell epitopes that could elicit antipeptide antibodies capable of cross-reacting with cognate proteins [3]. Such conflation of data on peptide–protein cross-reactions was motivated at least in part by the extreme paucity of available data in the earliest B-cell epitope databases; however, more recently published B-cell epitope data demonstrate the potential for inconsistencies arising from the practice (e.g., where antiprotein antibodies fail to cross-react with a peptide antigen, which nonetheless can elicit the production of antipeptide antibodies that cross-react with the cognate protein) [3].

Hence, the interrelated problems of defining negative data and conflating data on mechanistically distinct peptide–protein cross-reactions (i.e., involving antipeptide versus antiprotein antibodies) thus arose in the course of attempts to benchmark B-cell epitope prediction using extremely small datasets. The underlying paucity of available benchmark data has been partially mitigated with the accumulation of newer published B-cell epitope data, which have been curated for inclusion in various databases accessible via the Internet. These databases vary in their scope and purpose. For example, the database Bcipep [59, 60] explicitly focuses on supporting peptide-based vaccine design (with emphasis placed on epitope immunodominance, in line with prime-boost vaccination strategies), whereas CED (a conformational epitope database) [61] is specialized for storage and retrieval of detailed three-dimensional structural data on epitopes. Among all such databases, IEDB (Immune Epitope Database) [62–64] exemplifies both breadth and depth of coverage as regards epitope data (including both B-cell and T-cell epitope data, for peptidic and non-peptidic epitopes from biological and synthetic sources). With the notable exception of epitope data on human immunodeficiency virus (HIV), which are accessible via the Los Alamos National Laboratory HIV molecular immunology database (http://www.hiv.lanl.gov/content/immunology) rather than IEDB, IEDB comprises rigorously curated epitope data on an extremely wide variety of antigens including those of numerous pathogens, allergens, and autoantigens (i.e., self antigens targeted by autoimmune responses).

IEDB contains records that each pertain to a B-cell assay (typically an immunoassay to detect antibody–antigen binding) for a particular B-cell epitope structure (e.g., either a continuous or discontinuous peptidic epitope), which may be an individual B-cell epitope or an antigenic region containing one or more such epitopes [62, 65] and is denoted by "Epitope" in data-field names and values (*see* **Note 2**). Each of the said records contains the three minimal data components (i.e., structural data on an immunogen, structural data on an antigen used in an immunoassay, and data on the outcome of the immunoassay) required for benchmarking as outlined in Subheading 2.4. In particular, the records contain data fields defined in relation to the concepts of "1st Immunogen" (i.e., immunogen administered to produce antibodies) and "Antigen" (i.e., antigen used in the B-cell assay), such that each record contains two data fields named "1st Immunogen Epitope Relation" and "Antigen Epitope Relation" (hereafter referred to as the immunogen and antigen fields, respectively), at least one of which may be assigned the value "Epitope." Where both fields are assigned the value "Epitope," the B-cell assay aimed to detect reaction (e.g., of antipeptide antibodies with the peptide administered to elicit their production); otherwise, the B-cell assay aimed to detect cross-reaction (e.g., of antipeptide antibodies with a cognate protein, in which case the immunogen and antigen fields are assigned the values "Epitope" and "Source Antigen," respectively). As regards B-cell assay outcomes, each record contains a data field named "Qualitative Measurement" that is assigned a dichotomous outcome value of either "Positive" or "Negative" (indicating either detectable or undetectable binding, respectively); additionally, each record also contains data fields for capturing quantitative data (including numeric value and measurements units) on assay outcome where such data are available.

IEDB thus contains quantitative data suitable for benchmarking B-cell epitope prediction as outlined in Subheading 2.4. Records containing these data may be retrieved (and downloaded as comma-separated value [CSV] files) via the B Cell Search facility of IEDB, with filtering via appropriate restriction of data-field values (Fig. 1). For example, setting the epitope-type field to "Linear peptide" and both the immunogen and antigen fields to "Epitope" retrieves records on reaction of antipeptide antibodies with peptide antigens; but if the antigen field is set to "Source Antigen" rather than "Epitope," this retrieves records on cross-reaction of antipeptide antibodies with cognate protein antigens [5]. In this manner, undesirable conflation of benchmark data (e.g., on antipeptide antibodies cross-reacting with protein antigens versus antiprotein antibodies cross-reacting with peptide antigens) is avoided. Filtering on other data fields (e.g., representing immunized animal species and antibody clonality) may likewise be performed to control for possible confounding variables (e.g., interspecies variation in antibody responses, or potentially unrepresentative monoclonal-antibody data in the context of immunization with peptide-based vaccines [48]). Additional filtering with respect to B-cell assay type

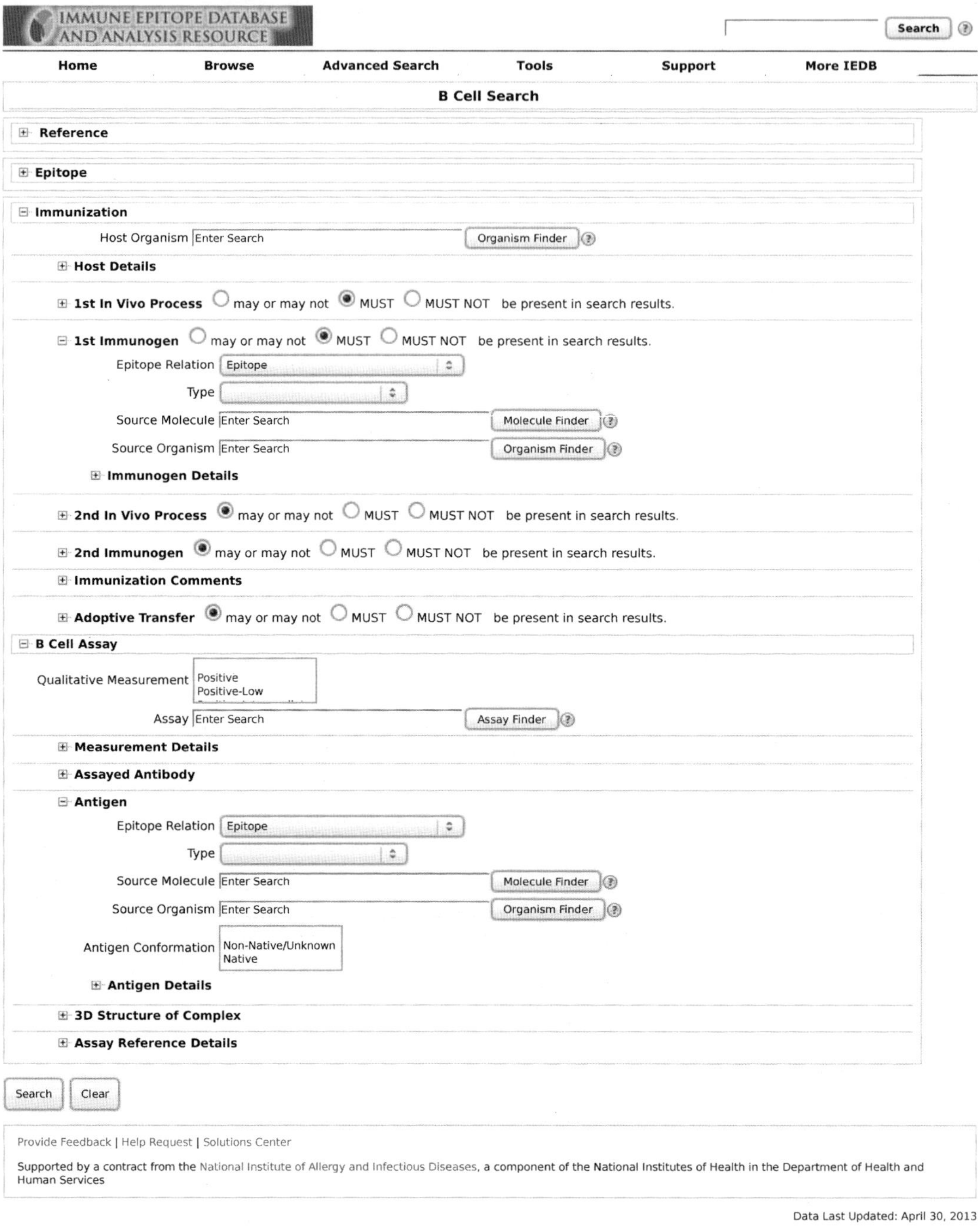

Fig. 1 IEDB B cell search facility web interface (directly accessible online through http://www.immuneepitope.org/advancedQueryBcell.php). Example shows user options selected from pull-down menus, for restricting searches by data fields of the type "Epitope Relation," particularly "1st Immunogen Epitope Relation" (within the Immunization section) and "Antigen Epitope Relation" (within the B cell assay section), with both fields set to "Epitope" (which is appropriate for retrieving records on B-cell assays wherein a short oligopeptide sequence was used both to elicit the production of antibodies and to detect the antibodies). Clicking on assay finder button (within the B cell assay section) launches assay finder popup window (Fig. 2)

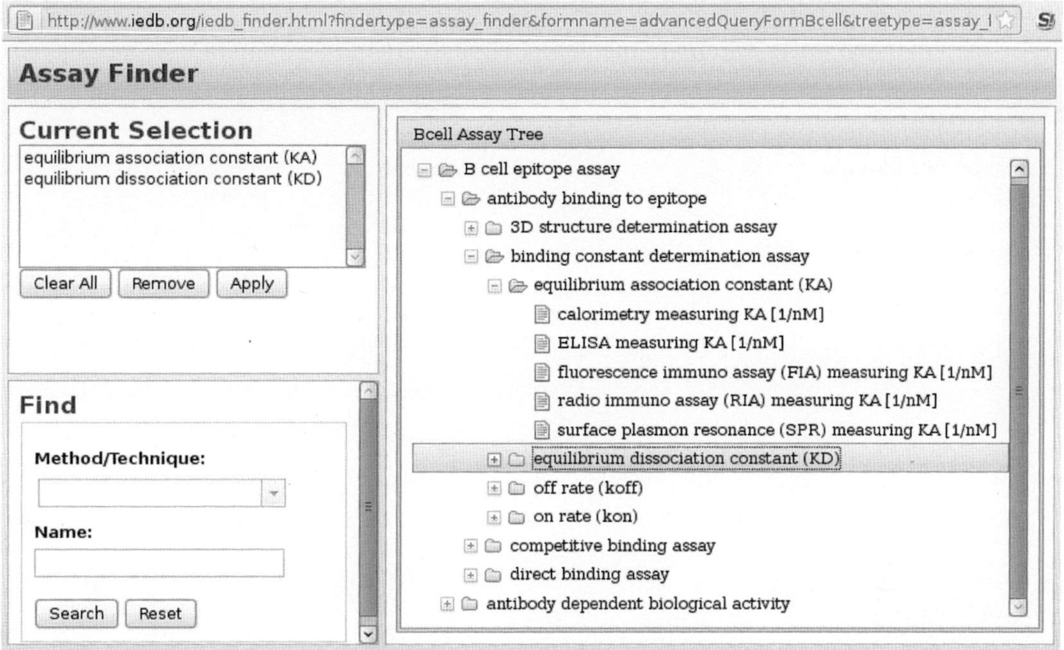

Fig. 2 IEDB assay finder popup window launched from B cell search facility web interface (Fig. 1), displaying B-cell assay tree. Example shows current selection comprising B-cell assays to obtain equilibrium association and dissociation constants. Within the assay tree, the node for equilibrium association constant is expanded to display its subsidiary assay types (calorimetry, ELISA, etc.) while the node for equilibrium dissociation constant, which has essentially the same subsidiary assay types, is collapsed; and the subsidiary assay types of the node for antibody-dependent biological activity (collapsed at bottom of assay tree panel) are presented in Table 1

(represented by the data field named "Assay") may be performed using the Assay Finder feature of the B Cell Search facility. Within the Assay Finder (popup) window, a B-cell assay tree is provided (Fig. 2); this tree can be navigated to view a hierarchical classification scheme of available assay types, which may be marked individually or by category (Table 1) to define appropriate selections of assay types for filtering in order to retrieve only those records matching one of the selected assay types. The two main assay-type categories are "antibody binding to epitope" and "antibody-dependent biological activity." "Antibody binding to epitope" comprises "binding constant determination assay," which in turn comprises "equilibrium association constant (KA)" and "equilibrium dissociation constant (KD)" (directly corresponding to the like-named quantities in Eq. 1); having measurement units of "[1/nM]" and "[nM]," respectively, these two categories are appropriate for retrieving quantitative data with which to benchmark paratope–epitope affinity predictions, noting that records thus retrieved should be checked for possible data duplication (*see* **Note 3**). "Antibody-dependent biological activity" comprises a wide variety

Table 1
IEDB B-cell assay categories and types for antibody-dependent biological activity

Assay category or type	Subsidiary assay types
Activation of additional immune response in vitro	Antibody-dependent cellular cytotoxicity Antibody-mediated histamine release Complement-dependent cytotoxicity Opsonization/phagocytosis
Efficacy of epitope-specific antibody intervention in vivo	Protection after challenge (e.g., survival after challenge) Decreased disease symptoms after treatment Disease symptom exacerbation Induction of hypersensitivity Induction of tolerance Reduction of fertility after treatment
Activation/enhancement of antigen activity	(Not applicable)
Antigen inhibition of antibody activity	(Not applicable)
Neutralization/inhibition of antigen activity	(Not applicable)

of assays (e.g., "neutralization/inhibition of antigen activity" and "survival after challenge"), all of which can yield quantitative data that are values of (or can be transformed into values of) the quotient q in Eqs. 16 and 17; hence, these assays are appropriate for retrieving quantitative data with which to benchmark predictions of antibody-mediated biological effects. In all cases (i.e., for predictions of affinities and biological effects), retrieved records should be checked for a valid numerical value assigned to the data field named "Quantitative measurement" [which may be empty (*see* **Note 4**)]; and also for any value assigned to the data field named "Measurement inequality" (which may be assigned a value of "$<$" or "$>$," indicating that the quantitative measurement represents a lower or upper limit rather than a point estimate) (*see* **Note 5**). For predictions on biological effects in particular, B-cell assay antibody concentrations (for use with Eqs. 7, 8, 10, and 11), which are typically absent from the IEDB records, may have to be either extracted or inferred (where possible) from the original literature references.

The most important aspect of IEDB and other actively maintained databases is their evolution over time to encompass ever greater amounts of epitope data (e.g., gleaned from published experimental work) for benchmarking prediction methods and to curate these data in ways that better facilitate the benchmarking process (e.g., by storing a wider variety of relevant data within readily searchable data fields). Apart from already published work, direct submission of yet unpublished data for inclusion in the databases is an increasingly significant driver of database growth, such that conventional published-literature searches may fail to retrieve newer data added to the databases. As database record structures and contents continue to evolve and expand, investigators may be compelled to retrieve additional data (e.g., immunoassay antibody concentrations) from primary sources (e.g., published manuscripts) where these data have yet to be captured within existing database records. Ultimately, the most crucial determinant of the availability of benchmark data is their actual generation in the first place. Experimentalists thus could greatly contribute to the further accumulation of informative benchmark data by generating dose-response data at or near half-maximal response levels, expressing antibody-mediated effects as apparent concentration-dependent changes in median effective doses of particular causative agents wherever possible, in line with the preceding Subheading 2.4. This demands explicit specification of antibody concentrations in molar or equivalent terms rather than incommensurable arbitrary units (e.g., based on titers operationally defined only for a particular immunoassay protocol) (*see* **Note 6**).

3.2 Prediction Methods

For the purpose of discussion, methods for B-cell epitope prediction are categorized herein as either classical or postclassical on the basis of their underlying computational approaches. The classical methods are exemplified by the earliest published epitope prediction algorithms, which generate simple sequence profiles using various amino-acid residue propensity scales. In contrast, the postclassical methods are more sophisticated in that they explicitly consider more structural detail (e.g., atomic coordinates of folded protein antigens) or otherwise employ more advanced computational techniques (e.g., machine learning [53, 66]); either way, the predictions are based on more structural information, although this may be only implicit (e.g., among machine learning approaches that utilize sequences without explicitly generating structural models). A recurrent theme in the development of both classical and postclassical methods is the trend towards hybrid methods (i.e., combinations of simpler methods to generate more reliable predictions), which is illustrated below using representative examples from the published literature.

Classical methods of B-cell epitope prediction entail sequence profiling of proteins. The earliest published B-cell epitope prediction method, namely that of Hopp and Woods [55], assigns numeric hydrophilicity values to each of the 20 canonical proteinogenic amino acids and evaluates the average hydrophilicity over a sliding window several residues in width along the entire sequence of a polypeptide chain, thus yielding sequence profiles whose peaks correspond to putative B-cell epitopes. The hydrophilicity values used were derived from free-energy changes associated with the transfer of model compounds from aqueous to organic phases [67], reflecting the notion that hydrophilic residues tend to be located on surface-exposed regions of proteins. Reasoning that B-cell epitopes must contain at least some residues that are physically accessible to paratopes, hydrophilicity is thus regarded as a surrogate for surface exposure, which in turn is regarded as a surrogate for antigenicity (i.e., the capacity to function as a B-cell epitope). This argument is open to the criticisms that surface exposure cannot be reliably inferred from hydrophilicity and that surface exposure is not even a sufficient condition for antigenicity [68]. Nevertheless, the method served as a provisional means for B-cell epitope prediction at a time when the best available biomolecular data were almost exclusively amino acid sequences deduced from nucleic acid sequences, as opposed to detailed three-dimensional protein structures, which are presently more readily obtained both experimentally [69] and computationally [70].

Notwithstanding the above mentioned limitations of the method of Hopp and Woods, it is the prototype of a large class of related methods whose common feature is the basic sliding-window algorithm [57, 71]. Members of this class are distinguished from one another by their chosen parameter of interest (e.g., hydrophilicity) as well as the mathematical averaging procedure applied to the parameter (i.e., simple computation of arithmetic means versus more elaborate procedures with weighting schemes biased in favor of more centrally located residues within the sliding window to yield smoother plots). Apart from the original hydrophilicity scale used by Hopp and Woods, alternative hydrophilicity scales have been developed, for example, based on retention times of model compounds in high-performance liquid chromatography [72]. Additionally, parameters other than hydrophilicity have been suggested as alternative bases for B-cell epitope prediction. These parameters include surface accessibility [73], segmental mobility (i.e., atomic mobility) [74, 75], and propensity for occurrence in aperiodic secondary structure exemplified by turns [58], all of which are based on observed statistical tendencies of amino-acid residues in three-dimensional protein crystal structures.

Given the diversity of parameters thus considered among classical methods for B-cell epitope prediction, the mathematical combination of multiple parameters (i.e., into hybrid methods for B-cell epitope prediction) has been proposed as a possible strategy for increasing the accuracy of B-cell epitope prediction [56, 57, 76, 77]. However, this introduces the problem of assigning relative weights to particular parameters, which is complicated by the fact that the parameters to be combined are strongly correlated with one another (i.e., they actually contain similar information); for example, the so-called antigenic index [56] was devised as the sum of multiple terms, each of which is the product of a parameter (secondary-structure propensity, hydrophilicity, flexibility, or surface probability) and a corresponding author-assigned numerical weight, such that secondary-structure propensity is assigned the greatest weight while both flexibility and surface probability are assigned the least weight (in view of their relatively poor correlation with antigenicity). The redundancy of information content among the parameters is a manifestation of the underlying physical interpretation of antigenicity in all cases, namely that antigenicity is primarily a function of surface exposure. For this reason, the limitations of the method of Hopp and Woods apply to all other methods that are based entirely on generation of sequence profiles using propensity scales for the 20 canonical proteinogenic amino acids.

Postclassical methods of B-cell epitope prediction transcend the simple sequence profiling of their classical predecessors, in the sense of exploiting higher-level structural information by means other than conventional residue-oriented propensity scales. The said structural information may pertain to rigid (e.g., folded) as well as flexible (e.g., unfolded) features, noting that rigidity and flexibility represent extremes on a continuum of possible conformational states. Although certain classical methods employ sequence-based secondary-structure and flexibility prediction (e.g., to predict turns and flexible loops as locations of putative B-cell epitopes), they nonetheless utilize residue-oriented propensity scales (e.g., based on statistical tendencies of residues to occur within turns [58] or on flexibility expressed as Debye–Waller temperature factors quantifying atomic mobility in crystal structures [75]).

With the increasing availability of detailed protein structural models [69, 70], protein atomic coordinates may be used instead of sequences as bases for B-cell epitope prediction via postclassical methods. For example, the CEP (conformational epitope prediction) server [78] uses protein atomic coordinates (specified as a user-supplied PDB ID or actual atomic coordinate file in PDB format) as input, explicitly considering amino-acid residue solvent accessibility (expressed as relative and absolute values per residue) and spatial proximity (grouping plausible epitope residues within 6 Å of one another) to delineate putative sequential (i.e., linear) and conformational (i.e., discontinuous) B-cell epitopes on proteins. Likewise, the DiscoTope server [79, 80] also uses protein atomic coordinates as input, although it is specifically intended for predicting discontinuous B-cell epitopes using a combination of residue contact number (i.e., count of neighboring residues in contact with a particular residue, inversely correlated with surface accessibility) and epitope log-odds ratio (expressing the statistical tendency of a residue to occur within known epitopes), thus employing a hybrid-method strategy to yield more accurate predictions than would be possible using either residue contact number or epitope log-odds ratio alone. The more recently developed ElliPro server [81] accepts either protein atomic coordinates or sequence as input; where a sequence is provided as input, this is used to generate a predicted structural model via homology modeling (with MODELLER [82–84]), as protein atomic coordinates are required to model the protein as an ellipsoid for which putative epitopes are delineated by spatial clustering of residues having a high protrusion index (i.e., degree of protrusion beyond the ellipsoid [85]).

Other postclassical methods utilize protein sequences as input without explicitly computing atomic coordinates, in which case structure is implicitly considered, likely in conjunction with other factors (e.g., biological correlates of structure such as immunodominance and immune protection). Such methods are typically based on machine learning approaches, including artificial neural networks and support vector machines, which entail training on empirical epitope data as may be obtained from the databases discussed in Subheading 3.1. For example, the ABCpred server [86] uses a recurrent neural network trained on data from the Bcipep database [59] for predicting immunogenic linear B-cell epitopes of fixed length, whereas the BCPREDS server [87] employs a support vector machine also trained on Bcipep data but for predicting immunogenic linear B-cell epitopes of variable length. Machine learning has also been applied to prediction of protective linear B-cell epitopes (i.e., eliciting protective antibody responses) among protein antigen sequences [38], employing a workflow that incorporates information on sequence variability and on conservation of patterns for post-translational modification and thus exemplifying a hybrid-method approach to B-cell epitope prediction.

More generally, machine learning approaches can be applied to B-cell epitope prediction where structure is either explicitly or implicitly considered. This is exemplified by the Epitopia server [88], which accepts either protein atomic coordinates or sequences and employs a naive Bayes classifier to predict immunogenicity of protein regions. As regards further development of such machine learning approaches, the main limiting factor is the availability of adequate amounts of suitable training data. These training data must be of the same type as the benchmark data for subsequent evaluation of predictive performance, but the training data must be kept strictly separate from the benchmark data so as to avoid the generation of misleadingly biased benchmark results that overestimate predictive performance. This is especially challenging where the training and benchmark data are quantitative rather than qualitative, considering the dearth of available quantitative data.

A collection of computational tools for B-cell epitope prediction using classical and postclassical methods is available online as part of the IEDB-AR (Immune Epitope Database and Analysis Resource) [63, 89], which also encompasses the database IEDB discussed in Subheading 3.1. The tools are under the heading of "B Cell Epitope Prediction Tools" (at http://tools.immuneepitope.org/main/html/bcell_tools.html) and listed with brief descriptions in Table 2, with sample output presented in Fig. 3. For more comprehensive accounts of various other published B-cell epitope methods, recent reviews [54, 90–92] may be consulted. Many of these tools are based on hybrid methods and may themselves be combined with each other and with other computational tools to create new hybrid methods (*see* **Note 7**).

4 Notes

1. Although minimally informative data correspond to H values of zero (*see* Eq. 17), they nonetheless point to the possibility of modifying experimental conditions (notably antibody concentration [Ab]) in order to yield new data that are more informative. In particular, q values of zero and unity, respectively, suggest that more informative data might be obtained by either increasing or decreasing [Ab] so as to bring q closer to half unity (e.g., in accordance with Eqs. 7, 8, 10, and 11), with the prospect of such improvement being more generally conceivable where [Ab] would be decreased rather than increased.
2. A curated IEDB B-cell epitope structure may comprise atoms or even entire amino-acid residues in addition to those forming part of any actual epitope relevant to a particular IEDB B-cell assay record. This is typically the case for linear peptide antigens, which may each contain one or more B-cell epitopes that have yet to be more precisely delineated by rigorous epitope mapping (e.g., by residue-wise incremental truncation of the peptide antigens from either or both amino and carboxy termini, until antibody–antigen binding becomes undetectable). Hence, investigators may opt to use IEDB data as starting points to design and perform additional experimental work to obtain more precise B-cell epitope data for benchmarking.
3. Paired IEDB B-cell assay records corresponding to equivalent association and dissociation constants (i.e., reciprocal values for exactly the same B-cell assay) may be curated because values of both constants were explicitly published (even though one had been mathematically derived from the other). Data duplication may thus occur where both records are erroneously included in analyses.
4. Where available, numerical values of quantitative data on B-cell assay outcomes are placed in an IEDB record data field named "Quantitative Measurement." However, this field may be left empty, for example, because the data are published only in graphical form (i.e., without printing the actual numerical values);

Table 2
IEDB-AR B-cell epitope prediction tools

Implemented method	Description/remarks
Chou and Fasman beta turn prediction	Classical; uses propensity scale based on statistics of residue occurrence in beta turns among protein crystals
Emini surface accessibility scale	Classical; uses propensity scale based on statistics of residue surface accessibility among protein crystals
Karplus and Schulz flexibility scale	Classical; uses propensity scale based on statistics of residue atomic mobility among protein crystals
Kolaskar and Tongaonkar antigenicity scale	Classical; uses propensity scale based on statistics of residue occurrence within known protein epitopes
Parker hydrophilicity scale	Classical; uses propensity scale based on chromatographic mobility of model compounds
BepiPred linear epitope prediction	Postclassical; uses recurrent neural network trained on immunogenic peptide epitope data
DiscoTope prediction of epitopes from protein structure	Postclassical; predicts discontinuous epitopes using residue contact number (inversely correlated with surface accessibility) and epitope log-odds ratio
ElliPro epitope prediction based upon structural protrusion	Postclassical; predicts linear and discontinuous epitopes using residue protrusion defined by modeling protein structure as an ellipsoid

still, the numerical values might be retrieved from the authors or inferred from graphical figures in certain cases (e.g., where the values are proportions of populations, as in Kaplan–Meier survival curves, especially where population sizes are relatively small such that values corresponding to discrete numbers of individuals may be more easily discerned).

5. The IEDB record data field named "Measurement inequality" may be either assigned a value of "=" or simply left empty (which is the usual case) to indicate that the quantitative measurement is a point estimate.
6. Quantitative data on B-cell assay outcome are often in the form of immunoassay signals (e.g., expressed as absorbance for enzyme immunoassays or fluorescence intensity for immunofluorescence assays) or corresponding titers

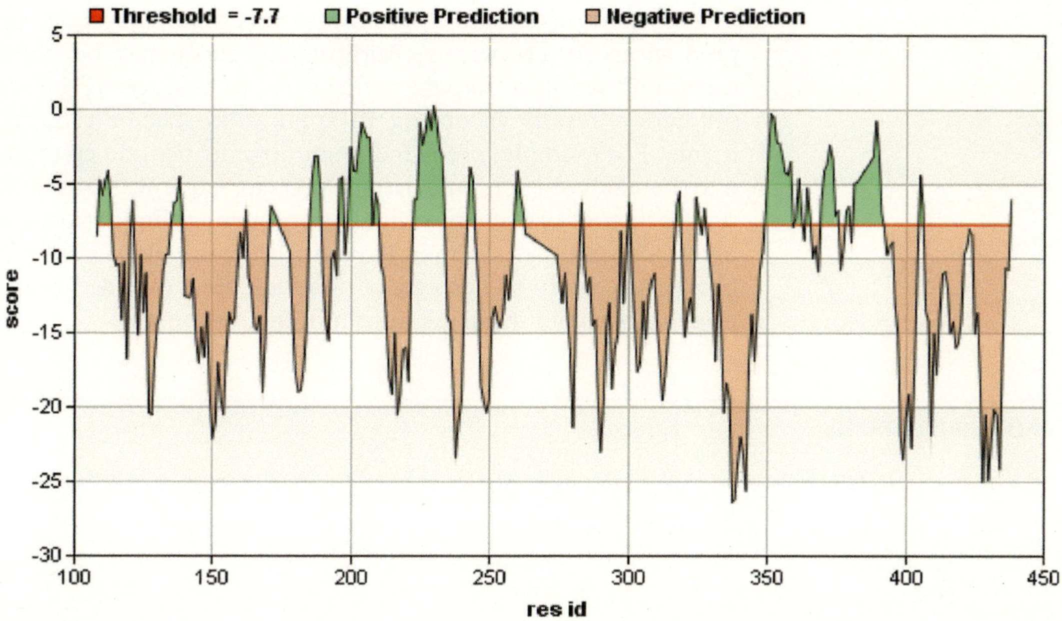

Fig. 3 IEDB-AR DiscoTope sample output for *Plasmodium falciparum* apical membrane antigen 1 (AMA1), using PDB ID 1Z40 chain A. The output is in the form of a sequence profile that superficially resembles one generated by means of a classical B-cell epitope prediction method. However, the score is calculated on the basis of a combination of residue contact number (i.e., count of neighboring residues in contact with a particular residue, inversely correlated with surface accessibility) and epitope log-odds ratio (expressing the statistical tendency of a residue to occur within known epitopes)

(e.g., maximum dilution factors beyond which signals become undetectable or comparable with background noise levels). These are typically difficult to quantitatively relate to absolute antibody concentrations. Hence, reporting of actual antibody concentrations (e.g., as dissociation constants and median inhibitory concentrations) along with other associated quantitative information (e.g., antigen concentrations and immunoassay temperatures) serves to provide more directly useful data for benchmarking B-cell epitope prediction methods.

7. For any given practical application (e.g., vaccine design), multiple methods may be used for B-cell epitope prediction, as consensus among these (e.g., one or more putative B-cell epitopes identified by all or most) may provide stronger decision support than predictions of each method taken separately. However, certain methods may be more appropriate than others for a given application (e.g., predicting immunogenic peptide epitopes is appropriate for designing peptide-based vaccine immunogens, whereas predicting epitopes recognized by antiprotein antibodies is appropriate for designing antigens as immunodiagnostic probes for antibody detection).

Hence, methods should be judiciously combined for rendering predictions by consensus. Additionally, tools may be combined into workflow pipelines that serve to progressively filter prediction results according to application-specific requirements. For example, predicted immunogenic peptide epitopes may be analyzed for similarity to protein sequences of a particular host organism, with subsequent elimination of predicted epitopes deemed likely to be epitopes of host self proteins (e.g., in the design of peptide-based vaccine immunogens, to avoid inducing autoimmune responses).

Acknowledgements

This work was supported by an Angelita T. Reyes Centennial Professorial Chair grant.

References

1. Van Regenmortel MH (2009) What is a B-cell epitope? Methods Mol Biol 524:3–20
2. Caoili SE (2006) A structural-energetic basis for B-cell epitope prediction. Protein Pept Lett 13:743–751
3. Caoili SE (2010) Immunization with peptide-protein conjugates: impact on benchmarking B-cell epitope prediction for vaccine design. Protein Pept Lett 17:386–398
4. Motte P, Alberici G, Ait-Abdellah M, Bellet D (1987) Monoclonal antibodies distinguish synthetic peptides that differ in one chemical group. J Immunol 138:3332–3338
5. Caoili SE (2012) On the meaning of affinity limits in B-cell epitope prediction for antipeptide antibody-mediated immunity. Adv Bioinformatics 2012:346765
6. Foote J, Eisen HN (1995) Kinetic and affinity limits on antibodies produced during immune responses. Proc Natl Acad Sci USA 92: 1254–1256
7. van Oss CJ (1997) Kinetics and energetics of specific intermolecular interactions. J Mol Recognit 10:203–216
8. Northrup SH, Erickson HP (1992) Kinetics of protein-protein association explained by Brownian dynamics computer simulation. Proc Natl Acad Sci USA 89:3338–3342
9. Raman CS, Jemmerson R, Nall BT, Allen MJ (1992) Diffusion-limited rates for monoclonal antibody binding to cytochrome c. Biochemistry 31:10370–10379
10. Watts C, Davidson HW (1988) Endocytosis and recycling of specific antigen by human B cell lines. EMBO J 7:1937–1945
11. Foote J, Eisen HN (2000) Breaking the affinity ceiling for antibodies and T cell receptors. Proc Natl Acad Sci USA 97:10679–10681
12. Ju ST, Nonogaki T, Bernatowicz MS, Matsueda GR (1993) The B cell immune response to an idiotype-inducing peptide epitope can be inhibited by immunodominance of a neighboring epitope. J Immunol 150:2641–2647
13. Novotny J, Handschumacher M, Haber E et al (1986) Antigenic determinants in proteins coincide with surface regions accessible to large probes (antibody domains). Proc Natl Acad Sci USA 83:226–230
14. Sanders RW, Venturi M, Schiffner L et al (2002) The mannose-dependent epitope for neutralizing antibody 2G12 on human immunodeficiency virus type 1 glycoprotein gp120. J Virol 76:7293–7305
15. Chothia C, Finkelstein AV (1990) The classification and origins of protein folding patterns. Annu Rev Biochem 59:1007–1039
16. Jones S, Thornton JM (1997) Prediction of protein-protein interaction sites using patch analysis. J Mol Biol 272:133–143
17. Murphy KP, Freire E (1992) Thermodynamics of structural stability and cooperative folding behavior in proteins. Adv Protein Chem 43:313–361
18. Edgcomb SP, Murphy KP (2000) Structural energetics of protein folding and binding. Curr Opin Biotechnol 11:62–66
19. Nakra P, Manivel V, Vishwakarma RA, Rao KV (2000) B cell responses to a peptide epitope. X. Epitope selection in a primary response is thermodynamically regulated. J Immunol 164:5615–5625

20. Francis T Jr (1960) On the doctrine of original antigenic sin. Proc Am Philos Soc 104: 572–578
21. Morens DM, Burke DS, Halstead SB (2010) The wages of original antigenic sin. Emerg Infect Dis 16:1023–1024
22. Tam JP (1988) Synthetic peptide vaccine design: synthesis and properties of a high-density multiple antigenic peptide system. Proc Natl Acad Sci USA 85:5409–5413
23. Posnett DN, Tam JP (1989) Multiple antigenic peptide method for producing antipeptide site-specific antibodies. Methods Enzymol 178: 739–746
24. Bainbridge J, Jones N, Walker B (2004) Multiple antigenic peptides facilitate generation of anti-prion antibodies. Clin Exp Immunol 137:298–304
25. Wang HW, Lin YC, Pai TW, Chang HT (2011) Prediction of B-cell linear epitopes with a combination of support vector machine classification and amino acid propensity identification. J Biomed Biotechnol 2011:432830
26. Herbst-Kralovetz M, Mason HS, Chen Q (2010) Norwalk virus-like particles as vaccines. Expert Rev Vaccines 9:299–307
27. Tobin GJ, Trujillo JD, Bushnell RV et al (2008) Deceptive imprinting and immune refocusing in vaccine design. Vaccine 26:6189–6199
28. Caoili SE (2013) Antidotes, antibody-mediated immunity and the future of pharmaceutical product development. Hum Vaccin Immunother 9:294–299
29. Laver WG, Air GM, Webster RG, Smith-Gill SJ (1990) Epitopes on protein antigens: misconceptions and realities. Cell 61:553–556
30. Schwab C, Bosshard HR (1992) Caveats for the use of surface-adsorbed protein antigen to test the specificity of antibodies. J Immunol Methods 147:125–134
31. Leder L, Wendt H, Schwab C et al (1994) Genuine and apparent cross-reaction of polyclonal antibodies to proteins and peptides. Eur J Biochem 219:73–81
32. Dunker AK, Oldfield CJ, Meng J et al (2008) The unfoldomics decade: an update on intrinsically disordered proteins. BMC Genomics 9:S1
33. Uversky VN, Dunker AK (2010) Understanding protein non-folding. Biochim Biophys Acta 1804:1231–1264
34. Wright PE, Dyson HJ (2009) Linking folding and binding. Curr Opin Struct Biol 19:31–38
35. Caoili SE (2010) Benchmarking B-cell epitope prediction for the design of peptide-based vaccines: problems and prospects. J Biomed Biotechnol 2010:910524
36. Van Regenmortel MH (2006) Immunoinformatics may lead to a reappraisal of the nature of B cell epitopes and of the feasibility of synthetic peptide vaccines. J Mol Recognit 19:183–187
37. Chen SW, Van Regenmortel MH, Pellequer JL (2009) Structure-activity relationships in peptide-antibody complexes: implications for epitope prediction and development of synthetic peptide vaccines. Curr Med Chem 16: 953–964
38. Sollner J, Grohmann R, Rapberger R et al (2008) Analysis and prediction of protective continuous B-cell epitopes on pathogen proteins. Immunome Res 4:1
39. Halstead SB, Mahalingam S, Marovich MA et al (2010) Intrinsic antibody-dependent enhancement of microbial infection in macrophages: disease regulation by immune complexes. Lancet Infect Dis 10:712–722
40. Lund O, Hansen J, Mosekilde E et al (1993) A model of enhancement and inhibition of HIV infection of monocytes by antibodies against HIV. J Biol Phys 19:133–145
41. Beck Z, Prohaszka Z, Fust G (2008) Traitors of the immune system—enhancing antibodies in HIV infection: their possible implication in HIV vaccine development. Vaccine 26: 3078–3085
42. Nelson S, Jost CA, Xu Q et al (2008) Maturation of West Nile virus modulates sensitivity to antibody-mediated neutralization. PLoS Pathog 4:e1000060
43. Cherrier MV, Kaufmann B, Nybakken GE et al (2009) Structural basis for the preferential recognition of immature flaviviruses by a fusion-loop antibody. EMBO J 28:3269–3276
44. Hill AV (1910) The nature of oxyhaemoglobin, with a note on its molecular weight. J Physiol 40:4–7
45. Weiss JN (1997) The Hill equation revisited: uses and misuses. FASEB J 11:835–841
46. Bounias M (1989) Algebraic potential of the Hill equation as an alternative tool for plotting dose (or time)/effects relationships in toxicology: a theoretical study. Fundam Clin Pharmacol 3:1–9
47. Casadevall A, Pirofski LA (2012) A new synthesis for antibody-mediated immunity. Nat Immunol 13:21–28
48. Caoili SE (2011) B-cell epitope prediction for peptide-based vaccine design: towards a paradigm of biological outcomes for global health. Immunome Res 7:2
49. Fedorov V, Mannino F, Zhang R (2009) Consequences of dichotomization. Pharm Stat 8:50–61

50. Shannon CE (1948) A mathematical theory of communication. Bell Syst Tech J 27:379–423, 623–656
51. Jaynes ET (1957) Information theory and statistical mechanics. Phys Rev 106:620–630
52. Jaynes ET (1957) Information theory and statistical mechanics II. Phys Rev 108:171–190
53. Sollner J, Mayer B (2006) Machine learning approaches for prediction of linear B-cell epitopes on proteins. J Mol Recognit 19:200–208
54. El-Manzalawy Y, Honavar V (2010) Recent advances in B-cell epitope prediction methods. Immunome Res 6(Suppl 2):S2
55. Hopp TP, Woods KR (1981) Prediction of protein antigenic determinants from amino acid sequences. Proc Natl Acad Sci USA 78: 3824–3828
56. Jameson BA, Wolf H (1988) The antigenic index: a novel algorithm for predicting antigenic determinants. Comput Appl Biosci 4: 181–186
57. Pellequer JL, Westhof E, Van Regenmortel MH (1991) Predicting location of continuous epitopes in proteins from their primary structures. Methods Enzymol 203:176–201
58. Pellequer JL, Westhof E, Van Regenmortel MH (1993) Correlation between the location of antigenic sites and the prediction of turns in proteins. Immunol Lett 36:83–99
59. Saha S, Bhasin M, Raghava GP (2005) Bcipep: a database of B-cell epitopes. BMC Genomics 6:79
60. Saha S, Raghava GP (2007) Prediction methods for B-cell epitopes. Methods Mol Biol 409:387–394
61. Huang J, Honda W (2006) CED: a conformational epitope database. BMC Immunol 7:7
62. Vita R, Vaughan K, Zarebski L et al (2006) Curation of complex, context-dependent immunological data. BMC Bioinform 7:341
63. Kim Y, Ponomarenko J, Zhu Z et al (2012) Immune epitope database analysis resource. Nucleic Acids Res 40:W525–W530
64. Salimi N, Fleri W, Peters B, Sette A (2012) The immune epitope database: a historical retrospective of the first decade. Immunology 137: 117–123
65. Vita R, Peters B, Sette A (2008) The curation guidelines of the immune epitope database and analysis resource. Cytometry A 73:1066–1070
66. Sollner J (2006) Selection and combination of machine learning classifiers for prediction of linear B-cell epitopes on proteins. J Mol Recognit 19:209–214
67. Levitt M (1976) A simplified representation of protein conformations for rapid simulation of protein folding. J Mol Biol 104:59–107
68. Van Regenmortel MH, Pellequer JL (1994) Predicting antigenic determinants in proteins: looking for unidimensional solutions to a three-dimensional problem? Pept Res 7:224–228
69. Berman H, Henrick K, Nakamura H, Markley JL (2007) The worldwide Protein Data Bank (wwPDB): ensuring a single, uniform archive of PDB data. Nucleic Acids Res 35:D301–D303
70. Zhang Y (2008) Progress and challenges in protein structure prediction. Curr Opin Struct Biol 18:342–348
71. Blythe MJ, Flower DR (2005) Benchmarking B cell epitope prediction: underperformance of existing methods. Protein Sci 14:246–248
72. Parker JM, Guo D, Hodges RS (1986) New hydrophilicity scale derived from high-performance liquid chromatography peptide retention data: correlation of predicted surface residues with antigenicity and X-ray-derived accessible sites. Biochemistry 25:5425–5432
73. Janin J (1979) Surface and inside volumes in globular proteins. Nature 277:491–492
74. Tainer JA, Getzoff ED, Alexander H et al (1984) The reactivity of anti-peptide antibodies is a function of the atomic mobility of sites in a protein. Nature 312:127–134
75. Karplus PA, Schulz GE (1985) Prediction of chain flexibility in proteins: a tool for the selection of peptide antigens. Naturwissenschaften 72:212–213
76. Alix AJ (1999) Predictive estimation of protein linear epitopes by using the program PEOPLE. Vaccine 18:311–314
77. Odorico M, Pellequer JL (2003) BEPITOPE: predicting the location of continuous epitopes and patterns in proteins. J Mol Recognit 16:20–22
78. Kulkarni-Kale U, Bhosle S, Kolaskar AS (2005) CEP: a conformational epitope prediction server. Nucleic Acids Res 33:W168–W171
79. Haste Andersen P, Nielsen M, Lund O (2006) Prediction of residues in discontinuous B-cell epitopes using protein 3D structures. Protein Sci 15:2558–2567
80. Kringelum JV, Lundegaard C, Lund O, Nielsen M (2012) Reliable B cell epitope predictions: impacts of method development and improved benchmarking. PLoS Comput Biol 8:e1002829
81. Ponomarenko J, Bui HH, Li W et al (2008) ElliPro: a new structure-based tool for the prediction of antibody epitopes. BMC Bioinformatics 9:514
82. Sali A, Potterton L, Yuan F et al (1995) Evaluation of comparative protein modeling by MODELLER. Proteins 23:318–326
83. Eswar N, Eramian D, Webb B et al (2008) Protein structure modeling with MODELLER. Methods Mol Biol 426:145–159

84. Yang Z, Lasker K, Schneidman-Duhovny D et al (2012) UCSF Chimera, MODELLER, and IMP: an integrated modeling system. J Struct Biol 179:269–278
85. Thornton JM, Edwards MS, Taylor WR, Barlow DJ (1986) Location of 'continuous' antigenic determinants in the protruding regions of proteins. EMBO J 5:409–413
86. Saha S, Raghava GP (2006) Prediction of continuous B-cell epitopes in an antigen using recurrent neural network. Proteins 65:40–48
87. El-Manzalawy Y, Dobbs D, Honavar V (2008) Predicting linear B-cell epitopes using string kernels. J Mol Recognit 21:243–255
88. Rubinstein ND, Mayrose I, Pupko T (2009) A machine-learning approach for predicting B-cell epitopes. Mol Immunol 46:840–847
89. Zhang Q, Wang P, Kim Y et al (2008) Immune epitope database analysis resource (IEDB-AR). Nucleic Acids Res 36:W513–W518
90. Reimer U (2009) Prediction of linear B-cell epitopes. Methods Mol Biol 524: 335–344
91. Costa JG, Faccendini PL, Sferco SJ et al (2013) Evaluation and comparison of the ability of online available prediction programs to predict true linear B-cell epitopes. Protein Pept Lett 20:724–730
92. Yao B, Zheng D, Liang S, Zhang C (2013) Conformational B-cell epitope prediction on antigen protein structures: a review of current algorithms and comparison with common binding site prediction methods. PLoS One 8: e62249

Chapter 15

Building Classifier Ensembles for B-Cell Epitope Prediction

Yasser EL-Manzalawy and Vasant Honavar

Abstract

Identification of B-cell epitopes in target antigens is a critical step in epitope-driven vaccine design, immunodiagnostic tests, and antibody production. B-cell epitopes could be linear, i.e., a contiguous amino acid sequence fragment of an antigen, or conformational, i.e., amino acids that are often not contiguous in the primary sequence but appear in close proximity within the folded 3D antigen structure. Numerous computational methods have been proposed for predicting both types of B-cell epitopes. However, the development of tools for reliably predicting B-cell epitopes remains a major challenge in immunoinformatics.

Classifier ensembles a promising approach for combining a set of classifiers such that the overall performance of the resulting ensemble is better than the predictive performance of the best individual classifier. In this chapter, we show how to build a classifier ensemble for improved prediction of linear B-cell epitopes. The method can be easily adapted to build classifier ensembles for predicting conformational epitopes.

Key words B-cell epitope prediction, Classifiers ensemble, Random forest, Epitope prediction toolkit

1 Introduction

Antigen-antibody interactions play a crucial role in the humoral immune response. Antibodies, a family of structurally related glycoproteins produced in membrane-bound or secreted form by B lymphocytes, serve as mediators of specific humoral immunity by engaging various effector mechanisms that serve to eliminate the bound antigens [1]. The part of the antigen recognized by antibodies is called B-cell epitope. B-cell epitopes often classified into two categories: (1) linear (continuous) B-cell epitopes consist of amino acid residues that are sequential in the primary structure of the protein and (2) conformational (discontinuous) B-cell epitopes consist of residues that are not sequential in the protein primary structure but come together in the protein 3D structure. Conformational B-cell epitopes form the majority of B-cell epitopes. Several experimental procedures for mapping both types of B-cell epitopes have been presented [2]. However, *in silico* methods

Rajat K. De and Namrata Tomar (eds.), *Immunoinformatics*, Methods in Molecular Biology, vol. 1184,
DOI 10.1007/978-1-4939-1115-8_15, © Springer Science+Business Media New York 2014

for identifying B-cell epitopes have the potential to dramatically decrease the cost and the time associated with the experimental mapping of B-cell epitopes [3].

Several computational methods have been proposed for predicting either linear or conformational B-cell epitopes [3–5]. Methods for predicting linear B-cell epitopes range from simple propensity scale profiling methods [6–9] to methods based on state-of-the-art machine learning predictors (e.g., [10–14]). Methods for predicting conformational B-cell epitopes (e.g., [15–19]) utilize some structure and physicochemical features derived from antigen-antibody complexes that could be correlated with antigenicity [3]. Despite the large number of B-cell epitope prediction methods proposed in literature, the performance of existing methods leaves significant room for improvement [4].

One of the promising approaches for improving the predictive performance of computational B-cell epitope prediction tools is to combine multiple classifiers. This approach is motivated by the observation that no single predictor outperforms all other predictors and that predictors often complement each other [20].

Against this background, we present a framework for developing classifier ensembles [21] and explain the procedure for building several variants of classifier ensembles based on the framework. Specifically, we describe a procedure for building classifier ensembles for predicting linear B-cell epitopes using Epitopes Toolkit (EpiT) [22]. We also show how to adapt the procedure for building classifier ensembles for predicting conformational B-cell epitopes (*see* **Note 1**). The procedures described in this chapter can be adapted for any other machine learning benchmark.

2 Materials

2.1 Data Set

We used the FBCPRED data set [11], a homology-reduced data set of variable-length linear B-cell epitopes extracted from Bcipep database [23]. The data set has 934 epitopes and non-epitopes (respectively) such that the length distribution of epitopes and non-epitopes is preserved.

2.2 Epitopes Toolkit (EpiT)

WEKA [24] is a machine learning workbench that is widely used by bioinformatics developers for developing prediction tools. Unfortunately, the vast majority of WEKA-implemented algorithms do not accept amino acid sequences as input. Hence, developers have to preprocess their sequence data for extracting useful features before using WEKA classification algorithms. Alternatively, developers of epitope prediction tools can use the Epitopes Toolkit (EpiT) [22] which is built on top of WEKA and provides a specialized set of useful data preprocessors (e.g., filters) and classification algorithms for developing B-cell epitope prediction tools.

A java implementation of EpiT is freely available at the project website, http://ailab.ist.psu.edu/epit. More information about how to install and use EpiT is provided in the project documentation.

3 Methods

In this section, we show how to use EpiT to build individual and classifier ensembles for predicting linear B-cell epitopes. The procedure can be easily adapted for any other machine learning workbench (e.g., RapidMiner [25] and KNIME [26]).

3.1 Building a Single Classifier with EpiT

Here, we show how to build a single predictor using FlexLenBCPred.nr80.arff, FBCPRED data in WEKA format available at http://ailab.ist.pdu.edu/red/bcell/FBCPred.zip, and a Random Forest classifier [27] with 50 trees (RF50).

1. Run EpiT.
2. Go to Application menu and select *model builder* application.
3. In the *model builder* window (WEKA explorer augmented with EpiT filters and prediction methods) click *open* and select the file *fbcprednr80.arff.*
4. Click *classify* tab.
5. In the *classifier* panel, click *choose* and browse for weka.meta.FilteredClassifier. The FilteredClassifier is a WEKA class for running an arbitrary classifier on data that has been passed through arbitrary filter.
6. Click on the *FilteredClassifier* in the classifier panel and specify the following classifier and filter. For the classifier, choose weka.classifiers.trees.RandomForest and set *numTrees* to 50. For the filter, choose epit.filters.unsupervised.attribute.AAP. The AAP filter implements the amino acid propensity scale features proposed in [28].
7. Having both the data set and the classification algorithm specified, we are ready to build the model and evaluate it using ten-fold cross-validation (*see* **Note 2**). Just click *start button* and wait for the ten-fold cross-validation procedure to finish. The *classifier output panel* shows several statistical estimates of the classifier using ten-fold cross-validation (*see* Fig. 1).

3.2 Building a Classifier Ensemble with EpiT

A classifier ensemble consists of a collection of individual (or base) classifiers that work together using a suitably designed fusion method (e.g., combination rule or second-level classifier) for optimally combining the outputs of the individual classifiers. This design process involves two basic steps: (1) design a set of complementary or diverse base classifiers: diversity of classifiers could be ensured by manipulating

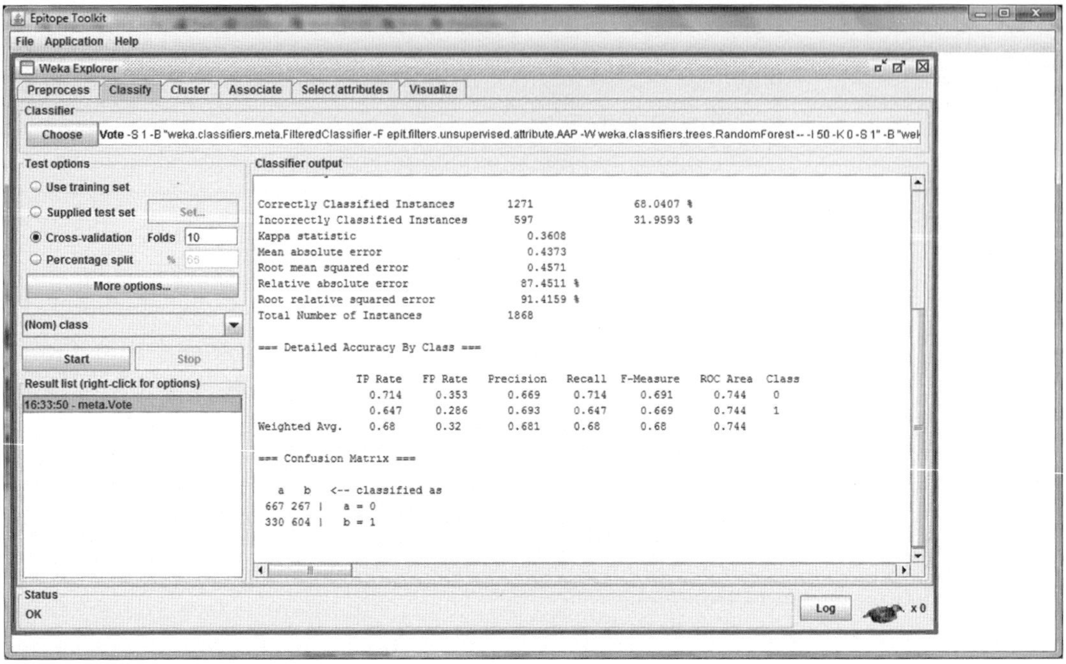

Fig. 1 Output statistics for ten-fold cross-validation experiment evaluating Vote classifier with average (AVG) of probabilities combination rule

the classifiers' inputs, outputs, or the training algorithms [21] (*see* **Notes 3** and **4**); (2) design a combination rule that exploits the behaviors of the individual classifiers to optimally combine them. Figure 2 shows a framework for constructing classifier ensembles using EpiT. In this framework, different classifier ensembles can be developed by using different combinations of choices of filters, base classifiers, and combination rules. In this example, we fix the base classifier to RF50 and use different filters for each individual classifier. We also experiment with different combination rules. To build a classifier ensemble for predicting flexible-length linear B-cell epitopes using EpiT, follow the following procedure:

1. Run EpiT.
2. Go to Application menu and select the *model builder* application.
3. In the *model builder* window (WEKA explorer augmented with EpiT filters and prediction methods) click *open* and select the file *fbcprednr80.arff.*
4. Click *classify* tab.
5. In the *classifier* panel, click *choose* and browse for weka.meta. Vote. The Vote classifier is a WEKA class for combining classifiers. Different combinations of probability estimates for classification are available.
6. Click on *classifiers* and enter four FilteredClassifiers. Set the *classifier* parameter for each FilteredClassifier to RF50 and set

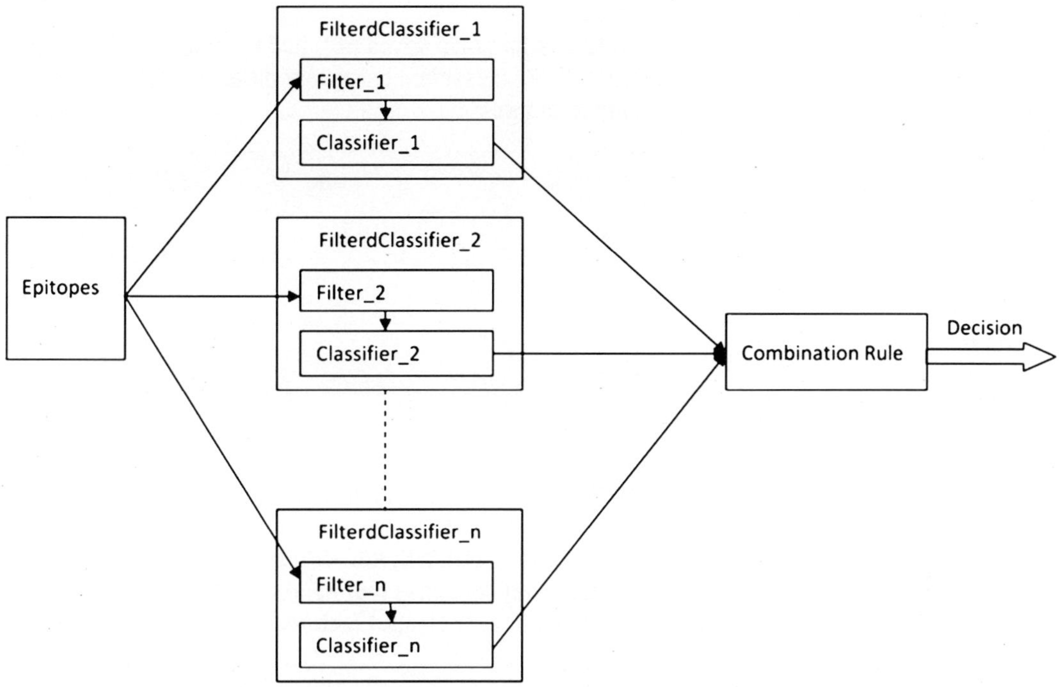

Fig. 2 Framework for building classifier ensembles using EpiT tool

the *filter* parameter to AAP, CTD, SequenceComposition, and SequenceDiCompositions, respectively.

7. Select one of the available combination rule options. In our experiment we used the WEKA default setting for this parameter, average of probabilities.
8. Click *start* button to start a ten-fold cross-validation experiment and wait for the output results (*see* Fig. 1).

A more sophisticated way for combining multiple classifiers according to the framework in Fig. 2 is to replace the simple combination rule used with Vote classifier with a meta-predictor, a second-stage classifier. The procedure for building such a classifier ensemble is as follows:

1. Run EpiT.
2. Go to Application menu and select the *model builder* application.
3. In the *model builder* window (WEKA explorer augmented with EpiT filters and prediction methods) click *open* and select the file *fbcprednr80.arff*.
4. Click *classify* tab.
5. In the *classifier* panel, click *choose* and browse for weka.meta. Stacking. The Stacking classifier is a WEKA class for combining several classifiers using the stacking method [29].

Table 1
AUC values for naïve Bayes (NB) and Random Forest (RF50) classifiers using four different sets of input features

Features	NB	RF50
AAP	0.67	0.72
CTD	0.65	0.65
AAC	0.66	0.71
DC	0.63	0.72

Table 2
AUC values for a classifier ensemble that combines four NB classifiers trained using the four sets of input features (AAP, CTD, AAC, DC) and a classifier ensemble that combines four RF50 constructed using the four sets of input features

Combination rule	NB	RF50
AVG	0.69	0.74
PROD	0.65	0.75
MIN	0.64	0.75
MAX	0.68	0.74

The classifier ensembles are obtained using the same base classifiers but different combination rules

6. Click on *classifiers* and enter four FilteredClassifiers. Set the *classifier* parameter for each FilteredClassifier to RF50 and set the *filter* parameter to AAP, CTD, SequenceComposition, and SequenceDiCompositions, respectively.
7. Click on *metaclassifier* and choose the naïve Bayes (NB) classifier, weka.classifiers.bayes.NaiveBayes.
8. Set *numFolds* to 3. This parameter sets the number of folds used for cross-validation experiment performed for training the meta-classifier. Click *OK*.
9. Click *start* button to start a ten-fold cross-validation experiment.

Table 1 compares the performance (in terms of AUC scores (*see* **Note 5**)) of two classifiers, NB and RF50, using four sets of input features: (1) amino acid pair (AA) propensities [28]; (2) composition-transition-distribution (CTD) [30]; (3) amino acid composition (AAC); and (4) dipeptide composition (DC). Table 2 compares the performance of a classifier ensemble that combines four NB classifiers

Table 3
AUC values for a classifier ensemble that combines four NB classifiers trained using the four sets of input features (AAP, CTD, AAC, DC) and a classifier ensemble that combines four RF50 constructed using the four sets of input features

Meta-predictor	NB	RF50
NB	0.69	0.75
Logistic	0.69	0.75

The classifier ensembles are obtained using the same base classifiers but different meta-predictors

trained using the four sets of input features (AAP, CTD, AAC, DC) and a classifier ensemble that combines four RF50 constructed using the four sets of input features. Four simple combination rules have been evaluated: AVG, PROD, MIN, and MAX which represent average, product, minimum, and maximum estimated probabilities from the four base classifiers for each input instance. Table 3 compares the performance of the NB- and RF50-based classifier ensembles (reported in Table 2) when the simple combination rule is replaced with a meta-classifier (second-stage classifier).

Table 1 shows that the predictive performance of each classifier seems to be highly dependent on the input features. For example, AUC scores of RF50 range from 0.65 to 0.72 for different choices of input features. Tables 2 and 3 show that combining individual classifiers constructed with different input features and using the same classification algorithm (e.g., NB and RF50) not only eliminate the dependency on the input features but also yields a classifier ensemble with performance higher than the best individual classifier performance obtained in Table 1.

It should be noted that the RF50 classifier, treated in our experiments as an individual classifier, is itself an ensemble of 50 different decision tree classifiers. The performance of RF50 might be improved using several approaches including (1) increasing the number of trees, (2) selecting a subset of the 50 trees using some criteria for eliminating redundant and poor tree predictors [31], and (3) building a multiple classifier system in which RF50 is treated as a base classifier.

4 Notes

1. The current implementation of EpiT does not support the extraction of evolutionary or structure-based features since most of these features require running third-party programs

(e.g., BLAST [32]). Building classifier ensemble that uses such features requires preprocessing the training data such that each epitope in the original data is represented with a combined set of extracted features (each set of features might be extracted using one or more third-party program (s)). The resulting combined set of features are used as inputs and the filter for each FilteredClassifier will select a range of attribute indices (corresponding to a set of features) to pass to the base classifier.

2. In ten-fold cross-validation experiments, the data set is randomly partitioned into ten equal subsets such that the relative proportion of epitopes to non-epitopes in each subset is preserved. Nine of the subsets are used for training the classifier and the remaining subset is used for testing the classifier. This procedure is repeated ten times, each time setting aside a different subset of the data for testing. The estimated performance of the classifier corresponds to an average of the results from the ten cross-validation runs.
3. Classifier ensembles can be developed using a single set of features and a single classification algorithm by training each base classifier with different training data (i.e., sampled instances or sampled subspace of the original training data). WEKA provides built-in classification algorithms for building such ensemble of classifiers (e.g., Bagging [33] and AdaBoost [34]).
4. For unbalanced data, an ensemble of classifiers system can be created by training each single classifier using all training instances from the minority class and an equal number of training instances (selected at random) from the majority class [21]. Such base classifiers can be created using EpiT Balanced Classifier (for more details please refer to EpiT documentation). The classifiers can be combined using a combination rule via Vote class or using a meta-classifier via Stacking class.
5. The receiver operating characteristic (ROC) curve is obtained by plotting the true positive rate as a function of the false-positive rate as the discrimination threshold of the binary classifier is varied. A widely used measure of classifier performance is the area under ROC curve (AUC). A perfect classifier will have an AUC=1, while a random guessing classifier will have an AUC=0.5, and any classifier performing better than random will have an AUC value that lies between these two values.

Acknowledgments

This work was supported in part by a grant from the National Institutes of Health (NIH GM066387) and by Edward Frymoyer Chair of Information Sciences and Technology at Pennsylvania State University.

References

1. Abbas AK, Lichtman AH, Pillai S (2007) Cellular and molecular immunology, 6th edn. Saunders Elsevier, Philadelphia
2. Reineke U, Schutkowski M (2009) Epitope mapping protocols, vol 524, 2nd edn, Methods in molecular biology. Humana Press, New York
3. Ansari HR, Raghava GP (2013) *In silico* models for B-cell epitope recognition and signaling. Methods Mol Biol 993:129–138
4. El-Manzalawy Y, Honavar V (2010) Recent advances in B-cell epitope prediction methods. Immunome Res 6(Suppl 2):S2
5. Yao B, Zheng D, Liang S et al (2013) Conformational B-cell epitope prediction on antigen protein structures: a review of current algorithms and comparison with common binding site prediction methods. PLoS One 8(4):e62249
6. Emini EA, Hughes JV, Perlow D et al (1985) Induction of hepatitis A virus-neutralizing antibody by a virus-specific synthetic peptide. J Virol 55(3):836–839
7. Karplus P, Schulz G (1985) Prediction of chain flexibility in proteins. Naturwissenschaften 72(4):212–213
8. Parker JM, Guo D, Hodges RS (1986) New hydrophilicity scale derived from high-performance liquid chromatography peptide retention data: correlation of predicted surface residues with antigenicity and X-ray-derived accessible sites. Biochemistry 25(19): 5425–5432
9. Pellequer J-L, Westhof E, Van Regenmortel MH (1993) Correlation between the location of antigenic sites and the prediction of turns in proteins. Immunol Lett 36(1):83–99
10. El-Manzalawy Y, Dobbs D, Honavar V (2008) Predicting linear B-cell epitopes using string kernels. J Mol Recognit 21(4):243–255. doi:10.1002/jmr.893
11. El-Manzalawy Y, Dobbs D (2008) Honavar V (3400678) Predicting flexible length linear B-cell epitopes. Comput Syst Bioinformatics, In, pp 121–132
12. Larsen JE, Lund O, Nielsen M (2006) Improved method for predicting linear B-cell epitopes. Immunome Res 2:2. doi:10.1186/1745-7580-2-2
13. Saha S, Raghava GP (2006) Prediction of continuous B-cell epitopes in an antigen using recurrent neural network. Proteins 65(1):40–48
14. Sweredoski MJ, Baldi P (2009) COBEpro: a novel system for predicting continuous B-cell epitopes. Protein Eng Des Sel 22(3):113–120
15. Haste Andersen P, Nielsen M, Lund O (2006) Prediction of residues in discontinuous B-cell epitopes using protein 3D structures. Protein Sci 15(11):2558–2567
16. Kringelum JV, Lundegaard C, Lund O et al (2012) Reliable B cell epitope predictions: impacts of method development and improved benchmarking. PLoS Comput Biol 8(12): e1002829
17. Ponomarenko J, Bui H-H, Li W et al (2008) ElliPro: a new structure-based tool for the prediction of antibody epitopes. BMC Bioinformatics 9(1):514
18. Sun J, Wu D, Xu T et al (2009) SEPPA: a computational server for spatial epitope prediction of protein antigens. Nucleic Acids Res 37(suppl 2):W612–W616
19. Sweredoski MJ, Baldi P (2008) PEPITO: improved discontinuous B-cell epitope prediction using multiple distance thresholds and half sphere exposure. Bioinformatics 24(12): 1459–1460
20. Resende DM, Rezende AM, Oliveira NJ et al (2012) An assessment on epitope prediction methods for protozoa genomes. BMC Bioinformatics 13:309
21. Wozniak M (2013) Hybrid Classifiers: Methods of Data, Knowledge, and Classifier Combination, vol 519. Studies in Computational Intelligence, Springer Heidelberg London
22. El-Manzalawy Y (2010) Honavar V A framework for developing epitope prediction tools. In: Proceedings of the First ACM International conference on bioinformatics and computational biology. ACM, pp 660–662
23. Saha S, Bhasin M, Raghava GP (2005) Bcipep: a database of B-cell epitopes. BMC Genomics 6:79
24. Frank E, Hall M, Holmes G, Kirkby R, Pfahringer B, Witten IH, Trigg L (2005) Weka: A machine learning workbench for data mining. In Data Mining and Knowledge Discovery Handbook (pp 1305–1314) Springer US
25. Jungermann F Information extraction with rapidminer. In: Proceedings of the GSCL Symposium'Sprachtechnologie und eHumanities, 2009. pp 50–61
26. Berthold MR, Cebron N, Dill F et al (2008) KNIME: The Konstanz information miner. Data Analysis, Machine Learning and Applications. Springer Berlin Heidelberg, In, pp 319–326
27. Breiman L (2001) Random forests. Mach Learn 45(1):5–32

28. Chen J, Liu H, Yang J et al (2007) Prediction of linear B-cell epitopes using amino acid pair antigenicity scale. Amino Acids 33(3): 423–428
29. Wolpert DH (1992) Stacked generalization. Neural Netw 5(2):241–259
30. Cai C, Han L, Ji ZL et al (2003) SVM-Prot: web-based support vector machine software for functional classification of a protein from its primary sequence. Nucleic Acids Res 31(13): 3692–3697
31. Bernard S, Heutte L, Adam S (2009) Towards a better understanding of random forests through the study of strength and correlation. Emerging Intelligent Computing Technology and Applications. With Aspects of Artificial Intelligence. Springer, In, pp 536–545
32. Altschul SF, Madden TL, Schäffer AA et al (1997) Gapped BLAST and PSI-BLAST: a new generation of protein database search programs. Nucleic Acids Res 25(17):3389–3402
33. Breiman L (1996) Bagging predictors. Mach Learn 24(2):123–140
34. Freund Y (1996) Schapire RE Experiments with a new boosting algorithm. ICML, In, pp 148–156

Chapter 16

Multiplex Peptide-Based B Cell Epitope Mapping

Sanne M.M. Hensen, Merel Derksen, and Ger J.M. Pruijn

Abstract

B cell epitope mapping is widely applied to determine antibody-binding sites. Several methods exist to map B cell epitopes and here we describe three methods that are characterized by the simultaneous analysis of multiple peptides. In the first approach a microarray of overlapping synthetic peptides derived from an antigenic protein is used and the binding of the antibodies is analyzed by fluorescently labeled secondary antibodies. This method is particularly suited for the identification of linear epitopes of an established target protein. In the second approach the binding of antibodies to a random synthetic peptide library immobilized on microbeads is determined by enzyme-conjugated secondary antibodies and the selection of antibody-bound beads by a light microscope. This method can be applied when information on the identity of the antigenic protein is lacking. In the third method an antigen is proteolytically digested and antibody binding to the resulting peptides is analyzed by surface plasmon resonance imaging (*i*SPR). The latter method can be applied when the purified antigenic protein is available.

Key words Antibody, Antigen, Peptide microarray, Microbeads, Antigen fragment library, *i*SPR

1 Introduction

B cell epitope mapping is essential for several biomedical applications, such as diagnostic peptide development and vaccine design. Numerous methods exist to identify either conformational B cell epitopes, in which amino acids in distinct parts of the protein contribute to antibody binding, or linear B cell epitopes, which are formed by a continuous stretch of amino acids in the antigenic protein. For the identification of a conformational epitope X-ray crystallography of the crystallized antibody-antigen complex represents one of the main methods. Although X-ray crystallography is a powerful method to characterize epitopes, a major drawback is the complicated and time-consuming crystallization process [1]. Another approach that is commonly used to determine conformational epitopes is based on mass spectrometry (MS). An antibody-antigen complex can be subjected to mild proteolysis and because the epitope is protected from cleavage by the bound antibody, the sequence of the region(s) bound by the antibody can be determined by MS. However, the resolution of

Rajat K. De and Namrata Tomar (eds.), *Immunoinformatics*, Methods in Molecular Biology, vol. 1184,
DOI 10.1007/978-1-4939-1115-8_16,

this method is rather low. A related approach is the so-called hydrogen/deuterium (H/D) exchange method for the identification of protein–protein interactions [2]. Antibody and antigen are separately labeled with deuterium and subsequently allowed to form a complex in D_2O. Next, the solution containing the complex is strongly diluted with water, allowing the exchange of deuterium for hydrogen, unless a part of the antigen is protected by a bound antibody. Finally, the complex is digested with a protease (e.g., pepsin) and deuterium retention can be measured with MS, resulting in the identification of amino acids at or close to the epitope.

Although most B cell epitopes may be conformational, the most commonly used methods are based on linear B cell epitopes, which is at least in part due to the time-consuming, expensive, and specific expertise-requiring properties of conformational epitope mapping methods. In this respect it is important to note that the characterization of linear epitopes generally suffices for the development of diagnostic and therapeutic molecules. A frequently used approach to map linear epitopes is the generation of a peptide library followed by the selection of antibody-binding peptides from this library. Peptide libraries can be composed of chemically synthesized peptides, biologically displayed peptides, or peptides obtained by proteolytic fragmentation of the antigen. Synthetic peptides can be produced on pins [3, 4] or on membrane supports [5, 6], but a major disadvantage of using pins and membrane supports is the limited number of peptides that can be screened simultaneously (96 to up to 2,000). Nowadays, it is more common to use glass slides (peptide microarrays) or microbeads. Peptide microarrays provide the ability to screen more than 100,000 peptides and require a significantly lower amount of reagents compared to pins and membrane supports [7]. Antibody binding can be detected by classical immunolabeling methods followed by visualization with a fluorescence scanning system [8] or by surface plasmon resonance imaging (*i*SPR), a label-free and easy-to-perform detection method. *i*SPR is based on the principle that the refractive index of a thin gold layer changes when molecules bind to the gold-coated surface [9, 10]. When beads are used, millions of peptides can be synthesized and screened for antibody binding [11]. Antibody-bound beads can be immunolabeled and selected by a fluorescence microscope or with a fluorescence-activated cell sorter. A disadvantage of the microbead-based approach is that the peptide sequence needs to be determined after antibody binding, unless "barcoded" beads are used [12].

Biological peptide libraries are generated using biological systems such as bacteriophages to produce and display the peptides and are characterized by the coupling of peptide and peptide-encoding nucleic acid sequence. The latter feature facilitates the elucidation of the amino acid sequence of targeted peptides. A major advantage of biologically displayed peptides is the high number (up to 10^9) and length of peptides that can be displayed. A disadvantage, on the other hand, is that only standard amino acids can be incorporated. If posttranslational modifications are important for epitope formation, these will not be detected. Bacteriophages are most frequently used as a biological system to display peptides (phage display) [13]. DNA fragments encoding the peptides are inserted in the phage genome, fused to the sequence encoding a surface protein, which results in the presentation of the peptides on the surface of the phage particles. Several selection rounds are performed to enrich for phages that bind with the highest affinity to the antibodies. Finally, the genome of the resulting phages can be sequenced to identify the epitope of the antibodies of interest.

Besides the synthetic and biological peptide libraries, antigen fragment libraries can be used for linear B cell epitope mapping. In this approach an antigenic protein is chemically or enzymatically cleaved, leading to the production of several fragments. An immunoprecipitation with the antibody of interest can be performed and bound peptides can be characterized by MS. An alternative possibility is the immobilization of the antigen fragments on a microarray and the subsequent analysis of antibody binding, e.g., by *i*SPR [14]. A major advantage of antigen fragment libraries is that it allows the detection of epitopes depending on posttranslational modifications.

In the following sections, we describe three multiplex methods that can be used for the mapping of B cell epitopes. The first is suitable for the characterization of epitopes of a defined antigenic protein, using a microarray of overlapping synthetic peptides. The second is the screening of a random peptide library using synthetic peptides immobilized on microbeads. The third starts with the generation of proteolytic fragments of a specific antigen and identifies antigenic fragments by *i*SPR analysis.

2 Materials

2.1 Overlapping Peptide Microarray Screening

This protocol is based on the PepStar™ microarrays (JPT Peptide Technologies) and was optimized for mapping epitopes of cytosolic 5′-nucleotidase IA targeted by antibodies in sera from sporadic inclusion body myositis (IBM) patients (Pluk, van Hoeve et al. [8]). The general aspects of the procedure are applicable to other microarrays. The protocol describes the incubation and washing steps by manual actions. However, the procedure described can also be performed using an incubation station.

1. Peptide microarray displaying an overlapping set of peptides derived from the antigen of interest (*see* **Note 1**), e.g., the PepStar™ microarray (JPT Peptide Technologies GmbH, Berlin, Germany), which contains the peptide array in triplicate.

2. Blank slide (dummy).
3. Two spacers per microarray.
4. Tris-buffered saline (TBS): 0.15 M NaCl, 0.05 M Tris–HCl, pH 7.6.
5. Blocking solution (MTBST): 5 % Nonfat dried milk, 0.05 % Tween-20 in TBS (*see* **Note 2**).
6. Antibody solution, e.g., serum 100- to 500-fold diluted in MTBST (*see* **Note 3**).
7. Deionized water.
8. Fluorescently labeled secondary antibody, e.g., 2 mg/mL Alexa Fluor-568-labeled goat-anti-human antibody (*see* **Note 3**).
9. If available, incubation station which can perform washing and incubation steps in a temperature-controlled environment.
10. Fluorescence scanner for microarrays, e.g., ProScanArray (PerkinElmer, *see* **Note 4**).
11. Software tool capable of assigning signal intensities to the individual spots on the microarray.

2.2 Microbead-Based Random Peptide Library Screening

This protocol describes the screening of a random peptide library (peptides immobilized on Tentagel beads) with IgG isolated from patient sera. Screening can be performed by the use of labeled secondary antibodies, though this requires the blocking with $F(ab')_2$ fragments generated from the healthy individuals' IgG. Alternatively, screening can be performed by directly "labeling" the patients' IgG, e.g., by conjugation to alkaline phosphatase (AP). In this method the peptide library beads are first blocked with IgG or $F(ab')_2$ from healthy individuals, followed by the binding of the labeled IgG and the detection of beads bound by the labeled antibodies. The protocol outlined below describes the screening of beads with AP-labeled patients' IgGs.

All solutions should be filtered through a 0.2 μm filter.

2.2.1 Materials for the Generation of $F(ab')_2$ Fragments

1. IgG from healthy individuals.
2. Tris/EDTA buffer: 50 mM Tris–HCl, pH 7.0, 2 mM EDTA.
3. Ficin, 36 U/mL in TE buffer.
4. L-Cysteine, 1 M in TE buffer.
5. *N*-ethylmaleimide, 100 mM in TE buffer.
6. PBS: 137 mM NaCl, 2.7 mM KCl, 10 mM Na_2HPO_4, 1.8 mM KH_2PO_4, pH 7.4.

2.2.2 Materials for IgG Isolation from Serum and AP Conjugation

1. Patient sera.
2. Protein A-agarose column.
3. PBS: 137 mM NaCl, 2.7 mM KCl, 10 mM Na_2HPO_4, 1.8 mM KH_2PO_4, pH 7.4.

4. PBS containing 0.5 M NaCl, 0.05 % NP40.
5. Elution buffer: 100 mM glycine-HCl, pH 2.5, 500 mM NaCl, 0.05 % NP-40.
6. 1 M Tris.
7. Reagents for protein concentration determination, e.g., Bradford reagents.
8. Dialysis tubing/devices (MW cutoff 3.5 kDa).
9. Alkaline phosphatase (e.g., activated alkaline phosphatase, Roche Applied Science Cat. No. 11464752001).

2.2.3 Peptide Bead Library Screening Components

1. Peptide bead library ($1–5 \times 10^6$ peptides) of 50,000–100,000 beads per microvial (*see* **Note 5**).
2. Buffer A: 50 mM Tris–HCl, pH 7.5, 150 mM NaCl. Before use add 0.5 % Tween-20.
3. 5 mg/mL $F(ab')_2$ from healthy individuals in PBS.
4. IgG isolated from patient sera.
5. AP buffer: 100 mM Tris, 100 mM NaCl, 5 mM $MgCl_2$, pH 7.5.
6. NBT: 50 mg/mL Nitro-blue tetrazolium in 70 % DMF.
7. BCIP: 25 mg/mL 5-Bromo-4-chloro-3′-indolyphosphate in 70 % DMF.
8. NBT/BCIP solution: 33 μL NBT, 33 μL BCIP, 10 mL AP.
9. 100 mM EDTA, pH 8.0.

2.3 Screening of Antigen Fragment Libraries by Microarray iSPR

When the antigenic protein is available in purified form and when posttranslational modifications may be involved in antibody binding, antigen fragment libraries might be used to determine B cell epitopes. The (modified) antigen is proteolytically digested and the resulting peptides are screened with the antibody for binding peptides by *i*SPR (Fig. 3a).

2.3.1 Preparation of Peptide Fragments

1. Protein of interest: 250 μg protease in 500 μL PBS (*see* **Note 6**).
2. 45 mM Dithiothreitol (DTT) in 50 mM NH_4HCO_3.
3. 100 mM Iodoacetamide in 50 mM NH_4HCO_3 (*see* **Note 7**).
4. Sequencing-grade trypsin (*see* **Note 8**).
5. Sequencing-grade chymotrypsin (*see* **Note 9**).
6. Lys-N protease (*see* **Note 10**).
7. Sep-Pak C18 cartridges (Waters Corporation, Milford, MA).
8. 100 mM Acetic acid (HAc).
9. 10 % Formic acid.

2.3.2 Separation of Peptide Fragments

1. Agilent 1100 HPLC system (Agilent Technologies, Santa Clara, California, USA).
2. Optilynx guard columns (Optimized Technologies, Oregon City, Oregon, USA).
3. Polysulfoethyl A strong cation exchange (SCX) column (PolyLC, Columbia, MD, USA; 200 mm × 2.1 mm inner diameter, 5 μm, 200 Å).
4. Water, pH 2.7.
5. 80 % Acetonitrile, pH 2.7.
6. Solvent A: 5 mM KH_2PO_4, 30 % acetonitrile, pH 2.7.
7. Solvent B: 5 mM KH_2PO_4, 30 % acetonitrile, 350 mM KCl, pH 2.7.
8. 10 % Formic acid.

2.3.3 Preparation of Microarray

1. *i*SPR sensor discs containing a dextran hydrogel with carboxylic acid groups (HC200, XantecBioanalytics GmbH, Dusseldorf, Germany) (*see* **Note 11**).
2. 50 mM MES buffer, pH 5.4.
3. 0.8 M *N*-hydroxysuccinimide (NHS) in 50 mM MES buffer, pH 5.4.
4. 0.2 M *N*-ethyl-*N*′-(dimethylaminopropyl)carbodiimide (EDC) in 50 mM MES buffer, pH 5.4.
5. 0.25 % Acetic acid (HAc), pH 4.5.
6. Microarray spotter.
7. 1 M Ethanolamine.

2.3.4 iSPR Analyses

1. *i*SPR apparatus (IBIS Technologies BV, Hengelo, The Netherlands).
2. PBS, 0.03 % Tween-20.
3. 10 mM Glycine-HCl, pH 1.5.
4. SPRint software (IBIS Technologies BV, Hengelo, The Netherlands).

3 Methods

3.1 Overlapping Peptide Microarray Screening

1. Block the microarray slide in MTBST at room temperature for 1 h (*see* **Note 12**).
2. Remove the MTBST and prevent the microarray slide from drying.
3. Assemble an incubation chamber: A small petri dish is placed upside down on a wet cloth in a larger petri dish. The cloth will prevent evaporation of the solution. A microarray slide, with

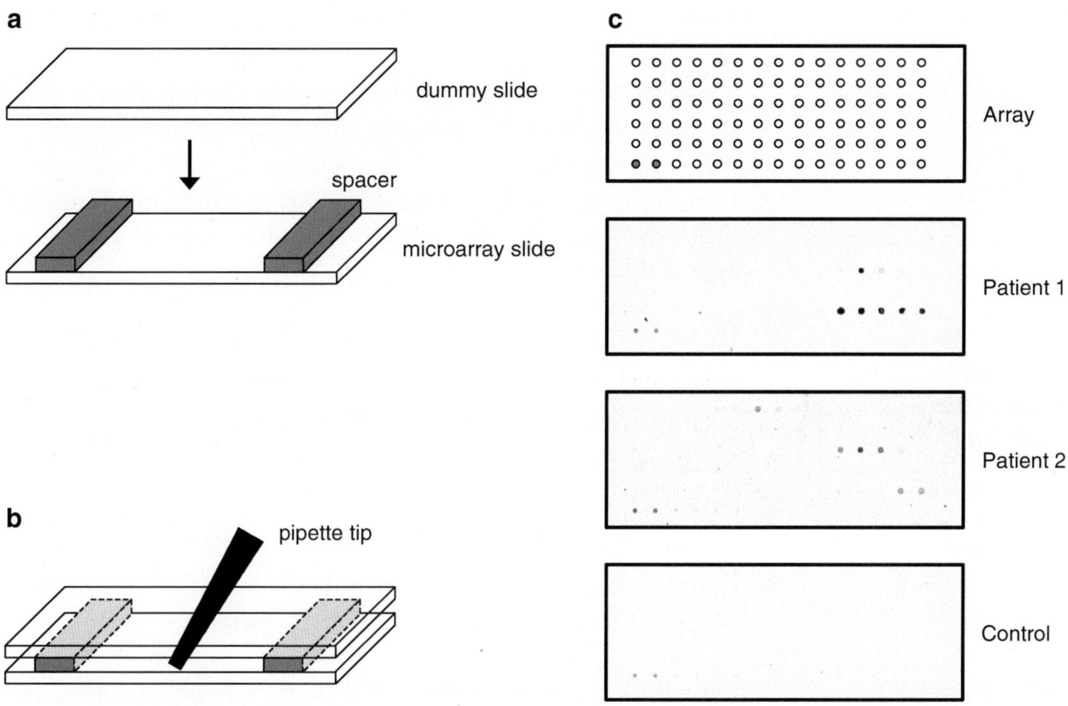

Fig. 1 Peptide microarray screening. (**a**) Assembly of the microarray incubation chamber. Spacers are placed on the slide containing the array and a second slide is placed on top of the spacers. The surface of the microarrays containing the peptides should be facing inwards. (**b**) The diluted patient serum is pipetted between the two slides. (**c**) Example of results obtained with a peptide microarray. A 90-spot array containing a set of overlapping peptides derived from the cytosolic 5′-nucleotidase IA protein was incubated with two patient sera and one healthy control serum. Bound antibodies were visualized by a fluorescently labeled secondary antibody and fluorescence scanning. The two spots in the lower left corner represent positive controls

the peptide surface upward, is placed on top of the small petri dish with the spacers on the ends of the slide. A second microarray slide, the dummy slide (*see* **Note 13**), is placed on top of the spacers (Fig. 1a).

4. Pipette 300 μL of the antibody solution (diluted serum) into the chamber (Fig. 1b).
5. Incubate the microarrays for 2 h at 37 °C.
6. Remove the antibody solution by suction and disassemble the incubation chamber.
7. Wash the microarray slide five times (5 min each) with MTBST.
8. Incubate the microarray slide in a closed petri dish for 1 h at 30 °C under agitation with the 2,000-fold diluted secondary antibody conjugate in MTBST (*see* **Note 14**).
9. Wash the microarray slide five times with MTBST buffer (5 min each).
10. Wash the microarray slide five times using deionized water (5 min each).

11. Blow a gentle stream of nitrogen on the microarray surface to dry the slide.
12. Visualize bound antibodies by scanning the microarray slides using a compatible fluorescence microarray scanner (*see* Fig. 1c for an example).
13. Determine the signal intensities for the individual peptide spots and calculate the average intensity of spots containing identical peptides.

3.2 Microbead-Based Random Peptide Library Screening

3.2.1 Generation of F(ab′)$_2$ Fragments

1. Dilute healthy control IgG in Tris/EDTA buffer to 4 mg/mL.
2. Add 70 μL Ficin solution to 500 μL of the diluted IgG.
3. Start the reaction by adding 2 μL L-cysteine.
4. Incubate the mixture at 37 °C for 24 h.
5. Stop the reaction by adding 60 μL of *N*-ethylmaleimide. Incubate the mixture for 15 min at room temperature.
6. Dialyze overnight against PBS at 4 °C.
7. The products can be analyzed by SDS-PAGE.

3.2.2 IgG Isolation from Serum and Conjugation to AP

1. Centrifuge serum at 2,000 × *g* at 4 °C for 15 min (*see* **Note 15**). Pass the supernatant through a 0.2 μm filter. The supernatant can be stored at 4 °C prior to use.
2. Wash the protein A-agarose column with PBS containing 0.5 M NaCl and 0.05 % NP40 for 2 h at 4 °C.
3. Apply 1 vol. of elution buffer to the column and subsequently pre-equilibrate the column with PBS.
4. Add 2 mL of serum, tenfold diluted in PBS, to the column. Circulate the diluted serum overnight through the column to ensure binding of all antibodies.
5. Wash the column with PBS containing 0.5 M NaCl and 0.05 % NP40.
6. Elute the bound IgG with elution buffer and collect fractions of 1 mL. Immediately add 70 μL 1 M Tris to each of the fractions.
7. Determine the protein concentration of the fractions, e.g., by using the Bradford reagent, and analyze the purity by SDS-PAGE and total protein staining.
8. Pool the IgG-containing fractions and dialyze against PBS.
9. Conjugate alkaline phosphatase to the isolated IgGs according to the instructions of the supplier.

3.2.3 Peptide-Bead Library Screening

All washing steps include mild vortexing of the samples followed by collecting the beads by brief centrifugation at 400 × *g*.

1. Wash the peptide beads four times with buffer A.
2. Divide beads in aliquots of approx. 10,000 beads per microvial.

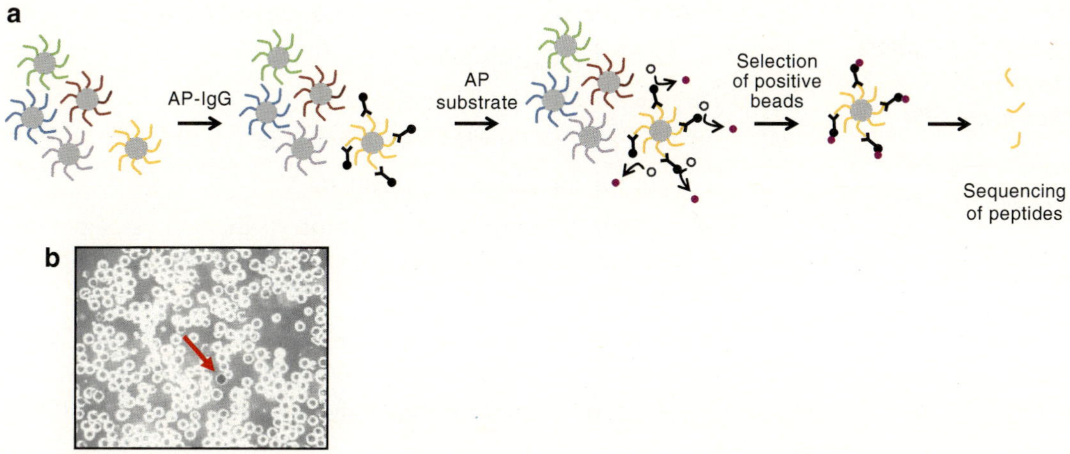

Fig. 2 Epitope mapping by randomized peptide library screening. (**a**) A randomized peptide library immobilized on beads (each bead contains many copies of the same peptide) is incubated with the antibody of interest conjugated to alkaline phosphatase (AP-IgG). Subsequently, bound antibodies are visualized by incubation with an AP substrate. Conversion of the substrate into a stained product, which precipitates on the beads, allows the isolation of the antibody-bound beads. Finally, the identity of the peptides can be determined by amino acid sequencing of the peptides, e.g., by Edman degradation. (**b**) Example of peptide library beads after the staining reaction. The *arrow* marks a stained bead

3. Block peptide beads that are recognized by common antibodies by adding 100 μL $F(ab')_2$ solution and 400 μL buffer A (*see* **Note 16**). Incubate the mixture by end-over-end rotation for 3 h at room temperature.
4. Centrifuge and wash the beads four times with 1 mL buffer A.
5. Incubate the beads with AP-conjugated IgGs in 0.5 mL buffer A. Incubate by end-over-end rotation at room temperature overnight.
6. Collect the beads by centrifugation and wash five times with buffer A.
7. Incubate the beads in 1 mL buffer A under agitation for 15 min.
8. Wash the beads twice with AP buffer.
9. Add 0.5 mL of NBT/BCIP solution and incubate for 30 min at room temperature.
10. Wash the beads with 100 mM EDTA to stop the AP reaction.
11. Transfer the beads to a petri dish and select the colored beads using a light microscope (Fig. 2).
12. The identity of the peptides on the selected beads can subsequently be analyzed by amino acid sequencing, e.g., by Edman degradation.

3.3 Screening of Antigen Fragment Libraries by Microarray iSPR

3.3.1 Preparation of Peptide Fragments

1. Dissolve 250 μg of protein per protease in 500 μL PBS.
2. Add 12.5 μL of 45 mM DTT.
3. Incubate for 30 min at 56 °C.
4. Add 12.5 μL of 100 mM iodoacetamide.
5. Incubate for 30 min at RT in the dark.
6. Digest protein with trypsin, chymotrypsin, or Lys-N by adding the enzyme at a 1:100 enzyme/substrate mass ratio by overnight incubation at 37 °C.
7. Desalt samples and exchange buffer to 100 mM HAc using Sep-Pak C18 cartridges, end volume 40 μL.
8. Add 20 μL of 10 % formic acid and freeze samples at −20 °C.

3.3.2 Separation of Peptide Fragments

1. Use 20 μL of samples to separate peptides by SCX chromatography [15].
2. Load peptides onto two C18 Optilynx guard columns using an Agilent 1100 HPLC system, with a flow rate of 100 μL/min using water, pH 2.7 as a solvent.
3. Elute peptides with 80 % acetonitrile, pH 2.7, and load onto a polysulfoethyl A SCX column for 10 min with a flow rate of 100 μL/min.
4. Separate different peptide populations using a nonlinear gradient at a flow rate of 200 μL/min: from 0 to 10 min, 100 % solvent A; from 10 to 15 min, 0–26 % solvent B; from 15 to 40 min, 26–35 % solvent B; from 40 to 45 min, 35–60 % solvent B; and from 45 to 49 min, 60–100 % solvent B.
5. Collect fractions in 1-min intervals.
6. Evaporate solvents.
7. Resuspend fractionated peptides in 60 μL 10 % formic acid.
8. Determine peptide composition of the fractions by LC-MS/MS.

3.3.3 Preparation of Microarray Chip

1. Mix equal volumes of 0.8 M NHS and 0.2 M EDC.
2. Add to the sensor disc surface and incubate for 20 min.
3. Wash the sensor disc with 0.25 % HAc, pH 4.5.
4. Dry sensor disc under nitrogen for 30 min.
5. Dilute peptides in 50 mM MES buffer (pH 5.4) to a final concentration of 1 ng/nL (*see* **Note 17**).
6. Spot 1 nL (or more, in case of a continuous flow microspotter) of the peptide solution on the surface of the sensor discs by the microarray spotter (*see* **Note 18**).
7. Incubate for 1 h at room temperature in a humidity chamber.
8. Block unreacted groups by incubation with 1 M ethanolamine for 10 min.
9. Rinse the sensor disc with PBS and keep wet until use.

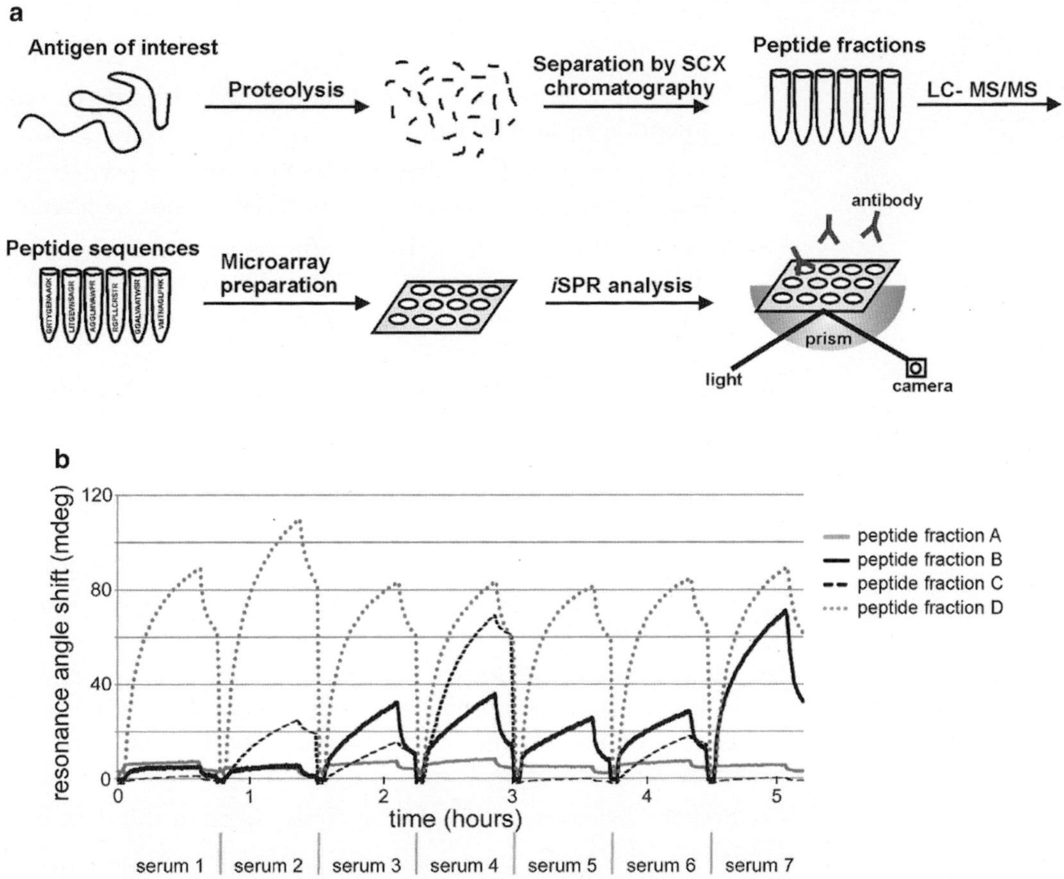

Fig. 3 Screening of antigen fragment libraries by microarray *i*SPR. (**a**) The antigen of interest is subjected to proteolysis and the resulting peptides are separated by SCX chromatography. Peptide fractions are sequenced with LC-MS/MS and spotted on a microarray chip. Antibody binding is subsequently determined by *i*SPR analysis. (**b**) Example of an *i*SPR sensorgram. Each curve represents one antibody-containing serum that is tested for its reactivity with different peptide fractions. The resonance angle shift corresponds to the amount of antibodies bound to a peptide spot on the array

3.3.4 *i*SPR Analyses

1. Incubation, washing, and regeneration are performed in an automated way using liquid-handling procedures in the *i*SPR apparatus.
2. Inject a sample plug of 80 μL containing the antibodies (in PBS, 0.03 % Tween-20) and pass 20 μL backward and forward over the array in a flow cell with a speed of 30 μL/s (*see* **Note 19**).
3. Rinse the flow cell with PBS and 0.03 % Tween-20.
4. Regenerate the array by two consecutive injections (30 s each) of 400 μL 10 mM glycine-HCl, pH 1.5.
5. Data can be analyzed with the SPRint software (*see* Fig. 3b for an example of an *i*SPR sensorgram).

4 Notes

1. Peptide microarray slides should be handled with care to keep the peptide surface intact.
2. It is recommended to filter the wash solutions used (using 0.4–2 μm filters). However, milk solutions cannot be filtered.
3. Serum dilutions in the 1:100–1:500 range and a secondary antibody concentration of 1 μg/mL are recommended. However, these concentrations might need optimization in order to reduce background signals.
4. Other fluorescence scanners with a resolution of at least 10 μm can be used.
5. A random peptide library can be generated by a mix-and-split approach, mixing and splitting the beads after each synthesis step. Peptide-containing Tentagel beads should be resuspended in acetonitrile:dichloromethane (82:18). The solvent should be evaporated using an exiccator and beads can be stored dry at 4 °C.
6. Method is suitable for the analysis of recombinant proteins as well as endogenous proteins isolated from a biological source. Posttranslationally modified proteins, modified either in vivo or in vitro, can be analyzed.
7. Prepare iodoacetamide solution fresh. Store in the dark until use.
8. Trypsin cleaves proteins at the C-terminal side of lysine and arginine residues, except when these residues are C-terminally flanked by a proline.
9. Chymotrypsin cleaves proteins at the C-terminal side of tyrosine, tryptophan, and phenylalanine residues.
10. Lys-N cleaves proteins at the N-terminal side of lysine residues, which means that the primary amines of the resulting fragments are located at the N-terminus.
11. Sensor discs containing other reactive groups can also be used.
12. The Pepstar™ microarrays are pretreated by the manufacturer to minimize nonspecific binding and blocking with protein solutions may cause high background signals. However, when using patient sera or plasma an additional blocking step is recommended. Besides blocking solutions from a commercial supplier, e.g., Pierce Biotechnology (Cat. No.37536), MTBST was found to be a suitable blocking solution [8].
13. The dummy slide is used during the procedure to facilitate the incubation and to prevent contamination of the solutions used.
14. To determine nonspecific binding of the secondary antibody, a parallel incubation of a microarray slide that is not exposed to the primary antibody can be performed.

15. If desired, patient sera can be pooled prior to IgG isolation.
16. When using AP-conjugated IgGs, blocking can also be performed using IgGs from control individuals. However, depending on the source of the AP-conjugated IgGs, blocking with $F(ab')_2$ fragments is preferred to reduce background staining. An example is the use of IgGs from rheumatoid arthritis patients which may contain antibodies directed to the Fc part of IgGs.
17. Fractions displaying a large overlap in peptide composition can be pooled if desirable.
18. As the peptides are relatively small, the contrast of the immobilized array to the background (visualized by *i*SPR) is low. To visualize the array, a protein, e.g., human IgG, can be spotted in parallel.
19. The sample plug is flanked by two air plugs to prevent the diffusion of sample components into the PBS and 0.03 % Tween-20.

Acknowledgements

We would like to thank Dr. Reinout Raijmakers for providing technical details and Dr. Joyce van Beers for providing an example of an *i*SPR sensorgram. Part of this work was financially supported by the Netherlands Proteomics Centre, a program embedded in the Netherlands Genomics Initiative.

References

1. Chayen NE, Saridakis E (2008) Protein crystallization: from purified protein to diffraction-quality crystal. Nat Methods 5(2):147–153
2. Mandell JG, Falick AM, Komives EA (1998) Identification of protein-protein interfaces by decreased amide proton solvent accessibility. Proc Natl Acad Sci USA 95(25):14705–14710
3. Geysen HM, Meloen RH, Barteling SJ (1984) Use of peptide synthesis to probe viral antigens for epitopes to a resolution of a single amino acid. Proc Natl Acad Sci USA 81(13):3998–4002
4. Rodda SJ (2001) Synthesis of multiple peptides on plastic pins. Curr Protoc Immunol Chapter 9:Unit 9–7
5. Frank R, Overwin H (1996) SPOT synthesis. Epitope analysis with arrays of synthetic peptides prepared on cellulose membranes. Methods Mol Biol 66:149–169
6. Frank R (2002) The SPOT-synthesis technique. Synthetic peptide arrays on membrane supports–principles and applications. J Immunol Methods 267(1):13–26
7. Katz C, Levy-Beladev L, Rotem-Bamberger S, Rito T, Rudiger SG, Friedler A (2011) Studying protein-protein interactions using peptide arrays. Chem Soc Rev 40(5):2131–2145
8. Pluk H, van Hoeve BJ, van Dooren SH, Stammen-Vogelzangs J, van der Heijden A, Schelhaas HJ, Verbeek MM, Badrising UA, Arnardottir S, Gheorghe K, Lundberg IE, Boelens WC, van Engelen BG, Pruijn GJ (2013) Autoantibodies to cytosolic 5′-nucleotidase 1A in inclusion body myositis. Ann Neurol 73(3):397–407
9. Rothenhäusler B, Knoll W (1988) Surface–plasmon microscopy. Nature 332:615–617
10. Lokate AM, Beusink JB, Besselink GA, Pruijn GJ, Schasfoort RB (2007) Biomolecular interaction monitoring of autoantibodies by scanning surface plasmon resonance microarray imaging. J Am Chem Soc 129(45):14013–14018

11. Lam KS, Salmon SE, Hersh EM, Hruby VJ, Kazmierski WM, Knapp RJ (1991) A new type of synthetic peptide library for identifying ligand-binding activity. Nature 354(6348):82–84
12. Jun BH, Kang H, Lee YS, Jeong DH (2012) Fluorescence-based multiplex protein detection using optically encoded microbeads. Molecules 17(3):2474–2490
13. Pande J, Szewczyk MM, Grover AK (2010) Phage display: concept, innovations, applications and future. Biotechnol Adv 28(6):849–858
14. van Beers JJ, Raijmakers R, Alexander LE, Stammen-Vogelzangs J, Lokate AM, Heck AJ, Schasfoort RB, Pruijn GJ (2010) Mapping of citrullinated fibrinogen B-cell epitopes in rheumatoid arthritis by imaging surface plasmon resonance. Arthritis Res Ther 12(6):R219
15. Gauci S, Helbig AO, Slijper M, Krijgsveld J, Heck AJ, Mohammed S (2009) Lys-N and trypsin cover complementary parts of the phosphoproteome in a refined SCX-based approach. Anal Chem 81(11):4493–4501

Chapter 17

Classification of Human Leukocyte Antigen (HLA) Supertypes

Mingjun Wang and Mogens H. Claesson

Abstract

Identification of new antigenic peptides, derived from infectious agents or cancer cells, which bind to human leukocyte antigen (HLA) class I and II molecules, is of importance for the development of new effective vaccines capable of activating the cellular arm of the immune response. However, the barrier to the development of peptide-based vaccines with maximum population coverage is that the restricting HLA genes are extremely polymorphic resulting in a vast diversity of peptide-binding HLA specificities and a low population coverage for any given peptide–HLA specificity. One way to reduce this complexity is to group thousands of different HLA molecules into several so-called HLA supertypes: a classification that refers to a group of HLA alleles with largely overlapping peptide binding specificities. In this chapter, we focus on the state-of-the-art classification of HLA supertypes including HLA-I supertypes and HLA-II supertypes and their application in development of peptide-based vaccines.

Key words Peptide, Vaccine, HLA-I supertypes, HLA-II supertypes

1 Introduction

The immune system, including the innate and adaptive as well as overlapping systems, plays a pivotal role in the defense against viral or bacterial infections, immune homeostasis, and cancer surveillance. Within the immune system, T lymphocytes are crucial for adaptive immune responses, and are activated upon recognition of peptides displayed by human leukocyte antigen class I (HLA-I) or class-II (HLA-II) molecules at the surfaces of antigen-presenting cells (APCs). T lymphocytes express the T cell receptor (TCR) that recognizes specific peptides, which have been processed and presented in combination with an HLA molecule. There are two major subtypes of T lymphocytes: $CD8^+$ cytotoxic T cells (CTLs) and $CD4^+$ helper T cells. CTLs recognize peptides in the context of HLA-I molecules, while $CD4^+$ helper T cells recognize peptides associated with HLA-II molecules. The functional activity of these

Rajat K. De and Namrata Tomar (eds.), *Immunoinformatics*, Methods in Molecular Biology, vol. 1184,
DOI 10.1007/978-1-4939-1115-8_17, © Springer Science+Business Media New York 2014

two subsets of T cells is said to be restricted by HLA-I and -II molecules, respectively.

It is known that CTLs play a major role in killing tumor cells [1, 2] and controlling viral or bacterial infections [3–7], while $CD4^+$ T cells are required for priming and expansion of naive $CD8^+$ T cells as well as secondary expansion of $CD8^+$ memory T cells [8–12]. It might therefore be of critical importance to incorporate both HLA-I- and -II-restricted epitopes in peptide-based vaccines to obtain participation of both $CD4^+$ and $CD8^+$ T cells for generation of strong and long-lasting immunity.

Thus, identification of new antigenic peptides, derived from infectious agents or tumor antigens, which may bind to HLA-I or HLA-II molecules in exchange with self-peptides normally occupying the HLA-binding site (*see* below), is important for developing new effective vaccines capable of activating the cellular arm of the immune responses. However, the barrier to development of peptide-based vaccines with maximum population coverage is that the restricting HLA genes are extremely polymorphic resulting in a vast diversity of peptide-binding HLA specificities and a low population coverage for any given peptide–HLA specificity. As of April 2013, it has been reported that there are 7,089 HLA-I alleles and 2,065 HLA-II alleles (http://hla.alleles.org). Undoubtedly, these numbers will be further increased in the future. To reduce this complexity, one option is to group thousands of different HLA molecules into clusters of several so-called HLA supertypes: a classification that refers to a group of HLA alleles with largely overlapping peptide binding specificities. In this chapter, we discuss the state-of-the-art classification of HLA-I and HLA-II supertypes and their application in development of peptide-based vaccines.

2 HLA-I Molecule and Assembly of HLA-I Peptide Complex

The major histocompatibility complex class I (MHC-I) antigens are referred to as the human leukocyte antigens class I (HLA-A, -B, and -C) and as H-2 class I antigens (K, D, and L) in mice. HLA-I antigens consist of three non-covalently associated components: a 45 kDa glycosylated amino acid (AA) heavy chain (HC), a 12 kDa light chain (beta 2 microglobulin, β2m), and a short 8–10 AA self-peptide. The heavy chain of HLA-I consists of about 340 AA residues, including a cytoplasmic region (about 30 AA residues), a transmembrane region (about 40 AA residues), and an extracellular region composed of three immunoglobulin-like domains (α1, α2, and α3), each consisting of approximately 90 AA. The α1 and α2 domains form a peptide-binding groove and contain the positions contributing to the binding pockets for the peptide and T cell receptors. The binding groove is divided into six distinct pockets (A–F) based on chemical and physical characteristics;

the most important pockets for peptide binding are the B and the F pockets. The membrane-proximal α3 domain of the HC contains a binding site for the co-stimulatory molecule CD8 [13] expressed by CTLs, which play an important enhancing role in killing virus-infected cells and cancer cells. The α1 and α2 domains consist of two segmented alpha helices forming the walls and eight antiparallel β strands forming the floor—together forming a unique peptide-binding groove, which is the site where the self (or foreign antigen-derived) peptide (8–10 AA) binds to the polymorphic parts of the HC and is presented to peptide-specific CTL for scrutiny. β2m is non-covalently associated with the extracellular region of the HLA-I heavy chain by non-covalent interactions with α2 and α3 domains [14]. β2m is essential for the correct conformation of the peptide-binding groove of the heavy chain and stabilizes the HLA-I antigen peptide complex on the cell surface. Thus, β2m indirectly participates in the antigen presentation to specific T-cell receptors of CTL [15–17].

The assembly of HLA-I peptide complex occurs in the endoplasmic reticulum (ER). Initially, the HLA-I HC associates with the chaperone calnexin (CNX) initiating an early folding and a disulfide bond formation within the HC. The newly synthesized HLA-I HC then associates with β2m to form heterodimer. This heterodimer is rapidly recruited into the peptide-loading complex (PLC) consisting of a transporter associated with antigen processing (TAP), and the chaperones tapasin, calreticulin (CRT), and ERp57. The HLA-I HC/β2m heterodimer is now ready for peptide loading. Peptides, both self- and pathogen-derived, are predominantly generated in the cytosol by the proteasome to degrade cytosolic proteins into short peptides, although a proteasome-independent peptide produced directly by insulin-degrading enzyme has been recently documented [18]. Thereafter, the peptides are transported into the ER by the TAP1 and TAP2. These peptides are further trimmed by aminopeptidase ERAAP1 and ERAAP2 to 8–10 AA, a length appropriate for HLA-I binding. Once HLA-I/HC-β2m dimers, physically associated with PLC, bind a subset of high-affinity peptides, the fully assembled MHC-I peptide complexes are released from PLC and transported via the Golgi apparatus to the cell surface, where the peptides are presented by HLA-I to CTL for scrutiny (*see* details in reviews [19, 20]).

3 HLA-II Molecule and Antigen-Presenting Pathway

The HLA-II molecule consists of two chains: α and β chain (each one with two domains: α1 and α2, β1 and β2) and a self-peptide with 13–25 AA located in a cleft formed by the α1 and β1 domains. Classical HLA-II molecules include HLA-DR, HLA-DQ, and

HLA-DP and are expressed mostly in the membrane of the professional antigen-presenting cells, where they present processed extracellular antigenic peptides to CD4⁺ T cells. In contrast to the antigen-binding groove of HLA-I molecule, which is closed at each end, the antigen-binding groove of HLA-II molecules is open at both ends and allows longer peptides (13–25 AA) to be loaded [21, 22]. During synthesis of HLA-II molecules in the ER, the α and β chains are produced and associate with an invariant chain, which stabilizes the HLA-II molecule and prevents it from binding of intracellular peptides or peptides from the endogenous pathway. The invariant chain directs transportation of HLA-II from the ER to the Golgi complex, followed by fusion with late endosomes which contain peptides derived from endocytosed, degraded proteins (self or foreign). The invariant chain is then cleaved by cathepsins to form a small fragment known as CLIP, which occupies the peptide-binding groove of the HLA-II molecules. HLA-DM facilitates CLIP removal and makes the peptide-binding groove of HLA-II ready for peptide loading before the HLA-II-peptide complex migrates to the cell surfaces to be scrutinized by CD4⁺ T cells [23].

4 Classification of Supertypes

4.1 HLA-I Supertypes

The concept of supertypes was firstly introduced by Alessandro Sette's group in 1995 [24, 25]. The definition of an HLA supertype is that HLA molecules with similar peptide binding features are grouped into one supertype; this means that if a peptide is able to bind to one allele within a supertype, it can also bind to all other alleles in this supertype. In practice, actually only a few peptides that are able to bind to one allele in a supertype can bind to all the other alleles within the supertype. To date, many methods have been used to define HLA-I supertypes, including structural similarities, shared peptide-binding motifs, and identification of cross-reacting peptides [26–29]. Based on motifs derived from binding data or sequencing of endogenously bound peptides, along with simple structural analyses, Sette and Sidney [30] defined nine supertypes (HLA-A1, -A2, -A3, -A24, -B7, -B27, -B44, -B58, -B62), which were reported to cover most of the HLA-A and -B polymorphisms. Subsequently, Ole Lund's group [26] constructed hidden Markov models (HMMs) [31] for HLA-I molecules using a Gibbs sampling procedure [32] and defined a similarity measure between these sequence motifs. By using this similarity to cluster alleles into supertypes, Ole Lund's group [26] further defined three new HLA-I supertypes (HLA-A26, -B8, and -B39), in addition to the nine supertypes described previously by Alessandro Sette's group [30], which was based on about 100 HLA-I peptide interactions. In the past few years, a lot of binding data have been generated; MHC-binding motif information is readily accessible (http://www.iedb.org), and MHC sequence data are also available in the IMGT (the international ImMunoGeneTics information system: http://www.imgt.org) database. In 2008 Alessandro Sette's group analyzed the updated list of alleles available through IMGT using a simple approach largely based on compilation of published motifs, binding data, and analyses of shared repertoires of binding peptides, in combination with clustering based on the primary sequence of the B and F peptide-binding pockets [29]. They provided updated supertype assignments, with new assignments for about 1,000 different HLA-I alleles, which is about a tenfold increase in the number of alleles compared to their original classification done in 1999 [30]. In the updated HLA-I classification, Alessandro Sette's group found that about 80 % of the 945 alleles examined were classified into one of the nine supertypes identified previously [30], and they did not suggest the existence of any other novel supertypes. However, they found that some alleles have specificities spanning two different supertypes, nine alleles share features of both the A01 and A03 supertypes, and another ten alleles have a specificity overlapping the A01 and A24 supertypes [29]. In addition, some alleles could not be assigned to any supertypes known today on the basis of the criteria mentioned above; thus these unclassified alleles remain to be addressed.

In summary, the updated HLA-I classification described by Alessandro Sette's group [29] is in agreement with those defined by other approaches from the other groups [26, 33, 34] including Ole Lund's group, and is now widely accepted and has been used for development of peptide-based vaccines [29, 35, 36].

4.2 HLA-II Supertypes

The structural composition between HLA-I and HLA-II molecules is fundamentally different, thus leading to very different binding characteristics. The binding groove is closed at both ends in an HLA-I molecule, while the peptide-binding groove of HLA-II molecules is open at both ends, which allow the binding of longer peptides (13–25 AA residues) than that for HLA-I molecules. A deeper understanding of the polymorphism of HLA-II molecules will contribute significantly to HLA-II-binding peptide prediction and classification of supertypes.

In contrast to HLA-I supertypes, HLA-II supertypes have been less intensively studied, although a few studies about HLA-II supertypes [26, 37–41] have been reported. One important reason is that peptide binding data for HLA-II molecules is less available than those for HLA-I molecules due to the complexity of HLA-II structure. Nevertheless, studies have suggested that many DR molecules [26, 37, 38] and many DP molecules [40, 42] can be grouped into supertypes. In 1998, Ou et al. [38] grouped HLA-DR molecules into seven different functional supertypes on the basis of their ability to bind and present antigenic peptides to T cells and their association with susceptibility or resistance to disease. In 2002, Castelli et al. [40] defined an HLA-DP4 supertype and supported the existence of three main binding supertypes among HLA-DP molecules. In 2005, Doytchinova et al. [37] applied a combined bioinformatics approach using both protein sequence and structural data, to 2,225 HLA-II molecules, to detect similarities in their peptide-binding sites for definition of HLA-II supertypes. They defined 12 HLA-II supertypes, including five DRs (DR1, DR3, DR4, DR5, and DR9), three DQs (DQ1, DQ2, and DQ3), and four DPs (DPw1, DPw2, DPw4, and DPw6). In 2011, Greenbaum et al. [41] determined the binding capacity of a large panel of non-redundant peptides for a set of 27 common HLA DR, DQ, and DP molecules. The measured binding data were then used to define class II supertypes on the basis of shared binding repertoires. Seven different supertypes (main DR, DR4, DRB3, main DQ, DQ7, main DP, and DP2) were defined. Subsequently, according to motif-based supertype classification [27], seven different supertypes were defined after the analysis of 27 HLA II proteins described in a previous report [41]. All the molecules belonging to the DP genetic locus (DPB1*0101, DPB1*0201, DPB1*0401, DPB1*0402, DPB1*0501, and DPB1*1401) were grouped into a single supertype; DQ proteins were grouped into two different supertypes, each containing three HLAs: (DQB1*0301, DQB1*0302, DQB1*0401) and (DQB1*0201, DQB1*0501, DQB1*0602). The motif-based classification of the DR proteins is less defined compared with the other loci. The HLA-DR can be grouped into four supertypes: (DRB1*0401, DRB1*0405, DRB1*0802, DRB1*1101), (DRB3*0101, DRB3*0202), (DRB1*0301, DRB1*1302), and the fourth containing the remaining DR proteins. Functional and motif-based clustering of 27 defined HLA-II molecules revealed the presence of proteins sharing both functional and structural properties, thus supporting the concept of HLA-II supertypes.

5 HLA Supertypes and Vaccines

To date, one of the major drawbacks of a peptide-based vaccine strategy is that the restricting HLA genes are extremely polymorphic resulting in a vast diversity of peptide-binding HLA specificities and a low population coverage for any given peptide–HLA specificity. To increase population coverage, one might include defined epitopes for each HLA-I allele; however, this would lead to a vaccine comprising hundreds of peptides. As mentioned above, one way to reduce this complexity is to group HLA molecules into HLA supertypes; a classification that as mentioned above refers to a group of HLA alleles with largely overlapping peptide binding specificities [24, 25, 30]. Ideally this means that a peptide, which binds to one allele within a supertype, has a high probability of binding to other allelic members of the same supertype. The concept of HLA supertypes has been successfully applied to characterize and identify T cell epitopes from a variety of different pathogens, including measles-mumps-rubella, SARS, EBV, HIV, HCV, HBV, HPV, influenza, LCMV, Lassa virus, *F. tularensis*, vaccinia, and cancer antigens as well [29].

HLA supertypes have been utilized as a component in several approaches and algorithms designed for predicting peptide candidates [43–48]. The technology behind "reverse immunology" is developing rapidly in order to identify T cell epitopes from tumor antigens and infectious microorganisms [44–51]. During the SARS epidemic back in 2003, the SARS genome was identified in a matter of weeks, and a complete CTL epitope scanning—just barely possible at that time—was completed a few months later [43]. Therefore, "reverse immunology" as a powerful tool to identify T cell epitopes has now reached the stage where genome-, pathogen-, and HLA-wide scanning for HLA-binding antigenic epitopes become feasible at a scale and speed that makes it possible to exploit the genome information as fast as it can be generated. Importantly, a large-scale dataset of measured HLA-II-binding affinities covering 26 allelic variants, including a total of 44541 affinity measurements for HLA-DR alleles as well as 11 HLA-DP and DQ molecules [52], are available to be used as training data for generating prediction tools utilizing several machine learning algorithms. To date, the computer-based algorithms for predicting peptides binding to HLA-I molecules are being developed for HLA-II-restricted peptide epitopes, a development, which is of pivotal importance for understanding the immune response and its effect on host-pathogen interactions [32, 52–55]. Those tools will definitely lead to fast identification of novel peptides restricted by HLA-I and HLA-II supertypes for use in vaccines against infectious agents as well as tumors. In this respect, individual peptides harboring both HLA-I and HLA-II binding potentials [46–48, 56] might be of particular importance.

In conclusion, classification of HLA supertypes reduces complexity of HLA polymorphisms and has a significant impact on the development of peptide-based vaccines with maximum population coverage. Since $CD4^+$ T cells are required for priming of naïve $CD8^+$ T cells as well as expansion of $CD8^+$ memory T cells [8–12], it is of critical importance to incorporate both HLA-I and -II supertype-restricted epitopes in peptide-based vaccines with maximum population coverage to obtain participation of both $CD4^+$ and $CD8^+$ T cells for generation of strong and long-lasting immunity.

Acknowledgements

This work was supported by National Institute of Allergy and Infectious Disease contracts HHSN266200400083C, HHSN266200400025C, EU 6FP 503231, National Institutes of Health contract HHSN266200400081C, and a grant from the Lundbeck Foundation, Copenhagen, Denmark.

References

1. Kanodia S, Kast WM (2008) Peptide-based vaccines for cancer: realizing their potential. Expert Rev Vaccines 7:1533–1545
2. Rosenberg SA, Dudley ME (2009) Adoptive cell therapy for the treatment of patients with metastatic melanoma. Curr Opin Immunol 21: 233–240
3. Woodworth JS, Wu Y, Behar SM (2008) Mycobacterium tuberculosis-specific CD8+ T cells require perforin to kill target cells and provide protection in vivo. J Immunol 181: 8595–8603
4. Woodworth JS, Behar SM (2006) Mycobacterium tuberculosis-specific CD8+ T cells and their role in immunity. Crit Rev Immunol 26:317–352
5. Kennedy R, Poland GA (2007) T-Cell epitope discovery for variola and vaccinia viruses. Rev Med Virol 17:93–113
6. McMichael AJ, Gotch FM, Noble GR et al (1983) Cytotoxic T-cell immunity to influenza. N Engl J Med 309:13–17
7. Bender BS, Croghan T, Zhang L et al (1992) Transgenic mice lacking class I major histocompatibility complex-restricted T cells have delayed viral clearance and increased mortality after influenza virus challenge. J Exp Med 175:1143–1145
8. Janssen EM, Droin NM, Lemmens EE et al (2005) CD4+ T-cell help controls CD8+ T-cell memory via TRAIL-mediated activation-induced cell death. Nature 434:88–93
9. Janssen EM, Lemmens EE, Wolfe T et al (2003) CD4+ T cells are required for secondary expansion and memory in CD8+ T lymphocytes. Nature 421:852–856
10. Shedlock DJ, Shen H (2003) Requirement for CD4 T cell help in generating functional CD8 T cell memory. Science 300:337–339
11. Sun JC, Bevan MJ (2003) Defective CD8 T cell memory following acute infection without CD4 T cell help. Science 300:339–342
12. Sun JC, Williams MA, Bevan MJ (2004) CD4+ T cells are required for the maintenance, not programming, of memory CD8+ T cells after acute infection. Nat Immunol 5:927–933
13. Salter RD, Benjamin RJ, Wesley PK et al (1990) A binding site for the T-cell co-receptor CD8 on the alpha 3 domain of HLA-A2. Nature 345:41–46
14. Bjorkman PJ, Saper MA, Samraoui B et al (1987) Structure of the human class I histocompatibility antigen, HLA-A2. Nature 329: 506–512
15. York IA, Rock KL (1996) Antigen processing and presentation by the class I major histocompatibility complex. Annu Rev Immunol 14: 369–396
16. Vitiello A, Potter TA, Sherman LA (1990) The role of beta 2-microglobulin in peptide binding by class I molecules. Science 250:1423–1426
17. Perarnau B, Siegrist CA, Gillet A et al (1990) Beta 2-microglobulin restriction of antigen presentation. Nature 346:751–754
18. Parmentier N, Stroobant V, Colau D et al (2010) Production of an antigenic peptide by insulin-degrading enzyme. Nat Immunol 11:449–454
19. Cresswell P, Ackerman AL, Giodini A et al (2005) Mechanisms of MHC class I-restricted antigen processing and cross-presentation. Immunol Rev 207:145–157
20. Kloetzel PM (2001) Antigen processing by the proteasome. Nat Rev Mol Cell Biol 2:179–187
21. Chicz RM, Urban RG, Lane WS et al (1992) Predominant naturally processed peptides bound to HLA-DR1 are derived from MHC-related molecules and are heterogeneous in size. Nature 358:764–768
22. Rudensky A, Preston-Hurlburt P, Hong SC et al (1991) Sequence analysis of peptides bound to MHC class II molecules. Nature 353:622–627
23. Watts C (2004) The exogenous pathway for antigen presentation on major histocompatibility complex class II and CD1 molecules. Nat Immunol 5:685–692
24. del Guercio MF, Sidney J, Hermanson G et al (1995) Binding of a peptide antigen to multiple HLA alleles allows definition of an A2-like supertype. J Immunol 154:685–693
25. Sidney J, del Guercio MF, Southwood S et al (1995) Several HLA alleles share overlapping peptide specificities. J Immunol 154:247–259
26. Lund O, Nielsen M, Kesmir C et al (2004) Definition of supertypes for HLA molecules using clustering of specificity matrices. Immunogenetics 55:797–810
27. Saha I, Mazzocco G, Plewczynski D (2013) Consensus classification of human leukocyte antigen class II proteins. Immunogenetics 65:97–105
28. Sidney J, Grey HM, Southwood S et al (1996) Definition of an HLA-A3-like supermotif demonstrates the overlapping peptide-binding repertoires of common HLA molecules. Hum Immunol 45:79–93

29. Sidney J, Peters B, Frahm N et al (2008) HLA class I supertypes: a revised and updated classification. BMC Immunol 9:1
30. Sette A, Sidney J (1999) Nine major HLA class I supertypes account for the vast preponderance of HLA-A and -B polymorphism. Immunogenetics 50:201–212
31. Eddy SR (1998) Profile hidden Markov models. Bioinformatics 14:755–763
32. Nielsen M, Lundegaard C, Worning P et al (2004) Improved prediction of MHC class I and class II epitopes using a novel Gibbs sampling approach. Bioinformatics 20:1388–1397
33. Tong JC, Tan TW, Ranganathan S (2007) In silico grouping of peptide/HLA class I complexes using structural interaction characteristics. Bioinformatics 23:177–183
34. Hertz T, Yanover C (2007) Identifying HLA supertypes by learning distance functions. Bioinformatics 23:e148–e155
35. Karosiene E, Lundegaard C, Lund O et al (2012) NetMHCcons: a consensus method for the major histocompatibility complex class I predictions. Immunogenetics 64:177–186
36. Thomsen M, Lundegaard C, Buus S et al (2013) MHCcluster, a method for functional clustering of MHC molecules. Immunogenetics 65(9):655–665
37. Doytchinova IA, Flower DR (2005) In silico identification of supertypes for class II MHCs. J Immunol 174:7085–7095
38. Ou D, Mitchell LA, Tingle AJ (1998) A new categorization of HLA DR alleles on a functional basis. Hum Immunol 59:665–676
39. Baas A, Gao X, Chelvanayagam G (1999) Peptide binding motifs and specificities for HLA-DQ molecules. Immunogenetics 50:8–15
40. Castelli FA, Buhot C, Sanson A et al (2002) HLA-DP4, the most frequent HLA II molecule, defines a new supertype of peptide-binding specificity. J Immunol 169:6928–6934
41. Greenbaum J, Sidney J, Chung J et al (2011) Functional classification of class II human leukocyte antigen (HLA) molecules reveals seven different supertypes and a surprising degree of repertoire sharing across supertypes. Immunogenetics 63:325–335
42. Sidney J, Steen A, Moore C et al (2010) Five HLA-DP molecules frequently expressed in the worldwide human population share a common HLA supertypic binding specificity. J Immunol 184:2492–2503
43. Sylvester-Hvid C, Nielsen M, Lamberth K et al (2004) SARS CTL vaccine candidates; HLA supertype-, genome-wide scanning and biochemical validation. Tissue Antigens 63: 395–400
44. Tang ST, Wang M, Lamberth K et al (2008) MHC-I-restricted epitopes conserved among variola and other related orthopoxviruses are recognized by T cells 30 years after vaccination. Arch Virol 153:1833–1844
45. Wang M, Lamberth K, Harndahl M et al (2007) CTL epitopes for influenza A including the H5N1 bird flu; genome-, pathogen-, and HLA-wide screening. Vaccine 25:2823–2831
46. Wang M, Larsen MV, Nielsen M et al (2010) HLA class I binding 9mer peptides from influenza A virus induce CD4 T cell responses. PLoS One 5:e10533
47. Wang M, Tang ST, Lund O et al (2009) High-affinity human leucocyte antigen class I binding variola-derived peptides induce CD4+ T cell responses more than 30 years post-vaccinia virus vaccination. Clin Exp Immunol 155: 441–446
48. Wang M, Tang ST, Stryhn A et al (2011) Identification of MHC class II restricted T-cell-mediated reactivity against MHC class I binding Mycobacterium tuberculosis peptides. Immunology 132:482–491
49. Sette A, Rappuoli R (2010) Reverse vaccinology: developing vaccines in the era of genomics. Immunity 33:530–541
50. Lundegaard C, Lund O, Buus S et al (2010) Major histocompatibility complex class I binding predictions as a tool in epitope discovery. Immunology 130:309–318
51. Kessler JH, Melief CJ (2007) Identification of T-cell epitopes for cancer immunotherapy. Leukemia 21:1859–1874
52. Wang P, Sidney J, Kim Y et al (2010) Peptide binding predictions for HLA DR, DP and DQ molecules. BMC Bioinformatics 11:568
53. Nielsen M, Lundegaard C, Blicher T et al (2008) Quantitative predictions of peptide binding to any HLA-DR molecule of known sequence: NetMHCIIpan. PLoS Comput Biol 4:e1000107
54. Nielsen M, Lund O, Buus S et al (2010) MHC class II epitope predictive algorithms. Immunology 130:319–328
55. Nielsen M, Justesen S, Lund O et al (2010) NetMHCIIpan-2.0—Improved pan-specific HLA-DR predictions using a novel concurrent alignment and weight optimization training procedure. Immunome Res 6:9
56. Wang M, Rasmussen S, Claesson MH (2012) HLA class II presentation of HLA class I binding antigenic 9mer peptides. In: Semiletova NV (ed) Major histocompatibility complex: biology, functions and roles in disease. Nova Science Publishers, Inc., New York, NY, pp 139–152

Chapter 18

Customized Predictions of Peptide–MHC Binding and T-Cell Epitopes Using EPIMHC

Magdalena Molero-Abraham, Esther M. Lafuente, and Pedro Reche

Abstract

Peptide binding to major histocompatibility complex (MHC) molecules is the most selective requisite for T-cell recognition. Therefore, prediction of peptide–MHC binding is the main basis for anticipating T-cell epitopes. A very popular and accurate method to predict peptide–MHC binding is based on motif-profiles and here we show how to make them using EPIMHC (http://imed.med.ucm.es/epimhc/). EPIMHC is a database of T-cell epitopes and MHC-binding peptides that unlike any related resource provides a framework for computational vaccinology. In this chapter, we describe how to derive peptide–MHC binding motif-profiles in EPIMHC and use them to predict peptide–MHC binding and T-cell epitopes. Moreover, we show evidence that customization of peptide–MHC binding predictors can lead to enhanced epitope predictions.

Key words MHC, HLA, PSSM, T-cell epitope, Prediction

Abbreviations

MHC Major histocompatibility complex
HLA Human leukocyte antigen
PSSM Position specific scoring matrices

1 Introduction

T cells play a key role in fighting infectious agents such as pathogenic viruses, bacteria and parasites, as well as in cancer immune surveillance, eliminating tumoral cells. T cells respond to foreign peptide antigens (T-cell epitopes) bound the cell surface expressed major histocompatibility complex (MHC) molecules [1–4]. There are two main classes of MHC molecules, MHC class I (MHCI) and MHC class II (MHCII) that are in turn recognized by $CD8^+$ T and $CD4^+$ T cells, respectively [4]. In humans, MHC molecules are known as Human Leukocyte Antigens (HLAs) and are

Rajat K. De and Namrata Tomar (eds.), *Immunoinformatics*, Methods in Molecular Biology, vol. 1184,
DOI 10.1007/978-1-4939-1115-8_18, © Springer Science+Business Media New York 2014

extremely polymorphic [5]. HLA polymorphisms are the basis for distinct peptide binding specificity of HLA allelic variants [5].

The relevance of T-cell epitopes for understanding disease pathology [6] and for epitope-based vaccines [7–10] has led to the identification of thousands of epitopes and MHC-peptide ligands from all kind of antigens. The availability of this vast amount of data has had two major intertwined consequences. On the one hand, it has given rise to comprehensive databases and resources to store the ever-increasing data. On the other hand, it has fueled the development of computational approaches for the prediction of T-cell epitopes.

Relevant examples of T-cell epitope and MHC-peptide ligand databases include SYFPEITHI [11], JenPep [12], and MHCBN [13], TEPIDAS [14], ImmuneEpitope Database [15], and EPIMHC [16]. These resources are instrumental for the analysis of peptide–MHC binding and T-cell epitope immunogenicity, primarily serving as source of data but also providing analytic and predictive tools. All these databases are based on relational databases and share a considerable amount of capabilities. Yet they also have unique features. Here we will work with EPIMHC [16], a highly curated database of T-cell epitopes and MHC-binding peptides that unlike any related resource enables tailored predictions of T-cell epitopes using custom-made peptide–MHC binding motif-profiles [17–19].

T-cell epitopes are determined by several molecular events [20–22], of which peptide–MHC binding is the most selective. Therefore, prediction of peptide–MHC binding is the main basis to anticipate T-cell epitopes [23]. Peptide–MHC binding predictions can be achieved through a great variety of methods [23], including peptide–MHC binding motif-profiles [17–19] which rank among the most successful and popular of them [24]. These motif-profiles consist of weighted position specific scoring matrices (PSSM) [25] created from sets of aligned peptide sequences known to bind to the relevant MHC molecules.

Prediction of T-cell epitopes using a large set of MHC-specific motif-profiles is readily available for free public use in at our RANKPEP site (http://imed.med.ucm.es/Tools/rankpep.html). We generated the peptide–MHC binding profiles available in RANKPEP from the largest non-redundant set of peptides that we could identify. In computational cross-validations, RANKPEP profiles exhibited a great performance [17]. However, they are not necessarily the best for all predictive matters. In fact there is no general consensus on what peptides should be included for peptide–MHC binding model building. Therefore, in this chapter, we illustrate how to use EPIMHC to derive custom-made peptide–MHC binding motif-profiles and produce tailored T-cell epitope predictions.

2 Materials

2.1 EPIMHC Database and Query Form

EPIMHC is a database with comprehensive information on MHC-restricted ligands and T-cell epitopes that are observed in real proteins from a great variety of sources including tumor antigens. Peptide data were collected from related databases [11, 26, 27] and the literature and was incorporated into the database upon computational curation (altered peptide ligands are not included). EPIMHC data is structured as a relational database with a set of related tables (Fig. 1). Entries in EPIMHC are unique with regard to the combination of two features: the sequence of the peptide and the MHC restriction element. Main annotations in EPIMHC include information on the ligand (sequence, length, MHC binding, T-cell activity, processing, protein source, protein name, and organisms), the MHC restriction element (CLASS, MHC molecule and MHC source), and publication reference. The "processing" field indicates whether the peptides are processed and presented from their protein sources in vivo (annotated as *natural*). "MHC binding" and "immunogenicity level" fields follow a qualitative annotation of four values (high, moderate, little, unknown). The immunogenicity level only applies to peptides with reported "T-cell activity." Immunogenicity and MHC-binding binding levels were obtained from the literature and translated onto the indicated qualitative values as previously reported [26]. If no information on peptide–MHC binding and/or Immunogenicity level was found, then such fields were annotated as unknown.

EPIMHC database is accessible online at http://imed.med.ucm.es/epimhc/ (Fig. 2) through an intuitive and user-friendly Web interface. This server allows for complex database queries,

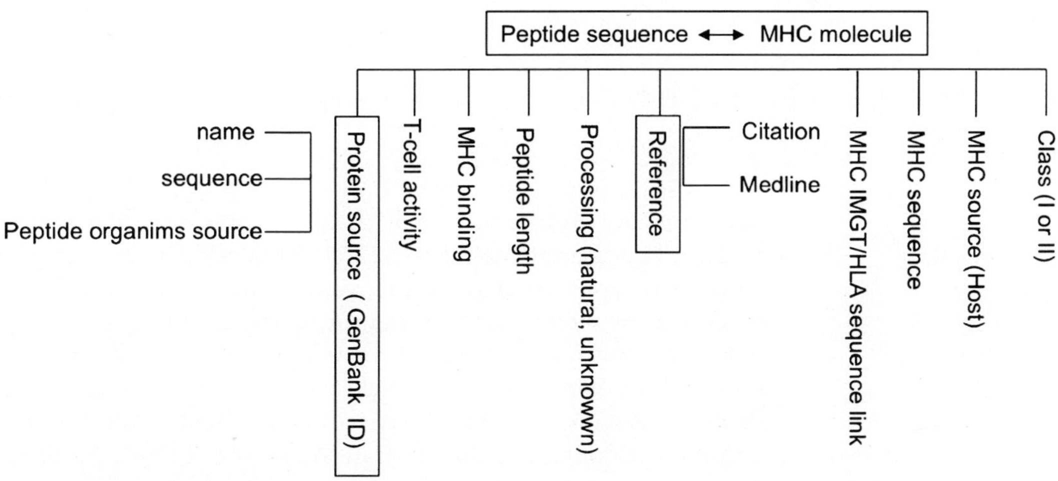

Fig. 1 EPIMHC database structure

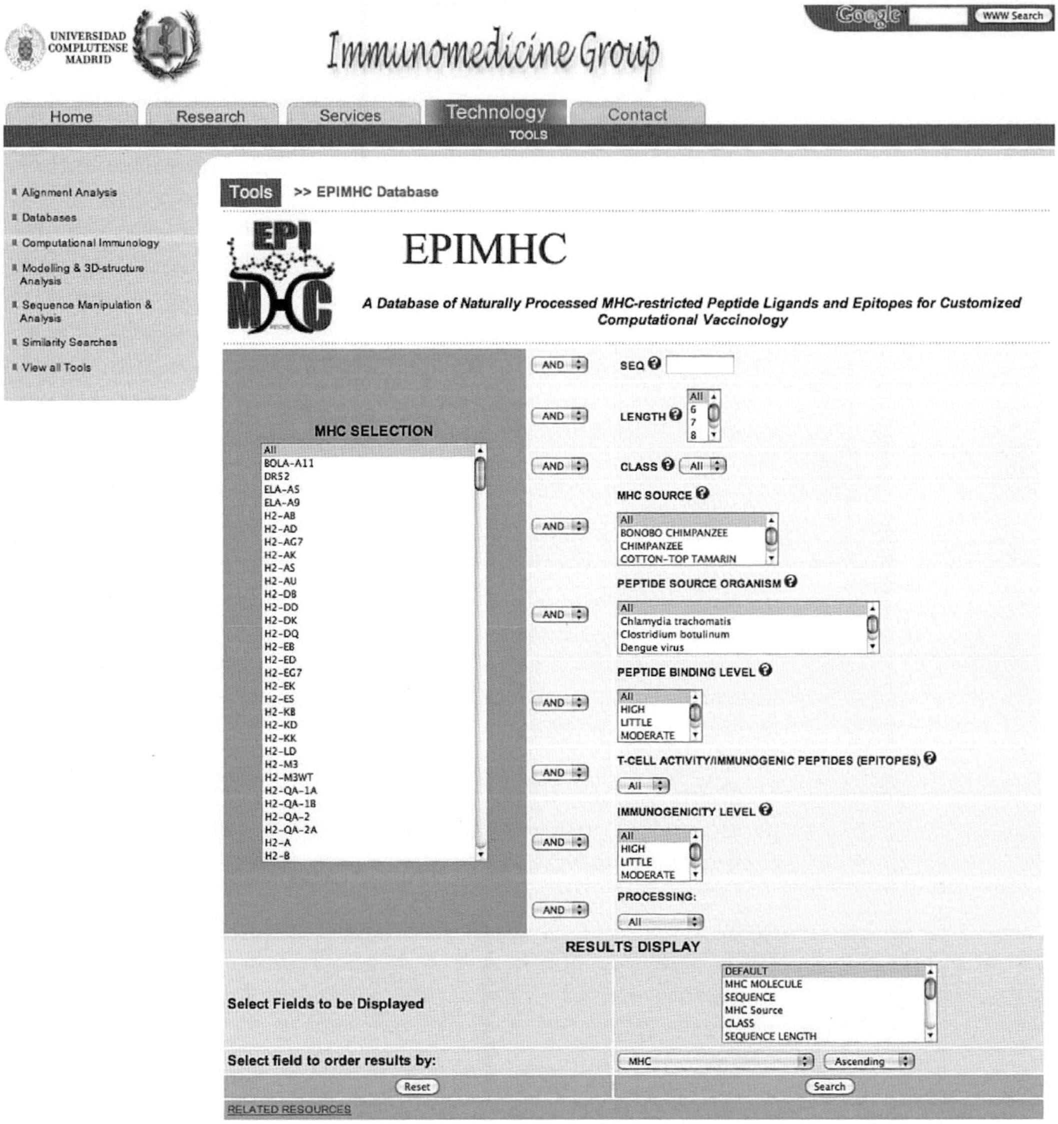

Fig. 2 EPIMHC database Web interface. EPIMHC resource is available for free public use at http://imed.med.ucm.es/epimhc/

combining any annotation field, thanks to the underlying SQL language. For example, users can search for peptide ligands and/or epitopes that are restricted by one, various or all MHC molecules (left side of the screen), and restrict the search according to various criteria like length and source of the peptide (right of the screen). Also, any field of interest can be included in the search output. The EPIMHC search output will be described in detail in the Method section in the context of the generation of custom-made profiles.

2.2 Prediction of a Peptide–MHC Binding and T-Cell Epitopes Using Profiles

As mentioned, motif-profiles consist of weighted PSSM [25] created from a set of aligned peptide sequences known to bind to a given MHC molecule. In order to predict peptide–MHC binding and T-cell epitope using motif-profiles, we used a search algorithm known as RANKPEP [17, 18]. The algorithm uses the profile coefficients to score all possible peptide fragments in a protein with the width of the PSSM and ranks them by score. The width of a PSSM is given by the number of residue sites in a multiple sequence alignment. Although rank per se is insufficient to assess whether a peptide is a potential binder, we have shown that T-cell epitopes score among the top 2 % ranking peptides [17, 18]. Motif-profiles assume that peptide residues contribute independently to MHC binding. This assumption is well supported by experimental data, although there are reported instances in which the contribution of peptide residues to MHC-binding is influenced by neighboring residues [28].

RANKPEP is accessible online for public use at the site http://imed.med.ucm.es/Tools/rankpep.html (Fig. 3). Currently, 88 and 50 different MHCI and MHCII molecules, respectively, can be targeted for peptide binding predictions in RANKPEP using the relevant motif-profiles. The profiles available in RANKPEP have been derived from large sets of non-redundant peptide–MHC binders, without taking in consideration their T-cell activity and source. These sets can include self-peptides eluted from MHC molecules. The RANKPEP Web server is flexible, intuitive and combines several interesting features. Notably, users can upload their own motif-profiles, such as those generated using EPIMHC (*see* Subheading 3). A simplified version of the RANKPEP input form can also be launched from EPIMHC to facilitate tailored prediction of T-cell epitopes using custom-made profiles (*see* Subheading 3).

3 Methods

In this section, we show a step-by-step guide to derive a specific peptide–MHC binding motif-profile in EPIMHC and produce tailored T-cell epitope predictions. In particular, we will target the prediction of A*0201-restricted CD8 T-cell epitopes from SARS coronavirus nucleoprotein (GI: 30173007). This protein contains 7 experimentally identified A*0201-restricted CD8 T-cell epitopes (Table 1) and we will use that knowledge to assess the predictive accuracy of various peptide–MHC binding motif-profiles.

3.1 Peptide Selection and Motif-Profile Building

We will build a motif-profile from all 9-mer peptides that are annotated in EPIMHC to bind with high affinity to the human MHC I molecule HLA-A*0201 (A*0201) (*see* **Notes 1–6**). To this end,

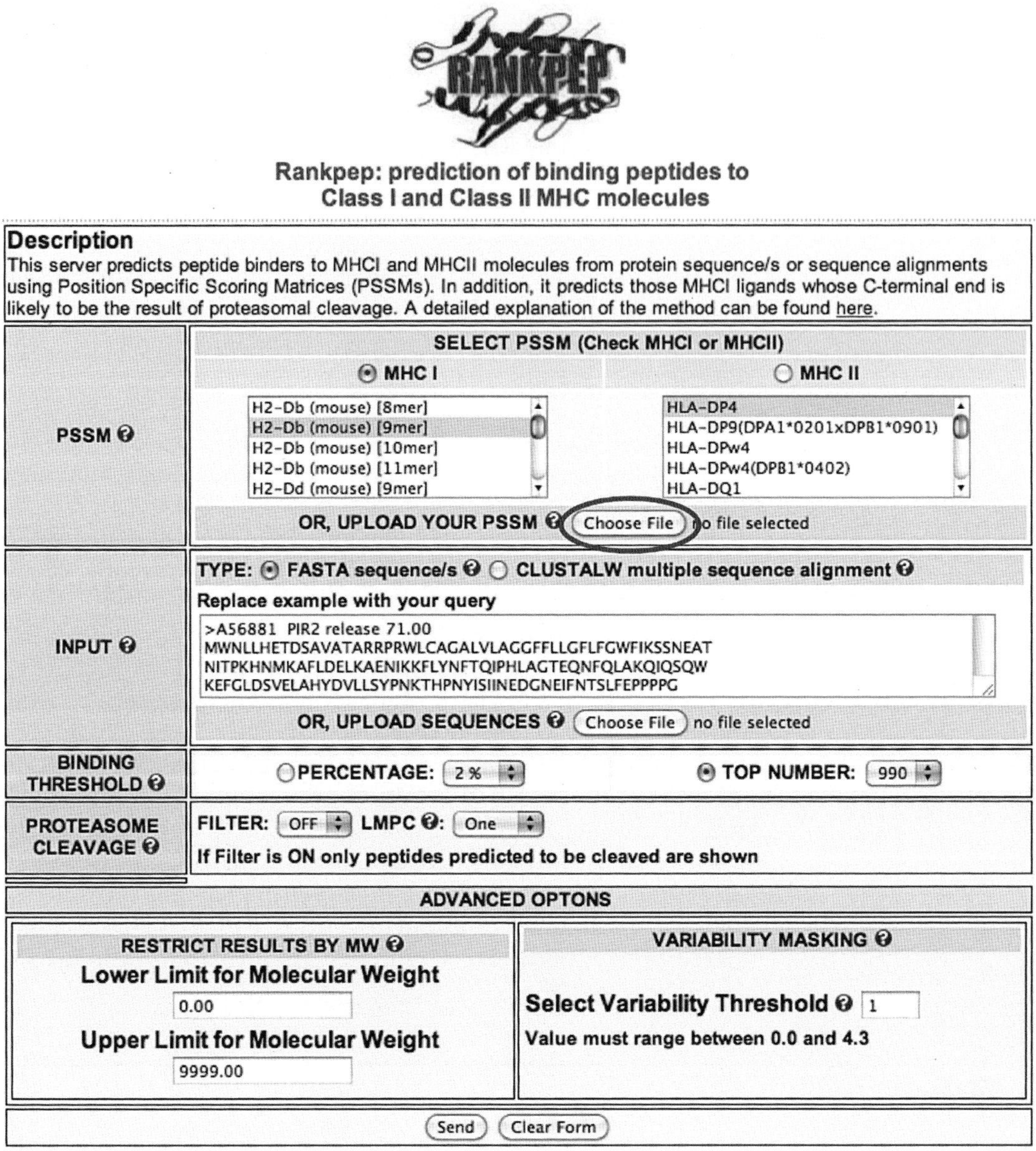

Fig. 3 RANKPEP Web server. The figure depicts a screenshot of the RANKPEP interface with the option of uploading custom-made profiles highlighted. RANKPEP is available for free public use at http://imed.med.ucm.es/Tools/rankpep.html

we first do a search in EPIMHC with the following selection criteria, leaving the remaining fields with default settings:

1. Select HLA-A*0201 in *MHC SELECTION*.
2. Select 9 in *LENGTH* (*see* **Note 2**).
3. Select high in *PEPTIDE BINDING* LEVEL.

Table 1
Known A*0201-restricted CD8 T-cell epitopes in SARS nucleoprotein

Epitope sequence	Location	References
ALNTPKDHI	139–147	[34]
LQLPQGTTL	160–168	[34]
LALLLLDRL	220–228	[34]
LLLDRLNQL	223–231	[34]
RLNQLESKV	227–235	[34]
GMSRIGMEV	317–325	[34]
ILLNKHIDA	352–360	[34]

None of these epitopes has been used for profile building in EPIMHC

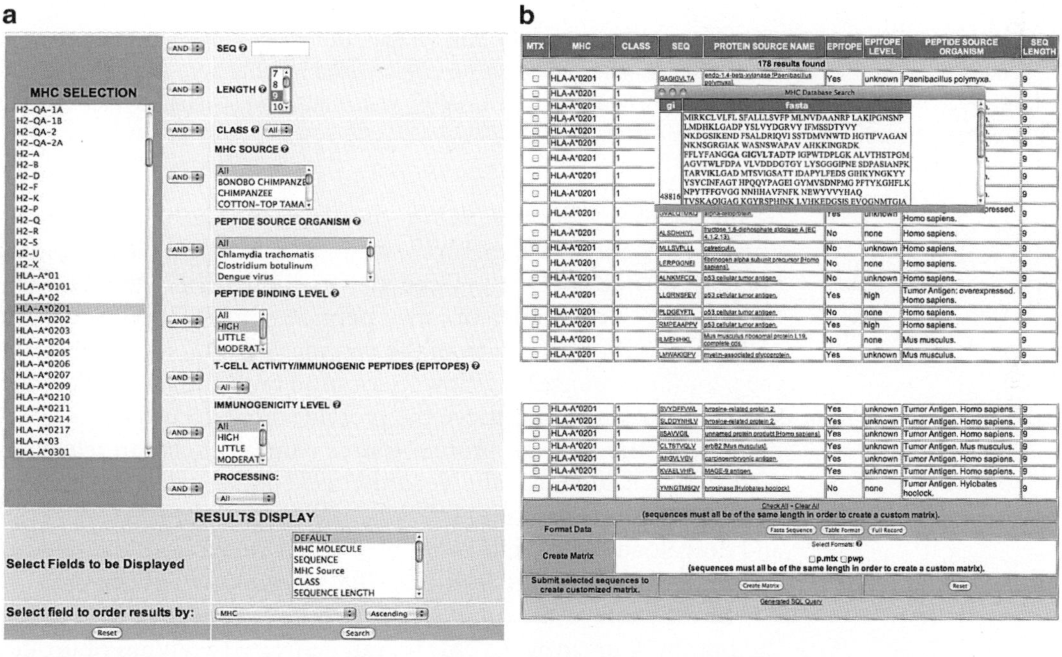

Fig. 4 EPIMHC search example and output. The figure illustrates a search example in the EPIMHC resource for peptides binding to HLA-A*0201 with high affinity (**a**) and the result of that specific search (**b**)

In Fig. 4a we show a screen capture of the selection described. Upon submitting the search, we get 178 peptides (Fig. 4b). EPIMHC search results consist of a tabulated list, rows, of records fitting the search criteria. The table columns provide the information fields selected by the users in the query form. The default fields are those shown in Fig. 4b and include the MHC restriction element (*MHC*), the MHC class (I or II) (*CLASS*), the sequence

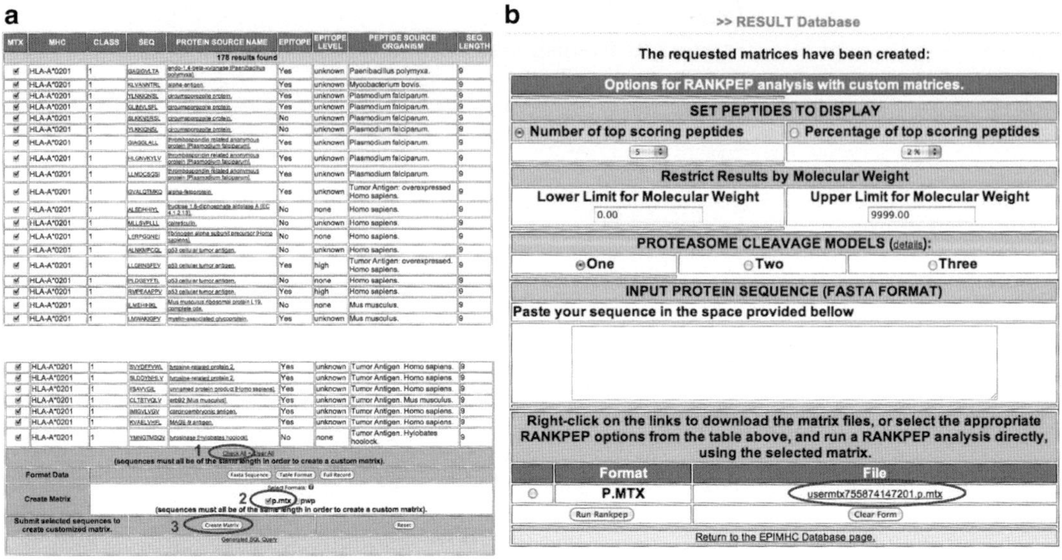

Fig. 5 Profile-building using peptides selected in the EPIMHC output. (**a**) Figure highlights the steps that are needed (labeled 1, 2, and 3) to make a motif-profile with position-based weights from all peptides in the EPIMHC search output (**b**) RANKPEP form launched by EPIMHC with ready-to-use custom-made profile (*highlighted*)

of the peptide (*SEQUENCE*), the name of the source protein (*PROTEIN SOURCE NAME*), whether the peptide is an epitope (Yes or NO) (*EPITOPE*), T-cell activity level (*EPITOPE LEVEL*), the source organisms of the peptide (*PEPTIDE SOURCE ORGANISMS*), and length of the peptide (*LENGTH*). Clicking on the peptide sequence will show its location onto the relevant protein source. Also, one can retrieve the protein source record in NCBI by clicking on the relevant protein source names. Users can select any record by clicking on the record-checkbox or select all records by clicking on the option *check all* at the bottom of the result page and download the data in a variety of text formats from the relevant action buttons (*FASTA Sequence*, *Table format*, or *Full Record*).

Motif-profiles are built from peptide sequences selected in the output. Currently, EPIMHC can generate two types of profiles that are specified by clicking on either the *p.mtx* or the *pwp* checkboxes. The first one uses a branch proportional weighting method [29], while the second uses position-based weights [30]. To make a motif-profile incorporating position-based weights from all peptide records, we follow the next sequential steps (highlighted in Fig. 5a):

1. Click on *check all* (all peptides will be selected) (*see* **Note 6**).
2. Click the *p.mtx* check box.
3. Click the *create matrix* action button.

A	C	D	E	F	G	H	I	K	L	M	N	P	Q	R	S	T	V	W	Y
1.242	0.511	-3.95	-6.57	0.868	1.588	1.939	0.437	0.183	-0.04	-0.47	-2.39	-5.13	-0.35	0.65	1.619	-2.53	0.18	0.974	2.959
-2.85	-7.01	-4.94	-5.05	-5.89	-10.8	-9.46	1.65	-4.4	7.78	4.35	-10.3	-11.4	-3.34	-9.33	-9.83	-0.81	1.236	-9.48	-8.34
-0.32	0.362	1.139	-3.5	0.567	0.459	2.033	0.959	-3.15	0.618	2.244	0.982	-3.94	-2.05	-0.83	-1.42	-3.04	1.318	5.311	2.123
-1.43	-1.02	3.157	2.219	-6.8	2.367	-5.67	0.641	0.4	-2.54	0.544	-0.58	2.026	-1.14	-0.65	-0.42	-0.61	-3.5	0.286	-1.77
-1.23	-0.58	-1.03	-1.3	1.308	2.357	0.944	0.333	-0.16	-0.64	-1.86	-0.22	-0.12	-0.3	-1.7	-2.67	-0.09	0.13	3.087	3.421
-2.24	3.381	-0.87	-3.12	0.527	-1.23	-1.19	3.077	-1.71	1.318	1.057	-1.49	-0.98	-1.86	-4.7	-0.17	0.915	2.648	-1.18	0.249
1.233	0.416	-3.41	-3.08	1.554	-1.54	2.554	1.719	-3.57	1.286	0.828	-0.21	2.362	-1.35	-4.06	-3.31	-0.16	1.889	3.005	-1.94
-1.8	-1.51	-2.67	1.279	0.087	-0.63	1.255	-0.88	-0.42	0.574	-0.17	-2.09	-1.91	1.059	0.347	1.467	1.8	-0.91	1.317	1.002
-0.77	-7.92	-12.2	-11.6	-7.63	-11.9	-11.1	3.887	-10.7	5.368	-0.05	-11.8	-12.6	-1.09	-11	-4.99	-9.15	7.339	-11.1	-9.67

Fig. 6 EPIMHC profile with position-based weights generated from 178 9-mer peptides binding to HLA-A*0201 with high affinity

Upon hitting the *create matrix* button, EPIMHC opens a simplified version of the RANKPEP Web interface that incorporates the custom-made motif-profile from the selected peptides (Fig. 5b). The profile appears under the *File* field of the form and can be downloaded with a mouse right click (PC) or by a mouse click-and-hold (MAC). The profile thus generated, shown in Fig. 6, has the format required by the MAST-motif search algorithm [31] and can be uploaded to the original RANKPEP Web server to produce custom predictions of peptide–MHC binding and T-cell epitopes. However, the RANKPEP interface launched by EPIMHC allows a more direct and simple way to achieve such a task (Fig. 5b). Under *SET DISPLAY OPTIONS,* users can select between two options to set the number of peptides to be returned by the algorithm: one is as a fixed *number of top scoring peptides* and the other as a *percentage of top scoring peptides.* Users can also restrict the peptides sorted by RANKPEP by molecular weight (MW) so that only peptides within a MW window will be returned. By default, MW filtering is not applied. The RANKPEP interface also provides three models for proteasomal cleavage predictions [17, 22]. By default, model *one* is selected. These models will be applied regardless of the class of the MHC targeted for predictions but the predictions are only meaningful for MHC I-restricted peptides (*see* **Note 4**).

3.2 Prediction of Peptide–MHC Binding and T-Cell Epitopes With EPIMHC Custom-Made Profiles

To target SARS nucleoprotein for T-cell epitope predictions using the RANKPEP form launched by EPIMHC with the custom-made profile we carry on as follows:

1. Set peptides to display to 2 % of top scoring peptides (*see* **Note 5**).
2. Paste the SARS nucleoprotein, FASTA format, in the text box INPUT section.
3. Click on the matrix check box.
4. Click on the action button *Run Rankpep.*

The indicated steps are highlighted in Fig. 7a and we describe next the RANKPEP output (Fig. 7b)

The top part of the RANKPEP output shows the matrix (profile) used for the predictions, a consensus sequence that would reach the largest score, optimal (largest) score and a binding threshold (BT). The later is an important feature. Large scores lead

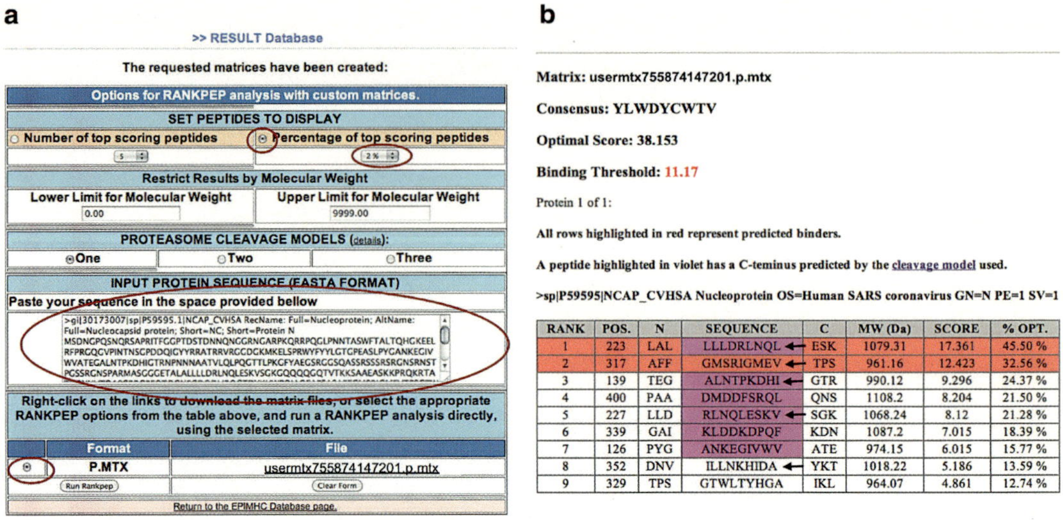

Fig. 7 Tailored prediction of peptide–MHC binding using the RANKPEP form launched by EPIMHC (**a**) The figure illustrate the steps to predict HLA-A*0201-restricted peptides from SARS nucleoprotein. (**b**) RANKEP output showing the prediction results

to top ranking peptides and are indicative of peptide–MHC binding. However, rank per se is insufficient to know whether a given peptide will bind to a particular MHC molecule, e.g., scoring a single peptide. Therefore, EPIMHC provides a profile-specific BT that serves to identify the most confident peptide–MHC binders and T-cell epitopes as those with a score ≥ BT. The profile-specific BT provided by EPIMC is obtained by scoring all the peptides used to make the relevant profile matching the 90 percentile value of all peptide scores [18]. The next part in the RANKPEP output consists of a list of peptides from the input protein ranked by the scores obtained with the relevant profile. In our case, RANKPEP shows only 9 peptides from SARS nucleoprotein because we selected to display only the 2 % of top scoring peptides. For every peptide, RANKPEP shows its rank (*RANK*), location in the protein sequence (*POS*), sequence (SEQUENCE), three N-terminal (*N*) and C-terminal (*C*) flaking residues, score (SCORE), and relative score, in percentage, with regard to the optimum score (*%OPT*). Peptides whose scores are equal or greater than the BT score are highlighted in red, and those containing a C-terminal end predicted to be the result of proteasomal cleavage are shown in violet.

As we made a profile from peptides binding with high affinity to A*0201, we are predicting peptides from SARS nucleoprotein that bind to A*0201 and hence potential A*0201-restricted CD8 T-cell epitopes. In fact, in the results shown in Fig. 7b, it is possible to identify 5 out of the 7 known A*0201-restricted CD8 T-cell epitopes from SARS nucleoprotein (Table 2).

Table 2
Description of custom-made profiles used in this study

	EPIMHC search selection					EPIMHC search result and profile building	
Profile name	MHC molecule	PEP. binding level	PEP. source organism	T-cell activity	Length	Peptides	Method
Profile #1	HLA-A*0201	High	All	All	9	178	p.mtx[b]
Profile #2	HLA-A*0201	High	All	YES	9	95	p.mtx[b]
Profile #3	HLA-A*0201	High	Viruses[a]	All	9	48	p.mtx[b]
Profile #4	HLA-A*0201	High	Viruses[a]	YES	9	32	p.mtx[b]

[a]The viruses selected in EPIMHC were the following: Dengue virus, Epstein-Barr virus, Hepatitis B virus, Hepatitis C virus, Human immunodeficiency virus 1, Human immunodeficiency virus 1 OPT, Human immunodeficiency virus 2, Influenza A virus, Measles virus, and West Nile virus. None of the CD8 T-cell epitopes from SARS nucleoprotein are included in any of the peptide selections used for making these profiles
[b]PSSMs generated using position-based weights [30]

3.3 Comparison of CD8 T-Cell Epitope Predictions Using Various Custom-Made Profiles

The goodness of peptide–MHC binding and T-cell epitope predictions provided by any predictive model, including motif-profiles, is tied to the data used for model building [32, 33]. To demonstrate such influence, here we will compare the predictions of A*0201-restricted CD8 T-cell epitopes from SARS nucleoprotein that are obtained with 4 distinct motif-profiles, including that described in the previous section (hereafter profile #1). The specific peptide selections that give rise to the different profiles used in this section are detailed in Table 2. Briefly, all profile-motifs are generated from peptides binding with high affinity to A*0201. In addition, profile #3 and #4 only include peptides from viruses and profile #2 and #4 only include peptides known to display T-cell activity (they are epitopes). To evaluate the predictive performance of these profiles, we scored and ranked all peptides from SARS nucleoprotein and compared the ranking achieved by the known SARS nucleoprotein A*0201-restricted CD8 T-cell epitopes shown in Table 1. These results are summarized in Table 3. All four motif-profiles produce related results, ranking the known CD8 T-cell epitopes among the top scoring peptides of SARS nucleoprotein. This is expected as A*0201-restricted CD8 T-cell epitopes and the peptides used for model building have in common the ability to bind to A*0201. However, there are also differences in the results. Thus, only the profiles derived from viral peptides are capable of predicting the known A*0201-restricted CD8 T-cell epitopes from SARS nucleoprotein among the top 11 scoring peptides (Table 3). Judging from the dispersion of the ranks (Table 3), the best overall epitope predictions are obtained

Table 3
Ranking and statistics of known SARS nucleoprotein A*0201-restricted CD8 T-cell epitopes using four custom-made motif-profiles

Epitope	Profile #1	Profile # 2	Profile #3	Profile # 4
ALNTPKDHI	3	3	4	4
GMSRIGMEV	2	1	1	3
ILLNKHIDA	8	7	5	5
LALLLLDRL	34	11	30	11
LLLDRLNQL	1	2	9	2
LQLPQGTTL	10	13	7	10
RLNQLESKV	5	4	3	1
		Rank	Statistics	
Median	5	4	5	4
Mean	9	5.8	5.4	5.1
Stdev	11.3	4.5	9.8	3.9
Range	1–34	1–13	1–30	1–11

Peptide ranks are obtained after scoring all peptides in SARS nucleoprotein with the relevant motif-profiles and depict the peptide score relative to that of all remaining peptides. Rank 1 means that the peptide has the largest score of all peptides

with a motif-profile derived from viral peptides with known T-cell activity (T-cell epitopes). It remains to be explored whether discarding peptide–MHC binders with no reported T-cell activity always improves the resulting T-cell epitope prediction models. It is important to note that none of the known epitopes used in these analyses have been used to derive any of the profiles. In fact, the current version of EPIMHC does not contain any SARS peptides at all.

In conclusion, profiles are very powerful at capturing nontrivial motifs and the results shown here support that epitope predictions can be improved using customized peptide–MHC binding profiles. EPIMHC is the only available resource readily suitable for that task.

4 Notes

1. In EPIMHC, users can make profiles from any peptide selection but the peptides must be related to some extend (e.g., binding to the same MHC) to produce profiles yielding meaningful predictions (*see* Subheading 3.1).
2. EPIMHC is better suited for making MHC I-specific profiles. Moreover, profiles can only be generated from peptides with the same length; otherwise EPIMHC returns an error

(*see* Subheading 3.1). There are practical and structure-based reasons for this limitation as discussed by Reche et al. [18].

3. EPIMHC can also produce profiles from peptides that have been selected to bind to MHC II molecules, provided that they have the same length (*see* Subheading 3.1). However, as data availability is limited, we recommend using a motif discovery program such as MEME [31] for making peptide–MHC II binding profiles from peptides of any length as described in previous reports [17, 18].
4. The proteasomal cleavage predictions should not be taken in consideration when predicting peptide binding to MHC II molecules: the proteasome is not involved in class II antigen processing. We are working in correcting this inconsistency.
5. For RANKPEP to return all peptides in a given protein sorted by score, users need to make the following selections in the RANKPEP input form: first set peptides to display by number and then select 990 from the pull-down menu.
6. To generate profiles that are capable of capturing the relevant peptide–MHC binding feature, we suggest using a minimum of five peptides.

Acknowledgements

This work was supported by grant SAF2009-08103 from Ministerio de Ciencia e Innovación to PAR.

References

1. Zinkernagel M, Doherty PC (1974) Restriction of *in vitro* T cell-mediated cytotoxicity in lymphocytic choriomeningitis within a syngeneic or semiallogeneic system. Nature 248:701–702
2. Margulies DH (1997) Interactions of TCRs with MHC-peptide complexes: a quantitative basis for mechanistic models. Curr Opin Immunol 9:390–395
3. Garcia KC, Teyton L, Wilson IA (1999) Structural basis of T cell recognition. Annu Rev Immunol 17:369–397
4. Wang J-H, Reinherz E (2001) Structural basis of T cell recognition of peptides bound to MHC molecules. Mol Immunol 38:1039–1049
5. Reche PA, Reinherz EL (2003) Sequence variability analysis of human class I and class II MHC molecules: functional and structural correlates of amino acid polymorphisms. J Mol Biol 331:623–641
6. Tchernev G, Orfanos CE (2006) Antigen mimicry, epitope spreading and the pathogenesis of pemphigus. Tissue Antigens 68:280–286
7. Reche PA, Keskin DB, Hussey RE, Ancuta P, Gabuzda D, Reinherz EL (2006) Elicitation from virus-naive individuals of cytotoxic T lymphocytes directed against conserved HIV-1 epitopes. Med Immunol 5:1
8. Akdis CA, Akdis M, Blesken T, Wymann D, Alkan SS, Muller U, Blaser K (1996) Epitope-specific T cell tolerance to phospholipase A2 in bee venom immunotherapy and recovery by IL-2 and IL-15 in vitro. J Clin Invest 98: 1676–1683
9. Stienekemeier M, Falk K, Rotzschke O, Weishaupt A, Schneider C, Toyka KV, Gold R, Strominger JL (2001) Vaccination, prevention, and treatment of experimental autoimmune neuritis (EAN) by an oligomerized T cell epitope. Proc Natl Acad Sci U S A 98: 13872–13877
10. Lazoura E, Apostolopoulos V (2005) Insights into peptide-based vaccine design for cancer immunotherapy. Curr Med Chem 12: 1481–1494

11. Rammensee HG, Bachmann J, Emmerich NPN, Bacho OA, Stevanovic S (1999) SYFPEITHI: database for MHC ligands and peptide motifs. Immunogenetics 50:213–219
12. Blythe MJ, Doytchinova IA, Flower D (2002) JenPep: a database of quantitative functional peptide data for immunology. Bioinformatics 18:434–439
13. Bhasin M, Singh H, Raghava GP (2003) MHCBN: a comprehensive database of MHC binding and non-binding peptides. Bioinformatics 19:665–666
14. Diez-Rivero CM, Garcia-Boronat M, Reche PA (2008) Integrating T-cell epitope annotations with sequence and structural information using DAS. Bioinformation 3:156–158
15. Zhang Q, Wang P, Kim Y, Haste-Andersen P, Beaver J, Bourne PE, Bui HH, Buus S, Frankild S, Greenbaum J et al (2008) Immune epitope database analysis resource (IEDB-AR). Nucleic Acids Res 36:W513–W518
16. Reche PA, Zhang H, Glutting JP, Reinherz EL (2005) EPIMHC: a curated database of MHC-binding peptides for customized computational vaccinology. Bioinformatics 21: 2140–2141
17. Reche PA, Glutting J-P, Reinherz EL (2004) Enhancement to the RANKPEP resource for the prediction of peptide binding to MHC molecules using profiles. Immunogenetics 56: 405–419
18. Reche PA, Glutting JP, Reinherz EL (2002) Prediction of MHC class I binding peptides using profile motifs. Hum Immunol 63:701–709
19. Reche PA, Reinherz EL (2007) Prediction of peptide-MHC binding using profiles. Methods Mol Biol 409:185–200
20. Jensen PE (2007) Recent advances in antigen processing and presentation. Nat Immunol 8: 1041–1048
21. Diez-Rivero CM, Chenlo B, Zuluaga P, Reche PA (2010) Quantitative modeling of peptide binding to TAP using support vector machine. Proteins 78:63–72
22. Diez-Rivero CM, Lafuente EM, Reche PA (2010) Computational analysis and modeling of cleavage by the immunoproteasome and the constitutive proteasome. BMC Bioinformatics 11:479
23. Lafuente EM, Reche PA (2009) Prediction of MHC-peptide binding: a systematic and comprehensive overview. Curr Pharm Des 15: 3209–3220
24. Gowthaman U, Agrewala JN (2008) In silico tools for predicting peptides binding to HLA-class II molecules: more confusion than conclusion. J Proteome Res 7:154–163
25. Gribskov M, McLachlan AD, Eisenberg D (1987) Profile analysis: detection of distantly related proteins. Proc Natl Acad Sci U S A 84: 4355–4358
26. Brusic V, Rudy G, Harrison LC (1994) MHCPEP: a database of MHC-binding peptides. Nucleic Acids Res 22:3663–3665
27. Korber BTM, Moor JP, Brander C, Walker BD, Haynes BF, Koup R (1998) HIV molecular immunology compendium. Los Alamos National Laboratory, Los Alamos, NM
28. Peters B, Tong W, Sidney J, Sette A, Weng Z (2003) Examining the independent binding assumption for binding of peptide epitopes to MHC-I molecules. Bioinformatics 19: 1765–1772
29. Thompson JD, Higgins DG, Gibson TJ (1994) Improved sensitivity of profile searches through the use of sequence weights and gap excision. Comput Appl Biosci 10:19–29
30. Henikoff S, Henikoff JG (1994) Position-based sequence weights. J Mol Biol 243: 574–578
31. Bailey TL, Gribskov M (1998) Combining evidence using p-values: application to sequence homology searches. Bioinformatics 14: 48–54
32. Martinez-Naves E, Lafuente EM, Reche PA (2011) Recognition of the ligand-type specificity of classical and non-classical MHC I proteins. FEBS Lett 585:3478–3484
33. Martínez-Naves E, Lafuente EM, Reche PA (2011) In: Liò P, Nicosia G, Stibor T (eds.), 10th International conference on artificial immune systems. Springer-Verlag Cambridge, England, Vol. 6825, 55–65
34. Cheung YK, Cheng SC, Sin FW, Chan KT, Xie Y (2007) Induction of T-cell response by a DNA vaccine encoding a novel HLA-A*0201 severe acute respiratory syndrome coronavirus epitope. Vaccine 25:6070–6077

Chapter 19

T-Cell Epitope Prediction Methods: An Overview

Dattatraya V. Desai and Urmila Kulkarni-Kale

Abstract

The scientific community is overwhelmed by the voluminous increase in the quantum of data on biological systems, including but not limited to the immune system. Consequently, immunoinformatics databases are continually being developed to accommodate this ever increasing data and analytical tools are continually being developed to analyze the same. Therefore, researchers are now equipped with numerous databases, analytical and prediction tools, in anticipation of better means of prevention of and therapeutic intervention in diseases of humans and other animals.

Epitope is a part of an antigen, recognized either by B- or T-cells and/or molecules of the host immune system. Since only a few amino acid residues that comprise an epitope (instead of the whole protein) are sufficient to elicit an immune response, attempts are being made to identify or predict this critical stretch or patch of amino acid residues, i.e., T-cell epitopes and B-cell epitopes to be included in multiple-subunit vaccines.

T-cell epitope prediction is a challenge owing to the high degree of MHC polymorphism and disparity in the volume of data on various steps encountered in the generation and presentation of T-cell epitopes in the living systems. Many algorithms/methods developed to predict T-cell epitopes and Web servers incorporating the same are available. These are based on approaches like considering amphipathicity profiles of proteins, sequence motifs, quantitative matrices (QM), artificial neural networks (ANN), support vector machines (SVM), quantitative structure activity relationship (QSAR) and molecular docking simulations, etc. This chapter aims to introduce the reader to the principle(s) underlying some of these methods/algorithms as well as procedural and practical aspects of using the same.

Key words T-cell epitope, Proteasomal cleavage, MHC–peptide binding, TAP transport, Quantitative matrix, Motif, MHC polymorphism, Epitope prediction algorithm, Vaccine design, Immunoinformatics, Bioinformatics

1 Introduction

1.1 Epitope: A Relational Entity

An epitope is the part of an antigen, recognized by the cells (B- and T-) and/or molecules (antibodies, MHCs, etc.) of the host immune system. A peptide epitope is a set of amino acid residues present either in continuity (linear or sequential) or as a surface patch (conformational or discontinuous) of a protein molecule. While B-cell epitopes are both sequential and conformational, the T-cell epitopes are linear.

Rajat K. De and Namrata Tomar (eds.), *Immunoinformatics*, Methods in Molecular Biology, vol. 1184, DOI 10.1007/978-1-4939-1115-8_19, © Springer Science+Business Media New York 2014

1.2 T-Cell Epitopes: Outcome of Multistep Processing

T-cell epitopes of length 8–11 and 13–17 bind to Major Histocompatibility Complex MHC class-I and MHC class-II molecules respectively and are presented on surface of Antigen Presenting Cells (APC). Cytotoxic T-cells are activated upon presentation of endogenous antigenic peptides by MHC class I molecules. The processing pathway of MHC class I restricted antigens involves three major steps: proteasomal cleavage, TAP transport, and MHC binding.

Antigenic proteins are subjected to cleavage by the proteasome to generate peptides in cells. The proteolytic activity of the proteasome is said to be trypsin-like, chymotrypsin-like, and peptidylglutamyl-peptide hydrolytic activity [1]. Proteolytic cleavage by the proteasome generates peptides with the correct C terminus, which is an essential requirement for their binding to MHC class I molecules [2, 3].

The transporters associated with antigen processing (TAP), transports some of these peptides to the endoplasmic reticulum (ER). It is known that the TAP has higher transport affinity for some peptides over others. The peptides of length 8–12 amino acids are transported with highest efficiency [4]. Three amino acids of N′ and C′ termini of the peptides have been found to be important for TAP binding [5]. There exists a correlation between TAP binding affinity and rate of transport of peptides [6].

The peptides bind to the MHC class I molecules in the ER. MHC class I molecules interact with the N′ terminal and C′ terminal amino acids of the peptide, thereby leaving a bulge in the middle. This restricts the length of peptide interacting with MHC class I molecules to 8–11 amino acids. The MHC class I–peptide complex is subsequently translocated to the cell surface, where it may activate cytotoxic T-cells.

The MHC class II molecules, expressed in APCs, bind to peptides derived primarily from degradation of endocytosed proteins and present them for recognition by the T-cell receptors of CD4+ T-helper cells. MHC class II molecules play a pivotal role in adaptive immune response [7].

1.3 MHC Diversity: Opportunities and Challenges

MHC molecules exhibit a high degree of polymorphism. The MHC molecules of humans are referred to as human leucocyte antigens (HLA). MHC class I molecules are encoded by the genes present at six loci, viz., HLA-A, HLA-B, HLA-C, HLA-E, HLA-F, and HLA-G, on Chromosome No. 6 of humans. MHC class II molecules are encoded by genes present at five loci on Chromosome No. 6 of humans. The loci are designated as HLA-DP, HLA-DQ, HLA-DR, HLA-DM, and HLA-DO. HLA alleles are highly polymorphic, in the sense, the number of alleles varies from as low as a single allele each, for HLA-DRB2, HLA-DRB8, and HLA-DRB9 to as high as 3,005 alleles for HLA-B, as reported in the current version of the IMGT/HLA database [as on July 26, 2013].

One may visit the IMGT/HLA database, available online for the detailed information and data on HLA molecules (http://www.ebi.ac.uk/ipd/imgt/hla/) [8].

Therefore, the MHC alleles of every human being would be unique. Although the MHC alleles of no two humans would be similar, many of the different MHC molecules have similar peptide binding specificities. MHC molecules with similar peptide binding specificities are grouped together to form MHC supertypes.

MHC supertypes have been defined based on sequential and structural similarities, presence of similar peptide binding motifs, amino acid pattern in the binding pocket, amino acid binding preferences, etc., wherein each MHC molecule of a particular supertype would bind to the same peptides.

Nine major MHC class I supertypes, viz., HLA-A1, A2, A3, A24,B7, B27, B44, B58, B62 have been recognized [9], while twelve MHC class II supertypes have been defined, viz., five DRs, three DQs, and four DPs. The HLA class II supertypes are DR1, DR3 , DR4, DR5, DR9, DQ1, DQ2, DQ3, DPw1, DPw2, DPw4, and DPw6 [10].

The existence of MHC polymorphism and MHC allelic distribution in one or more ethnic groups, diversity of the peptide repertoire together with the strength of MHC–peptide binding needs ample consideration in identification of T-cell epitopes and thereby multi-epitope vaccine design.

1.4 The Paradox: Data Deluge and Paucity

Owing to technological advancement, enormous amount of data has continually been pouring in, into the populist databases of sequences and structures of biological molecules in general and that of the immune system in particular. The dedicated databases containing data on the molecules of the immune system, per se are flooded with data. The IMGT/LIGM-DB [11] for example, contains 170,685 entries, corresponding to nucleotide sequences of immunoglobulin and T-cell receptors from 335 species (as on August 5, 2013), while the IMGT/3Dstructure-DB [12, 13], contains 2,802 entries (as on August 5, 2013) corresponding to the experimentally determined structures of molecules of the immune system, viz., immunoglobulins, T-cell receptors, MHC molecules, and related proteins of the immune system (RPI), etc. The IMGT/MH-DB [14] on the other hand has 9,719 allele sequences of HLA class I, HLA class II, and non-HLA alleles, along with detailed relevant information. Please note that the number of sequence and structure entries in the databases keeps increasing with time.

Paradoxical as it may sound, critical data on certain specific aspects is found to be woefully wanting. For instance, the experimental data on proteasomal cleavage sites is extremely limited. Furthermore, disparity also exists in the number of known peptide epitopes and non-epitopes. While the number of known peptide epitopes is much higher than that of non-epitopes, in any antigenic protein, the actual peptide epitopes would comprise of no more than a fraction of amino acid residues. There is a need to archive experimentally validated true positives as well as true negatives. This observation is true for MHC-binders and non-binders as well.

Development of immunoinformatics databases and tools is therefore necessitated by the voluminous increase in data and the need to predict with reasonable accuracy, the binding sites and the interactions of the molecules of immune system. T-cell epitope prediction, in particular, is a challenge owing to the high degree of MHC polymorphism, and disparity in the volume of data on various steps encountered in the generation and presentation of T-cell epitopes in living systems.

Development of new immunoinformatics tools to predict T-cell epitopes as well as improvising performance of existing tools is an ongoing process. Scientists have used various approaches, like considering amphipathicity profiles of proteins, motif-based methods, quantitative matrix based methods, methods based on Artificial neural networks, Support vector machines, QSAR and docking simulations to predict T-cell epitopes. This chapter provides details of some of these approaches.

2 Methods for T-Cell Epitope Prediction

Methods of T-cell epitope prediction are broadly categorized into two classes, viz., direct methods and indirect methods. Direct prediction methods are based on sequence and structure analyses of T-cell epitopes. They rely on features like the presence of amphipathicity, MHC-binding motifs, etc., but have rather limited accuracy and high false positive rate. Indirect methods, on the other hand, predict MHC–peptide binding, using some of the elegant techniques based on statistical learning theory, like the artificial neural networks (ANN), support vector machine (SVM), etc. Albeit prediction of MHC–peptide binding is obligatory, prediction of proteasomal cleavage and TAP transport are also essential components, in the realm of T-cell epitope prediction. Since MHC–peptide binding prediction is the subject matter of an earlier chapter, this chapter shall discuss the principles of some of the direct methods of T-cell epitope prediction as well as integrated methods that predict T-cell epitopes.

2.1 Amphipathicity Based Method

Amphipathicity refers to the presence of both hydrophilic (polar) and hydrophobic (apolar) regions in a single molecule. Membrane phospholipids for example are well known amphipathic molecules. Amphipathic regions may be displayed by proteins and peptides as well. The algorithm described below takes into account the amphipathicity present in proteins to predict T-cell epitopes.

2.1.1 AMPHI

One of the early studies to predict T-cell epitopes was the one carried out by DeLisi and Berzofski, published in 1985 [15], which led to the development of AMPHI algorithm in 1987 [16]. The AMPHI algorithm is based on the model of amphipathic helix, in which one face is predominantly hydrophilic (polar) while the other face is predominantly hydrophobic (nonpolar). Such amphipathic structures where hydrophilic and hydrophobic regions are formed when the polarity of residues along the sequence varies with a regular periodicity.

The developers of this tool divided the protein sequence into overlapping blocks of amino acid residues. For each block, average hydrophobicity and the extent of occurrence of hydrophobic residues in a pattern with a regular periodicity were determined. The periodicities of occurrence of hydrophobic residues corresponded to alpha-helical structure of proteins. Their studies showed that the T-cell epitopes show local secondary structural features, which reflect the periodicity in the hydrophobic profile of the segment of amino acid residues. T-cell epitopes possibly comprise of amphipathic structures, displaying periodicity in hydrophobic residues.

It must be kept in mind that even though T-cell epitopes comprise of contiguous stretch of blocks of amphipathic residues, it is not necessary that every such stretch would be a potential T-cell epitope. The developers of AMPHI algorithm reported a sensitivity of 75 %. This algorithm is currently not available to the users in the form of either an online or offline tool and has been included in the chapter as it was a pioneering effort.

2.2 Motif Based Methods

These methods follow an approach of searching protein sequences for regions that contain known MHC-binding amino acid motifs. The peptides that bind to MHC molecules contain certain amino acids at specific positions, which are called "anchor residues." The anchor residues facilitate peptide binding within the peptide-binding groove of the MHC molecule. The patterns of anchor residues are called "motifs." The motif present in peptides that bind to one MHC allele may differ from the motif present in peptides that bind to another MHC allele. Unusual anchor position and auxiliary anchor position for amino acid residues in the peptides have also been identified [17–20]. The motif based methods, thus utilize the knowledge of MHC binding motifs to identify T-cell epitopes.

2.2.1 EpiMer and OptiMer

The EpiMer algorithm searches for MHC-binding motifs in a given protein sequence. It generates a list of MHC-binding motif matches for a given protein. Clusters of MHC-binding motifs in a protein sequence are identified. T-cell epitopes are predicted based on the relative density of MHC-binding motifs. Therefore, EpiMer algorithm, predicts putative T-cell epitopes based on the clustering of MHC-binding motifs within a protein sequence. The developers of EpiMer algorithm have reported a sensitivity ranging from 53 to 71 %.

OptiMer algoritm takes into account both amphipathicity and presence of MHC binding motifs in a protein sequence. OptiMer algorithm generates a list of peptides from a protein sequence, which contain these MHC binding motifs. Using the AMPHI algorithm (discussed in Section 2.1.1), the OptiMer algorithm then identifies peptides that show amphipathicity and form a helix or beta strand. These amphipathic peptides are then compared with known (published) MHC binding motifs. The algorithm subsequently extends the predicted amphipathic peptides, to maximize the density of MHC binding motif matches per length of protein region. Therefore, OptiMer algorithm searches for MHC-binding motifs, as well as amphipathic secondary structural features. The developers of this algorithm have reported a sensitivity of 53–75 % in the various proteins tested by them. Both OptiMer and EpiMer can predict promiscuous T-cell epitopes for multiple MHC alleles [21].

2.2.2 SYFPEITHI

SYFPEITHI [22] is the name of a database of MHC ligands and peptide motifs of humans and other vertebrate species. The database facilitates search for peptides as well as prediction of T-cell epitopes. The prediction of T cell epitopes is based on an algorithm that takes into account the position of amino acids in the peptide, such as the anchor position, unusual anchor position, and auxiliary anchor position. Preferred amino acids as well as amino acids whose presence at particular positions is undesirable for peptide binding are also taken into account and are scored accordingly.

The scoring system of the algorithm evaluates every amino acid within a given peptide. The values are assigned to the amino acids at various positions in a peptide based on the frequency of occurrence of the respective amino acids in natural ligands, T-cell epitopes or binding peptides. The value of an amino acid can vary from a high positive value, say 15, the highest value that is attributed to ideal/optimal anchor residues to a low positive value of 1, which is attributed to amino acids that are only slightly preferred to a negative value which is attributed to amino acids that are disadvantageous to peptide binding at a particular position in the peptide. The values at each position are summed up to assign a final score for the peptide that acts as a T-cell epitope.

2.2.3 TEPITOPE and TEPITOPEpan

TEPITOPE is a tool, which implements an algorithm based on the prediction of HLA-II–peptide binding [23]. It consists of 11 position specific scoring matrices (PSSM) to represent MHC–peptide binding specificities. Each PSSM is a matrix where the binding pockets are represented by peptide binding specificity vectors. Since it covered only 51 HLA-DR alleles, it has limited usability in today's context.

TEPITOPEpan is a new method based on HLA-DR binding pocket similarity [24].

From the experimentally determined structures of MHC class II-peptide complexes available in the Protein Data Bank (PDB), HLA-DR binding pockets are identified. The residues that have close contact with one or more residues of peptide binding core represent these pockets. Then the pocket similarity between two HLA molecules is computed by the sequence similarity of the corresponding HLA residues. For an uncharacterized HLA-DR molecule, the binding specificity of each pocket is computed as a weighted average of pocket binding specificities over HLA-DR molecules characterized by TEPITOPE. Although TEPITOPEpan uses the library of specificity matrices obtained in TEPITOPE, it can be used for prediction of MHC class II binding peptides with over 700 HLA-DR alleles with known sequences.

2.3 Methods Based on Quantitative Matrices and/or Machine Learning Techniques

2.3.1 Quantitative Matrices

Quantitative matrix (QM) provides a means for quantitative representation of qualitative data/information. A quantitative matrix is essentially a position weight matrix, which contains the contribution of each amino acid located at every position in a peptide, towards the potential for MHC binding by the peptide. QMs are available for different MHC alleles. A QM for a particular MHC allele contains a value, which denotes the impact (favorable, neutral, or unfavorable) of presence of amino acid residues at various positions in a peptide on the binding of the peptide to a given MHC allele. As an example, QM for HLA-A2 [25–27], is given in Table 1, which is also available at http://www.imtech.res.in/raghava/nhlapred/matrices/a2.html.

The QM based methods hypothetically fragment the protein sequence into overlapping peptides of a chosen length, say 9-mer peptides for instance. Each amino acid of the peptides is assigned a coefficient value depending on the type of amino acid and its position in 9-mer peptide from the quantitative matrix. A score is then obtained for every peptide either by summation or multiplication of each of the coefficient values. Peptides having a score more than the threshold score are predicted to be MHC binders.

2.3.2 Artificial Neural Networks

An artificial neural network (ANN) is a computational model that emulates the neural circuitry of the brain. It is a "machine learning tool" that can be trained to learn the features of appropriate patterns and subsequently be used to recognize the similar patterns present in novel data. It is a network made up of a few to many neurons (not actual biological neurons!), also referred to as nodes or units, which are interconnected. These nodes may be present in multiple layers; viz., an input layer, one or more middle layers (hidden layers) and an output layer (Fig. 1). The nodes of the input layer receive the inputs and the nodes of the output layer give out the output. The middle layer(s), or hidden layer(s), consists of a network of neurons, whose connections are made and remade in such a manner, as to learn the pattern present in the data that is

Table 1
Quantitative matrix for HLA-A2 [25–27]

Amino acid/position	P1	P2	P3	P4	P5	P6	P7	P8	P9
A	0.52	−0.67	−0.25	−0.29	−0.35	−0.55	−0.1	−0.34	−0.05
C	0	−2	−0.4	0.29	1	1.67	1.33	0.67	1
D	−1.6	−2	0.08	0.34	−0.75	−0.86	−0.82	−0.4	−1.69
E	−1.41	−1.64	−1.48	−0.05	−0.43	−0.92	−1.08	−0.04	−2
F	0	−1.08	1.05	−0.4	1.28	0.27	1.39	−0.53	−2
G	0.91	−1.82	−0.47	1.18	0.3	−0.4	−0.11	0.13	−1.82
H	0.22	−2	0.22	0.22	−0.29	−0.5	0.93	−0.22	−2
I	−0.27	0.89	−0.62	−1.09	−0.62	0	−0.27	−0.07	0
K	0.25	−1.47	−1.14	−0.75	−0.77	−1.56	−1.2	−0.63	−1.43
L	0.51	1.62	1.24	−0.29	0.19	0.44	0.38	0.22	1.31
M	−0.67	1.47	0.29	1.43	1.33	1.67	0	0.4	1
N	−0.22	−2	0.29	−1	−1.11	−0.82	−0.22	−0.44	−2
P	−0.5	−2	−0.5	0.59	0.62	0.88	0.17	0.11	−2
Q	−0.75	−1.14	−1.64	0.26	−0.82	−0.35	−0.22	0.33	−1.33
R	0.17	−0.86	−0.29	0.32	−0.11	−1.11	−0.8	−0.15	−1.2
S	0.76	−2	0.4	0.5	0	0.11	−0.53	0.1	−1.08
T	−0.88	−0.75	−0.81	−0.92	−0.5	−0.67	−0.24	0.92	−0.71
V	−0.81	−0.88	0.22	−0.83	0	1.23	0.44	−0.5	1.38
W	−1.38	−1.6	−0.1	−1.64	−0.11	−1.47	−0.86	−1	−2
X	2	2	0	0	0	2	2	0	0
Y	−0.12	−2	0.09	−2	0.43	−0.12	−0.25	0	−1.43

Source: http://www.imtech.res.in/raghava/nhlapred/matrices/a2.html

used to train the network. The ANN is trained with the dataset called "training dataset" and tested with the dataset called "test dataset." As the ANN learns the inherent patterns present in the training dataset, it makes appropriate connections among the nodes (neurons) of the hidden layer, assigning them appropriate "weights." It is possible that the ANN may recognize correctly the patterns in the test dataset, but may commit errors as well. The error committed by the ANN can be corrected, wherein the network connections would be readjusted (accompanied by appropriate changes in the weights). Finally, a well learned ANN can be used to carry out an appropriate task. There are various types of ANNs, ranging from simple networks like "Perceptrons" to complex

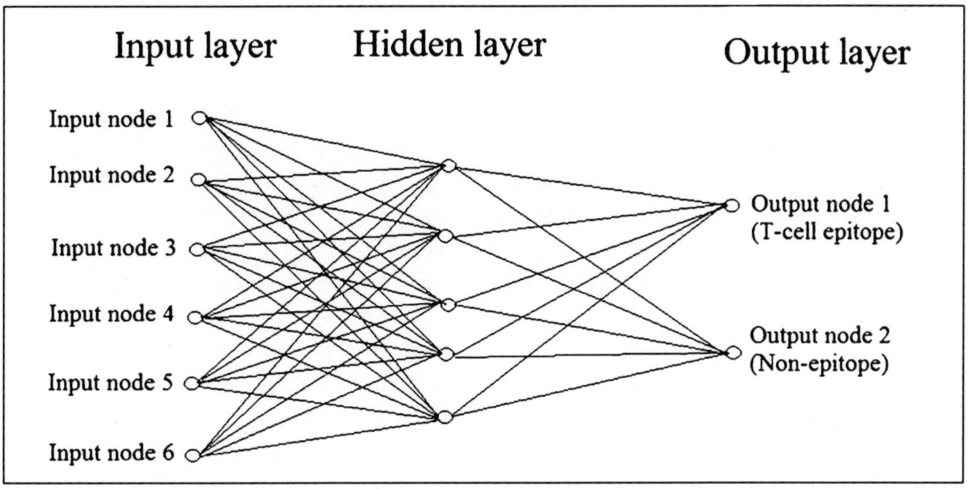

Fig. 1 Architecture of a simple artificial neural network (ANN)

networks like "Kohonen network," also called a "self-organizing map" [28].

ANNs are applied for various tasks, such as gene prediction, protein secondary structure prediction, B- and T-cell epitope prediction, etc and are extensively reviewed elsewhere [29–31]. For an ANN trained for T-cell epitope prediction, the typical input layer would have nodes for the amino acid sequence while the output layer would have nodes for "epitope" and "non-epitope."

2.3.3 Support Vector Machine (SVM)

A support vector machine (SVM) is a computational model, a machine learning tool that is based on statistical learning theory. It also learns and recognizes patterns present in the data and is widely used for classification of linear or non-linear data. Data points present in two-dimensional space may be divided into two categories by a line, while the data points present in three-dimensional space may be categorized into two categories by a plane. If data points are present in multi-dimensional space, a "hyperplane" would be required to classify such data. Essentially, a SVM classifies complex data present in multidimensional space (high-dimensional space) by constructing a hyperplane, which is defined by a kernel function. A SVM maps the data into a high-dimensional feature space, wherein every coordinate corresponds to a particular feature of the data. A suitable kernel function is used for classification of data. Linear kernel, polynomial kernel, radial basis function (RBF) kernel, string kernel, etc. are some of the types of kernel functions.

SVMs are widely used in modern biology [32–34]. For example, during analysis of microarray data on gene expression in normal cells and cancerous cells, a SVM may be used to separate out genes implicated in cancer from normal genes. Likewise, a SVM may be used to identify "T-cell epitopes" from among the many thousands of peptides which are non-epitopes.

2.3.4 CTLPred

It is a direct method for prediction of CTL epitopes. It employs quantitative matrices, ANN, and support vector machine for CTL epitope prediction. One can use any of these techniques separately, obtain a consensus of the methods, or have a combined approach. The quantitative matrices, as explained earlier, are matrices that quantify the impact of the presence of appropriate amino acid residues at particular positions in the peptides on immunogenicity of the peptides/ability to act as a CTL epitope. The feedforward neural network model was trained on curated dataset of CTL epitopes, non-epitopes, and MHC non-binders procured from MHCBN [35], a database of MHC binders and non-binders and subsequently tested using curated datasets as well as blind datasets. The support vector machine, employing a hyperplane defined by a polynomial kernel function was developed for classification of peptides into CTL epitopes and non-epitopes. CTLpred provides users with the option of using either a consensus of SVM and ANN or a combined approach. The tool also provides for MHC restriction for a number of MHC alleles, thereby providing a scope for either restricting the CTL epitope prediction to a particular MHC allele, or for multiple MHC alleles as per the need of the user. CTLPred performs with a reasonable accuracy of 62.0 %, 72.2 %, 75.4 %, 77.6 %, and 75.8 % for QM, ANN, SVM, consensus, and combined approaches respectively, as reported by the developers of this tool [26].

2.3.5 NetCTL

NetCTL is method that integrates predictions of MHC class I binding affinity, transporter associated with antigen processing (TAP) transport efficiency, and C-terminal proteasomal cleavage for prediction of CTL epitopes [36].

MHC class I affinity prediction is based on ANN. Each of the MHC supertypes is represented by an ANN trained on nonameric peptides with known binding affinity to a given MHC class I allele. Each peptide is assigned a value between 0 and 1, where 0 corresponds to low affinity for MHC class I binding and 1 to high MHC class I binding affinity.

C-terminal proteasomal cleavage prediction is carried out by four prediction methods which are used individually to assign a predicted cleavage value to the residues. These methods are C-term 2.0 and 20S networks of the NetChop 2.0 prediction server, NetChop C-term 3.0 and NetChop 20S-3.0 prediction servers [37, 38]. All these methods are based on artificial neural networks (ANN).

TAP transport efficiency prediction method is based on the matrix described by Peters et al. [39]. Predicted TAP transport efficiency of peptides with arbitrary length is calculated by scoring only the C terminus and the three N-terminal amino acid residues. The TAP transport efficiency score for a given nonamer is given as the average of the values for the nonamer and its decameric precursor.

Low TAP transport efficiency is indicated by a low predicted value while high TAP transport efficiency is indicated by a high predicted value.

When combining the predictions of MHC class I affinity, TAP transport efficiency, and proteasomal cleavage, the MHC class I affinities are rescaled to make the prediction values comparable between MHC class I supertypes using the approach given by Sturniolo et al. [40].

The predicted scores from proteasomal cleavage, MHC binding and TAP transport are integrated as a weighted sum with a relative weight on peptide–MHC binding of 1.

NetCTLpan is a pan-specific MHC class I pathway epitope prediction tool, which is customized to predict CTL epitopes for six vertebrate species, including humans [41].

2.3.6 WAPP

Whole antigen processing prediction (WAPP) is an integrated method, in which individual methods for prediction of proteasomal cleavage, TAP transport as well as MHC binding are combined to form a single prediction algorithm that emulates the whole processing pathway of MHC class I antigens [42].

The proteasomal cleavage prediction method in WAPP uses a probability-based model encoded by proteasomal cleavage matrices (PCMs). The developers of WAPP constructed proteasomal cleavage matrices from experimentally verified cleavage sites, together with four N-terminal and two C-terminal amino acids flanking each cleavage site, from three proteins, viz., beta-casein, enolase, and prion proteins. Using these small peptides, all of which contained a cleavage site between the fourth and fifth positions, position-specific scoring matrices (PSSM) were constructed, which are termed as PCMs.

The TAP transport affinity prediction method in WAPP is called SVMTAP, which is based on support vector regression (SVR). The SVM, optimized with a simple linear kernel, was trained and tested with data consisting of 9-mer peptides with experimentally verified ln IC50 values [43] and implemented using SVM-*light*, which is an implementation of SVM in the computing language "C". Detailed information on SVM-*light* is available at http://svmlight.joachims.org/.

The MHC binding prediction method in WAPP is SVMHC [44], a SVM-based method trained on verified MHC binding peptides from the SYFPEITHI and MHCPEP [45] databases. The SVM, optimized with linear, polynomial as well as radial basis function kernels is implemented using SVM-*light*.

In order to improve the prediction accuracy, the proteasomal cleavage method and SVMTAP were used as filters to remove peptides unlikely to be generated by proteasomal cleavage and/or transported by TAP, while MHC class I binding prediction by SVMHC is carried out due to high accuracy.

The output of WAPP is peptides that are predicted to possess a C terminus produced by proteasomal cleavage, good TAP affinity, and good affinity to MHC class I molecules. Peptides with a score below the threshold score of either proteasomal cleavage method or SVMTAP method were filtered out, to reduce the number of false positives.

2.3.7 EpiJen

It is an algorithm for T cell epitope prediction based on quantitative matrices. It carries out multiple steps, viz., prediction of proteasomal cleavage, TAP transport, MHC binding, and epitope selection successively, using QMs [46].

EpiJen mimics the cellular antigen processing pathway, working in a hierarchical or successive manner and not in parallel. Peptides that have been eliminated at any of the preceding steps do not continue to the successive steps.

Nonameric peptides derived from AntiJen [47] and SYFPEITHI databases were used to generate models for proteasomal cleavage, TAP binding, and MHC binding, trained them and tested them using "leave one out cross-validation," LOOCV using Receiver Operating Characteristic (ROC) curves.

Quantitative matrices of protein–peptide interaction were derived using additive method, wherein either multiple regression or discriminant analysis was used to derive the QMs, depending upon whether the dependent variable was continuous or discontinuous and solved using partial least squares (PLS) method, implemented in SYBYL 6.9 (http://www.tripos.com/).

EpiJen server can be used to identify epitopes from both protein sequences as well as nucleic acid sequences (which can be subjected to 3-frame or 6-frame translation).

Sensitivity and positive predictive value vary at varying thresholds used for prediction. The developers of this method recommend usage of a threshold of 5 % at which sensitivity of the method is reported to be 85 %.

2.3.8 EpiTOP

EpiTOP follows a proteochemometrics-based approach to MHC class II binding peptide prediction [48]. Proteochemometrics approach is an extrapolation of QSAR approach. Quantitative structure–activity relationship (QSAR) is a well-known approach which relates quantitative properties of ligands to their activity. It is popularly used in medicinal chemistry, drug design and discovery. In the proteochemometrics approach, a quantitative description of the protein is also considered in addition to the description of the ligands.

EpiTOP has developed and validated proteochemometric models to predict peptide binding to 12 HLA-DRB1 alleles using a quantitative matrix. The developers of EpiTOP extracted peptides binding to 12 HLA-DRB1 alleles from the Immune Epitope Database (IEDB) [49] to derive the QM.

They described the peptides using three *z*-scales broadly corresponding to volume, hydrophobicity, and polarizability for each of the constituent amino acid residues [50]. Nonameric peptides were encoded by a sequence of 27 *z*-descriptors (9 positions × 3 *z*-scales), while the HLA-DRB1 alleles were encoded by 54 descriptors (18 positions × 3 *z*-scales) corresponding to the polymorphic residues within the binding site that interact with the peptide. Cross-terms for adjacent peptide positions and peptide–protein cross-terms were also included in the models. The affinities of MHC binders were assessed as pIC50 values.

Iterative self-consistent (ISC) algorithm was used to derive the proteochemometric quantitative matrix. EpiTOP generates overlapping nonameric peptides from the input query protein sequence. Nonameric peptides bearing anchor residues at position 1 are assessed, while the rest are discarded. The binding affinities of the nonamers are predicted using the proteochemometric quantitative matrix. While the sensitivity of this tool, as reported by its developers is about 45 % for the top 5 % cutoff, it increases to about 95 % for the top 25 % cutoff.

2.3.9 PREDIVAC

It is a pan-specific method for CD4+ T-cell epitope prediction based on the specificity-determining residues (SDR) [51]. These are amino acid residues that are responsible for specific interactions between a given pair of interacting proteins, or between a protein and a peptide.

The developers of this tool studied crystal structures of peptide–MHC class II complexes involving HLA DQ and DR loci, as well as carried out quantum chemistry-based analyses of peptide–MHC class II interactions to identify the SDRs. They constructed a database called PredivacDB, which contains SDRs and nonameric high-affinity binding peptides derived from the Immune Epitope Database (IEDB), a database of MHC binders and nonbinders (MHCBN) and EPIMHC [52], a curated database of MHC binding peptides.

The PREDIVAC tool predicts MHC class II binding peptides by identifying the SDRs in the query protein and comparing them with the SDRs associated with HLA proteins of known specificity present in PredivacDB. It calculates amino acid frequencies and weights for peptides associated with allotypes sharing similar SDRs as the query protein sequence at each binding position. Subsequently, a position-weight matrix (PWM) is built based on the binding data. The query protein sequences are parsed into overlapping nonameric peptides, each of which is assigned a binding score using the PWM. Thereby, T-cell epitope mapping is carried out. The developers of this tool have reported identification of 75 % of immunodominant epitopes within the top 3 % scoring peptides. PREDIVAC covers over 95 % of HLA class II DR allotypes distributed in various geographical regions and ethnic groups across the globe.

3 Practical Considerations

There are many methods available for prediction of T-cell epitopes, some of which have been explained in Section 2. This section throws light on the practical considerations of some of the popular methods of T-cell epitope prediction.

3.1 SYFPEITHI

Availability: The graphical user interface for epitope prediction using SYFPEITHI is available online at http://www.syfpeithi.de/bin/MHCServer.dll/EpitopePrediction.htm
A screenshot of the same is given in Fig. 2. Offline version is also available for purchase.

Typical input: The server accepts protein sequence in single letter code in plain text format as input. The maximum sequence length accepted by the tool is 2,048 amino acid residues. A user needs to choose one, many or all MHC alleles provided by the tool. However, if the user chooses all alleles, the input sequence length accepted by the online tool is only 100 amino acid residues. If the sequence is longer than 100, it needs to be split. Else, purchase of offline version is recommended by the developers of this tool. Choice also exists for the length of peptide epitopes. One may choose octamers (8-mers), nonamers (9-mers), decamers (10-mers), endecamers (11-mers), pentadecamers (15-mers), and all mers. However, pentadecamer peptides are for MHC class II alleles only.

Typical output: The tool returns lists of peptide epitopes of the chosen length, corresponding to the respective MHC alleles chosen by the user. For a chosen peptide length and the MHC

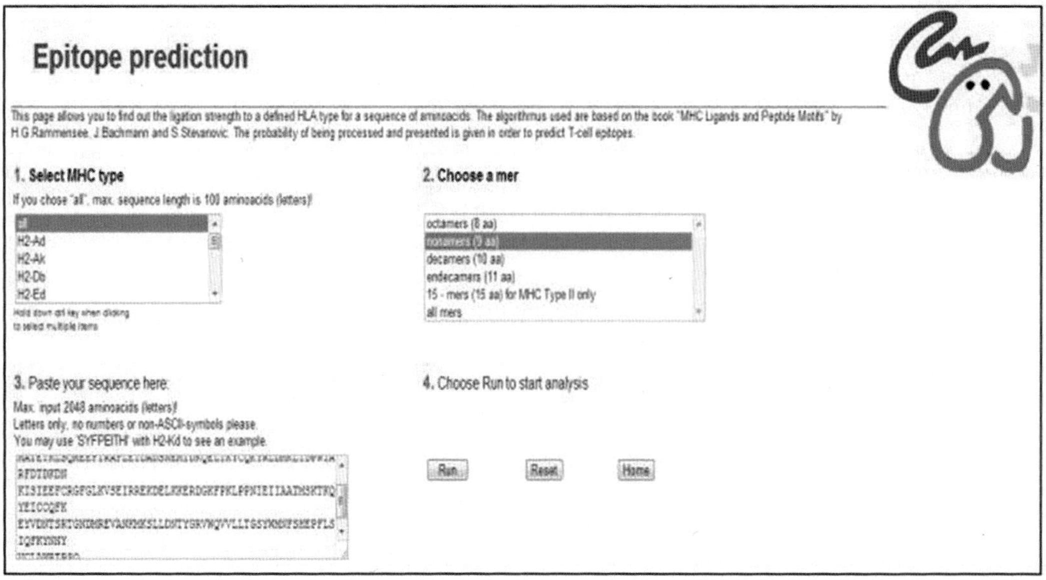

Fig. 2 Screenshot of the graphical user interface of SYFPEITHI epitope prediction tool

HLA-B*39:01 nonamers go to top

Pos	1 2 3 4 5 6 7 8 9	score
29	G R V W Q V V L L	26
28	Y G R V W Q V V L	15
7	S R T G N D M R E	12
15	E V A N K M K S L	12
45	F S H E P E L S I	12
16	V A N K M K S L L	11
43	M N F S H E P E L	11
13	M R E V A N K M K	10
23	L L D N T Y G R V	9
27	T Y G R V W Q V V	9
26	N T Y G R V W Q V	8
8	R T G N D M R E V	7
30	R V W Q V V L L T	7
31	V W Q V V L L T G	7
10	G N D M R E V A N	6
17	A N K M K S L L D	6

Fig. 3 Sample output of SYFPEITHI displaying the list of nonamer peptides for HLA-B*39:01 allele in the decreasing order of their score and peptide position

allele, the tool provides a list of peptides ranked in the descending order of their score along with the position of the first amino acid residue of the peptide in the input protein sequence. Position of each individual amino acid of the peptide is also indicated. Anchor amino acids are shown in bold face, while auxiliary amino acids are underlined. A sample output is given in Fig. 3. In case the user selects more than one MHC allele, and peptides of different lengths, an index of all the alleles along with the peptide lengths is provided followed by the list of peptides, as shown in Fig. 4.

Points to note:

- The developers of this tool state that the naturally presented epitope should be among the top-scoring 2 % of all peptides predicted for a particular MHC allele. For a 300 amino acid residue long input protein for example, of all the nonameric peptides predicted by the tool, the naturally presented epitope should be among the 6 top-scoring peptides.
- The preferred length of the peptide binding to different MHC molecules may be different. Some MHC molecules may prefer a nonamer, while some may prefer a decamer, for instance.
- Since the maximal scores of the peptide epitopes vary among different MHC alleles, the developers of SYFPEITHI advise users to include known epitopes in order to have an approximation of the scoring.
- SYFPEITHI predicts 15-mer peptides as MHC class II binding peptides. These 15-mer peptides contain three N-terminal flanking residues, the nonamer core residues located within the binding groove, and three C-terminal flanking residues.

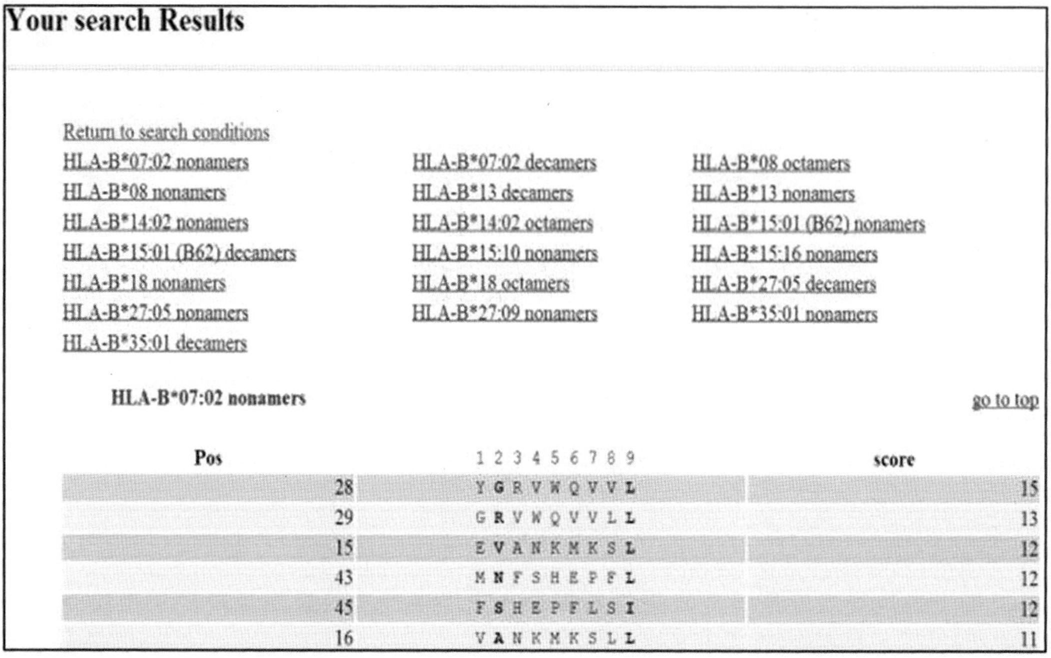

Fig. 4 Sample output of SYFPEITHI displaying the list of MHC alleles and peptides of appropriate lengths along with a partial list of peptide epitopes

3.2 CTLPred

Availability: CTLPred is available online at http://www.imtech.res.in/raghava/ctlpred.

A screenshot of the Web server is given in Fig. 5.

Typical input: The server accepts protein sequence in single letter code in plain text or any of the standard sequence formats as input. Minimum sequence length should be nine amino acid residues. Local sequence files may also be uploaded.

Prediction approaches: The user may choose any of the following approaches along with the cutoff score:

- QM based approach
- ANN based approach
- SVM based approach
- Consensus approach
- Combined approach

In the consensus approach as well as the combined approach, predictions are carried out using both ANN and SVM. However, in the consensus approach, only when both the ANN and SVM predict a peptide to be a T-cell epitope, does the CTLpred server report a peptide to be a T-cell epitope. If either of the methods predicts a peptide to be a non-epitope, the server reports the peptide to be a non-epitope. On the other hand, in the combined approach, even if one among ANN and SVM predicts a peptide to be an epitope, the CTLpred server reports the peptide to be an epitope.

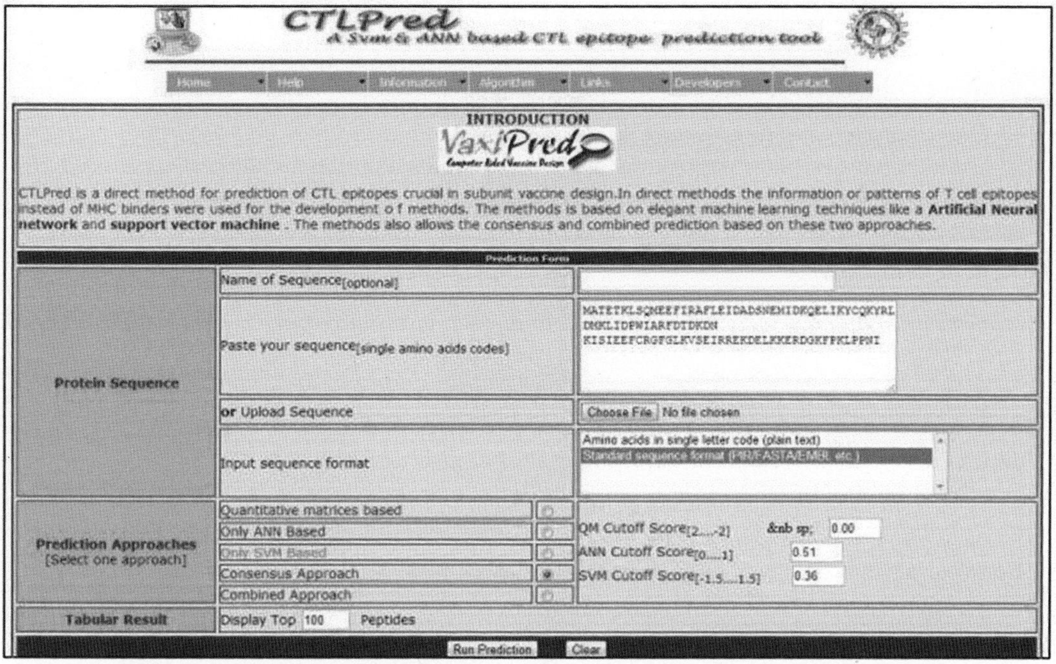

Fig. 5 Screenshot of the graphical user interface of CTLpred

Cutoff score: The cutoff score is used to differentiate between the epitopes and non-epitopes. The peptides achieving score greater than the cutoff score are predicted as epitopes. Default cutoff score of prediction methods will be used in case the user does not choose a cutoff score. The default cutoff score is the one at which the sensitivity and specificity of prediction methods are nearly equal.

Typical output: The output includes comprehensive information including the length of input sequence, prediction approach used, number of nonamers generated, cutoff score, date and time when the prediction was carried out, and the result in three formats, viz., color display, overlap display, and tabular display. A sample output is given in Fig. 6.

Color display: In this display, the amino acid sequence of the input protein is shown with 100 amino acid residues in each line. The first amino acid residue of the predicted CTL epitopes is colored red, the other amino acid residues of the epitope are colored in blue while the rest are colored black.

Overlap display: In this display, overlapping CTL epitopes are shown in separate lines. A scale indicating the position of the epitope in the protein sequence is given. The coloring scheme of the amino acid residues of the predicted CTL epitopes is as shown in the color display.

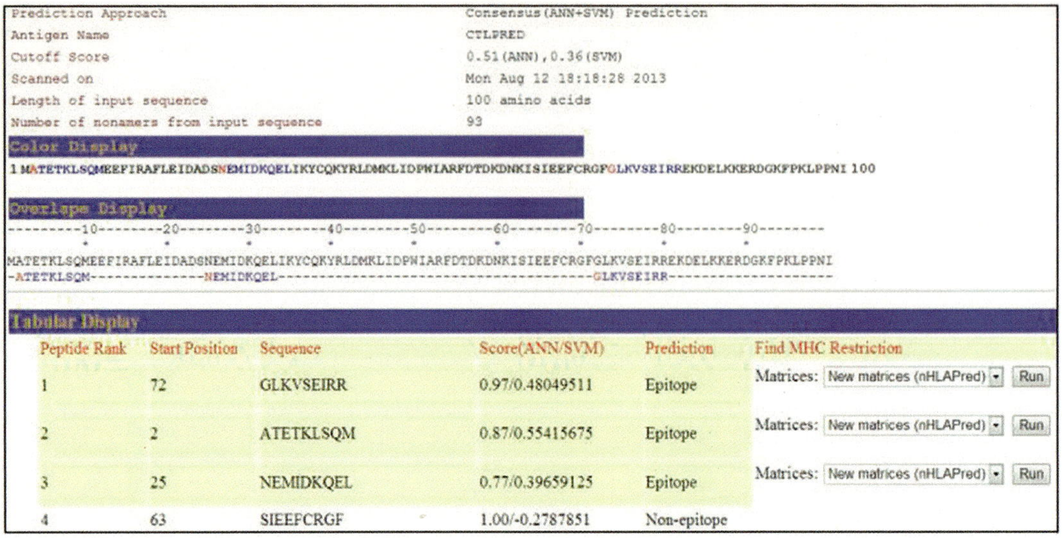

Fig. 6 Sample output of CTLpred, with consensus approach

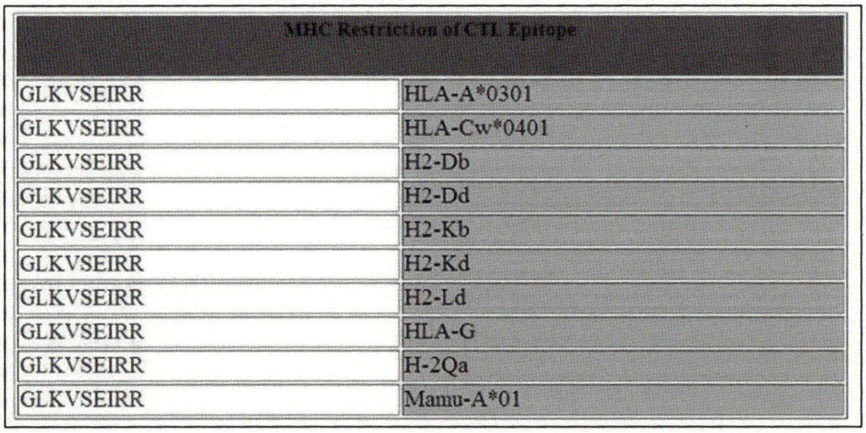

MHC Restriction of CTL Epitope	
GLKVSEIRR	HLA-A*0301
GLKVSEIRR	HLA-Cw*0401
GLKVSEIRR	H2-Db
GLKVSEIRR	H2-Dd
GLKVSEIRR	H2-Kb
GLKVSEIRR	H2-Kd
GLKVSEIRR	H2-Ld
GLKVSEIRR	HLA-G
GLKVSEIRR	H-2Qa
GLKVSEIRR	Mamu-A*01

Fig. 7 Sample output of CTLpred displaying information about the MHC alleles with which a particular epitope would interact

Tabular display: In this display, peptide rank, start position of the peptide, peptide sequence, score(s), and prediction—epitope or non-epitope are given in a tabular format. The peptides are displayed in the descending order of their score. For every peptide that is predicted to be an epitope, a set of matrices are provided to find MHC restriction. This option provides information about the MHC alleles for which the particular epitope is applicable (Fig. 7). Also, the user may choose the number of peptides to be displayed in the table.

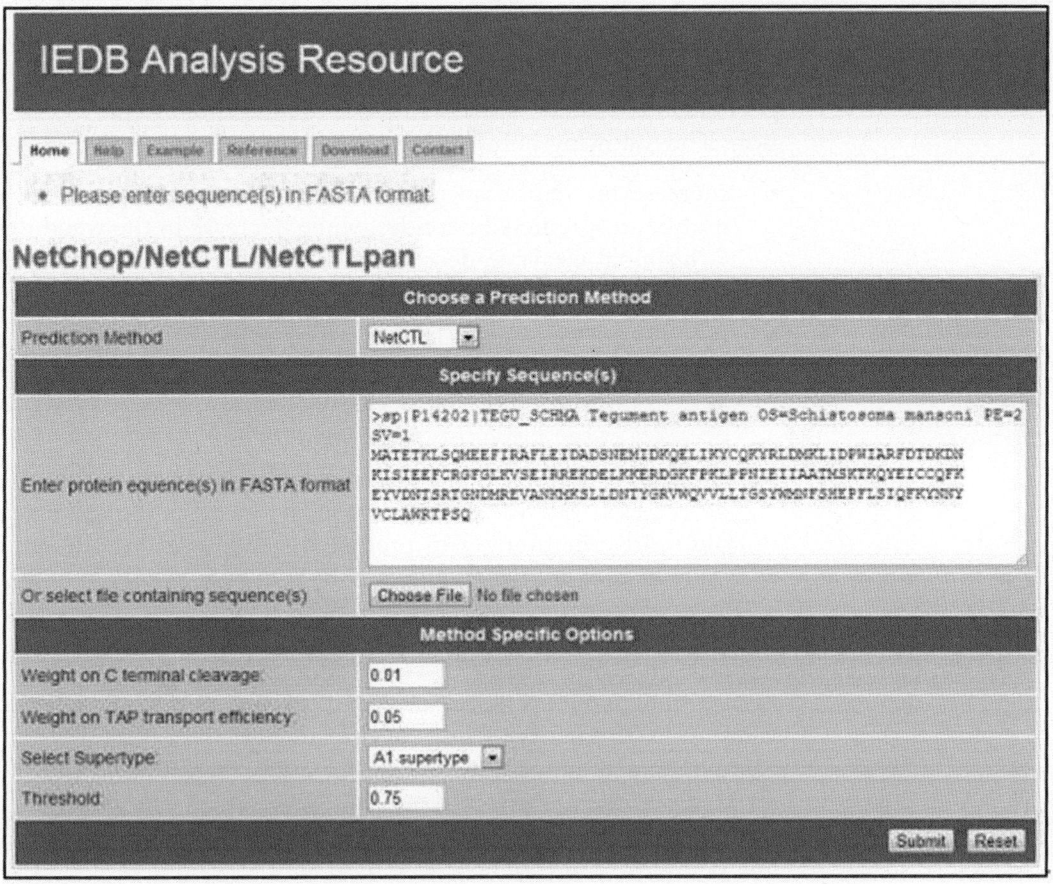

Fig. 8 Screenshot of the graphical user interface of NetCTL

3.3 NetCTL/NetCTLpan/NetChop

Availability: NetCTL is available online at http://tools.immuneepitope.org/stools/netchop/netchop.do. Please note that the same GUI provides the users the option to use NetCTLpan and NetCHOP as well. NetCTLpan is a tool that provides pan-specific CTL epitope predictions while NetCHOP predicts proteasome cleavage motifs using ANNs. A screenshot of the GUI is shown in Fig. 8. NetCTL alone is available online at http://www.cbs.dtu.dk/services/NetCTL/

Typical input: The server accepts protein sequence in FASTA format as input. Local sequence files in FASTA format may also be uploaded.

Choice of prediction methods: The user needs to choose one of the following prediction methods: NetCHOP/NetCTL/NetCTLpan.

3.3.1 NetCTL

Options available while using NetCTL: While the tool can be run with default parameters, there is a provision to alter the parameters by the user. The weights on the C terminal cleavage and TAP transport efficiency can be altered, and so is the threshold value of prediction. The following points need to be pondered over by the users:

- Increase in weight on the C terminal cleavage increases the number of predicted peptides, while decrease in the weight on C terminal cleavage decreases the number of predicted peptides at the given threshold score.
- Increase in weight on TAP transport efficiency increases the number of predicted peptides, while decrease in the weight on TAP transport efficiency decreases the number of predicted peptides at the given threshold score.
- Increase in threshold score increases specificity marginally but decreases sensitivity and vice versa. Therefore, at higher threshold score, the number of predicted epitopes may be lesser in number, but are highly specific.
- A user must choose any 1 of the 12 MHC class I supertypes.

Typical output for NetCTL: The prediction is shown as output in two formats, viz., graphical view and tabular view. The graphical view (Fig. 9) shows predicted epitopes in a graph of NetCTL score plotted against amino acid residue position. Peptides that have a score above the threshold score are predicted as binders and are shown in green, while the non-epitopes are shown in pink, the red line being the threshold score.

The tabular view (Fig. 10) of the output shows the following columns:

- # (amino acid residue position).
- Peptide Sequence (in single letter code).
- Predicted MHC Binding Affinity (given as 1 − log50k(aff), where log50k is the logarithm with base 50, and aff is the affinity in nM units).
- Rescale Binding Affinity (predicted binding affinity normalized by the 1st percentile score).
- C Terminal Cleavage Affinity (Predicted proteasomal cleavage score).
- TAP Transport Efficiency (Predicted TAP transport efficiency).
- Prediction Score (Overall prediction score).

The positive predictions are displayed in green, while the rest are shown in black. One can sort the predicted output by clicking on the respective column header.

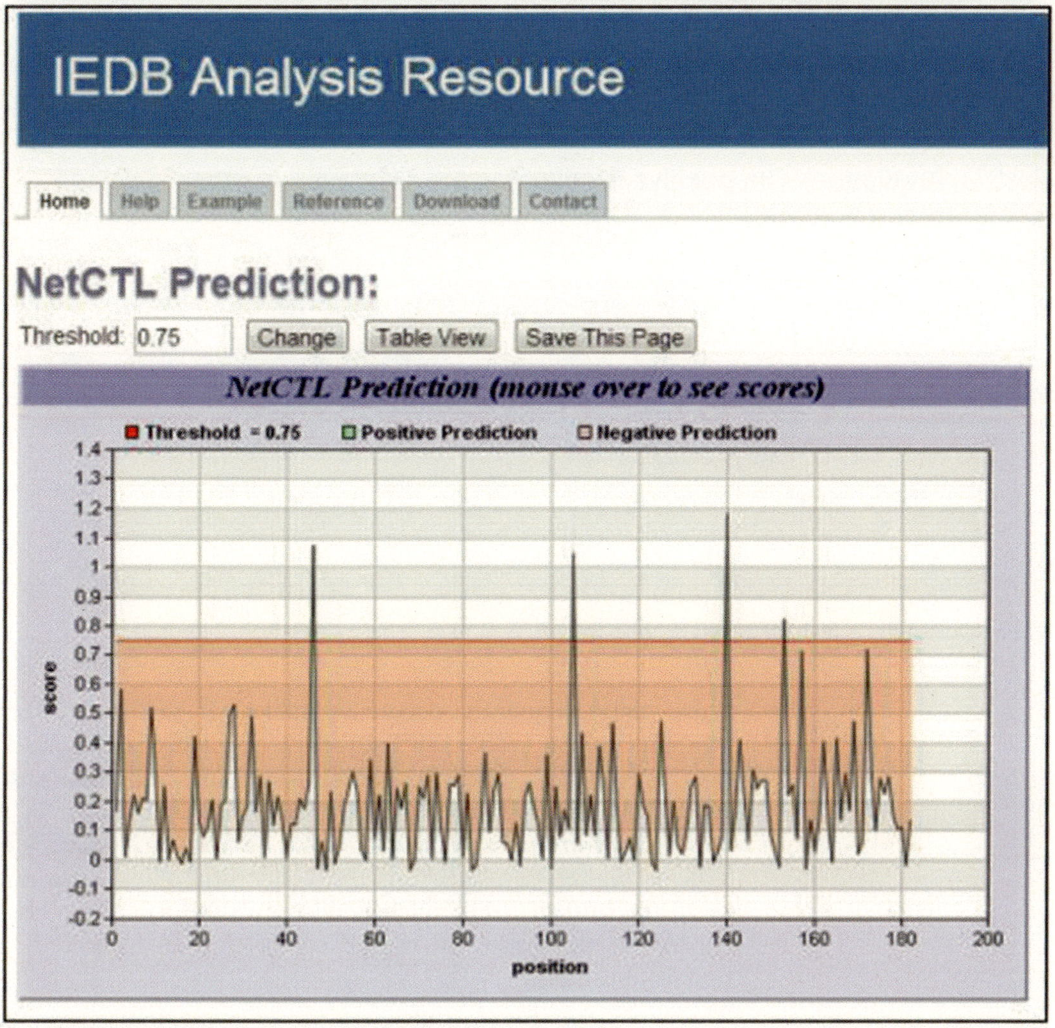

Fig. 9 Graphical view of sample output of NetCTL

3.3.2 NetCTLpan

A screenshot of the GUI for NetCTLpan is shown in Fig. 11. Options available while using NetCTLpan: NetCTLpan can be run using default parameters, but also provides users the provision to change the following parameters:

- Species: User may choose one of the six vertebrate species, including humans.
- Select allele: User may choose any of the enlisted MHC alleles of the selected species. If the user selects "human" as the species, the tool also provides for choice of length of the peptide (8-mer to 11-mer).

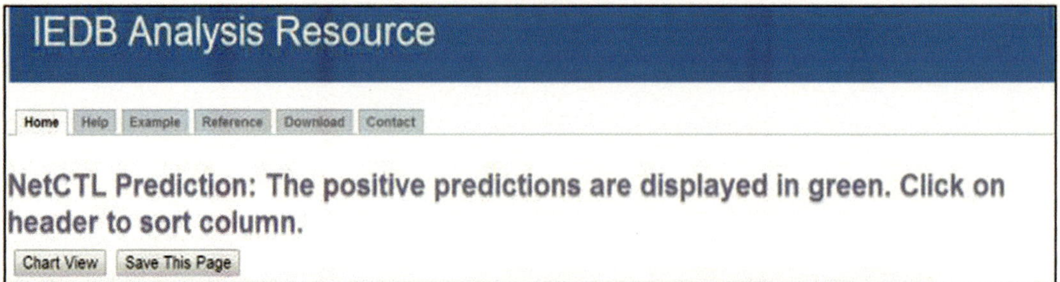

#	Peptide Sequence	Predicted MHC Binding Affinity	Rescale Binding Affinity	C Terminal Cleavage Affinity	TAP Transport Efficiency	Prediction Score
99	MKSLLDNTY	0.1306	0.8870	0.9455	3.1180	1.1847
5	LIDPWIARF	0.1167	0.7925	0.9990	2.5920	1.0720
64	ATMSKTKQY	0.1075	0.7302	0.9979	3.2660	1.0432
112	QVVLLTGSY	0.0765	0.5195	0.9948	3.0980	0.8236
131	SIQFKYNNY	0.0616	0.4184	0.9698	3.1220	0.7200
116	LTGSYWMNF	0.0687	0.4666	0.8526	2.4020	0.7146
128	PFLSIQFKY	0.0303	0.2060	0.9603	2.4630	0.4732
84	NTSRTGNDM	0.0635	0.4309	0.2198	0.1770	0.4728
73	EICCQFKEY	0.0317	0.2156	0.7392	2.8060	0.4668

Fig. 10 Tabular view of sample output of NetCTL

- Show only frequently occurring alleles: If the user selects "human" as the species, a "Frequently occurring alleles check-box" is provided, which is checked by default. This allows the selection of only those human MHC alleles that occur in at least 1 % of the human population or have allele frequency of 1 % or higher. Un-checking the check-box allows selection of all the human MHC alleles enlisted in the tool. The HLA supertype of some of the HLA alleles is indicated in parentheses.
- Threshold for showing predictions: It is the low combined prediction score threshold (ranging between −99.9 and 3) to filter with predictions to be displayed.
- Weight on C terminal cleavage: Increase in the weight on proteasomal cleavage will increase the number of predicted epitopes.
- Weight on TAP transport efficiency: Increase in the weight on TAP transport efficiency will increase the number of predicted epitopes.
- Threshold for epitope identification: It is the threshold to label predictions as epitopes. This threshold value is based on the % rank score.
- Percentile for positive prediction: It is the percentile cutoff value for positive prediction. Increase in the percentile cutoff value will increase the number of peptides labelled as epitopes.

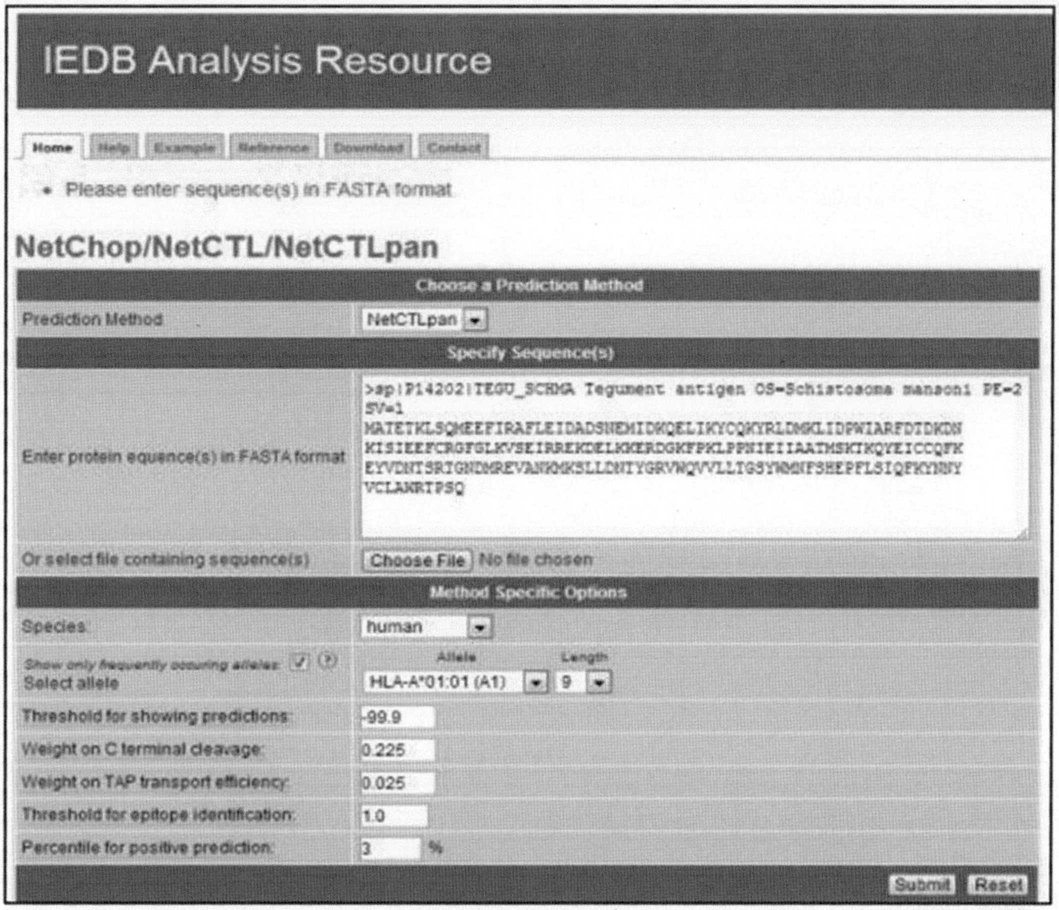

Fig. 11 Screenshot of the graphical user interface of NetCTLpan

Typical output for NetCTLpan: The output is in two formats, viz., graphical view and tabular view. The graphical view (Fig. 12) shows predicted epitopes in a graph of NetCTLpan score plotted against amino acid residue position. Peptides that have a score above the threshold score are predicted as binders and are shown in green, while the non-epitopes are shown in pink, the red line being the threshold score.

The tabular view (Fig. 13) of the output shows the following columns:

- # (amino acid residue position).
- MHC Prediction (MHC Prediction score given in 1-log50K(aff) where log50k is the logarithm with base 50, and aff is the affinity in nM units).
- TAP Prediction score (Predicted TAP transport efficiency).
- Cleavage Prediction score (Predicted proteasomal cleavage score).

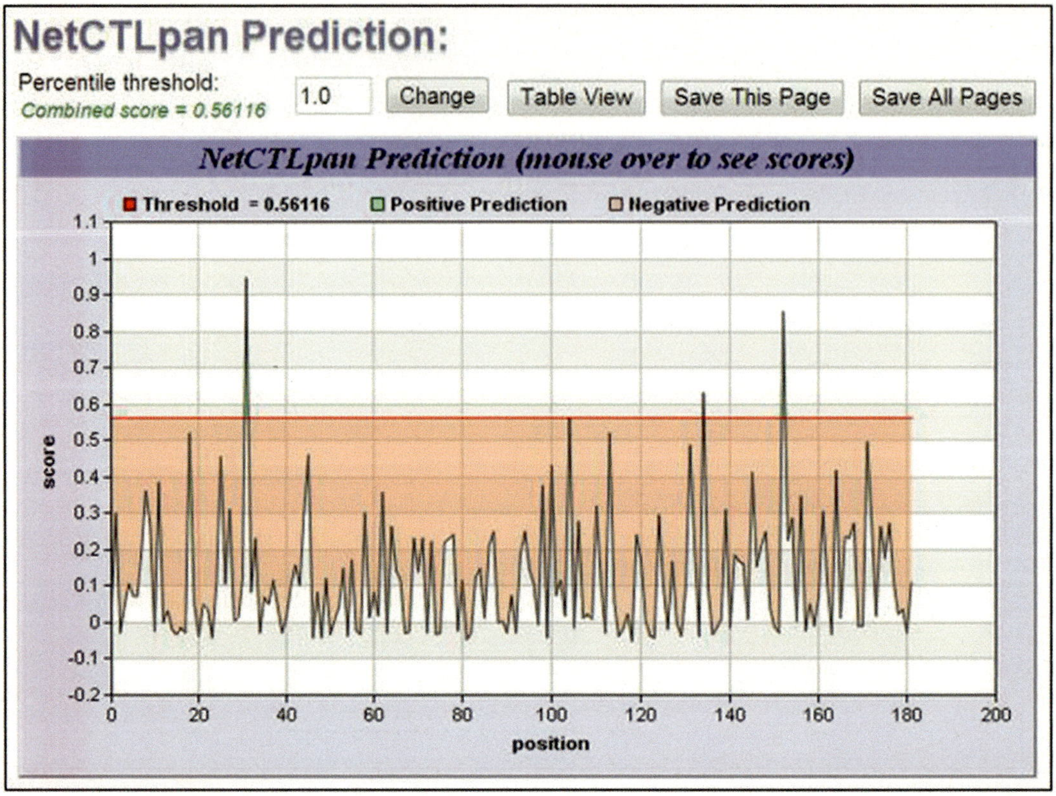

Fig. 12 Graphical view of sample output of NetCTLpan

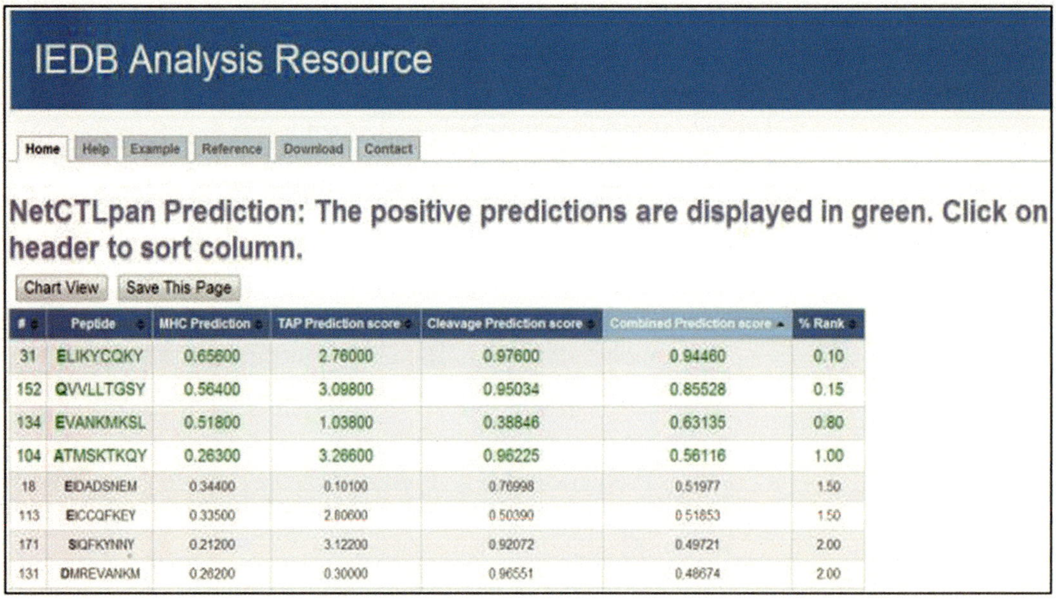

IEDB Analysis Resource

Home | Help | Example | Reference | Download | Contact

NetCTLpan Prediction: The positive predictions are displayed in green. Click on header to sort column.

Chart View | Save This Page

#	Peptide	MHC Prediction	TAP Prediction score	Cleavage Prediction score	Combined Prediction score	% Rank
31	ELIKYCQKY	0.65600	2.76000	0.97600	0.94460	0.10
152	QVVLLTGSY	0.56400	3.09800	0.95034	0.85528	0.15
134	EVANKMKSL	0.51800	1.03800	0.38846	0.63135	0.80
104	ATMSKTKQY	0.26300	3.26600	0.96225	0.56116	1.00
18	EIDADSNEM	0.34400	0.10100	0.76998	0.51977	1.50
113	EICCQFKEY	0.33500	2.80600	0.50390	0.51853	1.50
171	SIQFKYNNY	0.21200	3.12200	0.92072	0.49721	2.00
131	DMREVANKM	0.26200	0.30000	0.96551	0.48674	2.00

Fig. 13 Tabular view of sample output of NetCTLpan

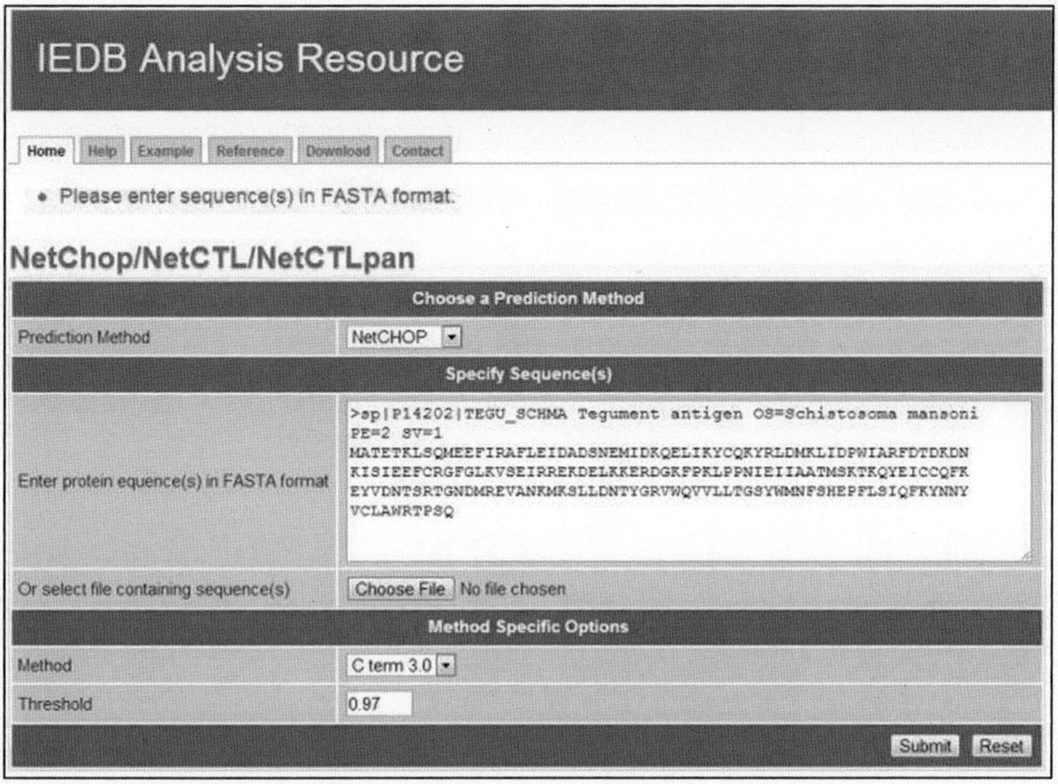

Fig. 14 A screenshot of the GUI for NetChop

- Combined Prediction score (Overall prediction score).
- %Rank (% Rank of prediction score to a set of 1,000 random natural 9-mer peptides).

The positive predictions are displayed in green, while the rest are shown in black. One can sort the predicted output by clicking on the respective column header.

3.3.3 NetChop

A screenshot of the GUI for NetCHOP is shown in Fig. 14.

Options available while using NetCHOP: While the tool can be run with default parameters, the user may choose either C term 3.0 or 20s 3.0 as the prediction servers. One may alter the threshold as well. Increase in threshold score increases specificity but decreases sensitivity and vice versa.

Typical output for NetCTLpan: The output is in two formats, viz., graphical view and tabular view. The graphical view (Fig. 15) shows predicted epitopes in a graph of NetCHOP score plotted against amino acid residue position. Peptides that have a score above the threshold score are predicted as binders and are shown in green, while the non-epitopes are shown in pink, the red line being the threshold score.

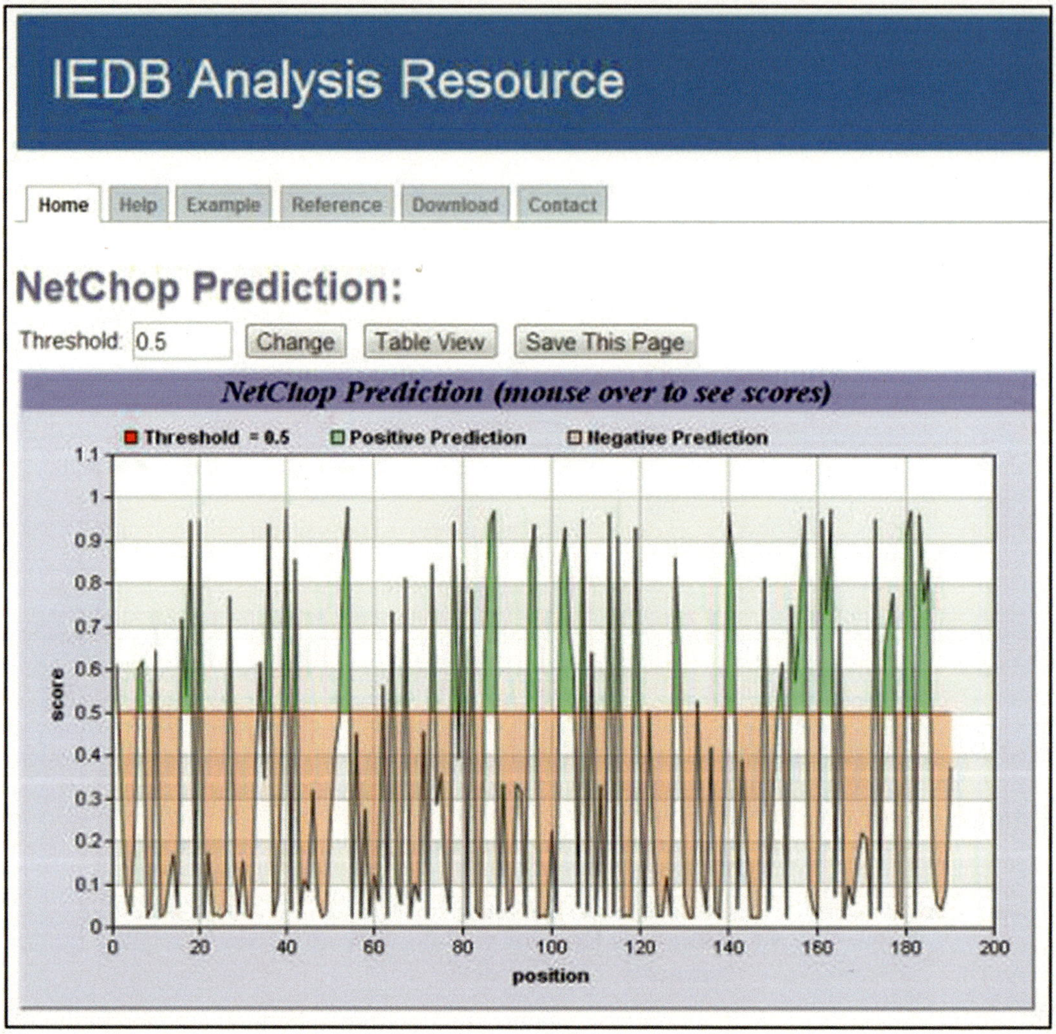

Fig. 15 Graphical view of output of NetChop

The tabular view (Fig. 16) of the output shows the following columns:

- # (amino acid residue position)
- Amino Acid (Amino acid residue)
- Prediction Score (NetChop prediction score)

3.4 EpiTOP

Availability: EpiTOP is available online at http://www.pharmfac.net/EpiTOP/. A screenshot of the GUI is shown in Fig. 17.

Typical input: The server accepts protein sequence in single letter code (raw format) as input. The user needs to choose the HLA class II allele from the drop-down menu provided. The output cutoff also needs to be chosen: 5 %, 10 %, 15 %, 20 %, 25 %, or all binders.

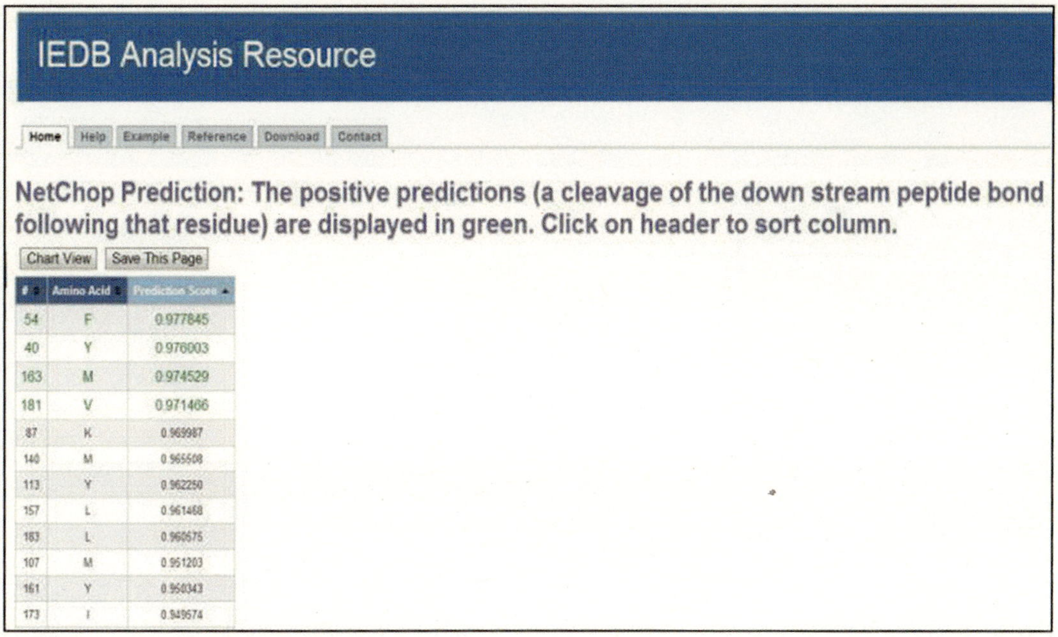

Fig. 16 Tabular view of output of NetChop

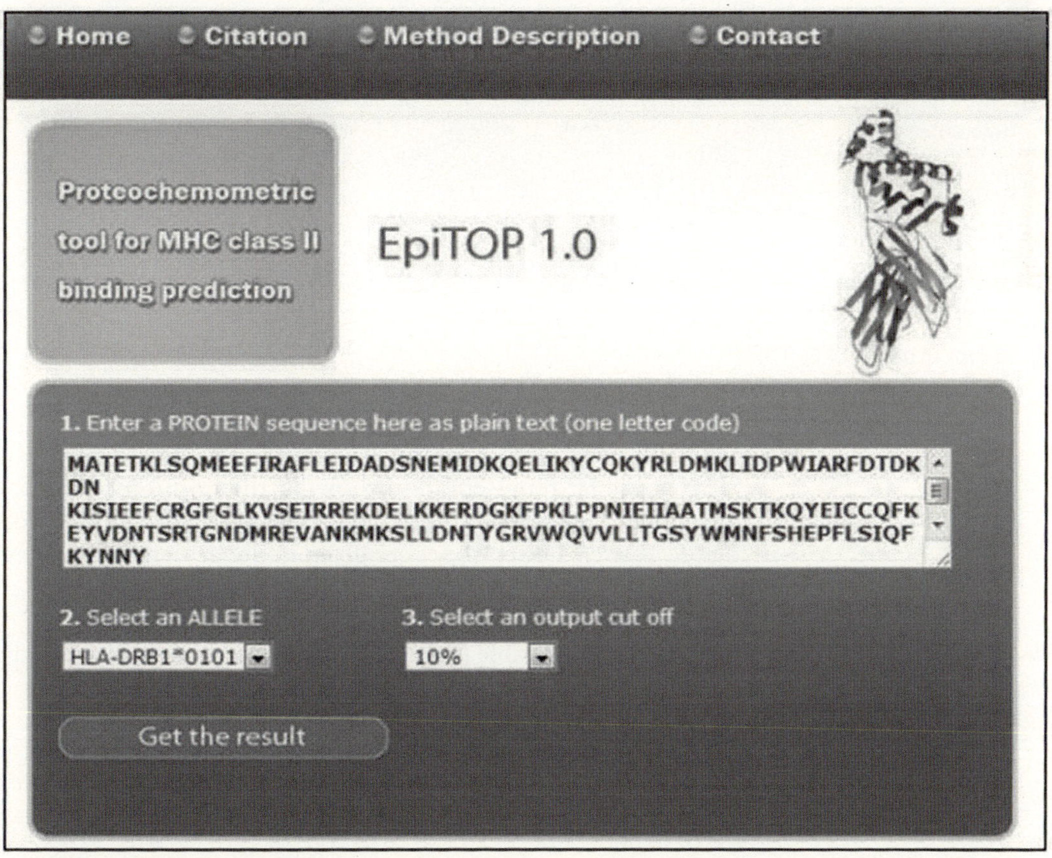

Fig. 17 Screenshot of GUI of EpiTOP

Fig. 18 Typical output of EpiTOP

Typical output: A sample output is shown in Fig. 18.

It is a tabular representation showing the position of the first amino acid of the peptide, the actual peptide epitope and the score in log (1/IC50). The epitopes are given in the descending order of scores.

3.5 PREDIVAC

Availability: It is available online at http://predivac.biosci.uq.edu.au/. Three options are provided to the users:

- Binding prediction.
- Population coverage prediction.
- Epitope prediction.

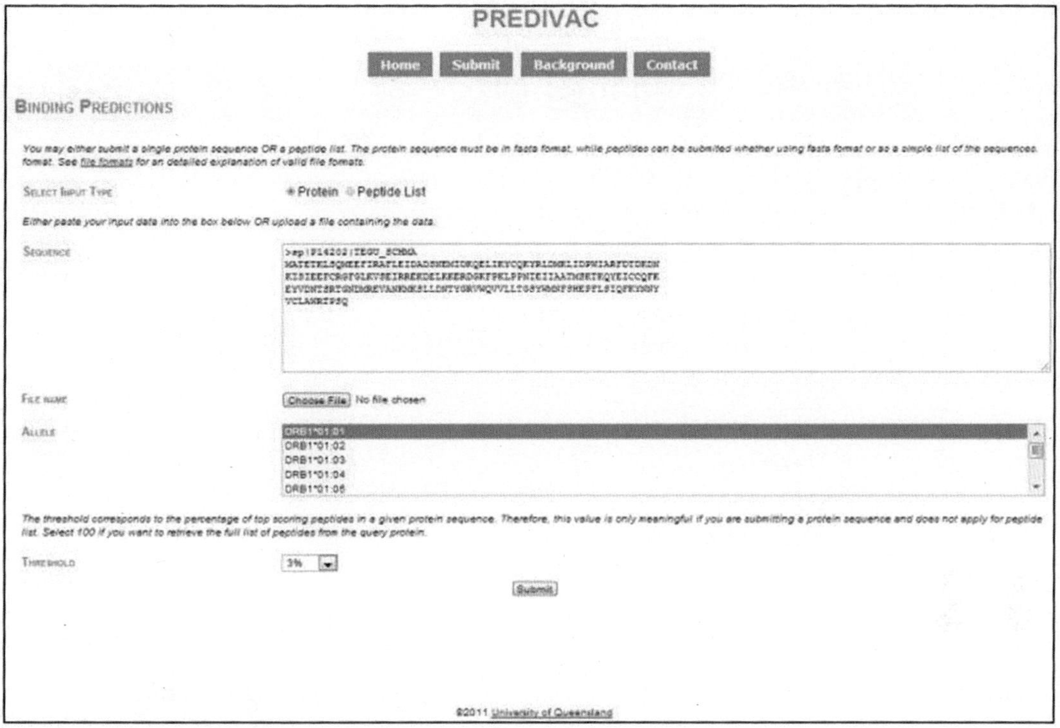

Fig. 19 A screenshot of the GUI of "Binding prediction" option of PREDIVAC

3.5.1 Binding Prediction

This predicts the MHC class II binders in the protein sequence. Also, it can evaluate whether a particular peptide can bind to a particular MHC class II allele.

Typical input: The server accepts a single protein sequence (in FASTA format) or a peptide list (in FASTA format or as a simple list of the peptide sequences) as input. Please note that the sequence submitted must not contain any non-standard amino acids. The user must specify the MHC class II allele from the drop-down menu as well as the threshold for prediction. The threshold refers to the percentage of top scoring peptides in the input protein sequence. Obviously, there is no need to specify the threshold value if the input is a peptide list. A screenshot of the GUI of "Binding prediction" option of PREDIVAC is shown in Fig. 19.

Typical output: It comprises of the list of peptides that would bind to the specified MHC class II allele, its start and end positions as well as the score. A frequency matrix and a scoring matrix are also given. Figure 20 shows the table of peptides in the output.

Results of "Population coverage prediction" and "Epitope prediction" using PREDIVAC were not received by the authors as on the date of submission this chapter.

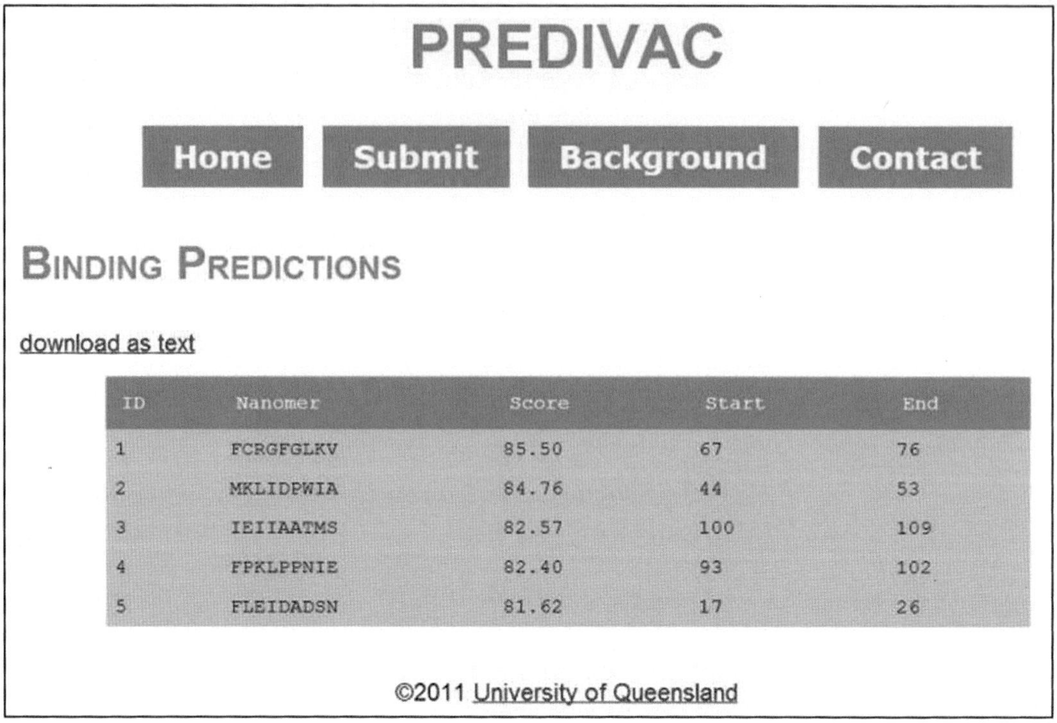

Fig. 20 A sample output of "Binding prediction" option of PREDIVAC showing the table of peptides

4 Note

Although many methods for "prediction of MHC class I and class II binding peptides" are available, for example, PropredI [53], Propred [54], MULTIPRED2 [55], etc., they have not been discussed here as discussion on the methods for "MHC class I and class II binding peptides" would form the subject matter of another chapter in the same book.

Acknowledgements

D.V.D. and U.K.K. gratefully acknowledge financial support under the aegis of Center of Excellence (CoE) grant from the Department of Biotechnology (DBT), Government of India.

References

1. Uebel S, Tampé R (1999) Specificity of the proteasome and the TAP transporter. Curr Opin Immunol 11:203–208
2. Niedermann G, King G, Butz S et al (1996) The proteolytic fragments generated by vertebrate proteasomes: structural relationships to major histocompatibility complex class I binding peptides. Proc Natl Acad Sci U S A 93:8572–8577
3. Craiu A, Akopian T, Goldberg A et al (1997) Two distinct proteolytic processes in the generation of a major histocompatibility complex class I-presented peptide. Proc Natl Acad Sci U S A 94:10850–10855

4. Koopmann JO, Post M, Neefjes JJ et al (1996) Translocation of long peptides by transporters associated with antigen processing (TAP). Eur J Immunol 26:1720–1728
5. Uebel S, Kraas W, Kienle S et al (1997) Recognition principle of the TAP transporter disclosed by combinatorial peptide libraries. Proc Natl Acad Sci U S A 94:8976–8981
6. Gubler B, Daniel S, Armandola EA et al (1998) Substrate selection by transporters associated with antigen processing occurs during peptide binding to TAP. Mol Immunol 35:427–433
7. Kindt TJ, Osborne BA, Goldsby RA (2006) Kuby immunology. W. H. Freeman & Company, New York
8. Robinson J, Halliwell JA, McWilliam H et al (2013) The IMGT/HLA Database. Nucleic Acids Res 41:D1222–D1227
9. Lund O, Nielsen M, Kesmir C et al (2004) Definition of supertypes for HLA molecules using clustering of specificity matrices. Immunogenetics 55:797–810
10. Doytchinova IA, Flower DR (2005) In silico identification of supertypes for class II MHCs. J Immunol 174:7085–7095
11. Giudicelli V, Duroux P, Ginestoux C et al (2006) IMGT/LIGM-DB, the IMGT® comprehensive database of immunoglobulin and T cell receptor nucleotide sequences. Nucleic Acids Res 34:D781–D784
12. Kaas Q, Ruiz M, Lefranc M-P (2004) IMGT/3Dstructure-DB and IMGT/Structural Query, a database and a tool for immunoglobulin, T cell receptor and MHC structural data. Nucleic Acids Res 32:D208–D210
13. Ehrenmann F, Lefranc M-P (2011) IMGT/3Dstructure-DB: Querying the IMGT Database for 3D Structures in Immunology and Immunoinformatics (IG or Antibodies, TR, MH, RPI, and FPIA). Cold Spring Harb Protoc 2011(6):750–761. doi:10.1101/pdb.prot5637
14. Robinson J, Waller MJ, Parham P et al (2003) IMGT/HLA and IMGT/MHC sequence databases for the study of the major histocompatibility complex. Nucleic Acids Res 31:311–314
15. DeLisi C, Berzofski JA (1985) T-cell antigenic sites tend to be amphipathic structures. Proc Natl Acad Sci U S A 82:7048–7052
16. Margalit H, Spouge JL, Cornette JL et al (1987) Prediction of immunodominant helper T cell antigenic sites from the primary sequence. J Immunol 138:2213–2229
17. Geluk A, Van Meijgaarden KE, Janson AA et al (1992) Functional analysis of DR17(DR3)-restricted mycobacterial T cell epitopes reveals DR17-binding motif and enables the design of allele specific competitor peptides. J Immunol 149:2864–2871
18. Malcherek G, Falk K, Rötzschke O et al (1993) Natural peptide ligand motifs of two HLA molecules associated with myasthenia gravis. Int Immunol 5:1229–1237
19. Geluk A, van Meijgaarden KE, Southwood S et al (1994) HLADR3 molecules can bind peptides carrying two alternative specific submotifs. J Immunol 152:5742–5748
20. Seeger FH, Schirle M, Keilholz W et al (1999) Peptide motif of HLA-B*1510. Immunogenetics 49:996–999
21. Meister GE, Roberts CG, Berzofsky JA et al (1995) Two novel T cell epitope prediction algorithms based on MHC-binding motifs; comparison of predicted and published epitopes from Mycobacterium tuberculosis and HIV protein sequences. Vaccine 13:581–591
22. Rammensee H, Bachmann J, Emmerich NP et al (1999) SYFPEITHI: database for MHC ligands and peptide motifs. Immunogenetics 50:213–219
23. Bian H, Hammer J (2004) Discovery of promiscuous HLA-II-restricted T cell epitopes with TEPITOPE. Methods 34:468–475
24. Zhang L, Chen Y, Wong H-S et al (2012) TEPITOPEpan: Extending TEPITOPE for Peptide Binding Prediction Covering over 700 HLA-DR Molecules. PLoS One 7:e30483. doi:10.1371/journal.pone.0030483
25. Parker KC, Bednarek MA, Coligan JE (1994) Scheme for ranking potential HLA-A2 binding peptides based on independent binding of individual peptide side-chains. J Immunol 152: 163–175
26. Bhasin M, Raghava GPS (2004) Prediction of CTL epitopes using QM, SVM and ANN techniques. Vaccine 22:3195–3201
27. Bhasin M, Raghava GPS (2007) A hybrid approach for predicting promiscuous MHC class I restricted T cell epitopes. J Biosci 32: 31–42
28. Rojas R (1996) Neural networks: a systematic introduction. Springer, Berlin
29. Narayanan A, Keedwell EC, Olsson B (2002) Artificial intelligence techniques for bioinformatics. Appl Bioinformatics 1:191–222
30. Yang ZR (2010) Neural networks. Methods Mol Biol 609:197–222. doi:10.1007/978-1-60327-241-4_12
31. Leman JK, Mueller R, Karakas M (2013) Simultaneous prediction of protein secondary structure and transmembrane spans. Proteins 81:1127–1140. doi:10.1002/prot.24258

32. Yang ZR (2004) Biological applications of support vector machines. Brief Bioinform 5: 328–338
33. Byvatov E, Schneider G (2003) Support vector machine applications in bioinformatics. Appl Bioinformatics 2:67–77
34. Kadam K, Sawant S, Kulkarni-Kale U et al. (2013) Prediction of protein function based on machine learning methods: an overview. In: Introduction to Sequence and Genome Analysis, iConcept Press Ltd., Hong Kong. (Accepted for publication)
35. Lata S, Bhasin M, Raghava GP (2009) MHCBN 4.0: A database of MHC/TAP binding peptides and T-cell epitopes. BMC Res Notes 2:61
36. Larsen MV, Lundegaard C, Lamberth K et al (2005) An integrative approach to CTL epitope prediction: A combined algorithm integrating MHC class I binding, TAP transport efficiency, and proteasomal cleavage predictions. Eur J Immunol 35:2295–2303
37. Kesmir C, Nussbaum AK, Schild H et al (2002) Prediction of proteasome cleavage motifs by neural networks. Protein Eng 15:287–296
38. Nielsen M, Lundegaard C, Lund O et al (2005) The role of the proteasome in generating cytotoxic T-cell epitopes: insights obtained from improved predictions of proteasomal cleavage. Immunogenetics 57:33–41
39. Peters B, Bulik S, Tampe R et al (2003) Identifying MHC class I epitopes by predicting the TAP transport efficiency of epitope precursors. J Immunol 171:1741–1749
40. Sturniolo T, Bono E, Ding J et al (1999) Generation of tissue-specific and promiscuous HLA ligand databases using DNA microarrays and virtual HLA class II matrices. Nat Biotechnol 17:555–561
41. Stranzl T, Larsen MV, Lundegaard C et al (2010) NetCTLpan: pan-specific MHC class I pathway epitope predictions. Immunogenetics 62:357–368
42. Dönnes P, Kohlbacher O (2005) Integrated modeling of the major events in the MHC class I antigen processing pathway. Protein Sci 14:2132–2140
43. Daniel S, Brusic V, Caillat-Zucman S et al (1998) Relationship between peptide selectivities of human transporters associated with antigen processing and HLA class I molecules. J Immunol 161:617–624
44. Donnes P, Elofsson A (2002) Prediction of MHC class I binding peptides, using SVMHC. BMC Bioinformatics 3:25
45. Brusic V, Rudy G, Harrsison LC (1998) MHCPEP, a database of MHC-binding peptides: Update. Nucleic Acids Res 26: 368–371
46. Doytchinova IA, Guan P, Flower DR (2006) EpiJen: a server for multistep T-cell epitope prediction. BMC Bioinformatics 7:131. doi: 10.1186/1471-2105-7-131
47. Toseland CP, Taylor DJ, McSparron H et al (2005) Anti-Jen: a quantitative immunology database integrating functional, thermodynamic, kinetic, biophysical, and cellular data. Immunome Res 1:4. doi:10.1186/1745-7580-1-4
48. Dimitrov I, Garnev P, Flower DR et al (2010) EpiTOP—a proteochemometric tool for MHC class II binding prediction. Bioinformatics 26:2066–2068
49. Vita R, Zarebski L, Greenbaum JA et al (2010) The immune epitope database 2.0. Nucleic Acids Res 38:D854–D862
50. Hellberg S, Sjöström M, Skagerberg B et al (1987) Peptide quantitative structure-activity relationships, a multivariate approach. J Med Chem 30:1126–1135
51. Oyarzún P, Ellis PJ, Bodén M et al (2013) PREDIVAC: CD4+ T-cell epitope prediction for vaccine design that covers 95% of HLA class II DR protein diversity. BMC Bioinformatics 14:52. doi:10.1186/1471-2105-14-52
52. Reche PA, Zhang H, Glutting JP et al (2005) EPIMHC: a curated database of MHC-binding peptides for customized computational vaccinology. Bioinformatics 21:2140–2141
53. Singh H, Raghava GP (2003) ProPred1: prediction of promiscuous MHC Class-I binding sites. Bioinformatics 19:1009–1014
54. Singh H, Raghava GPS (2001) ProPred: Prediction of HLA-DR binding sites. Bioinformatics 17:1236–1237
55. Zhang GL, Deluca DS, Keskin DB et al (2011) MULTIPRED2: A computational system for large-scale identification of peptides predicted to bind to HLA supertypes and alleles. J Immunol Methods 374:53–61. doi:10.1016/j.jim.2010.11.009

Chapter 20

Computational Antigenic Epitope Prediction by Calculating Electrostatic Desolvation Penalties of Protein Surfaces

Sébastien Fiorucci and Martin Zacharias

Abstract

The prediction of antigenic epitopes on the surface of proteins is of great importance for vaccine development and to specifically design recombinant antibodies. Computational methods based on the three-dimensional structure of the protein allow for the detection of noncontinuous epitopes in contrast to methods based on the primary amino-acid sequence only. A method recently developed to predict protein–protein binding sites is presented, and the application to predict putative antigenic epitopes is described in detail. The prediction approach is based on the local perturbation of the electrostatic field at the surface of a protein due to a neutral probe of low dielectric constant that represents an approaching binding partner. The calculated change in electrostatic energy corresponds to an energy penalty of desolvating a protein surface region, and antigenic epitope surface regions tend to be associated with a lower penalty compared to the average protein surface. The protocol to perform the calculations is described and illustrated on an example antigen, the outer surface protein A of *Borrelia burgdorferi*, a pathogenic organism causing lyme disease.

Key words Protein–protein interactions, Poisson Boltzmann calculation, Electrostatic properties, Epitope prediction

1 Introduction

The activation of the immune system typically involves the specific recognition of an antigen (Ag) by an antibody (Ab). Binding or recognition regions on the surface of an antigen for antibodies are called antigenic epitopes. If residues involved in an epitope are contiguous in the polypeptide chain, this epitope is called a continuous or linear epitope. A discontinuous or nonlinear epitope is composed of residues that are not necessarily continuous in the polypeptide sequence but have spatial proximity on the surface of a protein structure. The analysis of known epitope regions on proteins indicates that there are often characteristic protein surface regions which are preferentially recognized by antibody molecules, and hence are more suited as high-affinity epitopes compared to other surface regions.

Rajat K. De and Namrata Tomar (eds.), *Immunoinformatics*, Methods in Molecular Biology, vol. 1184,
DOI 10.1007/978-1-4939-1115-8_20, © Springer Science+Business Media New York 2014

The prediction of such antigenic epitopes is of major importance to specifically design new vaccines, new Abs and to possibly develop new therapeutic strategies of vaccine development. Several available epitope prediction methods are mainly developed to identify continuous epitope sequences of a protein based solely on the primary amino acid sequence of the protein. Such methods are based on amino acid physicochemical properties (e.g., hydrophilicity [1, 2], solvent accessibility [3]) or knowledge-based scoring functions derived from the analysis of Ag–Ab interaction databases using machine learning algorithms [4–7]. However, despite the use of consensus scoring functions the success of such approaches is limited [8, 9]. One reason for the limited performance is that a significant fraction of epitopes are discontinuous and the antigenicity of a linear peptide segment is also influenced by the surrounding surface regions. So far only a limited number of methods have been specifically developed to predict discontinuous B-cell epitopes [10–17] which is mainly due to the modest amount of available three-dimensional (3D) Ag–Ab complex structures. Structure-based methods often outperform more classical methods, based upon conservation and hydrophobicity of binding patches, that are often used to predict general protein binding sites [18]. Recent conformational B-cell epitope prediction algorithms have shown some successes, however, the level of prediction accuracy is not yet satisfactory [18].

The association of an antigen with its specific antibody partner follows general rules that drive protein–protein complex formation. Protein–protein interfaces are to a large extent well packed and are often composed of a buried hydrophobic core surrounded by a more hydrophilic ring partly exposed to solvent. Hydrophobic interactions and electrostatic complementarity are important driving forces for high affinity binding. The formation of the protein–protein complex requires the removal of water from the interface region. The removal of water molecules introduces a large desolvation penalty that needs to be overcome upon binding and which needs to be offset by attractive electrostatic and hydrophobic contributions. It is expected that the barrier to remove water (desolvation) is an important contribution that modulates the capacity of a surface region to interact with other proteins in general and with antibodies in particular.

A rapid method to calculate the solvation energy or a desolvation penalty is based on the accessible surface areas of atoms in the protein and on atomic solvation parameters derived from empirical vapor-to-water free energies of transfer of amino acid side-chain analogs. Such methods neglect the influence of the amino acid neighborhood and can lead to an incorrect prediction of the solvation energy. A more accurate method to calculate the electrostatic properties of a molecular system is to introduce a dielectric boundary between protein and water and treat the aqueous environment as a continuum. One can distinguish approaches based on solving the Poisson-Boltzmann equation numerically and approaches based on the Generalized Born formalism. We present here a method aimed at predicting protein–protein binding sites based on the electrostatic energy to remove water or to replace it with a region of low dielectric constant (electrostatic desolvation penalty, in the following: ESTADE). The method is based on the idea that preferred binding sites on protein surfaces may correspond to regions with a low electrostatic desolvation penalty. Notably, the present protocol has been successfully used to predict conformational antigenic epitopes in ref. 19 and in ref. 20.

2 Materials

The application of the approach for calculating electrostatic desolvation maps (ESTADE) on protein surfaces requires a 3D structure of the antigenic protein or at least a structural model. In the following the required steps for performing the calculations are outlined and explained.

2.1 Protein 3D Structure

1. The protein can be downloaded from Web servers like Protein Data Bank (http://www.rcsb.org/) that provide 3D coordinates of the protein.
2. In case where no crystallographic data is available, one can generate 3D structure of the protein by homology (MODELLER [21]: http://salilab.org/modeller/, SWISS-MODEL [22]: http://swissmodel.expasy.org/) or ab initio modeling (ROBETTA [23]: http://www.robetta.org/, I-TASSER [24]: http://zhanglab.ccmb.med.umich.edu/I-TASSER/).

2.2 Calculation of Protonation State of Residues

The electrostatic potential of a protein is influenced by the protonation states of surface residues. Several protocols can be used to predict the protonation state of residues: PROPKA [25] (http://propka.ki.ku.dk/) or H++ [26] (http://biophysics.cs.vt.edu/) for instance.

2.3 Electrostatic Potential Calculations

The electrostatic potential of the antigenic protein can be calculated by the finite-difference Poisson-Boltzmann (FDPB) method using programs such as APBS [27] (Adaptive Poisson Boltzmann Solver: http://www.poissonboltzmann.org/apbs) or PBEQ (Poisson Boltzmann Equation Solver: http://www.charmm-gui.org/?doc=input/pbeqsolver) or the Poisson-Boltzmann solvers implemented in several molecular modeling packages, like AMBER [28] (http://ambermd.org/) , CHARMM [29] (http://www.charmm.org/), NAMD [30] (http://www.ks.uiuc.edu/Research/namd/), or GROMACS [31] (http://www.gromacs.org/). Modifications of the approach employing electrostatic potential calculations based on the Generalized Born approach have also been developed [32].

3 Methods

3.1 System Setup

Before predicting antigenic epitopes at the surface of proteins based on electrostatic calculations, it might be necessary to first control the corresponding PDB files:

- Multiple conformations of a residue side chain must be eliminated because the calculations can only consider one side chain conformation for each residue. In such a case, the easiest solution is to retain only one of possible side chain conformation for the epitope mapping. If the residue is of critical importance for protein function (i.e., residue within the active site in the case of an enzyme), one may prepare several conformers of the protein and map desolvation properties for each of them.
- Missing atoms must be added before running the analysis. If only few atoms are missing, programs like PDB2PQR [33] (http://www.poissonboltzmann.org/pdb2pqr) or MMTSB [34] (http://mmtsb.org/) may be useful to add missing atoms. If a larger part of the structure is missing in the crystal structure, one can use homology or ab initio modeling software packages.
- As a next step, the protonation states of charged residues must be predicted. The issue can be important for Ag–Ab complexes since it is known that the propensity of charged or polar residues to be at the Ag–Ab interface is often higher than for other protein–protein complexes. PropKA is a very fast empirical method able to predict *p*Ka values of ionizable groups within a couple of seconds. Other software based on PB calculations (H++) may also help.
- Finally, atomic charges must be assigned to the protein atoms. Several force fields can be used and the most popular ones are related to well-known molecular modeling software packages: CHARMM, AMBER, GROMACS but some parameters were specifically designed to predict solvation properties of proteins (PARSE [35]). The Amber parm03 forcefield was used to assign atomic charges and radii in ref 19.

3.2 Parameters of the PB Equation Solver

- To obtain accurate electrostatic properties, the grid focusing technique is required during the FDPB calculations. In ref. 19, the coarse grid size equals twice the dimension of the finest grid. The latter encompassed the full protein (protein size plus 5 Å in XYZ dimension) and is centered on the center of mass of the protein.

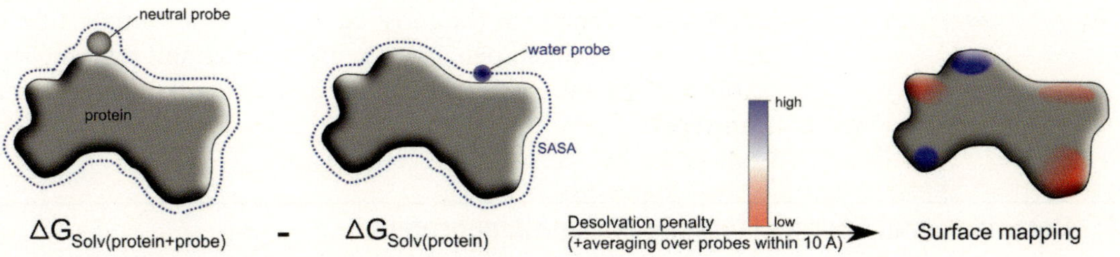

Fig. 1 Electrostatic desolvation mapping. Surface regions in *red* correspond to lowest electrostatic desolvation penalty (interpreted as high probability to be an antigenic epitope)

- A good compromise between speed of the calculation and accuracy of the electrostatic properties is to set the grid spacing of the final focused grid lower than 0.5 Å. 129 points in each space direction is generally sufficient for a small or medium sized system. In case of large proteins (>50 kDa) one may use a higher number of points to define the finest grid.
- The molecular surface is generated using a water probe with radius of 1.4 Å.
- A dielectric constant of 10 and 80 is used for protein and solvent, respectively. The choice for the dielectric constant of the protein is a compromise between estimates for the buried interior of proteins ($\varepsilon = 4$) and surface regions ($\varepsilon \sim 20$).

3.3 Desolvation Analysis

- The electrostatic desolvation free energy of a protein is calculated (once) and subtracted from the electrostatic energy of the protein with a neutral and low dielectric spherical probe ($\varepsilon = 10$) placed at many different positions at the protein surface. This calculation gives the electrostatic energy (penalty) of placing the neutral low dielectric probe at the particular surface position (*see* Fig. 1).
- The calculations can be performed systematically for various surface positions of the probe distributed approximately evenly at a distance of 3 Å from each other.
- The electrostatic desolvation penalty of a surface patch is then estimated as the average desolvation of all probes within a distance cutoff of 10 Å to a given surface point. The normalized desolvation penalty can then be mapped onto the surface defined by the probe adding the score in the bfactor column of a pdb file. The averaging procedure reduces the grid errors inherent to FDPB calculations. At the same time it has the advantage to include the effect of the shape of the surface on the desolvation free energy compared to using one probe with a larger radius [19].
- For the prediction of possible protein binding sites (the epitope in the case of an antigen) probe positions with the lowest (average) desolvation penalty must be considered (Cf. Subheading 5). Typically, one considers the patch with lowest electrostatic

desolvation penalty as the most likely antigenic epitope (it has the lowest associated penalty to remove water and replace it by a protein partner represented by a probe of low dielectric constant).

4 Computational Epitope Mapping of OspA Lipoprotein

The present protocol has previously been successfully tested on a set of 156 proteins in their bound and unbound conformations including a series of 27 Ag–Ab complexes with known structures [19]. The accuracy of the prediction was assessed using ROC curves (receiver operator curves) and the area under the curve reaches ~0.6. It needs to be emphasized that the approach can only predict a tendency of a surface region to be part of an antibody recognition region which may limit the prediction accuracy. However, by comparison with crystal structures of the bacterial protein lysozyme cocrystallized with different antibodies it could be demonstrated that the protocol is also able to predict multiple antigenic epitopes associated with a low electrostatic desolvation penalty [19].

We illustrate here on a test case, the outer surface protein A lipoprotein (OspA) of *Borrelia burgdorferi*, the results and performance which can be typically obtained using the computational epitope mapping based on the ESTADE approach. The OspA lipoprotein of the lyme disease causing *B. burgdorferi* is an important target and several recombinant monoclonal antibodies that bind to OspA have already been developed. The calculation of electrostatic desolvation penalties on the protein surface indicates several regions with a low associated desolvation penalty (Fig. 2). Surfaces in red correspond to low desolvation regions and they correlate well with the location of the Ag–Ab interface extracted from the 3D structure of OspA in complex with a monoclonal antibody (pdb 1FJ1). Among the five predicted low desolvation sites, three are located at the known Ag–Ab interface. For these putative epitopes, the accuracy of correctly predicted residues is roughly 80 %.

5 Notes

5.1 Implementation of the ESTADE Method in the Program Package Ptools

The ESTADE method described in ref. 19 has been implemented in the open source protein–protein docking program Ptools/ATTRACT [36, 37] (http://www.unice.fr/icn/fiorucci). Before running the analysis, the user should take care of the following recommendations: A "clean" PDB file must be provided (after removing multiple conformations of side chains and adding missing atoms as described above). Hydrogens must be removed from the original PDB file (the program PDB2PQR will add them automatically). The main script uses the programming language Python and several PTools script, so the library dependencies must be checked.

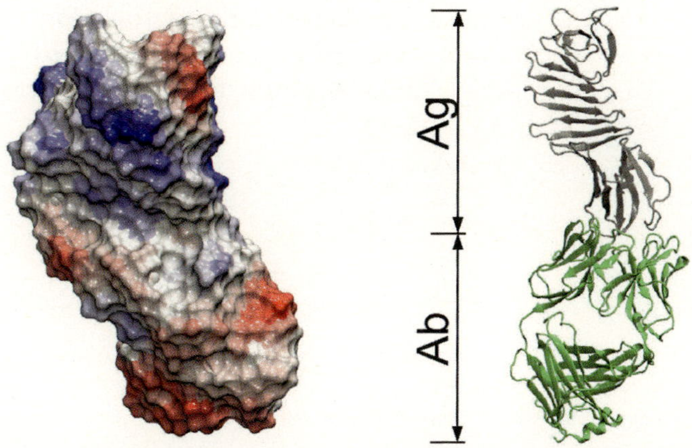

Fig. 2 Antigen–antibody complex structure (*right*, pdb1FJ1). Electrostatic desolvation mapping of OspA lipoprotein (*left*) to highlight the correlation between the epitope region and the predicted low desolvation surface regions (*red*). The surface desolvation mapping is shown in the same view as the Ag structure (*grey cartoon*) in the *right panel*

The script may also need extra Python programs/commands, not provided with PTools. For file preparation the following steps are implemented in the script:

- To generate pdb and pqr input files, the AMBERTOOLS and PDB2PQR tools are required.
- To calculate electrostatic properties and solve the FDPB equation, the APBS package has been chosen but alternative programs can also be used (*see* above). The AMBERTOOLS, PDB2PQR, and APBS are freely available software (current versions: Ambertools 13, pdb2pqr 1.8 and APBS 1.3).
- Environment variables ($AMBERHOME) must be set before running the script. PDB2PQR and APBS binaries must be in your PATH.

For instance, to run the electrostatic desolvation analysis on the protein coordinates stored in the file *1FJ1_l.pdb* the following main script can be used (part of the Ptools package):

$ e-static_profile.csh 1FJ1_l.pdb

5.2 Output Files

The program automatically generates a directory called *1FJ1_l* and stores all results in a series of files in this directory.

- The script produces pdb files called *1FJ1_l_Desolv.pdb* and *1FJ1_l_Desolv_av.pdb* which contain the desolvation energy of each probe position and the normalized desolvation energy, respectively.
- The desolvation energy per residue can also be obtained: The file *1FJ1_l_Desolv_res.txt* contains the residue name, residue

number and the desolvation energy per residue (NA means that the residue is not accessible to solvent), the first few rows of an example file are given below:

```
SER    1   6.51
LEU    2   5.47
ASP    3   5.06
GLU    4   6.52
LYS    5   5.66
ASN    6   4.31
SER    7     NA
VAL    8   3.40
SER    9   6.51
VAL   10   4.71
...
```

- The script produces also a pdb file called *1FJ1_l_Desolv_res.pdb* which stores in the bfactor column weights according to the binding site prediction: a weight of 2.0 means a high probability to be at the protein–protein interface and a weight of 1.0 means a low probability.
- A list of residues belonging to putative binding sites is also proposed in the file *1FJ1_l_Desolv_site.txt.* If two binding site are too close to each other (distance <10 Å), only the one with the lowest desolvation penalty will be retained. The output is also explained in the header of each script contained in the source directory. For visual inspection of the prediction, the file *1FJ1_l_Desolv_res.pdb* is most useful. The protein surface can be represented with a color code given by the B-factor that is a measure of the electrostatic desolvation penalty of the corresponding surface region.

Acknowledgements

The support by a grant (Multicomponent docking) from the Egide/DAAD (Deutscher Akademischer Austauschdienst) is gratefully acknowledged.

References

1. Hopp TP, Woods KR (1981) Prediction of protein antigenic determinants from amino acid sequences. Proc Natl Acad Sci U S A 78(6): 3824–3828
2. Parker JM, Guo D, Hodges RS (1986) New hydrophilicity scale derived from high-performance liquid chromatography peptide retention data: correlation of predicted surface residues with antigenicity and X-ray-derived accessible sites. Biochemistry 25(19):5425–5432
3. Emini EA, Hughes JV, Perlow DS, Boger J (1985) Induction of hepatitis A virus-neutralizing antibody by a virus-specific synthetic peptide. J Virol 55(3):836–839

4. El-Manzalawy Y, Dobbs D, Honavar V (2008) Predicting linear B-cell epitopes using string kernels. J Mol Recognit 21(4):243–255
5. Chen J, Liu H, Yang J, Chou KC (2007) Prediction of linear B-cell epitopes using amino acid pair antigenicity scale. Amino Acids 33(3): 423–428
6. Saha S, Raghava GP (2006) Prediction of continuous B-cell epitopes in an antigen using recurrent neural network. Proteins 65(1):40–48
7. Larsen JE, Lund O, Nielsen M (2006) Improved method for predicting linear B-cell epitopes. Immunome Res 2:2
8. Blythe MJ, Flower DR (2005) Benchmarking B cell epitope prediction: underperformance of existing methods. Protein Sci 14(1): 246–248
9. Greenbaum JA, Andersen PH, Blythe M, Bui HH, Cachau RE, Crowe J, Davies M, Kolaskar AS, Lund O, Morrison S, Mumey B, Ofran Y, Pellequer JL, Pinilla C, Ponomarenko JV, Raghava GP, van Regenmortel MH, Roggen EL, Sette A, Schlessinger A, Sollner J, Zand M, Peters B (2007) Towards a consensus on datasets and evaluation metrics for developing B-cell epitope prediction tools. J Mol Recognit 20(2):75–82
10. Zhang W, Xiong Y, Zhao M, Zou H, Ye X, Liu J (2011) Prediction of conformational B-cell epitopes from 3D structures by random forests with a distance-based feature. BMC Bioinformatics 12:341
11. Liang S, Zheng D, Standley DM, Yao B, Zacharias M, Zhang C (2010) EPSVR and EPMeta: prediction of antigenic epitopes using support vector regression and multiple server results. BMC Bioinformatics 11:381
12. Liang S, Zheng D, Zhang C, Zacharias M (2009) Prediction of antigenic epitopes on protein surfaces by consensus scoring. BMC Bioinformatics 10:302
13. Rubinstein ND, Mayrose I, Pupko T (2009) A machine-learning approach for predicting B-cell epitopes. Mol Immunol 46(5):840–847
14. Sun J, Wu D, Xu T, Wang X, Xu X, Tao L, Li YX, Cao ZW (2009) SEPPA: a computational server for spatial epitope prediction of protein antigens. Nucleic Acids Res 37(Web Server issue):W612–W616
15. Ponomarenko J, Bui HH, Li W, Fusseder N, Bourne PE, Sette A, Peters B (2008) ElliPro: a new structure-based tool for the prediction of antibody epitopes. BMC Bioinformatics 9:514
16. Sweredoski MJ, Baldi P (2008) PEPITO: improved discontinuous B-cell epitope prediction using multiple distance thresholds and half sphere exposure. Bioinformatics 24(12): 1459–1460
17. Haste Andersen P, Nielsen M, Lund O (2006) Prediction of residues in discontinuous B-cell epitopes using protein 3D structures. Protein Sci 15(11):2558–2567
18. Yao B, Zheng D, Liang S, Zhang C (2013) Conformational B-cell epitope prediction on antigen protein structures: a review of current algorithms and comparison with common binding site prediction methods. PLoS One 8(4):e62249
19. Fiorucci S, Zacharias M (2010) Prediction of protein–protein interaction sites using electrostatic desolvation profiles. Biophys J 98(9): 1921–1930
20. Soriani M, Petit P, Grifantini R, Petracca R, Gancitano G, Frigimelica E, Nardelli F, Garcia C, Spinelli S, Scarabelli G, Fiorucci S, Affentranger R, Ferrer-Navarro M, Zacharias M, Colombo G, Vuillard L, Daura X, Grandi G (2010) Exploiting antigenic diversity for vaccine design: the chlamydia ArtJ paradigm. J Biol Chem 285(39):30126–30138
21. Eswar N, Webb B, Marti-Renom MA, Madhusudhan MS, Eramian D, Shen MY, Pieper U, Sali A (2006) Comparative protein structure modeling using modeller. Curr Protoc Bioinformatics 15:5.6.1–5.6.30
22. Kiefer F, Arnold K, Kunzli M, Bordoli L, Schwede T (2009) The SWISS-MODEL repository and associated resources. Nucleic Acids Res 37:D387–D392
23. Raman S, Vernon R, Thompson J, Tyka M, Sadreyev R, Pei J, Kim D, Kellogg E, DiMaio F, Lange O, Kinch L, Sheffler W, Kim BH, Das R, Grishin NV, Baker D (2009) Structure prediction for CASP8 with all-atom refinement using Rosetta. Proteins 77(Suppl 9):89–99
24. Roy A, Kucukural A, Zhang Y (2010) I-TASSER: a unified platform for automated protein structure and function prediction. Nat Protoc 5(4):725–738
25. Rostkowski M, Olsson MH, Sondergaard CR, Jensen JH (2011) Graphical analysis of pH-dependent properties of proteins predicted using PROPKA. BMC Struct Biol 11:6
26. Anandakrishnan R, Aguilar B, Onufriev AV (2012) H++ 3.0: automating pK prediction and the preparation of biomolecular structures for atomistic molecular modeling and simulations. Nucleic Acids Res 40(Web Server issue): W537–W541
27. Baker NA, Sept D, Joseph S, Holst MJ, McCammon JA (2001) Electrostatics of nanosystems: application to microtubules and the ribosome. Proc Natl Acad Sci U S A 98(18): 10037–10041
28. Case DA, Darden TA, Cheatham TE III, Simmerling CL, Wang J, Duke RE, Luo R, Walker RC, Zhang W, Merz KM, Roberts B,

Hayik S, Roitberg A, Seabra G, Swails J, Goetz AW, Kolossváry I, Wong KF, Paesani F, Vanicek J, Wolf RM, Liu J, Wu X, Brozell SR, Steinbrecher T, Gohlke H, Cai Q, Ye X, Wang J, Hsieh M-J, Cui G, Roe DR, Mathews DH, Seetin MG, Salomon-Ferrer R, Sagui C, Babin V, Luchko T, Gusarov S, Kovalenko A, Kollman PA (2012) AMBER 12. University of California, San Francisco, CA

29. Brooks BR, Brooks CL 3rd, Mackerell AD Jr, Nilsson L, Petrella RJ, Roux B, Won Y, Archontis G, Bartels C, Boresch S, Caflisch A, Caves L, Cui Q, Dinner AR, Feig M, Fischer S, Gao J, Hodoscek M, Im W, Kuczera K, Lazaridis T, Ma J, Ovchinnikov V, Paci E, Pastor RW, Post CB, Pu JZ, Schaefer M, Tidor B, Venable RM, Woodcock HL, Wu X, Yang W, York DM, Karplus M (2009) CHARMM: the biomolecular simulation program. J Comput Chem 30(10):1545–1614
30. Phillips JC, Braun R, Wang W, Gumbart J, Tajkhorshid E, Villa E, Chipot C, Skeel RD, Kale L, Schulten K (2005) Scalable molecular dynamics with NAMD. J Comput Chem 26(16):1781–1802
31. Pronk S, Pall S, Schulz R, Larsson P, Bjelkmar P, Apostolov R, Shirts MR, Smith JC, Kasson PM, van der Spoel D, Hess B, Lindahl E (2013) GROMACS 4.5: a high-throughput and highly parallel open source molecular simulation toolkit. Bioinformatics 29(7):845–854
32. Schneider S, Zacharias M (2012) Combining geometric pocket detection and desolvation properties to detect putative ligand binding sites on proteins. J Struct Biol 180(3): 546–550
33. Dolinsky TJ, Czodrowski P, Li H, Nielsen JE, Jensen JH, Klebe G, Baker NA (2007) PDB2PQR: expanding and upgrading automated preparation of biomolecular structures for molecular simulations. Nucleic Acids Res 35(Web Server issue): W522–W525
34. Feig M, Karanicolas J, Brooks CL 3rd (2004) MMTSB tool set: enhanced sampling and multiscale modeling methods for applications in structural biology. J Mol Graph Model 22(5): 377–395
35. Sitkoff D, Sharp KA, Honig B (1998) Accurate calculation of hydration free energies using macroscopic solvent models. J Phys Chem 98(7):1978–1988
36. Schneider S, Saladin A, Fiorucci S, Prévost C, Zacharias M (2012) ATTRACT and PTools: open sources programs for protein-protein docking. In: Baron R (ed) Computational drug discovery and design. Methods in molecular biology, vol 819. Springer, Heidelberg, pp 221–232
37. Saladin A, Fiorucci S, Poulain P, Prevost C, Zacharias M (2009) PTools: an opensource molecular docking library. BMC Struct Biol 9:27

Chapter 21

In Silico Prediction of Allergenic Proteins

Gaurab Sircar, Bodhisattwa Saha, Swati Gupta Bhattacharya, and Sudipto Saha

Abstract

Currently, the prediction of new allergens is becoming important due to use of genetically modified (GM) foods and biopharmaceuticals. In this chapter, we describe how to use four popular allergenic prediction servers: (1) Structural Database of Allergenic Proteins (SDAP), (2) Allermatch, (3) Evaller 2, and (4) AlgPred. The first two prediction servers are based on traditional approaches, whereas Evaller 2 and AlgPred use sophisticated machine learning techniques.

Key words Codex bipartite test, IgE epitope, Sequence alignment, Support vector machine

1 Introduction

Here, we have described popular Web based tools, which allows users to predict the allergenicity of novel proteins. A number of different immunochemical, biochemical, and immunological methods have emerged and evolved over time, to predict the allergenicity and cross IgE reactivity of proteins causing hypersensitivity. A sensitized individual may respond similarly to proteins that share certain common structural and molecular features with the protein that elicited the initial immune reaction. This phenomenon is designated as cross reactivity and is tightly connected particularly to IgE epitopes which can be either linear or conformational. Pollen–fruit, latex–fruit are some common type of cross reactivity caused by promiscuous IgE epitope recognition due to protein structural similarity. The prediction tools are becoming important to assess the safety of GM crops, therapeutics, and biopharmaceuticals and also for the prediction of aeroallergens [1]. In the year 2001, World Health Organization (WHO) and the Food and Agriculture Organization (FAO) proposed guidelines to assess the potential allergenicity of proteins, in which partly similar bioinformatics testing is a mandatory introductive step. The bioinformatics part of the guidelines says that a protein is potentially allergenic

Rajat K. De and Namrata Tomar (eds.), *Immunoinformatics*, Methods in Molecular Biology, vol. 1184,
DOI 10.1007/978-1-4939-1115-8_21,

if it either shows a match of six consecutive amino acids or an identity of >35 %, across an 80 amino acid window. Subsequently, in 2003 the Codex Alimentarius Commission [2, 3] recognized some uncertainties in these tests and suggested weight of evidence approach that includes source of gene, sequence similarities with known allergens, stability of protein allergenicity, and IgE bindings. The relationships between amino acid sequence similarity of query proteins to known allergens and their type-I hypersensitivity potential have impels the development of bioinformatics tools for allergic risk assessment. The bioinformatics approaches for protein allergenicity assessment can be divided into following categories: (a) alignment based on Codex bipartite test; (b) alignment-based feature-extraction combined with statistical learning; (c) homology with allergen-derived motifs or reported IgE epitopes; (d) machine learning based approaches. Popular Web based tools that allow users to predict allergens using query amino acid sequences are SDAP [4], Allermatch [5], Evaller [6], and AlgPred [7]. In recent years, bioinformatics approaches have emerged as a relatively reliable and fast method to predict the allergenicity of new proteins.

2 Materials and Methods

2.1 Description of Structural Database of Allergenic Proteins (SDAP)

The SDAP is available from https://fermi.utmb.edu/SDAP/ and provides prediction tools along with database information of 1,526 allergens [4]. The menus are in the left side of the page. Following are the brief description of SDAP tools:

2.1.1 Description of SDAP

FAO/WHO Allergenicity Test

This page allows FAO/WHO allergenicity rules based on sequence homology as proposed in FAO/WHO report. It allows three types of tests: (1) perform an exact match search of contiguous amino acids; (2) perform FASTA alignments for 80 amino acids sliding window; and (3) perform full FASTA alignment.

FASTA Search in SDAP

This link allows users to find sequence similarity between query protein and all allergens from SDAP by using FASTA search.

Peptide Match

It allows users to search exact match of peptides in allergen sequences.

Peptide Similarity

It allows users to search property based peptide similarity search in allergen sequences.

Peptide-Protein PD Index

It allows users to search property based peptide similarity index property distance (PD) for two sequences.

Aller_ML_allergen Markup Language

This page describes about allergen markup language, AllerML. It is a tool to access data on allergens in multiple databases. AllerML is based on IUIS nomenclature and consists of a hierarchical set of tags that describes the information available in allergen databases including common names, sources, sequences, structures, IgE epitopes and cross-reactivity.

List SDAP

This page links to various lists available to download all allergens available in SDAP, including allergens with PDB structures, allergens with 3D models, and allergens with epitopes.

2.1.2 Usage of SDAP

FAO/WHO Allergenicity Rules Based on Sequence Homology

(a) Enter the name of the sequence (optional).

(b) Users can paste or type the query the protein sequence. The sequence must be written using one-letter amino acid code (*see* **Note 1**).

(c) Users can choose one out of three options: (1) Full FASTA alignment, in which default parameter for E-values is set at less than 0.01; (2) FASTA alignments for an 80 amino acids sliding window, in which sequence identity default cutoff is set at 35; (3) Exact match for contiguous amino acids, default number is set at 6.

2.1.3 FASTA Similarity Search in the SDAP Database

(a) Enter the name of the sequence (optional) as shown in Fig. 1a.

(b) Users can paste or type the protein sequence. The sequence must be written using one-letter amino acid code, and the maximum length of the sequence should be 1,000 (*see* **Note 2**).

Exact Match of Peptides in Allergen Sequences

(a) Users can select a sequence database out of two databases: SDAP allergens or SwissProt.

(b) Users can paste or type the query peptide sequence as a string of single-letter amino acid codes. The maximum length of the query peptide sequence is 30.

(c) Users can select the number of similar sequences in the output results. The default set for this, is and the maximum number for similar sequences in the output results is 100.

Property Based Peptide Similarity Search in Allergen Sequences

(a) Users can paste or type the query peptide sequence as a string of single-letter amino codes. The maximum length of the sequence is 30.

(b) Users can select the number of similar sequences in the output results, and in this case the default is set at 50, and the maximum number is 100.

2.1.4 Property Based Similarity Index Property Distance (PD) for Two Sequences

(a) Users can paste or type the first protein sequence in single-letter amino acids code. The maximum length of the sequence is 1,000. The first sequence should be shorter or equal to the second sequence.

(b) Secondly, users need to paste or type the second query protein sequence for computing the PD sequence similarity index.

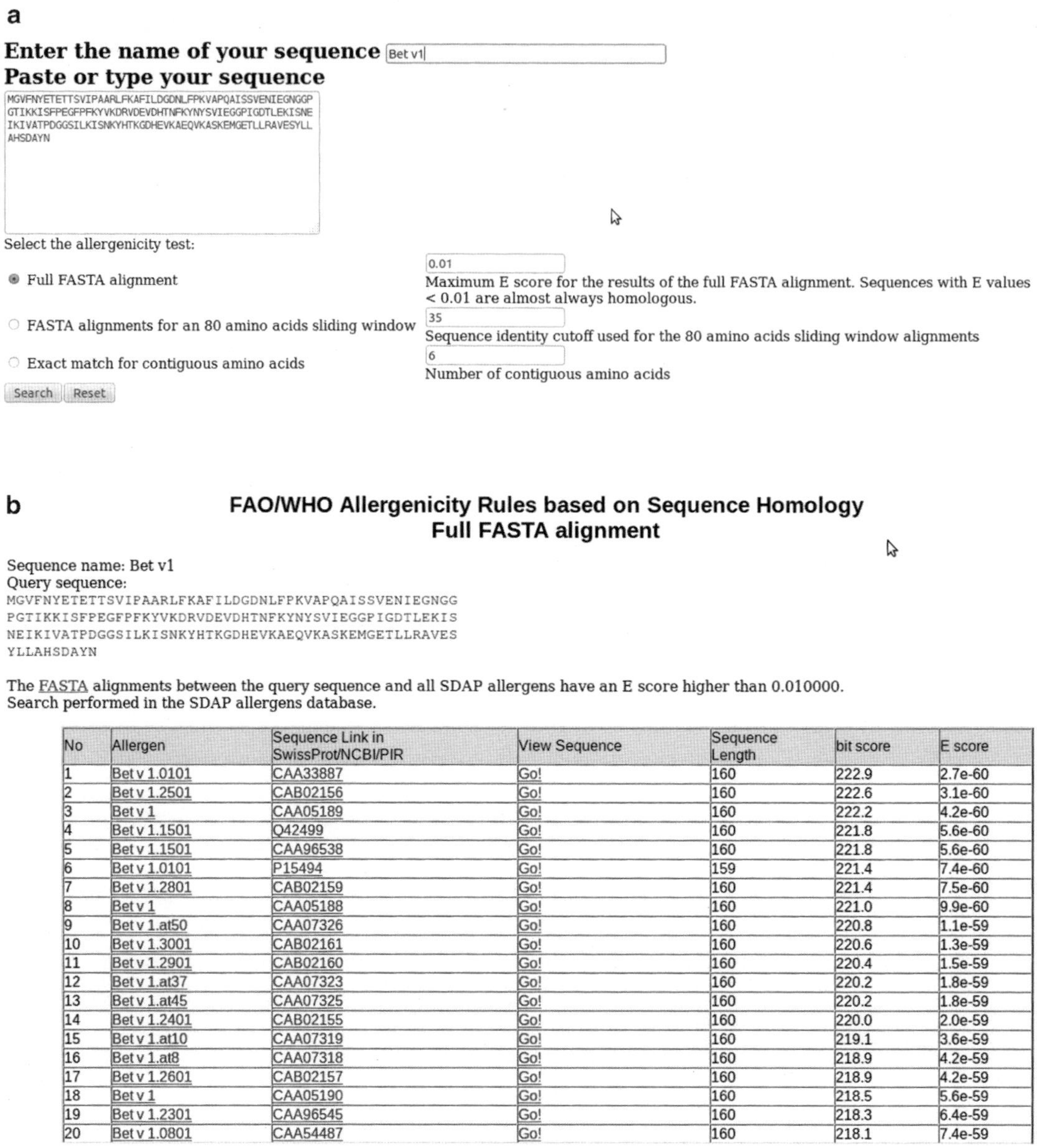

No	Allergen	Sequence Link in SwissProt/NCBI/PIR	View Sequence	Sequence Length	bit score	E score
1	Bet v 1.0101	CAA33887	Go!	160	222.9	2.7e-60
2	Bet v 1.2501	CAB02156	Go!	160	222.6	3.1e-60
3	Bet v 1	CAA05189	Go!	160	222.2	4.2e-60
4	Bet v 1.1501	Q42499	Go!	160	221.8	5.6e-60
5	Bet v 1.1501	CAA96538	Go!	160	221.8	5.6e-60
6	Bet v 1.0101	P15494	Go!	159	221.4	7.4e-60
7	Bet v 1.2801	CAB02159	Go!	160	221.4	7.5e-60
8	Bet v 1	CAA05188	Go!	160	221.0	9.9e-60
9	Bet v 1.at50	CAA07326	Go!	160	220.8	1.1e-59
10	Bet v 1.3001	CAB02161	Go!	160	220.6	1.3e-59
11	Bet v 1.2901	CAB02160	Go!	160	220.4	1.5e-59
12	Bet v 1.at37	CAA07323	Go!	160	220.2	1.8e-59
13	Bet v 1.at45	CAA07325	Go!	160	220.2	1.8e-59
14	Bet v 1.2401	CAB02155	Go!	160	220.0	2.0e-59
15	Bet v 1.at10	CAA07319	Go!	160	219.1	3.6e-59
16	Bet v 1.at8	CAA07318	Go!	160	218.9	4.2e-59
17	Bet v 1.2601	CAB02157	Go!	160	218.9	4.2e-59
18	Bet v 1	CAA05190	Go!	160	218.5	5.6e-59
19	Bet v 1.2301	CAA96545	Go!	160	218.3	6.4e-59
20	Bet v 1.0801	CAA54487	Go!	160	218.1	7.4e-59

Fig. 1 Screenshots of SDAP. (**a**) Query search of FAO/WHO allergenicity rules; (**b**) Output result of FAO/WHO allergenicity rules based on full FASTA alignment search

2.1.5 SDAP Query Result

(a) The output of FAO/WHO allergenicity rules based on sequence homology full FASTA alignment results in a tabular format as shown in Fig. 1b. The column names of the table are ordered based on best hits and its importance. The column names are (1) Allergen name, which on selecting gives total information on the allergen including its source, structure; (2) Sequence accession number and it links to SwissProt/NCBI/PIR databases (3) View Sequence; (4) Sequence length; (5) Bit score in descending order; (6) E score in ascending order.

(b) The output result of FASTA search in SDAP database is similar to FAO/WHO full FASTA alignment described above.

(c) The output of peptide match result shows mapping of a query peptide sequence with matched allergen sequences in SDAP database.

(d) The output of peptide similarity results in a tabular format with potential allergens hits are in sequential order based on PD sequence similarity index. The column names of the table are: (1) Allergen names; (2) Link to NCBI/PIR/Swiss Prot; (3) Property Distance (PD) sequence similarity index; (4) z (PD, min); (5) z (PD, all); (6) start residue; (7) matching region; and (8) end residue.

(e) The output result of Peptide-protein PD index provides a list of all the matched sequences and its PD values. At the bottom, it provides best matches with minimum PD score.

2.2 Allermatch

2.2.1 Description of Allermatch

Allermatch is available at www.allermatch.org/allermatch/. This Web tool allows users to predict allergenicity of proteins by bioinformatics approaches as recommended by the Codex Alimentarius and FAO/WHO [5]. The menus are in the left side and are interlinked. Following are the brief description of menus:

Home

It links to home page of Allermatch and users can directly go to search page from home page.

Search

This page links to input form, from where the users can make query search. More details about its usage are presented in Subheading 2.2.2.

Databases

This page links to the list of sequences of known allergenic proteins that have been hosted by UniProt Protein Knowledgebase and WHO-IUIS.

Publication

It links to publication references of Allermatch.

Introduction

This page links to brief introduction of bioinformatics approaches used for prediction of allergenic proteins.

Example

This page links to usage of Allermatch search page with query sequence and output result examples.

About Us

This page links to team members and contact information.

Feedback

This page provides e-mail link of contact person for feedback.

Disclaimer

It links to disclaimer information page.

Copyright

It links to copyright information.

Thanks

It links to acknowledgement page.

References

It links to references page.

2.2.2 Usage of Allermatch

(a) Users can go to search menu or directly go to search page from home page. The input form is shown in Fig. 2a (*see* **Note 1**). There are three options: (1) 80 amino acids sliding window alignment; (2) exact hit of 6 amino acids; (3) full FASTA alignment.

(b) For 80 amino acids sliding window search, the default cut off percentage value is set at 35 (*see* **Note 3**).

(c) For exact match search, the default wordmatch value is set at 6.

(d) Users can choose a database out of three databases: (1) UniProt and WHO-IUIS; (2) UniProt; and (3) WHO-IUIS. The default database set is Uniprot and WHO-IUIS.

2.2.3 Allermatch Query Result

Output Result of 80 Amino Acids Sliding Window Alignment

The search result is summarized into ten columns as shown in Fig. 2b. Each hit specific to allergenic protein is presented in a line, with the following information: (1) "Hit No." in ascending order; (2) "Db," database from which the allergen sequence has been retrieved; (3) "Allergen id," the Allermatch™ identifier for the allergenic protein; (4) "Best hit" (identity), ranked in descending order; (5) "No. of hits identity >35," the number of 80-amino acids subsequences (windows) of the query sequence that showed hits above the cut-off value; (6) "% of hits identity >35," the fraction (percentage, %) of the total number of analyzed subsequences (windows) of the input sequence that showed hits above the cutoff value with the allergenic protein; (7) "Full identity," percentage of identical amino acids in the FASTA alignment against the complete input sequence; (8) "External link," links to external protein databases; (9) "Species name"; (10) "Detailed information," links to more information about the allergenic protein hit as shown in Fig. 2c.

Output Result of Wordmatch

The output result is similar to 80 amino acids sliding window approach. The query result with at least one hit is listed and users are allowed to retrieve more detailed information from the link.

Output Result of Full FASTA Alignment

The output results provide a list of allergens showing significant homology with the query protein and the top hit with lowest E-value. The FASTA alignment of query and matched sequence are also shown in the result.

2.3 Evaller 2

2.3.1 Description of the Evaller

Evaller 2 is a Web tool which allows users to predict allergens based on its amino acid sequence [6]. It is available from http://www.slv.se/en-gb/Group1/Food-Safety/e-Testing-of-protein-allergenicity/. This Web tool is based on the core algorithm named 'Detection based on Filtered Length-adjusted Allergen Peptides' (DFLAP) and uses support vector machine (SVM) for training the FLAPs and testing the query protein sequence. The main menus are in the left side and are interlinked. In addition, there are more menus about Evaller on the right side as well. At the bottom of the page, there are some links including contact details, information about developers, and previous reports. Following are the brief description of the main menus:

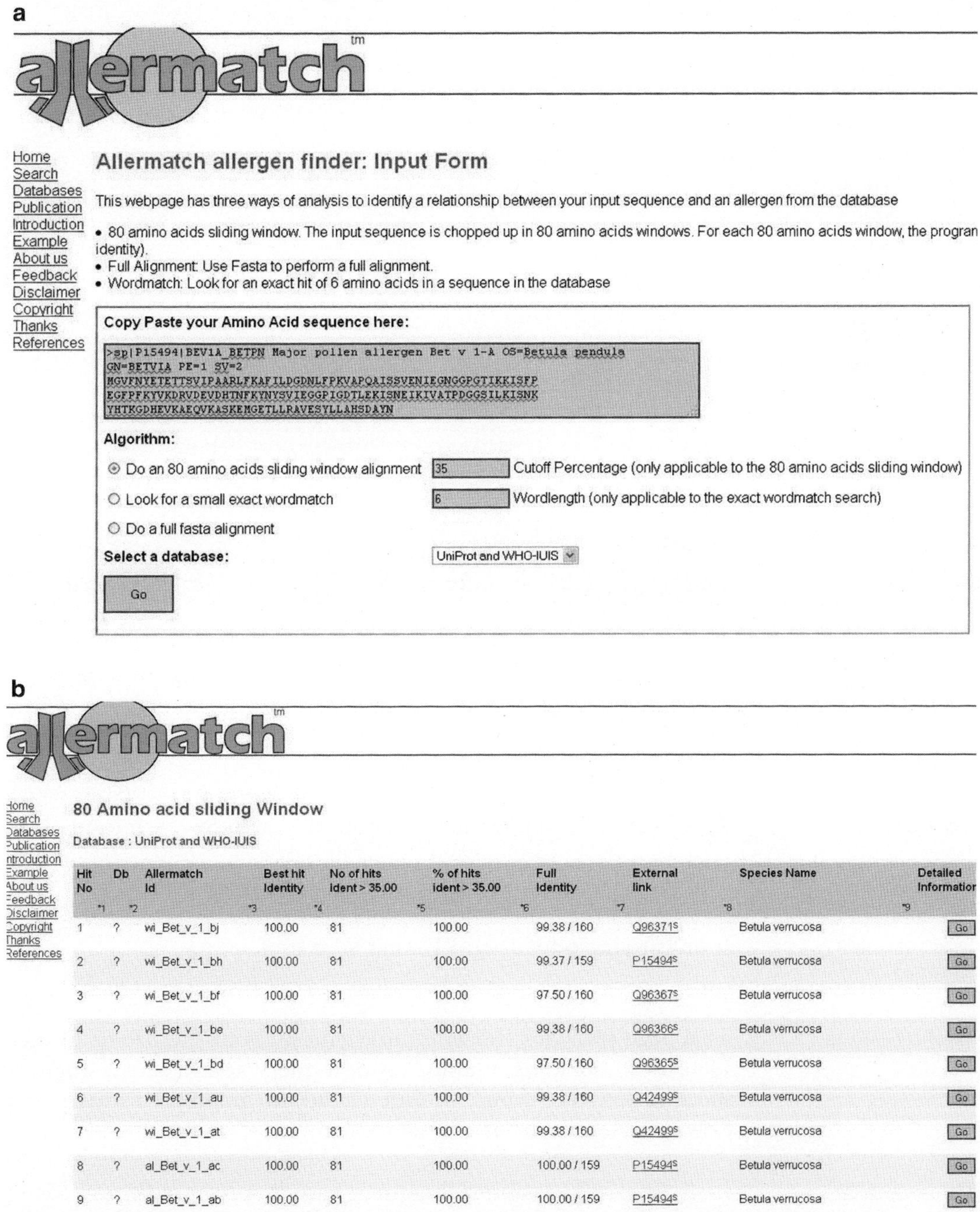

Fig. 2 Screenshots of Allermatch. (**a**) Query search page with Bet v 1 amino acid sequence; (**b**) Output result of 80 amino acids sliding window against Bet v 1. (**c**) Results of 80 amino acid sliding window against wi_Bet_v_1_bj

c

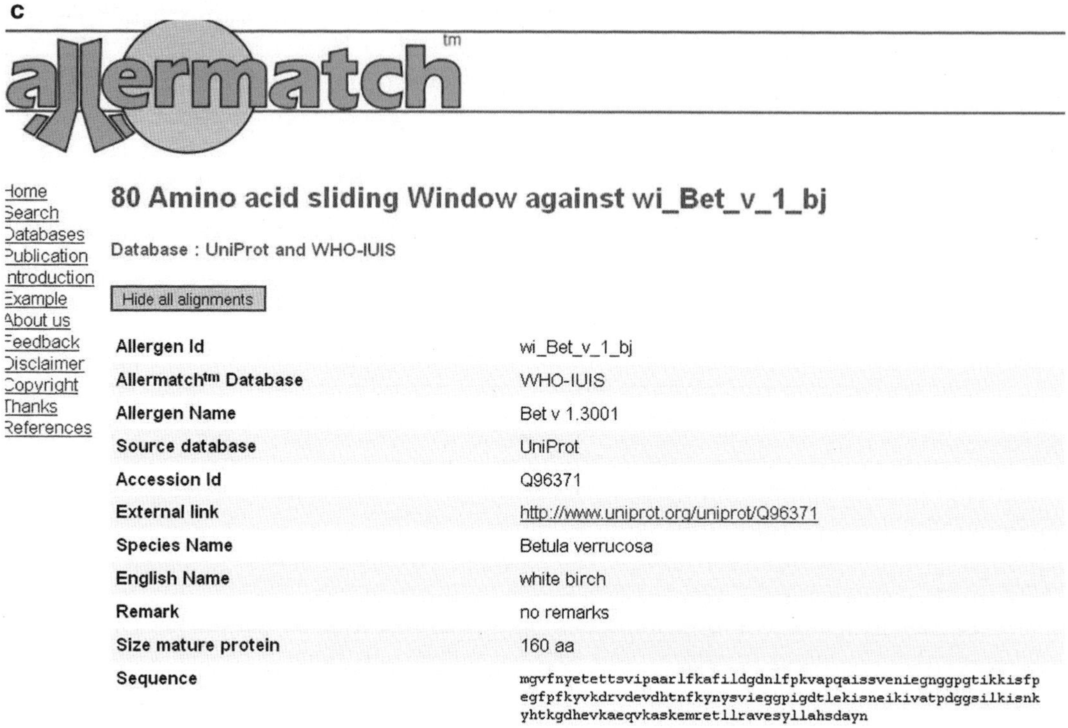

Fig. 2 (continued)

Food Safety	This page links to information about food safety specific to allergenicity.
Acrylamide	This page links to information about toxicological effects of acrylamide.
Allergens	This page links to lists food allergens.
Dioxin in Swedish food	This page links to information about dioxin in Swedish food.
Energy Drinks	This page links to comments on usage of energy drinks.
e-Testing of Protein Allergenicity	It links to introduction of Evaller Web tool.
How EVALLER Works	This page provides stepwise information about the use of Evaller Web tool, including submission of a query amino acid sequence, specification of output, output results in graphical format, and downloading results as text file.
e-Test Allergenicity	It allows users to submit the query sequence for prediction. More details about its usage are available in Subheading 2.3.2.

Heavy Metals and Minerals in Foods for Children

This page provides facts about the usage of heavy metals and minerals in foods for children.

List of Plants and Plant Parts Unsuitable for Use in Food

It links to a reference about plants and plant parts that are unsuitable for use in food.

Mushrooms and Mushroom Toxins

This page links to some advice to reduce the risk of being poisoned by mushrooms.

Pine Nuts with a Strange Metallic Taste

This page links to facts about pine nuts.

2.3.2 Usage of EVALLER

1. Users can query the protein sequence from the left menu, "e-Test allergenicity" as shown in Fig. 3a (*see* **Note 1**). The Web tool allows users to copy or paste the query sequence in FASTA format and the minimum sequence length allowed is 40. In addition, there is an option to browse the input sequence from local computer.
2. The user can specify the number of best matching FLAPs within a range from 1 to 7 from a drop down menu and the default value set is 5 (*see* **Note 3**).

2.3.3 EVALLER Query Result

1. In the graphical output result, the query amino acid appears as a bar either red (presumably an allergen) or green (presumably not an allergen). Below the bar, the query matching FLAPs come into view as short bars in descending darkness of grey, where the darkest bar represents the highest scoring FLAP as shown in Fig. 3b.
2. It also allows users to download results as a text file, by clicking on the clickable links.
3. In the textual assignment of the output result, the query sequence is represented as being either of the two assessment categories "Presumably an allergen" or "Presumably not an allergen."

2.4 AlgPred

2.4.1 Description of AlgPred

AlgPred is available at http://www.imtech.res.in/raghava/algpred/. It uses different approaches to predict allergens including mapping of IgE epitopes, motif based modeling search, classifiers based on support vector machine (SVM) and blast search on allergen representative proteins (ARPs) [7]. The menus are in the left side and are interlinked. Following are the brief description of menus:

Home

It links to the home page of AlgPred. It describes about salient features of the Web server and publication reference.

Help	This page allows users to query a protein sequence. Its detail usage is available at Subheading 2.4.2.
Submission	This page provides information about the usage of AlgPred server.
Algorithm	It links to information about datasets and different methods/algorithm used in developing AlgPred. It also provides results (sensitivity, specificity, and accuracy) on various thresholds of SVM based classifications, so that users can change the default parameters set on the submission accordingly (*see* **Note 3**).
Supplementary	This page provides supplemental data including dataset, performance of SVM modules, and hybrid approaches.
Related Links	It links to other related information about allergens.

a

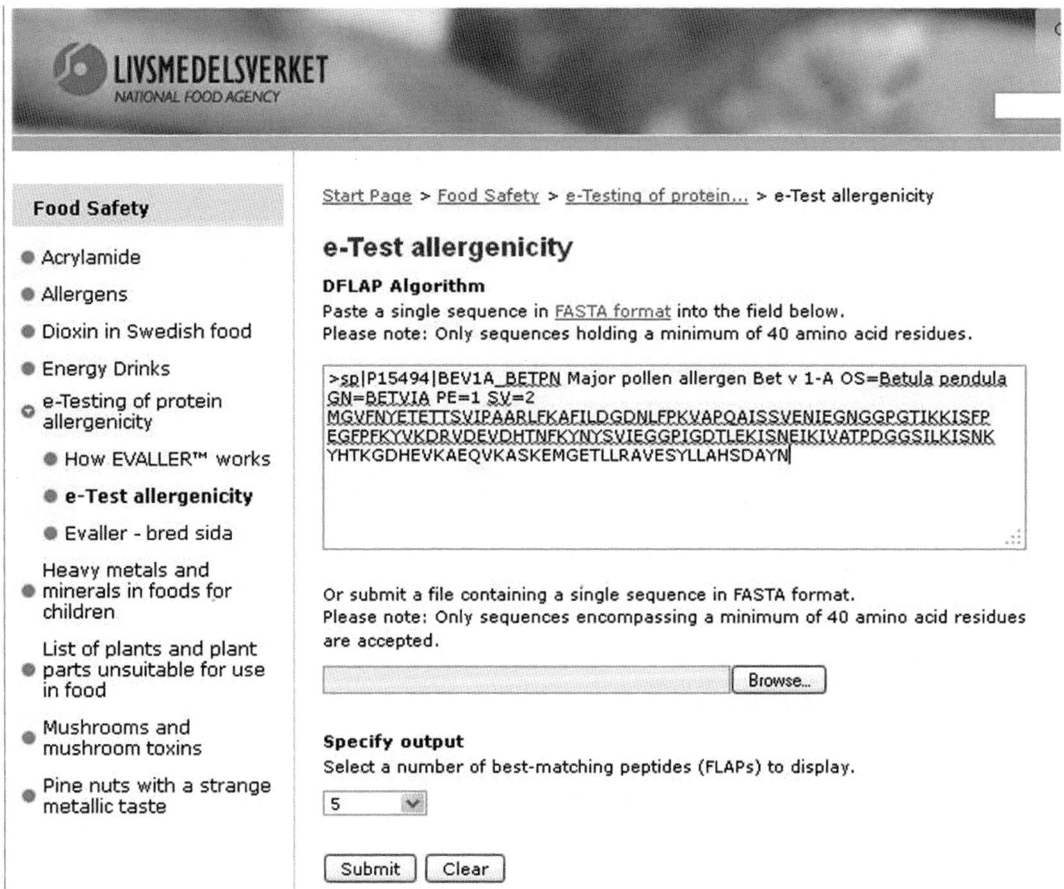

Fig. 3 Screenshots of Evaller. (**a**) Query search page with Bet v 1 amino acid sequence. (**b**) Output results of Evaller

b

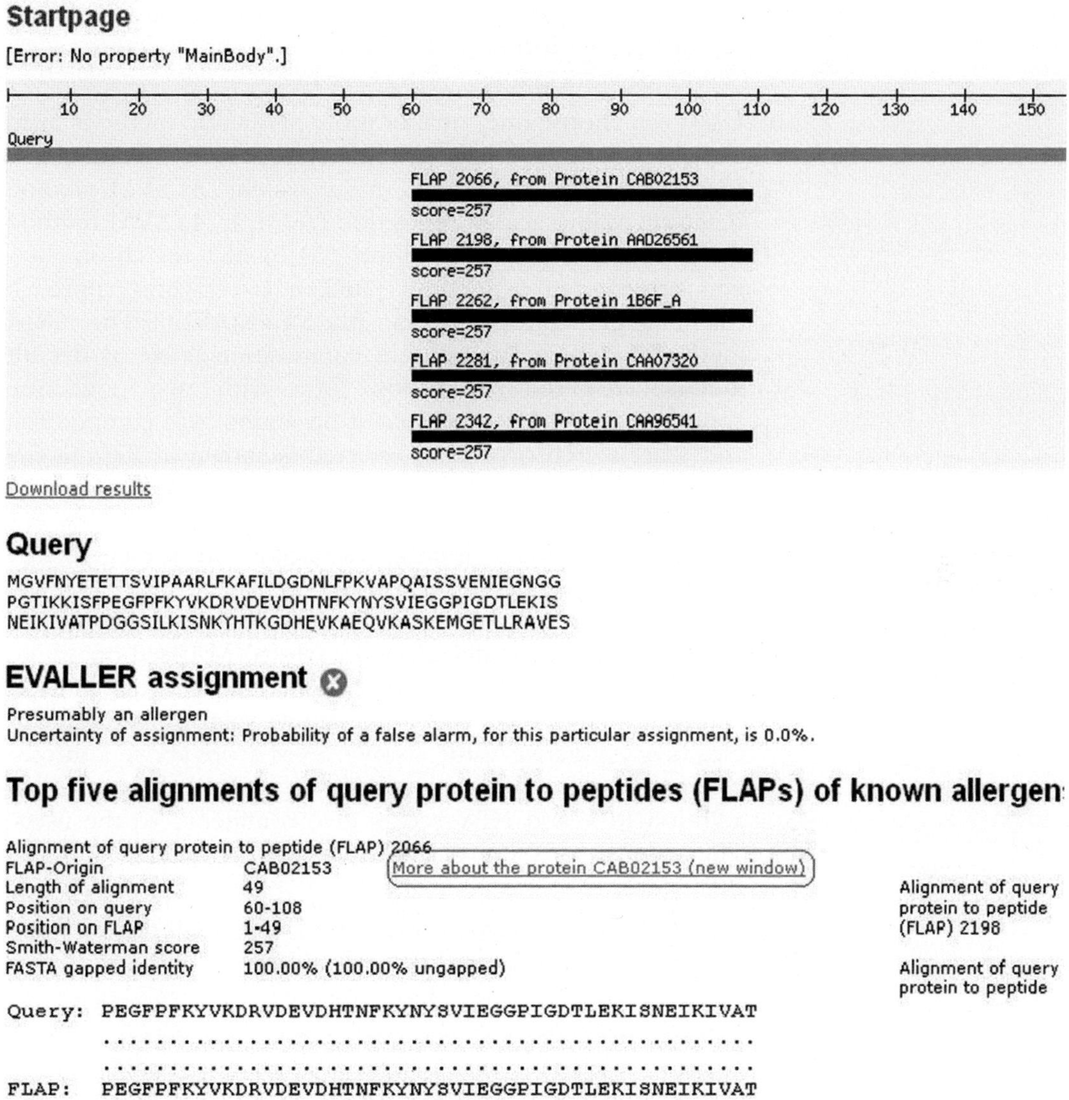

Fig. 3 (continued)

Acknowledgements	It links to acknowledgement page.
Developers	It links to contact address of the developers of AlgPred.
Contact	It links to the contact address for feedback information.

2.4.2 Usage of AlgPred Server

(a) Users can input the query protein sequence from the "submission" menu. It allows users to paste or type the amino acid sequence in single-letter code or upload the sequence file from local computer (*see* **Note 1**). Users are allowed to select any of the two formats: (1) plain format (single-letter code), (2) standard format like FASTA or PIR.

(b) Users can choose one, two, or more prediction methods from available six approaches. (1) IgE epitope and Percentage of Identity; (2) MEME/MAST motif search; (3) SVM method based on amino acid composition (SVMc); (4) SVM method based on dipeptide composition; (5) BLAST search on allergen representative peptides (ARPs); (6) Hybrid approach (SVMc + IgE epitope + ARPs BLAST + MAST). The SVM method based on amino acid composition is set as default method. AlgPred submission form with two prediction approaches (SVM module based on amino acid composition and Blast search on ARPs) selected is shown in Fig. 4a (*see* **Note 4**).

2.4.3 AlgPred Query Result

(a) The results of all the approaches selected are presented in a single output page as shown in Fig. 4b.

(b) Each red box represents the output result of the selected approach. In case of default approach (SVM based on amino acid composition), there are three confidence scores: (1) SVM predicted scores, (2) Positive Predictive Value (PPV), and (3) Negative Predictive Value (NPV). SVM score above a threshold cut off value (−0.4) is predicted as potential allergen and in addition PPV and NPV scores give confidence to the users. The accuracy of the prediction depends on higher SVM and PPV scores.

(c) The BLAST result displays hits found with ARPs database and a single hit is predicted as potential allergen.

3 Notes

1. The input query protein sequence should be in single-letter amino acid code and users need to select whether it is in standard format.
2. In case of SDAP Web tool, the input sequence length more than 1,000 were not used for prediction.
3. The default parameters set were the best performance values based on accuracy, sensitivity, and specificity on datasets used to develop the methods. However, users may change the default parameter according to their need.
4. It is advisable to search in two or more allergen prediction servers based on different approaches for reliability.

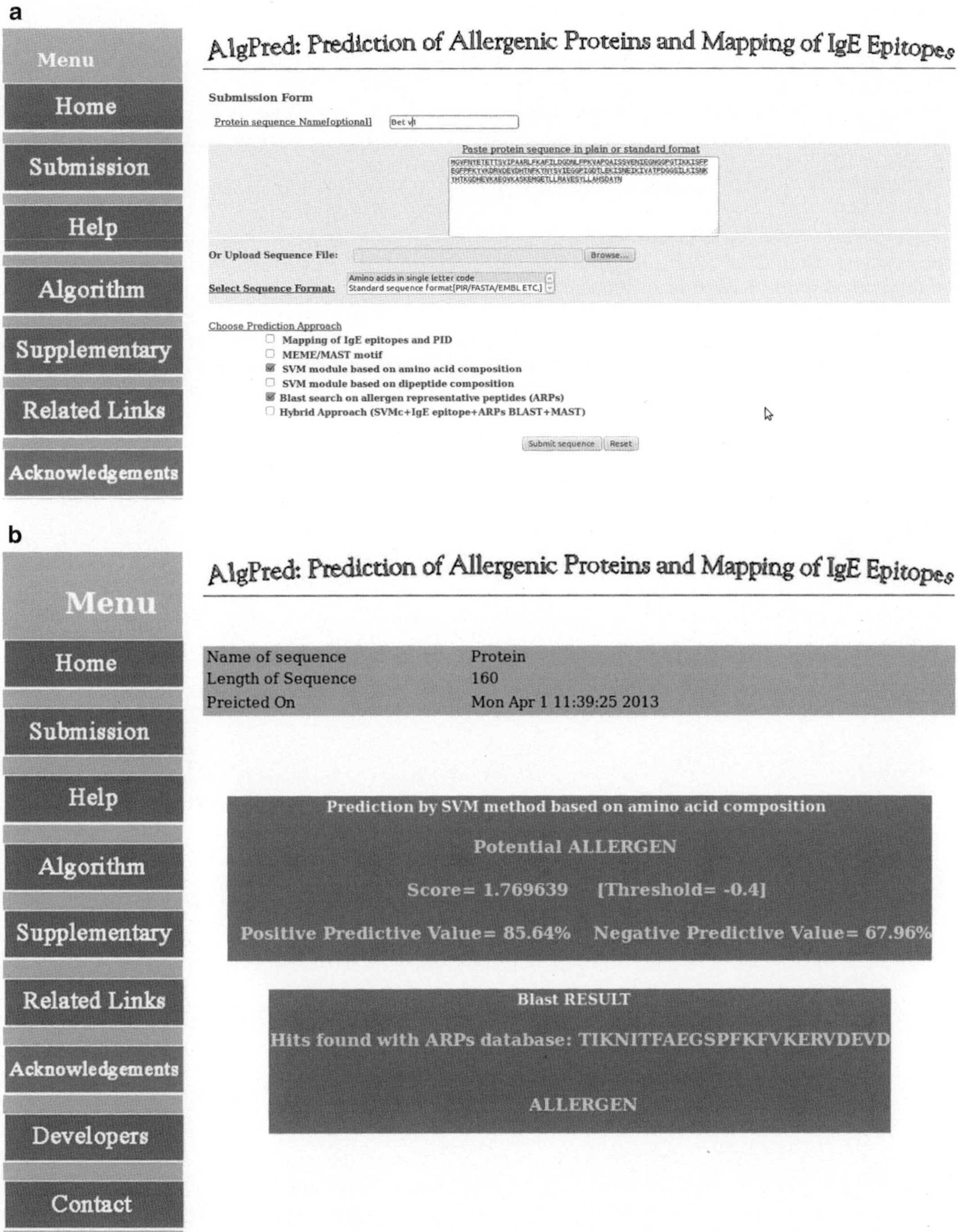

Fig. 4 Screen shots of AlgPred. (**a**) Query search page with Bet v 1 amino acid sequence. (**b**) Output results of AlgPred

References

1. Ghosh D, Gupta Bhattacharya S (2011) Allergen bioinformatics: recent trends and developments. In: Xia X (ed) Selected works in bioinformatics, InTechOpen, Croatia
2. Stadler MB, Stadler BM (2003) Allergenicity prediction by protein sequence. FASEB J 17(9):1141–1143
3. WHO (2003) Joint FAO/WHO food standards programme. Codex Ad Hoc intergovernmental task force on foods derived from biotechnology. World Health Organization, Yokohama (Japan) Accessed 18 Sept 2006
4. Ivanciuc O, Schein CH, Braun W (2003) SDAP: database and computational tools for allergenic proteins. Nucleic Acids Res 31(1): 359–362
5. Fiers MWEJ, Gijs KA, Nijland H (2004) Allermatch™, a webtool for the prediction of potential allergenicity according to current FAO/WHO Codex alimentarius guidelines. BMC Bioinformatics 5:133
6. Martinez Barrio A, Soeria-Atmadja D, Nistér A (2007) EVALLER: a web server for in silico assessment of potential protein allergenicity. Nucleic Acids Res 35:W694–W700
7. Saha S, Raghava GPS (2006) AlgPred: prediction of allergenic proteins and mapping of IgE epitopes. Nucleic Acids Res 34:W202–W209

Chapter 22

Prediction of Virulence Factors Using Bioinformatics Approaches

Rupanjali Chaudhuri and Srinivasan Ramachandran

Abstract

Virulence factors produced by a pathogen are essential for causing disease in the host. They enable the pathogen to establish itself within the host thus enhancing its potential to cause disease and in some instances underlie evasion of host defense mechanisms. Identification of these molecules, especially those of immunological interest and their use in vaccine development are attractive and are among the initial steps of reverse vaccinology. Surface localized virulence factors such as adhesins serve as excellent immunogenic candidates in this regard. In this chapter we have described the bioinformatics approaches for adhesin prediction, which include specific adhesin prediction algorithms.

Key words Virulence factors, Host, Pathogen, Immunogenic, Adhesins, Vaccine

1 Introduction

Despite advances in technologies to combat infections, infectious diseases continue to challenge humans. This may be attributed to the rise in drug-resistant strains of pathogens such as *Mycobacterium tuberculosis* and new emerging infectious pathogens such as SARS coronavirus and influenza virus. A key step in the establishment of infectious disease is microbial virulence, which has been described as an emergent property of host–microbe interaction [1]. At the molecular level, entities like proteins, carbohydrates, or lipids enable the pathogens to establish themselves in a susceptible host. These molecules form inherent part of the pathogen cellular system and are collectively termed "virulence factors" [2, 3]. Virulence factors in various pathogens play diverse roles in the establishment of disease. These include colonization of the host, evasion of host defense mechanisms, immunosuppression, acquisition of nutrients from host cell, mediation of entry and exit into host cell in intracellular pathogens, and sensing change of environment [4, 5]. These factors enable colonization of host niche and eventually cause damage to host tissues [2, 4, 6].

Rajat K. De and Namrata Tomar (eds.), *Immunoinformatics*, Methods in Molecular Biology, vol. 1184,
DOI 10.1007/978-1-4939-1115-8_22, © Springer Science+Business Media New York 2014

It was therefore realized that targeting these microbial molecules by identifying their immunogenicity and use in vaccine formulations could serve as efficient anti-infective strategy. Vaccinologists are therefore preparing vaccine formulations with these molecules for priming the immune system in order to neutralize their activity in the event of a host–pathogen contact [5, 7].

A diverse array of molecules is involved during host–pathogen interaction and the prominent players vary between the pathogens. These include adhesins, toxins, enzymes, and capsules (polysaccharides or polypeptides).

Adhesins have attracted interest from immunological perspective because they are located on the cell surface and are likely to be accessible to the molecules of the immune system [8]. In the subsequent sections we provide an overview of these molecules and describe their prediction using Bioinformatics.

1.1 Adhesins

Adhesins enable adherence of the pathogen to host cells and constitutes the initial major step in the process of infection. This role of adhesins qualifies them for vaccine candidates as targeting adhesins could arrest infection at the initial stage [8]. Even though adhesins exhibit sequence polymorphisms, the conserved regions may serve for potential vaccine especially those containing receptor binding domain [9]. Recently, a potent combination of adhesins of *Plasmodium falciparum* has been identified, which could transcend strain variations [10].

Examples include FimH adhesin of uropathogenic *Escherichia coli*. Vaccination with this protein proved effective against urinary tract infection caused by *E. coli* in both mice and in nonhuman primates [11]. Filamentous hemagglutinin (FHA) and pertactin adhesins of gram-negative bacteria *Bordetella pertussis* elicits long-lasting cell mediated respiratory immune response [12]. These adhesins are components of the approved acellular pertussis licensed vaccine [13]. Another adhesin *Neisseria meningitidis* adhesin A (NadA) is part of a multicomponent meningococcal serogroup B vaccine named Bexero, which is capable of eliciting a robust immune response. This vaccine has cleared all clinical trials and awaiting license for use [14].

2 Materials and Methods

2.1 Bioinformatics Approaches of Adhesin Characterization

The advent of genomics technologies has revolutionized biological research. The complete genome sequence of a pathogen provides an abundance of opportunities to identify putative virulence factors through sequence analysis. These investigations are being aided by the development of new computational algorithms in this area.

In the sections below, we discuss and outline the methods used in several investigations:

1. Sequence Similarity Search: Sequence similarity search is very popular and is among the first to be applied in sequence analysis. The goal here is to obtain orthologous sequences corresponding to a given query. This approach has been used to identify orthologues of known adhesins characterized in other pathogens (*see* **Note 1**). The best known algorithm is the Basic Local Alignment Search Tool (BLAST) algorithm [15]. Examples include application of BLAST algorithms in screening for potential adhesins in *Mycoplasma agalactiae*, *Escherichia coli*, *Mycoplasma conjunctivae*, *Mycoplasma pneumonia*, *Rickettsial species* [16–21]. In addition BLAST can be used to identify orthologues of enzymes from pathogens involved in virulence: Hyaluronidase, Neuraminidase, Phospholipases, Proteases, Collagenase, Kinase, Coagulase, Leukocidins, Hemolysins.
2. Sequence Motif search: Sequence motif refers to a particular arrangement or pattern of amino acids within a protein sequence, or nucleotides within a DNA sequence, which is characteristic of a specific biochemical function [22]. In particular, majority of protein sequence motifs, provide unique detectable sequence features for a set of protein sequences and thus act as signatures of protein families. Such motifs indicate similar functional roles.

   For example, in fungi, many Glycosylphosphatidylinositol-modified (GPI) proteins linked to plasma membrane via preformed GPI anchor play role in adhesion and virulence [23, 24]. These proteins have C-terminal GPI-motif described as follows: "[GNSDAC]-[GASVIETKDLF]-[GASV]-X(4,19)-[FILMVAGPSTCYWN](10)>" in Prosite format, where ">" indicates the C-terminal end of the protein [26]. Algorithm based on identifying sequences having a C-terminal, fungus-specific, consensus sequence for GPI modification (GPI-motif) helps screen a set of potential fungal adhesins [25]. Table 1 lists the motifs identified in several adhesins.
3. Signal Peptide: Signal Peptide (SP) is a short stretch of sequence present in the N-terminus of the protein directing it to the secretory pathway [31]. Adhesins being membrane attached proteins usually posses N-terminal signal peptide for translocation across the membrane of the endoplasmic reticulum [32, 33]. Therefore, algorithms using this information to screen for proteins having N-terminal signal peptide may help identifying potential adhesins (*see* **Note 2**). However, there are adhesins called "anchorless adhesins," which do not have Signal peptide or Transmembrane domain. These "anchorless adhesins" cannot be identified through these approaches.

Table 1
Motifs in adhesins and other virulence factors

Motif	Description	Reference
Beta-helix motif	These are right-handed parallel beta-helix supersecondary structural motifs in primary amino acid sequences. Present in toxins, virulence factors, adhesins, and surface proteins of Chlamydia, Helicobacteria, Bordetella, Leishmania, Borrelia, Rickettsia, Neisseria, and *Bacillus anthracis*	[26]
FxxN, GGA(I,L,V)	These are tetrapeptide motifs FxxN and GGA(I, L, V) present in polymorphic membrane protein family (Pmp) of *Chlamydia pneumonia*. They are required as duplicate copies for adhesion to host cells	[27]
RGD, SGxG	These are arginine-glycine-aspartic acid (RGD) and glycosaminoglycan binding site (SGXG) motifs present in autotransporter family proteins of *Bordetella pertussis*—pertactin (Prn), Bordetella resistance to killing (BrkA) and Bordetella autotransporter protein-C (BapC). The arrangement of motifs confer BapC adhesive property to binding sites on the macrophages and epithelial cells	[28]
PARF motif (A/T/E) xYLxx(LYF)N	This is a (A/T/E)XYLXXLN amino acid sequence motif referred to as PARF (peptide associated with rheumatic fever). It is located in the N-terminal hypervariable region of the collagen binding M protein type 3 of *Streptococcus pyogenes* and *Streptococcus dysgalactiae ssp. equisimilis* (SDSE)	[29]
HExxH containing metalloprotease adhesins	This is a zinc binding sequence motif His-Glu-Xaa-Xaa-His. It is present in certain adhesins like *Treponema pallidum* extracellular matrix binding adhesin Tp0751	[30]

4. Transmembrane domain: Transmembrane domains are the regions of membrane proteins which traverse in and out, looping through the membrane. They are characteristics of integral membrane proteins. Since adhesins are mostly membrane proteins, the prediction of proteins having transmembrane domain would contribute to the set of putative adhesins (*see* **Note 3**). However, this approach would lead to large number of false positives as not all proteins possessing transmembrane domains are adhesins.
5. Domain Search: Domains are conserved autonomously folding functional unit of a protein [34]. The domains of a protein together define the function of the protein. The domain information of an unannotated protein sequence can be used to predict its function (*see* **Note 4**).

Some adhesin domains are known. Examples include GLEYA adhesin domain, PA14 domain, ALS_N domain in fungal species,

YadA adhesin protein domain, fibrinogen-binding domain, Gingipain adhesin domains forming part of cleaved adhesin domain in bacterial species [35–40]. Sequence analysis to study the presence of such adhesin related domains in the query protein sequence may help predicting potential adhesins.

2.2 Challenges in Bioinformatics Characterization of Adhesins

Although the computational methods described in preceding section permit identifying potential adhesins they are limited in their scope. Unlike many families of proteins, adhesins lack a well defined common sequence pattern or signatures, rendering their identification using the general signature sequence search or unique motif search difficult. This is mainly because adhesins include diverse proteins. Even adhesins belonging to same species include diverse molecular types and lack a common specific pattern in sequence. For example, the adhesins- M proteins in *Streptococcus pyogenes*, Gal/GalNAc lectin in *Entamoeba histolytica*, Fimbrial adhesins in *Escherichia coli*, Blood group antigen binding adhesin (BabA) in *Helicobacter pylori*, YadA collagen binding adhesin in *Yersinia enterolitica* [41–45] lack significant similarity among each other.

However, in certain cases like in fungal species where many adhesins possess fungal specific GPI-motif, sequence motif search algorithm can be used to screen for potential fungal adhesins. However, identification methods solely based on motif searches such as GPI-anchor searches could return several false positives because all GPI-anchored proteins are not adhesins. Similar concerns apply to other identification methods such as Signal peptide search. The basic principles and limitations of various bioinformatics approaches used to characterize adhesins are summarized in Fig. 1.

These limitations formed the foundation for developing non-homology group of algorithms, which use a large number of compositional properties.

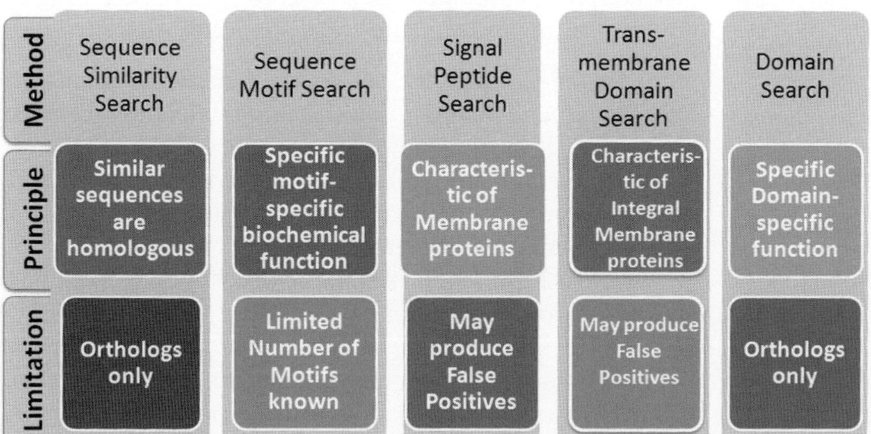

Fig. 1 Advantages and limitations of different sequence and motif based approaches for prediction of potential virulence factors

2.3 Specialized Algorithms for Adhesin Prediction

2.3.1 SPAAN: A Software Program for Prediction of Adhesins and Adhesin-Like Proteins

SPAAN is an adhesin prediction tool developed using artificial neural network trained on compositional properties of known adhesins and non-adhesins. The algorithm is trained to predict adhesins and adhesin-like proteins solely from the sequence data. It is a non-homology method. SPAAN was trained using 105 compositional properties including 20 amino acid frequencies, 20 selected dipeptide frequencies, 20 multiplet frequency, 20 charge compositions, and 25 hydrophobic compositions. It showed an optimal sensitivity of 89 % and specificity of 100 % on a defined test set and could identify 97.4 % of known adhesins at high Pad value from a wide range of bacteria. Though SPAAN was trained on datasets dominated by bacterial adhesins, it can be used for general purpose to identify adhesins from a wide spectrum of species belonging to diverse phyla. Many novel adhesins in diverse species have been characterized using SPAAN [46]. It is one of the most widely used adhesin prediction tool available. The standalone software package of SPAAN can be downloaded from http://sourceforge.net/projects/adhesin/files/.

System Requirement: Red Hat Linux version 7.3 or above.

Other requirements: C compiler

Instruction for usage

1. SPAAN is provided as a tar-gzipped file. Post download, it should be unzipped and untarred by the command "tar xvzf SPAAN.tar.gz."
2. The query sequences should be in FASTA format. Multiple sequences can be present in the input file.
3. The input file should be named as "query.dat."
4. The command to run the software SPAAN is "./askquery."
5. The output data is stored in "query.out."
6. If the existing binary files are not compatible to the system, the source C codes provided need to recompiled using the following example command-"gcc –lm standard.c –o standard.o."

List of C source codes to be compiled—standard.c, filter.c, annotate.c, and finalp1.c in the main SPAAN directory; recognize.c, AAcompo.c, hdr.c, multiplets.c, querydipep.c, and charge.c in their respective directories: AAcompo, hdr, multiplets, dipep, and charge: recognize.c needs to be compiled individually in each of the five mentioned directories.

Figure 2 describes an example of a run of SPAAN output result file "query.out."

2.3.2 MAAP: Malarial Adhesin and Adhesin-Like Proteins Predictor

MAAP was developed using Support Vector Machine (SVM) trained through compositional properties for classifying malarial adhesins and adhesin-like proteins [47]. The SVM^{light} package [48] of Support Vector Machine was used for this purpose. A total of 420 compositional properties including amino acid frequencies of

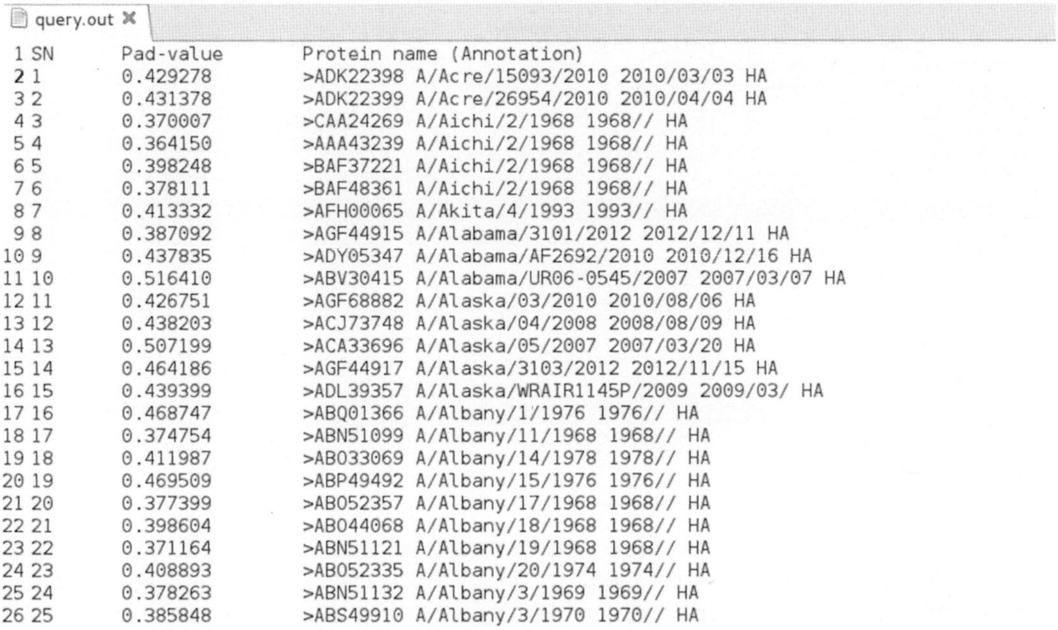

```
query.out
1 SN       Pad-value      Protein name (Annotation)
2 1        0.429278       >ADK22398 A/Acre/15093/2010 2010/03/03 HA
3 2        0.431378       >ADK22399 A/Acre/26954/2010 2010/04/04 HA
4 3        0.370007       >CAA24269 A/Aichi/2/1968 1968// HA
5 4        0.364150       >AAA43239 A/Aichi/2/1968 1968// HA
6 5        0.398248       >BAF37221 A/Aichi/2/1968 1968// HA
7 6        0.378111       >BAF48361 A/Aichi/2/1968 1968// HA
8 7        0.413332       >AFH00065 A/Akita/4/1993 1993// HA
9 8        0.387092       >AGF44915 A/Alabama/3101/2012 2012/12/11 HA
10 9       0.437835       >ADY05347 A/Alabama/AF2692/2010 2010/12/16 HA
11 10      0.516410       >ABV30415 A/Alabama/UR06-0545/2007 2007/03/07 HA
12 11      0.426751       >AGF68882 A/Alaska/03/2010 2010/08/06 HA
13 12      0.438203       >ACJ73748 A/Alaska/04/2008 2008/08/09 HA
14 13      0.507199       >ACA33696 A/Alaska/05/2007 2007/03/20 HA
15 14      0.464186       >AGF44917 A/Alaska/3103/2012 2012/11/15 HA
16 15      0.439399       >ADL39357 A/Alaska/WRAIR1145P/2009 2009/03/ HA
17 16      0.468747       >ABQ01366 A/Albany/1/1976 1976// HA
18 17      0.374754       >ABN51099 A/Albany/11/1968 1968// HA
19 18      0.411987       >AB033069 A/Albany/14/1978 1978// HA
20 19      0.469509       >ABP49492 A/Albany/15/1976 1976// HA
21 20      0.377399       >AB052357 A/Albany/17/1968 1968// HA
22 21      0.398604       >AB044068 A/Albany/18/1968 1968// HA
23 22      0.371164       >ABN51121 A/Albany/19/1968 1968// HA
24 23      0.408893       >AB052335 A/Albany/20/1974 1974// HA
25 24      0.378263       >ABN51132 A/Albany/3/1969 1969// HA
26 25      0.385848       >ABS49910 A/Albany/3/1970 1970// HA
```

Fig. 2 An example of a run of SPAAN output result file "query.out." The results are output under three column heads, Serial No. (SN), Probability of adhesin (Pad-value), Protein name (Annotation)

20 and 400 dipeptide frequencies were used to characterize the sequences of known adhesins and nonadhesins of *Plasmodium* species. MAAP runs on complete proteomes of *Plasmodium* species revealed that in *Plasmodium falciparum* at P*maap* scores above 0.0, a sensitivity of 100 % was observed with two false positives. In *P. vivax* and *P. yoelii* an optimal threshold P*maap* score of 0.7 was found optimal with very few false positives (upto 5). The MAAP Web server provides users with an interface where they can paste or upload their query sequences and predict whether the protein sequence is an adhesin (*see* **Note 5**). Users have the facility to set their own desired threshold cutoff value. The result can be exported as tab delimited text file by the users. The standalone version can be downloaded from the "Download" tab of MAAP Web server or http://sourceforge.net/projects/adhesin/files/.

Figure 3 describes the output result obtained using MAAP Web server.

2.3.3 FungalRV adhesin Predictor

In pathogenic fungi, adhesins play major roles as virulence factors mediating the interaction of the pathogens to variety of host cell types. In addition, adhesins in fungi aid in biofilm formation contributing to increased drug resistance and persistence of infections [49]. It has been established that differences in adhesion are responsible for greater virulence of one strain compared to other in fungi [50]. The fungal pathogens represent a diverse group of species.

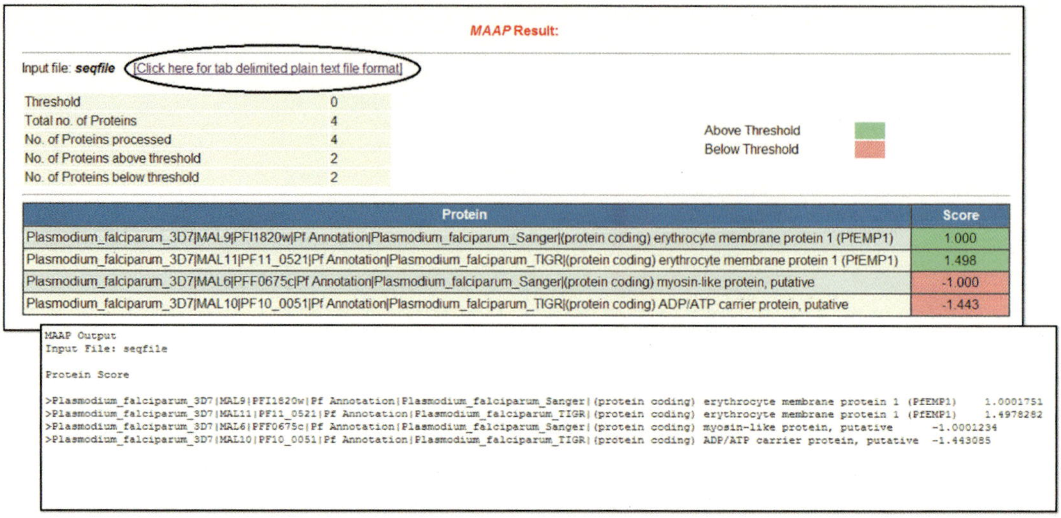

Fig. 3 Screenshot of output result obtained using MAAP Web server. The protein sequences scoring above threshold are highlighted in *green* color, whereas those scoring below the threshold are highlighted in *red* color. The result can be saved in a tab delimited plain text file format by clicking on the *purple* colored link (*encircled*)

FungalRV adhesin predictor was developed using Support Vector Machine (SVM) trained through compositional properties for classifying human pathogenic fungal adhesins and adhesin like proteins [51]. This tool was developed using SVM^{light} package of Support Vector Machine trained through 3,945 compositional properties including amino acid frequencies of 20 from amino acids, 247 selected dipeptide frequencies, 3,653 selected tripeptide frequencies, 20 amino acid multiplets frequencies, frequency of the hydrophobic amino acids and four moments of hydrophobic amino acid distribution of order 2–5. This is a non-homology based prediction tool. We obtained an overall MCC value of 0.8702 considering all 8 pathogens, namely, *Candida albicans*, *Candida glabrata*, *Aspergillus fumigatus*, *Coccidioides immitis*, *Coccidioides posadasii*, *Histoplasma capsulatum*, *Blastomyces dermatitidis*, and *Paracoccidioides brasiliensis* thus showing high sensitivity and specificity at a threshold of 0.511. In case of *P. brasiliensis* the algorithm achieved a sensitivity of 66.67 %. This tool was made into FungalRV Web server available at http://fungalrv.igib.res.in. The "Adhesin Predictor" tab of the FungalRV Web server provides users with an interface where they can paste or upload their query sequences and predict whether the protein sequence is a fungal adhesin (*see* **Note 6**). Users have been provided the facility to set their own desired threshold cutoff value. This facility has been provided to allow users to optimize the threshold for other fungi for which "FungalRV adhesin predictor" was not trained. The result can be exported as tab delimited text file by the users. The facility to search for fungal specific GPI

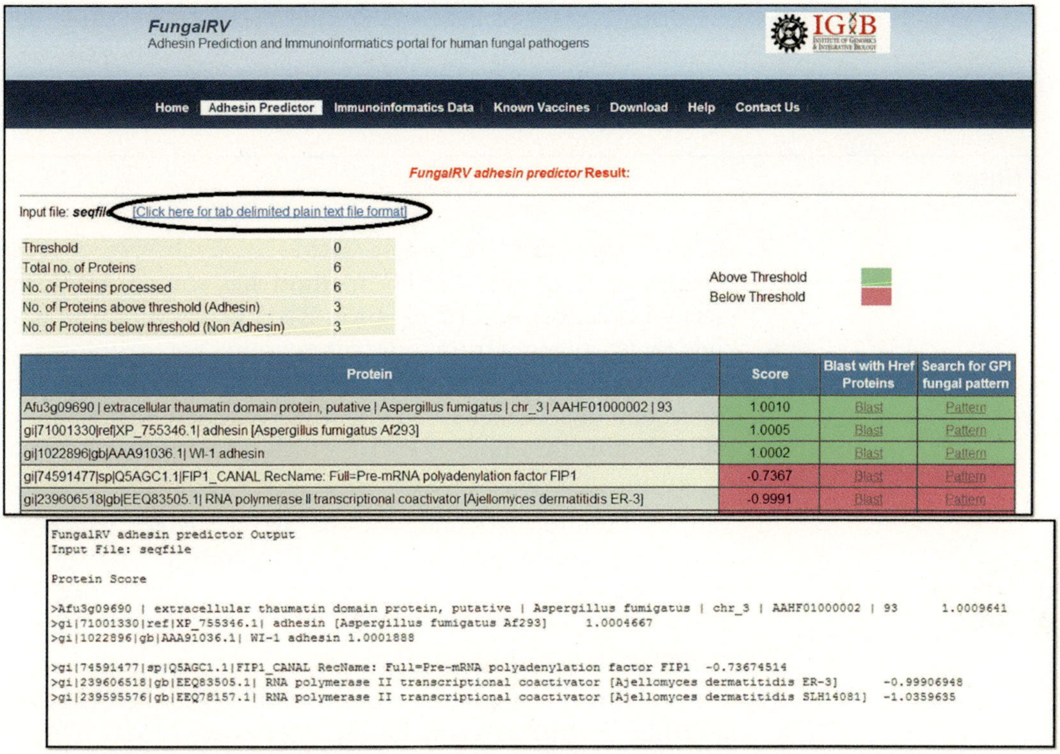

Fig. 4 Screenshot of output result obtained using FungalRV Web server. The protein sequences scoring above threshold are highlighted in *green* color, whereas those scoring below the threshold are highlighted in *red* color. The result can be saved in a tab delimited plain text file format by clicking on the *purple* colored link (*encircled*). Additional data on BLAST with Href proteins and GPI patterns are also displayed

pattern in the predicted adhesins and adhesin like proteins using fuzzpro program of EMBOSS has been provided. Users also have been provided the facility to conduct BLAST search with human reference proteins (*see* **Note 7**). The standalone version can be downloaded from the "Download" tab of FungalRV Web server or http://sourceforge.net/projects/adhesin/files/. Figure 4 describes the output adhesin prediction results obtained using FungalRV Web server.

2.3.4 Faapred

In addition to FungalRV, another Support Vector Machine (SVM) based algorithm named Faapred for prediction of fungal adhesins and adhesin-like proteins is available [52]. The SVM models for Faapred development were trained with compositional features- amino acid, dipeptide, multiplet fractions, charge and hydrophobic compositions, as well as PSI-BLAST derived PSSM matrices. The best classifiers were screened based on high MCC and accuracy. The amino acid composition model (ACHM), PSSM-a, and PSSM-b came out as the best classifiers with ACHM providing the highest MCC value of 0.610. Thus the prediction of Faapred uses classifiers based on compositional properties as well as PSSM. Faapred provides overall accuracy of 86 %. The prediction method is freely available as a World Wide Web based server at http://bioinfo.icgeb.res.in/faap.

3 Notes

1. BLAST algorithm is widely used to fetch orthologues. Reciprocal Best Hits (RBH) method has shown good efficiency in identifying orthologues. RBH is based on the principle that two genes from different genomes are orthologous if they find each other as the best hit in BLAST search in the other genome. Here BLASTP is usually carried out at a maximum E-value threshold of 1×10^{-6}, including Smith–Waterman algorithm and Soft-filtering.
2. Various bioinformatics algorithms are available, which aid identifying signal peptides. SignalP algorithm available at http://www.cbs.dtu.dk/services/SignalP/ is widely used. The query sequences input in FASTA format can be submitted to predict presence of signal peptides.
3. Transmembrane prediction algorithms for example TMHMM available at http://www.cbs.dtu.dk/services/TMHMM/ is generally used to predict presence of transmembrane regions.
4. Conserved Domains can be predicted using domain prediction algorithms for example CDD search available at http://www.ncbi.nlm.nih.gov/Structure/cdd/cdd.shtml. The presence of known adhesin related domains in the query sequences can be predicted.
5. The query proteins in FASTA format can be uploaded in the MAAP Web server. The server can be used to analyze the whole genome in one run.
6. Query protein sequences in FASTA format can be uploaded in FungalRV Web server. This Web server can be used to analyze the whole genome.
7. An adhesin vaccine should ideally not have similarity to human reference proteins to avoid cross-reactivity. The facility to conduct BLAST search with human reference proteins has therefore been provided in the FungalRV Web server. The cutoff E-value used here is 0.01, which borders on the limits of threshold similarity.

Acknowledgements

RC thanks The Indian Council of Medical Research for fellowship. This work was funded through grants "GENESIS" BSC0121 to SR from CSIR.

References

1. Casadevall A, Fang FC, Pirofski LA (2011) Microbial virulence as an emergent property: consequences and opportunities. PLoS Pathog 7:e1002136
2. Weiss RA (2002) Virulence and pathogenesis. Trends Microbiol 10:314–317
3. Poulin R, Combes C (1999) The concept of virulence: interpretations and implications. Parasitol Today 15:474–475
4. Cross AS (2008) What is a virulence factor? Crit Care 12:196
5. Casadevall A, Pirofski LA (2009) Virulence factors and their mechanisms of action: the view from a damage-response framework. J Water Health 7(Suppl 1):S2–S18
6. Rothy A, James L, Ed. (1988) Virulence mechanisms of bacterial pathogens. American Society for Microbiology, ISBN 0-914826-99-9
7. Casadevall A, Pirofski LA (1999) Host-pathogen interactions: redefining the basic concepts of virulence and pathogenicity. Infect Immun 67:3703–3713
8. Wizemann TM, Adamou JE, Langermann S (1999) Adhesins as targets for vaccine development. Emerg Infect Dis 5:395–403
9. Ofek I, Hasty DL, Sharon N (2003) Anti-adhesion therapy of bacterial diseases: prospects and problems. FEMS Immunol Med Microbiol 38:181–191
10. Pandey AK, Reddy KS, Sahar T, Gupta S, Singh H, Reddy EJ, Asad M, Siddiqui FA, Gupta P, Singh B, More KR, Mohmmed A, Chitnis CE, Chauhan VS, Gaur D (2013) Identification of a potent combination of key *Plasmodium falciparum* merozoite antigens that elicit strain-transcending parasite-neutralizing antibodies. Infect Immun 81:441–451
11. Langermann S, Mollby R, Burlein JE, Palaszynski SR, Auguste CG, DeFusco A, Strouse R, Schenerman MA, Hultgren SJ, Pinkner JS et al (2000) Vaccination with FimH adhesin protects cynomolgus monkeys from colonization and infection by uropathogenic *Escherichia coli*. J Infect Dis 181:774–778
12. Ausiello CM, Lande R, Urbani F, la Sala A, Stefanelli P, Salmaso S, Mastrantonio P, Cassone A (1999) Cell-mediated immune responses in four-year-old children after primary immunization with acellular pertussis vaccines. Infect Immun 67:4064–4071
13. Halperin SA, Scheifele D, Mills E, Guasparini R, Humphreys G, Barreto L, Smith B (2003) Nature, evolution, and appraisal of adverse events and antibody response associated with the fifth consecutive dose of a five-component acellular pertussis-based combination vaccine. Vaccine 21:2298–2306
14. Gorringe AR, Pajón R (2012) Bexsero: a multicomponent vaccine for prevention of meningococcal disease. Hum Vaccin Immunother 8: 174–183
15. Altschul SF, Gish W, Miller W, Myers EW, Lipman DJ (1990) Basic local alignment search tool. J Mol Biol 215:403–410
16. Pichel M, Binsztein N, Viboud G (2000) CS22, a novel human enterotoxigenic *Escherichia coli* adhesin, is related to CS15. Infect Immun 68:3280–3285
17. Fleury B, Bergonier D, Berthelot X, Peterhans E, Frey J, Vilei EM (2002) Characterization of P40, a cytadhesin of *Mycoplasma agalactiae*. Infect Immun 70:5612–5621
18. Belloy L, Vilei EM, Giacometti M, Frey J (2003) Characterization of LppS, an adhesin of Mycoplasma conjunctivae. Microbiology 149: 185–193
19. Nakane D, Adan-Kubo J, Kenri T, Miyata M (2011) Isolation and characterization of P1 adhesin, a leg protein of the gliding bacterium Mycoplasma pneumoniae. J Bacteriol 193: 715–722
20. Renesto P, Samson L, Ogata H, Azza S, Fourquet P, Gorvel JP, Heinzen RA, Raoult D (2006) Identification of two putative rickettsial adhesins by proteomic analysis. Res Microbiol 157:605–612
21. Palaniappan RU, Chang YF, Jusuf SS, Artiushin S, Timoney JF, McDonough SP, Barr SC, Divers TJ, Simpson KW, McDonough PL, Mohammed HO (2002) Cloning and molecular characterization of an immunogenic LigA protein of *Leptospira interrogans*. Infect Immun 70:5924–5930
22. D'haeseleer P (2006) What are DNA sequence motifs? Nat Biotechnol 24:423–425
23. De Groot PW, Hellingwerf KJ, Klis FM (2003) Genome-wide identification of fungal GPI proteins. Yeast 20:781–796
24. Richard ML, Plaine A (2007) Comprehensive analysis of glycosylphosphatidylinositol-anchored proteins in *Candida albicans*. Eukaryot Cell 6:119–133
25. Weig M, Jänsch L, Gross U, De Koster CG, Klis FM, De Groot PW (2004) Systematic identification in silico of covalently bound cell wall proteins and analysis of protein-polysaccharide linkages of the human pathogen Candida glabrata. Microbiology 150:3129–3144
26. Bradley P, Cowen L, Menke M, King J, Berger B (2001) BETAWRAP: successful prediction of parallel beta -helices from primary sequence reveals an association with many microbial pathogens. Proc Natl Acad Sci U S A 98: 14819–1424

27. Mölleken K, Schmidt E, Hegemann JH (2010) Members of the Pmp protein family of *Chlamydia pneumoniae* mediate adhesion to human cells via short repetitive peptide motifs. Mol Microbiol 78:1004–1017
28. Bokhari H, Bilal I, Zafar S (2012) BapC autotransporter protein of Bordetella pertussis is an adhesion factor. J Basic Microbiol 52:390–396
29. Reissmann S, Gillen CM, Fulde M, Bergmann R, Nerlich A, Rajkumari R, Brahmadathan KN, Chhatwal GS, Nitsche-Schmitz DP (2012) Region specific and worldwide distribution of collagen-binding M proteins with PARF motifs among human pathogenic streptococcal isolates. PLoS One 7:e30122
30. Houston S, Hof R, Francescutti T, Hawkes A, Boulanger MJ, Cameron CE (2011) Bifunctional role of the *Treponema pallidum* extracellular matrix binding adhesin Tp0751. Infect Immun 79:1386–1398
31. Blobel G, Dobberstein B (1975) Transfer of proteins across membranes. I. Presence of proteolytically processed and unprocessed nascent immunoglobulin light chains on membrane-bound ribosomes of murine myeloma. J Cell Biol 67:835–851
32. Lodish H, Berk A, Zipursky SL et al. (2000) Molecular cell biology, 4th edn. W. H. Freeman, New York, NY. Section 17.4, translocation of secretory proteins across the ER membrane. http://www.ncbi.nlm.nih.gov/books/NBK21532/
33. Lee VT, Schneewind O (2001) Protein secretion and the pathogenesis of bacterial infections. Genes Dev 15:1725–1752
34. Phillips DC (1966) The three-dimensional structure of an enzyme molecule. Sci Am 215: 78–90
35. Rigden DJ, Mello LV, Galperin MY (2004) The PA14 domain, a conserved all-beta domain in bacterial toxins, enzymes, adhesins and signaling molecules. Trends Biochem Sci 29:335–339
36. Linder T, Gustafsson CM (2008) Molecular phylogenetics of ascomycotal adhesins—a novel family of putative cell-surface adhesive proteins in fission yeasts. Fungal Genet Biol 45:485–497
37. Phan QT, Myer CL, Fu Y, Sheppard DC, Yeaman MR, Welch WH, Ibrahim AS, Edwards JE Jr, Filler SG (2007) Als3 Is a *Candida albicans* invasin that binds to cadherins and induces endocytosis by host cells. PLoS Biol 5:e64
38. Nummelin H, Merckel MC, Leo JC, Lankinen H, Skurnik M, Goldman A (2004) The *Yersinia adhesin* YadA collagen-binding domain structure is a novel left-handed parallel beta-roll. EMBO J 23:701–711
39. Li N, Yun P, Nadkarni MA, Ghadikolaee NB, Nguyen KA, Lee M, Hunter N, Collyer CA (2010) Structure determination and analysis of a haemolytic gingipain adhesin domain from *Porphyromonas gingivalis.* Mol Microbiol 76: 861–873
40. Konkel ME, Christensen JE, Keech AM, Monteville MR, Klena JD, Garvis SG (2005) Identification of a fibronectin-binding domain within the Campylobacter jejuni CadF protein. Mol Microbiol 57:1022–1035
41. Ellen RP, Gibbons RJ (1972) M protein-associated adherence of *Streptococcus pyogenes* to epithelial surfaces: prerequisite for virulence. Infect Immun 5:826–830
42. Schembri MA, Kjaergaard K, Sokurenko EV, Klemm P (2001) Molecular characterization of the *Escherichia coli* FimH adhesin. J Infect Dis 183(Suppl 1):S28–S31
43. Mann BJ (2002) Structure and function of the *Entamoeba histolytica* Gal/GalNAc lectin. Int Rev Cytol 216:59–80
44. Wen S, Moss SF (2009) *Helicobacter pylori* virulence factors in gastric carcinogenesis. Cancer Lett 282:1–8
45. Nummelin H, Merckel MC, Leo JC, Lankinen H, Skurnik M, Goldman A (2004) The *Yersinia adhesin* YadA collagen-binding domain structure is a novel left-handed parallel beta-roll. EMBO J 23:701–711
46. Sachdeva G, Kumar K, Jain P, Ramachandran S (2005) SPAAN: a software program for prediction of adhesins and adhesin-like proteins using neural networks. Bioinformatics 21:483–491
47. Ansari FA, Kumar N, Bala SM, Gnanamani M, Ramachandran S (2008) MAAP: malarial adhesins and adhesin-like proteins predictor. Proteins 70:659–666
48. Joachims T (1999) Making large-scale SVM learning practical. In: Scholkopf B, Burges C, Smola A (eds) Advances in Kernel methods—support vector learning. MIT, Cambridge, MA, pp 169–185
49. Hawser SP, Douglas LJ (1994) Biofim formation by *Candida* species on the surface of catheter materials in vitro. Infect Immun 62:915–921
50. Verstrepen KJ, Klis FM (2006) Flocculation, adhesion and biofilm formation in yeasts. Mol Microbiol 60:5–15
51. Chaudhuri R, Ansari FA, Raghunandanan MV, Ramachandran S (2011) FungalRV: adhesin prediction and immunoinformatics portal for human fungal pathogens. BMC Genomics 12:192
52. Ramana J, Gupta D (2010) FaaPred: a SVM-based prediction method for fungal adhesins and adhesin-like proteins. PLoS One 5:e9695

Part IV

Systems Biology Approaches in Immunoinformatics

Chapter 23

A Systems Biology Approach to Study Systemic Inflammation

Bor-Sen Chen and Chia-Chou Wu

Abstract

Systemic inflammation needs a precise control on the sequence and magnitude of occurring events. The high throughput data on the host–pathogen interactions gives us an opportunity to have a glimpse on the systemic inflammation. In this article, a dynamic *Candida albicans*–zebrafish interactive infectious network is built as an example to demonstrate how systems biology approach can be used to study systematic inflammation. In particular, based on microarray data of *C. albicans* and zebrafish during infection, the hyphal growth, zebrafish, and host–pathogen intercellular PPI networks were combined to form an integrated infectious PPI network that helps us understand the systematic mechanisms underlying the pathogenicity of *C. albicans* and the immune response of the host. The signaling pathways for morphogenesis and hyphal growth of *C. albicans* were 2 significant interactions found in the intercellular PPI network. Two cellular networks were also developed corresponding to the different infection stages (adhesion and invasion), and then compared with each other to identify proteins to gain more insight into the pathogenic role of hyphal development in the *C. albicans* infection process. Important defense-related proteins in zebrafish were predicted using the same approach. This integrated network consisting of intercellular invasion and cellular defense processes during infection can improve medical therapies and facilitate development of new antifungal drugs.

Key words Host–pathogen interaction network, Infection, Hyphal development, Dynamic PPI network, Immune response, Host defense

1 Introduction

Candida albicans is an opportunistic fungal pathogen responsible for various mucosal infections, such as candidiasis and other potentially life-threatening diseases [1]. It is also the species most frequently responsible for hospital-acquired fungal infections. This pathogen can colonize various biomaterials, such as ventricular assist devices and urinary and vascular catheters, forming dense biofilms that are resistant to most antifungal drugs [2]. *C. albicans* infections and candidiasis are difficult to treat and create very serious therapeutic challenges. Mortality rates among patients with candidiasis can be as high as 40–60 % [3]. Therefore, knowledge of

Rajat K. De and Namrata Tomar (eds.), *Immunoinformatics*, Methods in Molecular Biology, vol. 1184,
DOI 10.1007/978-1-4939-1115-8_23, © Springer Science+Business Media New York 2014

the molecular mechanisms underlying the pathogenicity of *C. albicans* and the defense of host could improve medical therapy and facilitate new antifungal drugs development.

Under normal circumstances, *C. albicans* lives in 80 % of the human population with no harmful effects, although its overgrowth, often observed in immunocompromised (e.g., HIV-positive) individuals, results in candidiasis, [4, 5]. *C. albicans* can grow in a variety of morphological forms, ranging from yeast form, pseudohyphae form, to true tubular hyphae form, depending on the growth conditions in the host environment [6]. A number of molecules have been implicated as associated with the virulence of *C. albicans*, such as host recognition biomolecules, secreted aspartyl proteases, and phospholipases, as well as life cycle factors like adhesion and morphogenesis [7]. Among those factors, the transition from yeast to hyphal form is considered to be critical for *C. albicans* pathogenesis [6, 8]. Although previous studies have provided some hints, the detailed molecular mechanisms responsible for morphological forms remain to be elucidated.

The *C. albicans* (strain SC5314) genome, used in this article, has already been sequenced , revealing that almost two-thirds of its ~6,000 open reading frames are orthologous to genes of *Saccharomyces cerevisiae*, the most intensively studied eukaryotic model organism to have its entire genome sequenced [9, 10]. Compared to *C. albicans*, *S. cerevisiae* has abundant high-throughput screening data and it is closely related to *C. albicans* (i.e., both fall within the hemiascomycetes class), the information from *S. cerevisiae* could be usefully adapted for our understanding biology and pathogenesis of *C. albicans* [9].

The zebrafish (*Danio rerio*) has emerged as a powerful new vertebrate model for human disease. Numerous studies have already utilized the zebrafish system to study the pathogenesis of various human infectious diseases, including those caused by bacteria or viruses [11, 12]. The zebrafish immune system shows remarkable similarities to mammalian counterparts. As a demonstration of the zebrafish's utility as a model organism for human disease, in 49 cases of a zebrafish mutant gene being cloned based on a forward genetic screen, the genes were found to have homologs in human disease [13]. Overall, the zebrafish genetic map demonstrates highly conserved synteny with the human genome [14]. Chao et al. have also demonstrated that *C. albicans* can colonize and invade the fish host at multiple anatomical sites and prove fatal in a dose-dependent manner [15]. Therefore, a zebrafish infection model could be used to investigate the details of the *C. albicans* invasive process and infectious mechanisms.

In this article, we construct an infectious *C. albicans* and zebrafish intercellular PPI network by mining and integrating microarray data, PPI information, and host–pathogen intercellular interactions to investigate how the infectious behaviors of *C. albicans* on host tissue are regulated. Consequently, we discovered that all major hyphae-related pathways are visible in our hyphal PPI network, confirming the reliability and accuracy of our methods. From a systems perspective, we were able to predict the proteins with the largest changes in the number of interactions and the hub proteins for morphological switching processes. We identified several important hyphae growth-related proteins–e.g., Ubi4, Act1, Kex2, Hsl1, and Tsa1–and some proteins worth further exploration for pathogenicity research such as Hht21, Kre1, and Orf19.5438. Moreover, three noteworthy functions in *C. albicans* infection—cellular iron ion homeostasis, glucose transport, and cell wall molecular biosynthesis—were named as pathogen invasion mechanisms from analysis of the integrated intercellular protein interaction networks. Further, several functions such as apoptosis and immune response were also found to be involved in host defense mechanisms.

2 Materials

2.1 Ethics Statement

Manipulation of the animal model was approved by the Institutional Animal Care and Use Committee of National Tsing Hua University.

2.2 Simultaneous Time Course Microarray Experiments During C. albicans Infection

C. albicans (strain SC5314) and adult zebrafish (strain AB) were used in the experiments. Their maintenance and preparation were performed according to procedures described previously [15]. Zebrafish were anesthetized by immersion in water containing 0.17 g/ml of tricaine (Sigma) and then intraperitoneally injected with 1×10^8 CFU *C. albicans* cells suspended in 10 μl sterile phosphate-buffered saline (PBS). The infected fish were sacrificed by immersion in ice water at 0.5, 1, 2, 4, 6, 8, 12, 16, and 18 h post-injection (hpi) and frozen in liquid nitrogen. *C. albicans-infected* zebrafish were treated with Trizol® Reagent (Invitrogen, USA), pulverized in liquid nitrogen using a small mortar and pestle, and then disrupted using a MagNA Lyser System (Roche) with glass beads (cat. no. G8772-100G, Sigma) by shaking at 5,000 rpm for 15 s. After phase separation by adding chloroform, the total RNA was purified using an RNeasy Mini Kit (Qiagen, Germany). Purified RNA was quantified at OD260 nm using a ND-1000 spectrophotometer (Nanodrop Technology, USA) and analyzed using a Bioanalyzer 2100 (Agilent Technologies, USA) with RNA 6000 Nano LabChip kit (Agilent Technologies, USA). 1 μg of the total RNA was amplified using a Quick-Amp Labeling kit (Agilent Technologies, USA) and labeled with Cy3 (CyDye, PerkinElmer, USA) during the in vitro transcription process. 0.625 μg of Cy3 cRNA for the *C. albicans* array and 1.65 μg of Cy3 cRNA for the zebrafish array, were fragmented to an average size of 50–100 nucleotides by incubation with fragmentation buffer at 60 °C for 30 min. The fragmented labeled cRNA was then hybridized to both *C. albicans* and zebrafish oligo microarrays (Agilent Technologies, USA) at 60 °C for 17 h. After washing and drying using a nitrogen gun, microarrays were scanned using an Agilent microarray scanner (Agilent Technologies, USA) at 535 nm for Cy3. For each time point, three biological replicates were done for both organisms. The raw data were processed with Loess normalization and the results have been deposited in Gene Expression Omnibus (accession number GSE32119).

3 Methods

3.1 Overview of the Data Processing

The global screening method for infection-related proteins was divided into three key steps: (1) data preprocessing and selection, (2) dynamic *C. albicans* hyphal growth and zebrafish networks construction, (3) intercellular PPI network between pathogen and host construction. The flowchart of the method is shown in Fig. 1. After constructing the overall network, consisted of the hyphal growth PPI network for *C. albicans*, the PPI network for zebrafish and the intercellular PPI network between pathogen and host, we search for potential infection-related proteins and immune response pattern recognition molecules in both *C. albicans* and zebrafish.

3.2 Data Preprocessing and Selection

Several types of data were mined and integrated to construct the integrated cellular network. In *C. albicans*, the required data included its microarray gene expression profiles, PPIs from *S. cerevisiae*, gene orthologs data between *C. albicans* and *S. cerevisiae*, and gene annotations for *C. albicans*. There are 9 time points in the *C. albicans* microarray data spanning from 0.5 to 18 hpi (i.e., 0.5, 1, 2, 4, 6, 8, 12, 16, and 18 hpi). The gene ortholog data were acquired from the Candida Genome Database (CGD), and the *C. albicans* gene annotations were retrieved from the Gene Ontology. The PPI data of *S. cerevisiae* were extracted from the Biological General Repository for Interaction Datasets (BioGRID).

In zebrafish, the required data included gene expression profiles, PPIs from *Homo sapiens*, data on human and zebrafish gene orthologs and functional gene annotations for zebrafish. There were also 9 time points in the zebrafish microarray data spanning from 0.5 to 18 hpi (i.e., 0.5, 1, 2, 4, 6, 8, 12, 16, and 18 hpi). The gene ortholog data were acquired using the ZFIN and InParanoid. The zebrafish gene annotations were retrieved from the Gene Ontology. The PPI data for *Homo sapiens* were extracted from BioGRID and the Human Protein Reference Database (HPRD).

3.3 Protein Pool Selection for Candidate PPI Networks

Because of the lack of PPI databases between *C. albicans* and zebrafish at present, gene orthology data between *C. albicans* and *S. cerevisiae* as well as between zebrafish and human were utilized to set up protein data pools for our candidate *C. albicans* and zebrafish PPI networks, respectively. *C. albicans* PPIs can be

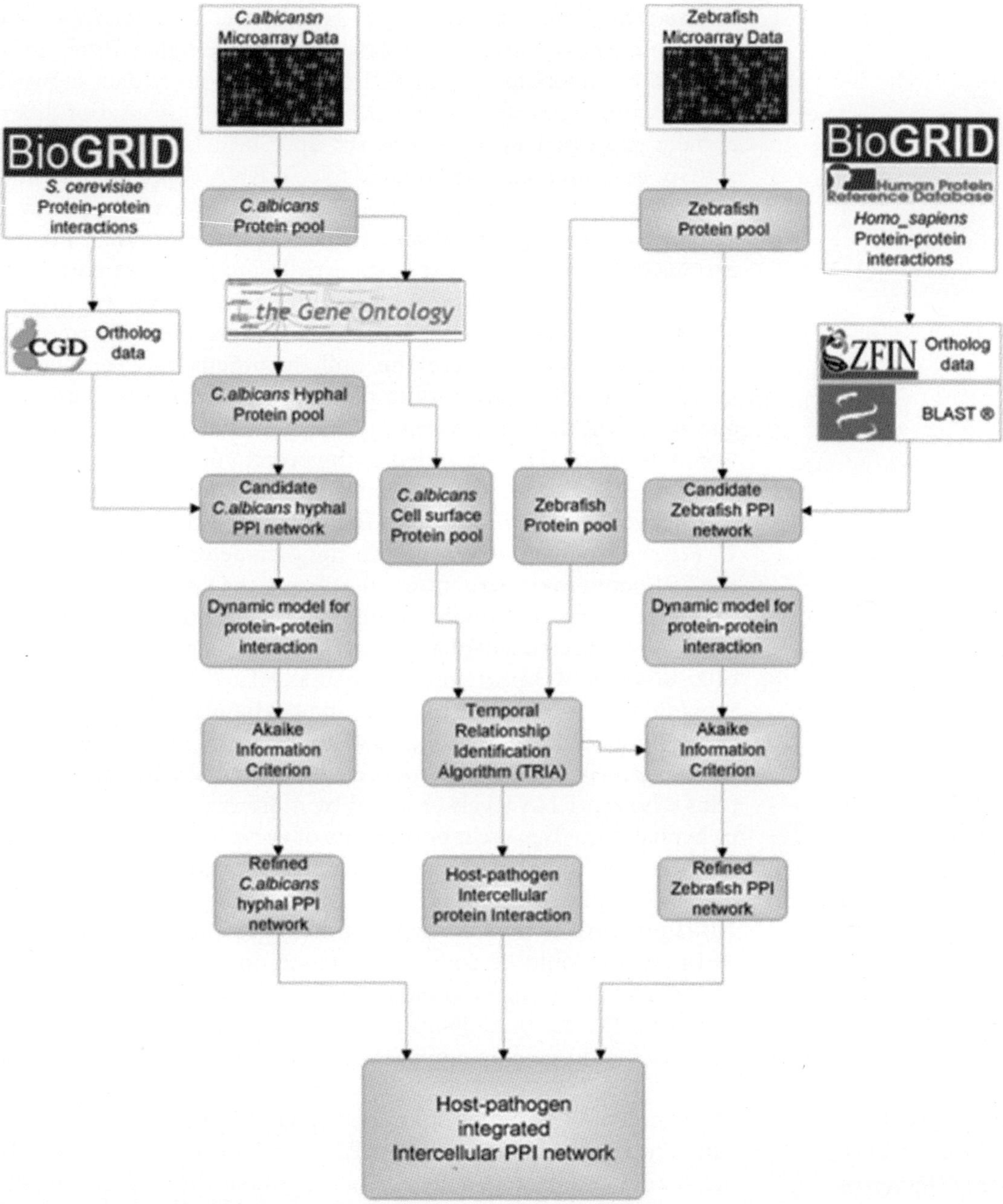

Fig. 1 Flowchart for construction of the integrated infectious PPI network. The construction of our integrated intercellular PPI network was performed by database mining and network identification. The network construction combines DNA microarray data with different types of information from various databases, as shown in the *white boxes. Blue boxes* show the steps of candidate subnetwork construction. The next, bottom part (*orange boxes*) of the flowchart illustrates the steps of network identification and the subsequent construction of the integrated cellular network

inferred by applying orthology data between *C. albicans* and *S. cerevisiae* to the latter's PPI data; similarly, zebrafish PPIs can be inferred by mapping human PPI data to orthology data between humans and zebrafish. Then, the protein pool consisting of differentially expressed proteins was set up. Since large-scale protein activity measurements are unavailable, mRNA expression profiles were used as a substitute. Although the mRNA expression levels cannot be completely representative of the corresponding protein expression levels, they are at least partially and positively correlated [16, 17]. The mRNA expression level for each protein was used to select differentially expressed proteins using one-way analysis of variance (ANOVA), where the null hypothesis was the average expression levels at every time point being equal. In *C. albicans* and zebrafish, the proteins with p-values returned by ANOVA that were less than 0.01 were added to the protein pool. In this step, we found that a set of 4,820 proteins is too large for constructing the PPI network for *C. albicans* and then narrowed the protein pool of *C. albicans* to avoid overfitting in the parameter identification for the PPI network construction. So utilizing the GO database to further select a hyphal growth protein pools within the 4,820 proteins set, we constructed a hyphal PPI network for *C. albicans* consisting of a subset of 403 proteins identified as related to hyphal growth. In addition, we were able to locate the beginning of hyphal growth in the body of the zebrafish at 2–4 hpi from microscopy images of the experiment (Fig. 2). Therefore, we selected 598 additional proteins whose mRNA levels changed by more than twofold in 1–6 h to hyphal growth protein pool. Most of these 598 proteins had not yet been confirmed as associated with hyphal growth. Combining the 403 hyphae-related proteins with the 598 proteins yielded 1,001 proteins for the final hyphal growth protein pool. A candidate PPI network could be constructed based on this protein pool and PPI information. Since candidate PPI networks contain many false positive PPIs, the candidate PPI network was pruned using microarray data and based on a dynamic interaction model (*see* **Note 1**).

3.4 Dynamic Interaction Model for PPI Networks During Infection

The candidate PPI network can be depicted as a dynamic system in which interactive proteins and mRNA are considered as inputs of the system and protein activities as outputs of the system. All proteins in the candidate PPI network can be considered as target proteins. For a target protein p in the candidate PPI network with N interacting proteins, a dynamic model of the protein's activity can be represented as follows:

$$y_p[t+1] = y_p[t] + \sum_{q=1}^{Q_p} b_{pq} y_p[t] y_q[t] + \alpha_p x_p[t] - \beta_p y_p[t] + \omega_p[t] \quad \text{for } p = 1, 2, \ldots, N \tag{1}$$

where $y_p[t]$ represents the activity level of p at time t, b_{pq} denotes the interaction ability of the qth interactive protein to p, $y_q[t]$ represents the protein activity level of the q-th protein interacting with

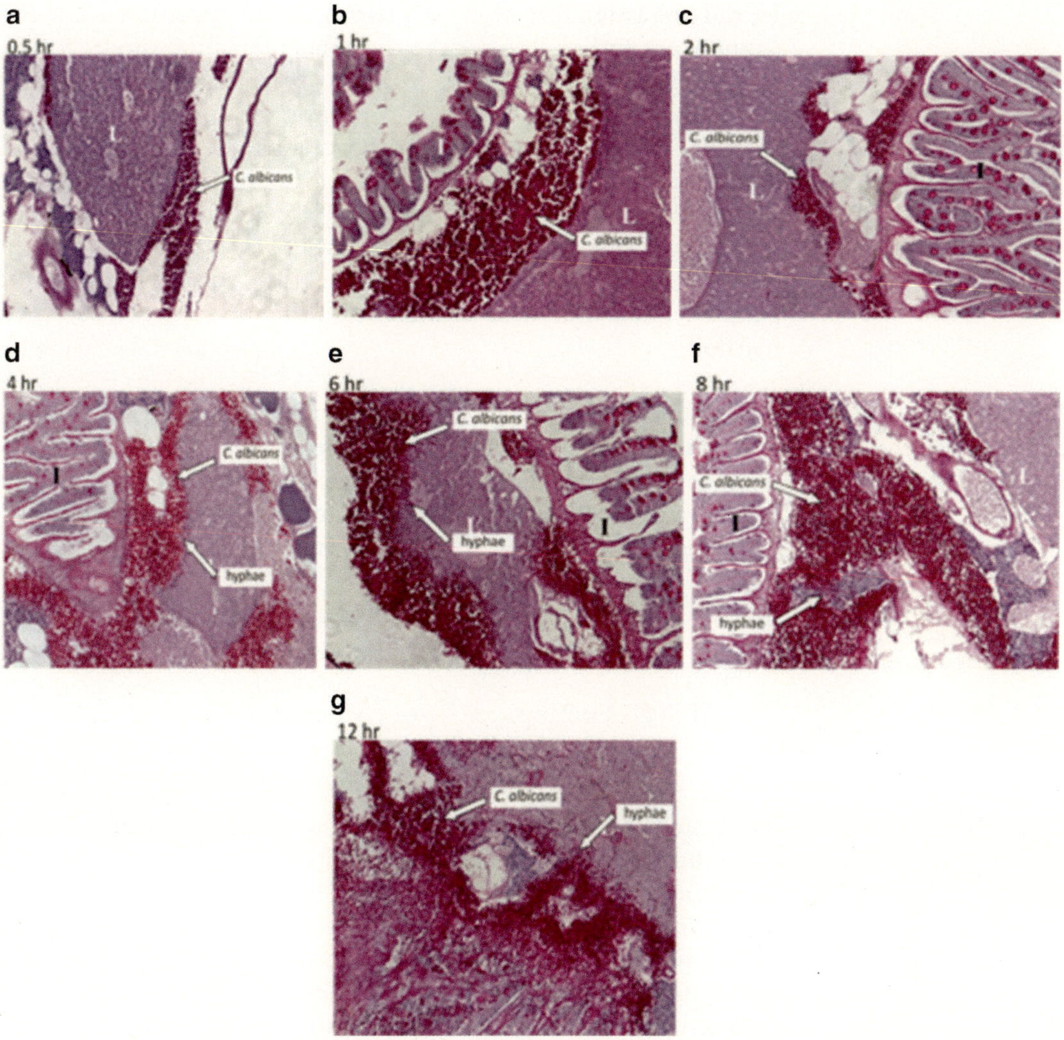

Fig. 2 Microscopy images of the infection process of *C. albicans* in zebrafish tissue. Infection of zebrafish with *C. albicans*. The respective time points are 0.5 (**a**), 1 (**b**), 2 (**c**), 4 (**d**), 6 (**e**), 8 (**f**), and 12 h (**g**). "L" indicates liver and "I" indicates intestines. It is apparent that hyphae began to grow between the 2 and 4 h

p, α_p denotes the translation rate from mRNA to protein, $x_p[t]$ represents the mRNA expression level of p, β_p indicates the decay rate of the protein, and $\omega_p[t]$ is stochastic noise. The PPI rate is proportional to the product of two proteins' concentrations [18], and thus the interaction is modeled as a nonlinear multiplication scheme. The biological interpretation of Eq. 1 is that the protein activity level of target protein p at time $t+1$ is a function of the present protein activity level plus regulatory interactions with Q_p interactive proteins, plus additive translation effects from mRNA, minus present protein degradation effects, and plus some stochastic noise. Because of the undirected nature of protein interactions, we did not assign direction for a two-protein interaction in the PPI subnetwork in Eq. 1. After the dynamic interaction model for the *p*th protein is constructed as in Eq. 1, the interaction parameters b_{pq}, translation parameter α_p and decay rate β_p can be estimated from microarray data by least square parameter estimation method. Since the number of interactions in a candidate PPI network is dependent on the biological situation or condition used instudies, there exist many false positives and several interactions may not be relevant for our purposes. Therefore, the estimated interaction parameters $\hat{b}_{pq}$ should be pruned using the model order detection method.

3.5 Determination of Significant Interaction Pairings

After identifying the regulatory interaction parameters $\hat{b}_{pq}$, Akaike Information Criterion (AIC) [19] is then employed for both model order selection and determination of significant interactions in the infection PPI networks, i.e., to determine the number of interactions Q_P in Eq. 1. The AIC, which includes both the estimated residual error and model complexity in one statistical measure, decreases as the residual error decreases and increases as the number of interactions (i.e., complexity) increases [20].

$$\mathrm{AIC}(Q_P) = 2\log\varepsilon_P + \frac{2Q_P}{L}, \quad \text{where } \varepsilon_P = \Upsilon_P - \Phi_p\hat{\theta}_P \tag{2}$$

As the expected residual error decreases with increasing interactions for inadequate model complexities, there should be a minimum located near the correct interaction number [19, 20]. Therefore, *AIC* can be used to select model order (i.e., the number of interactions) based on the protein interaction abilities $\hat{b}_{pq}$ identified above. After constructing the PPI networks for host and pathogen, we constructed a network for the protein interactions between pathogen and host in the following (*see* **Note 2**).

3.6 Construction of an Intercellular PPI Networks Between Pathogen and Host

To identify the intercellular PPIs between pathogen and host during infection of zebrafish with *C. albicans*, we utilized the Temporal Relationship Identification Algorithm (TRIA), which uses gene expression data to identify a given transcription factor's regulatory targets from its binding targets as inferred from ChIP-chip data [21]. The first step was to build a pool of *C. albicans* cell surface proteins. We used the GO database to select 195 cell surface proteins to build the resultant protein pool for *C. albicans*. Because host resistance against *C. albicans* infections is mediated predominantly by phagocytes, namely neutrophils and macrophages [22, 23], we assumed that cell surface proteins of *C. albicans* may interact with any protein of zebrafish in the infectious process. So we let $\overline{x} = (x_1,,\ldots,,x_N)$ denote the gene expression time profile of *C. albicans* cell surface protein *x* and $\overline{y} = (y_1,,\ldots,,y_N)$

denote the gene expression time profile of zebrafish protein y. We constructed the protein interactions between *C. albicans* and zebrafish via cross-correlation calculations of their time series microarray data.

We compute the cross-correlation between $\overline{x}$ and $\overline{y}$ with a lag of k time points as follows:

$$c(k) = \frac{\sum_{i=1}^{N-k}(y_{i+k} - \overline{y})(x_i - \overline{x})}{\sqrt{\sum_{i=1}^{N-k}(y_{i+k} - \overline{y})^2} \cdot \sqrt{\sum_{i=1}^{N-k}(x_i - \overline{x})^2}}, \quad k = 0,1,\ldots,T \tag{3}$$

where $\overline{y} \triangleq \sum_{i=1}^{N-k} y_{i+k} \Big/ N-k$, $\overline{x} \triangleq \sum_{i=1}^{N-k} x_i \Big/ N-k$ and T is the maximal time lag after the *C. albicans* infection. We interpolated the 9 time points available for both *C. albicans* and zebrafish into 36 time points. The interval between each time point was half an hour. In this study, we set $T=8$, meaning that we computed the cross-correlation between a *C. albicans* cell surface protein and a zebrafish protein for all possible time lags less than 4 h. Although the beginning of hyphal growth in the body of the zebrafish occurs at 2–4 h post-infection, we assumed the hyphae-related proteins of *C. albicans* might influence zebrafish proteins ahead of 4 h post-infection. Then, we tested the null hypothesis $H_0{:}c(k)=0$ (i.e., the cell surface proteins of *C. albicans* and zebrafish proteins are uncorrelated) and the alternative hypothesis $H_a{:}c(k)\neq 0$ by the bootstrap method [24] to obtain a p-value. After all cross-correlations were calculated, we set the constraint that cross-correlation levels must be higher than 0.95. The PPIs satisfying this constraint were considered as potential intercellular PPIs between *C. albicans* and zebrafish (*see* **Note 3**).

4 Notes

1. Our aim is to construct the integrated intercellular interaction network between the hyphal proteins of *C. albicans* and zebrafish proteins during the infection process. The flowchart for construction is shown in Fig. 1, and has three main routes, among which two separately construct the hyphal PPI network of *C. albicans* and the PPI network of zebrafish. The third constructs the host–pathogen intercellular PPI network. Based on the microarray data, we selected 4,820 and 9,665 proteins for inclusion in the source protein pools of *C. albicans*

and zebrafish respectively. In addition, we selected 1,001 proteins for inclusion in the hyphal growth protein pool from the *C. albicans* protein pool due to the need to investigate what factors are behind the transition from yeast form to hyphal form in the infection process. In the candidate *C. albicans* hyphal PPI network, there were 3,604 protein–protein interactions; in the candidate zebrafish PPI network, there were 1,129.

2. We utilize the 9-time-point *C. albicans* time series microarray data to construct two dynamic networks for different infection stages. Since hyphae appear to begin to grow in the zebrafish body from 2 to 4 hpi in the experimental microscopy images (Fig. 2), we collected two groups of data at different stages of infection to construct two separate networks. With the *C. albicans* microarray data spanning 0.5–4 h, we constructed a network called the 'adhesive stage network', which represents *C. albicans* cells in the adhesion stage. Since cubic spline interpolation requires at least four data points to solve a cubic polynomial [25], we included the 4 h data point to construct this network. With the *C. albicans* microarray data spanning 2–12 h, we constructed another network called the 'hyphal stage network', which represents *C. albicans* cells transitioning to the hyphal form. Similarly, we collected two groups of data at different stages of infection to construct two separate PPI networks for zebrafish as well: one for microarray data from 0.5 to 4 h, and another for data from 2 to 12 h, named the zebrafish stage 1 network and zebrafish stage 2 network, respectively. By estimating the system parameters using the time course microarray data and selecting model order using the AIC [19, 20], the likelihood of false positive interactions in the potential PPI network for the infection process was reduced. Network refinement yielded 550 proteins with 2,725 PPIs in the adhesive stage network and 555 proteins with 3,171 PPIs in the hyphal stage network: these two networks could then be combined into the *C. albicans* dynamic hyphal PPI network for the infection process (Fig. 3). Similar refinements in the zebrafish data returned 1,248 proteins with 2,344 PPIs in the zebrafish stage 1 network and 1,265 proteins with 2,379 PPIs in the zebrafish stage 2 network, and these two networks could then be combined into the zebrafish dynamic PPI network for the defensive process (Fig. 4). The *C. albicans* dynamic hyphal PPI network, the zebrafish dynamic PPI network, and the host–pathogen intercellular PPI network could be merged into an integrated infection intercellular PPI network.

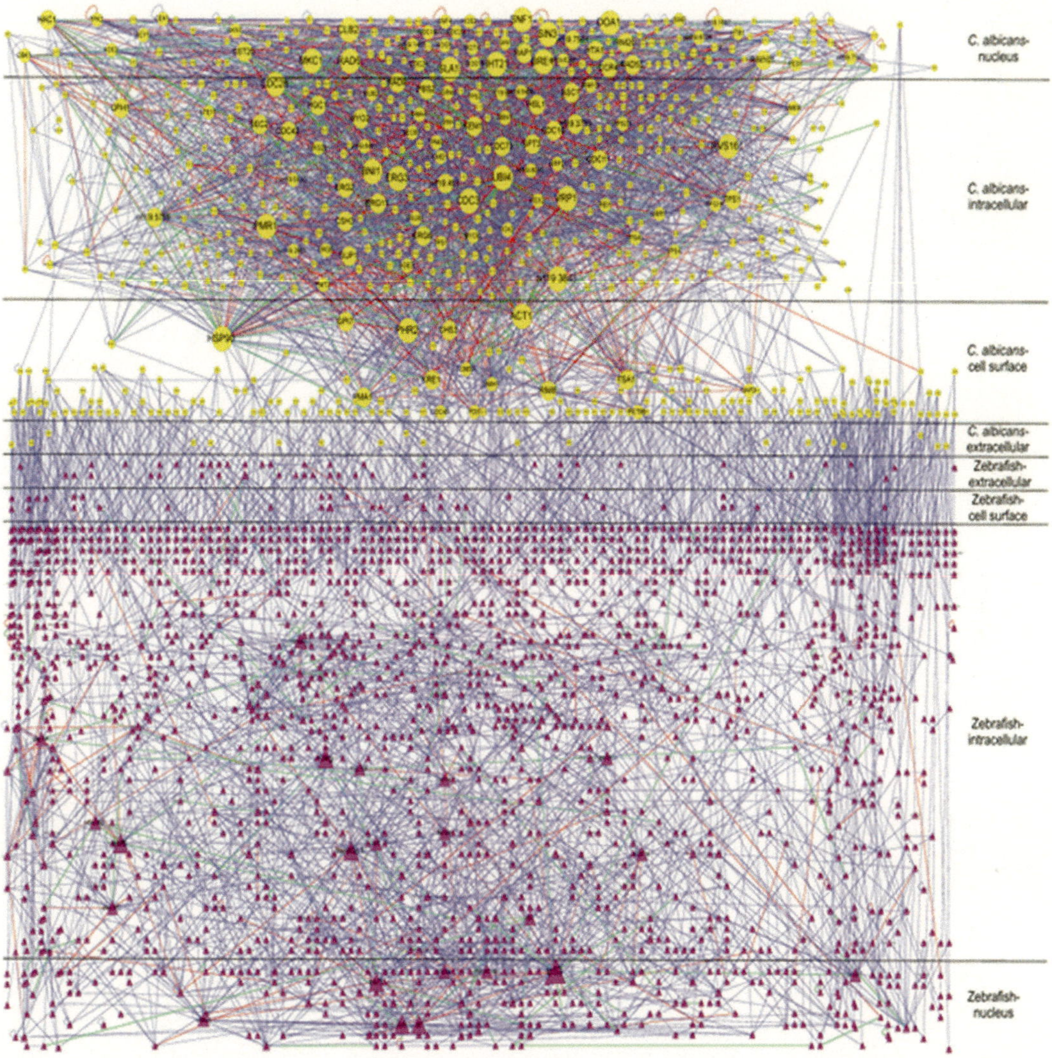

Fig. 3 Dynamic hyphal PPI network of *C. albicans* during infection. The dynamic hyphal PPI network of *C. albicans* contains 3,452 PPIs among 557 proteins. This network contains three different color lines. The *red lines* denote PPIs that did not appear in the adhesive stage network but did in the hyphal stage network. The *green lines* indicate PPIs that appeared in the adhesive stage network but did not appear in the hyphal stage network. The *blue lines* and *yellow nodes* indicate the PPIs and proteins (respectively) that appeared in the adhesive stage network and the hyphal stage network. The node size denotes the connectivity degree

3. The global system view of the *C. albicans*- and zebrafish-integrated infection intercellular PPI network is illustrated in Fig. 5. The entire integrated infection intercellular network can be divided into eight levels according to the location of protein action (i.e., nucleus, intracellular, cell surface, or extracellular) and species (i.e., *C. albicans* or zebrafish), and is composed

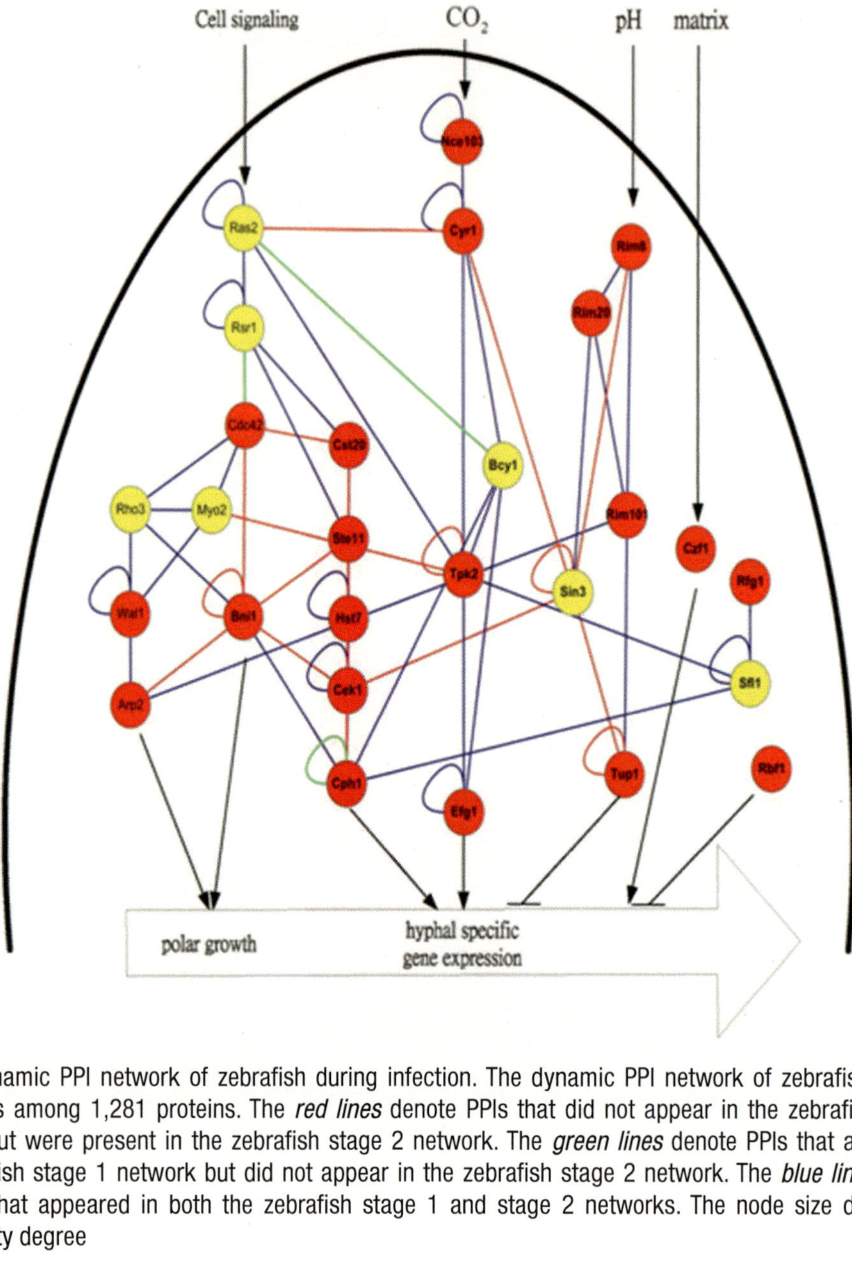

Fig. 4 Dynamic PPI network of zebrafish during infection. The dynamic PPI network of zebrafish contains 2,500 PPIs among 1,281 proteins. The *red lines* denote PPIs that did not appear in the zebrafish stage 1 network but were present in the zebrafish stage 2 network. The *green lines* denote PPIs that appeared in the zebrafish stage 1 network but did not appear in the zebrafish stage 2 network. The *blue lines* indicate the PPIs that appeared in both the zebrafish stage 1 and stage 2 networks. The node size denotes the connectivity degree

of three subnetworks. The upper subnetwork is the dynamic hyphal PPI network of *C. albicans*. The middle subnetwork shows the host–pathogen intercellular interaction network. For simplicity, only the top five correlated interactions of the *C. albicans* cell surface proteins are listed. The bottom subnetwork is the dynamic defensive protein interaction network of zebrafish.

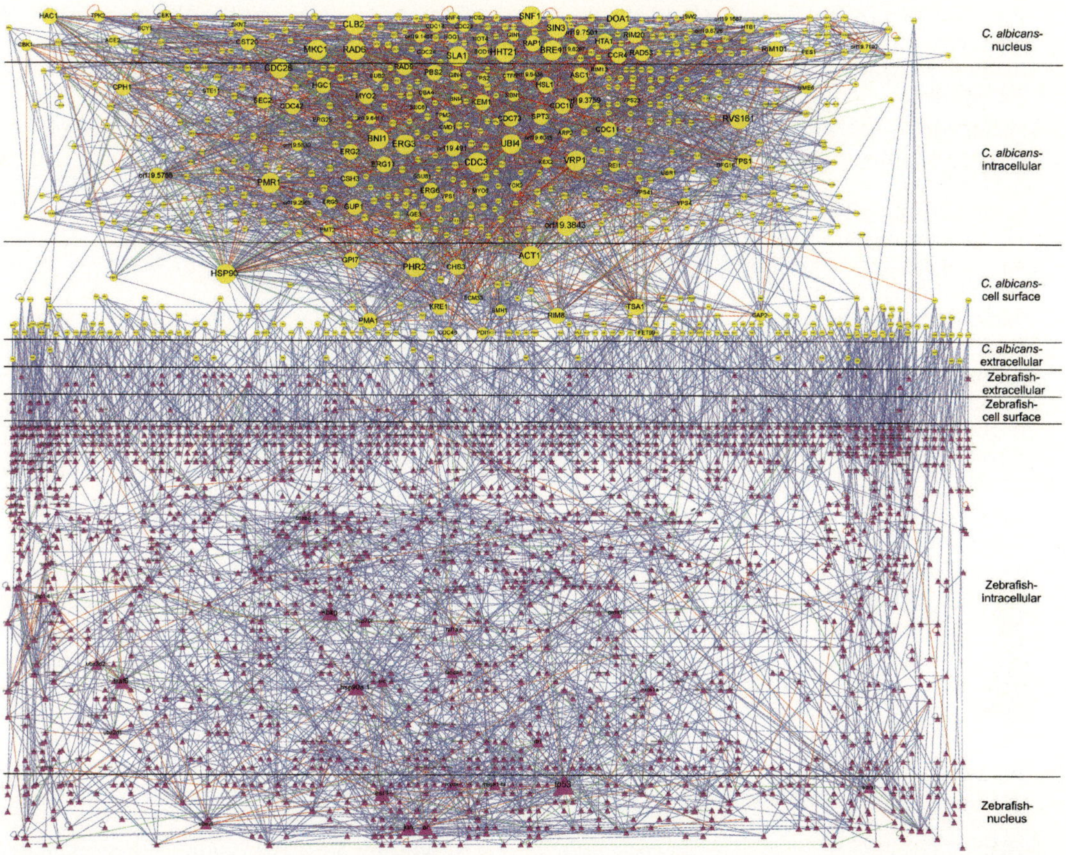

Fig. 5 Integrated intercellular dynamic PPI network during *C. albicans* infection. The infectious intercellular network is composed of three subnetworks. The *upper* subnetwork is the dynamic hyphal PPI network of *C. albicans*. The *middle* subnetwork shows the host–pathogen intercellular interaction network. For simplicity, only the top five correlated interactions of the *C. albicans* cell surface proteins are listed. The *bottom* subnetwork is the dynamic defensive protein interaction network of zebrafish. This infectious intercellular PPI network contains lines and nodes of three different colors. The *red lines* denote PPIs that did not appear in the stage 1 network but did in the stage 2 network. The *green lines* denote PPIs that appeared in the stage 1 network but did not in the stage 2 network. The *blue lines* denote PPIs that appeared in both the stage 1 and 2 networks. The node size denotes connectivity degree

References

1. Leroy O, Gangneux JP, Montravers P, Mira JP, Gouin F, Sollet JP et al (2009) Epidemiology management, and risk factors for death of invasive Candida infections in critical care: a multicenter, prospective, observational study in France (2005–2006). Crit Care Med 37: 1612–1618
2. Kojic EM, Darouiche RO (2004) Candida infections of medical devices. Clin Microbiol Rev 17:255–267
3. Seneviratne CJ, Jin L, Samaranayake LP (2008) Biofilm lifestyle of Candida: a mini review. Oral Dis 14:582–590
4. Pfaller MA, Diekema DJ (2007) Epidemiology of invasive candidiasis: a persistent public health problem. Clin Microbiol Rev 20:133–163
5. Olorode OA, Okpokwasili GC (2012) The efficacy of disinfectants on abattoirs' Candida albicans isolates in Niger Delta region. Mycoses 55:323–323

6. Lo HJ, Kohler JR, DiDomenico B, Loebenberg D, Cacciapuoti A, Fink GR (1997) Nonfilamentous C. albicans mutants are avirulent. Cell 90:939–949
7. Calderone RA, Fonzi WA (2001) Virulence factors of Candida albicans. Trends Microbiol 9:327–335
8. Leberer E, Ziegelbauer K, Schmidt A, Harcus D, Dignard D, Ash J et al (1997) Virulence and hyphal formation of Candida albicans require the Ste20p-like protein kinase CaCla4p. Curr Biol 7:539–546
9. Ihmels J, Bergmann S, Berman J, Barkai N (2005) Comparative gene expression analysis by differential clustering approach: application to the Candida albicans transcription program. PLoS Genet 1:e39
10. Goffeau A, Barrell BG, Bussey H, Davis RW, Dujon B, Feldmann H et al (1996) Life with 6000 genes. Science 274:546, 563–567
11. Meeker ND, Trede NS (2008) Immunology and zebrafish: spawning new models of human disease. Dev Comp Immunol 32: 745–757
12. Sullivan C, Kim CH (2008) Zebrafish as a model for infectious disease and immune function. Fish Shellfish Immunol 25:341–350
13. Amsterdam A, Hopkins N (2006) Mutagenesis strategies in zebrafish for identifying genes involved in development and disease. Trends Genet 22:473–478
14. Postlethwait J, Amores A, Force A, Yan YL (1999) The zebrafish genome. Methods Cell Biol 60:149–163
15. Chao CC, Hsu PC, Jen CF, Chen IH, Wang CH, Chan HC et al (2010) Zebrafish as a model host for Candida albicans infection. Infect Immun 78:2512–2521
16. Orntoft TF, Thykjaer T, Waldman FM, Wolf H, Celis JE (2002) Genome-wide study of gene copy numbers transcripts, and protein levels in pairs of non-invasive and invasive human transitional cell carcinomas. Mol Cell Proteomics 1:37–45
17. Newman JRS, Ghaemmaghami S, Ihmels J, Breslow DK, Noble M, DeRisi JL et al (2006) Single-cell proteomic analysis of S-cerevisiae reveals the architecture of biological noise. Nature 441:840–846
18. Alon U (2007) An introduction to systems biology: design principles of biological circuits. Chapman & Hall/CRC, Boca Raton, FL
19. Akaike H (1974) New look at statistical-model identification. IEEE Trans Automat Contr Ac19:716–723
20. Johansson R (1993) System modeling and identification. Prentice Hall, Englewood Cliffs, NJ
21. Wu WS, Li WH, Chen BS (2007) Identifying regulatory targets of cell cycle transcription factors using gene expression and ChIP-chip data. BMC Bioinformatics 8:188
22. Edwards JE Jr, Rotrosen D, Fontaine JW, Haudenschild CC, Diamond RD (1987) Neutrophil-mediated protection of cultured human vascular endothelial cells from damage by growing Candida albicans hyphae. Blood 69:1450–1457
23. Hummert S, Hummert C, Schroter A, Hube B, Schuster S (2010) Game theoretical modelling of survival strategies of Candida albicans inside macrophages. J Theor Biol 264: 312–318
24. Efron B, Tibshirani R (1993) An introduction to the bootstrap. Chapman & Hall, New York, NY
25. Dyer SA, Dyer JS (2001) Cubic-spline interpolation: part 1. IEEE Instrum Meas Mag 4:44–46

Chapter 24

Procedures for Mucosal Immunization and Analyses of Cellular Immune Response to Candidate HIV Vaccines in Murine and Nonhuman Primate Models

Shailbala Singh, Pramod Nehete, Patrick Hanley, Bharti Nehete, Guojun Yang, Hong He, Scott M. Anthony, Kimberly S. Schluns, and K. Jagannadha Sastry

Abstract

Sampling the mucosal tissues and analyses of immune responses are integral to vaccine-development strategies against human immunodeficiency virus (HIV), which is transmitted predominantly across the oro-genital mucosa. While immune assay development and standardization attempts employ mouse models, immunogenicity and protective efficacy that can be extrapolated to humans are realized only from experiments in nonhuman primates. Here, we describe commonly used practices for immunizations in rhesus macaques (*Macaca mulatta*) along with procedures for obtaining important mucosal tissues samples from macaques and mice. We also describe detailed protocols for two important assays applicable in mouse as well as primate experiments for determining antigen-specific T cells responses induced after vaccination.

Key words HIV–AIDS, Vaccines, Animal models, Nonhuman primates, Rhesus macaques, Mucosal immunity, T cells, ELISPOT, Cytokine, Flow cytometry

1 Introduction

Vaccination in general may be the most cost-effective strategy against global infectious diseases. This cannot be emphasized enough in case of the acquired immunodeficiency syndrome (AIDS) induced by human immunodeficiency virus (HIV) infection, an epidemic with enormous monitory and human resources being expended [1–3]. Despite worldwide efforts over the past few decades a vaccine against HIV–AIDS is still not a reality. Incredible amounts of variations among the strains prevalent around the world have been formidable obstacles [3, 4]. However, great strides have been made in the understanding of the biology and pathology of HIV infection mainly due to the research involving

Rajat K. De and Namrata Tomar (eds.), *Immunoinformatics*, Methods in Molecular Biology, vol. 1184,
DOI 10.1007/978-1-4939-1115-8_24,

nonhuman primate (NHP) models. Investigations employing a variety of NHP species created a wealth of knowledge mainly because of the close genetic links between NHP and humans, more specifically the similarities with respect immune and hematopoietic organizations [5, 6]. Nonhuman primates have been a critical component to the successes in molecular biology, largely in part due to the advances in procedure development in techniques for vaccination and immune function monitoring [7]. Over the past 20+ years our laboratory has investigated the suitability of synthetic peptides, recombinant proteins, or immunogens expressed from viral vectors as HIV vaccine candidates for their efficiency to induce immune responses in multiple systemic and mucosal tissues [8–10]. We describe in this chapter key methods for immunization, harvesting tissues or biopsies of mucosal tissues, immune assays to detect T cell responses. Detailed protocols covering these techniques from HIV vaccine studies mainly in rhesus macaques and some in mice are described below.

2 Procedures Describing Immunization by Various Routes in Rhesus Macaques

Nonhuman primates are routinely restrained with ketamine, xylazine, telazol, or domitor for any procedures that require handling outside the cage [11, 12]. When appropriate, administration of biohazardous substances should be conducted with the primate located inside a biosafety cabinet (designated Class IIB) in the biocontainment laboratory or, alternatively, biosafety level 3 (BSL 3) practices should be utilized. Prior to experimental inoculation, the base level immune responses must be determined for future comparison with the post-immunization values such that each animal can serve as its own control in addition to having separate group of animals that are either unimmunized or mock-immunized with control reagents. Usually blood and tissue biopsy samples are collected from each animal and single cell suspensions are prepared for either immediate analysis of T cells responses or banked along with serum samples for later batch analyses. Experimental animals can receive the vaccine formulation by any of multiple routes. The route as well as concentration and volume of the vaccine administration must be conducted according to the protocols approved by the institutional animal care and use committee (IACUC). Administration of the vaccine can be accomplished in a variety of different routes that include different mucosal tissues as well as topical, intramuscular, intravenous, intradermal, subcutaneous, and intraperitoneal [13–16]. Beyond these more traditional routes, the use of a gene gun and electroporation are also options for delivering vaccines [17].

2.1 Intravaginal (IVAG) Immunization

2.1.1 Materials

1. Visualization device (depending on size of vaginal opening).
   (a) Vaginal speculum.
   (b) Colposcope.
   (c) Otoscope with appropriate sized ear cone.
2. Appropriate syringes (1–3 ml) and needles (normally 23–25 G).
3. Catheter (22–25 G×2–3 in.) for topical administration or feeding tube.
4. Inoculum.

2.1.2 Methods

1. Animal is anesthetized using standard anesthesia techniques [11, 12].
2. Animal is placed in ventral recumbency.
3. Perineum is cleaned using a betadine solution or chlorhexidine solution.
4. Visualization device is inserted and inoculation site identified.
5. If required, the animal may be placed in a biosafety cabinet.
6. If the inoculum is to be injected, then a small gauge needle attached to a syringe is used (*see* **Note 1**).
7. If the inoculum is to be applied topically, then either a feeding tube or catheter without the stylet is used to infuse the inoculum slowly (*see* **Notes 2** and **3**).
8. Following infusion, the animal should be observed immediately post infusion, 1 h post infusion, and 24 h post infusion for any signs of toxicity.

2.2 Intrarectal (IR) Immunization

2.2.1 Materials

1. Visualization device (depending on depth of inoculation).
   (a) Flexible endoscope.
   (b) Anoscope.
   (c) Otoscope with appropriate sized ear cone.
2. Catheter 22–25 G with 2–3 in. length.
3. Inoculum.

2.2.2 Methods

1. Animal is anesthetized using standard anesthesia techniques.
2. Animal is placed in ventral recumbency.
3. Perineum is cleaned using a betadine solution or chlorhexidine solution.
4. If required, the animal may be placed in a biosafety cabinet.
5. Depending on the needed depth of the infusion, then a flexible endoscope, anoscope, or otoscope may be used to visualize inoculation site (*see* **Note 2**).
6. Once identified, the inoculum can be administered using either the biopsy channel in the endoscopes or a catheter with the stylet removed.

7. Following infusion, the animal should be observed immediately post infusion, 1 h post infusion, and 24 h post infusion for any signs of toxicity.

2.3 Intranasal (IN) Immunization

2.3.1 Materials

1. Catheter without stylet 22–25 G with 1–2 in. length.
2. Inoculum.

2.3.2 Methods

1. Animal is anesthetized using standard anesthesia techniques.
2. Animal is placed in either right or left lateral recumbency depending on the side to be inoculated.
3. If required, the animal may be placed in a biosafety cabinet.
4. The inoculum is infused slowly (over 1 min) into the nasal cavity of choice (*see* **Note 4**).
5. Following infusion, the animal should be observed immediately post infusion, 1 h post infusion, and 24 h post infusion for any signs of toxicity.

2.4 Intratracheal Immunization

2.4.1 Materials

1. Flexible rhinoscope or endoscope (3.5 mm) or feeding tube.
2. Endotracheal tube (size dependent on monkey).
3. Inoculum.

2.4.2 Methods

1. Animal is anesthetized using standard anesthesia techniques.
2. Animal is intubated via standard procedure.
3. If required, the animal may be placed in a biosafety cabinet.
4. The inoculum is infused slowly (over 1 min) into the trachea either through the biopsy channel on the flexible rhinoscope/endoscope or feeding tube.
5. Following the procedure, the animal should be extubated via standard procedure.
6. Following infusion, the animal should be observed immediately post infusion, 1 h post infusion, and 24 h post infusion for any signs of toxicity.

2.5 Intratonsillar Immunization

2.5.1 Materials

1. Laryngoscope.
2. Appropriate sized needles/syringes.
3. Catheter without stylet 22–25 G with 1–2 in. length.
4. Inoculum.

2.5.2 Methods

1. Animal is anesthetized using standard anesthesia techniques.
2. Animal may need to be intubated via standard procedure.
3. Animal is placed in either dorsal or ventral recumbency depending upon the injector's preference.

4. If required, the animal may be placed in a biosafety cabinet.
5. Use the laryngoscope to visualize the tonsils.
6. Using the appropriate length needle with a small gauge inject the inoculum directly into the tonsils.
7. Monitor for hemostasis following the injection.
8. Extubate the animal via standard procedure.
9. Following infusion, the animal should be observed immediately post infusion, 1 h post infusion, and 24 h post infusion for any signs of toxicity.

2.6 Topical Immunization (Skin)

2.6.1 Materials

1. Hair Clippers.
2. Exfoliating pad.
3. Adhesive (Stripping) Tape.
4. Pipettes.
5. Transcutaneous immunization device.
6. Inoculum.

2.6.2 Methods

1. Animal is anesthetized using standard anesthesia techniques.
2. Animal is placed in a manner to give the best access to the area to be inoculated.
3. If required, the animal may be placed in a biosafety cabinet.
4. The area to be inoculated is clipped and cleaned with 70 % isopropyl alcohol or other appropriate cleaning agent.
5. Exfoliation method:
   (a) Rub the area approximately 50 times with the exfoliation pad.
   (b) Apply adhesive tape to area and remove.
   (c) Turn the tape 90°, reapply, and then remove.
6. For topical immunization use a pipette to apply inoculum to the exfoliated area. Ensure that the inoculum is spread over the entire area.
7. For transcutaneous immunization, use a specialized device to inject the inoculum into the site of interest.
8. There may be erythema at the site of inoculation following immunization but it should resolve within 48 h.
9. Following infusion, the animal should be observed immediately post infusion, 1 h post infusion, and 24 h post infusion for any signs of toxicity.

2.7 Intramuscular (IM) Immunization

2.7.1 Materials

1. Appropriate syringes and needles.
2. Squeeze back cage.
3. Inoculum.

2.7.2 Methods

1. If elected, intramuscular injections may be administered to anesthetized animals. Anesthetize the animal according to standard procedures. When the animal is fully anesthetized, proceed by following **steps 3** through **6**.
2. To administer injections to conscious animals, immobilize the monkey into a lateral profile by pulling the back of the squeeze cage forward (*see* **Note 5**).
3. Preferred areas for injection are the hamstrings, quadriceps, and/or triceps muscles.
4. Insert the needle with the attached syringe into the muscle.
5. Pull back slightly on the syringe plunger. If there is no evidence of blood, slowly advance plunger to inject the drug. If there is evidence of blood, withdraw the needle slightly to reposition it and repeat **step 4**.
6. Remove the syringe unit and deposit into a sharps container (*see* **Note 6**).
7. Return the squeeze-back device to its normal position to release the animal.
8. In case of an anesthetized animal, return the squeeze-back device to the normal position only when the animal is sitting unsupported in an upright position and appears alert.
9. Following injection, the animal should be observed immediately post injection, 1 h post injection, and 24 h post injection for any signs of toxicity.

2.8 Intravenous (IV) Immunization

2.8.1 Materials

1. Appropriate syringes and needles.
2. Appropriate sized catheter.
3. Inoculum.

2.8.2 Methods

1. Anesthetize animals per standard procedure.
2. Shave the hair around the cephalic vein (dorsum of the forearm) or the saphenous vein (back of the leg).
3. Cleanse the area with either a betadine scrub or Nolvasan scrub.
   (a) Scrub the area in a circular motion, begin with small circles and work outwards.
   (b) Wipe clean with 70 % isopropyl alcohol.
   (c) Repeat **steps 3a**, **b** at least three times.
4. Using aseptic technique, with the bevel of the needle up introduce the catheter with stylet through the skin and into the vein.
5. Identify a flash of blood.
6. Slowly thread the catheter into the vein.
7. Remove the stylet.

8. Attach an injection plug or a stopcock to the catheter.
9. If administering test article or medication, flush the catheter with saline or heparinized saline for longer catheter patency (*see* **Note 7**).
10. Check for perivascular leaks (*see* **Note 8**).
11. If no leaks are apparent, tape catheter in place and attach syringe or infusion set.
12. If a leak develops, remove the catheter and repeat **steps 4** through **8** proximal to the original site of the catheter insertion or in a different arm/leg.
13. Once the catheter is in place, infuse the inoculum through the catheter (*see* **Note 9**).
14. Following injection, the animal should be observed immediately post injection, 1 h post injection, and 24 h post injection for any signs of toxicity.

2.9 Intradermal (ID) Immunization

2.9.1 Materials

1. Appropriate syringes and needles.
2. Squeeze back cage.
3. Inoculum.

2.9.2 Methods

1. If elected, intradermal injections may be administered to anesthetized animals. Anesthetize animals via standard protocol and once the animal is fully anesthetized, proceed by following **steps 3** through 7.
2. To administer injections to conscious animals, immobilize the monkey by pulling the back of the squeeze cage forward (*see* **Note 5**).
3. Shave, or use a pre-shaved area to inject (*see* **Note 10**).
4. By holding the skin taut, insert the needle bevel up just under the surface of the skin at an angle of 15–20° until the bevel is covered and inject slowly. A distinct bleb must form at the site of inoculation.
5. Remove the syringe unit and deposit it into a sharps container (*see* **Note 6**).
6. Return the squeeze-back device to its normal position.
7. Following injection, the animal should be observed immediately post injection, 1 h post injection, and 24 h post injection for any signs of toxicity.

2.10 Subcutaneous (SC) Immunization

2.10.1 Materials

1. Appropriate syringes and needles.
2. Squeeze back cage.
3. Inoculum.

2.10.2 Methods

1. If elected, subcutaneous injections may be administered to anesthetized animals. Anesthetize per standard procedures and once the animal is fully anesthetized, proceed by following **steps 3** through 7.
2. To administer injections to conscious animals, immobilize the monkey into a lateral profile by pulling the back of the squeeze cage forward (*see* **Note 5**).
3. Preferred areas for subcutaneous injections are the lateral flanks.
4. Shave, or use a pre-shaved area to inject (*see* **Note 10**).
5. If possible, grasp the skin and pull it away from the body slightly.
6. At the end of the skin closest to body, insert the needle with attached syringe through the skin approximately at an angle of 45°.
7. Pull back the syringe plunger slightly to ensure that the needle has not penetrated any blood vessel. If there is no evidence of blood, slowly advance the plunger to inject the drug. If there is evidence of blood, withdraw the needle slightly to reposition it and repeat **step 6**.
8. Remove the syringe unit and deposit it into a sharps container (*see* **Note 6**).
9. Return the squeeze-back device to its normal position.
10. Following injection, the animal should be observed immediately post injection, 1 h post injection, and 24 h post injection for any signs of toxicity.

Notes

1. The inoculum is injected and a small "bleb" should be identified.
2. The perineum of the animal should be elevated prior to the infusion and approximately 10–20 min following the procedure to prevent any leakage.
3. If necessary, a vaginal speculum, otoscope, or colposcope may be used to visualize the cervix.
4. The head should remain as lateral as possible to avoid leakage out of the nostril or leakage into the throat. Maintain the head in lateral position for at least 3–5 min if possible.
5. Use caution not to injure the monkey during the squeeze operation.
6. Do not recap the needle.
7. A saline flush is not required for IV fluids.
8. Leaks are noted if the fluid begins to form a subcutaneous bump on the skin around the injection site.

9. If necessary, an infusion pump may be used to control the rate of infusion.
10. When performing on conscious animals, shave the injection site at a prior sedation.

3 Collection of Peripheral Blood and Body Fluids in Rhesus Macaques

Vaccine-mediated induction of humoral and cellular immune responses is generally determined in the blood and tissue samples, and in case of HIV a variety of mucosal tissues are also sampled to realize the effectiveness in inducing immunity at these portals of viral entry [15, 16]. Although venipuncture is the normal route of blood collection arterial puncture may be necessary for certain parameters. Size of the animal is an important consideration for blood collection as it determines the blood volume that can be safely removed without any adverse consequences for the animal. Additional samples collected in relation to mucosal routes of vaccination, specifically oral and intranasal include saliva and secretions/washes from vaginal, rectal, bronchial, and nasal tissues. Tissue collection is also an important need for most experiments. Most commonly, during the course of the experiment that does not involve sacrificing the animals, either pinch or punch biopsies are used to obtain the tissue. If the anatomic location of tissue is internal such as liver or kidney, then visualization through either ultrasound or directly via laparoscopy or open surgery is preferred. Tissue sample from the lumen of an organ (e.g., stomach, colon) can be retrieved via a natural orifice by using flexible endoscopy. Collection of biopsies from lymph nodes is a common practice for most vaccination studies. Normally these biopsies are performed on peripheral lymph nodes (e.g., inguinal or axillary); however, sometimes specific lymph nodes may also be required due to their association with a certain tissue and/or route of immunization. In those cases, rigid endoscopy or open surgery may be necessary.

3.1 Blood

3.1.1 Materials

1. Appropriate syringes and needles.
2. Vacutainer blood collection device.
3. Appropriate blood collection tubes.

3.1.2 Methods

1. Anesthetize animals per standard procedure.
2. Position the monkey to allow for venous access.
3. Prepare the site of the blood collection by rubbing with either 70 % isopropyl alcohol-soaked gauze or cotton balls, or alcohol wipes.
4. Palpate the pulse if the targeted vein is not visible.

5. For collection of blood sample using a syringe, proceed as described in **steps a–e**.
   (a) Penetrate the skin with syringe/needle unit at an appropriate angle.
   (b) Apply slight negative pressure to plunger to draw out the blood from the vein.
   (c) Allow the blood to fill the syringe to the desired amount by maintaining negative pressure on the plunger.
   (d) Remove the syringe/needle unit and dispense blood into the appropriate blood collection tube.
   (e) Immediately following the removal of the needle, apply pressure to the area to maintain hemostasis (*see* **Note 1**).
6. For collection of blood sample using a vacutainer, proceed as described in **steps a–f**.
   (a) Align the vacutainer with the needle at the same angle of entry as used with the syringe/needle method.
   (b) Following complete entrance of the needle into the skin, attach a blood collection tube to the vacutainer and advance the needle to the correct depth to adequately fill the tube.
   (c) Once the blood collection tube is filled, remove the tube from the vacutainer.
   (d) If other tubes are needed, insert another blood collection tube into the vacutainer and repeat step c.
   (e) When the blood is collected in the last tube, gently remove the vacutainer system with needle from the animal.
   (f) Apply pressure to the area to maintain hemostasis (*see* **Note 1**).

3.2 Bronchoalveolar Lavage (BAL)

3.2.1 Materials

1. Flexible rhinoscope or endoscope (3.5 mm) or feeding tube.
2. Endotracheal tube (size dependent on monkey).
3. Sterile saline (amount dependent on monkey size).

3.2.2 Methods

1. Animal is anesthetized using standard anesthesia techniques.
2. Animal is intubated via standard procedure.
3. If required, the animal may be placed in a biosafety cabinet.
4. The sterile saline is rapidly infused into the bronchi through either the biopsy channel on the flexible rhinoscope/endoscope or the feeding tube.
5. After infusion, negative pressure is applied on either the feeding tube or biopsy channel to extract bronchial fluid.
6. Place the collected fluid into an appropriate container.

7. It may be necessary to reposition the feeding tube or scope for better fluid extraction (*see* **Note 2**).
8. Following the procedure, the animal should be extubated via standard procedure
9. The animal should be observed immediately post infusion, 1 h post infusion, and 24 h post infusion for any sign of respiratory distress.

3.3 Vaginal Secretions

3.3.1 Materials

1. Appropriate syringes (1–10 ml).
2. Catheter (22–25 G × 2–3 in.) or feeding tube (6–8 fr).
3. Sterile phosphate buffered saline (1× PBS).

3.3.2 Methods

1. Animal is anesthetized using standard anesthesia techniques.
2. Animal is placed in ventral recumbency.
3. Perineum is cleaned using a betadine solution or chlorhexidine solution.
4. If required, the animal may be placed in a biosafety cabinet.
5. Place the feeding tube or catheter into the vaginal vault and rapidly infuse approximately 3–5 ml of sterile 1× PBS into the vagina.
6. Aspirate the fluid into syringe and place into appropriate container.

3.4 Rectal Secretions

3.4.1 Materials

1. Appropriate syringes (1–10 ml).
2. Feeding tube (6–8 fr).
3. Sterile phosphate buffered saline (1× PBS).

3.4.2 Methods

1. Animal is anesthetized using standard anesthesia techniques.
2. Animal is placed in ventral recumbency.
3. Perineum is cleaned using a betadine solution or chlorhexidine solution.
4. If required, the animal may be placed in a biosafety cabinet.
5. It may be necessary to manual evacuate feces near the entrance of the rectum.
6. Place the feeding tube into the rectum up to a depth of approximately 4–5 cm and rapidly infuse approximately 3–5 ml of sterile 1× PBS into rectum.
7. Aspirate fluid into syringe and place into appropriate container.

3.5 Nasal Lavage

3.5.1 Materials

1. Catheter without stylet 22–25 G with 1–2 in. length.
2. Sterile saline (3–5 ml).
3. 15 or 50 ml conical tube.

3.5.2 Methods

1. Animal is anesthetized using standard anesthesia techniques.
2. Animal is placed in either right or left lateral recumbency depending on the side chosen for collection of nasal lavage fluids.
3. If required, the animal may be placed in a biosafety cabinet
4. With the head either tilted forward or hanging off table saline (3–5 ml) is infused using the catheter into the upper nostril and the fluid is captured into a 15 or 50 ml conical tube from the lower nostril.
5. The procedure can be repeated on the other side if necessary.
6. Fluid can also be aspirated from the nostril using the catheter, but be cautious about disrupting the nasal mucosa (*see* **Note 3**).

Notes

1. Always monitor the animals for hemostasis.
2. Only a fraction of the fluid infused will be able to be extracted.
3. If the nasal mucosa is disrupted, then blood can contaminate the fluid being collected. In the event of bleeding from nasal cavity either push cotton into the nose or pinch the nostrils with gauze to achieve hemostasis.

4 Harvesting of Mucosal and Lymphoid Tissues from Rhesus Macaques

In rhesus macaques, because of the large size of the animals and the cost of the animals over the course of an experiment biopsies are conducted to evaluate the immune responses in the tissues of interest.

4.1 Rectal Pinch Biopsies

4.1.1 Materials

1. Visualization device (depending on depth of inoculation).
   (a) Flexible endoscope.
   (b) Anoscope.
   (c) Otoscope with appropriate sized ear cone.
2. Biopsy device (1.8 or 5 mm cup biopsy).
3. Hanks Balanced Salt Solution (HBSS).

4.1.2 Methods

1. Animal is anesthetized using standard anesthesia techniques.
2. Animal is placed in ventral recumbency.
3. Perineum is cleaned using a betadine solution or chlorhexidine solution.
4. If required, the animal may be placed in a biosafety cabinet.
5. Depending on the depth required for the collection of biopsy, a flexible endoscope or anoscope or otoscope may be used to visualize the biopsy site.

6. Once identified, use the biopsy device to obtain samples needed (*see* **Note 1**).
7. Place biopsies in HBSS.
8. Following procedure, the animal should be observed immediately post procedure, 1 h post procedure, and 24 h post procedure for any signs of hemorrhage or pain. Give analgesia and antibiotics as necessary.

4.2 Vaginal/Cervical Pinch Biopsies

4.2.1 Materials

1. Visualization device (depending on size of vaginal opening).
   (a) Vaginal speculum.
   (b) Colposcope.
   (c) Otoscope with appropriate sized ear cone.
2. Biopsy device (1.8 or 5 mm cup biopsy).
3. HBSS.

4.2.2 Methods

1. Animal is anesthetized using standard anesthesia techniques.
2. Animal is placed in ventral recumbency.
3. Perineum is cleaned using a betadine solution or chlorhexidine solution.
4. Visualization device is inserted and biopsy site identified.
5. Once identified, use the biopsy device to obtain samples needed (*see* **Note 1**).
6. Place biopsies in the HBSS.
7. Following procedure, the animal should be observed immediately post procedure, 1 h post procedure, and 24 h post procedure for any signs of hemorrhage or pain. Give analgesia and antibiotics as necessary.

4.3 Gastric/Duodenal Pinch Biopsies

4.3.1 Materials

1. Flexible endoscope.
2. Endotracheal tube.
3. Biopsy device (1.8 mm).
4. HBSS.

4.3.2 Methods

1. Animal is anesthetized using standard anesthesia techniques.
2. Intubate animal per standard procedure.
3. Animal is placed in left lateral recumbency.
4. Endoscope is placed into the esophagus and advanced into the stomach or duodenum (*see* **Note 2**).
5. Identify biopsy location for each area to be sampled.
6. Once identified, use the biopsy device to obtain samples needed (*see* **Note 1**).
7. Place biopsies in the appropriate medium.

8. After gastroscopy, remove scope and extubate the animal.
9. Following procedure, the animal should be observed immediately post procedure, 1 h post procedure, and 24 h post procedure for any signs of nausea, hemorrhage, or pain. Give analgesia and antibiotics as necessary.

4.4 Peripheral Lymph Node Biopsies

4.4.1 Materials

1. Small surgical pack.
   (a) Scalpel (#10 or #15 blade).
   (b) Mosquito hemostats.
   (c) Brown–Adson tissue forceps.
   (d) Olsen-Hegar or Mayo-Hegar needle holders.
   (e) Mayo scissors.
   (f) Metzenbaum scissors.
   (g) Surgical towels.
   (h) Surgical drape.
2. Sterile Gauze.
3. Suture.
4. HBSS.

4.4.2 Methods

1. Animal is anesthetized and intubated using standard anesthesia techniques.
2. Animal is placed in dorsal recumbency.
3. Lymph nodes are identified either in the axilla or inguinal area.
4. Hairs covering the identified area are clipped and the site is prepared for surgery.
5. Use a scalpel to make a small incision over the lymph node.
6. Use blunt dissection to identify the lymph node and to separate it from the underlying tissues.
7. If excising the entire lymph node, then prior to removal use an encircling suture around the lymph node vessels. Place the excised lymph node in the appropriate medium.
8. If only removing a section of the lymph node, then use sharp scissors or scalpel to remove the section. Place the section of lymph node in HBSS and hold the gauze on the remaining lymph node to maintain hemostasis.
9. Close the incision using a two-layer closure with the appropriate sized suture.
10. Extubate the animal.
11. Following procedure, the animal should be observed immediately post procedure, 1 h post procedure, and 24 h post procedure for any signs of hemorrhage or pain. Give analgesia and antibiotics as necessary.

4.5 Colon Biopsy

4.5.1 Materials

1. Flexible endoscope.
2. Biopsy device (1.8 mm).
3. Bowel cleansing solution.
4. HBSS.

4.5.2 Methods

1. Prior to performing a colon biopsy by using colonoscopy, the colon of the animal must be cleansed to allow for proper evaluation. This colonic cleanse is normally done using a bowel cleansing solution such as NuLYTELY over the course of 2 days.
2. Animal is anesthetized and intubated using standard anesthesia techniques.
3. Animal is placed in dorsal recumbency.
4. Endoscope is placed into the rectum and advanced into the colon (*see* **Notes 3** and **4**).
5. Identify biopsy location for each area to be sampled.
6. Once identified, use the biopsy device to obtain samples needed (*see* **Note 1**).
7. Place biopsies in HBSS.
8. After colonoscopy, remove the scope and extubate the animal.
9. Following procedure, the animal should be observed immediately post procedure, 1 h post procedure, and 24 h post procedure for any signs of nausea, hemorrhage, or pain. Give analgesia and antibiotics as necessary.

4.6 Buccal Biopsies

4.6.1 Materials

1. Visualization device.
   (a) Oral speculum.
   (b) Tongue Depressor.
2. Biopsy device (1.8 or 5 mm cup biopsy).
3. HBSS.

4.6.2 Methods

1. Animal is anesthetized using standard anesthesia techniques.
2. Animal is placed in dorsal or ventral recumbency.
3. Visualization device is inserted and biopsy site identified.
4. Once identified, use the biopsy device to obtain samples needed (*see* **Note 1**).
5. Place biopsies in HBSS.
6. Following procedure, the animal should be observed immediately post procedure, 1 h post procedure, and 24 h post procedure for any signs of hemorrhage or pain. Give analgesia and antibiotics as necessary.

4.7 Tonsillar Biopsies

4.7.1 Materials

1. Visualization device.
   (a) Oral speculum.
   (b) Tongue Depressor.
2. Endotracheal Tube.
3. Laryngoscope.
4. Small surgical pack.
   (a) Scalpel (#10 or #15 blade).
   (b) Mosquito hemostats.
   (c) Brown–Adson tissue forceps.
   (d) Olsen-Hegar or Mayo-Hegar needle holders.
   (e) Mayo scissors.
   (f) Metzenbaum scissors.
   (g) Surgical towels.
   (h) Surgical drape.
5. Gauze.
6. HBSS.

4.7.2 Methods

1. Animal is anesthetized and intubated using standard anesthesia techniques.
2. Animal is placed in dorsal or ventral recumbency.
3. Visualization device is inserted and biopsy site identified (*see* **Note 5**).
4. Once identified, use the sharp scissors or scalpel to obtain sample needed.
5. Place biopsies in the HBSS.
6. Place gauze on the tonsil to maintain hemostasis.
7. Extubate the animal and monitor for excessive hemorrhaging from the biopsy site.
8. Following procedure, the animal should be observed immediately post procedure, 1 h post procedure, and 24 h post procedure for any signs of hemorrhage or pain. Give analgesia and antibiotics as necessary.

Notes

1. If multiple biopsies from the same organ are required over a period of time make sure not to take them from the exact same location to avoid possible trauma and perforation of the organ of interest.
2. If duodenal biopsies are needed, then advance the endoscope into the duodenum prior to taking stomach biopsies.
3. Depending on the needs and the experience of the endoscopist, the endoscope can be advanced to the ileocecal junction.

4. Be sure to take biopsies on removal of the scope not during advancement of the scope to avoid possible colonic perforation.
5. Laryngoscope may also be used to provide light.

5 Harvesting of Mucosal and Lymphoid Tissues from Mice

Unlike rhesus macaques, where biopsies are conducted to collect tissue samples for evaluation of immune response, in case of mice collection of tissues is usually done by performing necropsy on the animal. The size and age of mice along with the vaccination regimen determines the yield of lymphocytes from the different organs.

5.1 Necropsy and Collection of Mouse Mucosal and Lymphoid Tissue

5.1.1 Materials

1. CO_2 tank and chamber for euthanasia.
2. Dissection board (thin Styrofoam or cork board).
3. Hypodermic needles—27 G.
4. Syringe—10 ml.
5. 70 % Ethanol.
6. Pair of curved scissors.
7. Pair of straight scissors.
8. Forceps.
9. Petri dishes (60 mm).
10. HBSS containing HEPES, L-glutamine, gentamicin, and penicillin/streptomycin (HGPG).
11. Phosphate Buffered Saline (1× PBS).

5.1.2 Methods

1. Prior to collection of tissue, euthanize the mouse using CO_2 (*see* **Note 1**). Prepare one mouse at a time for collection.
2. Spray down the mouse with 70 % ethanol to disinfect as well as limit the contamination of tissues with mouse hair/fur (*see* **Note 2**).
3. For good visualization and cleaner collection of tissues secure the mouse on its back on a dissecting board by pinning down its fore and hind limbs such that the abdomen is stretched.
4. Using a blunt ended or curved scissor, make a small vertical incision on the abdomen. Through the incision, insert the scissors between the skin and peritoneum and proceed to bluntly dissociate the skin from underlying peritoneum from neck to pelvic region. Extend the blunt separation laterally in axillae and inguinal regions (*see* **Note 3**).
5. Use a different sterile pair of curved scissors to cut open the peritoneal cavity from the base of abdomen to the ribcage. Pin down the fragments of peritoneum laterally with the skin.

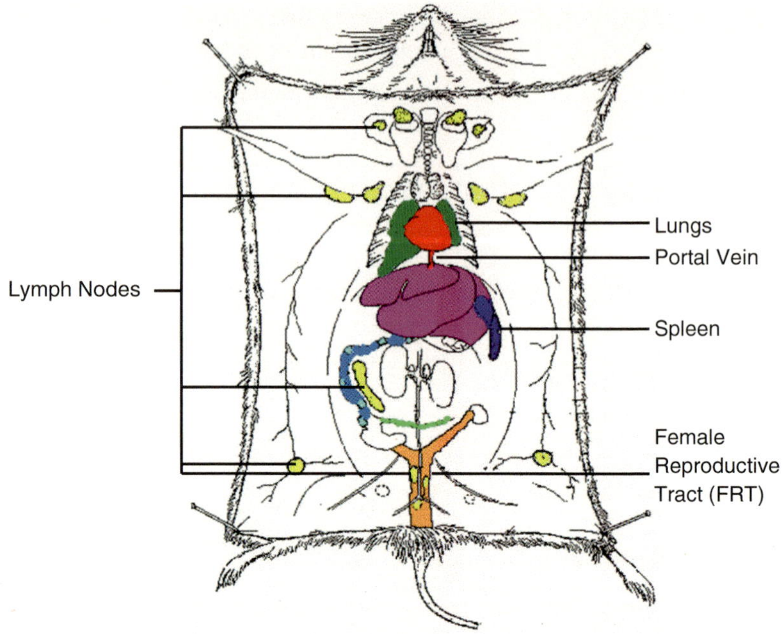

Fig. 1 Ventral view of mouse during necropsy showing the anatomic locations of lymph nodes, spleen, lungs, and the female reproductive tract (modified from Dunn T.B. 1954 J. Nat. Cancer Inst.14: 1281–1434)

This will expose the organs of abdomen such as intestines, reproductive tract and liver. To access the lungs open the thoracic cavity by cutting the ribs and lift the sternum with fine forceps and cut the diaphragm carefully (*see* Fig. 1) [18, 19].

6. *Lungs*: For collection of lymphocytes from the lungs, the organs must be first perfused to remove any circulating cells and then excised according to the following steps.
   (a) For perfusion 10–20 ml of cold 1× PBS is injected into the right ventricle of the mouse heart until the lungs are cleared of blood and turn white in color. A slit in the left ventricle or severing of portal vein allows the blood to leave circulation.
   (b) Once the lungs are white, gently excise them and collect them in a petri dish with 5 ml of HBSS taking care not to remove any peritracheal lymphoid tissue.
7. *Female Reproductive Tract* (*FRT*): Female reproductive tract of mice comprises of bicornuate uterus where the two horns merge into the body of uterus. Between the body of uterus and highly muscular vagina lies the cervix. The uterine horns are connected to the ovaries that lie caudal to the kidneys. FRT is embedded in the fat and lies close to the dorsal body wall (*see* Fig. 1) [19]. For collection of FRT proceed according to the following steps.

(a) Upon removal of peritoneum lining the abdomen, FRT can be accessed by moving the GI tract to the right. Insert the scissors between colon and pelvis and cut the bone.

(b) For excising the FRT, commence by cutting away the two horns from the ovaries. Gently hold the horns together with forceps and using the scissors remove all the associated fat and connective tissue.

(c) Proceed with removal of connective tissue till the end of vagina and cut out FRT to remove it from the abdominal cavity.

8. *Lymphoid tissues* such as lymph nodes and spleen can be collected and lymphocytes isolated directly without any digestion (*see* **Note 4**) [18–20].

Notes

1. To ensure humane euthanasia, the flow of CO_2 from the gas cylinder must start slowly so that the air in the chamber is displaced at a rate of 10–30 % per minute. At this rate, in about 1 min, the animal becomes unconscious and there is an absence of righting reflex when the CO_2 concentration in the chamber is about 50 %. The flow rate of CO_2 can be increased at this time and once the animal has ceased breathing, the flow of CO_2 must be maintained for another minute. Observe the animal for any muscle activity for another 30 s before proceeding with the dissection.
2. When the cells from the mice tissues need to be cultured for immune assay such as ELISPOT or the mice are immunized using viral vectors, the collection of tissues must be conducted in a tissue culture hood and the instruments to be used for collection must be sterile.
3. Care must be taken to avoid puncturing of the peritoneal cavity or incision of any blood vessels in the neck region. Once the two layers are separated, extend the incision in the skin to the neck region anteriorly and till the anal region posteriorly. Reflect the skin laterally and pin it down on the dissecting board on each side.
4. While collecting lymphoid tissues, remove most of the connective tissue and fat because their presence negatively affects the recovery and viability of the lymphocytes

6 Cell Isolation

Mononuclear cells from blood, body fluids, and tissues are purified for immune assays and these methods are common for mice and rhesus macaques as described below (any unique differences are mentioned at appropriate places).

6.1 Preparation of Peripheral Blood Mononuclear Cells (PBMC)

Heparinized or citrated venous blood samples of rhesus macaques are obtained as described above and PBMC are isolated by density gradient sedimentation using Ficoll Histopaque-1077 separation solution according to the protocol described below (*see* **Note 1**) [21].

6.1.1 Materials

1. Swinging Rotor Centrifuge.
2. Ficoll Histopaque-1077.
3. Phosphate Buffered Saline (1× PBS).
4. Complete RPMI medium (RPMI-1640 medium supplemented with 10 % heat-inactivated FBS, 2 mM L-glutamine, 100 U/ml penicillin/streptomycin, 25 μg/ml gentamicin, and 50 μM β-mercaptoethanol).
5. Trypan blue solution.
6. Conical centrifuge tubes—15 and 50 ml.

6.1.2 Methods

1. A 1:2 dilution of whole blood is made by mixing with equal volume of 1× PBS. In a 50 ml conical tube, carefully layer the blood on a cushion of Ficoll Histopaque-1077 (*see* **Note 2**).
2. Centrifuge the tube at 2,700 RPM or 1,565 ×*g* for 20 min in a swinging bucket rotor without brakes.
3. After centrifugation, discard the upper layer without disturbing the band of PBMC present at the interface and then carefully collect the band of PBMC into another 50 ml conical tube.
4. Wash the cells twice by mixing with 20 ml of 1× PBS and centrifuging the tubes at 1,800 RPM or 700 ×*g* for 10 min.
5. Resuspend the cell pellet in complete RPMI 1640 medium and determine the number of viable PBMC using Trypan blue exclusion method (*see* **Notes 3** and **4**).

Notes

1. For ideal separation of mononuclear cells from the blood, Ficoll must be brought to room temperature before overlaying of blood.
2. The ratio of volume of diluted whole blood to Ficoll Histopaque-1077 can range from 3:1 to 4:1.
3. Purified PBMC can either be used directly for the immune assays or be stored frozen (freezing medium is a mixture of 90 % FBS and 10 % DMSO) in liquid nitrogen for later use. When using the cryopreserved PBMC, the vials of frozen PBMC are removed from liquid nitrogen and rapidly thawed in a 37 °C water bath, gently mixed, washed with complete RPMI medium to remove the freezing medium and resuspended in complete RPMI medium.

4. PBMC are used directly or further processed to purify into lymphocytes subsets such as CD4+ T cells and CD8+ T cells by using specific isolation kits such as Dynal T cell negative selection kit (Invitrogen, Carlsbad, CA).

6.2 Percoll Gradient Enrichment

In case of mucosal tissues such as colon and vagina, a Percoll density gradient is used for isolation and purification of lymphocytes from the other cells such as epithelial cells and fibroblasts [22].

6.2.1 Materials

1. Swinging Rotor Centrifuge.
2. Buffered Percoll.
3. Phosphate Buffered Saline (1× PBS).
4. RPMI 1640 medium.
5. Complete RPMI medium (RPMI-1640 medium supplemented with 10 % heat-inactivated FBS, 2 mM L-glutamine, 100 U/ml penicillin/streptomycin, 25 μg/ml gentamicin, and 50 μM β-mercaptoethanol.
6. Syringe-30 ml.
7. Blunt ended needle-18 G, 6 in. long.
8. Trypan blue solution.
9. Conical centrifuge tubes—15 and 50 ml.

6.2.2 Methods

1. In 15 ml conical centrifuge tubes, prepare the Percoll gradients by underlaying 4 ml of 35 % Percoll with 4 ml of 60 % Percoll, both diluted from concentrated stock using serum free RPMI (*see* **Note 1**). One tube can accommodate 6 ml of cell suspension on top of this gradient.
2. Refrigerate the gradients at 4 °C for 1 h before using.
3. Gently layer 6 ml of cell suspension on top of each gradient. Final volume in each 15 ml conical tube will be 14 ml.
4. Centrifuge the tubes in at 1,800 RPM or 700×*g* for 20 min, 4 °C in a swinging bucket rotor *without* brakes.
5. Collect the epithelial cells such enterocytes present in the top interface layer (between media and 35 % Percoll) and the lymphocytes that are located primarily at the lower interface (between 35 and 60 % Percoll) separately in sterile 50 ml conical tubes.
6. Add 1× PBS to the collected cells and achieve a final volume of 50 ml.
7. Invert the tubes several times to mix the cells with 1× PBS.
8. Centrifuge the tubes for 10 min at 1,800 RPM or 700×*g*, 25 °C.
9. Discard the supernatant and repeat the washing step by resuspending the cell pellet in 50 ml of fresh 1× PBS.

10. After the second wash, discard the supernatant and resuspend the cell pellet in 5 ml of complete RPMI medium.
11. Count viable cells in a hemocytometer using trypan blue.

Notes

1. The 60 % Percoll solution, which will be the lower solution in the tube, is tinted red in order to readily distinguish it from the 35 % Percoll.

6.3 Isolation of Mononuclear Cells from Mouse Lungs

6.3.1 Materials

1. Petri dish—60 mm.
2. Scalpel.
3. Forceps.
4. Conical tubes—15 and 50 ml.
5. Nylon mesh cell strainer—70 μm pore.
6. Collagenase (from *Clostridium histolyticum*) solution.
7. Rocking platform.
8. Incubator—37 °C.
9. Ammonium chloride–Potassium (ACK) Lysing Buffer.
10. Complete RPMI medium (RPMI-1640 medium supplemented with 10 % heat-inactivated FBS, 2 mM L-glutamine, 100 U/ml penicillin/streptomycin, 25 μg/ml gentamicin, and 50 μM β-mercaptoethanol).
11. Trypan blue solution.
12. Swinging bucket rotor centrifuge.

6.3.2 Methods

1. Cut the lungs into 100–300 mm^2 pieces using a scalpel and transfer the tissue along with HBSS to a 15 ml conical tube.
2. Centrifuge the tubes at 1,500 RPM or 483 × *g* for 5 min at room temperature. Since lung tissue is spongy in nature and does not form a pellet at the bottom of the tube, the supernatants must be removed carefully so that no tissue pieces are discarded.
3. Digest the tissue with gentle agitation by incubating in 5 ml of collagenase solution (RPMI-1640 containing 10 % FBS, 100 U Penicillin/streptomycin, and 125 U/ml of Collagenase (Type II)) for 60 min at 37 °C.
4. After 1 h, strain the cell suspension through a disposable 70 μm disposable cell strainer to remove any undigested tissue and large cells. Wash the strainer by pipetting 10 ml of complete RPMI medium through it to recover any cells trapped in the mesh of the strainer.
5. Centrifuge the tubes for 5 min at 1,500 RPM or 483 × *g* and discard the cell supernatants.
6. Lyse the red blood cells by resuspending the cell pellet in 1 ml of ACK Lysing Buffer (Invitrogen) and incubating for 5 min at room temperature.

7. After 5 min incubation, add 5 ml of complete RPMI and centrifuge the cells for 5 min at 1,500 RPM or 483 ×*g*.
8. Discard the supernatants and resuspend the cell pellets in 5 ml of complete RPMI medium and count the cells by trypan blue exclusion method.

6.4 Isolation of Mononuclear Cells from Mouse Lymphoid Tissues

6.4.1 Materials

1. Petri dish—60 mm.
2. Frosted glass slides.
3. Forceps.
4. Conical tubes—50 ml.
5. Nylon mesh cell strainer—70 μM pore.
6. ACK Lysing Buffer.
7. Complete RPMI medium (RPMI-1640 medium supplemented with 10 % heat-inactivated FBS, 2 mM L-glutamine, 100 U/ml penicillin/streptomycin, 25 μg/ml gentamicin, and 50 μM β-mercaptoethanol)
8. Trypan blue solution.
9. Swinging bucket rotor centrifuge.

6.4.2 Methods

1. Lymphoid tissues such as lymph nodes and spleen can be homogenized between frosted ends of glass slides to form single cell suspension.
2. Pass the cell suspension through a 70 μm cell strainer to remove any large cells and pieces of connective tissue such as splenic capsule.
3. Rinse the strainer by pipetting 10 ml of complete RPMI to recover all the cells trapped in the mesh of the strainer.
4. Centrifuge the tube at 1,500 RPM or 483×*g* for 5 min and discard the supernatants.
5. For cells isolated from spleens, lyse RBC by resuspending the cell pellet in 1 ml of ACK lysis buffer and incubate for 5 min to room temperature.
6. After the incubation, add 5 ml of complete RPMI and centrifuge the cells for 5 min at 1,500 RPM or 483×*g*.
7. Discard the supernatants and resuspend the cell pellets in 10 ml of complete RPMI and count the cells.

6.5 Isolation of Mononuclear Cells from Mouse Female Reproductive Tract (FRT)

6.5.1 Materials

1. Petri dish—60 mm.
2. Forceps.
3. Scissors.
4. Conical tubes—15 and 50 ml.
5. Nylon mesh cell strainer—40 μm pore.
6. HBSS containing HGPG [21, 22].

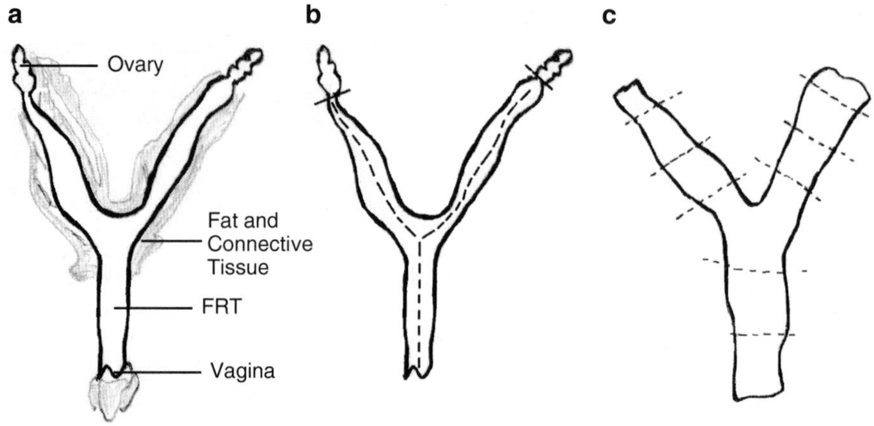

Fig. 2 Cartoon depicting typical processing procedures for isolating lymphocytes from the FRT tissue from mice. Panel (**a**) shows the female reproductive tract (FRT) along with the associated connective tissue excised from a mouse. Panel (**b**) shows procedures for making longitudinal incision on the FRT to open the lumen of the organ. Panel (**c**) shows procedures for making lateral sections of the FRT prior to digestion of tissue to isolate lymphocytes

7. HBSS containing HGPG and 1.3 mM EDTA (*see* **Note 1**) [23, 24].
8. Collagenase IV solution (100 ml RPMI, 1 mM $MgCl_2$, 1 mM $CaCl_2$, 2 ml HGPG, and 15,000 U of Collagenase IV) (*see* **Note 1**).
9. Incubator—37 °C.
10. Erlenmeyer Flask—25 ml.
11. Magnetic stir bar.
12. Magnetic stirrer.
13. RPMI 1640 medium (HyClone laboratories, Logan, UT).
14. Complete RPMI medium (RPMI-1640 medium supplemented with 10 % heat-inactivated FBS, 2 mM L-glutamine, 100 U/ml penicillin/streptomycin, 25 μg/ml gentamicin, and 50 μM β-mercaptoethanol).
15. Trypan blue solution.
16. Swinging bucket rotor centrifuge.

6.5.2 Methods

1. Place the organ on a moist paper and gently trim off any remaining fat, connective tissue, and blood vessels that may be attached to the reproductive tract (*see* Fig. 2a).
2. Using fine tipped pair of scissors cut open the lumen of the reproductive tract. Commence from the vaginal end and proceed anteriorly towards the reproductive horns (*see* Fig. 2b).
3. Lay the tissue flat and use the blunt edge of scissors to scrape the mucosa to remove any mucus.

4. Cut the flattened FRT laterally into 0.5 cm long pieces and place in a 50 ml conical tube with 10 ml of HBSS containing HGPG (*see* Fig. 2c).
5. Invert the tubes several times to rinse off any mucus attached to pieces of FRT. Allow the tubes to sit for a minute till the pieces settle down at the bottom of the tube and then pour off the HBSS carefully (*see* **Note 2**).
6. Add 20 ml of HBSS + EDTA solution (pre-warmed to 37 °C) to the tube with the tissue. Place a stir bar in a 25 ml Erlenmeyer flask and transfer the contents of the conical tube to the flask (*see* **Note 3**). Incubate the flask with constant stirring for 60 min at 37 °C (*see* **Note 4**). Retain the conical tubes for later use.
7. After incubation for 60 min, transfer the content of the flask back to the conical tube. Gently pour off HBSS–EDTA (*see* **Note 2**).
8. Add 20 ml of RPMI to the tube with tissues and invert several times to rinse off any HBSS–EDTA (*see* **Note 5**).
9. Take out the pieces of FRT in a petri dish and add about 2–3 ml of collagenase solution to it. Use a sharp pair of scissors to cut the pieces into further smaller size (*see* **Note 6**).
10. Tilt the petri dish to collect the pieces of FRT and the collagenase on the side of the dish and pour them back into the conical tubes. Make up the volume of the contents in the tube to 20 ml with collagenase solution. Stir the tubes and pour the contents into flask with stir bar. Incubate the flask for another 60 min at 37 °C with constant stirring. Continue to retain 50 ml conical tubes for further use.
11. After an hour, gently add the cell suspension from the flask into a 40 μm disposable cell strainer placed in the 50 ml conical tubes. Use the plunger of a 1 ml tuberculin syringe to release any cells that might be trapped in the strainer. Rinse the strainer by pipetting additional 10 ml of complete RPMI.
12. Centrifuge the tubes at 1,600 RPM or 550 × *g* for 6 min. Remove the cell supernatants and resuspend the cell pellet in 5 ml HBSS + HGPG.
13. Transfer the cell suspension to a 15 ml conical tube and centrifuge the tube at 1,600 RPM or 550 × *g* for 6 min.
14. Remove the supernatants and resuspend the cell pellet in 1 ml of HBSS + HGPG.
15. Proceed to counting and staining the cells for flow cytometry.

Notes

1. Prepare EDTA and collagenase solutions in the morning on the day of isolation [24].
2. To ensure that no FRT segments are lost during the washing process, gently tilt the tubes on a beaker and hold a pair of

forceps inside the tube to catch any floating pieces of FRT. Monitor the beaker as well for the presence of any lost tissue pieces. Repeat the wash step once and try to remove as much HBSS as possible.

3. Confirm that all the pieces of FRT have been transferred to the flask.
4. Retain the conical tubes for later use.
5. Since EDTA is a chelating agent and may interfere with the activity of collagenase during digestion, it is important to remove all the traces of HBSS–EDTA solution. Besides the tissue, the Erlenmeyer flask must also have no traces of EDTA, therefore, use vacuum to suction it out of the flask [25].
6. Smaller pieces of tissue are digested better because of increased surface area for collagenase action.

7 Immune Function Analyses

In principle vaccine studies, including many in the NHP models, are designed to improve adaptive immune responses in order to prepare the host to fight against diseases by priming humoral and cellular immune responses specific to a variety of HIV antigens. Since viruses by definition are obligate intracellular pathogens that upon entering the host will quickly invade the host cells to secure shelter and resources for propagation, T cell responses (helper and CTL) against viral antigens have special importance in aiming towards potential elimination of the virus-infected/producing cells [26–29]. The T cells responses particularly at the genital mucosal portals of virus entry such as the oral, vaginal and rectal tissues are highly relevant and critical for providing barrier protection [15, 28]. There are many methods available to measure the T cell mediated immune responses that include (a) assaying the cytolytic activity of $CD8^+$ T cells termed cytotoxic T lymphocytes (CTL), (b) quantitative determination of cytokine production by $CD8^+$ T cells as well as $CD4^+$ T helper cells (T_h) by employing fluorescence tagged antibodies for intracellular cytokine staining (ICS), and (c) enumerating the numbers of cytokine producing T cells using the enzyme-linked spot-forming (ELISPOT) assay [30, 31]. All these T cell assays involve in vitro activation of the cells for varying lengths of time with peptide or protein antigens corresponding to viral sequences. The ICS and ELISPOT assays have advantage over classical CTL assay to bypass the need for MHC-matched cell lines and the time consuming prior expansion of effector cells. Accurate and sensitive methodology for measuring the function of antigen specific T cells is important for determining the strength and breadth of cell-mediated immunity induced by vaccine candidate. Since NHP models are expensive resources with limited availability, detailed murine studies are performed first for selecting and optimizing not only the type and quantity of vaccine candidates but also the route of immunization, tissues to be analyzed, and methods/assays to be utilized before testing in the NHP models. Detailed protocols for the cytokine ELISPOT and ICS assays to measure antigen-specific T cells responses are described below and where applicable differences as they pertain to analyzing mouse versus macaque samples are noted.

7.1 Enzyme Linked Immuno Spot (ELISPOT) Assay

The ELISPOT assay employs a quantitative sandwich enzyme-linked immune adsorbing assay methodology for enumerating the number of cells secreting the cytokine of interest in response to specific stimulation [32, 33]. Monoclonal or polyclonal antibody specific for the cytokine of interest is coated onto an ELISPOT plate, which is generally a 96-well microtiter plate with polyvinylidene difluoride (PVDF) membrane bottom. The coated antibody captures the cytokine(s) secreted by the cells seeded into the wells when incubated with the antigen of interest, usually at 37 °C for 24–48 h. After washing off the cells, biotinylated polyclonal second antibody specific to the cytokine being determined is added. This is followed by treatment with Streptavidin-enzyme (ALP or HRP) and chromogen AEC substrate or BCIP/NBT substrate to visualize the signal in terms of spots representing the individual cytokine secreting cells. The spots are counted manually using a stereomicroscope or automated systems. Alternately, for unbiased interpretation of the data the counting of the spots can be contracted to third party commercial sources (e.g., KS ELISPOT, from Carl Zeiss, Inc., Thornwood, NY). The data is presented as cytokine spot forming cells (SFC) per total number of cells in the well [34].

ELISPOT assay is performed using either commercially available kits (e.g., MABTECH) or selecting an antibody pair specific to the cytokine to be determined. In general, when using the commercially available kits, the manufacturers provide detailed directions to be followed. The following are the required materials and the step-by-step methodology for the ELISPOT assay as performed in our laboratory for enumerating IFN-γ producing cells within the mononuclear cells prepared from blood or tissue specimens of mice and rhesus macaques in vaccine studies [35]:

7.1.1 Materials

1. 96-well PVDF-bottomed plates (EMD Millipore, Billerica, MA).
2. IFN-γ ELISPOT Set (BD Biosciences, San Jose, CA) (*see* **Note 1**).
3. Concanavalin A (Con A) (Sigma Aldrich, St. Louis, MO).
4. Dulbecco's PBS (DPBS, Ca_2/Mg_2-free; Life technologies, Rockville, MD).
5. HBSS (Sigma Aldrich, St. Louis, MO).

6. RPMI 1640 culture medium (HyClone laboratories, Logan, UT).
7. Complete RPMI medium [Complete RPMI medium [RPMI-1640 containing 10 % heat-inactivated FBS, 2 mM L-glutamine, 100 U/ml penicillin/streptomycin, 25 μG/ml gentamicin, and 50 μM β-mercaptoethanol]
8. Tween 20 (Sigma Aldrich, Saint Louis, MO).
9. 3-Amino-9-ethyl-cardazole (AEC; Sigma Aldrich, St. Louis, MO).
10. *N,N*-Dimethyl-Formamide (DMF).
11. 0.1 M sodium acetate buffer.
12. 30 % hydrogen peroxide (H_2O_2) (protect from light and store at 4 °C).
13. 70 % Ethanol.
14. Avidin Peroxidase.
15. 1× Phosphate Buffered Saline–Tween (0.5 % Tween 20).

7.1.2 Methods

Coating Plate with capture antibody: Day 1

1. Pre-wet the PVDF membrane of ELISPOT plate with either 15 μl of 35 % ethanol/well for 1 min or 50 μl of 70 % ethanol/well for 30 s (*see* **Note 1**). Discard the ethanol and wash the plate twice with 200 μl of sterile water/well followed by three washes with 200 μl of sterile 1× PBS for each well.
2. Coat the wells with 100 μl of diluted purified anti-IFN-γ capture antibody. For making the dilutions of purified antibody in 1× PBS either follow the manufacturer's recommendation or start with a stock solution of 1–10 μg/ml (*see* **Notes 2** and **3**).
3. Cover the plate and seal it with parafilm to prevent evaporation and incubate the plates overnight at 4 °C (*see* **Note 4**).

Setting up ELISPOT Assay: Day 2

1. Discard any unbound capture antibody by washing the plate thrice with 1× PBS. Block the plate with complete RPMI medium (RPMI medium supplemented with 10 % FBS) for at least 2 h at room temperature (*see* **Notes 5** and **6**).
2. Adjust the single suspension of PBMC or lymphocytes recovered from different tissues (isolated by protocol described previously) to a final concentration of $1–2 \times 10^6$ cells/ml in complete RPMI medium. Add 100 μl of the cell suspension to each well (equivalent to $1–2 \times 10^5$ cells/well, *see* **Notes 7** and **8**).
3. Depending upon the vaccine administered, the lymphocytes are stimulated with the test antigen (single or pools of peptides corresponding to the antigen) (*see* **Notes 9** and **10**). Mitogens such as Concanavalin A or PHA are used as positive control

and complete RPMI medium as negative control reagents for stimulation of the cells (*see* **Note 11**). There should be either duplicate or triplicate wells for determining the response for each stimulating agent.

4. The plates are incubate for 36–48 h in a humidified 37 °C, 5 % CO_2 incubator (*see* **Notes 12** and **13**).

Developing ELISPOT plate: Day 4/5

1. Discard the cells and supernatants from the ELISPOT plate and wash the plate five times with 1× PBS-T (wash buffer). Each well must be soaked with 200 μl of wash buffer for 3–5 min during each wash (*see* **Note 14**).
2. After discarding the wash buffer, add 100ul of diluted anti-IFN-γ biotinylated detection antibody (dilution buffer is 1× PBS + 10 % FBS) to each well (*see* **Note 2**). Cover the ELISPOT plate and incubate at room temperature for 2 h.
3. Discard the detection antibody and wash the plate thrice with 200 μl of wash buffer/well. Allow wells to soak in the wash buffer for 1–3 min each time.
4. Following the washes, add 100 μl of streptavidin-conjugated horseradish peroxidase or alkaline phosphatase (diluted in 1× PBS-10 % FBS according to manufacturer's recommendation) and incubate for 1 h at room temperature (*see* **Note 15**).
5. At the end of the incubation, discard the enzyme solution and wash the wells four times with 1× PBS-T. Soak the wells with 200 μl of wash buffer for 1–3 min for each wash.
6. Discard wash buffer and wash the plates twice with 1× PBS using 200 μl/well. Allow the membrane to soak for 1–2 min each time to remove any trace amounts of 1× PBS-T that may interfere with the enzyme reaction.
7. Dispense 100 μl of the substrate solution/well of the ELISPOT plate and monitor for color development of spots. It usually takes 5–60 min for the spots to develop (*see* **Note 16**). Stop the reaction by discarding the substrate solution and washing the plate with DI water (*see* **Note 17**).
8. Allow the plates to air-dry at room temperature in the dark overnight.
9. Enumerate the spots by counting manually under a dissection microscope or using an automated ELISPOT reader system (Carl Zeiss Microimaging, Thornwood, NY) (*see* Fig. 3a for sample images of wells with IFN-γ spot forming cells after treatment with test and positive/negative control reagents, and Fig. 3b typical data from the analyses of cells from macaques in a vaccine study).

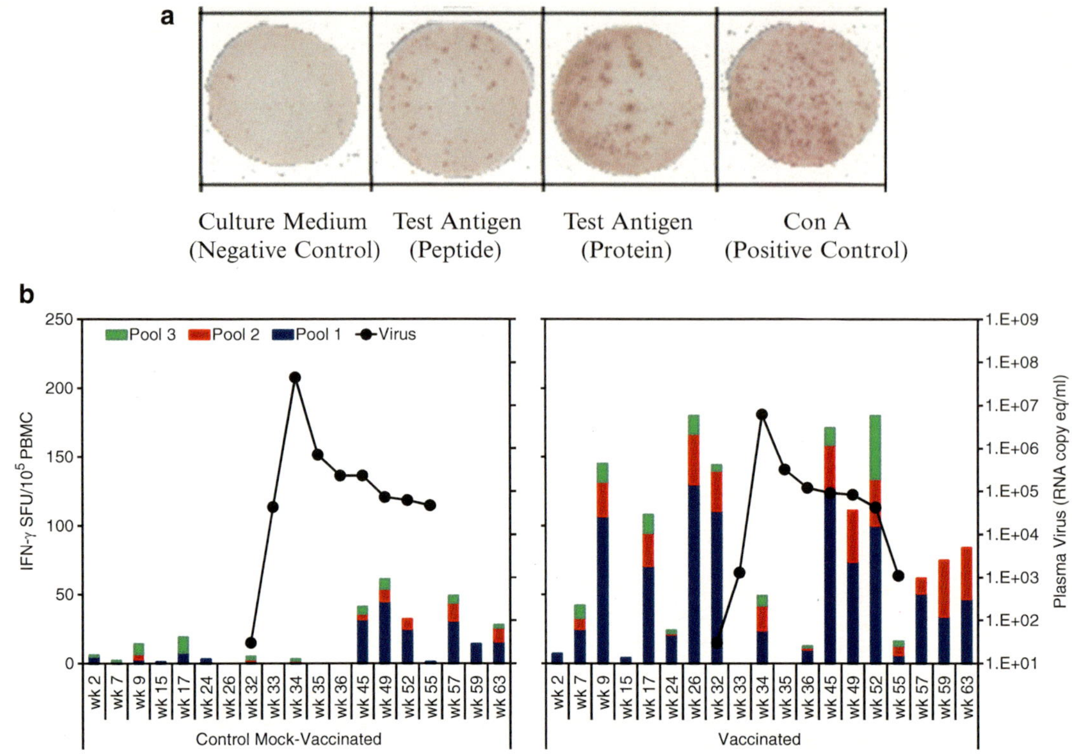

Fig. 3 Representative data from the IFN-γ ELISPOT assay. Panel (**a**) shows typical appearance of IFN-γ spot forming cells in the different wells of an ELISPOT assay plate where the lymphocytes were incubated with test antigen, Con A, and culture medium (the latter two are positive and negative control treatments). Panel (**b**) shows data for a typical assay determining the protective efficacy of adenoviral vectors expressing the HIV-1 envelope protein against pathogenic challenge with the simian human immunodeficiency virus (SHIV). The vaccine-induced immune responses at different time points before and after immunization and virus challenge, in terms of IFN-γ producing cells in response to stimulation with three overlapping peptide pools corresponding to HIV-1 envelope protein (stacked columns and data values shown on *left vertical axis*), and viral loads in terms of plasma RNA copy equivalents (*solid line* and data values shown on *right vertical axis*) in a representative mock-vaccinated control monkey (panel on the *left*) and a vaccinated monkey (panel on the *right*)

Notes

1. Pre-wetting PVDF membrane improves the efficiency of coating with the antibody. However, prolonged exposure or large volume of ethanol leads to leakage of the membrane therefore time and volume requirement for ethanol must be adhered to strictly. Once the membrane has been pre-wet, it must be ensured that the membrane does not dry out.
2. It is critical to follow the recommended dilution mentioned in the Certificate of Analysis included with the kit because the quality of the antibody may change with different lots of reagents.
3. The Human IFN-γ ELISPOT Kit from BD Biosciences shows cross-reactivity with nonhuman primates. The ELISPOT reagents (capture antibody and detection antibody) and color

development reagents as individual reagents or kits along with pre-coated plates are also available from R&D System.

4. Incubation can be done at room temperature for 4 or 2 h at 37 °C. For convenience plates can be coated few days prior to assay performance and stored at 4 °C.
5. Blocking before addition of PBMC or lymphocytes to the plate is necessary for reducing any nonspecific background.
6. In case of mice, since the sacrifice of the animals and setting up of the ELISPOT assay are usually done on the same day, it is useful to block the plates prior to counting and making dilutions of the lymphocytes.
7. The number of responder cells in each well as 2×10^5 is an optimum number because in our experience $1–2 \times 10^5$ cells yielded decreased number of positive spots whereas $>2 \times 10^5$ cells gave higher background.
8. Inclusion of a step of stimulation of PBMC with the antigens for a defined period of time prior to setting up of the ELISPOT assay is called short-term culture (STC) ELISPOT assay. This pre-stimulation step can be used to increase the sensitivity of the assay where there is a concern that the direct ELISPOT assay may fail to detect the immune response. In addition, in experiments where the assay is performed with cells cryopreserved for longer periods of time, inclusion of a pre-stimulation step can reduce the variations that may exist with different sample collections. Smith et al. [36] have noted that pre-stimulation step prior to transfer of cells to ELISPOT plate leads to an increase in the spot forming units being detected.
9. In case of direct stimulation of lymphocytes in the ELISPOT plates, to avoid drying out of the PVDF membrane of ELISPOT plates while adding cells + antigenic stimulant, pre-mix 100 μl of the diluted lymphocytes with 100 μl of antigen solution in a 96-well cell culture plate and then transfer this mixture to the ELISPOT plates using a multichannel pipettor. This approach also minimizes the time required for transfer of cells.
10. For evaluating immune responses to multiple peptides, a cocktail composed of 2 μg/ml of each peptide is used for stimulation.
11. In our laboratory, whenever possible, we also routinely include similarly stimulated PBMC from naïve animals as additional negative controls
12. The incubation period for the detection of cytokine producing cells in the ELISPOT assay can range from 36 to 72 h depending upon the stimulating agent and the cytokine response being evaluated.

13. Ensure that the ELISPOT plate is not disturbed during the incubation period. Also do not stack up the plates during incubation. Movement of plates can result in the movement of cells and as a result streaks will be observed instead of spots. Stacking of plates can result in uneven distribution of temperature and cells.
14. Adherence to recommended soaking time is critical for limiting background in the wells containing cells treated with cell medium alone.
15. During the incubation with enzyme solution, commence preparation of the working substrate solution according to the manufacturer's protocol. Solution of either AEC or TMB is used for HRP while BCIP/NBT is used as substrate for alkaline phosphatase.
16. Development of spots must be monitored carefully. Over development leads to staining of the membrane making it tough to distinguish spots from the background. The color development must be stopped when the brown spots in the wells containing Con A start turning green.
17. Remove the plastic tray under the plate carefully and wash the plate with DI water to remove any residues of the substrate.

7.2 Intracellular Cytokine Analyses

Intracellular cytokine (ICC) analysis is widely used to assess cell mediated immune responses to various antigens and infectious agents [30, 31]. This assay is commonly used in many laboratories to evaluate the T cells responses in nonhuman primates immunized with vaccines against SIV or HIV. Besides requiring relatively less time for conducting the procedure, this assay also enables accurate determination of the frequency of cytokine-producing cells in different T cell subsets within the mononuclear cells from peripheral blood as well as tissues [31]. Using commercially available anti-human monoclonal antibodies to human immune cells surface markers and cytokines that can also cross-react with their simian analogues, we routinely perform ICC assays for determining the frequency of IFN-γ and IL-2 secreting rhesus macaque CD4+ and CD8+ T cells responding to in vitro stimulation with either proteins or peptides corresponding to the vaccine(s) administered to the animals.

7.2.1 Materials

1. 12 mm×75 mm polystyrene test tube (Falcon, Lincoln Park, NJ).
2. Dulbecco's PBS (DPBS, Ca_2/Mg_2-free; Life technologies, Rockville, MD).
3. Phorbol 12-myristate 13-acetate (PMA) and Ionomycin (Sigma-Aldrich, St. Louis, MO).

4. 96-well round-bottom cell culture plate (BD Biosciences, Franklin Lake, NJ).
5. Cytofix/Cytoperm™ Fixation/Permeabilization Solution Kit with BD GolgiPlug solution (BD Biosciences; San Jose, CA).
6. FACS Buffer: DPBS supplemented with 2 % FBS.
7. Complete RPMI medium [RPMI-1640 containing 10 % heat-inactivated FBS, 2 mM L-glutamine, 100 U/ml penicillin/streptomycin, 25 μg/ml gentamicin, and 50 μM β-mercaptoethanol].
8. Aqua LIVE/DEAD® Fixable Dead Cell Stain Kits (Invitrogen; Carlsbad, CA).
9. Antigenic proteins/peptides.
10. Centrifuge with Swinging Bucket Rotor.
11. Antibodies (*see* **Note 1**): such as
    (a) CD3 PE-Cy7 (SP34-2) (BD Biosciences; San Jose, CA).
    (b) CD8 Alexa Fluor700 (RPA-T8) (BD Biosciences; San Jose, CA).
    (c) CD28 PerCP-cy5.5 (L293) (BD Biosciences; San Jose, CA).
    (d) CD95 APC (DX2) (BD Biosciences; San Jose, CA).
    (e) IFN-γ FITC (B27) (BD Biosciences; San Jose, CA).
    (f) IL-2 PE (MQ1-17H12) (BD Biosciences; San Jose, CA).
    (g) CD4 eFluor4 (OKT4) (eBiosciences, San Diego, CA).

7.2.2 Methods

Stimulation of cells:

1. Single cell suspensions from PBMC or tissues are first resuspended at a concentration of 10×10^6 cells/ml, in complete RPMI medium and then100 μl (1×10^6 cells) of the cell suspension is added to each well of a 96-well round-bottomed cell culture plate for stimulation with the antigen. Both freshly isolated as well as cryopreserved cells can be used for this protocol (*see* **Note 2**).
2. The choice of use of either peptide pools or proteins as stimulating antigens is dictated by the immunogen(s) used in the vaccine study. Unstimulated cells serve as negative and cells stimulated with PMA and ionomycin serve as positive controls, respectively (*see* **Note 3**).
3. Cells are first incubated for 1.5 h in a humidified 37 °C, 5 % CO_2 incubator with the appropriate antigen. To allow for the intracellular accumulation of cytokines, 10 μg/ml of the protein transport inhibitor, brefeldin A (*see* **Note 4**) is then added to the cell suspension and the incubation is continued further for 4.5 h.

Staining for surface markers:

1. After 6 h of stimulation, centrifuge the plate at 1,600 RPM or 550 × *g* for 2 min and discard the cell culture medium.
2. Wash the cell pellet twice with FACS buffer (200 μl/well) and stain the cells with live/dead fixable Aqua LIVE/DEAD fluorescent reactive dye at 4 °C for 30 min in the dark.
3. Wash the cells once with cold FACS wash buffer and proceed for staining of surface markers.
4. Prepare a cocktail of fluorescently labeled antibodies against cell surface markers such as CD3, CD4, CD8, CD28, and CD95 by diluting them in FACS buffer and add 100 μl of this cocktail to each well. Incubate the cells with the antibodies in the dark for 30 min at 4 °C (*see* **Note 5**). For each experiment both compensation controls (*see* **Note 6**) and fluorescence minus one (FMO) controls (*see* **Note 7**) must also be included.
5. After incubation, wash the stained cells twice with cold FACS wash buffer as described previously, and fix with fixation buffer (3 % buffered paraformaldehyde solution, 200 μl/well) for at least 20 min (*see* **Note 8**).
6. After fixing, cells can be resuspended in either FACS buffer or fixation solution and stored overnight at 4 °C in the dark before proceeding for intracellular cytokine staining.

Staining for intracellular cytokines:

1. Fixed cells are first permeabilized by washing twice with permeabilization buffer and then incubated with 50 μl/well of permeabilization buffer for 20 min at 4 °C.
2. Add 50 μl/well of appropriately diluted antibodies against cytokines such as IFN-γ (B27) and IL-2 (MQ1-17H12) to the permeabilized cells and incubate them in the dark cells for 60 min at 4 °C (*see* **Note 9**).
3. Following staining, wash the cells twice with the permeabilization buffer and then fix with 200 μl of 3 % buffered paraformaldehyde solution/well before proceeding for acquisition on a flow cytometer (*see* **Notes 10** and **11**).

Flow cytometry analysis:

The fixed samples can be acquired on a flow cytometer such as LSR II or BD LSRFortessa (BD Bioscience) and FACS data can be analyzed by using software such as FlowJo (TreeStar, OR, USA).

An example of the gating strategy and analysis of different cell surface markers is illustrated in Fig. 4. Lymphocyte population is gated by generating forward (FSC) and side (SSC) scatter dot plots. Live lymphocytes are identified by gating on cells negative

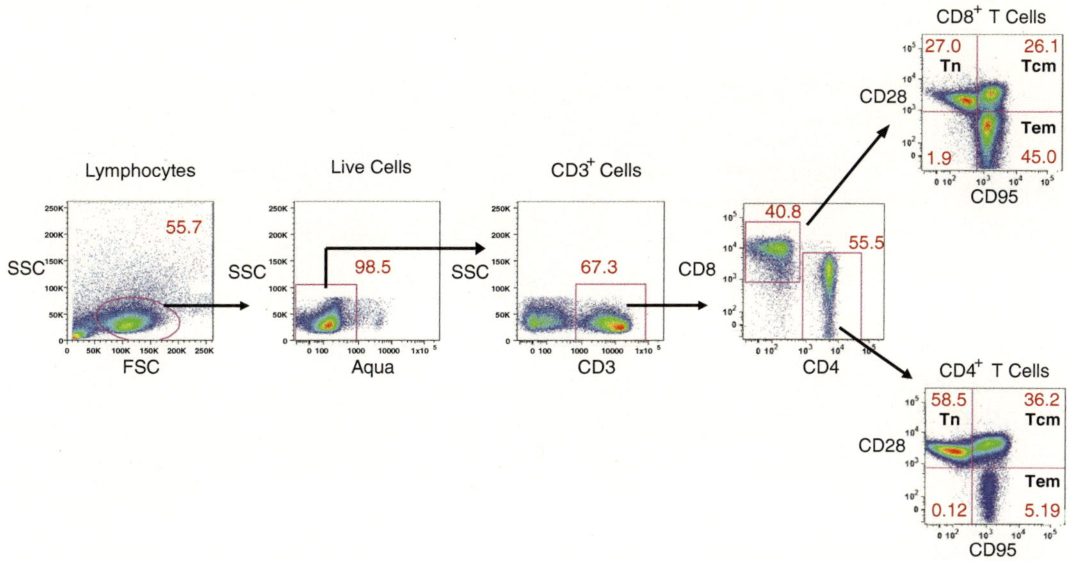

Fig. 4 Gating scheme utilized for the analyses of the different T cell subsets from a representative monkey. The lymphocytes were first gated using a dot pot with FSC versus SSC, and then live lymphocytes were identified based on SSC and aqua-negative population. The T cells were then identified by CD3 expression. The CD4+ CD8– and CD4– CD8+ populations within the CD3+ T cell population were also determined. On the basis of CD28 and CD95 expression, the CD4+ and CD8+ T cells were further differentiated into naıve (Tn CD28+ CD95−), central memory (Tcm CD28+ CD95+) and effector memory (Tem CD28− CD95+) subsets

for aqua stain. The T lymphocytes are identified by gating on CD3 positive live cells and then CD8 and CD4 T lymphocytes can be identified based on the expression of the two markers. Further characterization of the different effector and memory CD8 and CD4 subsets in case of macaque samples is done based on the expression of CD28 and CD95 wherein naïve cells are CD28+ CD95–, central memory cells are CD28+ CD95+ and effector memory cells are CD28– CD95+. An example of intracellular staining for production of cytokines is shown in Fig. 5. Expression of cytokines IL-2 and IFN-γ is determined for the different subsets on T lymphocytes defined on the basis on surface expression of cellular makers.

The protocol for activating and staining the cells is similar for the assay with cells from mice. An example of the gating strategy and analysis of different cell surface markers in murine FRT is illustrated in Fig. 6. The lymphocytes are identified by gating on cells positive for CD45. Activated CD8 T lymphocytes are identified by gating on cells which are positive for both CD8 and CD44 expression.

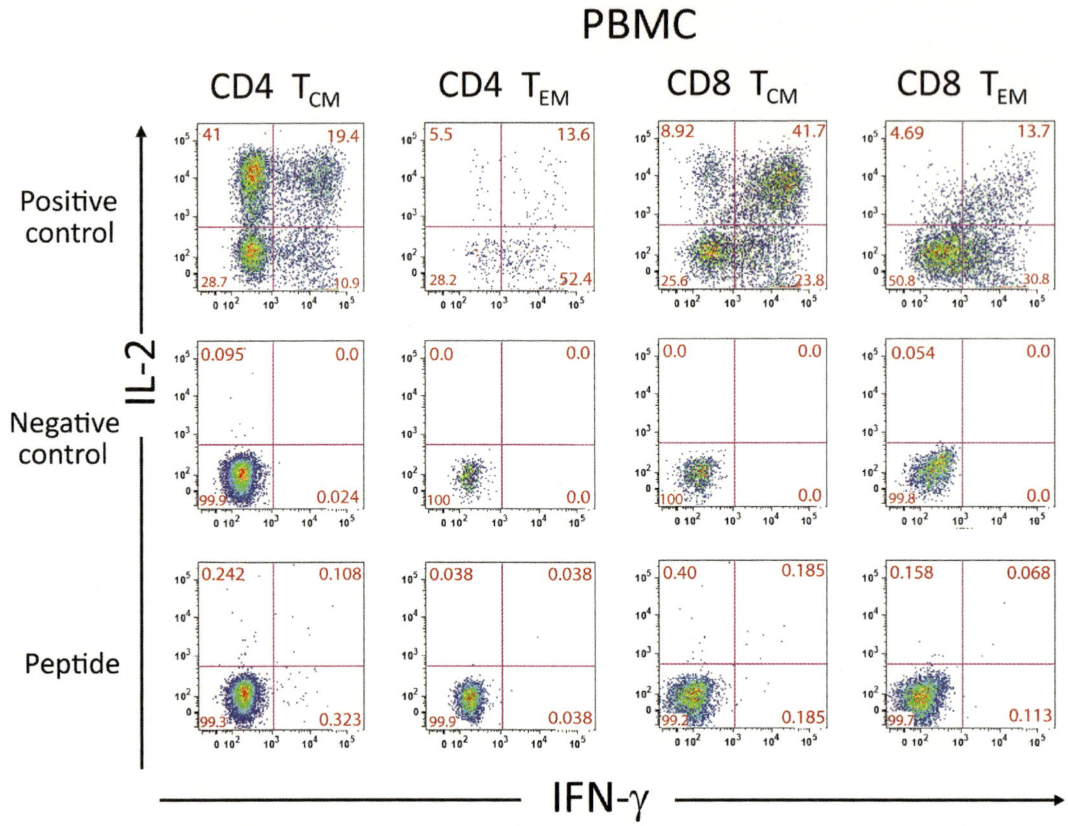

Fig. 5 Typical plots showing the INF-γ and IL-2 profiles of CD4+ and CD8+ memory subsets after stimulation with PMA+ionomycin (positive control) or antigen-specific peptides or culture medium (negative control) for monkey PBMC

Notes

1. All antibodies used in this study are cross-reactive to rhesus monkeys as reported in NIH Nonhuman Primate Reagent Resource core facility.
2. When using the cryopreserved PBMC, the vials of frozen PBMC are removed from liquid nitrogen and rapidly thawed in a 37 °C water bath, gently mixed, washed with complete RPMI medium to remove the freezing medium, and resuspended in complete RPMI medium.
3. For the stimulation with peptide pools, a cocktail of peptides containing 2 μg/ml/peptide is prepared in complete RPMI medium and 100 μl/well is used for treatment of cells. For positive control a mixture of PMA and ionomycin is used and the final concentration of 10 ng and 100 ng/well, respectively. For negative control, cells are treated with 100 μl of complete RPMI medium. The final total volume for 96-well tissue culture plate is 0.2 ml/well (100 μl cell suspension and 100 μl of

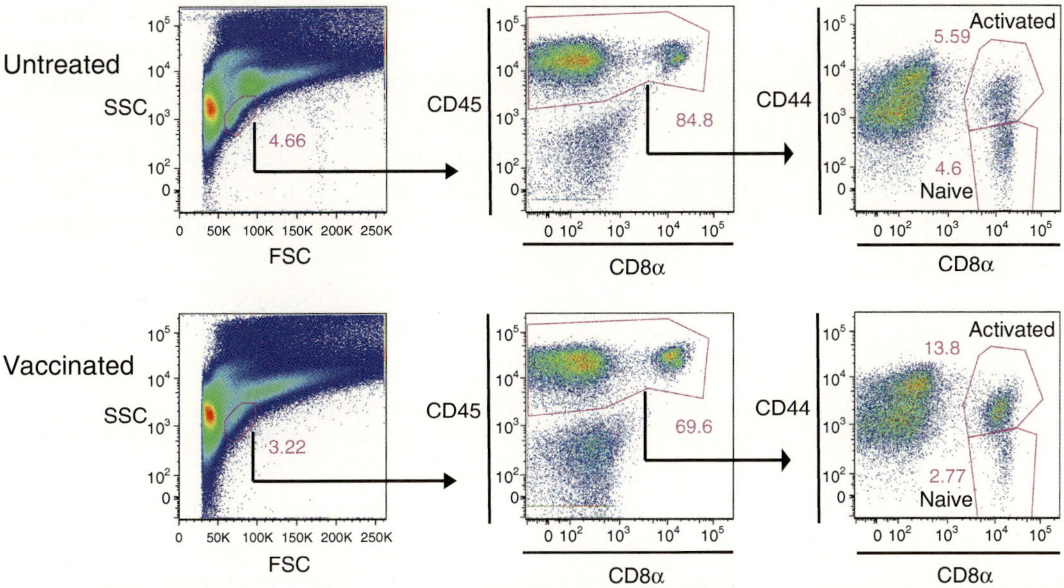

Fig. 6 Gating scheme utilized for the analyses of the different T lymphocyte subsets from FRT of mice. The lymphocytes from unvaccinated and vaccinated mice are first gated using a dot pot with FSC versus SSC. Cell surface marker CD45 is used to identify the leukocytes among the gated cells. To determine the activation status of CD8 T lymphocytes, expression of CD44 and CD8 is determined on CD45 positive cells. Activated cells can be identified as those positive for the expression of both CD44 and CD8, while naïve CD8 T cells are negative for the expression of CD44

the stimulating antigen). In addition, we routinely include PBMC obtained from naive macaques as an additional negative control.

4. Since it inhibits any secretory function of the cells, brefeldin is toxic to the cells if present for longer than 6 h. Therefore, duration of treatment with brefeldin must be monitored carefully.
5. A properly titrated antibody will achieve the optimal separation between positive and negative staining and thereby preventing wastage of reagents.
6. For the compensation controls, a complete set of tubes containing cell suspensions from one of monkey are stained with fluorescent conjugated monoclonal Antibody individually as single color stains. Compensation controls are utilized to eliminate false signal that results from spectral overlap between fluorescent dyes.
7. Fluorescence minus one (FMO) is a multicolor staining combination that contains all reagents but the one of interest. FMO and is used to determine the boundary between a positive and negative population by duplicating autofluorescence

level and data present in fully stained sample. It is critical to eliminate nonspecific background to observe minor and subtle changes in the lymphocyte functions which can be obscured by the presence of this auto fluorescence.

8. Permeabilization buffer contains detergent such as saponin or Triton™ X; therefore, it is important to fix the antibodies on the surface of the cells before proceeding to the permeabilization step.
9. For intracellular staining, the antibodies must be diluted in permeabilization buffer and cells must be maintained in this buffer for the pores of the cell membrane to remain open.
10. While fixation for 30 min is enough for lymphocytes isolated from immunized but uninfected animals, to ensure safe handling of lymphocytes from SIV or SHIV-infected macaques, FACS analysis must be performed only after overnight fixation in 3 % buffered paraformaldehyde
11. Always protect stained cells from light during staining, storage as well as acquisition stages to prevent bleaching of fluorescence. Store the stained cell suspensions at 4 °C in the dark.

References

1. Gamble LJ, Matthews QL (2010) Current progress in the development of a prophylactic vaccine for HIV-1. Drug Des Devel Ther 5: 9–26
2. Girard MP, Osmanov SK, Kieny MP (2006) A review of vaccine research and development: the human immunodeficiency virus (HIV). Vaccine 19:4062–4081
3. Kim JH, Rerks-Ngarm S, Excler JL et al (2010) HIV vaccines: lessons learned and the way forward. Curr Opin HIV AIDS 5:428–3
4. Esparza J, Osmanov S (2003) HIV vaccines: a global perspective. Curr Mol Med 3:183–193
5. Hu SL (2005) Non-human primate models for AIDS vaccine research. Curr Drug Targets Infect Disord 2:193–201
6. Duerr A (2010) Update on mucosal HIV vaccine vectors. Curr Opin HIV AIDS 5:397–403
7. Nehete PN, Singh S, Sastry KJ (2013) Lessons on non-progression of HIV disease from monkeys. Front Immunol 4:64
8. Nehete PN, Nehete BP, Hill L et al (2008) Selective induction of cell-mediated immunity and protection of rhesus macaques from chronic SHIV(KU2) infection by prophylactic vaccination with a conserved HIV-1 envelope peptide-cocktail. Virology 370:130–141
9. Sastry KJ, Nehete PN, Venkatnarayanan S et al (1992) Rapid in vivo induction of HIV-specific CD8+ cytotoxic T lymphocytes by a 15-amino acid unmodified free peptide from the immunodominant V3-loop of GP120. Virology 188: 502–509
10. Weaver EA, Nehete PN, Nehete BP et al (2009) Protection against mucosal SHIV challenge by peptide and helper-dependent adenovirus vaccines. Viruses 1:920
11. Murphy KL, Baxter MG, Flecknell PA (2012) Anesthesia and analgesia in nonhuman primates. In: Abee CR, Mansfield K, Tardif SD et al (eds) Nonhuman primates in biomedical research: biology and management. Academic, San Diego, CA, pp 403–436
12. Popilskis SJ, Lee DR, Elmore DB (2011) Anesthesia and analgesia in nonhuman primates. In: Fish R, Danneman PJ, Brown M, Karas A (eds) Anesthesia and analgesia in laboratory animals. Academic, San Diego, CA, pp 335–363
13. Bellanti JA, Zeligs BJ, Mendez de Inocencio J et al (2001) Alternative routes of immunization for prevention of infectious diseases: a new paradigm for the 21st century. Allergy Asthma Proc 22:173–176
14. Scherliess R (2011) Delivery of antigens used for vaccination: recent advances and challenges. Ther Deliv 2:1351–1368
15. Meeusen EN (2011) Exploiting mucosal surfaces for the development of mucosal vaccines. Vaccine 29:8506–8511

16. Demberg T, Robert-Guroff M (2009) Mucosal immunity and protection against HIV/SIV infection: strategies and challenges for vaccine design. Int Rev Immunol 28:20–48
17. Pachuk CJ, McCallus DE, Weiner DB, Satishchandran C (2000) DNA vaccines—challenges in delivery. Curr Opin Mol Ther 2: 188–189
18. Dunn TB (1954) Normal and pathological anatomy of the reticular tissue in laboratory mice. J Natl Cancer Inst 14:1281–1434
19. Covelli V (2013) Guide to the necropsy of the mouse. http://eulep.pdn.cam.ac.uk/Necropsy_of_the_Mouse. Accessed 28 July 2013
20. Reeves JP, Reeves PA (2001) Removal of lymphoid organs. Curr Protoc Immunol 1:1.9.1–1.9.3
21. Fuss IJ, Kanof ME, Smith PD, Zola H (2009) Isolation of whole mononuclear cells from peripheral blood and cord blood. Curr Protoc Immunol 85:7.1.1–7.1.8
22. Ulmer AJ, Scholz W, Ernst M et al (1984) Isolation and subfractionation of human peripheral blood mononuclear cells (PBMC) by density gradient centrifugation on Percoll. Immunobiology 166:238–250
23. Schluns KS, Nowak EC, Cabrera-Hernandez A et al (2004) Distinct cell types control lymphoid subset development by means of IL-15 and IL-15 receptor alpha expression. Proc Natl Acad Sci U S A 101:5616–5621
24. Sheridan BS, Lefrançois L (2012) Isolation of mouse lymphocytes from small intestine tissues. Curr Protoc Immunol 99:3.19.1–3.19.11
25. Seifter S, Gallop PM, Klein L et al (1959) Studies on collagen, part II. Properties of purified collagenase and its inhibition. J Biol Chem 234:285
26. Koup RA, Douek DC (2011) Vaccine design for CD8 T lymphocyte responses. Cold Spring Harb Perspect Med 1(1):a007252. doi:10.1101/cshperspect.a007252
27. Appay V, Douek DC, Price DA (2008) CD8+ T cell efficacy in vaccination and disease. Nat Med 14:623–628
28. Swain SL, McKinstry KK, Strutt TM (2012) Expanding roles for CD4+ T cells in immunity to viruses. Nat Rev Immunol 12:136–148
29. Perrin H, Canderan G, Sékaly RP, Trautmann L (2010) New approaches to design HIV-1T-cell vaccines. Curr Opin HIV AIDS 5: 368–376
30. Schmittel A, Keilholz U, Thiel E, Scheibenbogen C (2000) Quantification of tumor-specific T lymphocytes with the ELISPOT assay. J Immunother 23:289–295
31. De Rosa SC (2012) Vaccine applications of flow cytometry. Methods 57:383–391
32. Lehmann PV, Zhang W (2012) Unique strengths of ELISPOT for T cell diagnostics. Methods Mol Biol 792:3–23
33. Kalyuzhny AE (2005) Chemistry and biology of the ELISPOT assay. In: Kalyuzhny AE (ed) Handbook of ELISPOT: methods and protocols. Humana Press, Totowa, pp 15–31
34. Lehman PV (2005) Image analysis and data management of ELISPOT assay results. In: Kalyuzhny AE (ed) Handbook of ELISPOT: methods and protocols. Humana Press, Totowa, pp 117–132
35. Weaver EA, Nehete PN, Nehete BP et al (2013) Comparison of systemic and mucosal immunization with helper-dependent adenoviruses for vaccination against mucosal challenge with SHIV. PLoS One 8(7):e67574. doi:10.1371/journal.pone.0067574
36. Smith SG, Joosten SA, Verscheure V et al (2009) Identification of major factors influencing ELISpot-based monitoring of cellular responses to antigens from *Mycobacterium tuberculosis*. PLoS One 4(11):e7972. doi:10.1371/journal.pone.0007972

Chapter 25

Immunoinformatics and Systems Biology in Personalized Medicine

Guillermo Lopez-Campos, Jesús F. Bermejo-Martin, Raquel Almansa, and Fernando Martin-Sanchez

Abstract

Every year new databases and tools for the storage and analysis of biological data are developed, updated, and discontinued. For this reason it is very important to have a clear picture of the major repositories providing information about the availability of these databases and tools as well as a brief description of them. This chapter provides an overview of the most important information sources which can guide researchers through the process of selecting databases and tools of interest for immunoinformatics and systems biology in personalized medicine.

As an example of a particular resource of interest that combines a curated database and tools for data analysis, this chapter also includes a description of InnateDB. This database offers access to curated information relative to the innate immune response in a systems biology context.

Key words Immunoinformatics, Databases, Systems biology, Bioinformatics, Immunology, Standards, Web servers

1 Introduction

In the last decade, biology, immunology, and their applications in medicine have witnessed a revolution in the methods and tools available for research. The development of new laboratory techniques has been supported by the maturity of bioinformatics tools, databases and their subsequent applications. These new techniques and methods have enabled the systematic collection and analysis of large amounts of data in global perspectives in what have been called "-omics" approaches and thus the use of the "-omics" suffix has expanded across the study of the different molecular levels (genomics, transcriptomics, proteomics, metabolomics) and disciplines (immunomics). A common aspect across all these approaches is the ability to generate unprecedented amounts of data that require strong support from bioinformatics for their processing (data storage and retrieval in databases, standards).

Rajat K. De and Namrata Tomar (eds.), *Immunoinformatics*, Methods in Molecular Biology, vol. 1184,
DOI 10.1007/978-1-4939-1115-8_25, © Springer Science+Business Media New York 2014

The term immunoinformatics was coined in 2001 [1] referring to the use of an informational network used to model and understand the regulatory elements and feedback processes associated with the immune system. Nowadays, the term can be broadly considered as the area of bioinformatics specifically devoted to the analysis and management of information related with the immune system. Although by June 2013, a PubMed search using the term "immunoinformatics" yields a total of 120 articles it must be noted that this term is not comprehensive enough to cover all the published articles that describe tools and applications of bioinformatics in the area of immunology. The number of tools and databases available in this field has increased during the last decade and several of these resources and methods have been oriented to specific aspects of immunology, such as the MHC or T-cell receptors, and provide single-level analysis or information (e.g., analysis of epitope recognition, antigen binding regions, sequences, or three-dimensional structures). In some cases some of the resources combine a few of these aspects but either remaining at the same molecular level (e.g., protein sequences and structures) or combining nucleotide and protein sequences.

In their work "Immunology in the post genomic area" A. Aderem and L. Hood [2] described the pioneering role of immunology in the analyses of complex regulatory networks, the use of informatics and the key role of integrating all this multilevel information within the concept of systems biology. The aim of systems biology is to provide a "system" or "multilevel" understanding of biological processes through the integration and modeling of different data sources. Therefore, immunology, with all its complex interaction between different cell types, different regulatory and signaling pathways, and different molecules and genes, provides a perfect environment for the development and use of approaches based on systems biology.

Systems biology combined with the data gathered from "-omics" methods are key to understand the trends towards precision medicine, which aims at defining diseases not only by its traditional signs and symptoms but also by its underlying molecular causes and other factors such as environmental risk factors [3, 4]. In this context the immune system plays a central role as a protective element, including protection against infections. The immune system is also responsible for the development of some diseases or the exacerbation of other processes such as allergies or autoimmune phenomena. Therefore, it provides a good model for the study of the equation "Genome*Exposome = Phenome" which is the key for supporting the current investigations towards personalized and precision medicine. Achieving these goals exceeds the scope of immunoinformatics since it requires the use of other data types and information sources, such as for example electronic clinical records, as well as other different techniques provided by other areas of Biomedical Informatics.

2 Finding Relevant Databases for Immunoinformatics

Along the last years there have been many publications describing databases which provide data and information associated with immunology. In this chapter we have considered that rather than providing a list with a brief description of the different databases, it is more useful to provide a description of two of the major resources containing information about available databases and tools associated with immunoinformatics, namely, the Nucleic Acids Research Database Annual Issue and the Canadian Bioinformatics Links Directory [5]. The reason to describe these resources, rather than the most traditional approach based on listing a set of databases of interest with a brief description for each of them, is that these resources provide access to up-to-date annotated lists of immunoinformatics resources which ensures the quality and relevance of these databases and tools.

2.1 Molecular Biology Database Collection

Since 1993, the Nucleic Acids Research (NAR) journal publishes every year two special issues devoted to the broader field of bioinformatics, one focuses on Databases and the other describes Web Services.

The first of these yearly issues is released every January and it is the one devoted to biological databases. In this issue it is possible to find articles associated with the publication of new biological data repositories as well as updates and major changes from previously published databases. Associated with this publication, the journal also maintains an online repository, the Molecular Biology Database Collection (MBDC) (http://www.oxfordjournals.org/nar/database/c), with all the databases published in the previous years as well as their original articles where the resources were described.

On July 2013 the collection included more than 1,500 databases, organized under a series of major topics such as nucleotide databases, RNA sequence databases or cell biology (Fig. 1). Among those various topics it is possible to find one annotated as immunological databases, which contains 31 different databases (Table 1). These databases are specifically focused on immunology and therefore can be considered as reference immunoinformatics databases.

The MBDC also includes a number of resources associated with systems biology under the category "Metabolic and Signaling Pathways," where many resources are stored in its four subcategories (Table 2).

The collection provides a category where human genes and disease databases are grouped. In this category it is possible to find interesting databases for the development of personalized medicine from an immunological perspective. In this category and subcategories (Table 1) it is possible to find polymorphisms databases

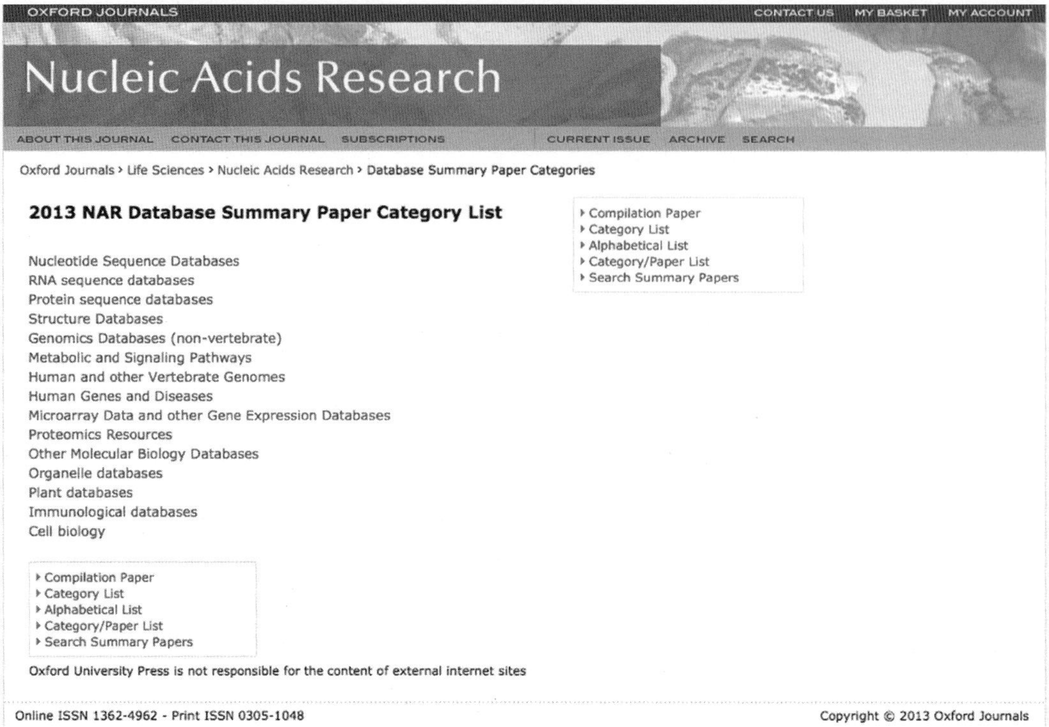

Fig. 1 Screenshot showing the major topics used to classify the resources in the Molecular Biology Database Collection. It is also possible to see in the *upper-right* and *bottom-left* corners the different options to access and retrieve the information

Table 1
List of the databases found in the category "Immunological" at the MBDC

Name	URL
ALPSbase [22]	http://www.niaid.nih.gov/topics/alps/Pages/default.aspx
AntigenDB [23]	http://www.imtech.res.in/raghava/antigendb
AntiJen [24]	http://www.ddg-pharmfac.net/antijen/AntiJen/antijenhomepage.htm
BCIpep [25]	http://bioinformatics.uams.edu/mirror/bcipep/
dbMHC [26]	http://www.ncbi.nlm.nih.gov/gv/mhc/
DIGIT [27]	http://www.biocomputing.it/digit
Epitome [28]	http://www.rostlab.org/services/epitome/
GPX-Macrophage Expression Atlas [29]	http://www.gpxmea.gti.ed.ac.uk/

(continued)

Table 1
(continued)

Name	URL
HaptenDB [30]	http://www.imtech.res.in/raghava/haptendb/
HPTAA [31]	http://www.bioinfo.org.cn/hptaa/
IEDB-3D [32]	http://www.immuneepitope.org/bb_structure.php
IL2Rgbase [33]	http://www.ncbi.nlm.nih.gov/lovd/home.php?select_db=IL2RG
IMGT [34]	http://www.imgt.org/
IMGT/GENE-DB [35]	http://www.imgt.org/IMGT_GENE-DB/GENElect?livret=0/
IMGT/HLA [36]	http://www.ebi.ac.uk/ipd/imgt/hla/
IMGT/LIGM-DB [37]	http://www.imgt.org/ligmdb/
InnateDB [16]	http://www.innatedb.com/
IPD-ESTDAB [8]	http://www.ebi.ac.uk/ipd/estdab/
IPD-HPA—Human Platelet Antigens [8]	http://www.ebi.ac.uk/ipd/hpa/
IPD-KIR—Killer-cell Immunoglobulin-like Receptors [8]	http://www.ebi.ac.uk/ipd/kir/
IPD-MHC [8]	http://www.ebi.ac.uk/ipd/mhc/
MHCBN [38]	http://www.imtech.res.in/raghava/mhcbn/
MHCPEP [39]	http://bio.dfci.harvard.edu/DFRMLI/
MPID-T2 [40]	http://biolinfo.org/mpid-t2/
Protegen [41]	http://www.violinet.org/protegen/
SuperHapten [42]	http://bioinformatics.charite.de/superhapten/
The Immune Epitope Database (IEDB) [43]	http://www.iedb.org/
VBASE2 [44]	http://www.vbase2.org/
FIMM [45]	http://www.research.i2r.a-star.edu.sg/fimm/ Not operative in June 2013
MUGEN Mouse Database [46]	http://www.mugen-noe.org/database/ Not operative in June 2013
Interferon Stimulated Gene Database [47]	http://www.lerner.ccf.org/labs/williams/xchip-html.cgi Not operative in June 2013

Three out of the 31 were not operative in June 2013

Table 2
Category and subcategories used in the molecular biology database collection compiled by NAR

Category	Subcategory
Nucleotide sequence databases	International nucleotide sequence database collaboration Coding and noncoding DNA Gene structure, introns and exons, splice sites Transcriptional regulator sites and transcription factors
RNA sequence databases	
Protein sequence databases	General sequence databases Protein properties Protein localization and targeting Protein sequence motifs and active sites Protein domain databases; protein classification Databases of individual protein families
Structure databases	Small molecules Carbohydrates Nucleic acid structure Protein structure
Genomics databases (non-vertebrate)	Genome annotation terms, ontologies, and nomenclature Taxonomy and identification General genomics databases Viral genome databases Prokaryotic genome databases Unicellular eukaryotes genome databases Fungal genome databases Invertebrate genome databases
Metabolic and signaling pathways	Enzymes and enzyme nomenclature Metabolic pathways Protein–protein interactions Signaling pathways
Human and other vertebrate genomics	Model organisms, comparative genomics Human genome databases, maps, and viewers Human ORFs
Human genes and diseases	General human genetics databases General polymorphism databases Cancer gene databases Gene-, system-, or disease-specific databases
Microarray data and other gene expression databases	
Proteomics resources	
Other molecular biology databases	Drugs and drug design Molecular probes and primers

(continued)

Table 2
(continued)

Category	Subcategory
Organelle databases	Mitochondrial genes and proteins
Plant databases	General plant databases Arabidopsis thaliana Rice Other plants
Immunological databases	
Cell biology	

In some the listed resources might appear in the subcategories, whereas in other cases they might be appear directly under the main category

such as ClinVar [6], databases from the Human Genome Variation Society, PharmGKB [7] or the reference of IPD—Immuno Polymorphisms database [8].

The collection provides links to different sets of interesting databases from an immunological perspective but because of their generalistic approach they are placed under different subjects. Some examples of such resources include gene expression databases, such as the Gene Expression Omnibus (GEO) from the National Center for Biotechnology Information (NCBI) [9] or the European Bioinformatics Institute's [10] ArrayExpress database [11], or structure databases, such as the Protein Data Bank (PDB) [12].

Alternatively to the searches based on the categories, the collection also provides an option to search for a specific database using an alphabetic listing of the contents as well as the possibility of using a search tool that can be found as "Search Summary Papers." This tool is very useful since it provides a text box where the user can type the search term of interest and select the field (among "title," "author," "affiliation," "paper," "references," or "all") where the search of interest will be carried out (Fig. 2).

Both approaches, search by category and search by term, should be combined when looking for a database of interest since they provide different insight and different results. An interesting example is the comparison between the resources grouped under the subject immunological databases and a text search strategy using the terms "immunological" or "immunology." The results (Table 3) show that different search strategies present different results with some degree of overlapping (*see* **Note 1**).

Fig. 2 Screenshot showing the search box and possible search topics in the Molecular Biology Database Collection when the "Search Summary Papers" option is used to query the database

2.2 The Canadian Bioinformatics Links Directory

Bioinformatics Links Directory (http://www.bioinformatics.ca/links_directory/) is another important resource providing information about databases and tools. It is a community driven resource and includes contents selected from the recommendation of experts in bioinformatics, and where registered members can suggest and submit reviews of links, resources, databases, and tools. It is important to remark that there is a close relationship between the NAR special issues and the Bioinformatics Links Directory, because part of the content is directly extracted from those NAR special issues. This relationship is more visible for the tools contained in the Links Directory because the contents of the NAR Web Services Special issue have been feeding the directory since 2003.

The elements contained in the Links Directory are contextually annotated, and there are tags based on MeSH (Medical Subject Headings) terms to describe the resources. This annotation is an important and characteristic feature of the Links Directory and can be used for searching purposes. Part of this annotation includes a rating of each of the resources. This rating is presented as the "Links Directory Index" which is based on the citations of that particular resource in PubMed and Google Scholar or, when applicable, in social media such as Twitter and Google+. These rates give an idea of how often the resource is used and can be used to rank the results of the searches carried out in the site. Another important characteristic is the curation of the contents, eliminating "dead" contents and links that are not available any longer.

The Bioinformatics Links Directory contained 620 databases, 164 links and 1,459 tools in June 2013, and it is being continuously updated with new contents.

The information in the directory is structured around three categories, Resources, Databases and Tools, and 11 concepts, Computer Related, DNA, Education, Expression, Human Genome, Literature,

Table 3
Comparison of the results between a search using category list "immunological databases" and search terms "immunological" and the related term "immunology"

Category list "immunological databases"	Search term "immunological"
ALPSbase	**IPD-ESTDAB**
AntigenDB	**IPD-HPA—Human Platelet Antigens**
AntiJen	**IPD-KIR—Killer-cell Immunoglobulin-like Receptors**
BCIpep	**IPD-MHC**
dbMHC	**The Immune Epitope Database (IEDB)**
DIGIT	**VBASE2**
Epitome	
FIMM	
GPX-Macrophage Expression Atlas	
HaptenDB	
HPTAA	
IEDB-3D	
IL2Rgbase	
IMGT	
IMGT/GENE-DB	
IMGT/HLA	
IMGT/LIGM-DB	
InnateDB	
Interferon Stimulated Gene Database	
IPD-ESTDAB	
IPD-HPA—Human Platelet Antigens	
IPD-KIR—Killer-cell Immunoglobulin-like Receptors	
IPD-MHC	
MHCBN	
MHCPEP	
MPID-T2	
MUGEN Mouse Database	
Protegen	
SuperHapten	
The Immune Epitope Database (IEDB)	
VBASE2	

(continued)

Table 3 (continued)

Category list "immunological databases"	Search term "immunological" / Search term "immunology"
	AGRIS—*Arabidopsis* Gene Regulatory Information Server
	AntiJen
	FIMM
	HAGR—Human Ageing Genomic Resources
	HemoPDB—Hematopoiesis Promoter Database
	HIV Molecular Immunology Database
	IEDB-3D
	NMPDR—National Microbial Pathogen Data Resource
	Protegen
	RECODE
	SOURCE
	SysZNF
	The Immune Epitope Database (IEDB)
	VBASE2
	VFDB—Virulence Factors Database

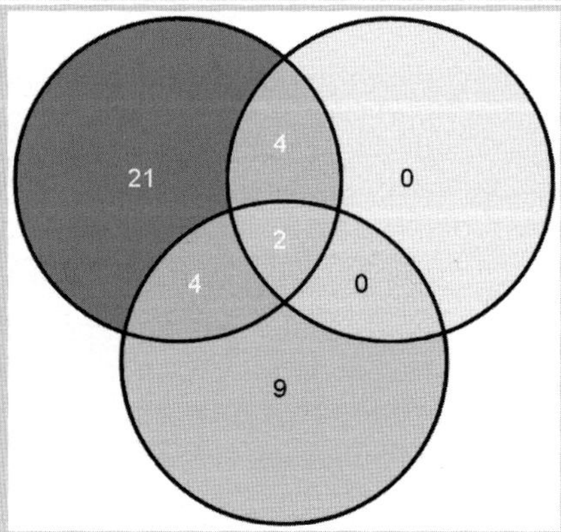

It can be seen that the three strategies lead to different results with a different degree of overlapping among them. Common elements are shown in bold and the Venn diagram shows the shared elements in the three searches

Model Organisms, Other Molecules, Proteins, RNA, and Sequence Comparison. Each of these concepts contains in turn different subcategories (Fig. 3).

Each link entry includes the name and link of the described element, the concepts to which is associated, a brief description of the contents, a link to the PubMed citation and Directory Index, a link to the NAR Issue when available, a report on users feedback, and finally the set of tags used for annotating the entry.

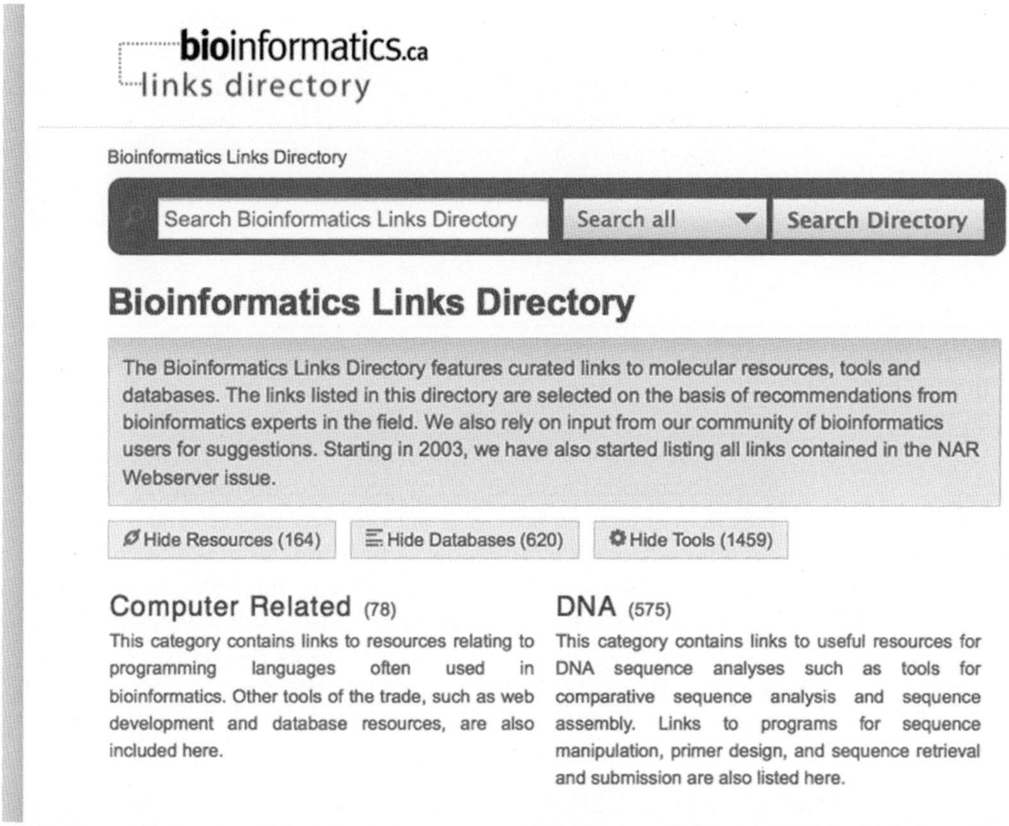

Fig. 3 Screenshot of the Home page of the Bioinformatics Links Directory. It is possible to see the different elements that enable the data retrieval from this resource. On the *upper* part the "text-box" search and in the *lower* part of this screenshot just two of the different topics used to classify the resources. Above these two topics are the buttons used to select between the contents category (resource, database or tool) and the number of elements contained on each of them in June 2013

Accessing the information contained in the resources can be performed in two different ways. The first one is based on browsing by concepts and sub-concepts. The sub-concepts are shown when the cursor is over each of the main categories, then it is possible to click on them and then all the resources are shown. The results are then presented showing all the resources, databases, and tools annotated under that sub-concept in alphabetical order. On top of the page there are several buttons that allow the user to subscribe to a RSS channel associated with that search to receive the latest news related with changes in those contents, change the results presentation to see them in a compact way or sort them by the Directory Index rank. It is also possible to download the results in a variety of formats and filter them by hiding the undesired categories (resources, databases or tools). In the bottom of the screen the associated Tag Cloud represents the tags used in the annotation of those links.

Alternatively it is possible to query the directory using a text box placed on the top of the page. These searches can be performed on the titles, the descriptions, the tags or by default in all these fields at once (Fig. 4).

3 Finding Relevant Tools for Immunoinformatics, Systems Biology, and Personalized Medicine

The main resource to search for bioinformatics tools is the Bioinformatics Links category, since it contains all the tools published in the NAR Special Issue on Web services since 2003. The search methodology for tools is the same described in the previous section, with the only difference that in this case the user should press the buttons "Hide Databases" and "Hide resources" in order to retrieve only the tools. A useful strategy for the retrieval of tools is to use the "Tags" used to annotate the links, in Table 4 there is a list of possible tags of interest and Table 5 shows some of the results of a search performed using the "Tag" *Health and Disease*.

4 Sharing Data in Immunoinformatics

The widespread use of knowledge representation standards and ontologies in biomedical informatics has greatly simplified the data integration and sharing processes. Immunoinformatics has not been different and there have been some important developments not only in the integration of ontologies in existing resources (for example in IEDB [13]) but also in the development of some specific ontologies focused on immunology. A good example of this specific ontologies is the IMGT ontology [14], firstly developed in 1999 as part of the efforts of the Immunogenetics international collaboration (Table 6). Another important aspect related with data sharing and data integration is the development and use of reporting guidelines. These guidelines enable the reporting of minimal information related with biological investigations [15]. It is possible to access to the list of available reporting guidelines through the biosharing.org Web portal (http://www.biosharing.org/standards/reporting_guideline). There were 62 guidelines in June 2013 and several of them are recommended or associated with the contents of this chapter (Table 7).

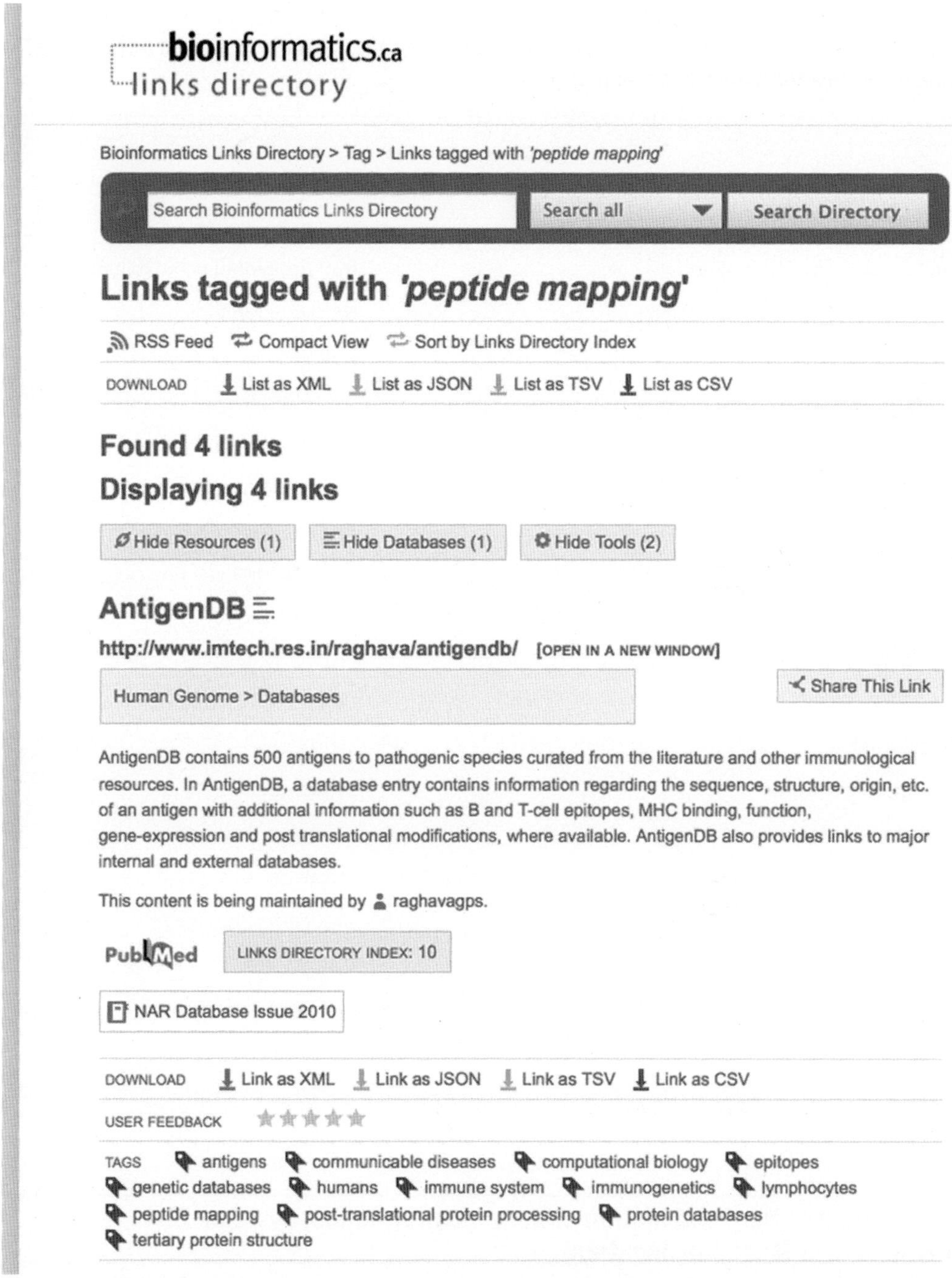

Fig. 4 Screenshot showing the details of an entry in the Bioinformatics Links Directory from the results of a search using the Tag "peptide mapping". In the *upper* part of the image it is possible to see the sorting and download options available. In the *lower* part of the screenshot it is possible to appreciate the contents of the entries with the link to the resource, its description, ranking and finally the associated tags assigned during the annotation process

Table 4
Tags for the retrieval of links associated with immunoinformatics, systems biology or personalized medicine in the Bioinformatics links directory

Algorithms	Histocompatibility antigens class II
Alleles	Immune system
Amino acid sequence	Immunogenetics
Amino acid sequence homology	Immunoglobulin variable region
Amino acids	Immunoglobulins
Antibodies	Kinetics
Antibody specificity	Lymphocytes
Antigen receptors	Major histocompatibility complex
Antigen–antibody complex	Molecular structure
Antigens	Peptide mapping
b-Lymphocyte epitopes	Post-translational protein processing
Binding sites	Protein binding
Biological evolution	Protein databases
Chemical models	Protein interaction mapping
Chemical models	Proteins
Communicable diseases	Proteomics
Computational biology	Reference standards
Computer simulation	Sensitivity and specificity
Epitope mapping	Sequence alignment
Epitopes	Systems biology
Genetic databases	Systems integration
Histocompatibility antigens	t-Lymphocyte epitopes
Histocompatibility antigens class I	Tertiary protein structure

5 An Example of the Combination of Immunoinformatics, Systems Biology, and Personalized Medicine: InnateDB

Nowadays, with the development of new "-omics" based technologies, the analysis of different sets of biological molecules represents a key aspect of most biomedical research projects. An important aspect related with the use of bioinformatic resources and tools consists of the need to provide a context for the interpretation of the results. In this sense the protocol described in this section describe the use of the database InnateDB [16]

Table 5
Example of some of the tools that can be found at the Links directory when the search is done using the Tag "Health and Disease"

Tools	Description
IEDB-AR. The Immune Epitope Database Analysis Resource (IEDB-AR) [48]	Analysis of immune epitopes
DyNaVacS [49]	DNA vaccine design tool
DigSee—Disease Gene Search Engine with Evidence Sentences [50]	Tool for understanding the relationship between genes and diseases from the literature
EpiToolKit [51]	Suite of tools for immunological research
NetMHC3.0 [52]	Prediction of MHC Class I peptide binding
OptiTope [53]	Identification of epitopes for vaccine design
PEPVAC [54]	Multiple epitope vaccines design tool
PGMRA [55]	Phenotype–genotype association tool

Table 6
Example of some of the ontologies found at OBO Foundry (http://www.obofoundry.org/) that are relevant for Immunoinformatics

Ontology name	URL
Cell Type Ontology	http://www.cellontology.org/
Foundational Model of Anatomy	http://www.fma.biostr.washington.edu/
Protein Ontology	http://www.pir.georgetown.edu/pro/
Human Disease Ontology	http://www.disease-ontology.org
Gene Ontology	http://www.geneontology.org
Human Phenotype Ontology	http://www.human-phenotype-ontology.org
Infectious Disease Ontology	http://www.infectiousdiseaseontology.org
Ontology for Biomedical Investigations	http://www.obi-ontology.org/
Systems Biology ontology	http://www.ebi.ac.uk/sbo/
Vaccine Ontology	http://www.violinet.org/vaccineontology/

(http://www.innatedb.com) (*see* **Note 2**), a curated resource developed in Canada and focused on the innate immune response. This system provides access to a unified interface for accessing curated information (at different levels) about pathways and molecules, and their interactions, associated with the innate immune response of human (*Homo sapiens*), mouse (*Mus musculus*), and cow (*Bos taurus*).

Table 7
Examples of some of the relevant minimum information reporting guidelines

Reporting guideline name	Purpose
MIAME [56]	Minimum information about a microarray experiment http://www.fged.org
MIATA [57]	Minimal information about T cell assays http://www.miataproject.org
MIMIx [58]	Minimum information about a molecular interaction experiment http://www.psidev.info/mimix
MISFISHIE [59]	Minimum information specification for in situ hybridization and immunohistochemistry experiments http://www.mged.sourceforge.net/misfishie/

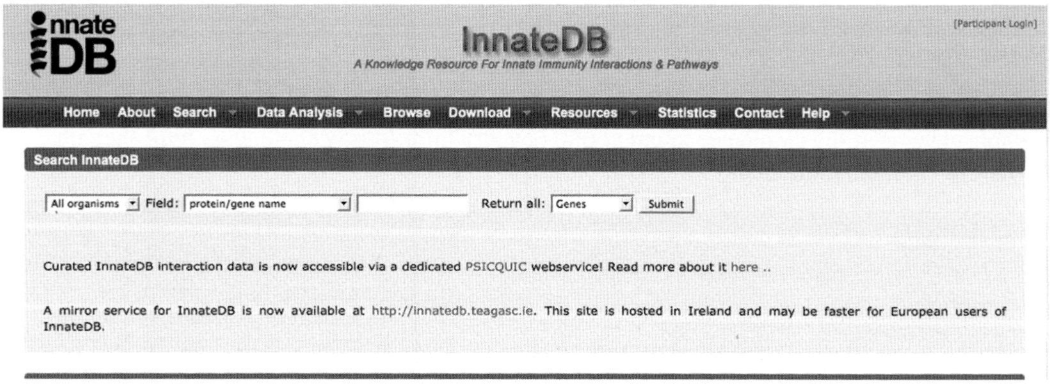

Fig. 5 Screenshot of the home page of InnateDB showing the menu as well as the simplified query interface

The information retrieval process at this resource can be performed in different ways. The simplest query system is based on the selection among the different organisms available in the database, the selection of the identifier used for the search (which could be a protein or a gene name, an Ensembl ID, or a RefSeq ID) and finally the type of results of interest, choosing among genes, proteins or interactions (Fig. 5). Under the tab named "Search" the database provides the user with a more complex query interface for queries based on the molecules, the interactions or the pathways. The contents of the database are grouped together in different categories depending on the type of interactions, the pathways or the different immune gene lists, and this grouping is also provided as a way to browse the database.

A very interesting characteristic of InnateDB is that it also offers some tools for the analysis of user's data (*see* **Notes 3** and **4**). The tab "Data Analysis" gives access to a set of tools to analyze pathways, the gene ontology, networks, interactions and transcription binding factor sites (TFBSs). Each of these analytical tools allows the users to upload their own data using a tabular format, providing at least the name of the genes or proteins of interest and depending on the tool also some other numerical data such as the gene expression intensities or *p*-values.

In the pathway analysis the system uses all the pathways from a selection of major public databases such as KEGG [17], Reactome [18], NetPath [19], or PID [20] and makes the tool one of the most comprehensive in terms of the number of resources used for pathway analysis. In the analysis of gene ontology, interactions, and networks, the analysis exploits the curated information stored in the database to enrich the results of these analyses.

6 Notes

1. The molecular Biology Database Collection is a very powerful resource to identify databases and it is updated on a yearly basis on January. An important characteristic of the MBDC is that it provides an historical perspective of those resources that have been previously published in the database special issue and are still available at the date of each annual update in January. For this reason those resources that are removed after January and therefore are not available any longer might still be listed in Web site.
2. An interesting feature of InnateDB is that it has a mirror in Ireland (http://www.innatedb.teagasc.ie) that might be faster for European users and helps to reduce the load of the Canadian server.
3. InnateDB interface is a mouse over interface so it is necessary to put the mouse pointer over the tags for the different tabs for the different services available avoiding clicking on them because it would generate an error.
4. An important characteristic associated with the use of the analytical tools included in InnateDB is that they require the use of accession numbers from any of the following databases: Ensembl (preferred), RefSeq, Entrez, UniProt, or InnateDB. For this reason in the case of working with gene symbols these must be transformed into any of the previously cited accession numbers. This gene ID conversion step might be carried out using a Web conversion tool such as the one offered at DAVID [21] (http://www.david.abcc.ncifcrf.gov/conversion.jsp).

References

1. Segel LA (2001) Controlling the immune system: diffuse feedback via a diffuse informational network. Novartis Found Symp 239:31–40, discussion 40–51
2. Aderem A, Hood L (2001) Immunology in the post-genomic era. Nat Immunol 2(5):373–375
3. Committee on a Framework for Development a New Taxonomy of Disease (2011) Toward precision medicine: building a knowledge network for biomedical research and a new taxonomy of disease. The National Academies, Washington, DC
4. Desmond-Hellmann S (2012) Toward precision medicine: a new social contract? Sci Transl Med 4(129):129ed3
5. Brazas MD et al (2012) A decade of Web Server updates at the Bioinformatics Links Directory: 2003–2012. Nucleic Acids Res 40(Web Server issue):W3–W12
6. Wheeler DL et al (2013) Database resources of the National Center for Biotechnology Information. Nucleic Acids Res 41(Database issue):D8–D20
7. Hernandez-Boussard T et al (2008) The pharmacogenetics and pharmacogenomics knowledge base: accentuating the knowledge. Nucleic Acids Res 36(Database issue):D913–D918
8. Robinson J et al (2013) IPD—the Immuno Polymorphism Database. Nucleic Acids Res 41(Database issue):D1234–D1240
9. Barrett T et al (2013) NCBI GEO: archive for functional genomics data sets—update. Nucleic Acids Res 41(Database issue):D991–D995
10. Huebinger RM et al (2013) Examination with next-generation sequencing technology of the bacterial microbiota in bronchoalveolar lavage samples after traumatic injury. Surg Infect (Larchmt) 14(3):275–282
11. Rustici G et al (2013) ArrayExpress update—trends in database growth and links to data analysis tools. Nucleic Acids Res 41(Database issue):D987–D990
12. Rose PW et al (2013) The RCSB Protein Data Bank: new resources for research and education. Nucleic Acids Res 41(Database issue):D475–D482
13. Vita R et al (2013) Query enhancement through the practical application of ontology: the IEDB and OBI. J Biomed Semantics 4(Suppl 1):S6
14. Giudicelli V, Lefranc MP (2012) Imgt-Ontology 2012. Front Genet 3:79
15. Taylor CF et al (2008) Promoting coherent minimum reporting guidelines for biological and biomedical investigations: the MIBBI project. Nat Biotechnol 26(8):889–896
16. Lynn DJ et al (2008) InnateDB: facilitating systems-level analyses of the mammalian innate immune response. Mol Syst Biol 4:218
17. Kanehisa M et al (2012) KEGG for integration and interpretation of large-scale molecular data sets. Nucleic Acids Res 40(Database issue): D109–D114
18. Croft D et al (2011) Reactome: a database of reactions, pathways and biological processes. Nucleic Acids Res 39(Database issue): D691–D697
19. Kandasamy K et al (2010) NetPath: a public resource of curated signal transduction pathways. Genome Biol 11(1):R3
20. Schaefer CF et al (2009) PID: the Pathway Interaction Database. Nucleic Acids Res 37(Database issue):D674–D679
21. da Huang W, Sherman BT, Lempicki RA (2009) Systematic and integrative analysis of large gene lists using DAVID bioinformatics resources. Nat Protoc 4(1):44–57
22. Price S et al (2014) Natural history of autoimmune lymphoproliferative syndrome associated with FAS gene mutations. Blood 123(13): 1989–1999
23. Ansari HR, Flower DR, Raghava GP (2010) AntigenDB: an immunoinformatics database of pathogen antigens. Nucleic Acids Res 38(Database issue):D847–D853
24. Toseland CP et al (2005) AntiJen: a quantitative immunology database integrating functional, thermodynamic, kinetic, biophysical, and cellular data. Immunome Res 1(1):4
25. Saha S, Raghava GP (2007) Searching and mapping of B-cell epitopes in Bcipep database. Methods Mol Biol 409:113–124
26. Thanh HD et al (2013) Development of a 16S-23S rRNA intergenic spacer-based quantitative PCR assay for improved detection and enumeration of Lactococcus garvieae. FEMS Microbiol Lett 339(1):10–16
27. Chailyan A, Tramontano A, Marcatili P (2012) A database of immunoglobulins with integrated tools: DIGIT. Nucleic Acids Res 40(Database issue):D1230–D1234
28. Schlessinger A et al (2006) Epitome: database of structure-inferred antigenic epitopes. Nucleic Acids Res 34(Database issue):D777–D780
29. Grimes GR et al (2005) GPX-Macrophage Expression Atlas: a database for expression profiles of macrophages challenged with a variety of pro-inflammatory, anti-inflammatory, benign and pathogen insults. BMC Genomics 6:178
30. Singh MK et al (2006) HaptenDB: a comprehensive database of haptens, carrier proteins

and anti-hapten antibodies. Bioinformatics 22(2):253–255
31. Wang X et al (2006) HPtaa database-potential target genes for clinical diagnosis and immunotherapy of human carcinoma. Nucleic Acids Res 34(Database issue):D607–D612
32. Ponomarenko J et al (2011) IEDB-3D: structural data within the immune epitope database. Nucleic Acids Res 39(Database issue): D1164–D1170
33. Puck JM (1996) IL2RGbase: a database of gamma c-chain defects causing human X-SCID. Immunol Today 17(11):507–511
34. Lefranc MP (2011) IMGT, the International ImMunoGeneTics Information System. Cold Spring Harb Protoc 2011(6):595–603
35. Giudicelli V, Chaume D, Lefranc MP (2005) IMGT/GENE-DB: a comprehensive database for human and mouse immunoglobulin and T cell receptor genes. Nucleic Acids Res 33(Database issue):D256–D261
36. Robinson J et al (2013) The IMGT/HLA database. Nucleic Acids Res 41(Database issue):D1222–D1227
37. Giudicelli V et al (2006) IMGT/LIGM-DB, the IMGT comprehensive database of immunoglobulin and T cell receptor nucleotide sequences. Nucleic Acids Res 34(Database issue):D781–D784
38. Bhasin M, Singh H, Raghava GP (2003) MHCBN: a comprehensive database of MHC binding and non-binding peptides. Bioinformatics 19(5):665–666
39. Brusic V, Rudy G, Harrison LC (1998) MHCPEP, a database of MHC-binding peptides: update 1997. Nucleic Acids Res 26(1):368–371
40. Khan JM et al (2011) MPID-T2: a database for sequence-structure-function analyses of pMHC and TR/pMHC structures. Bioinformatics 27(8):1192–1193
41. Yang B et al (2011) Protegen: a web-based protective antigen database and analysis system. Nucleic Acids Res 39(Database issue): D1073–D1078
42. Gunther S et al (2007) SuperHapten: a comprehensive database for small immunogenic compounds. Nucleic Acids Res 35(Database issue):D906–D910
43. Vita R et al (2010) The immune epitope database 2.0. Nucleic Acids Res 38(Database issue):D854–D862
44. Retter I et al (2005) VBASE2, an integrative V gene database. Nucleic Acids Res 33(Database issue):D671–D674
45. Schonbach C et al (2005) An update on the functional molecular immunology (FIMM) database. Appl Bioinformatics 4(1):25–31
46. Aidinis V et al (2008) MUGEN mouse database; animal models of human immunological diseases. Nucleic Acids Res 36(Database issue):D1048–D1054
47. de Veer MJ et al (2001) Functional classification of interferon-stimulated genes identified using microarrays. J Leukoc Biol 69(6):912–920
48. Kim Y et al (2012) Immune epitope database analysis resource. Nucleic Acids Res 40(Web Server issue):W525–W530
49. Harish N et al (2003) DyNAVacS: an integrative tool for optimized DNA vaccine design. Nucleic Acids Res 34(1):W264–W266
50. Kim J et al (2013) DigSee: disease gene search engine with evidence sentences (version cancer). Nucleic Acids Res 41(Web Server issue): W510–W517
51. Feldhahn M et al (2008) EpiToolKit–a web server for computational immunomics. Nucleic Acids Res 36(Web Server issue):W519–W522
52. Lundegaard C et al (2008) NetMHC-3.0: accurate web accessible predictions of human, mouse and monkey MHC class I affinities for peptides of length 8-11. Nucleic Acids Res 36(Web Server issue):W509–W512
53. Toussaint NC, Kohlbacher O (2009) OptiTope—a web server for the selection of an optimal set of peptides for epitope-based vaccines. Nucleic Acids Res 37(Web Server issue):W617–W622
54. Reche PA, Reinherz EL (2005) PEPVAC: a web server for multi-epitope vaccine development based on the prediction of supertypic MHC ligands. Nucleic Acids Res 33(Web Server issue):W138–W142
55. Arnedo J et al (2013) PGMRA: a web server for (phenotype x genotype) many-to-many relation analysis in GWAS. Nucleic Acids Res 41(Web Server issue):W142–W149
56. Brazma A et al (2001) Minimum information about a microarray experiment (MIAME)-toward standards for microarray data. Nat Genet 29(4):365–371
57. Hoos A, Janetzki S, Britten CM (2012) Advancing the field of cancer immunotherapy: MIATA consensus guidelines become available to improve data reporting and interpretation for T-cell immune monitoring. Oncoimmunology 1(9):1457–1459
58. Orchard S et al (2007) The minimum information required for reporting a molecular interaction experiment (MIMIx). Nat Biotechnol 25(8):894–898
59. Deutsch EW et al (2006) Development of the minimum information specification for in situ hybridization and immunohistochemistry experiments (MISFISHIE). OMICS 10(2):205–208

Part V

Applications of Immunoinformatics

Chapter 26

The Role of Small RNAs in Vaccination

Ajeet Chaudhary and Sunil Kumar Mukherjee

Abstract

The concept of vaccination came to light following Edward Jenner's classical observation on milkmaids who were protected against smallpox. However, plants lack the cellular based immunity system and thus it was not appreciated earlier that plants can also be protected from their pathogens. But phenomena like cross-protection, pathogen derived resistance (PDR), viral recovery, etc. in plants suggested that plants have also evolved immunity against their pathogens. The further advances in the field revealed that an endogenous defense system could have multiple prongs. With the advent of RNAi, it was clear that the antiviral immune responses are related to the induction of specific small RNAs. The detection of virus specific small RNAs (vsiRNA) in immunized plants confirmed their roles in the immunity against pathogens. Although many issues related to antiviral mechanisms are yet to be addressed, the existing tools of RNAi can be efficiently used to control the invading viruses in transgenic plants. It is also possible that the microRNA(s) induced in infected plants impart immunity against viral pathogens. So the small RNA molecules play a vital role in defense system and these can be engineered to enhance the immunity against specific viral pathogens.

Key words Cross-protection, PDR, RNA interference, Endogenous siRNA, Artificial microRNA

1 Introduction

Cross-protection is a phenomenon known for developing immunity against various diseases like smallpox, influenza, measles, and tetanus in human beings in the form of vaccination (*see* **Note 1**). The earliest documented examples of vaccination were from India and China in the seventeenth century, where vaccination with powdered scabs from people infected with smallpox was used to protect against the disease [1]. This kind of immunity essentially depends upon the B-cells and T-cells in the animal kingdom [2]. However, plants lack this cellular immune system; and instead they have evolved different mechanisms to deal with the protection against the invading pathogens, especially viruses. Previously, a few defense mechanisms have been known in plants like PAMP-triggered immunity (PTI) and effector triggered immunity (ETI) and others like systemic acquired response (SAR), etc. [3, 4] (*see* **Note 2**).

Rajat K. De and Namrata Tomar (eds.), *Immunoinformatics*, Methods in Molecular Biology, vol. 1184,
DOI 10.1007/978-1-4939-1115-8_26, © Springer Science+Business Media New York 2014

Later, a more effective RNA-dependent antiviral immunity was discovered in plants which is now more commonly known as RNA interference (RNAi). RNAi mediated antiviral defense mechanism is similar in some way to cell based immunity of the animal system [5]. In plants, instead of antibodies against pathogens, the prevailing RNAi mechanisms produce small RNAs which specifically target the viral genome sequence for inactivation. This RNAi mechanism can also be artificially directed against viral pathogen to induce immunity against the pathogens in transgenic plants. The subsequent part of this article focuses on the plant defense response against the model pathogens, i.e., viruses only.

2 Cross-Protection

Plant scientists have been trying to develop resistance through prophylactic inoculation with attenuated viral strains in plants for decades. This kind of immunization is known as cross-protection. It originated from the classical observation that many plants did not show secondary infection of the virus if they had previously been infected by same or closely related non-virulent viruses (Fig. 1).

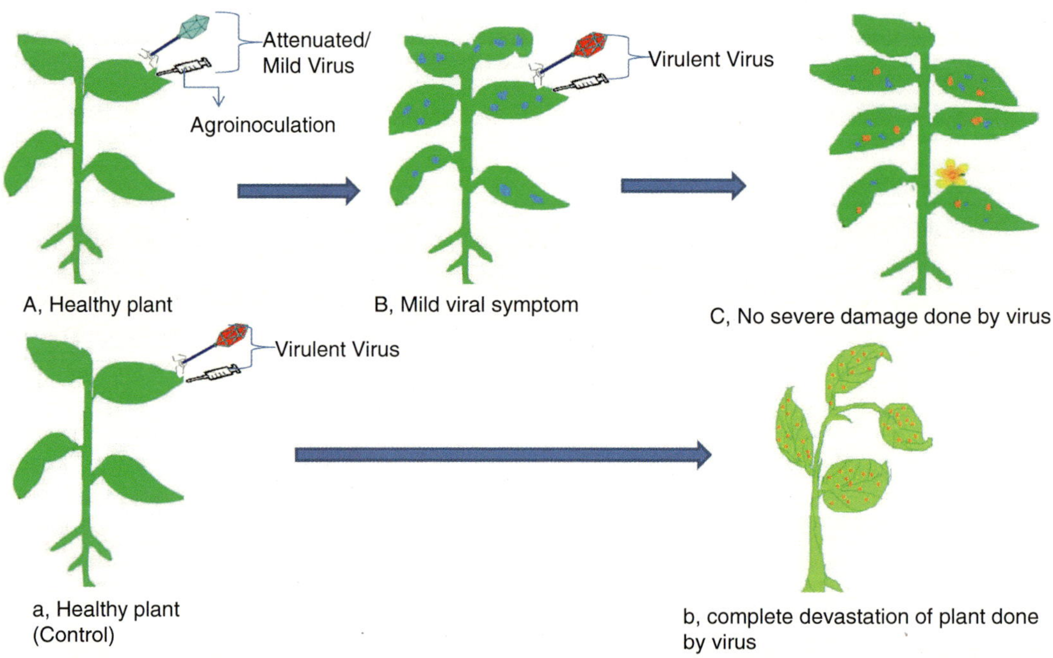

Fig. 1 Cross-protection in tobacco plants: (**A**) A healthy tobacco plant was first inoculated with mild or attenuated virus by using agrobacteria or white fly. (**B**) After few days when the plant started showing mild viral symptoms, the second dose of virus was given with the virulent strain of the same virus. (**C**) The plant did not show any severe symptoms and instead was healthy with normal flowering and fruiting. (**a**) In control experiment a healthy plant was directly inoculated with a severe strain of virus. (**b**) The plant was severely affected by the virus, and showed severe symptoms with delayed flowering and fruiting

Table 1
List of plants treated with cross-protection against different viruses worldwide [144]

Virus system	Host plant	Country
Tobacco mosaic virus (TMV)	Tomato	Netherlands, UK, Japan, France, USA
Tomato mosaic virus (ToMV)	Tomato, pepper	China
Citrus tristeza virus (CTV)	Citrus	Brazil, Australia, India
Apple mosaic virus	Apple	New Zealand
Passion fruit woodiness virus (PFWV)	Passion fruit	Australia
Cacao swollen shoot virus (SSV)	Cacao	West Africa
Cucumber mosaic virus (CMV) + RNA-5 versus CMV	Pepper	China, India

The earliest evidence for the cross-protection was recorded by Mckinney [6]. He observed that tobacco plants were no longer vulnerable to subsequent infection of TMV strain causing green mosaic symptom if previously they had been infected with the mild strain of TMV causing mild green mosaic symptom. Shortly thereafter, the similar results were also observed by Thung [7], using TMV strains, and Salaman [8] with strains of potato virus X (PVX). They both independently demonstrated that simultaneous inoculation of a mild and a severe virus strain into a plant led to less severe symptoms than when the severe strain was inoculated alone. Since then cross-protection has been used against many viral diseases in various parts of the world to protect the crop plants against devastating viral disease (Table 1).

In China, an experiment was performed in the year 1964; using symptom-less mutant of tomato mosaic virus (ToMV) and significant increase in yield of tomato was observed [9]. In the same manner, the mild strain of Citrus tristeza virus (CTV) was used to save several million citrus trees in Brazil and Australia from more severe strains of CTV [10]. A successful operation aimed for the cross-protection of apple orchards against several strains of Apple mosaic virus (AMV) was also completed in New Zealand [11].

Passion fruit vines are generally productive for as long as 8 years, but productivity drops to 2 years in areas heavily infected with the passion fruit woodiness virus. Protection from severe strains of the virus has been achieved in Australia, using field isolates obtained from fruit plants displaying mild symptoms [12, 13]. In West Africa, virulent strains of cacao swollen shoot virus (SSV) cause typical symptoms of swollen shoot disease in cacao plantations. Mild strains of the virus obtained in the field effectively protected the plant from attack by a severe strain [14] after field trial periods of 3 years. Only 8 % of protected plants developed severe symptoms as opposed to 70 % among non-protected plants [15]. Papaya ringspot virus (PRV) causes extensive damage to papaya trees, and limits papaya production in several tropical and subtropical areas. No resistant papaya cultivars have been obtained so far. To diminish the economic loss caused by the virus, symptom-less strain of PRV obtained by nitrous acid treatment had been used that protected papaya trees from infection with severe strains [16].

During the worldwide practice of cross-protection in crop plants, it was observed that in some cases no resistant plants could be obtained, or it has adversely affected crop plants. To explain this ambiguity of cross-protection, many hypotheses were given, but no single hypothesis could account for all data obtained. Later it was found that, for plant protection, whole virus genome is not required, only part of the viral genome is sufficient to confer the resistance.

3 Pathogen Derived Resistance (PDR)

In 1985, Sanford and Johnston [17] proposed the concept of pathogen derived resistance (PDR). PDR is a method to induce resistance in plants against viral pathogen by introducing gene(s) of pathogen into the susceptible host. The PDR approach is based on the fact that in all pathogen–host interaction, there are certain pathogen encoded cellular functions that are essential for pathogens but not for host. These functions are mostly indispensable for pathogens. If one of such functions is compromised in host, pathogens will not survive. These essential cellular functions which are under control of pathogen's gene might be altered by the presence of a corresponding gene product in dysfunctional form or in excess or appearance of the same at a wrong developmental phase of the pathogen life cycle in the host. Therefore, resistance to a particular pathogen can be attained by transforming a plant with an appropriate pathogen's gene. The validity of the concept of PDR was first demonstrated by Powell Abel. In his work, transgenic plant overexpressing the coat protein (CP) of tobacco mosaic virus (TMV) showed resistance to TMV. In these experiments, transgenic tobacco plants expressing high levels of TMV-CP were more resistant to the TMV virions [18]. Subsequently, there have been numerous attempts to generate virus resistance in transgenic plants based on this concept, i.e., through the expression of virus derived genes or genome fragments [19–23]. Furthermore, in a quest to find the mechanism for resistance, John A. Linbdo performed an experiment using Tobacco etch virus (TEV) coat protein. The transgenic tobacco plant expressing TEV coat protein recovered from TEV infection after 3–5 weeks of inoculation. The transgene mRNA levels in recovered tissue were 12–22 times less than in un-inoculated transgenic tissue of the same developmental stage [24]. It was concluded that virus infection induced destruction of RNAs related to the invading viral genome, conferring resistance against the virus and suggesting the existence of RNA-directed antiviral defense mechanism [25]. The RNA (and not the viral protein in the transgenic) directed antiviral defense was demonstrated by many different means including the transgenic introduction of the non-translatable version of RNA of the pathogen gene. Moreover, the discovery of virus encoded suppressor protein of gene silencing made it clear that small RNAs are a key player behind the antiviral immunity of plants [26–28].

4 Viral Recovery

Another form of cross-protection is also visible in virus infected plant which is known as viral recovery. Viral recovery is a phenomenon in which, virus infected plant tends to recover from viral infection after some time and becomes immune upon reinfection with same or similar virus. During 1920s, many researchers in USA made the observation that in the ringspot virus infected plants, the ringspot symptoms gradually failed to develop on the newly emerging leaves, but the sap from the new growth continued to be infectious and would readily produce the disease on healthy plants [29]. Later in the year 1939, W. M. Stanley had experimentally showed the viral recovery phenomenon in tobacco plant [30]. He observed that the inoculation of young Turkish tobacco plants with tobacco ringspot virus is followed by the appearance of marked systemic lesions. As the plant grows and the disease progresses, the new leaves which the plant produces show less and less severe symptoms until after about 2 weeks, following which the new emerging leaves appear quite normal in comparison to the leaves of healthy plants. The plant is then considered to have recovered, for the leaves produced thereafter look healthy and cuttings grow into normal looking plants. Since then, this recovery phenomenon has been observed in many combinations of plants and viruses. For example, the shock disease of Blue berry was observed in 1950. The plant stayed in shock (diseased) condition for 1–2 years, after which it recovered from disease though it carried the virus. Many other similar observations of viral recovery have been made in different crops. Like cross-protection, viral recovery is also strain specific. It works only for closely related strains of viruses. However, there is no need of prior infection with mild or attenuated virus. The discovery of the "gene silencing system" in 1990s resolved the long debate over the mechanism of cross-protection and viral recovery. Molecular virologists proved that cross-protection (or PDR or viral recovery) is a consequence of RNA silencing mechanism operating at the transcript level.

5 Homology Dependence

Both cross-protection and viral recovery phenomena are homology dependent (Fig. 2). They work for the homologous or nearly homologous viruses only. In the beginning researchers tried with many different combinations of viruses to induce resistance, but they could not get desired results [31–33]. Attempts were made to inoculate unrelated virus like PVX into same plants immunized with different viruses such as TMV a priori, but the TMV-immunized plants could not resist the development of PVX viral symptoms. Sequence homology dependence nature of viral protection was more descriptively revealed by Ratcliff et al. [34]. In their experiment, when recovered leaves of tomato black ring nepovirus strain W22 infected tobacco plants were inoculated with progressively less related virus PVX or closely related BUK strain, the plants did not resist the occurrence of disease. This analysis confirmed that the resistance associated with the recovery was specific to strains that were related in genomic sequence to the recovery-inducing virus. Of the viruses used for the secondary infection, BUK is the most closely related to W22, having 68 % nucleotide identity in RNA2. Tomato ringspot nepovirus RNA1 and PVX RNA have no long

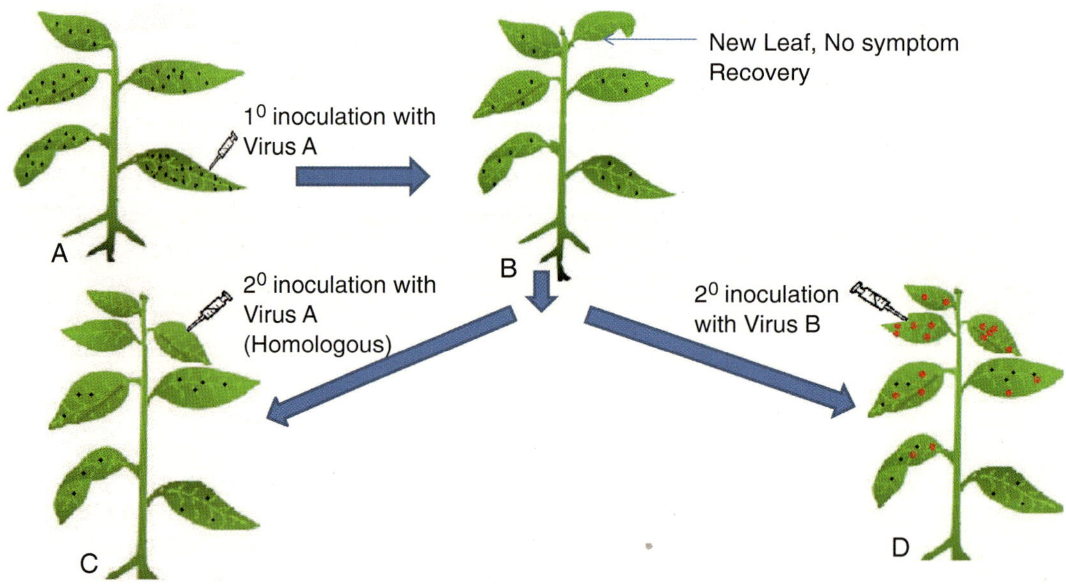

Fig. 2 Homology-dependent protection and recovery: (**a**) The plant was inoculated with Virus A, leading to the development of viral symptoms (Black ringspot). (**b**) After 2–3 weeks of primary inoculation, in the *upper* part of plant the new leaf showed no symptoms of black ringspot disease. The plant is said to be recovered. (**c**) When recovered leaves were again inoculated with same virus (Virus A), plant leaves failed to develop any symptoms. This is homology-dependent cross-protection. (**d**) However, if secondary inoculation is with different strains of the virus (Virus B), the plant leaves develop viral symptom in usual manner. Adopted from Hamilton et al. (1999)

stretches of sequence identity with W22 RNA. Therefore, resistance in the recovered leaves is specific for viruses that have RNA sequence that are similar to the virus used for primary inoculation. In Fig. 2, the homology-dependent viral protection phenomenon is elaborated.

6 Protein or RNA?

The mechanism for all three phenomena (PDR, cross-protection, viral recovery) took some time to come out. To resolve the dilemma over what is more important; Protein or RNA of the inoculating viral ORF for the protection against virus, researchers looked at the consequences with both the translatable and non-translatable form of viral ORFs. When the researchers tried to express the viral protein in plants, they surprisingly noticed that the plants expressing the lowest or even undetectable level of protein often displayed the highest resistance [35]. Furthermore, attempts were made to express the untranslatable viral mRNA or viral transcripts which cannot synthesis functional protein in plants. In one such experiment the expression of untranslatable viral mRNA demonstrated the resistance in plants, thus confirming the involvement of RNA in resistance [36]. The other examples of PDR directed at the RNA level involve the expression of non-structural protein gene sequences resulting from frameshift mutation [35, 37] or sequence present in untranslated regions [38] or various antisense RNAs [39, 40]. In all these cases the transgenic plants were unable to generate functional protein for corresponding mRNA. So it was concluded that RNA plays a key role in the PDR.

7 Small RNA (sRNA)

In the recent discovery it has been shown that whole RNA complement is not required, only fragment(s) of RNA is sufficient to implement the viral recovery. For the first time, presence of small RNA corresponding to the virus in the infected plant was shown by Andrew J. Hamilton and Baulcombe [41]. They detected the ~25 nucleotide long double-stranded RNA (ds-RNA) corresponding to the potato virus X (PVX) positive strand in tobacco plant after 4 days of inoculation with PVX virus. Twenty-five-nucleotide PVX RNA accumulated to the similar extent in systemic leaves but not in mock inoculated leaves. The presence of these small RNAs in PVX infected tobacco plant is linked with the role of small RNAs in the antiviral immunity. These small dsRNAs were found from both the virus-infected and -recovered leaves. Long before the discovery of small RNA in antiviral immunity, it was made clear that the gene silencing mechanisms control both antiviral immunity and

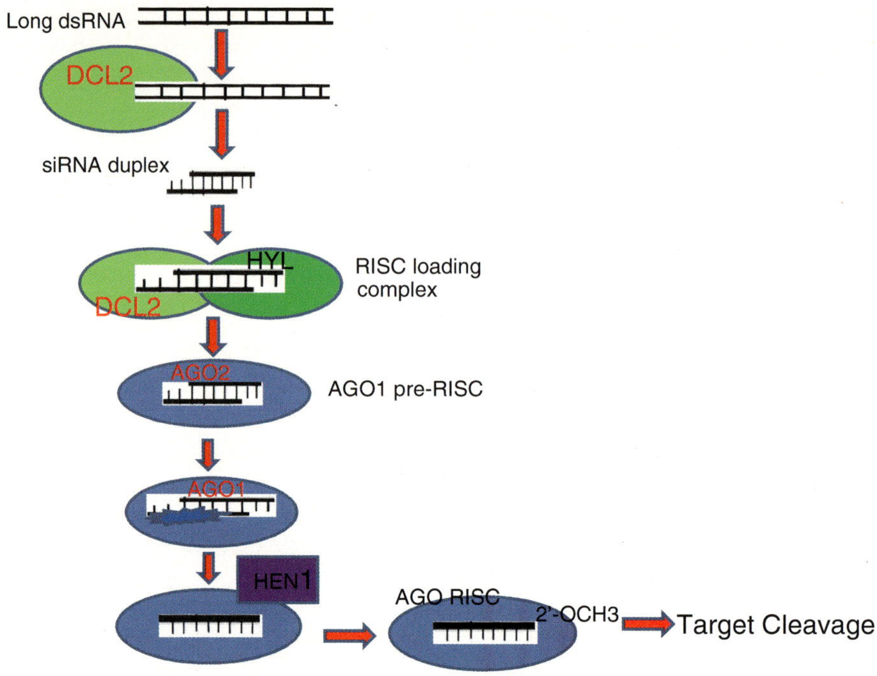

Fig. 3 siRNA Pathway: dsRNA precursors (viral DNA, structural Loci) are processed by Dicer (DCL-2 or DCL-4) to produce siRNA duplexes containing guide and passenger strands. The plant dicer and the cognate dsRNA-binding protein HYL (which together form the RISC-loading complex, RLC) load the duplex into Argonaute1 (AGO1) the passenger strand is later destroyed and the guide strand directs AGO1 to the target RNA. Adopted from Nature Reviews by Megha Ghildiyal and Phillip D. Zamore (2009)

transgene silencing in the homology-dependent manner [42–46]. In the subsequent years, when RNA based gene silencing mechanism more descriptively explained in various model organisms, it was found that the small RNAs are sole executioners for all type of silencing and antiviral immunity.

The small RNAs (sRNAs) are 21–24 nucleotides long non-coding RNA and are involved in sequence specific regulation of gene expression at transcriptional and posttranscriptional level. RNA silencing is an evolutionary conserved, sequence-specific mechanism that regulates gene expression and chromatin states and defends against invasive nucleic acids such as transposons, transgenes, and viruses [47–49]. Silencing is directed by 21–24 ntsRNAs, processed from the double-stranded (ds) RNA precursors by Dicer or Dicer-like (DCL) enzymes (*see* **Note 3**). The sRNAs associate with Argonaute (AGO) (*see* **Note 4**) proteins and guide the resulting RNA-induced silencing complexes (RISC) (*see* **Note 5**) to silence complementary RNA or DNA (Fig. 3). In plants, silencing pathways generate two types of sRNAs: microRNAs (miRNAs) and short interfering (si)RNAs. The miRNAs are produced by DCL1 from hairpin dsRNA containing precursors transcribed by Pol II from the endogenous *MIR* genes. They silence target genes through mRNA cleavage and/or translational repression. The miRs are present in all forms of eukaryotic life except the fungi, and about 30,000 of them are known till today. The siRNAs of distinct size classes are processed by DCL4 (21-nt), DCL2 (22-nt) or DCL3 (24-nt) enzymes from dsRNA precursors in plants. These precursors are produced by plant encoded RNA-dependent RNA polymerases (RDRs) (*see* **Note 6**) or from the overlapping sense and antisense Pol II transcripts. RDR6-dependent, 21-nt *trans*-acting siRNAs (tasiRNAs) and secondary siRNAs silence genes posttranscriptionally (like miRNAs), while the RDR2-dependent 24-nt heterochromatic siRNAs (hcsiRNAs) silence repetitive DNA transcriptionally through RNA-dependent DNA methylation (RdDM).

As discussed in the paragraph above, among other small RNAs, small interfering RNAs (siRNAs) are regarded as main player of the antiviral immunity. There are many strong evidences to support the exclusive role of siRNA in antiviral activity in plants. The detection of siRNA specific to the viral coat protein gene in transgenic plants resistant to Papaya ringspot virus (PRSV) support the involvement siRNAs in the establishment of resistance against viral pathogens [50]. The virus infected plant elicits the RNA silencing through biogenesis of siRNA to target virus and homologous RNAs for degradation [51]. Discovery of viral suppressor proteins (*see* **Note** 7) of RNA silencing [26–28, 52], and mutation studies of RNA silencing gene like *SGS2/SDE1*, *SGS3*, *SDE3*, and *AGO1 in Arabidopsis thaliana* [53] further validate the role of siRNA in plant immunity against viral pathogens. *SGS2/SDE1*, *SGS3*, *SDE3*, and *AGO1* gene were identified as important for transgene induced RNA silencing in *Arabidopsis thaliana*. The products of these genes are involved in the production of dsRNA which acts as a substrate for downstream machinery of RNA silencing. Moreover, it was observed that the mutant *A. thaliana* for above genes were hypersensitive to CMV infection as well as susceptible for some other viruses [54–58].

8 Exogenous siRNA

To develop a better understanding about plant antiviral immunity, it is imperative to go into the details of siRNA. In the broader sense, on the basis of origin of dsRNA, siRNAs are classified as exogenous and endogenous. For exogenous siRNA, the source of dsRNA may be transgenes or viral nucleic acids. However, in the case of endogenous siRNA it may be transposons, repeat sequences, or convergently transcribed RNAs. The siRNA pathway in Arabidopsis consists of mainly two families of ribonuclease protein, Dicer and Argonaute. The member of dicer like (DCL1, DCL2,

DCL3, DCL4) protein family possesses RNaseIII type activity and involve in the processing of dsRNA to generate siRNA. DCL2, DCL3, DCL4 generate 22-, 24-, 21-bp siRNA respectively while DCL1 produces 21-nt small RNA and act in the production of miRNAs. The viral nucleic acids (DNA/RNA viruses) are mainly processed by DCL4 protein and produce 21 nt viral siRNA (vsiRNA). In the absence of DCL4, DCL2 and DCL3 act to process viral dsRNA. However, DCL1 is also a player but a minor contributor to vsiRNA production. It has been observed that a triple mutant (loss of function for dcl2, dcl3, dcl4 gene) Arabidopsis plant produces low amount of vsiRNA upon infection with Turnip mosaic virus (TuMV). So it suggests that DCL1 is also capable of producing vsiRNA when other DCL-activities are hampered. The Argonaute (AGO) is the RNaseH nuclease type protein which cleaves the single-stranded (ss) RNA. The AGO protein bound with ssRNA of the ds-siRNA forms the RNA-induced silencing complex (RISC) that targets the complementary mRNA [59]. There are ten types of AGO proteins (AGO1-10) present in Arabidopsis plant with their distinct functions.

Two models have been proposed to explain the silencing trigger on virus attack. In the first model, the genomic components of the incoming virus may have an inbuilt capability to form intramolecular dsRNA due to the presence of extensive complementary regions in their genome. These folded viral RNAs are then directly recognized and processed by DCL to vsiRNA [60]. In second model, dsRNA can be produced by viral RNA polymerase from DNA or RNA viruses as a converging bidirectional transcript (DNA virus) or an intermediate in genome replication and transcription [61]. The systemic spread of virus induced silencing gives rise to another type of pathway for dsRNA production which requires the host RdRPs (RNA-dependent RNA polymerases, also called RDRs). The Arabidopsis plant has 6 RDRs with specialized but interconnected function in production of different dsRNAs [62–64]. Only RDR1 and RDR6 are involved in targeting viral genome for vsiRNA synthesis. Some lines of evidence also support the involvement of RDR2 in the production of vsiRNA [65]. The combination of RDR1, RDR2, and RDR6 produces nearly 90 % of vsiRNA in the virus infected Arabidopsis plant [60]. The vsiRNA strand that guides silencing through RISC is called the guide strand, while the other strand, which is eventually destroyed, is known as the passenger strand. In addition to AGO protein and small RNA complex, RISC also contains some other proteins that direct the RISC to the site of mRNA degradation [66]. The thermodynamic stability of 5′ end of two siRNA strand in the duplex determines the guide and passenger strand [67, 68]. Then the RISC loading complexes (RIC) recruit AGO2 protein which cleaves the passenger strand at the phosphodiester bond between the nucleotide at position 10 and 11 nucleotide of paired guide strand [69]. The release of passenger strand from pre-RISC complex converts it into mature RISC. The guide strand is subsequently 2′-O-methylated at 3′ end by s-adenosylmethionine-dependent methyltransferase HEN1 (*see* **Note 8**) [70, 71]. Following this, the matured RISC can bind with complementary mRNA transcript and inactivate it.

9 Use of Exogenous siRNA to Engineer Plant Antiviral Immunity: Transgenic Use

The RNA silencing phenomenon has provided a wonderful weapon to combat viral pathogens. Various strategies have been adopted by researchers to augment the plant antiviral immunity by mimicking the natural RNA silencing pathway. In past RNA silencing phenomenon was inadvertently evoked in quest to make virus resistance plant by introducing virus derived sequences into plants. As the siRNA based RNA silencing phenomenon is better understood, it has been widely used for engineering virus resistance [72]. The idea to engineer virus resistance is based on the expression of artificial dsRNAs, homologous to viral sequences in plants [73, 74]. As dsRNA is a substrate for the Dicer, plant-encoded RdRps will not be necessary to turn on RNAi effects. So it is possible to target the wide range of virus by expressing artificial dsRNAs homologous to viral sequences [75–78]. The transgenic plant expressing dsRNAs has been successfully employed against various plant viruses like tomato golden mosaic virus (TGMV) [79], Tomato yellow leaf curl Sardinia virus (TYLCSV) [80], Tomato yellow leaf curl virus (TYLCV) [81]. Artificial dsRNAs in plants can be generated by two methods: (1) Hairpin constructs in which virus sequence is cloned in sense and antisense manner and separated by an intron [82–84]; (2) Independent expression of transgene in the sense and/or antisense manner, resembling the mechanism of co-suppression [85, 86]. Pooggin et al. obtained the improved resistance in transgenic plants against *Vigna mungo* yellow mosaic virus (MYMV) using IR (inverted repeat) construct containing the common region of the MYMV [73]. Similarly, Noris et al. [87] and Ribeiro et al. [88] produced transgenic plants expressing siRNAs against TYLCSV and Tomato chlorotic mottle virus (ToCMoV), respectively. The artificial dsRNA technology has no limitation in the choice of targeted sequences. But mostly conserved and viability related nucleotide sequences in viruses are opted for targeting. For example Rep gene (Replication gene, AC1 gene from bogomovirues) is strictly required for replication, so it is a high-priority RNAi target for almost all plant viruses [89]. Various researchers have obtained promising results by targeting Rep gene of ACMV [90, 91]. This approach definitely can be aimed to any viral coding gene, until the decrease in viral mRNA has negative effect in virus life cycle. Some researchers also tried to target the viral silencing suppressor gene.

The viral silencing suppressors are the viral proteins to counteract the RNAi of plants. They play a significant role in the accumulation of viral transcripts [75, 92, 93]. Fagoaga et al. [94] have engineered *Citrus tristeza* virus resistance by targeting the p23 gene, a viral silencing suppressor. In the midst of cat and mouse game between viruses and plants, nature has bestowed plants with many endogenous siRNA which provides spontaneous response to biotic and abiotic stresses.

10 Endogenous siRNA

Endogenous siRNAs (endo-siRNAs) are small RNA which are encoded by own genome of an organism. The first endo-siRNA was discovered in plants and *C. elegans* [95, 96], and in recent years they have been reported from mammal and flies too. In plants they arise from various sources like transposons, repetitive elements, and tandem repeats such as 5S ribosomal gene [97]. They are also called *cis*-acting siRNA (casiRNAs) and comprise the bulk of endo-siRNA in cellular milieu. The production of casiRNA requires DCL3, RDR2, Pol IV, and AGO6 or AGO4 proteins [97–105]. These 24-nt long casiRNAs are methylated by HEN1 and promote the heterochromatin formation by histone modification and DNA methylation at loci from which they emerge [95, 97, 106–108]. The other class of plant endo-siRNAs includes tasiRNA, nat-siRNA, lsiRNA, etc. The *trans*-acting siRNa are produced by convergence of miRNa and siRNA pathway in plants [109–113]. Sometimes the cleavage of certain transcripts by miRNA directed pathways recruits RDR6, which copies the cleaved transcript and converts it to dsRNA and provides the substrate for DCL4. DCL4 splices the dsRNA into 21-nt long tasiRNA [113]. In response to biotic stress plants also produce natural antisense transcript-derived siRNAs (natsiRNAs) [114, 115]. The nat-siRNAs are produced from a pair of convergently transcribed RNA. In such case one transcript is expressed constitutively while the complementary RNA is transcribed only under stress conditions such as pathogen attack or abiotic stresses. They are 21 and 24-nt long (or even could be longer), require DCL2/DCL1, RDR6, SGS3 (Suppressor of Gene Silencing 3, an RNA binding protein), and RNA Pol IV for production from an overlapping region of two transcripts [114–116]. The cleavage of one mRNA from the pair is then directed by the same nat-siRNA, and it is another example of secondary siRNA production. In addition to natsiRNA, there is another stress responsive class of siRNA that is atypical in size (39–41-nt long) called long siRNA. It is also produced from natural antisense transcript pairs with the help of DCL1, DCL4, AGO7, RDR6, and Pol IV [117]. Taken together, endogenous siRNAs have evolved with diverse functions at various levels of defense. They can induce transcriptional gene silencing via DNA methylation or histone modification, or they can posttranscriptionally silence the gene by mRNA degradation [118]. The natural antisense siRNA, "nat-siRNA ATGB2" and long siRNA, "lsiRNA-1" both were observed to be induced specifically in *Arabidopsis thaliana* upon recognition of *Pseudomonas syringae* effector AvrPt2 by the cognate *Arabidopsis* disease resistance (R) protein RPS2 (Katiyar-Agarwal et al. [114]). The mutation in the small RNA biogenesis components like RDR6 and HYL1 (*see* **Note 9**) but not in the silencing components hampered the RPS2 mediated resistance in Arabidopsis. It suggests the role for nat-siRNAATGB2 and lsiRNA-1 in AvrPt2-specified effector triggered immunity (ETI) upon bacterial infection [117]. It is also claimed that endo-siRNAs are responsible for transgeneration-systemic acquired resistance in *Arabidopsis thaliana* through chromatin modification against bacterial pathogen *P. syringae* [119–121]. Although the significant role of endogenous siRNAs in pathogen triggered immunity against bacterial pathogens, in biotic and abiotic stresses, has been documented, their role in antiviral immunity has not been explored yet.

The virus induced endo-siRNA can be screened by blocking the generation of ds RNAs from exogenous sources in plants.

11 MicroRNA (miRNA)

MicroRNAs (miRs) are another important class of endogenous small RNA, but they are different from endo-siRNA on the basis of origin, biogenesis, and function. Unlike exo-siRNAs, they have their own genes from which they are generated by serial trimming of their precursor structures. They play a vital role in developmental process, pathogen response, abiotic stress, gene regulation, etc. The miRs are 21–22 nt noncoding small RNA, transcribed from their own gene present in intergenic region or some time from intron region (*mirtrons*) by RNA Pol II [122, 123]. The MIRNA genes are transcribed as primary transcripts (pri-miRNAs) with hairpin structure, which are cleaved by DCL1 along with HYPONASTIC LEAVES 1 (HYL1) and SERRATE (SE) (*see* **Note 10**), producing pre-miRNAs. These pre-miRNAs are in turn processed by DCL1 and HEN1 producing a duplex comprising the mature miRNA imperfectly base paired with a miRNA* strand. The newly formed duplex is then methylated by the methyltransferase HUA ENHANCER 1 (HEN1) proteins [124]. This process occurs in the nucleus from where the methylated duplex is exported out by HASTY proteins and incorporated into AGO1 containing RISC complex. In the RISC complex the passenger strand is cleaved off while guide strand remains attached. The mi-RISC then binds to the cognate target mRNA usually at 3′ UTRs (mostly in animals and insects) or within the protein coding region (mostly in plants) by exact or near-exact complementary base pairing, following which the mRNA target is cleaved or translationally repressed [109, 125, 126]. This miRNA directed RNAi machinery also can be exploited to develop antiviral immunity in plants. Recently many virus induced novel miRNAs have been identified in various virus infected plants. For example, miR156 and miR164 have been identified, which are induced upon infection with Turnip Mosaic Virus (TuMV) in *Arabidopsis* plant [92]. The miR158 and miR1885 are also identified as virus induced miRNA in *Brassica* against Cucumber Mosaic Virus (CTV) [127]. The identification of novel miRNAs in virus infected plants has some bearing on the development of antiviral strategies in terms of over-expressing the virus responsive host miRNAs in plants.

12 Transgenic Use

Artificial miRNA can be generated *in planta* to target a gene of interest by mimicking the intact secondary structure of endogenous miRNA precursor [128–132]. The precursor miRNA selectively produces the sRNA duplex of miRNA–miRNA* in vivo. The change in precursor sequences are allowed until the structural integrity is maintained. The precisely designed miRNAs often result in high level accumulation of miRNA with desired consequence of silencing the target mRNA. The first amiRNAs were designed and used in human cell lines [132] and later in Arabidopsis [130] where they suppressed the reporter gene. Very soon it was realized that amiRNAs can be used for various purposes like silencing of endogenous plant gene(s) or to develop antiviral immunity in plants with some obvious advantage over hairpin construct (*see* **Note 11**). Since then, various resistant transgenic plants have been generated by using amiRNA constructs specifically designed to silence viral pathogenic ORFs, leading to resistance against viruses. The resistance was observed in Arabidopsis against turnip yellow mosaic virus (TYMV) by targeting gene silencing suppressor gene, P69 of TYMV with amiRNA construct [129]. In another experiment 2b gene of cucumber mosaic virus was targeted using the same strategy of amiRNA, and significant resistance was observed [48]. More recently in our laboratory, transgenic tomato lines expressing amiRNA against Tomato Leaf Curl Virus New Delhi (ToLCNDV) were shown to resist/tolerate the virus [133, 134].

13 Conclusion

Various methods are available to induce resistance against pathogens since ancient time. But now the journey of vaccination/immunization has reached to a new horizon. In the incessant marathon to develop immunity against pathogens, researchers have made surmounting progresses from classical observation of cross-protection to RNA interference. The concept of PDR has opened a complete new scope for induced resistance. Though it has been in practice for a long time, only recently we have more grips on this technique as we understand more of the phenomenon in mechanistic terms. The detailed study of RNA gene silencing mechanism revealed small RNAs as the chief executioner molecule responsible for antiviral immunity in an organism. Ultimately researchers have started to exploit the RNAi mechanism against viral pathogen by devising various strategies, like hairpin construct for production of dsRNA, artificial miRNA, etc. The discovery of endogenous siRNAs in the pathogen infested plants has laid the foundation for a new approach to develop antiviral immunity.

14 Notes

1. *Vaccine*. The term vaccine is derived from Latin *vaccīn-us* (means from *vacca*, cow) and was used first for Edward Jenner's preparation from cowpox to prevent smallpox [135]. It is a biological preparation that contains mild or attenuated form of pathogens, its toxin, or one of its surface proteins and elicits an immune response against virulent form of the same pathogens. The vaccine catalyzes the body's defense response by being recognized as a foreign material, and the body neutralizes it by secreting antibodies against it. Moreover, these immune responses are also memorized for later time-periods.
2. *PAMP-triggered immunity* (*PTI*). Pathogen-associated molecular patterns (PAMPs) are conserved molecules associated with nonviral pathogens like the bacterial flagellin-derived peptide flg22 [4]. These PAMPs are recognized by diverse pattern recognition receptors (PRRs) in plants and elicit the first line of defense. These PRRs are transmembrane in nature and include receptor-like kinase (RLKs) and receptor-like proteins. The bacterial flagellin flg22 is recognized by complex of receptor-like kinase Flagellin Sensing 2 (RLK FLS2) and regulatory kinase BAK1 which trigger a set of pathogen-related responses. The pathogen-related response includes production of reactive oxygen species (ROS), activation of calcium-dependent kinases (CDK) and mitogen-activated kinases (MAPKs), etc. [4].

   Effector triggered immunity (*ETI*). To neutralize the primary defense response (viz., PTI) of a host organism, a pathogen uses effector proteins. The effector proteins counteract the defense response of the host by blocking the PTI associated signalling cascade. However, to overcome the effector proteins, plants have evolved resistance gene (R gene).

The R gene mediates the recognition of effector proteins and results in the effector triggered immunity (ETI). The ETI leads to the hypersensitive responses and programmed cell death (PCD) to confine the pathogen to limited regions only [4].

Systemic-acquired resistance (SAR). SAR is a hormonally controlled immune response which induces defenses in distal non-infected tissues after PTI and ETI. The key molecules involved in SAR are salicylic acid (SA) and ethylene/jasmonic acid (ET/JA). Salicylic acid stimulates the hypersensitive response and programmed cell death through ETI, while ethylene/jasmonic acid controls the spread of PCD [4].

3. *Dicer or Dicer-like (DCL) enzymes.* Dicer is very important ribonuclease protein of RNase III family, involved in the processing of double-stranded RNA (dsRNA) to form short double-stranded fragment of RNA (20–25 bp) with two base-pair overhang at 3′ end [136]. The Dicer contains one PAZ domain and two RNase III domains, and the distance between these two domains influences the length of siRNAs it produces [136]. The Dicer protein ultimately facilitates the formation of RISC. In plants, they are called Dicer-like (DCL) proteins. The Arabidopsis mainly encode four different DCL genes, DCL 1–4. The DCL 2, 3, 4 are the main players in viral siRNA formation, while DCL1 engages itself in producing microRNAs. The DCL4 is a chief sensor of viral dsRNAs and produces 21 nt vsiRNA, while DCL2 and DCL3 act in the absence of DCL4 and generate 22 nt and 24 nt vsiRNA respectively [46].
4. *Argonaute (AGO) proteins.* Argonaute proteins are specialized small-RNA binding protein which constitute the catalytic component of RISC. Typically they have a molecular weight of ~100 kDa and are characterized by a Piwi-Argonaute-Zwille (PAZ) domain and a PIWI domain [137]. It is named after an AGO knockout in Arabidopsis, which shows typical phenotype resembling to tentacle of octopus *Argonauta argo* [138]. AGO proteins specifically bind with different classes of small non-coding RNAs viz, miRNA, siRNA, etc. Small RNAs direct Argonaute proteins to their specific targets through sequence complementarity, and lead to silencing of the target. On the basis of sequence homology, Argonaute protein has been classified in two subclasses: (1) Arabidopsis Ago subfamily which resemble with AGO1 and (2) Drosophila PIWI protein called Piwi subfamily [139]. Ago proteins are conserved throughout species and many organisms express multiple family members, ranging from 27 in *C. elegans*, 10 in *Arabidopsis*, 8 in humans, 5 in *Drosophila*, to 1 in *Schizosaccharomyces pombe* [139].
5. *RNA-induced silencing complexes (RISC).* The RNA-induced silencing complex is effector molecule of gene silencing. It is a multiprotein complex containing AGO proteins bound with

guide siRNA or miRNA molecule [46]. RISC uses the siRNA or miRNA as a template to recognizing target mRNA. When the RISC encounters the complementary region, it binds to target mRNA, activates RNaseH function of AGO, and cleaves/inactivates the mRNA [46]. The whole process of gene silencing through RISC is very crucial for growth and development of an organism as well as defense against invading viral pathogens.

6. *RNA-dependent RNA polymerases (RDRs)*. RNA-dependent RNA polymerase is required for synthesis of RNA strand from RNA template. They possess a conserved RNA-dependent RNA polymerase catalytic domain. RDRs are widely present in plants, fungi, protists as well as in RNA viruses, but are absent in humans, mice, and *Drosophila* [140]. However, the viral RDRs are more recently named as RdRps to differentiate them from eukaryotic RDRs. The *Arabidopsis thaliana* possesses six RDRs (RDR1-6) [140]. RDR1, RDR2, and RDR6 are more ubiquitously involved in the production of dsRNA molecules that are eventually generated into different types of siRNAs targeting respective endogenous loci [139]. Beside their role in antiviral activities through gene silencing mechanism, plant RDRs also have important functions in growth and development [140].
7. *Viral RNAi suppressor protein*. Suppression of RNAi mechanism is a common strategy employed by viruses to suppress the antiviral effects of the host's RNAi mediated defense system against the viruses.

   These viral suppressor proteins interact with components of host RNA silencing machinery and block their immediate action. For example, the p19 protein is a known suppressor of RNA silencing and encoded by tombusviruses. It sequesters small RNA duplex molecules and blocks the initiation of RNAi pathways against viral genome [141]. About 70 such suppressors are known till date, and the function and crystal structures of many of those are known. However, they lack a broad consensus suppression motif.
8. *Methyl-transferase HUA ENHANCER 1 (HEN1)*. The HUA ENHANCER 1 (HEN1) is a methyltransferase protein that adds methyl group to the ribose moiety of 3′-most nucleotide of miRNAs and siRNA to increase the stability against 3′–5′ degradation and 3′ uridylation [142]. The plant specific HEN1 contains two double-stranded RNA binding domains (dsRBD1 and dsRBD2) and a La-motif-containing domain (LCD). The substrate recognition is accomplished through both dsRBDs. However, the length of the substrate is influenced by the distance between the MTase domain and the LCD, each interacting with one end of the small RNA duplex. The methylation process is Mg^{2+} dependent [142].

9. *HYPONASTIC LEAVES 1 (HYL1)*. The HYPONASTIC LEAVES1 (HYL1) is a dsRNA-binding protein that involves the processing of primary miRNAs into microRNAs along with SERRATE and DCL1. It has two tandem double-stranded RNA binding domains (dsRBDs). The C terminus contains the putative protein–protein interaction domain, while the N terminus has the putative nuclear localization signals. The N terminus domain containing dsRBDs is indispensable for its function; however, the C terminus can be compromised. In Arabidopsis, HYL1 plays a critical role in processing of miRNA from pri-miRNA through DCL1 [143].
10. *SERRATE (SE)*. SERRATE (SE) protein is another important protein, along with DCL1 and HYL1, which is involved in the processing of miRNA from long transcripts (pri-miRNAs). It is a zinc finger protein with N terminus domain that used to bind with RNA, while both domains (zinc finger and N terminus) are required for binding with DCL1 and stimulate the cleavage of dsRNA in an ionic strength-dependent manner [119].
11. *Advantages of amiRNAs*. Various strategies have been adopted for silencing genes of interest in plants. Some of these approaches were/are based on the generation of siRNAs derived from dsRNAs or hairpin constructs. The large inserts used in these approaches produce a diverse set of siRNAs. Because of amplifiable nature of siRNAs, very often multiple species of siRNAs are generated in the form of transitive siRNAs. Hence, the chances of silencing of undesired genes (off-targets), resulting from fortuitous binding, are also increased. In some extreme situations, transgenes might become less stable due to autosilencing, and loss of the silencing activity on target sequences in subsequent progenies is also observed. The artificial miRNA approach has offered a new alternative way to target genes of interest, circumventing these above difficulties.

References

1. Lombard M, Pastoret PP, MoulinA M (2007) A brief history of vaccines and vaccination. Rev Sci Tech 26(1):29–48
2. Janeway C, Paul T, Mark W, Mark S (2001) Immunobiology, 5th edn. Garland Science, New York
3. Jones JDG, Dangl JL (2006) The plant immune system. Nature 444:323–329
4. Zvereva AS, PoogginM M (2012) Silencing and innate immunity in plant defense against viral and non-viral pathogens. Viruses 4(11):2578–2597. doi:10.3390/v4112578
5. Iriti M, Faoro F (2009) Chitosan as a MAMP, searching for a PRR. Plant Signal Behav 4(1): 66–68
6. Mckinney HH (1929) Mosaic diseases in the Canary island, West Africa and Gibraltar. J Agric Res 39:557–578
7. Thung TH (1931) Handel. VI Ned. Ind. Natuur Congr Bandoeng, Java. p 450
8. Salaman RN (1933) Nature 131:468
9. Tien P, Chang XH (1983) Seed Sci Technol 11:969–972
10. Frazer LR, Long K, Cox J (1968) Proc 4th Conf Int Organis of citrus virologists. pp 27–31
11. Chamberlin EE, Atkinson JD, Hunter JA (1964) N Z J Agric Res 7:480–490
12. Simmonds JH (1959) Queensl J Agric Sci 16: 371–380

13. Greber RS (1966) Queensl J Agric Anita Sci 23:533–536
14. Posnette AF, Todd JM (1951) Ann Appl Biol 38:785–800
15. Posnette AF, Todd JM (1955) Ann Appl Biol 43:433–453
16. Yeh SD, Gonsalves D (1984) Evaluation of induced mutants of papaya ringspot virus for control by cross protection. Phytopathology 74(9):1086–1091
17. Sanford JC, Johnston SA (1985) The concept of parasite-derived resistance—deriving resistance genes from the parasite's own genome. J Theor Biol 113:395–405
18. Abel PP, Nelson RS, De B et al (1986) Delay of disease development in transgenic plants that express the tobacco mosaic virus coat protein gene. Science 232(4751):738–743
19. Beachy RN (1993) Transgenic resistance to plant viruses. Semin Virol 4:327–416
20. Wilson TMA (1993) Strategies to protect crop plants against viruses: pathogen-derived resistance blossoms. Proc Natl Acad Sci U S A 90:3134–3141
21. Baulcombe DC (1994) Replicase-mediated resistance: a novel type of virus resistance in transgenic plants? Trends Microbiol 2:60–63
22. Baulcombe DC (1994) Novel strategies for engineering virus resistance in plants. Curr Opin Biotechnol 5:117–124
23. Lomonossoff GP (1995) Pathogen-derived resistance to plant viruses. Annu Rev Phytopathol 33:323–343
24. Lindbo JA, Silva-Rosales L, Proebsting WM et al (1993) Induction of a highly specific antiviral state in transgenic plants: implications for regulation of gene expression and virus resistance. Plant Cell 5:1749–1759
25. Mlotshwa S, Pruss GJ, Vance V (2008) Small RNAs in viral infection and host defense. Trends Plant Sci 13(7):375–382. doi:10.1016/j.tplants.2008.04.009
26. Anandalakshmi R, Pruss GJ, Ge X et al (1998) A viral suppressor of gene silencing in plants. Proc Natl Acad Sci U S A 95(22): 13079–13084
27. Brigneti G, Voinnet O, Li WX et al (1998) Viral pathogenicity determinants are suppressors of transgene silencing in Nicotiana benthamiana. EMBO J 17(22):6739–6746
28. KasschauK D, Carrington JC (1998) A counter defensive strategy of plant viruses: suppression of posttranscriptional gene silencing. Cell 95:461–470
29. Wingard SA (1928) J Agric Res 37:127
30. StanleyW M (1939) Isolation of virus from plants recovered from the tobacco ring spot disease. J Biol Chem 129:429–436
31. HarrisonB D (1958) Further studies on raspberry ringspot and tomato black ring, soil-borne viruses that affect raspberry. Ann Appl Biol 46(4):571–584
32. Jedlinski H, Brown CM (1965) Cross protection and mutual exclusion by three strains of barley yellow dwarf virus in Avenasativa L. Virology 4:613–621
33. Kohler E, Panjan M (1943) Das Paratabakmosaikvirus der Tabakpflanze. Ber Deut Botan Ges 61:175–180
34. Ratcliff F, Harrison BD, Baulcombe DC (1997) A similarity between viral defense and gene silencing in plants. Science 276(5318): 1558–1560. doi:10.1126/science.276.5318.1558
35. Golemboski DB, Lomonossoff GP, Zaitlin M (1990) Plants transformed with a tobacco mosaic virus nonstructural gene sequence are resistant to the virus. Proc Natl Acad Sci U S A 87(16):6311–6315
36. Smith HA, Swaney SL, Parks TD et al (1994) Transgenic plant virus resistance mediated by untranslatable sense RNAs: expression, regulation, and fate of nonessential RNAs. Plant Cell 6:1441–1453
37. Carr JP, Zaltlln M (1991) Resistance in transgenic tobacco plants expressing a nonstructural gene sequence of tobacco mosaic virus is a consequence of markedly reduced virus replication. Mol Plant-Microbe Interact 4: 579–585
38. Morch MD, Joshi RL, Denial TM et al (1987) A new 'sense' RNA approach to block viral RNA replication in vitro. Nucleic Acids Res 15(10):4123–4130
39. Hemenway CL, Fang RX, Kaniewski WK et al (1988) Analysis of the mechanism of protection in transgenic plants expressing the potato virus X coat protein or its antisense RNA. EMBO J 7:1273–1280
40. Powell PA, Stark DM, Sanders PR et al (1989) Protection against tobacco mosaic virus in transgenic plants that express tobacco mosaic virus antisense RNA. Proc Natl Acad Sci U S A 86:6949–6952
41. Hamilton AJ, Baulcombe DC (1999) A species of small antisense RNA in posttranscriptional gene silencing in plants. Science 286(5441):950–952. doi:10.1126/science.286.5441.950
42. Ingelbrecht I, Van Houdt H, Van Montagu M et al (1994) Posttranscriptional silencing of

reporter transgenes in tobacco correlates with DNA methylation. Proc Natl Acad Sci U S A 91:10502–10506
43. de CarvalhoNiebel F, Frendo P, Van Montagu M et al (1995) Post-transcriptional cosuppression of β-13. Glucanase genes does not affect accumulation of transgene nuclear mRNA. Plant Cell 7:347–358
44. Mueller E, Gilbert JE, Davenport G et al (1995) Homology-dependent resistance: transgenic virus resistance in plants related to homology-dependent gene silencing. Plant J 7:1001–1013
45. Goodwin J, Chapman K, Swaney S et al (1996) Genetic and biochemical dissection of transgenic RNA-mediated virus resistance. Plant Cell 8:95–105
46. Agrawal N, Dasarathi PVN, Mohmmed A, Malhotra P, Bhatnagar RK, Mukherjee S (2003) RNA interference: biology, mechanism and applications. Microbiol Mol Biol Rev 67:657–685
47. Vaucheret H (2006) Post-transcriptional small RNA pathways in plants: mechanisms and regulations. Genes Dev 20(7):759–771
48. Ding SW, Voinnet O (2007) Antiviral immunity directed by small RNAs. Cell 130(3): 413–426
49. Matzke M, Kanno T, Daxinger L et al (2009) RNA-mediated chromatin-based silencing in plants. Curr Opin Cell Biol 21(3):367–376. doi:10.1016/j.ceb.2009.01.025
50. Krubphachaya P, Jurícek M, Kertbundit S (2007) Induction of RNA-mediated resistance to papaya ringspot virus type W. J Biochem Mol Biol 40(3):404–411
51. Ding SW, Li H, Lu R et al (2004) RNA silencing: a conserved antiviral immunity of plants and animals. Virus Res 102(1):109–115
52. Karjee S, Islam MN, Mukherjee SK (2008) Screening and identification of virus encoded RNA silencing suppressors. Methods Mol Biol 442:187–203
53. Vance V, Vaucheret H (2001) RNA silencing in plants—defense and counterdefense. Science 292(5525):2277–2280
54. Boutet S, Vazquez F, Liu J (2003) Arabidopsis HEN1: a genetic link between endogenous miRNA controlling development and siRNA controlling transgene silencing and virus resistance. Curr Biol 13:843–848
55. Dalmay T, Hamilton A, Rudd S (2000) An RNA-dependent RNA polymerase gene in Arabidopsis is required for posttranscriptional gene silencing mediated by a transgene but not by a virus. Cell 101(5):543–553
56. Dalmay T, Hamilton A, Mueller E (2001) Potato virus X amplicons in Arabidopsis mediate genetic and epigenetic gene silencing. Plant Cell 12(3):369–379
57. Morel JB, Mourrain P, Béclin C et al (2000) DNA methylation and chromatin structure affect transcriptional and post-transcriptional transgene silencing in Arabidopsis. Curr Biol 10(24):1591–1594
58. Mourrain P, Béclin C, Vaucheret H (2000) Are gene silencing mutants good tools for reliable transgene expression or reliable silencing of endogenous genes in plants? Genet Eng (N Y) 22:155–170
59. Hammond SM, Bernstein E, Beach D et al (2000) An RNA-directed nuclease mediates post-transcriptional gene silencing in Drosophila cells. Nature 404:293–296. doi:10.1038/35005107
60. Donaire L, Wang Y, Gonzalez-Ibeas D et al (2009) Deep-sequencing of plant viral small RNAs reveals effective and widespread targeting of viral genomes. Virology 392(2):203–214. doi:10.1016/j.virol.2009.07.005
61. Ahlquist P (2006) Parallels among positive-strand RNA viruses, reverse-transcribing viruses and double-stranded RNA viruses. Nat Rev Microbiol 4(5):371–382
62. Xie Z, Johansen LK, Gustafson AM et al (2004) Genetic and functional diversification of small RNA pathways in plants. PLoS Biol 2(5):E104
63. Kasschau KD, Fahlgren N, Chapman EJ et al (2007) Genome-wide profiling and analysis of Arabidopsis siRNAs. PLoS Biol 5(3):e57
64. Howell MD, Fahlgren N, Chapman EJ et al (2007) Genome-wide analysis of the RNA-DEPENDENT RNA POLYMERASE6/DICER-LIKE4 pathway in Arabidopsis reveals dependency on miRNA- and tasiRNA-directed targeting. Plant Cell 19(3):926–942
65. Qi X, Bao FS, Xie Z (2009) Small RNA deep sequencing reveals role for Arabidopsis thaliana RNA-dependent RNA polymerases in viral siRNA biogenesis. PLoS One 4(3):e4971
66. Liu J, Valencia-SanchezM A, HannonG J et al (2005) Micro RNA-dependent localization of targeted mRNAs to mammalian P-bodies. Nat Cell Biol 7:719–723. doi:10.1038/ncb1274
67. Schwarz DS, Hutvágner G, Du T et al (2003) Asymmetry in the assembly of the RNAi enzyme complex. Cell 115(2):199–208
68. Khvorova A, Reynolds A, Jayasena SD (2003) Functional siRNAs and miRNAs exhibit strand bias. Cell 115(2):209–216
69. Elbashir SM, Martinez J, Patkaniowska A et al (2001) Functional anatomy of siRNAs for mediating efficient RNAi in Drosophila melanogaster embryo lysate. EMBO J 20(23): 6877–6888

70. Horwich MD, Li C, Matranga C et al (2007) The Drosophila RNA methyltransferase, DmHen1, modifies germline piRNAs and single-stranded siRNAs in RISC. Curr Biol 17(14):1265–1272
71. Pélisson A, Sarot E, Payen-Groschêne G et al (2007) A novel repeat-associated small interfering RNA-mediated silencing pathway downregulates complementary sense gypsy transcripts in somatic cells of the Drosophila ovary. J Virol 81(4):1951–1960
72. Tenllado F, Llave C, Díaz-Ruíz JR (2004) RNA interference as a new biotechnological tool for the control of virus diseases in plants. Virus Res 102(1):85–96
73. Pooggin MM, Hohn T (2004) Fighting geminiviruses by RNAi and vice versa. Plant Mol Biol 55(2):149–152
74. Vanitharani R, Chellappan P, Fauquet CM (2005) Geminiviruses and RNA silencing. Trends Plant Sci 10(3):144–151
75. Chen J, Li WX, Xie D et al (2004) Viral virulence protein suppresses RNA silencing-mediated defense but upregulates the role of microrna in host gene expression. Plant Cell 16(5):1302–1313
76. Hily JM, Scorza R, Webb K et al (2005) Accumulation of the long class of siRNA is associated with resistance to Plum pox virus in a transgenic woody perennial plum tree. Mol Plant Microbe Interact 18(8):794–799
77. Kalantidis K, Psaradakis S, Tabler M (2002) The occurrence of CMV-specific short RNAs in transgenic tobacco expressing virus-derived double-stranded RNA is indicative of resistance to the virus. Mol Plant Microbe Interact 15(8):826–833
78. Nomura K, Ohshima K, Anai T et al (2004) RNA silencing of the introduced coat protein gene of turnip mosaic virus confers broad-spectrum resistance in transgenic Arabidopsis. Phytopathology 94(7):730–736. doi:10.1094/PHYTO.2004.94.7.730
79. Day AG, Bejarano ER, BuckK W et al (1991) Expression of an antisense viral gene in transgenic tobacco confers resistance to the DNA virus tomato golden mosaic virus. Proc Natl Acad Sci U S A 88:6721–6725
80. Bendahmane M, Gronenborn B (1997) Engineering resistance against tomato yellow leaf curl virus (TYLCV) using antisense RNA. Plant Mol Biol 33:351–357
81. Yang Y, Sherwood TA, Patte CP et al (2004) Use of tomato yellow leaf curl virus (TYLCV) Rep gene sequences to engineer TYLCV resistance in tomato. Phytopathology 94: 490–496
82. Smith NA, Singh SP, Wang MB et al (2000) Gene expression—total silencing by intron-spliced hairpin RNAs. Nature 407:319–320
83. Wesley SV, Helliwell CA, Smith NA et al (2001) Construct design for efficient, effective and high-throughput gene silencing in plants. Plant J 27:581–590
84. Helliwell CA, Waterhouse PM (2005) Constructs and methods for hairpin RNA-mediated gene silencing in plants. Methods Enzymol 392:24–35
85. Napoli CA, Lemieux C, Jorgensen RA (1990) Introduction of a chimeric chalcone synthase gene into petunia results in reversible co-suppression of homologous genes in trans. Plant Cell 2:279–289
86. Van der Krol AR, Mur LA, Beld M et al (1990) Flavonoid genes in petunia: addition of a limited number of gene copies may lead to a suppression of gene expression. Plant Cell 2:291–299
87. Noris E, Lucioli A, Tavazza R et al (2004) Tomato yellow leaf curl Sardinia virus can overcome transgene-mediated RNA silencing of two essential viral genes. J Gen Virol 85: 1745–1749
88. Ribeiro SG, Lohuis H, Goldbach R et al (2007) Tomato chlorotic mottle virus is a target of RNA silencing but the presence of specific siRNAs does not guarantee resistance in transgenic plants. J Virol 81:1563–1573
89. Hanley-Bowdoin L, Settlage SB, Orozco BM et al (1999) Geminiviruses: models for plant DNA replication, transcription, and cell cycle regulation. Crit Rev Plant Sci 18:71–106
90. Vanitharani R, Chellappan P, Fauquet CM (2003) Short interfering RNA-mediated interference of gene expression and viral DNA accumulation in cultured plant cells. Proc Natl Acad Sci U S A 100:9632–9636
91. Chellappan P, Vanitharani R, Fauquet CM (2004) Short interfering RNA accumulation correlates with host recovery in DNA virus infected hosts, and gene silencing targets specific viral sequences. J Virol 78:7465–7477
92. Kasschau KD, Xie Z, Allen E et al (2003) P1/HC-Pro, a viral suppressor of RNA silencing, interferes with Arabidopsis development and miRNA function. Dev Cell 4:205–217
93. Chellappan P, Vanitharani R, Ogbe F et al (2005) Effect of temperature on geminivirus-induced RNA silencing in plants. Plant Physiol 138:1828–1841
94. Fagoaga C, Lopez C, de Mendoza AH et al (2006) Post-transcriptional gene silencing of the p23 silencing suppressor of citrus tristeza virus confers resistance to the virus in

transgenic Mexican lime. Plant Mol Biol 60: 153–165
95. Hamilton A, Voinnet O, Chappell L et al (2002) Two classes of short interfering RNA in RNA silencing. EMBO J 21:4671–4679
96. Ambros V, Lee RC, Lavanway A et al (2003) MicroRNAs and other tiny endogenous RNAs in C. elegans. Curr Biol 13(10): 807–818
97. Chan SW, Zilberman D, Xie Z et al (2004) RNA silencing genes control de novo DNA methylation. Science 303:1336
98. Zilberman D, Cao X, Jacobsen SE (2003) ARGONAUTE4 control of locus-specific siRNA accumulation and DNA and histone methylation. Science 299:716–719
99. Zheng X, Zhu J, Kapoor A, Zhu JK (2007) Role of Arabidopsis AGO6 in siRNA accumulation. DNA methylation and transcriptional gene silencing. EMBO J 26:1691–1701
100. El-Shami M, Pontier D, Lahmy S et al (2007) Reiterated WG/GW motifs form functionally and evolutionarily conserved ARGONAUTE-binding platforms in RNAi-related components. Genes Dev 21:2539–2544
101. Herr AJ, Jensen MB, Dalmay T et al (2005) RNA polymerase IV directs silencing of endogenous DNA. Science 308:118–120
102. Kanoh J, Sadaie M, Urano T et al (2005) Telomere binding protein Taz1 establishes Swi6 heterochromatin independently of RNAi at telomeres. Curr Biol 15:1808–1819
103. Lee SK, Dykxhoorn DM, Kumar P et al (2005) Lentiviral delivery of short hairpin RNAs protects CD4 T cells from multiple clades and primary isolates of HIV. Blood 106:818–826
104. Onodera Y, Haag JR, Ream T et al (2005) Plant nuclear RNA polymerase IV mediates siRNA and DNA methylation-dependent heterochromatin formation. Cell 120:613–622
105. Pontier D, Yahubyan G, Vega D et al (2005) Reinforcement of silencing at transposons and highly repeated sequences requires the concerted action of two distinct RNA polymerases IV in Arabidopsis. Genes Dev 19:2030–2040
106. Tran RK, Henikoff JG, Zilberman D et al (2005) DNA methylation profiling identifies CG methylation clusters in Arabidopsis genes. Curr Biol 15:154–159
107. Llave C, Kasschau KD, Rector MA et al (2002) Endogenous and silencing-associated small RNAs in plants. Plant Cell 14: 1605–1619
108. Mette MF, Aufsatz W, van der Winden J et al (2000) Transcriptional silencing and promoter methylation triggered by double-stranded RNA. EMBO J 19:5194–5201
109. Vazquez F, Vaucheret H, Rajagopalan R et al (2004) Endogenous trans-acting siRNAs regulate the accumulation of Arabidopsis mRNAs. Mol Cell 16:69–79
110. Peragine A, Yoshikawa M, Wu G et al (2004) SGS3 and SGS2/SDE1/RDR6 are required for juvenile development and the production of trans-acting siRNAs in Arabidopsis. Genes Dev 18:2368–2379
111. Yoshikawa M, Peragine A, Park MY et al (2005) A pathway for the biogenesis of trans-acting siRNAs in Arabidopsis. Genes Dev 19:2164–2175
112. Williams L, Carles CC, Osmont KS et al (2005) A database analysis method identifies an endogenous trans-acting short-interfering RNA that targets the Arabidopsis ARF2, ARF3, and ARF4 genes. Proc Natl Acad Sci U S A 102:9703–9708
113. Allen E, Xie Z, Gustafson AM et al (2005) microRNA-directed phasing during trans-acting siRNA biogenesis in plants. Cell 121: 207–221
114. Katiyar-Agarwal S, Morgan R, Dahlbeck D et al (2006) A pathogen-inducible endogenous siRNA in plant immunity. Proc Natl Acad Sci U S A 103:18002–18007
115. Borsani O, Zhu J, Verslues PE et al (2005) Endogenous siRNAs derived from a pair of natural cis-antisense transcripts regulate salt tolerance in Arabidopsis. Cell 123:1279–1291
116. Zhang D, Trudeau VL (2008) The XS domain of a plant specific SGS3 protein adopts a unique RNA recognition motif (RRM) fold. Cell Cycle 7:2268–2270
117. Katiyar-Agarwal S, Gao S, Vivian-Smith A et al (2007) A novel class of bacteria-induced small RNAs in Arabidopsis. Genes Dev 21:3123–3134
118. Brodersen P, Voinnet O (2006) The diversity of RNA silencing pathways in plants. Trends Genet 22(5):268–280
119. Rasmann S, De Vos M, Casteel CL et al (2012) Herbivory in the previous generation primes plants for enhanced insect resistance. Plant Physiol 158(2):854–863. doi:10.1104/pp. 111.187831
120. Luna E, Bruce TJ, Roberts MR et al (2012) Next-generation systemic acquired resistance. Plant Physiol 158(2):844–853. doi:10.1104/pp. 111.187468

121. Slaughter A, Daniel X, Flors V et al (2012) Descendants of primed Arabidopsis plants exhibit resistance to biotic stress. Plant Physiol 158(2):835–843. doi:10.1104/pp. 111.191593
122. Bartel DP, Chen CZ (2004) Micromanagers of gene expression: the potentially widespread influence of metazoan microRNAs. Nat Rev Genet 5:396–400
123. Jones-Rhoades MW, Bartel DP, Bartel B (2006) MicroRNAS and their regulatory roles in plants. Annu Rev Plant Biol 57:19–53
124. Chapman EJ, Carrington JC (2007) Specialization and evolution of endogenous small RNA pathways. Nat Rev Genet 8(11): 884–896
125. Ghildiyal M, Zamore PD (2009) Small silencing RNAs: an expanding universe. Nat Rev Genet 10(2):94–108. doi:10.1038/nrg2504
126. Gustafson AM, Allen E, Givan S et al (2005) ASRP: the Arabidopsis small RNA project database. Nucleic Acids Res 33(Database issue):D637–D640
127. He S, Yang Z, Skogerbo G et al (2008) The properties and functions of virus encoded microRNA, siRNA, and other small noncoding RNAs. Crit Rev Microbiol 34(3–4):175–188. doi:10.1080/10408410802482008
128. Alvarez JP, Pekker I, Goldshmidt A et al (2006) Endogenous and synthetic microRNAs stimulate simultaneous, efficient, and localized regulation of multiple targets in diverse species. Plant Cell 18:1134–1151
129. Niu QW, Lin SS, Reyes JL et al (2006) Expression of artificial microRNAs in transgenic Arabidopsis thaliana confers virus resistance. Nat Biotechnol 24:1420–1428
130. Parizotto EA, Dunoyer P, Rahm N et al (2004) In vivo investigation of the transcription, processing, endonucleolytic activity, and functional relevance of the spatial distribution of a plant miRNA. Genes Dev 18:2237–2242
131. Schwab R, Ossowski S, Riester M et al (2006) Highly specific gene silencing by artificial microRNAs in Arabidopsis. Plant Cell 18: 1121–1133
132. Zeng Y, Wagner EJ, Cullen BR (2002) Both natural and designed micro RNAs can inhibit the expression of cognate mRNAs when expressed in human cells. Mol Cell 9:1327–1333
133. Vu TV, Choudhury NR, Mukherjee SK (2013) Transgenic tomato plants expressing artificial microRNAs for silencing the precoat and coat proteins of a begomovirus, Tomato leaf curl New Delhi virus, show tolerance to virus infection. Virus Res 172:35–45
134. Yadava P, Mukherjee SK (2012) Artificial microRNA and its applications. In: Bibekanand M, Zhumur G (eds) Regulatory RNAs: basics, methods and applications. Springer, Berlin, pp 505–521
135. Riedel S (2005) Edward Jenner and the history of smallpox and vaccination. Proc (Bayl Univ Med Cent) 18(1):21–25
136. Macrae IJ, Zhou K, Li F et al (2006) Structural basis for double-stranded RNA processing by Dicer. Science 311(5758): 195–198
137. Ender C, Meister G (2010) Argonaute proteins at a glance. J Cell Sci 123(Pt 11):1819–1823. doi:10.1242/jcs.055210
138. Bohmert K, Camus I, Bellini C et al (1998) AGO1 defines a novel locus of Arabidopsis controlling leaf development. EMBO J 17(1):170–180
139. Seo J-K, Wu J, Lii Y et al (2013) Contribution of small RNA pathway components in plant immunity. Mol Plant Microbe Interact 26(6):617–625. doi:10.1094/MPMI-10-12-0255-IA
140. Willmann MR, Endres MW, Rebecca T et al (2011) The functions of RNA-dependent RNA polymerases in Arabidopsis. Arabidopsis Book 9:e0146. doi:10.1199/tab.0146
141. Rodrigo G, Carrera J, Jaramillo A et al (2011) Optimal viral strategies for bypassing RNA silencing. J R Soc Interface 8(55):257–268. doi:10.1098/rsif.2010.0264
142. Huang Y, Ji L, Huang Q et al (2009) Structural insights into mechanisms of the small RNA methyltransferase HEN1. Nature 461:823–827. doi:10.1038/nature08433
143. Iwata Y, Takahashi M, Fedoroff NV et al (2013) Dissecting the interactions of SERRATE with RNA and DICER-LIKE 1 in Arabidopsis microRNA precursor processing. Nucleic Acids Res 41(19):9129–9140
144. Valle RPC, Skrzeczkowski J, Morch MD et al (1988) Plant viruses and new perspectives in cross-protection. Biochimie 70(5): 695–703

Chapter 27

Structure-Based Clustering of Major Histocompatibility Complex (MHC) Proteins for Broad-Based T-Cell Vaccine Design

Joo Chuan Tong, Tin Wee Tan, and Shoba Ranganathan

Abstract

Structure-based clustering technique is useful for identifying superfamilies of major histocompatibility complex (MHC) proteins with similar binding specificities. The resolved MHC superfamilies play an important role in vaccine development, from discovering new targets for broad-based vaccines and therapeutics to optimizing the affinity and selectivity of hits. Here, we describe a protocol and provide a summary for grouping MHC proteins according to their structural interaction characteristics.

Key words Bioinformatics, Immunoinformatics, Clustering, MHC superfamily, Virtual screening, Computer-aided vaccine design

1 Introduction

The identification of major histocompatibility complex (MHC) superfamilies with similar antigen-binding specificities has a tremendous impact in the field of vaccine design. The epitopes that bind these MHC proteins are useful for broad-based T-cell vaccine design, based on their population coverage with the maximum number of MHC proteins [1]. The classification of MHC proteins into superfamilies has been achieved in three ways: (a) conservation of peptide sequences [2], (b) conservation of amino acid residues in MHC-binding pockets [3], and (c) structural interaction profiles of MHC proteins and their binding peptides [4]. The procedure for clustering MHC proteins using their interaction profiles was demonstrated in 2007, when Tong et al. [4] investigated the structural interaction patterns of 68 peptide/HLA complexes spanning 13 class I alleles. Here, we share our experience with the structure-based classification technique and implementing the protocol presented.

Rajat K. De and Namrata Tomar (eds.), *Immunoinformatics*, Methods in Molecular Biology, vol. 1184,
DOI 10.1007/978-1-4939-1115-8_27, © Springer Science+Business Media New York 2014

2 Materials

2.1 Data Sources

Retrieve all required information to your computer. Diligently check to ensure that all duplicate sequences are removed.

1. Protein sequences of MHC alleles: The international ImMunoGeneTics information system (IMGT)/HLA database (http://www.ebi.ac.uk/imgt/hla/) [5].
2. Sequences of MHC-bound ligands and their restricting alleles: The Immune Epitope Database (IEDB; http://www.immuneepitope.org/) [6].
3. 3-D structures of MHC–peptide complexes: The Protein Data Bank (PDB; http://www.rcsb.org/) [7].
4. Refer Table 1 for other data sources.

2.2 Software

1. Sequence similarity searches: The Basic Local Alignment Search Tool (BLAST; http://blast.ncbi.nlm.nih.gov/) [8].
2. Multiple sequence alignments: Clustal Omega (http://www.clustal.org/omega/) [9].
3. Homology modeling: MODELLER version 9.11 (http://www.salilab.org/modeller/) [10].
4. Optimizing side chains of protein structures: SCWRL [11].
5. Clustering of MHC–peptide interaction parameters: MATLAB Statistics Toolbox functions [12].

Table 1
Publicly available immunological databases

Name	URL
IEDB	http://www.iedb.org
IMGT	http://imgt.org
IPD	http://www.ebi.ac.uk/ipd/
The HIV Molecular Immunology Database	http://www.hiv.lanl.gov/content/immunology/
SYFPEITHI	http://www.syfpeithi.de/
AntiJen	http://www.ddg-pharmfac.net/antijen/
MHCBN	http://www.imtech.res.in/raghava/mhcbn/
MPID-T	http://biolinfo.org/mpid-t2/
BEID	http://datam.i2r.a-star.edu.sg/BEID/
CED	http://immunet.cn/ced/

6. Calculating hydrogen bonding potential in proteins: HBPLUS (http://www.biochem.ucl.ac.uk/bsm/hbplus/) [13].
7. Calculating accessible surface areas and gap volumes: SURFNET (http://www.cgl.ucsf.edu/chimera/docs/ContributedSoftware/surfnet/surfnet.html) [14].
8. Assessing the stereochemical quality of protein structures: PROCHECK (http://www.ebi.ac.uk/thornton-srv/software/PROCHECK/) [15].
9. Optimizing side-chain positions of proteins: ICM-Pro (http://www.molsoft.com/icm_pro.html) [16].

3 Methods

3.1 Homology Modeling

Perform protein homology modeling on MHC alleles with no experimentally solved 3-D structures. Diligently follow the below procedure when building the models.

3.1.1 Template Selection

1. Perform a BLAST search against the PDB to identify homologous proteins to a target MHC sequence based on sequence identity.
2. Refer Fig. 1 on how to select appropriate structure as template based on the percentage of identical residues in the alignment [17].
3. Choose a template that contains a ligand in the solved structure.

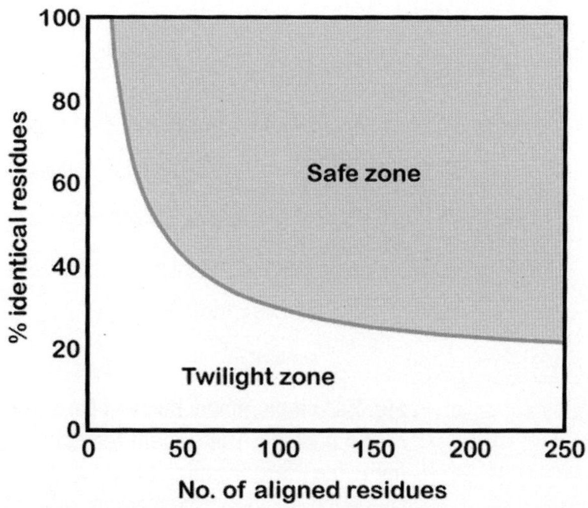

Fig. 1 Zones for protein homology modeling. Two sequences of similar length will most likely fold into the same structure if their pairwise sequence identity falls within the safe zone [17]

4. Where two or more representative sequences with comparable pairwise sequence identity are available, choose the highest resolution homologue with minimal missing residues as template structure.

3.1.2 Target-Template Alignment

1. Input the target and template sequences in FASTA format into the Clustal Omega program [9].
2. Set the "Output Alignment Format" to FASTA.
3. Click "Submit" to run the job.
4. Manually inspect and edit to minimize the number of misaligned residues (*see* **Note 1**).
5. Next, check the initial alignment in view of the template structure(s).
6. Avoid gaps in secondary structure elements and in buried regions.

3.1.3 Model Building

1. Use the MODELLER program version 9.11 [10] for homology modeling.
2. The program takes in as inputs one or more PDB files, an alignment file, and a model building file and generates as output homology models of the target sequence.
3. Create a model building file as shown in Fig. 2.

```
from modeller import *

from modeller.automodel import &


env = environ()

a = automodel(env, alnfile='target-template.ali',

        knowns=' 1hhh', sequence='hla-a3',

        assess_methods=(assess.DOPE, assess.GA341))

a.starting_model = 1

a.ending_model = 3

a.make()
```

Fig. 2 Example model building file used as input for the MODELLER program for model building. The system reads in a template structure "1hhh" and an alignment file "target-template.ali" and generates three homology models of the sequence "hla-a3." The starting_model and ending_model define the number of models to be calculated (i.e., in this case, it will run from indices 1 to 3). The make method in the last line will compute the models

4. Run MODELLER by typing "mod9v1 <model_building_file_name>" in the command prompt at the working directory containing the input files (*see* **Note 2**).
5. Next, use the ICM-Pro software [16] to refine the side-chain positions of proteins.
6. Input the target structure in PDB format into the ICM-Pro software.
7. Right click on the PDB name in the ICM Workspace and choose the option "Convert PDB" to convert the PDB file into an ICM object.
8. Select the side chains you wish to optimize in the graphical display.
9. From the right click menu, select "Advanced/Optimize Side Chains" to call the data entry box.
10. Set the number of calls per variable and press "OK." In general, the simulation length increases with increasing number of calls per variable. The default number was recommended as an appropriate simulation length.
11. The software will output a table displaying a list of energy conformations. View each conformation by clicking the entries in the table.

3.1.4 Model Validation

1. Use the PROCHECK program [15] to check for normality of bond lengths and bond angles.
2. The system reads a file in PDB format and generates as output the following plots:
   (a) Ramachandran plot—to show the φ and ψ torsion angles for all residues in the structure.
   (b) Ramachandran plots by residue type—to show individual Ramachandran plots for each of the 20 different types of amino acids.
   (c) Chi1-Chi2 plots—to show the χ_1 and χ_2 side-chain torsion angles for all amino acid residue types.
   (d) Main-chain parameters—to compare the main-chain parameters of the target structure with that of well-defined structures at a similar resolution.
   (e) Side-chain parameters—to compare the side-chain parameters of the target structure with that of well-defined structures at a similar resolution.
   (f) Residue properties—to show how the protein's geometrical properties change along its primary sequence.
   (g) Main-chain bond length distributions—to show how each of the different main-chain bond lengths is distributed in the structure.

(h) Main-chain bond angle distributions—to show how each of the different main-chain bond angles is distributed in the structure.

(i) RMS distances from planarity—to show the RMS distances from planarity for the different planar groups in the structure.

(j) Distorted geometry plots—to show all the distorted main-chain bond lengths, main-chain bond angles, and planar groups.

3. Run PROCHECK by typing the following in command prompt: procheck <pdb_file> <chain_id> <resolution>.

3.2 Interaction Parameters

3.2.1 Intermolecular Hydrogen Bonds

1. The HBPLUS program reads a file in PDB format and generates as output a list of potential hydrogen bonds [13].
2. HBPLUS requires all atoms in the input PDB file to be correctly labeled and ordered, and no atoms have alternate locations. Use the "Clean" program that comes bundled with HBPLUS to check and correct inconsistencies in the PDB file.
3. Run HBPLUS on the "clean" PDB file by typing "hbplus <cleaned filename> <uncleaned filename>" in the command prompt at the working directory containing the cleaned and uncleaned PDB files (*see* **Note 3**).

3.2.2 Interface Area

1. The SURFNET program [14] is used to compute the accessible surface area (ASA) between an MHC molecule and its bound peptide.
2. For MHC class I complexes, the ASA between an MHC molecule and its bound peptide is defined as the mean ΔASA on complexation when going from a monomeric MHC protein to a dimeric MHC–peptide complex state.
3. The program takes in as inputs a PDB file, atom range of MHC and peptide in the file, output format, and grid spacing of the surface maps and generates as output the interface area of the protein.
4. The parameter file surfnet.par is used to configure all parameters for the plot.
5. Create a parameter file for the MHC molecule.
6. In the section "OUTPUT FILES," the first line of the text will contain information of the MHC molecule (*see* Fig. 3).
7. Assign the first column name as "mhc."
8. Set the atom range of the MHC molecule in the third (i.e., start atom) and fourth (i.e., last atom) columns.
9. Run SURFNET by typing "surfnet" in the command prompt at the working directory containing the input PDB file and the parameter file.

```
<----SURFNET----><-SURFACE-><---------SURFPLOT----------->
                    Foregmd Backgrnd
          Atom range        colour  colour   Line
     File   1st  Last Contour /shade  /shade   width SOLID
Filename type[1] atom  atom level[2] (0-10)[3] (0-10)[3]  [4] /WIRE

OUTPUT FILES
mhc       S    1  2248  100.0     1     1     0.1
peptide   S  2249  2332  100.0     8     8     0.1
```

Fig. 3 Example parameter file used as input for the SURFNET program to compute interface area of MHC molecule

```
<----SURFNET----><-SURFACE-><---------SURFPLOT----------->
                    Foregmd Backgrnd
          Atom range        colour  colour   Line
     File   1st  Last Contour /shade  /shade   width SOLID
Filename type[1] atom  atom level[2] (0-10)[3] (0-10)[3]  [4] /WIRE

OUTPUT FILES
Mhc-peptide S    1  2332  100.0     1     1     0.1
```

Fig. 4 Example parameter file used as input for the SURFNET program to compute interface area of MHC–peptide complex

10. An output file "mhc.srf" will be generated containing the interface area of the MHC molecule.
11. Now, create another parameter file for the MHC–peptide complex (*see* Fig. 4).
12. Assign the first column name as "mhc-peptide."
13. Set the atom range of the MHC–peptide complex in the third (i.e., start atom) and fourth (i.e., last atom) columns.

14. Half the difference between the ASA of the MHC–peptide complex and the ASA of the MHC molecule. This will give you the ASA between an MHC class I molecule and its bound peptide.
15. The ASA between an MHC class II molecule and its bound peptide can be computed in a similar manner as above.

3.2.3 Gap Volume

1. The gap volume ($Å^3$) or the gap region between the bound peptide and MHC protein is an indicator of the goodness of fit between the two molecules. This is computed using the SURFNET program [14].
2. The parameter file surfnet.par is used to configure all parameters for the plot.
3. Set the maximum radius to 5.00 Å and minimum radius to 1.00 Å.
4. Set the option "Calculate gap volume" to "Y."
5. Use the default values for all other parameters.
6. Run SURFNET by typing "surfnet" in the command prompt at the working directory containing the input PDB file and the parameter file.
7. An output file "gaps.srf" will be generated containing the gap volume between the MHC and its bound ligand.

3.2.4 Gap Index

1. The gap index [18] is an indicator of the electrostatic and geometric complementarity the bound peptide and MHC molecule.
2. Divide the gap volume of the bound complex (refer Subheading 3.2.3) with its accessible surface area (refer Subheading 3.2.2). This will give you the gap index of the MHC–peptide complex.

3.3 Hierarchical Clustering

1. The MATLAB Statistics Toolbox functions are used to cluster the derived MHC–peptide interaction data set.
2. Calculate the distance between objects using the "pdist" function: Y=pdist(X) where X is the data set.
3. Group the objects into binary clusters using the "linkage" function: Z=linkage(Y).
4. Use the "dendrogram" function to plot the hierarchical cluster tree: dendrogram(Z).
5. Use the "cluster" function to prune the bottom of the hierarchical tree and partition the data into groups: cluster(Z, "cutoff," c) where c is a threshold for cutting Z into clusters.
6. Alternatively, use the "clusterdata" function to perform all the above steps at one go: T=clusterdata(X, c) where X is the data set and c is a threshold for cutting Z into clusters.

4 Notes

1. Sometimes it may be difficult to align two sequences where the percentage sequence similarity is low. To arrive at a better alignment, we find that it is best to visually inspect the results and manually edit where necessary.
2. The homology modeling program may not generate the best model in a single run. To arrive at a better model, we find that it is best to generate several models and select the model with the lowest "MODELLER objective function" value.
3. All hydrogen bond potentials are defined in accordance to standard geometric parameters using HBPLUS [13].

References

1. Sidney J, Peters B, Frahm N, Brander C, Sette A (2008) HLA class I supertypes: a revised and updated classification. BMC Immunol 9:1
2. Brusic V, Petrovsky N, Zhang G, Bajic B (2002) Prediction of promiscuous peptides that bind HLA class I molecules. Immunol Cell Biol 80:280–285
3. Doytchinova IA, Guan P, Flower DR (2004) Identifying human MHC supertypes using bioinformatic methods. J Immunol 172: 4314–4323
4. Tong JC, Tan TW, Ranganathan S (2007) In silico grouping of peptide/HLA class I complexes using structural interaction characteristics. Bioinformatics 23:177–183
5. Lefranc M-P, Giudicelli V, Ginesoux C, Jabado-Michaloud J, Folch G, Bellahcene F et al (2009) IMGT®, the international ImMunoGeneTics information system®. Nucleic Acids Res 37: D1006–D1012
6. Vita R, Zarebski L, Greenbaum JA, Emami H, Hoof I, Salimi N et al (2010) The immune epitope database 2.0. Nucleic Acids Res 38: D854–D862
7. Berman HM, Westbrook J, Feng Z, Gilliland G, Bhat TN, Weissig H et al (2000) The Protein Data Bank. Nucleic Acids Res 28:D235–D242
8. Altschul SF, Gish W, Miller W, Myers EW, Lipman DJ (1990) Basic local alignment search tool. J Mol Biol 215:403–410
9. Sievers F, Wilm A, Dineen D, Gibson TJ, Karplus K, Li W et al (2011) Fast, scalable generation of high-quality protein multiple sequence alignments using Clustal Omega. Mol Syst Biol 7:539
10. Sali A, Blundell TL (1993) Comparative protein modelling by satisfaction of spatial restraints. J Mol Biol 234:779–815
11. Wang Q, Canutescu AA, Dunbrack RL Jr (2008) SCWRL and MollDE: computer programs for side-chain conformation prediction and homology modeling. Nat Protoc 3:1832–1847
12. Gilat A (2004) MATLAB: an introduction with applications, 2nd edn. Wiley, Canada
13. McDonald IK, Thornton JM (1994) Satisfying hydrogen bonding potential in proteins. J Mol Biol 238:777–793
14. Laskowski RA (1995) SURFNET: a program for visualizing molecular surfaces, cavities, and intermolecular interactions. J Mol Biol 13(323–330): 307–308
15. Laskowski RA, MacArthur MW, Moss DS, Thornton JM (1993) PROCHECK—a program to check the stereochemical quality of protein structures. J Appl Cryst 26:283–291
16. Abagyan RA, Totrov MM, Kuznetsov DA (1994) ICM: a new method for protein modeling and design: applications to docking and structure prediction from the distorted native conformation. J Comput Chem 15:488–506
17. Rost B (1999) Twilight zone of protein sequence alignments. Protein Eng 12:85–94
18. Jones S, Thornton JM (1996) Principles of protein-protein interactions. Proc Natl Acad Sci U S A 93:13–20

Chapter 28

Immunoinformatics, Molecular Modeling, and Cancer Vaccines

Seema Mishra and Subrata Sinha

Abstract

Cancer vaccines are a natural way of fighting the development and progression of cancer as they harness the power of immune system to tweak it into killing cancerous cells. One of the most important agents in an immune system, the cytotoxic T cells (CTL), play a major role and the CTL epitopes in the form of an immunotherapeutic product have been shown to help mount an immune response towards tumor cell destruction. Immunoinformatics and molecular modeling tools have proven powerful towards the prediction of plausible CTL epitopes as well as other epitopes, cutting short the time and cost. We focus on the sequential methodology using these tools as well as some databases to generate a succinct list of enterprising subtype-specific or promiscuous peptide epitopes.

Key words Cancer vaccine, Immunoinformatics, Cytotoxic T cell, Peptide epitopes, MHC-binding epitopes, Proteasomal cleavage prediction, TAP transporter-binding epitopes, Molecular modeling

1 Introduction

Immunoinformatics tools are a powerful means of designing cancer vaccines for the purpose of finding potential, effective immunotherapy candidates. Peptide epitopes in the form of B-cell and T-cell antigens from proteins that are either present uniquely or overexpressed in tumors are capable of eliciting an immune response towards tumor cells. These peptide or protein epitopes can function as plausible immunotherapy candidates and one of the ways to harvest these is to utilize the power of immunoinformatics.

It has been estimated that experimental vaccine discovery and development to registration takes about 10–20 years and a cost amounting to US$ 200–900 million [1]. Immunoinformatics tools cut both the time and costs involved and have been found to be reliable in terms of accuracy of results as well as effectiveness in several studies [2–5].

This chapter focuses on the methodology adopted in harnessing several different sequence- and structure- or molecular modeling-based

Rajat K. De and Namrata Tomar (eds.), *Immunoinformatics*, Methods in Molecular Biology, vol. 1184, DOI 10.1007/978-1-4939-1115-8_28, © Springer Science+Business Media New York 2014

prediction tools to harvest a list of peptide epitopes from tumor-specific or tumor-associated antigens (TSA or TAA) that can be studied further in experiments to deduce their activity as good binders to major histocompatibility complex (MHC) molecules, to T cells and subsequent elicitation of immune response [6].

Before we begin, the authors believe that a succinct refresher in T cell antigen processing and presentation will be helpful in understanding the overall methodology. Nature designed CD8+ cytotoxic T lymphocytes (CTLs) to function as Shiva, the lord of destruction of evil. It is precisely the CTLs that target the molecules or the cells for destruction. CD4+ helper T lymphocytes (HTLs) serve to prime and maintain these CTLs. Generation of an integrated CD8+ and CD4+ T cell immune response may prove a more effective immunotherapeutic procedure as opposed to using either one alone [7].

T cell epitopes in the form of peptide antigens are presented on the cell surface. After cleavage of relevant protein/s within the cells, the individual peptides bind to the surface of the proteins of the major histocompatibility complex (MHC) class I molecules (also known as the human leukocyte antigen (HLA) system in humans—in this chapter, the two terms are used interchangeably). In case of tumor antigens which are generated within the cells endogenously, antigen presentation by MHC class I pathway is required. Tumor cells, in order to evade the immune system, may vary the expression of MHC class I on their surface, but that is another story.

Virtually all tumor-specific or tumor-associated antigens are processed through MHC class I antigen-processing pathway in three major steps (as mentioned in Fig. 1):

1. Generation of antigenic peptides by hydrolysis of the protein antigen by constitutive proteasome or immunoproteasome
2. Transport of peptides from cytosol to endoplasmic reticulum (ER) by transporter associated with antigen-processing (TAP) protein.
3. Binding of peptides to human leukocyte antigen (HLA) class I molecules assisted by several chaperones
4. Transport of peptide-MHC complex through Golgi bodies to cellular surface for presentation to CD8+ T cells.

MHC class II antigen-processing pathway for presentation to CD4+ T cells involves assembly of HLA class II molecules in ER with invariant chain, followed by degradation of this invariant chain in MHC class II compartments to generate class II-associated invariant peptide (CLIP) which remains bound in the MHC groove. CLIP is then exchanged for antigenic peptides that are derived from exogenous protein molecules internalized *via* endosomes. This process is facilitated by HLA-DM molecules and then presented to CD4+ T cells. This is the main pathway for exogenous antigens. While MHC class I molecules are found on the surface of

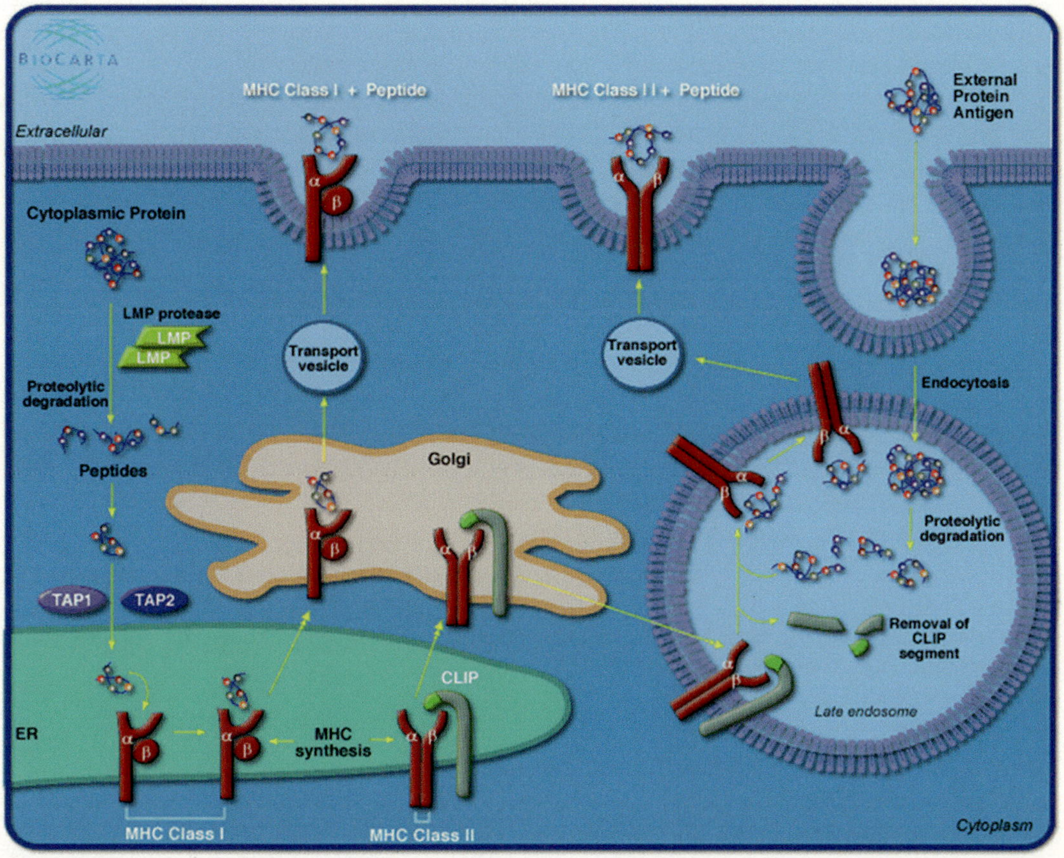

Fig. 1 Figure depicting a clear representation of antigen processing and presentation pathway in the case of MHC class I and MHC class II molecules (image taken with permission from BioCarta pathways website, antigen processing and presentation, http://www.biocarta.com/pathfiles/h_mhcpathway.asp) (reproduced with kind permission)

nucleated cells, MHC class II molecules are present only on the antigen-presenting cells and lymphocytes.

Tools developed for CD8+ T cell epitope prediction occur in greater number than those developed for CD4+ T cell epitope prediction, because of the difficulty in accurately identifying CD4+ T cell epitopes owing to greater length of the epitopes extending out of the binding groove of MHC class II molecules [8]. Further, the former prediction softwares utilize more diverse algorithmic approaches. This algorithmic diversity leads towards more accurate prediction of results generated through a consensus. In literature, antigen processing and presentation pathway has been explored more in the context of CD8+ T cell epitope prediction, primarily due to their main role in cellular immune response. A sequential methodological pathway towards promiscuous and/or subtype-based T cell epitope prediction for use in cancer vaccine design is explained in the following sections.

2 Materials

1. A list of tumor-specific or tumor-associated protein/s can be retrieved from literature search or from the following databases:
   (a) TANTIGEN: Tumor T Cell Antigen database from the website http://cvc.dfci.harvard.edu/tadb/.
   (b) T cell-defined tumor antigens from the website http://cancerimmunity.org/peptide/ [9].
2. Amino acid sequence and known structure, if any, is obtained from National Center for Biotechnology Information (NCBI) or from Swiss-Prot or Protein Data Bank (PDB) databases.
3. Sequence-based prediction tools based on a variety of algorithms are used for the studies:
   (a) MHC-binding epitope prediction: BIMAS (http://www-bimas.cit.nih.gov/molbio/hla_bind/) [10], SYFPEITHI (http://www.syfpeithi.de/Scripts/MHCServer.dll/EpitopePrediction.htm) [11, 12], Propred-I (http://www.imtech.res.in/raghava/propred1/) [13], Propred (http://www.imtech.res.in/raghava/propred/) [14], NetMHC 3.4 (http://www.cbs.dtu.dk/services/NetMHC/) [15, 16].
   (b) Proteasomal cleavage prediction: PAProC (http://www.paproc.de/) [17], NetChop3.1 (http://www.cbs.dtu.dk/services/NetChop/) [18], MAPPP (based on FRAGPREDICT, http://www.mpiib-berlin.mpg.de/MAPPP/cleavage.html) [19].
   (c) TAP transporter-binding peptide prediction: TAPPred (http://www.imtech.res.in/raghava/tappred/) [20], TAP Hunter (http://datam.i2r.a-star.edu.sg/taphunter/index.html).
   (d) Combination tools: Propred-I (http://www.imtech.res.in/raghava/propred1/), MAPPP (http://www.mpiib-berlin.mpg.de/MAPPP/expertquery.html) [21], NetCTL1.2 (http://www.cbs.dtu.dk/services/NetCTL/) [22].
4. Structure-based molecular modeling software: Discovery Studio (http://www.accelrys.com/), Sybyl (http://www.tripos.com/), Swiss PDB Viewer (http://www.expasy.ch/spdbv/), AMBER (http://amber.scripps.edu/).

3 Methods

A combination of different algorithms for MHC-binding T cell epitope prediction lends greater accuracy. These algorithms span across experimentally derived matrices, virtual matrices, binding

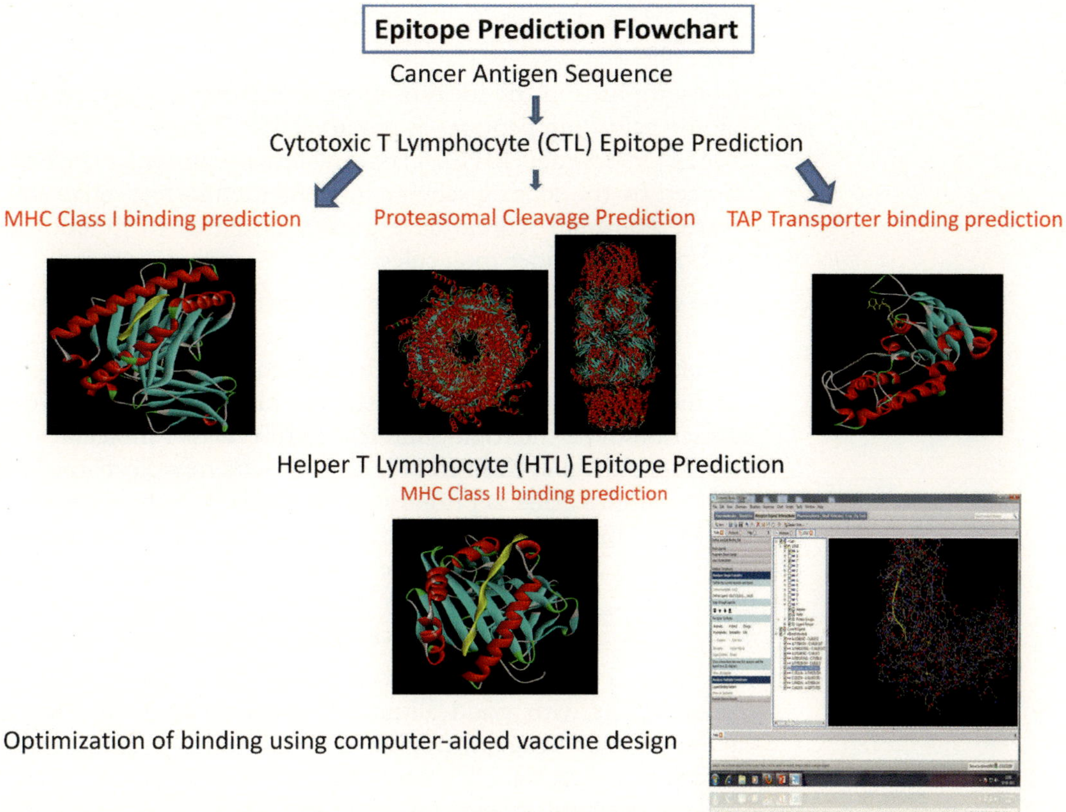

Fig. 2 A flow chart depicting key steps during T cell epitope prediction and shortlisting of potential T cell epitopes. Images (in a generalized form) were created from downloaded structure files from RCSB PDB website with PDB IDs 1G7Q (MHC class I molecule in complex with octameric peptide from tumor-associated antigen MUC1), 1FNT (20S proteasome), 1JJ7 (C-terminal ATPAse domain of human TAP1 molecule), 3L6F (MHC class II molecule in complex with 15 amino acid residue-long phosphopeptide derived from melanoma antigen recognized by T cells 1), and 1DUZ (MHC class I molecule in complex with nonameric peptide from Tax protein of HTLV-1)

motif-based algorithm, profile hidden Markov model, and artificial neural networks to generate a repertoire of potential epitopes that can generate an immune response. When combined with proteasomal cleavage prediction and TAP transporter-binding prediction, the picture may present the events happening within real cellular milieu. A pictorial representation in the form of a flow chart of these steps is shown in Fig. 2.

The methodologies outlined below are generalized and can be used with any software of choice as mentioned in Subheading 2.

3.1 Sequence-Based Prediction for Peptide-MHC Binding

1. The complete sequence of TSA or TAA, or if it is unavailable, a partial one, is used as an input in the web server of the selected prediction tool, mostly in FASTA format.
2. In one of the main parameters present in web servers, one or more of the HLA alleles are chosen to represent each supertype,

e.g., "HLA-A," "HLA-B," or "HLA-C" alleles for MHC class I molecules.

3. The output is in the form of mostly nonamer (9-mer) or nine amino acid-long peptides as potential MHC class I-binder epitopes or 15-mer MHC class II-binder epitope sequences, ranked by the scoring scheme implemented in respective software. The high scorers have greater potential to function as good MHC-binder molecules.
4. The scoring scheme reflects the binding affinity and a variety of scoring schemes are implemented. As an example, BIMAS algorithm provides score for the subsequence which is a predicted epitope, according to estimated half time ($T_{1/2}$) of dissociation of β2 microglobulin from MHC class I molecules. In another example, SYFPEITHI scoring scheme reflects the role of the amino acids as primary and secondary anchors involved in MHC binding (primary anchor residues occur at positions 2 and 9 in a nonamer peptide, while secondary anchors are at first, third, fifth, and seventh positions) and thereby the frequency of respective amino acid in T cell epitopes.
5. To harvest promiscuous epitopes that can bind to multiple HLA alleles with good affinity in order to provide a larger population coverage, the epitopes common in the output produced using several alleles are enlisted through manual search among the high scorers.

3.2 Structure- or Molecular Modeling-Based Prediction

While the sequence-based predictions can be validated using structure-based predictions using computer-aided vaccine design approach, an insight into intermolecular interactions provided by this approach is also valuable. Modifications of peptide epitope sequence, or of T cell receptor sequences, have been studied for enhanced binding [7] and modeling can provide useful insights into experimentation. Several modeling software such as Schrodinger's molecular modeling, Discovery Studio ((DS), http://www.accelrys.com/), Sybyl (http://www.tripos.com/), Swiss PDB Viewer (http://www.expasy.ch/spdbv/), and AMBER (http://amber.scripps.edu/) among others can be implemented. One of the methods to study MHC-peptide epitope-binding affinity using Accelrys' Discovery Studio 3.5 is explained below.

1. The three-dimensional coordinates of HLA molecules are retrieved from PDB, or if a particular HLA allele/subtype is not available in PDB, then a model can be generated via homology modeling. The HLA molecules are polymorphic and their binding site is like a cleft. This cleft is closed at both ends in case of class I MHC and open at both ends in case of class II MHC molecules.

2. The list of peptide epitopes predicted by sequence-based prediction tools is used to model these peptides onto the HLA molecules.
3. The structures are checked to make sure that they are of the lowest possible X-ray crystallographic resolution good enough to be used further and that there are no missing residues in the groove region where peptide epitope is to be modelled. If there are missing residues, they should be added first before proceeding further.
4. Hydrogens are added to the crystallographic complexes, which are energy minimized to reduce structural strain. During energy minimization, care should be taken so that the peptide-binding groove region does not get distorted extensively.
5. The predicted epitopes are threaded onto the nonamer peptides present in the complex by mutating/replacing the original peptide residues using Macromolecules tool in DS.
6. After the predicted peptides have been threaded, the entire complex is energy minimized using Simulation tool. A 13 Å non-bonded cutoff and a distance-dependent dielectric constant of $4r$, in order to simulate the solvent in an approximate way, are applied. Potentials and charges are assigned using default CHARMM/CFF force-field parameters. A full minimization of the complex is achieved using 1,000 steps of steepest descent algorithm followed by 1,000 steps of conjugate gradient algorithm or till the value of RMS gradient of potential energy becomes less than 0.4 kcal/mol/Å.
7. After minimization, rigid-body docking of the peptide to HLA molecules using LigandFit followed by intermolecular energy calculations is done. This provides an insight into the binding affinity. Further intermolecular interaction analyses for the number of hydrogen bonds and interaction contacts among others can be done using receptor-ligand interaction protocol in DS (*see* **Notes 1–4**).

3.3 Sequence-Based Prediction for Proteasome Cleavage

1. The complete sequence of TSA or TAA, or if it is unavailable, a partial one, is used as an input in the web server of the selected prediction tool, mostly in FASTA format (*see* **Note 3**).
2. Specify the length of fragments, which is mostly 9-mer epitope for MHC class I molecules, and 10- to 15-mer epitope for MHC class II molecules.
3. Keeping all other parameters at default value, the input sequence is submitted for processing. The output mainly consists of fragments with probabilities of cleavage. The fragments with highest probabilities of cleavage on the carboxy-terminal side are selected and enlisted.

3.4 Sequence-Based Prediction for TAP-Transporter Binding

1. The complete sequence, in FASTA format or any acceptable format, of TSA or TAA, or if a full sequence is unavailable, a partial sequence, is used as an input in the web server of the selected prediction tool (*see* **Note 3**).
2. Keeping all other parameters default, output is generated in terms of binders with higher or lower affinities or non-binders.

4 Notes

1. As an alternative, peptide-MHC docking and binding affinity studies can also be done using the Dock Proteins protocol, implementing ZDOCK and RDOCK software, in Discovery Studio.
2. Users are requested to use the most recent versions of software tools available, as these are in stages of continuous development as more and more datasets and improved algorithms become available.
3. There is almost always a help/FAQ section on the web server pages, which users can consult for proper usage and to dispel any doubts. Before starting to use the software/web server, reading these sections first and understanding the basic principles behind the software/web server development and usage for several parameters provided therein are highly recommended.
4. We found a good correlation between sequence-based prediction scores and intermolecular interaction energy values [6]; that is, peptide epitopes with a high score in the sequence-based prediction methods (translating into high binders to MHC class I molecules) also had the lowest interaction energy values obtained *via* molecular docking to same MHC molecules.

References

1. Pronker ES, Weenen TC, Commandeur H, Claassen EHJHM, Osterhaus ADME (2013) Risk in vaccine research and development quantified. PLoS One 8(3):e57755
2. Tu SH, Huang HI, Lin SI, Liu HY, Sher YP, Chiang SK, Chong P, Roffler S, Tseng GC, Chen HW, Liu SJ (2012) A novel HLA-A2-restricted CTL epitope of tumor-associated antigen L6 can inhibit tumor growth in vivo. J Immunother 35(3):235–244
3. Bellone S, Anfossi S, O'Brien TJ, Cannon MJ, Silasi DA, Azodi M, Schwartz PE, Rutherford TJ, Pecorelli S, Santin AD (2009) Induction of human tumor-associated differentially expressed gene-12 (TADG-12/TMPRSS3)-specific cytotoxic T lymphocytes in human lymphocyte antigen-A2.1-positive healthy donors and patients with advanced ovarian cancer. Cancer 115(4):800–811
4. Neumann F, Kubuschok B, Ertan K, Schormann C, Stevanovic S, Preuss KD, Schmidt W, Pfreundschuh M (2011) A peptide epitope derived from the cancer testis antigen HOM-MEL-40/SSX2 capable of inducing CD4+ and CD8+ T-cell as well as B-cell responses. Cancer Immunol Immunother 60(9):1333–1346
5. Gritzapis AD, Fridman A, Perez SA, La Monica N, Papamichail M, Aurisicchio L, Baxevanis CN (2009) HER-2/neu (657-665) represents an immunogenic epitope of HER-2/neu oncoprotein with potent antitumor properties. Vaccine 28(1):162–170

6. Mishra S, Sinha S (2006) Prediction and molecular modeling of T cell epitopes derived from placental alkaline phosphatase for use in cancer immunotherapy. J Biomol Struct Dyn 24(2):109–121
7. Mishra S, Sinha S (2009) Immunoinformatics and modeling perspective of T cell epitope-based cancer immunotherapy: a holistic picture. J Biomol Struct Dyn 27(3):293–306
8. Jørgensen KW, Buus S, Nielsen M (2010) Structural properties of MHC class II ligands, implications for the prediction of MHC class II epitopes. PLoS One 5(12):e15877
9. van der Bruggen P, Stroobant V, Vigneron N, Van den Eynde B (2013) Peptide database: T cell-defined tumor antigens. Cancer Immun 13:15, http://cancerimmunity.org/peptide/
10. Parker KC, Bednarek MA, Coligan JE (1994) Scheme for ranking potential HLA-A2 binding peptides based on independent binding of individual peptide side-chains. J Immunol 152:163
11. Rammensee H-G, Friede T, Stevanovic S (1995) MHC ligands and peptide motifs: 1st listing. Immunogenetics 41:178–228
12. Rammensee, H-G. Bachmann, J., Stevanovic, S. (1997) MHC ligands and peptide motifs. Landes Bioscience (International distributor—except North America). Springer, Heidelberg
13. Singh H, Raghava GP (2003) ProPred1: prediction of promiscuous MHC class-I binding sites. Bioinformatics 19:1009–1014
14. Singh H, Raghava GPS (2001) ProPred: prediction of HLA-DR binding sites. Bioinformatics 17(12):1236–1237
15. Lundegaard C, Lamberth K, Harndahl M, Buus S, Lund O, Nielsen M (2008) NetMHC-3.0: accurate web accessible predictions of human, mouse and monkey MHC class I affinities for peptides of length 8–11. Nucleic Acids Res 36(Web Server issue):W509–W512
16. Lundegaard C, Lund O, Nielsen M (2008) Accurate approximation method for prediction of class I MHC affinities for peptides of length 8, 10 and 11 using prediction tools trained on 9mers. Bioinformatics 24(11):1397–1398
17. Nussbaum AK, Kuttler C, Hadeler KP, Rammensee H-G, Schild H (2001) PAProC: a prediction algorithm for proteasomal cleavages available on the WWW. Immunogenetics 53:87–94
18. Nielsen M, Lundegaard C, Lund O, Kesmir C (2005) The role of the proteasome in generating cytotoxic T cell epitopes: Insights obtained from improved predictions of proteasomal cleavage. Immunogenetics 57(1–2):33–41
19. Holzhütter HG, Kloetzel P-M (2000) A kinetic model of vertebrate 20S proteasome accounting for the generation of major proteolytic fragments from oligomeric peptide substrates. Biophys J 79:1196–1205
20. Bhasin M, Raghava GPS (2004) Analysis and prediction of affinity of TAP binding peptides using cascade SVM. Protein Sci 13(3): 596–607
21. Hakenberg J, Nussbaum A, Schild H, Rammensee H-G, Kuttler C, Holzhütter H-G, Kloetzel P-M, Kaufmann SHE, Mollenkopf H-J (2003) MAPPP—MHC-I antigenic peptide processing prediction. Appl Bioinformatics 2(3):155–158
22. Larsen MV, Lundegaard C, Lamberth K, Buus S, Lund O, Nielsen M (2007) Large-scale validation of methods for cytotoxic T-lymphocyte epitope prediction. BMC Bioinformatics 8:424

Chapter 29

Investigating Host–Pathogen Behavior and Their Interaction Using Genome-Scale Metabolic Network Models

Priyanka P. Sadhukhan and Anu Raghunathan

Abstract

Genome Scale Metabolic Modeling methods represent one way to compute whole cell function starting from the genome sequence of an organism and contribute towards understanding and predicting the genotype–phenotype relationship. About 80 models spanning all the kingdoms of life from archaea to eukaryotes have been built till date and used to interrogate cell phenotype under varying conditions. These models have been used to not only understand the flux distribution in evolutionary conserved pathways like glycolysis and the Krebs cycle but also in applications ranging from value added product formation in *Escherichia coli* to predicting inborn errors of *Homo sapiens* metabolism. This chapter describes a protocol that delineates the process of genome scale metabolic modeling for analysing host–pathogen behavior and interaction using flux balance analysis (FBA). The steps discussed in the process include (1) reconstruction of a metabolic network from the genome sequence, (2) its representation in a precise mathematical framework, (3) its translation to a model, and (4) the analysis using linear algebra and optimization. The methods for biological interpretations of computed cell phenotypes in the context of individual host and pathogen models and their integration are also discussed.

Key words Genome scale metabolic modeling, Network reconstruction, Host–pathogen, Flux balance analysis, Constraints-based modeling

1 Introduction

Metabolites form the link between high-level physiological function and molecular components at both the protein and DNA level. Metabolites are centrally placed in the reductionist causal chain, interconnected closely to the proteins (enzymes) that transform them (and the genes that encode them) on the lower end and higher levels of biological organization in the cells, tissues, and organs of the organism. Thus capturing metabolism in a model not only allows us to probe downward causation, as it is connected to gene/transcript/protein components but also simultaneously link them to higher level properties of the organism [1]. The causal

Rajat K. De and Namrata Tomar (eds.), *Immunoinformatics*, Methods in Molecular Biology, vol. 1184, DOI 10.1007/978-1-4939-1115-8_29,

link for infection lies in the critical balance and dynamics of the pathogen and the host and their potential interaction. Such interactions can typically manifest as clearance, latency, symbiosis, death of the invader or the host. The success of a pathogen lies in the availability of several virulence mechanisms and components encoded by the genome. These include but are not limited to quorum sensing (QS) siderophores-based iron uptake systems, cable pili for adhesion, motility, hemolysin, proteases, phospholipases, secretion systems, lipopolysaccharides (LPS), toxins, and extracellular capsules [2]. It is critical to understand metabolism (the biochemical engine directly related to proliferative potential) including metabolic virulence factors like QS, LPS, and rhamnolipids to unravel mechanisms of pathogenesis. Probing such host–pathogen interactions using metabolic models can be considered a distinct three-step process involving:

1. Metabolic modeling of the pathogen.
2. Metabolic modeling of the host.
3. Integrating host and pathogen metabolism together to understand their interaction.

Constraints-based modeling techniques define biological systems by a set of constraints that characterize all feasible cell behaviors rather than precise phenotypes [3]. Flux balance analysis (FBA) calculates the flow of metabolites through a network making it possible to predict growth rates of organisms or biosynthesis rates of specific metabolites. FBA represents all metabolic reactions in the network as a mathematical formalism—a numerical matrix of stoichiometric coefficients of each reaction. Classically, constraints-based FBA has been used to compute cell function using a genome scale metabolic network as a starting point, but can be applied to regulatory and signaling networks as well. Constraints are typically represented as bounds and balances [4]. The stoichiometry of each reaction imposes a physicochemical constraint on the flow of metabolites through the network [5, 6]. The stoichiometric matrix imposes balance (mass) constraints on the system and each reaction is given bounds (the maximum and minimum allowable fluxes of the reaction). Together, they define the space of allowable flux distributions of a system—that is, the rates at which every metabolite is consumed or produced by each reaction at steady state. The dependence on "hard to measure" parameters in vivo is thus minimal and the method is scalable [7]. The prediction in terms of fluxes or reaction rates rather than concentrations is an outcome of applying the steady state assumption and optimality criterion. The power of the FBA predictions and elegance of the constraints-based approach have led to accurate in silico representation of organisms [8–10]. This approach has become extremely popular for understanding the cell phenotypes in varying environmental

(C-source, N-source, O_2) conditions, minimal media and also in the event of genetic perturbation (gene deletions). These applications can be extended to understand host and pathogen behavior in isolation and their dynamics together.

This chapter summarizes steps that have almost become standard operating procedures in the area of network modeling of metabolism over the last decade that can be applied while investigating host and pathogen behavior. It covers briefly metabolic network reconstruction, translation to mathematical models their simulation techniques. The analysis of the solution and their biological relevance are stated using examples in literature. In a concise form this illustrates the paradigm of metabolic systems biology in the context of host–pathogen behavior and dynamics.

2 Materials

The materials can be broadly categorized into two categories. The first category includes tools for reconstruction that mainly consist of varied databases and the second category includes software and programs to actually convert the networks to models and analyze them.

2.1 Databases

Metabolic network reconstruction of a specific species is generally built using information from biological databases and literature sources. Aiding the process, are several online public resources. Some representative bioinformatics resources utilized for metabolic reconstructions are discussed below:

1. *BRENDA* (http://www.brenda-enzymes.info/) provides enzyme related data and includes organism-specific information on localization. These databases also cite literature references making it easy to look up the source and evaluate the information.
2. *MetaCyc* (http://metacyc.org/), *KEGG* (http://www.genome.jp/kegg/) and *NCBI* (http://www.ncbi.nlm.nih.gov/) include gene, protein, and reaction information for several organisms. *KEGG* also delineates pathways with detailed maps that are a great resource while reconstructing a metabolic network. *NCBI* has a comprehensive literature database that provides access to all the legacy data relating to the organism of interest.
3. Model SEED is an online tool for available for semi-automated model generation.
4. The NCBI, Gene Expression Omnibus (GEO: www.ncbi.nlm.nih.gov/geo/) provides gene expression data.

5. Recon X (http://humanmetabolism.org/) provides information on human metabolism and the recent version of reconstruction of human genome (Recon 2, discussed in detail below). The preferred format for representing a network reconstruction is the SBML format.

2.2 Software and Programs

All mathematical simulations for the toy network and the genome-scale model are worked out using the COBRA Toolbox [11, 12], available for Python or MATLAB® (MathWorks, Inc.) platforms. Although in Subheading 4 details are provided for functions from the COBRA Toolbox (MATLAB) to simulate host–pathogen behavior and interaction, the functions for Python based toolbox are similar and work on the same logic.

An optimization toolbox that works with MATLAB is recommended. Examples include Gurobi, Tomlab, and LINDO. All simulation results in Subheading 4 are based on using Tomlab as the solver.

3 Methods

There are four major steps for metabolic modeling of any organism (host or pathogen). Following this procedure (Fig. 1) allows computation of biological phenotype of the in silico organism and extends the analysis to interacting systems. The four steps are as follows:

1. Network Reconstruction.
2. Translation into Mathematical Model.
3. Simulation and Analysis.
4. Biological Interpretation.

3.1 Network Reconstruction

Genome-Scale Metabolic Network Reconstructions have become increasingly popular for studying metabolism and have been built for organisms across the three kingdoms of life. Multiple versions also exist for the same organism. Quality Controlled, Quality Assured reconstructions are thus indispensible yet critical to ensure accuracy of predictions of cell function. Reconstructed networks (discussed here specifically for metabolic networks), but applicable to regulatory [13] and signaling [14] can be used for computational modeling. The bottom-up approach is the method of choice and based on genomic and legacy data. The outcome is a biochemically, genetically, and genomically structured database: a knowledge-base of the organism of interest [15]. Mathematical translation into a stoichiometric matrix allows simulation of biological phenotypes like growth rate or biosynthesis of metabolites. Strict adherence to the rules of network reconstruction as discussed

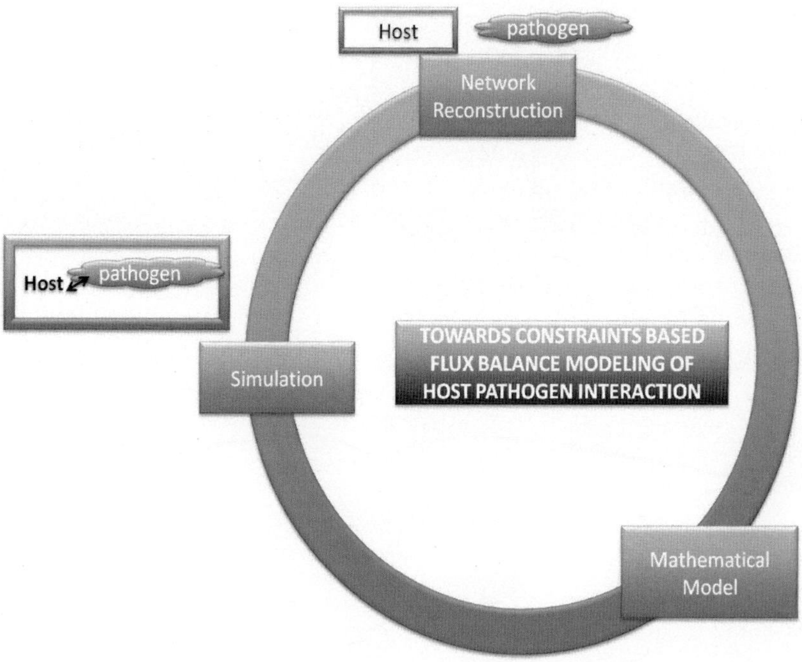

Fig. 1 Schematic representation of the iterative process for modeling host–pathogen behavior and interaction

in this section result in mathematical models that can accurately predict cell function (also refer **Notes 1–3** and 7). A detailed description of the process is delineated elsewhere [16].

3.1.1 Generating a Draft Reconstruction for a Genome Scale Network

The very first step involved is the generation of a draft reconstruction based on the most current version of the genome annotation of the organism of interest. This is generally done computationally from the genome sequence by identifying open reading frames (ORFs), genes and coding sequences, using a family of algorithms for similarity search like FASTA and BLAST. Candidate metabolic functions are assigned based on the similarity scores in conjunction with several biochemical databases like KEGG [17] and EcoCyc [18]. The draft reconstruction is primarily a collection of genome encoded metabolic functions. Incorrect inclusions and exclusions of existing function are expected at this point due to errors in annotation of the genome or gene function in the databases used. The similarity scores represent the confidence level of a given gene function assignment and can be used subsequently to include or preclude reactions in the model. Currently, a few tools including MODEL SEED [19] and SCRUMPY [20] exist to build genome-scale metabolic draft reconstructions. A metabolic reconstruction primarily would include all genes coding enzymes that are involved in metabolism and small molecule transport. This would amount

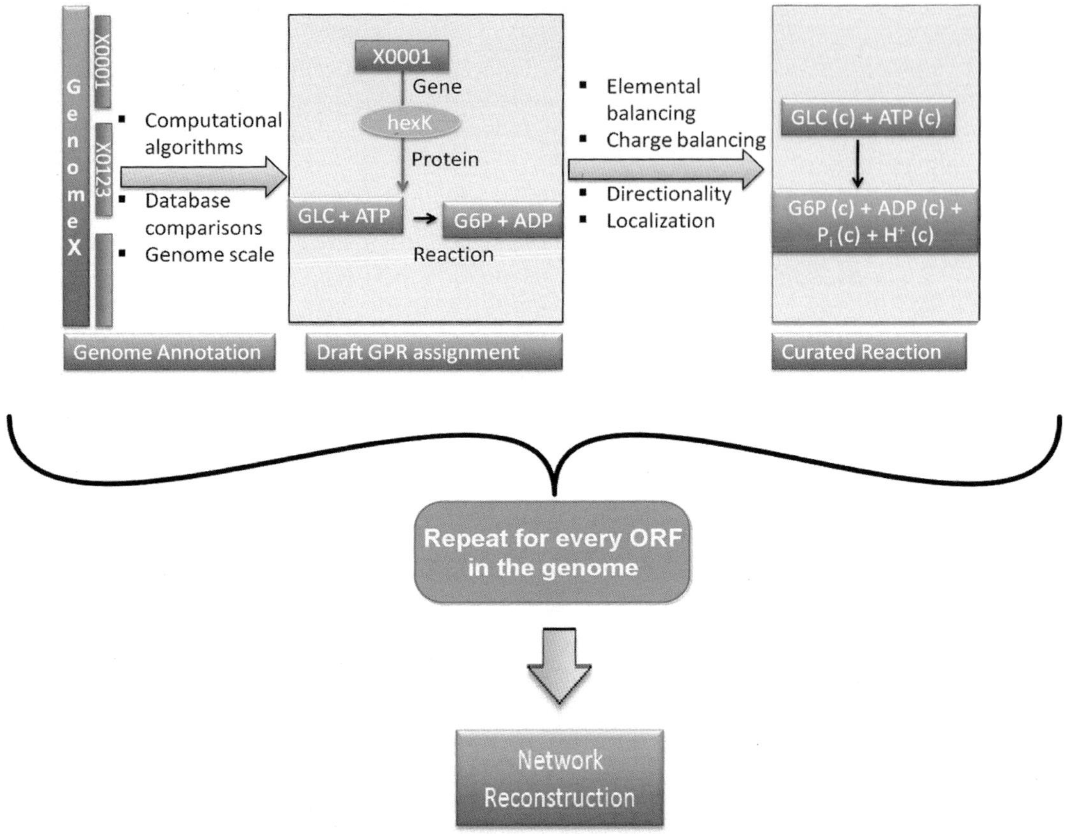

Fig. 2 The network reconstruction process. The network reconstruction process starts with the genome annotation and identification of ORFs, the assignment of gene, protein and reactions based on similarity search algorithms. Once the GPR is assigned as seen for hexokinase in the figure, the curation steps begin. Curation involves checking reaction content including elemental and charge balances, directionality of the reaction and the localization. Once the curation of the GPR is completed on genome-wide basis, one has a genome-scale curated network reconstructed ready for translation to mathematical matrix model

to several hundred gene–protein–reaction (GPR) relationships that need refinement through a process of manual curation. The curation process determines the validity of each gene's functional assignment. A quick search on PubMed with the name of the organism name should give an idea of how extensively the organism under consideration has been studied.

3.1.2 Refining the Network Reconstruction

Manual refinement of the content in the draft reconstruction (Fig. 2) is required to build a model that predicts cell function accurately [16]. The ease of the curation process and the time for refinement depend on the quality and accuracy of the genome annotation and the extent of experimental data that exists on the organism for which the network is being reconstructed.

3.1.3 Assigning Correct Gene–Protein–Reaction Relationships

The accuracy of the network reconstruction relies on correct assignment of Gene–Protein–Reaction Relationships (GPRs). Candidate metabolic functions identified in the draft reconstruction need to be verified through literature search. Incorrect genome annotations are generally the cause for errors in the draft and legacy data based on biochemistry, genetics, or physiology of the organism should be used extensively to validate the metabolic function assigned. The two driving questions that need to be answered for every GPR in the reconstruction [16] are: (1) Is the GPR really present in the organism? and (2) How is this GPR connected to the rest of the network? Since this is a tedious job, it is a good recommendation to assemble the reconstruction one biochemical pathway after the other in order to understand gaps in knowledge as well. If organism-specific information is absent in literature, phylogenetically close organisms can be used to make connections. However, a note regarding that should be included along with the confidence score for assigning that function. Generic databases like KEGG, BRENDA, and METACYC are useful for this process.

GPRs are written as Boolean statements (*see* Fig. 3c for TCA cycle) and define the genetics behind the functional protein. GPRs mainly comprise of two operators, to define the associations, "AND" and "OR." The AND operator is applied when two or more genes are required together for a particular protein (e.g., protein isocitrate lyase is a hetero-dimer encoded by two genes "*aceAa*" and "*aceAb*"). A protein can be coded by more than one gene, and the expression of either of these genes may result in a functional protein, this relation is defined with the OR operator. Also this operator can be applied where two or more proteins are required for one function or reaction (Fig. 3c). There are four types of functional protein GPRs that can be defined as Boolean relations: (1) multifunctional proteins, (2) isozymes, (3) multimeric protein, and (4) multiprotein complex. Mis-assignments in the GPR associations affect results of in silico gene deletion studies in the future. However, these discrepancies that are essentially failure modes of the model, can be used later to refine reconstructions iteratively.

3.1.4 Curating the "Reaction" in the GPR

Once GPR assignments are made, the reaction needs to be further curated. Curating a reaction involves validating the content of a reaction in the GPR [16]. The content of the reaction includes (1) use of substrates and cofactors, (2) metabolite charge, (3) directionality, and (4) localization.

3.1.5 Use of Substrates and Cofactors

In biochemical reactions, in addition to the primary substrate and product, cofactors are important. Since these vary with organisms, specific substrate and cofactor utilization needs to be ascertained from organism-specific databases as changes in the cofactor usage affect the overall prediction potential of the model. Since KEGG

a

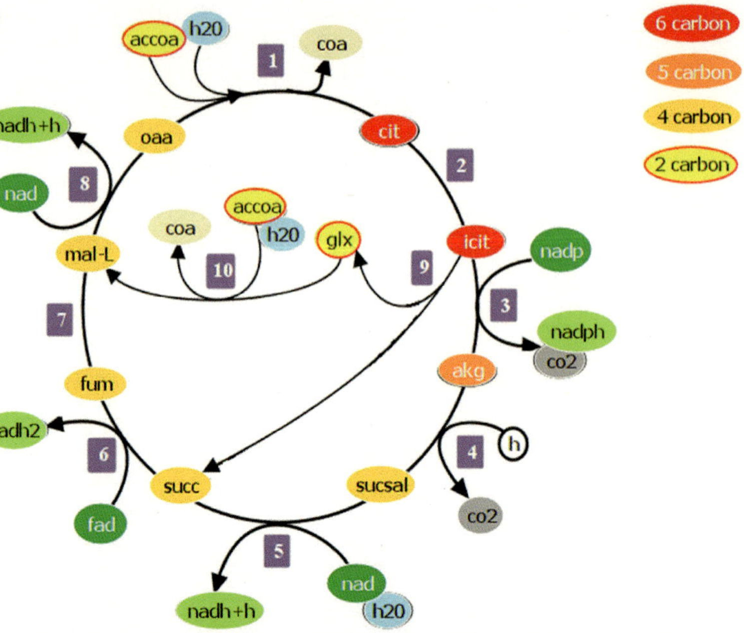

b

Rxn name	Rxn description	Equation
ACONT	aconitase	cit[c] <=> icit[c]
CS	citrate synthase	accoa[c] + h2o[c] + oaa[c] -> cit[c] + coa[c] + h[c]
FUM	fumarase	h2o[c] + fum[c] <=> mal-L[c]
ICDHy	isocitrate dehydrogenase (NADP)	icit[c] + nadp[c] -> akg[c] + co2[c] + nadph[c]
ICL	Isocitrate lyase	icit[c] -> glx[c] + succ[c]
MALS	malate synthase	accoa[c] + h2o[c] + glx[c] -> coa[c] + h[c] + mal-L[c]
MDH	malate dehydrogenase	mal-L[c] + nad[c] <=> oaa[c] + h[c] + nadh[c]
OXGDC	2-oxoglutarate decarboxylase	h[c] + akg[c] -> co2[c] + sucsal[c]
SSALx	succinate-semialdehyde dehydrogenase (NAD)	h2o[c] + nad[c] + sucsal[c] -> 2 h[c] + succ[c] + nadh[c]
SUCD1i	succinate dehydrogenase	succ[c] + fad[c] -> fum[c] + fadh2[c]

Fig. 3 (**a**) The TCA cycle in *M. tuberculosis*. The TCA cycle reactions are as follows: (*1*) citrate synthase, (*2*) aconitase, (*3*) isocitrate dehydrogenase, (*4*) 2-oxoglutarate decarboxylase, (*5*) succinate-semialdehyde dehydrogenase, (*6*) succinate dehydrogenase, (*7*) fumarase, (*8*) malate dehydrogenase, (*9*) isocitrate lyase, and (*10*) malate synthase. The intermediates of the TCA cycle are color coded based on the number of carbon atoms. The reduced form of the cofactors involved is represented in *light green* and the oxidized form in *dark green*. (**b**) Elementally balanced equations for every reaction are derived for the model. (**c**) Boolean representation of Gene protein reaction associations. The four different types of protein associations are illustrated here. All figures represent GPR associations of different types in the model. The first level (in *sky-blue*) is the gene level, the second (in *dark blue*) represents the corresponding translated peptide, the third level (in *orange*) represents a functional protein and the last level (in *yellow*) is the reaction. (*1*) RxnAbbr: *Acont*. This reaction is carried out

c

Rxn abbr.	Protein name	Protein abbr.	Gene associated	Data derived from
ACONT	aconitate hydratase	Acn	(Rv1475c)	KEGG
CS	citrate synthase	CitA or PrpC or GltA2	(Rv0889c) or (Rv1131) or (Rv0896)	KEGG
FUM	fumarase (fumarate hydratase)	Fum	(Rv1098c)	KEGG
ICDHy	isocitrate dehydrogenase (NADP)	Icd1 or Icd2	(Rv3339c) or (Rv0066c)	KEGG
ICL	Isocitrate lyase	(AceAb and AceAa) or Icl	(Rv1916 and Rv1915) or (Rv0467)	KEGG
MALS	malate synthase	GlcB	(Rv1837c)	KEGG
MDH	malate dehydrogenase	Mdh	(Rv1240)	KEGG
OXGDC	2-oxoglutarate decarboxylase	Multifunctional alpha-ketoglutarate metabolic enzyme (Two components: 2-oxoglutarate dehydrogenase E1 component, Dihydrolipoyllysine-residue succinyltransferase component of 2-oxoglutarate dehydrogenase complex)	(Rv1248c)	KEGG + UniprotKB
SSALx	succinate-semialdehyde dehydrogenase (NAD)	GabD2 and GabD1	(Rv1731 and Rv0234c)	KEGG
SUCD1i	succinate dehydrogenase	SdhB, SdhD, SdhC, SdhA, succinate dehydrogenase flavoprotein subunit, SdhD, SdhC, SdhA, SdhD, SdhC, succinate dehydrogenase iron-sulfur subunit, SdhA	(Rv3319 and Rv3317 and Rv3316 and Rv3318 or Rv0248c and Rv3317 and Rv3316 and Rv3318 or Rv3317 and Rv3316 and Rv0247c and Rv3318)	KEGG

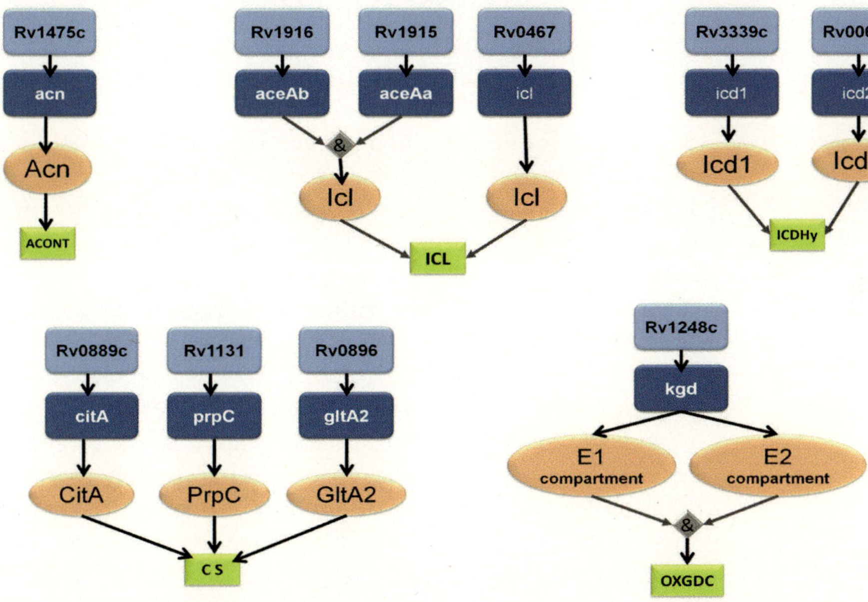

Fig. 3 (continued) by one protein, namely, *Acn*. It has a simple 1-to-1 gene–protein relation. (*2*) Rxn Abbr.: *CS*. This reaction is carried out by three isozymes, from three different ORFs. (*3*) Rxn Abbr.: *ICDHy*. This reaction is carried out by isozymes, from two different ORFs. (*4*) Rxn Abbr.: *ICL*. This reaction is carried out by two isozymes, from three different ORFs. Peptides *AceAa* and *AceAb*, *translated* from their respective ORFs, form the subunits of the protein *Icl*, that catalyzes the isocitrate lyase reaction. This reaction is also catalyzed by another isozyme, which is translated from only one gene, as represented in the figure. (*5*) Rxn Abbr.: *SSALx*. This reaction is carried out by one protein, which has two distinct domains that are responsible for specific functions. Both domains are necessary to catalyze this reaction

and BRENDA [21] list all possible transformations of an enzyme that have been identified in any organism, if there is only one reaction associated with that enzyme in the database, the reaction does not require organism based refinement. At this stage each network reaction should be assigned a confidence score (normally 0–4) reflecting the availability of information and strength of evidence.

3.1.6 Metabolite Charge

Metabolites are generally listed with their uncharged formula in most databases. However, many metabolites are protonated or deprotonated in cells or in medium conditions. The protonation state represents the charged formula and is related to the pH. As metabolic networks are reconstructed assuming an intracellular pH of 7.2, the protonated formula calculated based on the *p*Ka value of the functional groups of the metabolite should be used in the model for accuracy.

3.1.7 Directionality

Biochemical data for reaction directionality for the target organism help assign correct directionality. Whenever experimental data are not available, new approaches use the estimation of the standard Gibbs free energy of formation (ΔfGo) and of reaction (ΔrGo) to assign the direction [22]. Other methods combine thermodynamic information with network topology and heuristic rules to assign reaction directionality [23] also have been developed.

3.1.8 Localization

Subcellular localization of a protein or compartmental localization of a reaction needs to be included as they affect certain functions. PSORT [24] and PASUB [25] are nucleotide/amino acid sequence-based algorithms that can be used to predict the cellular localization. Novel Web-based methods [26] have recently been developed to predict the subcellular location of eukaryotic and prokaryotic proteins.

3.1.9 Macromolecule Biosynthesis

Reactions for biosynthesis of macromolecules or polymers (LPS, peptidoglycan, lipids, phospholipids) are difficult to curate. These reactions are generally not represented accurately in a database as their molecular formula is different in every organism. These reactions generally result in stoichiometric inconsistencies. Translation of structural data of peptidoglycan or LPS to molecular formulae and elementally balanced biosynthesis reactions for macromolecules and polymers is a challenge. Few algorithms have been developed to identify all the reactions that are stoichiometrically inconsistent [27]. This step is particularly important for quality control of networks generated by community efforts. Community approaches towards network reconstruction are becoming increasing popular [28, 29] and the use of automated and computational approaches towards ensuring the quality and accuracy is imminent. These components are very important as in some pathogens they could potentially differentiate serovars or be implicated in differential

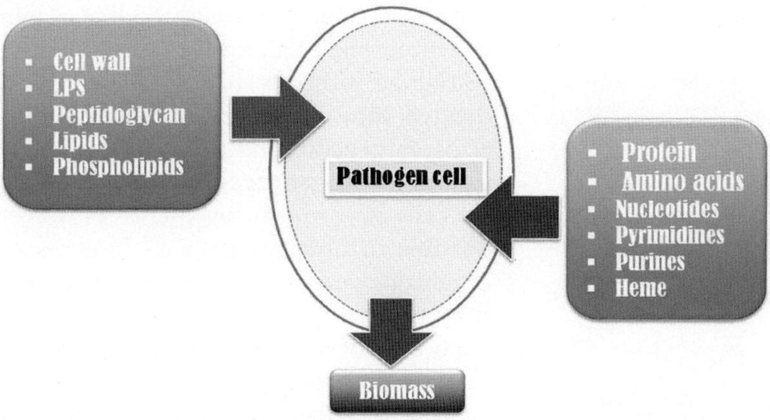

Fig. 4 Formulation of biomass composition. The biomass composition reaction is represented typically in the metabolic models as a linear combination of many inputs and a balanced output that corresponds to the biomass

mechanisms of pathogenesis like *F. tularensis*, *P. gingivalus*, and *H. pylori* [7, 30, 31].

3.1.10 Reconstruction of Small Networks

Although an automated procedure of draft reconstruction is convenient for building large models, small networks can be assembled one biochemical pathway after the other until the required network is developed. Here, we use the assembly of the TCA cycle (Fig. 3) for the pathogen *Mycobacterium tuberculosis* H37Rv as an example. The proteins representations are also shown and include examples of isozymes, protein complexes, and the Boolean representation of their association (refer Fig. 3c).

The TCA cycle plays essential roles in cell metabolism, providing reducing equivalents for energy generation and biosynthetic reactions, along with precursors for lipids, amino acids, and heme. Many variants of TCA cycle operate in pathogens based on the diversity of their metabolic niches. Although the Mtb genome is annotated to encode a complete TCA cycle, Tian et al. have shown that the conversion of α-KG to succinate is via a succinic semialdehyde intermediate instead of the classical succinyl CoA [32]. The reconstruction of the TCA Cycle network for Mtb and its curation using the COBRA Toolbox is delineated stepwise (refer **Note 2**) and methods for saving and exporting the model is also discussed (**Note 13**).

3.1.11 Additional Reactions

To have a complete representation of an organism and be able to simulate its fundamental phenotype of growth, one has to represent biomass composition in a network reconstruction. Demand reactions are also artificial reactions that are added to the network to represent metabolite accumulation properties of the cell.

3.1.12 Formulating the Biomass Reaction

The biomass of an in silico organism is derived from the macromolecular components that make up cells (Fig. 4). The biomass reaction is thus depicted as a weighted linear combination of the fractional contributions of all known cell constituents required to make 1 g of cellular biomass [33]. Each cellular biomass macromolecule is divided into its corresponding building blocks, such as amino acids, fatty acids, and nucleotides. The detailed biomass composition of the target organism needs to be experimentally determined for cells growing in log phase. However, in cases where it is not possible to obtain a detailed biomass composition for the target organism, one can estimate the relative fraction of the precursors from the genome data in a database. Since this reaction is often the main phenotype computed by the model, care should be taken to build an accurate representation.

The biomass reaction is also referred to as a "Biomass objective function" (BOF) and directly influences growth rate calculations [34]. BOFs allow differentiating phenotypic states of a cell. For example, in *Y. pestis* 91001, the biomass composition differs at 25 °C versus 37 °C [35]. The composition also determines the potential host (insect or mammalian). It is the weightage or the coefficients of the terms for the four fatty acid acyl chains 14:0, 16:1, 16:0, and 18:1 that differ between the two BOFs [36].

3.1.13 Demand and Sink Reactions

Demand reactions are unbalanced network reactions that allow the accumulation of a compound, that are otherwise not allowed due to mass-balance principles in steady-state models. Sink reactions are similar to demand reactions but are defined to be reversible and thus provide the network with metabolites. Demand reactions need to be added for compounds like cofactors, lipopolysaccharide, and antigens that the organism is known to produce [16]. These reactions should be used carefully and be biologically relevant if used while computing cell phenotypes. A brief description is given in **Note 8**.

3.1.14 Exchange Reactions

Exchange reactions typically allow specifications of media or environment using uptake rates. Growth conditions are crucial to defining the biological question or hypothesis addressed and the success of the simulation.

Growth conditions would ideally identify (1) essential carbon and nitrogen sources, (2) known auxotrophies, (3) composition of media base (water, protons, metal ions), (3) different environments the pathogen thrives in, (4) crucial metabolites in the host that attenuate or increase the growth and survival of pathogen, (5) specific host niches the pathogen resides in. To define a media constraint file, the reader is referred to **Note 5**.

3.2 Conversion of Network to Mathematical Model

The network reconstruction is converted into a mathematical format by extraction of the S-matrix. The extraction of the S-matrix is illustrated (Fig. 5) with an example of the TCA cycle from *M. tuberculosis* H37Rv. Several condition specific models can be defined for testing, by application of systems boundaries

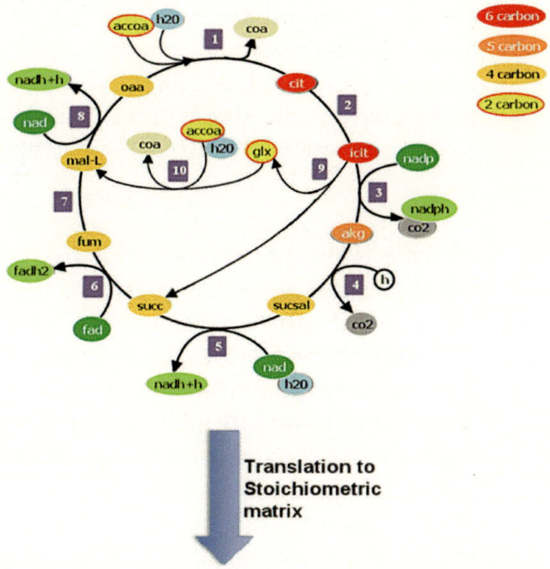

	ACONT	CS	FUM	ICDHy	ICL	MALS	MDH	OXGDC	SSALx	SUCD1i
cit[c]	-1	1	0	0	0	0	0	0	0	0
icit[c]	1	0	0	-1	-1	0	0	0	0	0
accoa[c]	0	-1	0	0	0	-1	0	0	0	0
h2o[c]	0	-1	-1	0	0	-1	0	0	-1	0
oaa[c]	0	-1	0	0	0	0	1	0	0	0
coa[c]	0	1	0	0	0	1	0	0	0	0
h[c]	0	1	0	0	0	1	1	-1	2	0
fum[c]	0	0	-1	0	0	0	0	0	0	0
mal-L[c]	0	0	1	0	0	1	-1	0	0	1
nadp[c]	0	0	0	-1	0	0	0	0	0	0
akg[c]	0	0	0	1	0	0	0	-1	0	0
co2[c]	0	0	0	1	0	0	0	1	0	0
nadph[c]	0	0	0	1	0	0	0	0	0	0
glx[c]	0	0	0	0	1	-1	0	0	0	0
succ[c]	0	0	0	0	1	0	0	0	1	-1
nad[c]	0	0	0	0	0	0	-1	0	-1	0
nadh[c]	0	0	0	0	0	0	1	0	1	0
sucsal[c]	0	0	0	0	0	0	0	1	-1	0
fad[c]	0	0	0	0	0	0	0	0	0	-1
fadh2[c]	0	0	0	0	0	0	0	0	0	1

Fig. 5 Translation of the TCA metabolic network to stoichiometric matrix format. The set of reactions that represent the TCA cycle is translated into a mathematical model, represented as the S matrix. The reactions represented here are from *M. tuberculosis*: (*1*) citrate synthase, (*2*) aconitase, (*3*) isocitrate dehydrogenase, (*4*) 2-oxoglutarate decarboxylase, (*5*) succinate-semialdehyde dehydrogenase, (*6*) succinate dehydrogenase, (*7*) fumarase, (*8*) malate dehydrogenase, (*9*) isocitrate lyase, and (*10*) malate synthase

that defines the successful conversion of a network reconstruction into a condition-specific model. These are added as exchange reactions. Multiple iterations of validation allow definition of a refined model that can accurately simulate phenotypic behavior that is biologically relevant.

3.2.1 Analyzing the Metabolic Network: The Rest Is Math

FBA is a method for assessing the systemic properties and cell behaviors of a metabolic genotype. The fundamentals of FBA have been reviewed [3, 4, 37, 38]. In short, the matrix equation (Eq. 1) that describes the steady-state mass balances of the biochemical reaction network [39, 40] is the lynchpin of FBA.

$$\mathbf{S.v} = \mathbf{b} \tag{1}$$

where S is the stoichiometric matrix ($m \times n$), v is the vector of n metabolic fluxes, and b is the vector representing m transport fluxes (i.e., known consumption rates, by-product production rates, and uptake rates). Mathematically, the S matrix acts as a linear transformation between the vector that defines fluxes through **n** reactions in the biochemical network and the vector of the time derivatives of the concentrations of **m** metabolites involved in these reactions.

For example, the stoichiometric matrix, derived directly from the genome-scale metabolic network for Mtb, *i*NJ66 [41] has a dimension of $826 \times 1{,}025$ ($m=826$; $n=1{,}025$). Based on matrix theory, the row space and null space of the S-matrix define the flux solution spaces while the column and left null spaces are related to the concentration space. Based on the definition of the null space, all steady state solutions to the flux balance equations are contained in this space. The solution space of interest to us is actually the intersection of the region that satisfy the mass balance constraints as defined by Eq. 1 and the linear inequalities defined by maximum and minimum rates ($\alpha_i \leq v_i \geq \beta_i$) or any other physicochemical constraints [42] (defined through physiological data or OMICS data). The feasible set of solutions defines the boundaries and capabilities of the in silico cell and represent what a cell can theoretically do [43]. Since Eq. 1 is an underdetermined system of linear equations having infinite number of solutions, defining biological objectives (as linear combinations of fluxes) for optimization is a critical factor allowing convergence to only one relevant solution (the one that results in the maximum or minimum value of the objective function). Thus designing objective functions is critical to posing the right biological question and also in interpretation of computed cell function.

3.3 Simulation and Analysis

For small models it is possible to extract the stoichiometric matrix manually but genome-scale stoichiometric models, need efficient editing to translate into a model, run simulations, and visualize results. Since the mathematical methods are a combination of linear algebra and optimization, a program in addition to the linear solver is needed to interpret the solver's output. Currently, tools like Model SEED are available for automated model generation. OptFlux [44] and CellNetAnalyzer [45] mainly feature editing and visualization. PySCeS [46], YANASquare [47], MEGU [48],

BioMet Toolbox [49], and Cytoscape [50] are limited in for network modeling capabilities. None of these however are scalable to genome level and allow simulations for interrogation and prediction of cell function and behavior. FAME: the Flux Analysis and Modeling Environment [51] a Web-based, python scripted environment for FBA and some related functions, that doesn't require proprietary software is also slowly gaining popularity as a "one stop shop" for FBA. The most versatile and popular tool for mathematical simulation is the COBRA Toolbox [11, 12], a matlab, based tool for model solving and command-line manipulation. This chapter discusses methods and objective functions from the COBRA Toolbox to simulate host–pathogen behavior and interaction.

Analysis methods

A model is only as powerful as the questions posed. The success of a model is based on the defined objective functions. Choice and design of the right mathematical objective makes way for validating biological hypotheses and driving biological discovery. Most COBRA methods depend on physicochemical and biological constraints to delineate the allowable phenotypic space under specific conditions. These constraints are hard physicochemical constraints including compartmentalization, mass conservation, molecular crowding, and thermodynamic directionality. Designing constraints that represent the actual organism pathology using known experimental data is critical to exploiting the power of CBM. Methods have also been described recently to use transcriptomic, proteomic data to reduce the size of the set of computed feasible states [52–54]. Although existent, more integrative methods to translate OMICs data types into constraints are needed to expand the repertoire of analyses related to host–pathogen interactions. Although studying host–pathogen interactions using constraints-based FBA has its limitations, it is probably one of the few methods that allows for system level analyses. The COBRA Toolbox gives access to several COBRA methods. Detailed descriptions of COBRA methods and protocols can be found elsewhere [11, 12]. The following examples listed here discuss protocols for in silico experiments to probe pathogen or host metabolic behavior and understand their interaction.

3.3.1 Structural Analysis of the Metabolic Network

Understanding the topology of the network by identifying gaps and dead ends is advisable before proceeding with other functional analysis. The steady-state assumption in FBA prevents accumulation of any metabolite in the network, and hence, the reactions that produce them are never used in computing of the phenotypic state of the cell. Incomplete networks can be assessed, the gaps delineated and filled by this set of functions in COBRA.

Gap Analysis

Gaps in network structure due to missing information and functional characterization of putative genes limit phenotypic outcomes. A gap is the link between two reactions that is missing in the network. Filling these gaps makes it possible to reconcile failure modes of the model.

Only dead-ends with strong genetic evidence should be included in the model. Three potential hypotheses can be extended towards identified dead ends. These include (1) reactions required to produce or consume the metabolite are absent either from the reconstruction or genome annotation (2) the reaction that causes the dead-end may not actually occur in the organism, or (3) the dead-ends actually exists in the organism.

The two COBRA functions that identify gaps are (A) detectDeadEnds or (B) gapFind. (Refer **Note 4**).

Find All Gaps in a Model

The GapFind algorithm [15] allows one to find all gaps in a model and all metabolites that are downstream from a model gap.

```
>>[allGaps,rootGaps,downstreamGaps]
=gapFind(model,findNCgaps,verbFlag)
```

where array: **allGaps** is a list of the metabolite indices for a metabolite at a gap; **rootGaps** is a list of metabolites that cannot be produced; and **downstreamGaps** is a list of metabolites that are produced in a reaction that requires a metabolite that cannot be produced.

This function is run in an interactive and iterative fashion to guarantee that all gaps are identified. It is necessary to set the upper and lower bounds of the exchange to relative large or small magnitudes to get accurate results. If the bound magnitudes are too small, the algorithm will incorrectly identify many metabolites as gaps; if this occurs, increase the bound magnitudes by tenfold (refer **Note 4**).

Detect Dead Ends in a Model

The detectDeadEnds function searches the model.S matrix for metabolites that participate in only one reaction (can only be produced or only be consumed) and returns the corresponding indices for the metabolites in the model.mets field. Setting removeExternalMets to true removes external metabolites from the results. Not all gaps can be identified by simply inspecting the model.S matrix (refer **Note 4**).

For example, 108 reactions have been identified as dead ends with high confidence in *S. aureus* and are included in the mode [55]. Subsequent additions to the model will likely close some of these gaps. Thus Gap analysis is the first step towards iterative model building with refinements coming from filling these gaps.

GapFill

The function GapFill [56] can be used to bridge network gaps after their identification. GapFill can achieve this by adding metabolic reactions, transport pathways and relaxing irreversibilities of reactions. Reactions known not to be present in an organism (e.g., an incomplete TCA cycle in *M. genitalium*, *H. Pylori*, incomplete glycolysis in *F. tularensis*) although identified through the program, should be excluded as gap filling candidates. Thus it is essential to study the literature carefully for the organism whose reconstruction is being developed to interpret the GapFill results.

GapFill can be used efficiently to unblock synthesis of constituents of biomass guided by the known components in the growth medium. For example, in *M. genitalium* the authors [57] unblocked biomass production by addition of 65 reactions, for which more than 60 % were involved in metabolite transport (uptake of amino acids, folate, riboflavin, metal ions, cofactors such as CoA), the rest involving hydrolysis of dipeptides and some biotransformations.

Integrating GapFill predictions, homology search (BLASTp), and structural fold (PFAM) identification algorithms, genes encoding three subunits (MG098, MG099, and MG100) in the *M. gentalium* genome have been connected to the glutamyl-tRNA(Gln) amidotransferase protein [57].

growthExpMatch is another optimization-based algorithm that identifies the minimum number of reactions from a universal reaction database that are required for the in silico organism to grow in specified media conditions [15].

3.4 Functional Analysis of the Cell

Computing cell function or phenotype is critical to hypothesis generation and biological discovery. Objective functions in an optimization problem are used to define the exact biological function one wants to compute. Insights into mechanisms of infection and pathogenesis from a metabolic stand point that can be extended to the catalyst enzymes and encoding proteins can be obtained using COBRA methods. These methods allow computing phenotypes like growth in an environmental niche, oxygen sensitivity, energy requirements to identification of virulent genes and drug targets (Fig. 6). The following protocols discuss the set up of the computation via a mathematical function that would provide the correct biological interpretation.

3.4.1 Optimal Growth

Simulating maximal growth of the organism using FBA is one of the most fundamental calculations consistent with genome scale behavior. The growth rate on a variety of carbon sources can be predicted by fixing the uptake of nutrients and defining the composition of 1 g of biomass.

The function **optimizeCbModel(model)**, discussed in **Note 9**, in COBRA toolbox can be used. This function runs FBA on the selected model for maximization of the set objective function. The data structure solution that results contains an optimal solution where “f” gives the value of the objective function. This is essentially the unique optimal growth rate predicted in silico. The vector “x” gives a non-unique optimal flux distribution through the network. The shadow prices and reduced costs are also calculated by

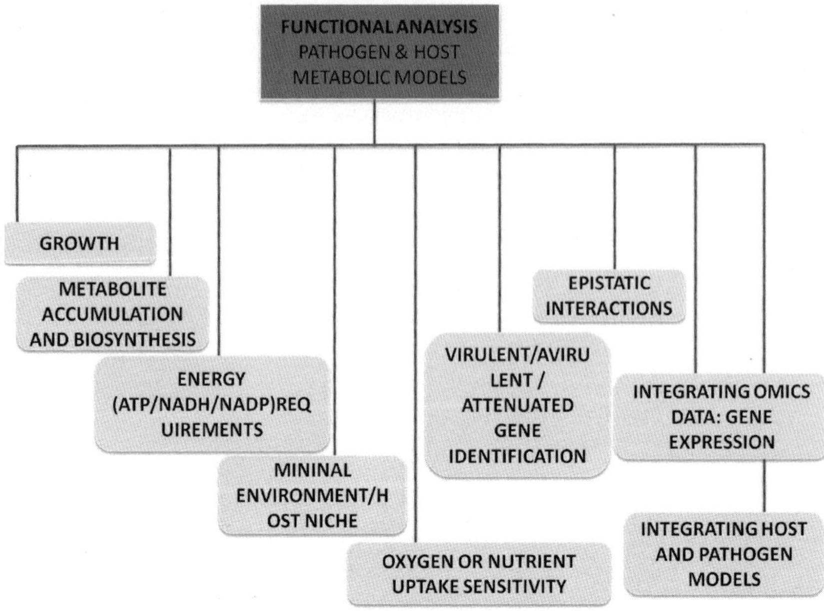

Fig. 6 Functional analysis of host–pathogen behavior and interaction. The flow chart delineates functional cell phenotypes that can be computed using COBRA methods

optimizeCbModel function. The vector of *m* shadow prices is **FBAsolution.y** and the vector of *n* reduced costs is FBAsolution.w.

The sensitivity of the FBA solution is indicated by either shadow prices or reduced costs. The sensitivity analysis of the optimal value of the objective function allows one to understand how constraints affect its value. Since constraints in FBA are defined by bounds on reaction rates or flux values, shadow prices for them determine how the cell is limited by a particular constraint and indicate how much the addition of that metabolite will increase or decrease the objective. Shadow prices are the derivative of the objective function with respect to the exchange flux of a metabolite.

The reduced costs on the other hand represent the amount by which the objective function will be reduced if the corresponding reaction is forced to carry a flux (i.e., if the gene is expressed or "turned on"). By understanding what set of reduced costs are zero, alternate equivalent flux distributions can be analyzed to get redundancies built into the cell. Reduced costs are the derivatives of the objective function with respect to an internal reaction with 0 flux, indicating how much each particular reaction affects the objective.

Sensitivity analysis of the objective functions helps interpret understanding decisions a cell makes regarding metabolic options. The reader is referred to a Linear Programming textbook to understand the concepts of optimization.

The function described above can be used for optimal growth predictions in a network model of any organism. What makes it amenable to study host–pathogen interactions is the ability to define several alternate environments the pathogen can reside in as constraints and calculate the optimal growth possible. Of course, the identification of these microenvironments depend on specific niches in the host, the stage of infection etc. and are nontrivial to define. However, multiple simulations in different environments allow us to identify not only the proliferative potential of the pathogen but also the accuracy of defining the host milieu as discussed in the protocol that follows.

3.4.2 Optimal Metabolite Biosynthesis in the Pathogen and Host

Using the same protocol as above, FBA can be used to calculate the maximum yields of cofactors like ATP, NADH, and NADPH. Proton balancing limits production of cofactors ATP, NADH, and NADPH. Sensitivity analysis on maximal ATP yield objective for *E. coli* core model shows that the shadow price of cytosolic protons (h[c]) is −0.25 [4]. What this means is that one needs to add 4 mol protons/mol glucose for the ATP yield to drop by a unit, i.e., 1 mol ATP/mol glucose. As discussed previously, the cap on the ATP production is a direct outcome of the steady-state assumption. Any increase would result in import of additional protons that have no way to leave the cell and change the intracellular pH.

3.4.3 Predicting the Host Niche (Environment) the Pathogen Encounters

This protocol generally is used to determine the minimal media required to produce biomass in silico can be extended to understand the host niche or nutrient environment the pathogen may encounter. Each potential environmental (media) component that is defined by existence of a transport mechanism in the real organism is taken out of the network and then interrogated for possibility of growth. If the network is not capable of producing biomass without taking up that specific metabolite, the metabolite is deemed "essential." Generally during this process, the uptake rates for protons and water are not constrained and the uptake rates of all other media components are set to be maximally 20 mmol/h/g (dry weight)[43]. The oxygen uptake rate is set based on aerobicity required anywhere from 0 for anaerobic to the maximum value after which biomass is not affected in silico.

Using these methods minimal media of several organisms have been designed. One example is *Helicobacter pylori* [31], where in it was predicted that amino acids are alone sufficient to form the bulk carbon requirements. Minimal media or environmental calculations also allow us to identify any major problems with the reconstructions, e.g., if growth is possible in the absence of a carbon source that is a major discrepancy unless noted so in literature. Generally agreements between the in silico-predicted and the in vitro/in vivo-determined requirements indicate validity of the network reconstruction and consistency of predicted conditions dependent in silico behavior. Further, this function can be used to delineate the minimum nutritional requirements for wild type and mutant strains of any organism [58]. For example, in the *S. aureus* reconstruction [55] discrepancies were detected in the computationally predicted, amino acid requirements for growth of the pathogen. Kuroda and colleagues report that the six amino acids are specifically required for strain N315 to grow despite presence of pathways for the synthesis of all amino acids. Flux balance models in general and specifically iSB619 do not account for regulatory effects and predictions made assume gene to be expressed at the needed levels. The authors have explored the discrepancy between experimental results and computational predictions by studying the in silico effect of adding each amino acid individually to the predicted minimal media listed (always with arginine). They found on average, providing one of amino acids noted as essential from experimental data led to more biomass production than providing one amino acid not listed as essential. A Wilcoxon rank sum test can be further used to test if the results are likely to be the result of random chance. *S. aureus* is said to require an organic source of nitrogen provided by amino acids, so the requirement of at least one amino acid is not surprising. The authors predicted that *S. aureus* can grow more efficiently by taking up certain amino acids rather than synthesizing them, even though the genome encodes that functionality [55].

Differentiating serovars of a pathogen can also be accomplished using FBA when the major difference is the ability to ferment carbon sources. For example, the ability or inability to ferment rhamnose and melibiose is typically used to classify strains of *Y. pestis*. YP CO92 is known to be lethal to humans and is characterized by the inability to ferment rhamnose as is predicted in silico. A simple simulation can also identify the reason for this outcome (potential over accumulation of L-lactaldehyde) [35]. Metabolic models can also recapitulate common amino acid auxotrophies seen in strains like the epidemic strains of *Y. pestis*.

3.4.4 Robustness Analysis

The effect of reducing flux through any reaction in the network on optimal function (growth, energy or biosynthetic capability) is very important in host–pathogen studies. Robustness analysis allows the computation of how an objective of interest (e.g., growth rate) changes as the flux through a specific reaction of interest varies in magnitude. One could systematically study the effect of varying flux through each gene on growth and virulence of the pathogen in a host environment. The interdependence of oxygen or h+ flux on multiple objectives of the cell can be predicted to understand how microenvironments affect growth, function and pathogenicity. Predicting haplo-insufficient phenotypes in eukaryotes has also been proposed by studying the effect of decreasing the expression level of a specific metabolic enzyme on the growth rate [11]. This function (*see* **Note 10**) is used to compute and plot the value of the model objective function as a function of flux values for a reaction of interest (**controlRxn**) as a means to analyze the network robustness with respect to that reaction [11].

Aerobicity: Effect of oxygen on metabolism and growth

The sensitivity of growth rate to oxygen uptake on a variety of different carbon sources can be calculated using FBA. The carbon source uptake rate must be restricted to the same molar maximum for all calculations. Under these conditions, biomass production would intuitively increase with oxygen uptake until there is no longer an oxygen limitation. Robustness has also been used to understand glutamate mediated acid resistance and growth for the live vaccine strain of *Francisella tularensis* in the chamberlain media [30].

3.4.5 Single Gene Deletion Analysis

Gene deletion phenotypes can be simulated [11, 12] by using the same **optimizeCbModel(model)** function used for calculating optimal growth. Perturbation parameter however is the removal of the gene and its corresponding reaction in the network. The gene–reaction relationships defined by Boolean rules are critical to this protocol. The upper and lower flux bounds for the reaction(s) corresponding to the deleted gene are both set to zero. The result obtained in the data structure f: now describes growth objective under the influence of the gene perturbation, i.e., deletion. To look at genome wide effects it is also possible to use the functions that performs a model-wide single gene deletion study (*see* **Note 11**).

Biological interpretation of gene deletion data has far reaching implications in understanding essentiality of genes and also potential drug targets. The three categories of results that can be obtained include (1) unchanged maximal growth, (2) reduced maximal growth, or (3) no growth. The interpretation has significant implications in understanding pathogen genetics and host–pathogen interactions. These simulations can thus differentiate virulent, avirulent and attenuated (growth of mutant in vivo at a lower rate than the wild type) genes. The first category indicates a nonlethal gene or avirulent genes while the third category result indicates a lethal gene or virulent gene function. The second category of genes would be identified attenuated genes in pathogens that could translate to vaccine strains. For example, in silico gene deletion predictions for salmonella compared to experimentally identified in vivo essential and nonessential genes [59]. And the growth rates predicted were consistent with physiologically identified virulent, avirulent, and attenuated strains in literature [60]. Using in silico gene deletion study, potential new drug targets can be identified with low sequence identity to human proteins. One of these antimalarial targets discovered, nicotinate mononucleotide adenylyltransferase (NMNAT), was experimentally tested in a growth inhibition assay using a recently discovered small molecule inhibitor [61].

The effect of deleting two genes simultaneously on growth can be simulated. Here, reactions bounds of all the reactions corresponding to these two genes are set to zero and the effect of that perturbation is observed in the value of the objective function. These results may be useful in explaining potential epistatic interactions. Theoretically double deletion of every gene pair in the genome-scale network is a simple task, although experimental validation is quite daunting. However, identifying microenvironments to probe the effects in silico of these gene perturbations allows one to narrow down the experimental targets by several orders of magnitude and prioritize the most relevant epistatic interactions. These interacting pairs suggest combination targets in treatment strategies and can be classified as condition-dependent or independent.

Pairwise double gene deletions can be performed using the function **doubleGeneDeletion**. This function calculates the growth rates and relative growth ratios for every two-gene combination in the model. Relative growth rate data can be used to identify epistatic (synthetic lethal or synthetic sick) interactions between genes in the model. A synthetic sick interaction is one in which the growth rate ratios of the double deletion and each single gene deletion are less than 0.01 [12]. The function in COBRA is **findEpistaticInteractions(model,grRatio)**.

3.4.6 Flux Variability Analysis (FVA)

The metabolic capacity of an organism, and hence its robustness, is determined by all alternate routes it can use to achieve an objective [30, 59]. FVA is a derivative of FBA allows one to examine the redundancies by calculating the full range of numerical values for each reaction flux in a network [62]. In silico prediction of metabolic pathways utilized during infection allows us to identify redundant metabolism that the pathogen could exploit in order to survive and replicate. This is carried out by optimizing for a particular objective, while still satisfying the given constraints set on the system. One can thus also determine the minimum and maximum flux value that each reaction in the model can take up while satisfying all constraints on the system for a specific objective (refer **Note 12**).

For example, for Salmonella given the possible nutrients provided in the host-cell environment, FVA identified a reactome of 417 reactions when biomass production is optimal. Flux variability analysis (FVA) was used to identify such metabolic reactions that might be operational in different environments including that during infection. This allows us to identify the host metabolism that *Francisella* could exploit in order to survive and replicate. By setting the **optPercentage** variable (refer **Note 12**) one can further investigate flux variability at suboptimal values of the objective function [30, 59].

3.4.7 Metabolite Essentiality for Drug Design

The concept of metabolite essentiality has been introduced recently for studying the cellular robustness and to complement the classical reaction (gene)-centric approach [63, 64]. Metabolite essentiality analysis predicts whether the removal of a metabolite from the cell causes death. Structural analogs of metabolites that are essential can be potential drug candidates [64]. Simulations involve deleting all reactions that either consume or produce the metabolite and calculating optimal growth [63]. All metabolites that result in no cell growth are deemed essential to the organism.

Further use of this data to identify potential drug candidates is possible using the EMFilter framework [65]. This EMFilter is a four step procedure that eliminates metabolites from the essential list by removing (1) evolutionarily conserved currency metabolites with high connectivity, e.g., ATP and NADH. (2) metabolites that form nodes with triconnectivity (more than three reactions), and an out degree of two (metabolite-consuming), (3) metabolites that are common to human metabolism, and (4) metabolites whose connected enzymes possess homologs in humans. The metabolites that remain are potential candidates for designing drug candidates.

3.4.8 Building Tissue Models Using OMICS Data

A lot of methods and protocols focus on building bacterial pathogen and host models. The human genome sequence annotation provided the appropriate foundation for human metabolic reconstructions and have resulted in four genome-scale reconstructions, the HumanCyc [66], the Edinburgh Human Metabolic Network [67, 68], Recon 1[69], and Recon2[29].

The genome-scale human network models can serve as host for most pathogens. Use of generic metabolism models of hosts develops understanding of mechanisms of either the pathogen or the host a whole. However, it is also necessary to build models for individual organelles as well to understand the different stages of infection. For example interactions between macrophages and the pathogen can be understood better if the host model is tailored to represent macrophage phenotypes rather than just overall metabolism. Several OMICS data types can be used for this purpose [52, 70, 71]. An example is the incorporation of gene expression data [72] to develop a human alveolar macrophage model [53]. The GIMME [73] algorithm retains those present in the high-throughput data and orphan reactions in Recon 1. The reactions with no detected expression are minimized and those not required to retain flux through the objective reaction are removed. Recon1 is the first version of the genome-scale metabolic network for *Homo sapiens* [69]. As an example, Bordbar and colleagues have developed a macrophage model using Recon1 [53]. The function **createTissueModel** in the COBRA toolbox can be used to build these tissue specific models. The details of this protocol for iAB-AMØ-1410 (refer **Note 15**) delineates the development of a host–pathogen interaction model in the context of alveolar macrophage and mTB.

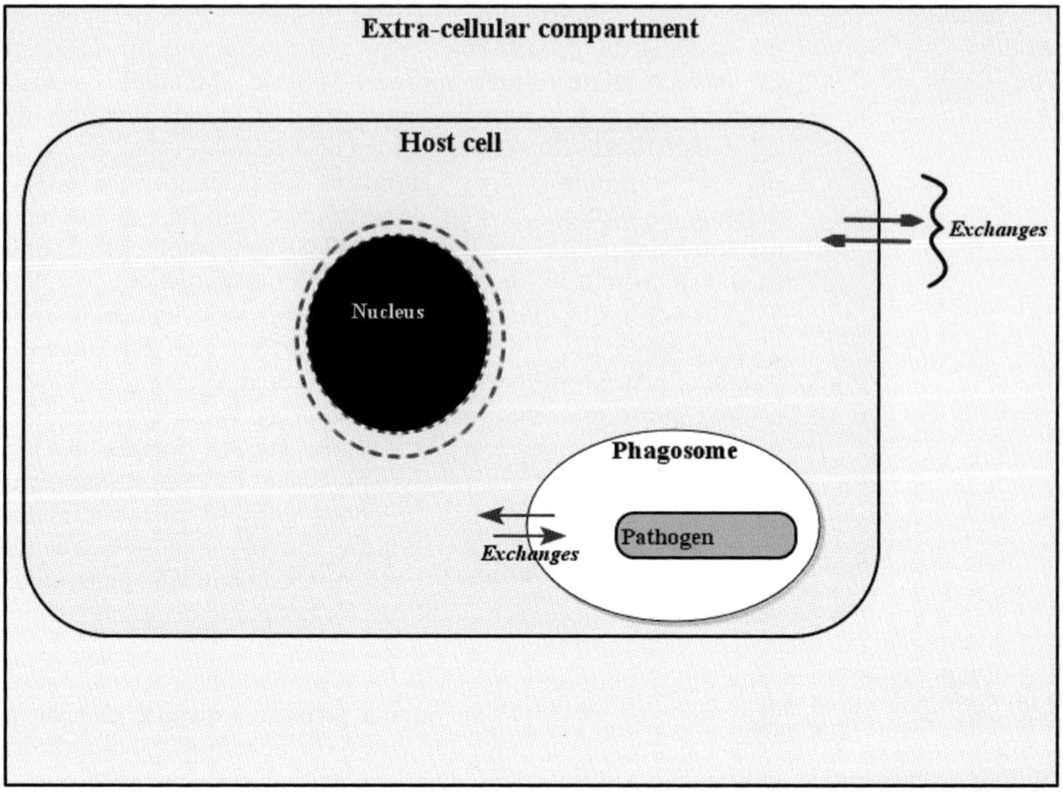

Fig. 7 Schematic of compartmentalization in host–pathogen interaction model. Re-compartmentalization process while integrating a host and a pathogen model is critical to understanding the infection process. In the given *Human macrophage–M. tuberculosis* example, the pathogen engulfed is within a phagosome in the macrophage cell. The phagosome environment mainly consists of lipids, and thus, the metabolite pool available within the phagosome is different from the host cytosolic metabolite pool. Hence, an additional phagosome compartment is added to the model

3.5 Host–Pathogen Interaction Model

The integration of the refined host and pathogen models is critical in developing a host–pathogen interaction model. Since most metabolic pathways are evolutionarily conserved across species, both the individual models have several metabolites and reactions in common. It is essential to rename metabolites and reactions to delineate between organisms while integrating the host and pathogen models [53]. As pathogens reside in a phagosome, the models need to be re-compartmentalized and phagosome environment and necessary transport across the macrophage cytoplasm needs to be delineated (Fig. 7). As an example, the macrophage model is developed here with the latest version of the genome-scale human metabolic network and is then integrated with iNJ661 as discussed (**Notes 14–16**).

4 Conclusions

This chapter discussed the many protocols developed for the constraints-based flux balance modeling of genome-scale metabolic networks that can be used to probe host and pathogen behavior and their interaction. The philosophy of this systems approach as applied to genome-scale metabolic networks results in three important insights. First, accurate in silico organism-scale models can now be reconstructed that help drive experiments and generate novel hypotheses. Reconciliation of failure modes of these models allows iterative model-building. Second, cell function and biological phenotype is a result of the many hard constraints (physicochemical) that organisms operate under. Third, availability of impressive computational toolboxes driving in silico interrogation of cell behavior makes analysis of complex interacting systems achievable in the near future.

5 Notes

This section outlines the actual codes and results for using functions in the COBRA toolbox for reconstruction, simulation and analysis of metabolic models. The corresponding subsection in the main text that references these notes is listed here. These are helpful to the beginner and the already initiated user alike and will help in connecting mathematical simulations to biological phenotypes. The thought process behind formulating the mathematical problem for the biological question to be answered is the holy grail of modeling host–pathogen behavior and their interactions for novel hypothesis generation and biological discovery.

1. *Basic functions for metabolic network reconstruction and analysis*:
   Before using any COBRA functions, check whether COBRA toolbox is working by using the command:
   ```
   >>initCobraToolbox
   ```
   To change the cobra solver:
   ```
   >>changeCobraSolver('tomlab_cplex', 'LP')
   ```
2. *Toy network of Mycobacterium tuberculosis TCA (citric acid cycle): Network reconstruction and implementation.*
   The reader is referred to the TCA cycle of *M. tuberculosis* (Fig. 3) to understand the reactions involved (Fig. 3b) and the GPR Boolean logic (Fig. 3c) before proceeding to the reconstruction of the network. This network essentially consists of ten steps of the TCA cycle and a glyoxylate bypass. The reactions have been adopted from the iNJ661 *M. tuberculosis* model.

3. *To build a model in COBRA usable format*:
   The network and stoichiometric model (referred as **my TCAmodel** below) can be built from scratch using the **createModel** function. Reaction abbreviations, protein and gene names, metabolites and equations for all reactions involved in TCA need to be assembled prior to creating then network in a text or excel file.

```
>>rxnAbrList={'ACONT';'CS'; 'FUM'; 'ICDHy'; 'ICL';
'MALS'; 'MDH'; 'OXGDC'; 'SSALx'; 'SUCD1i'};
```

```
>>rxnNameList={'aconitase'; 'citrate synthase';
'fumarase'; 'isocitrate dehydrogenase (NADP)';
'Isocitrate lyase'; 'malate synthase'; 'malate
dehydrogenase'; ' 2-oxoglutarate decarboxylase';
'succinate-semialdehyde dehydrogenase (NAD)';
'succinate dehydrogenase'};
```

```
>rxnList={'cit[c]<=>icit[c] ';
'accoa[c]+h2o[c]+oaa[c] ->cit[c]+coa[c]+h[c] ';
'h2o[c]+fum[c]<=>mal-L[c] ';
'icit[c]+nadp[c] ->akg[c]+co2[c]+nadph[c] ';
'icit[c] ->glx[c]+succ[c] ';
'accoa[c]+h2o[c]+glx[c] ->coa[c]+h[c]+mal-L[c] ';
'mal-L[c]+nad[c]<=>oaa[c]+h[c]+nadh[c] ';
'h[c]+akg[c] ->co2[c]+sucsal[c] ';
'h2o[c]+nad[c]+sucsal[c]->2h[c]+succ[c]+nadh[c]';
'succ[c]+fad[c] ->fum[c]+fadh2[c] '}
```

```
>>myTCAmodel=createModel(rxnAbrList,rxnName
List,rxnList);
```

   The function **createModel** automatically creates the S matrix from the network reconstruction based on the information derived from the elementally balanced equations that define the reaction.

   The directionality of the reactions in the network decide the value of the upper bound value (10^6) and the lower bound (0 for irreversible and -10^6 for irreversible).

   The network made using the above reactions is incomplete, as the system is isolated in the cytosolic compartment, denoted by "[c]" in the model. The TCA cycle cannot function in isolation in the pathogen. The genome-scale *Mycobacterium tuberculosis* metabolic model, iNJ661 (for detailed description, refer **Note 6**) is used, to describe the COBRA toolbox utilities. This toy TCA network is used only to describe structural analysis functions (Gap analysis).

   The toy network can also be imported from a COBRA acceptable format excel sheet (discussed in detail in **Note 6**).

```
>>myTCAmodel=xls2model('modelFile')
```

```
allGaps =                downstreamGaps =            rootGaps =

    'cit[c]'                 'cit[c]'                    'accoa[c]'
    'icit[c]'                'icit[c]'                   'nadp[c]'
    'accoa[c]'               'h2o[c]'                    'fad[c]'
    'h2o[c]'                 'oaa[c]'
    'oaa[c]'                 'coa[c]'
    'coa[c]'                 'h[c]'
    'h[c]'                   'fum[c]'
    'fum[c]'                 'mal-L[c]'
    'mal-L[c]'               'akg[c]'
    'nadp[c]'                'co2[c]'
    'akg[c]'                 'nadph[c]'
    'co2[c]'                 'glx[c]'
    'nadph[c]'               'succ[c]'
    'glx[c]'                 'nad[c]'
    'succ[c]'                'nadh[c]'
    'nad[c]'                 'sucsal[c]'
    'nadh[c]'                'fadh2[c]'
    'sucsal[c]'
    'fad[c]'
    'fadh2[c]'
```

Fig. 8 Gapfind Simulation Output For A Network. GapFind simulation output lists all the gaps (blocked metabolites) including rootgaps and downstream gaps for the TCA cycle network discussed

4. *Identification of gaps and dead-ends in the network*: **gapFind** function identifies all blocked metabolites, downstream of a gap in a model. MILP algorithm is required to solve this function. Accordingly, the COBRA solver must be changed.

```
>>changeCobraSolver('tomlab_cplex', 'MILP')
```

It is preferable to change the reaction bounds (rxnbounds) on exchange reactions appropriately to allow uptake of every metabolite, to find every gap in a model.

```
>>[allGaps,rootGaps,downstreamGaps]=gapFind
(myTCAmodel);
```

The output of the function gives three arrays described below:

- **allGaps**: all blocked metabolites (metabolites with no in-flux).
- **rootGaps**: all root no production (and consumption) gaps.
- **downstreamGaps**: array of all downstream gaps.

In this example, **gapFind** predicts all 20 metabolites in the TCA cycle as blocked metabolites or gaps (Fig. 8).

This is expected and true as no metabolite transport reactions are provided to the system. In addition to the transport reactions that are biologically relevant for symport or antiport and exchange reactions must be added to the model. This concept is further discussed in detail (*see* **Note 6**), while explaining how cell growth medium is provided in the model. Transport and exchange reactions can either be added for all the metabolites in the network based on legacy data.

"Deadend" metabolites participate in only one reaction or are either only produced or only consumed. Deadends can be identified using the **detectDeadEnds** function. This function checks if the *S matrix* values are all −1 or all +1 and also check if the lower bound is zero for each metabolite.

```
>>deadends=detectDeadEnds(myTCAmodel);
```

The function outputs indices of dead end metabolites. In the TCA network;

```
accoa[c],coa[c],nadp[c],co2[c],nadph[c],fad[c]
,fadh2[c]; are identified as dead ends.
```

As discussed in the reaction content curation sub-protocol, all reactions need to be elementally and charge balanced. The function **checkMassChargeBalance** checks if reactions are mass-balanced. The function compares sum of elements on left side with the sums of elements on the right hand side of each reaction.

```
>>[massImb,imBaMass,imBaCharge,imbaBool,elements]=checkMassChargeBalance(myTCAmodel);
```

The output:

- **massImb**: Gives an *ExN* sparse matrix with mass imbalance, if the reaction is elementally balanced, value is zero.
- **imBaMass**: Cell array of size *N* with mass imbalance. e.g.: −1H represents 1 hydrogen missing in the reaction.
- **imBaCharge**: Array of size *N* with charge imbalance for each reaction.
- **imbaBool**: Boolean vector indicating imbalanced reactions.

The function checks for the elements "H," "C," "O," "P," "S," "N," "Mg," "X," "Fe," "Zn," "Co," "R"; in each reaction for imbalances.

5. *Adding transport reactions and exchanges to the network*: Transport reactions represent the exchange of metabolites between any two compartments. In the case of prokaryotes, exchange between cytosol and extracellular compartment is required. In the model it is represented as the cytosolic metabolite going out to the extracellular environment and becoming an extracellular metabolite and vice versa (i.e., [c]⟺[e]). Transport can be by diffusion, where nutrients like H_2O or O_2 are freely exchanged or via antiport or symport of an ion like H^+ along with the metabolite. Examples for free diffusion of water, oxygen, and carbon dioxide are shown below.

```
>>myTCAmodel=addReaction(myTCAmodel,'Tr_
h2o(c)','h2o[c]<=>h2o[e]');

>>myTCAmodel=addReaction(myTCAmodel,'Tr_
co2(c)','co2[c]<=>co2[e]');

>>myTCAmodel=addReaction(myTCAmodel,'Tr_
o2(c)','h[c]<=>o2[e]');
```

In order to complete the TCA cycle network, transport reactions are necessary for all intermediates in the pathway, but the equation has to be based on literature data.

Metabolite uptake and secretion to and from the environment is defined by addition of exchange reactions in the stoichiometric matrix. Exchange reactions represent the flow of metabolites across the boundary of the cell. Also exchange reactions, associated with media components, can be assigned reaction bounds accordingly to measured uptake rates; for example, glucose exchange fixed at a value of −1, would mean uptake of glucose from the media into the cell at a rate of 1 mmol/gDW/h.

Exchange reactions need to be added for metabolites that can exchange across the system boundary. For example to add these reactions for cofactors nad[e], nadh[e], fad[e], fadh2[e], nadp[e], nadph[e] an easy function **addExchangeRxn** is used.

```
>>myTCAmodel=addExchangeRxn(myTCAmodel,{'na
d[e]','nadh[e]','fad[e]','fadh2[e]','nadp[e]',
'nadph[e]'});
```

The reaction bounds can be adjusted to

(a) input certain metabolites (media) into the cell at a particular rate

```
>>myTCAmodel=changeRxnBounds(myTCAmodel,'Ex_
glc[e]',-1,'b');
```

(b) force certain metabolites only into the cell at varying rates

```
>>myTCAmodel=changeRxnBounds(myTCAmodel,'Ex_
glc[e]',-1,'l');
```

```
>>myTCAmodel=changeRxnBounds(myTCAmodel,'Ex_
glc[e]',0,'u');
```

(c) force certain metabolites (fermentation product) to be only secreted out of the cell

```
>>myTCAmodel=changeRxnBounds(myTCAmodel,'Ex_
lac-D[e]',0,'l');
```

```
>>myTCAmodel=changeRxnBounds(myTCAmodel,'Ex_
lac-D[e]',10,'u');
```

(d) allow a metabolite to freely go in and out of the cell

```
>>myTCAmodel=changeRxnBounds(myTCAmodel,'Ex_
h2o [e]',-1000,'l');
```

```
>>myTCAmodel=changeRxnBounds(myTCAmodel,'Ex_
h2o[e]',1000,'u');
```

6. *Flux balance and constraints-based analysis of the genome-scale Mycobacterium tuberculosis metabolic model, Inj661*:
The TCA cycle does not function in isolation inside the cell. Thus, it is not fruitful to convert it into a complete model that can be subject to meaningful simulations and analyses using COBRA methods. Thus, the genome-scale *Mycobacterium*

tuberculosis metabolic model, iNJ661 [41] is used in this note to demonstrate the use of COBRA toolbox; the SBML format of the model is provided in the BIGG database (http://bigg.ucsd.edu/). The biomass reaction is provided in the model itself [41].

The following notes include details on how to import an already reconstructed network model in SBML format, run FBA and constraints-based analysis simulations.

7. *Importing the reconstruction file into COBRA toolbox*:
To import a constraints-based model from SBML format file into MATLAB, the following COBRA toolbox function is used:

```
>>myModel=readCbModel
```

This function opens a prompt box, which asks the user to select the model file to be imported.

One can also import model from excel file:

```
>>myModel=xls2model('modelFile')
```

COBRA requires a specific format for 'modelFile'. It must consist of two tabs, '*reactions*' and '*metabolites*'.
'*reactions*' tab format:

```
col 1    Abbreviation     HEX1
col 2    Name  Hexokinase
col 3    Reaction    1 atp[c]+1 glc-D[c] -->1
         adp[c]+1 g6p[c]+1 h[c]
col 4    GPR   b0001
col 5    Genes b0001 (optional: column can be
         empty)
col 6    Protein AlaS (optional:column can be
         empty)
col 7    Subsystem  Glycolysis
col 8    Reversible  0 (irreversible)
col 9    Lower bound 0 (irreversible reactions
         have lower bound 0)
col 10   Upper bound 1000,
col 11   Objective   0 (optional: column can
         be empty)
col 12   Confidence Score  0,1,2,3,4
col 13   EC. Number  1.1.1.1
col 14   Notes N/A  (optional:  column  can  be
         empty)
col 15   References  PMID: 1111111 (optional:
         column can be empty)
```

'metabolites' tab format:

```
col 1  Abbreviation
col 2  Name
col 3  Formula (neutral)
```

```
col 4  Formula (charged)
col 5  Charge
col 6  Compartment
col 7  KEGG ID
col 8  PubChem ID
col 9  ChEBI ID
col 10 InChI string
col 11 Smiles
```

'***metabolites***' tab must consist of complete list of metabolites occurring in all compartments, i.e., the same metabolite appearing in multiple compartments must be represented distinctly as two metabolites. For example, ATP occurring in mitochondria and in the cytosolic compartments must be represented in two separate rows as 'atp[m]' and 'atp[c]', respectively. Abbreviations for metabolites must be the same as used in the Reactions.

8. *Defining the in silico media or environmental constraints file*: To represent the environment or the media in silico, the exchange reactions must be set accordingly. This is also necessary for combining experimentally derived data into the model. This process allows a more accurate representation of the real system. The default flux unit used is *mmol* $gDW^{-1} h^{-1}$ (*millimoles per gram dry cell weight per hour*). The exchange bounds can be changed for various exchange reactions. In the following example glucose exchange is set to an uptake of *1 mmol* $gDW^{-1} h^{-1}$

```
>>myModel = changeRxnBounds(model,'EX_
glc(e)', -1,'l');
>>myModel = changeRxnBounds(model,'EX_
glc(e)', -1,'u');
```

In certain cases, one may need provide sink or demand reactions into the model, to compensate for production of certain compounds (lipopolysaccharides, toxins, etc.) (refer Subheading 3.1.13).

To add sink or demand reactions, the following functions are used:

```
>>myModel=addSinkRxn(myModel, {metabolite
List});
>>myModel=addDmdReaction(myModel, {metabolite
List});
```

Sink and demand reactions also help in debugging the model. These reactions are critical to integrating two models (**Note 16** discusses the host–pathogen interaction model). It is recommended to simply write a list of **changeRxnBounds** function for all required exchanges into a Matlab script file that can be used numerous times for any simulations. Or save the script as a function, which can be used for various models (*see* Fig. 9).

```
EDITOR    PUBLISH    VIEW

function model = minimalMedia(model)

% Function provides minimal media to the cell model
% Changes the bounds of exchange reactions

% Priyanka Sadhukhan May 14 2013

model = changeRxnBounds(model,'EX_acac(e)',-1,'l');
model = changeRxnBounds(model,'EX_acetone(e)',-1,'l');
model = changeRxnBounds(model,'EX_ala_L(e)',-1,'l');
model = changeRxnBounds(model,'EX_arg_L(e)',-1,'l');
model = changeRxnBounds(model,'EX_asn_L(e)',-1,'l');
model = changeRxnBounds(model,'EX_asp_L(e)',-1,'l');
model = changeRxnBounds(model,'EX_bhb(e)',-1,'l');
model = changeRxnBounds(model,'EX_cys_L(e)',-1,'l');
model = changeRxnBounds(model,'EX_glc(e)',-1,'l');
model = changeRxnBounds(model,'EX_gln_L(e)',-1,'l');
model = changeRxnBounds(model,'EX_glu_L(e)',-1,'l');
model = changeRxnBounds(model,'EX_gly(e)',-1,'l');
model = changeRxnBounds(model,'EX_hdca(e)',-1,'l');
model = changeRxnBounds(model,'EX_his_L(e)',-1,'l');
model = changeRxnBounds(model,'EX_ile_L(e)',-1,'l');
model = changeRxnBounds(model,'EX_leu_L(e)',-1,'l');
model = changeRxnBounds(model,'EX_lnlc(e)',-1,'l');
model = changeRxnBounds(model,'EX_lnlnca(e)',-1,'l');
model = changeRxnBounds(model,'EX_lys_L(e)',-1,'l');
model = changeRxnBounds(model,'EX_met_L(e)',-1,'l');
model = changeRxnBounds(model,'EX_phe_L(e)',-1,'l');
model = changeRxnBounds(model,'EX_pro_L(e)',-1,'l');
model = changeRxnBounds(model,'EX_ser_L(e)',-1,'l');
model = changeRxnBounds(model,'EX_thr_L(e)',-1,'l');
model = changeRxnBounds(model,'EX_trp_L(e)',-1,'l');
model = changeRxnBounds(model,'EX_tyr_L(e)',-1,'l');
model = changeRxnBounds(model,'EX_val_L(e)',-1,'l');
```

Fig. 9 Constraint File defining in silico media or environmental niche. The in silico media or environment for the cell in the simulation is defined by changing the bounds of the exchanges

9. *Setting up the FBA problem*
   Choice of objective functions is at the discretion of the modeler. Typically, biomass growth rate, fermentation product production rate, cofactor yield for respiration (NAD/NADH balance) or energy (ATP/ADP balance) can be used. Linear combination of metabolites can also be aggregate objectives; minimization of sum of fluxes is a mathematical representation of optimizing resources for a cell and is also a frequently used objective function. In this example the biomass growth rate is set as the objective function (Fig. 10).

   The function to define the objective reaction is

   ```
   >>myModel = changeObjective(model, rxnName, objective Coeff).
   ```

   rxnName is a cell array consisting reaction(s) to be set as objective, and **objectiveCoeff** is the value of objective coefficient for each reaction.

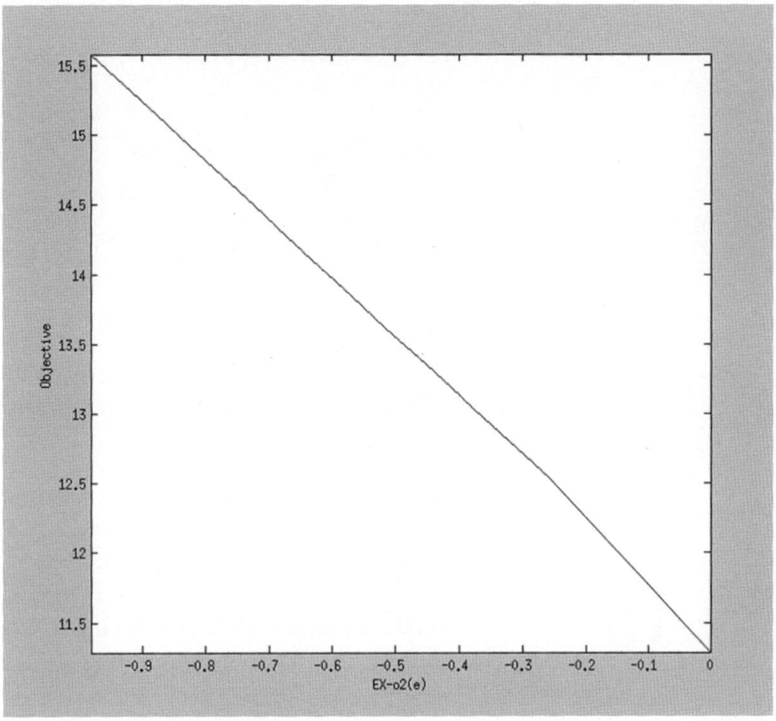

Fig. 10 Graphical representation of Robustness Analysis. The Robustness diagram shows the sensitivity of cell growth rate to oxygen availability

To set biomass as objective in the iNJ661 model:

```
>>myModel=changeObjective(iNJ661, 'biomass_
Mtb_9_60atp', 1)
```

optimizeCbModel solves the FBA problem

This function solves LP problems of the form:

max/min c'*v

$$\{\text{subject to } S \cdot v = b\}$$

$$lb \le v \le ub$$

```
>>Solution=optimizeCbModel(myModel)
```

This output of the function is a matlab structure with six fields given below:

- f: Objective value
- x: Primal, i.e., flux through each reaction
- y: Dual
- w: Reduced costs
- s: Slacks
- stat: Solver status in standardized form (1 = Optimal solution, 2 = Unbounded solution, 0 = Infeasible -1 = No solution reported).

```
Solution =

            x: [1027x1 double]
            f: 15.5822
            y: [826x1 double]
            w: [1027x1 double]
         stat: 1
     origStat: 1
       solver: 'tomlab_cplex'
         time: 0.0179
```

Fig. 11 Flux Balance Analysis (FBA) Solution Output. The Solver output for a FBA simulation summarizes the value of the objective function and other sensitivity parameters of the optimization like reduced costs and shadow prices

The time taken to run the optimization is also given in the output. The output is described in depth in Subheading 3.3.1. Similarly, to set the FBA problem to maximize respiration using balanced NAD/NADH production as objective, an artificial reaction, can be added using the following function:

```
>>myModel=addReaction(myModel,'nadRxn','nad
h[c]->nad[c]+h[c]');
```

The demand reaction added, '**nadRxn**' is set as objective for FBA

```
>>myModel=changeObjective(myModel, 'nadRxn', 1)
>>Solution=optimizeCbModel(myModel)
```

A result is shown in Fig. 11:

Similarly, ATP production can also be predicted by adding an artificial reaction that couples ATP and ADP in the cell.

10. *Robustness analysis*:
Robustness analysis can be performed for a control reaction against any objective. The following example shows the sensitivity of growth rate to oxygen uptake rate 'EX_o2(e)'.

```
>>myModel = changeObjective(myModel,
'biomass_Mtb_9_60atp');
>[controlFlux,objFlux]=robustnessAnalysis
(myModel,'EX_o2(e)',20);
```

A plot of growth rate (objective function) on the ordinate and oxygen uptake (control reaction) on the abscissa axes is automatically generated, the number of points on the plot is being user defined (the above example produces 20 points). A snapshot of the plot produced is given (Fig. 10).

11. *Gene deletion analysis*:
Gene Deletion can be simulated using the **singleGeneDeletion** and **doubleGeneDeletion** functions in the COBRA toolbox.

```
>>[grRatio,grRateKO,grRateWT,hasEffect,delR
xns]=singleGeneDeletion(myModel);
```

The function iteratively blocks the reaction(s) affected by each gene (depending on the Boolean relationships) and runs an FBA to check for viability. This function is useful for identifying gene essentiality.
The output is explained below:

- **grRateWT**: growth rate of the Wild Type (1/h).
- **grRateKO**: respective growth rates of the Knock outs (1/h).
- **hasEffect**: Indicates the effect of corresponding gene deletion {1 = reactions removes as result of deletion; 0 = no effect}.
- **delRxns**: Lists the reactions affected by the respective gene deletion.
- **grRatio**: Computed growth rate ratio between deletion strain and wild type {grRateKO/grRateWT}.

The simulations predict that genes 'Rv1436', 'Rv1437', 'Rv1438' are most essential due to their low grRatio.

```
>>[grRatioDble,grRateKO,grRateWT]=doubleGene
Deletion(myModel);
```

The function by default computes deletion analysis for each gene in the model, which takes a long time to compute. Depending on the goal of the study a list of genes to be deleted can also be provided by the user which also reduces the computation time required.

```
>>[grRatioDble,grRateKO,grRateWT]=doubleGene
Deletion(myModel, list1,list2);
```

The output for *doubleGeneDeletion* is similar to *singleGene Deletion* function:

- **grRateWT**: growth rate of the Wild Type (1/h).
- **grRateKO**: respective growth rates of the Knock outs (1/h).
- **grRatio**: Computed growth rate ratio between deletion strain and wild type {grRateKO/grRateWT}.

12. *To set up an FVA*:
FVA can be set up in COBRA toolbox using the function:

```
>>myModel=changeObjective(myModel, 'biomass_
Mtb_9_60atp');
[minFlux, maxFlux]=fluxVariability(myModel)
```

The function can also perform FVA by considering solutions that give a certain percentage of the optimal solution.

```
>>[minFlux,maxFlux]=fluxVariability(myModel, 50)
```

The value 50 defines the flux variability at half the value of the optimal solution.

13. *Saving and exporting the model*:
Any modified model with exchanges and additional constraints can be saved and exported in different formats.

The following command line functions in COBRA toolbox allow saving the file in sbml, mat, excel, and text formats.

(a) SBML: A model can be imported from SBML format to MATLAB structure, using COBRA toolbox. Model structures can also be exported as SBML file, using the following function:

```
>>writeCbModel(model, 'sbml')
```

(b) mat file: The standard COBRA model structure in MATLAB saved as mat file.

```
>>writeCbModel(model, 'mat')
```

(c) Excel sheet: Model can be saved in COBRA acceptable format excel sheet.

```
>>writeCbModel(model, 'xls')
```

(d) Text file: Similarly, the model can be saved in text file, with tab separated columns.

```
>>writeCbModel(model, 'text')
```

14. *Building host–pathogen interaction models*:
Coupling individual host and pathogen metabolic models, helps better understand host–pathogen interaction and pathogenesis. This example builds a combined model of the human alveolar macrophage and *M. tuberculosis*, using the Recon 2.02 (refer Subheading 3.4.8), the latest version of the human metabolic network reconstruction [29] and the iNJ661 model discussed previously (Bordbar et al. [53]) have shown the use of this protocol for converting Recon1 into an alveolar macrophage. This method delineates the use of ReconX, a general metabolic reconstruction, and gene expression data for building tissue specific models.

15. *Building the macrophage model*:
The following function uses the Recon 2.02 model as a template to build the alveolar macrophage model using gene expression data.

```
>>[macrophageModel,Rxns]=createTissueSpecific
Model(recon, GE);
```

The command uses the GIMME [11] and Shlomi [54] algorithms to build tissue specific models. The algorithm desired to be used can be specified in the solver while calling the function. The input to the function, namely, '**recon**' is the Recon 2.02 model. It can be downloaded in the toolbox acceptable format from ReconX (refer Subheading 2). The other input "GE" is a matlab structure consisting of two arrays *.Locus* and *.Data*, that represent a given transcriptome data set. *.Locus* is an array containing the gene identifier and *.Data* can have two values (1 = presence calls) and (0 = absence calls). In this example, only human macrophage gene expression data from ref. 74 was retrieved from the GEO. Transcriptome data of various stages of disease compared with undiseased controls can be used for delineating host–pathogen interactions better [53]. Information on exchanges required to define exchange of metabolites between pathogen and host requires additional literature survey and is critical to interpretation of simulation data in the context of host–pathogen interaction.

16. *Building the human macrophage–M. Tuberculosis integrated model*:
After building and curating the macrophage model, the iNJ661 pathogen model is integrated. For this, each species and reaction in the models must be represented uniquely, e.g., 'adp[c]' represents cytosolic ADP in both the models. Hence, the compartmentalization would be erroneous. The Recon 2 model has six compartments including the extracelluar compartment '[e]' and the iNJ661 has two compartments'[e]&[c]'. In addition to this a “phagosome” compartment, with appropriate exchanges between macrophage cytosol and phagosome, must be added (*see* Fig. 7). The new reaction names and metabolites names can be reassigned to the existing arrays.

```
>>myModel.mets(i)='newMetaboliteName';{where 'i'
is the array index}
```

Another way to change the names and abbreviation is by exporting the models into excel or text format and renaming the compartments, using an editor. One can add a short string to the existing names to differentiate between various compartments. For instance, 'adp[c]' can be reassigned as 'adp[H_c]' and 'adp[P_c]' the H referring to the host and the P referring to the pathogen models respectively. The reaction formula is updated automatically, on changing the *myModel.mets*. The two models can be integrated by using the aforementioned '**addReaction**' function.

```
>>myModel.mets={ };{where 'myModel' is the pathogen
model}
>>myModel.rxns={ };
>>myRxns=myModel.rxns
>>rxnFormulas=printRxnFormula(myModel, myRxns);
{array of reaction formulas}
>>intModel=macrophageModel; {to build an integrated
model}
>>for i=1:numel(myRxns)
intModel=addReaction(intModel, myRxns{i}, rxns
Formulas{i});
end
```

In addition, the exchanges for iNJ661 must be specified accordingly to represent exchange between the pathogen and the phagosome within the macrophage. Sink reactions may have to be added for species absent in the host model. Transcriptome data can also be incorporated in the iNJ661 (pathogen) model, for a specific condition-tailored model that could be useful for certain experiments (refer [53] for detailed study).

References

1. Noble D (2008) Genes and causation. Philos Trans R Soc A 366:3001–3015
2. Fang K, Zhao H, Sun C, Lam CMC et al (2011) Exploring the metabolic network of the epidemic pathogen Burkholderia cenocepacia J2315 via genome-scale reconstruction. BMC Syst Biol 5:83
3. Price ND, Reed JL, Palsson BØ (2004) Genome-scale models of microbial cells: evaluating the consequences of constraints. Nat Rev Microbiol 2(11):886–897
4. Jeffrey DO, Thiele I, Palsson BØ (2010) What is flux balance analysis? Nat Biotechnol 28(3): 245–248
5. Edwards J, Palsson BØ (2000) Metabolic flux balance analysis and the in silico analysis of Escherichia coli K-12 gene deletions. BMC Bioinformatics 1:1
6. Kauffman KJ, Prakash P, Edwards JS (2003) Advances in flux balance analysis. Curr Opin Biotechnol 14:491–496
7. Mazumdar V, Snitkin ES, Amar S et al (2009) Metabolic network model of a human oral pathogen. J Bacteriol 191(1):74–90
8. Famili I, Forster J, Nielsen J et al (2003) Saccharomyces cerevisiae phenotypes can be predicted by using constraint-based analysis of a genome-scale reconstructed metabolic network. Proc Natl Acad Sci U S A 100: 13134–13139
9. Ibarra RU, Edwards JS, Palsson BØ (2002) Escherichia coli K-12 undergoes adaptive evolution to achieve in silico predicted optimal growth. Nature 420:186–189
10. Segre D, Vitkup D, Church GM (2002) Analysis of optimality in natural and perturbed metabolic networks. Proc Natl Acad Sci U S A 99:15112–15117
11. Becker SA, Feist AM, Mo ML et al (2007) Quantitative prediction of cellular metabolism with constraint-based models: the COBRA Toolbox. Nat Protoc 2(3):727–738
12. Schellenberger J, Que R, Fleming RM et al (2011) Quantitative prediction of cellular metabolism with constraint-based models: the COBRA Toolbox v2.0. Nat Protoc 6:1290–1307
13. Thiele I, Jamshidi N, Fleming RM, Palsson BØ (2009) Genome-scale reconstruction of Escherichia coli's transcriptional and translational machinery: a knowledge base, its mathematical formulation, and its functional characterization. PLoS Comput Biol 5(3):e1000312
14. Papin JA, Palsson BO (2004) The JAK-STAT signaling network in the human B-cell: an extreme signaling pathway analysis. Biophys J 87:37–46
15. Schellenberger J, Park J, Conrad T et al (2010) BiGG: a biochemical genetic and genomic knowledgebase of large scale metabolic reconstructions. BMC Bioinformatics 11:213
16. Thiele I, Palsson BØ (2010) A protocol for generating a high-quality genome-scale metabolic reconstruction. Nat Protoc 5:93–121
17. Kanehisa M, Goto S, Hattori M et al (2006) From genomics to chemical genomics: new developments in KEGG. Nucleic Acids Res 1 (34(Database Issue)):D354–D357
18. Karp PD et al (2002) The EcoCyc database. Nucleic Acids Res 30:56–58
19. Devoid S, Overbeek R, DeJongh M et al (2013) Automated genome annotation and metabolic model reconstruction in the SEED and Model SEED. Methods Mol Biol 985: 17–45
20. Poolman MG (2006) ScrumPy: metabolic modelling with Python. Syst Biol 153(5): 375–378
21. Barthelmes J, Ebeling C, Chang A et al (2007) BRENDA, AMENDA and FRENDA: the enzyme information system in 2007. Nucleic Acids Res 35:D511–D514
22. Fleming RMT, Thiele I, Nasheuer HP (2009) Quantitative assignment of reaction directionality in constraint-based models of metabolism: application to Escherichia coli. Biophys Chem 145:47–56
23. Kümmel A, Panke S, Heinemann M (2006) Systematic assignment of thermodynamic constraints in metabolic network models. BMC Bioinformatics 7:1–12
24. Gardy JL et al (2005) PSORTb v. 2.0: expanded prediction of bacterial protein subcellular localization and insights gained from comparative

proteome analysis. Bioinformatics (Oxford) 21:617–623

25. Lu Z et al (2004) Predicting subcellular localization of proteins using machine-learned classifiers. Bioinformatics (Oxford) 20:547–556
26. Emanuelsson O, Brunak S, von Heijne G et al (2007) Locating proteins in the cell using TargetP, SignalP and related tools. Nat Protoc 2:953–971
27. Gevorgyan A, Poolman MG, Fell DA (2008) Detection of stoichiometric inconsistencies in biomolecular models. Bioinformatics 24(19): 2245–2251
28. Zengler K, Palsson BØ (2012) A road map for the development of community systems (CoSy) biology. Nat Rev Microbiol 10(5):366–372
29. Thiele I, Swainston N, Fleming RM et al (2013) A community-driven global reconstruction of human metabolism. Nat Biotechnol 31(5):419–425
30. Raghunathan A, Shin S, Daefler S (2010) Systems approach to investigating host–pathogen interactions in infections with the biothreat agent Francisella. Constraints-based model of Francisella tularensis (2010). BMC Syst Biol 4:118
31. Schilling CH, Covert MW, Famili I et al (2002) Genome scale metabolic model of Helicobacter pylori 26695. J Bacteriol 184(16): 4582–4593
32. Tian J, Bryk R, Itoh M et al (2005) Variant tricarboxylic acid cycle in Mycobacterium tuberculosis: identification of alpha-ketoglutarate decarboxylase. Proc Natl Acad Sci U S A 102(30): 10670–10675
33. Feist AM, Palsson BØ (2010) The biomass objective function. Curr Opin Microbiol 13(3): 344–349
34. Liao Y, Huang T, Chen F et al (2011) An experimentally validated genome-scale metabolic reconstruction of Klebsiella pneumoniae MGH 78578, iYL1228. J Bacteriol 193(7): 1710–1717
35. Charusanti P, Chauhan S, McAteer K et al (2011) An experimentally-supported genome-scale metabolic network reconstruction for Yersinia pestis CO92. BMC Syst Biol 5:163
36. Abbas CA, Card GL (1980) The relationships between growth temperature, fatty acid composition and the physical state and fluidity of membrane lipids in Yersinia enterocolitica. Biochim Biophys Acta 602:469–476
37. Feist AM, Herrgård MJ, Thiele I et al (2009) Reconstruction of biochemical networks in microorganisms. Nat Rev Microbiol 7(2): 129–143
38. Reed JL, Famili I, Thiele I et al (2006) Towards multidimensional genome annotation. Nat Rev Genet 7(6):130–141
39. Varma A, Palsson BØ (1994) Stoichiometric flux balance models quantitatively predict growth and metabolic by-product secretion in wild-type Escherichia coli W3110. Appl Environ Microbiol 60(10):3724–3731
40. Edwards JS, Palsson BØ (2000) The Escherichia coli MG1655 in silico metabolic genotype: its definition, characteristics, and capabilities. Proc Natl Acad Sci U S A 97(10): 5528–5533
41. Jamshidi N, Palsson BØ (2007) Investigating the metabolic capabilities of Mycobacterium tuberculosis H37Rv using the in silico strain iNJ661 and proposing alternative drug targets. BMC Syst Biol 1:26
42. Schilling CH, Palsson BØ (2000) Assessment of the metabolic capabilities of Haemophilus influenzae Rd through a genome-scale pathway analysis. J Theor Biol 203:249–283
43. Thiele I, Vo TD, Price ND et al (2005) Expanded metabolic reconstruction of Helicobacter pylori (iT341 GSM/GPR): an in silico genome scale characterization of single and double deletion mutants. J Bacteriol 187(16): 5818
44. Rocha I, Maia P, Evangelista P et al (2010) OptFlux: an open-source software platform for in silico metabolic engineering. BMC Syst Biol 4:45
45. Klamt S, Saez-Rodriguez J, Gilles E (2007) Structural and functional analysis of cellular networks with Cell NetAnalyzer. BMC Syst Biol 1:2
46. Olivier B, Rohwer J, Hofmeyr J (2005) Modelling cellular systems with PySCeS. Bioinformatics 21:560–561
47. Schwarz R, Liang C, Kaleta C et al (2007) Integrated network reconstruction, visualization and analysis using YANAsquare. BMC Bioinformatics 8:313
48. Kono N, Arakawa K, Tomita M (2006) MEGU: pathway mapping web-service based on KEGG and SVG. In Silico Biol 6:621–625
49. Cvijovic M, Olivares-Hernández R, Agren R et al (2010) BioMet toolbox: genome-wide analysis of metabolism. Nucleic Acids Res 38: W144–W149
50. Shannon P, Markiel A, Ozier O et al (2003) Cytoscape: a software environment for integrated models of biomolecular interaction networks. Genome Res 13:2498–2504
51. Boele J, Olivier BG, Teusink B (2012) FAME, the flux analysis and modeling environment. BMC Syst Biol 30:6–8
52. Blazier AS, Papin JA (2012) Integration of expression data in genome-scale metabolic network reconstructions. Front Physiol 3:299
53. Bordbar A, Lewis NE, Schellenberger J et al (2010) Insight into human alveolar macrophage

and M. tuberculosis interactions via metabolic reconstructions. Mol Syst Biol 6:422

54. Shlomi T, Cabili MN, Herrgard MJ et al (2008) Network-based prediction of human tissue-specific metabolism. Nat Biotechnol 26: 1003–1010
55. Becker SA, Palsson BØ (2005) Genome-scale reconstruction of the metabolic network in Staphylococcus aureus N315: an initial draft to the two-dimensional annotation. BMC Microbiol 5:8
56. Kumar SV, Dasika MS, Maranas CD (2007) Optimization based automated curation of metabolic reconstructions. BMC Bioinformatics 8:212
57. Suthers PF, Dasika MS, Kumar VS et al (2009) A genome-scale metabolic reconstruction of Mycoplasma genitalium, iPS189. PLoS Comput Biol 5(2):e1000285
58. Navid A, Almaas E (2009) Genome-scale reconstruction of the metabolic network in Yersinia pestis, strain 91001. Mol BioSyst 5: 368–375
59. Raghunathan A, Reed J, Shin S et al (2009) Constraint-based analysis of metabolic capacity of Salmonella typhimurium during host–pathogen interaction. BMC Syst Biol 3:38
60. Fields PI, Swanson RV, Haidaris CG et al (1986) Mutants of Salmonella typhimurium that cannot survive within the macro-phage are avirulent. Proc Natl Acad Sci U S A 83: 5189–5193
61. Germán P, Hsiao T, Olszewski KL et al (2010) Reconstruction and flux-balance analysis of the Plasmodium falciparum metabolic network. Mol Syst Biol 6:408
62. Mahadevan R, Schilling CH (2003) The effects of alternate optimal solutions in constraint-based genome-scale metabolic models. Metab Eng 5(4):264–276
63. Kim PJ, Lee DY, Kim TY et al (2007) Metabolite essentiality elucidates robustness of Escherichia coli metabolism. Proc Natl Acad Sci U S A 104:13638–13642
64. Dobson PD, Patel Y, Kell DB (2009) 'Metabolite-likeness' as a criterion in the design and selection of pharmaceutical drug libraries. Drug Discov Today 14:31–40
65. Kim HU, Kim TY, Lee SY (2010) Genome-scale metabolic network analysis and drug targeting of multi-drug resistant pathogen Acinetobacter baumannii AYE. Mol Biosyst 6:339–348
66. Romero P, Wagg J, Green ML et al (2005) Computational prediction of human metabolic pathways from the complete human genome. Genome Biol 6:R2
67. Hao T, Ma H, Zhao X et al (2010) Compartmentalization of the Edinburgh Human Metabolic Network. BMC Bioinformatics 11:393
68. Ma H, Sorokin A, Mazein A et al (2007) The Edinburgh human metabolic network reconstruction and its functional analysis. Mol Syst Biol 3:135
69. Duarte NC, Becker SA, Jamshidi N et al (2007) Global reconstruction of the human metabolic network based on genomic and bibliomic data. Proc Natl Acad Sci U S A 104:1777–1782
70. Raghunathan A, Price ND, Galperin MY et al (2004) In silico metabolic model and protein expression of haemophilus influenzae strain Rd KW20 in rich medium. OMICS 8(1): 25–41
71. Schmidt BJ, Ebrahim A, Metz TO et al (2013) GIMME: condition-specific models of cellular metabolism developed from metabolomics and expression data. Bioinformatics 29(22): 2900–2908
72. Kazeros A, Harvey BG, Carolan BJ et al (2008) Overexpression of apoptotic cell removal receptor MERTK in alveolar macrophages of cigarette smokers. Am J Respir Cell Mol Biol 39:747–757
73. Becker SA, Palsson BØ (2008) Context-specific metabolic networks are consistent with experiments. PLoS Comput Biol 4:e1000082
74. Thuong NT, Dunstan SJ, Chau TT et al (2008) Identification of tuberculosis susceptibility genes with human macrophage gene expression profiles. PLoS Pathog 4(12):e1000229

Chapter 30

Mathematical Models of HIV Replication and Pathogenesis

Dominik Wodarz

Abstract

This review outlines how mathematical models have been helpful, and continue to be so, for obtaining insights into the in vivo dynamics of HIV infection. The review starts with a discussion of a basic mathematical model that has been frequently used to study HIV dynamics. Some crucial results are described, including the estimation of key parameters that characterize the infection, and the generation of influential theories which argued that in vivo virus evolution is a key player in HIV pathogenesis. Subsequently, more recent concepts are reviewed that have relevance for disease progression, including the multiple infection of cells and the direct cell-to-cell transmission of the virus through the formation of virological synapses. These are important mechanisms that can influence the rate at which HIV spreads through its target cell population, which is tightly linked to the rate at which the disease progresses towards AIDS.

Key words Mathematical models, Virus dynamics, Evolution, Multiple infection of cells, Cell-to-cell transmission, Virological synapse

1 Introduction

Human immunodeficiency virus infection results in a complex disease process. Following the acute phase, characterized by relatively high viral loads, the infection enters an asymptomatic phase where viral loads are significantly reduced. The asymptomatic phase lasts on average between 5 and 10 years, but the duration is highly variable. As virus load rises and the immune system becomes progressively impaired, AIDS eventually develops, which is the end stage of the disease. HIV infects different immune cell types, including CD4+ T helper cells, dendritic cells, and macrophages. The CD4+ T cell count is an important measure for how well the immune system functions, and HIV infection depletes the CD4+ T helper cell population, eventually leading to a collapse of immune function.

Much research has been done into the mechanisms that drive HIV pathogenesis [1–4]. In this respect, mathematical models have played a crucial role in complementing experimental work in

Rajat K. De and Namrata Tomar (eds.), *Immunoinformatics*, Methods in Molecular Biology, vol. 1184,
DOI 10.1007/978-1-4939-1115-8_30, © Springer Science+Business Media New York 2014

order to improve our understanding about the mechanisms of HIV pathogenesis. This has been reviewed extensively e.g. in refs. 5–8. Importantly, the collaboration between mathematical modelers and experimentalists lead to a detailed quantification of viral and immunological parameters. In addition, various types of mathematical models have been used to explore mechanisms that could lead to the progression from the asymptomatic period of the infection to AIDS. Much of this work was based on evolutionary dynamics [9], because virus evolution in vivo is thought to be a crucial driving force underlying disease progression.

This work on HIV infection is part of the larger field of virus dynamics [5, 10], and research on HIV dynamics has been a very prominent topic in this field. Much of the classic work on this topic has been very well reviewed in books and papers, with excellent summaries provided in e.g. refs. 5–8. This review will start by summarizing some of this work. The rest of the article, however, will focus on more recent topics that are relevant for understanding the replication and evolution of HIV in vivo. These include the multiple infection of cells and the direct transmission of the virus from cell to cell via the formation of virological synapses. Although these topics are only beginning to be explored in detail, they are of great interest for improving our understanding of HIV dynamics, evolution, and pathogenesis.

2 The Basic Model of Virus Dynamics

The basic model of virus dynamics [5–7, 11] (Fig. 1) has three variables: the population sizes of susceptible, uninfected cells, T; infected cells, I; and free virus particles, V. These quantities can either denote the total abundance in a host, or the abundance in a given volume blood or tissue. Free virus particles infect uninfected cells at a rate proportional to the product of their abundances, βTV. The rate constant, β, describes the efficacy of this process, including the rate at which virus particles find uninfected cells, the rate of virus entry, and the rate and probability of successful infection. Infected cells produce free virus at a rate k. Infected cells die at a rate a, and free virus particles are removed from the system at rate u. Therefore, the average life-time of an infected cell is $1/a$, whereas the average life-time of a free virus particle is $1/u$. The total amount of virus particles produced from one infected cell, the "burst size", is k/a. Uninfected cells are produced at a constant rate, λ, and die at a rate d. The average life-time of an uninfected cell is $1/d$. In the absence of infection, the population dynamics of host cells is given by $dT/dt = \lambda - dT$. This is a simple linear differential equation. Without virus, the abundance of uninfected cells converges to the equilibrium value λ/d. Combining the

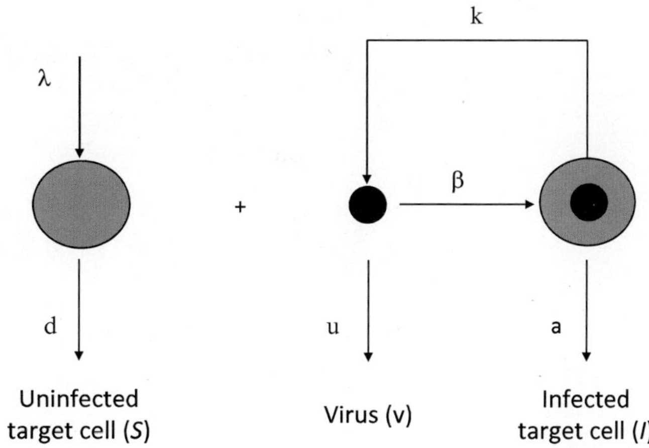

Fig. 1 Schematic diagram explaining the basic model of virus dynamics. See text for explanations

dynamics of virus infection and host cells, we obtain the basic model of virus dynamics:

$$\frac{dT}{dt} = \lambda - dT - \beta TV$$
$$\frac{dI}{dt} = \beta TV - aI \qquad (1)$$
$$\frac{dV}{dt} = kI - uV$$

This is a system of nonlinear differential equations. An analytic solution of the time development of the variables is not possible, but we can derive various approximations and thereby obtain a complete understanding of the system. Before infection, we have $I = 0$, $v = 0$, and uninfected cells are at equilibrium $T = \lambda/d$. Denote by $t = 0$ the time when infection occurs. Suppose infection occurs with a certain amount of virus particles, v_0. Thus the initial conditions are $T_0 = \lambda/d$, $I_0 = 0$, and V_0. Whether or not the virus can grow and establish an infection depends on a condition very similar to the spread of an infectious disease in a population of host individuals [12]. The crucial quantity is the basic reproductive ratio of the virus, R_0, which is defined as the number of newly infected cells that arise from any one infected cell when almost all cells are uninfected (Fig. 2). The rate at which one infected cell gives rise to new infected cells is given by $\beta kT/u$. If all cells are uninfected then $T = \lambda/d$. Since the life-time of an infected cell is $1/a$, we obtain $R_0 = \beta\lambda k/(adu)$.If $R_0 < 1$ then the virus will not spread, since every infected cell will on average produce less than one other infected cell. The chain reaction is sub-critical. On average we expect $1/(1 - R_0)$ rounds of infection before the virus population dies out.

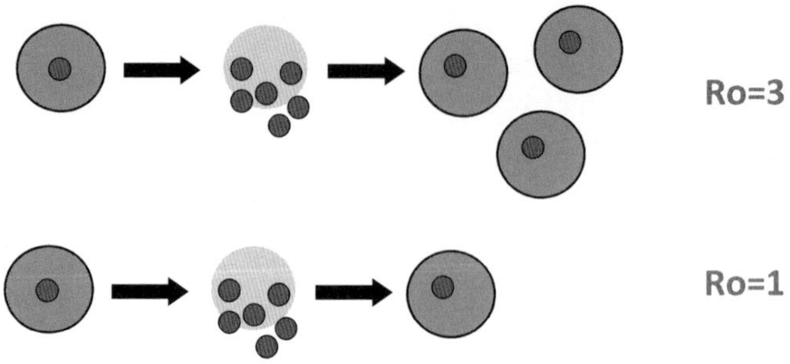

Fig. 2 Schematic diagram that illustrates the basic reproductive ratio of the virus. See text for explanations

If on the other hand $R_0 > 1$, then every infected cell will on average produce more than one newly infected cell. The chain reaction will generate an explosive multiplication of virus. Virus growth will not continue indefinitely, because the supply of uninfected cells is limited. There will be a peak in virus load and subsequently damped oscillations to an equilibrium. The equilibrium abundance of uninfected cells, infected cells, and free virus is given by $T^* = T_0/R_0$, $I^* = (R_0 - 1)du/(\beta k)$, $V^* = (R_0 - 1)d/\beta$. At equilibrium, any one infected cell will on average give rise to one newly infected cell.

If the virus has a basic reproductive ratio much larger than one, then T^* will be greatly reduced compared to T_0, which means that during infection the equilibrium abundance of uninfected cells is much smaller than before infection. In other words, the above simple model cannot explain a situation where during a persistent virus infection almost all infectable cells remain uninfected, except in the case when R_0 is only slightly bigger than unity (which is unlikely in general). Furthermore, if $R_0 \gg 1$, then the equilibrium abundance of infected cells and free virus is approximately given by $I^* \approx \lambda/a$ and $V^* \approx (\lambda k)/(au)$. Interestingly, both quantities do not depend on the infection parameter β [13]. The reason is that a highly infectious virus (large β) will rapidly infect uninfected cells, but at equilibrium there will only be few uninfected cells in the system. A less infectious virus (smaller β) will take longer to infect uninfected cells, but the equilibrium abundance of uninfected cells is higher. For both viruses the product βT will be the same at equilibrium, resulting in a constant rate of production of new infected cells, and therefore in similar equilibrium abundances of infected cells and free virus. For a highly cytopathic virus (a much larger than d), the equilibrium abundance of infected cells will be small compared to the abundance of cells prior to infection. In fact, the larger a, the smaller the abundance both of infected cells and of free virus. For a non-cytopathic virus ($a \approx d$), the equilibrium abundance of infected cells will be roughly equivalent to the total abundance of susceptible cells prior to infection.

3 Insights into the Kinetics and Evolutionary Dynamics of HIV Infection

One of the first major insights generated by the collaboration between mathematical modelers and experimentalist was the quantification of the turnover rates of infected cells and free virus during the chronic phase of the infection [14–19]. Patients were treated with drugs, preventing the infection of new cells. The resulting exponential decline of the virus population could be fit with the type of mathematical model described in the previous section, which allowed quantification of the turnover rate of free virus particles and infected cells. This work showed for the first time that HIV turns over rapidly during the chronic phase of the infection, which implies an enormous potential of the virus to evolve during this time, allowing it to escape immune responses and to acquire drug resistance. Specifically, the half-life of the infected cell population was found to be between 1 and 3 days, and the half-life of free virus particles is of the order or hours. Subsequent work showed that the decline of the virus population for longer periods of time is characterized by several phases: a first fast phase followed by is followed by a slower second phase. As virus load declines, the rate of decline slows down even more. The fast phase was due to the decline of productively infected T cells, while the slower phases were due to the decline of longer lived infected antigen presenting cells, and latently infected T cells that provide long lived viral reservoirs [20–22]. This quantitative work provided a much improved understanding of the natural history of HIV infection. Another important measure that was quantified was the basic reproductive ratio of HIV [23–25]. This was done by analyzing the dynamic of virus growth and decline during the acute phase of the infection, and was performed both in SIV-infected macaques and in HIV-infected humans. Other important kinetic estimates concerned immune response such as the rates at which cytotoxic T lymphocytes (CTL) or CD8 T cells kill infected cells—an important branch of the immune system for fighting the virus [26, 27]. This also leads to an important discussion about the role of lytic and non-lytic CTL responses for the control of the infection during the asymptomatic phase.

While obtaining a variety of parameter estimates that characterize HIV infection has been of central importance to the field, mathematical models also made conceptual advances. Theories were developed about the processes that contribute to the transition from the asymptomatic period towards the development of AIDS [5]. A central theme in these models has been that viral evolution in vivo is a crucial driving force of disease progression. One of the more prominent theories argued that viral evolution of escape from immune responses can gradually weaken the immune system and lead to its collapse once the virus population had diversified sufficiently [28–30]. Other theories looked at different evolutionary processes, considering virus evolution towards faster replication rates, increased cytopathicity, or broadened cell tropism [31, 32], and theories about the in vivo evolution of HIV have been developed further e.g. [33–36]. While it is difficult to validate these theories, partly because an array of mechanisms is at work in HIV infection, it has become clear that viral evolution does indeed play a pivotal role in the disease process. In an elegant set of experiments, monkeys were infected with SIV, and the virus was sampled and characterized at an early, intermediate, and late stage post infection [37]. This showed that over time, the virus became faster replicating, more cytopathic, and less recognized by the immune system. When the later virus isolates were injected into a new monkey host, the rate of initial virus growth was faster, the set-point virus load higher, and the rate of disease progression faster compared to a scenario where the earlier virus isolates were used to infect a new monkey host. This set of experiments showed that the virus evolves to more virulent phenotypes, and that this evolution enables the virus to be more aggressive and to cause faster disease progression. More recent work investigated viral evolutionary processes in more detail, especially the escape of HIV from immune responses [38–44], analyzing the fitness cost of escape mutants and the dynamics of escape during the disease.

In the following, the review will switch gear and discuss some more recent data and concepts that have important implications for our understanding of the principles that govern virus growth through its target cell population, the evolution of the virus, and thus the mechanisms that contribute to pathogenesis. In particular, the review will focus on multiple infection of cells and on the direct cell-to-cell transmission of the virus through virological synapses. Both data and mathematical modeling concepts will be explored.

4 Multiple Infection of Cells

Until fairly recently, the concept of coinfection has not played a prominent role in HIV research. This likely stems at least in part from the early observation that infection leads to the down-modulation of the CD4 receptor (reviewed in refs. 45, 46), and the more recent observation that HIV also down modulates the CCR5 and CXCR4 viral coreceptors [47] from the cell surface, reducing the susceptibility of cells to reinfection over time. In fact, three separate HIV proteins, Nef, Vpu, and Env, mediate CD4 down modulation [48, 49], emphasizing its biological significance. Furthermore, it was also observed quite early in the epidemic that infection frequency of cells in blood is low, on the order of one in one thousand to one in one hundred thousand, leading to the incorrect assumption that the probability of two infection events in the same cell must be exponentially lower.

Over time it has become clear that this picture is not correct and that coinfection with two or more viruses, i.e. multiple infection of cells, is a frequent phenomenon that plays an important role in the natural history of HIV. First and perhaps foremost, HIV-1 replication occurs predominantly in the lymphoid tissues, where the concentration of target cells is higher than in the blood, and cell to cell contact facilitates transmission of multiple virions between cells. Even when few infected cells can be identified in mucosal tissues, they are observed to be infected with multiple viruses [50], and in situ staining in splenocytes of HIV-1 patients observed an average of 3–4 integrated proviruses per cell and sequencing of HIV-1 nucleic acids in these cells confirms multiple infection with divergent viruses and recombination between them [51]. During acute infection of macaques with a pathogenic strain of SIV, an average of 1.5 viruses per cell was observed, indicating coinfection of a large fraction of cells [52]. The recent description of cell to cell transmission of HIV via virological synapse formation [53–55] dramatically illustrates how multiple infection of cells can be locally generated.

Although CD4 loss from the cell surface is a consequence of HIV-1 infection, it is not clear that its primary function is to prevent reinfection (superinfection) of cells prior to virion production. Instead removal of CD4 from the cell surface has been shown by several groups to increase the infectivity of the newly produced virions [56, 57], allowing more Env protein to associate with virions and increasing viral pathogenesis [58]. Further, there is an 18–24 h delay between infection of a cell and production of viral proteins which modulate CD4 expression, during which the cell remains susceptible to reinfection, so inhibition of superinfection is only operative during the productive phase of infection, when superinfection might be more toxic to the already stressed cell, causing premature cell death through apoptosis, lowering virus production (reviewed in refs. 45, 59). Since the lifespan of a productively infected T cell in vivo is only about ½ to 1 day, once virus production is underway in the cell, superinfection at this late stage would most likely be unhelpful to the virus.

Experimental systems to study the dynamics of multiple infection have frequently utilized recombinant viruses bearing different reporter genes, allowing quantification of cells infected with one or both viruses [60–62]. These studies, carried out in tissue culture or in vivo within human thymic tissue in SCID mice (SCID-hu Thy/Liv mice) have made it abundantly clear that multiple infection is a natural consequence of HIV-1 replication. Over many rounds of replication in tissue culture or in the SCID-hu Thy/Liv system multiple infection proceeds without apparent inhibition, despite the ability of HIV-1 to inhibit reinfection, resulting in frequent recombination [62]. The inference is that the pace of HIV-1 replication exceeds the inhibition effect, and that fostering multiple infection, rather than inhibiting it, may be to the benefit of the virus.

Recombination is the best studied outcome of multiple infection. It can have important implications for the evolution of HIV in vivo. Recombination can potentially speed up the rate of evolution by bringing together different advantageous alleles into a single genome. On the negative side (from the virus' standpoint), recombination can also break up existing advantageous allelic combinations or it can lead to the inactivation of viable viruses if they are coinfected and recombine with defective viruses. The effect of recombination on the evolutionary dynamics in vivo is complex, and can depend on several population genetic phenomena, such as the degree of epistasis. This has been studied in a variety of theoretical papers [63–68].

There are other important consequences of coinfection for virus dynamics. Viruses defective in vital functions can be phenotypically complemented during coinfection, resulting in chimeric virions bearing mixtures of genes and proteins from more than one parental strain [61, 69], and recombination can repair the defect at the genetic level [61, 69]. Viruses with essentially zero fitness can replicate as a result of complementation during coinfection [61]. In addition, it is likely that other evolutionary processes are influenced by the occurrence of multiple infection of cells [70–73].

5 Does Multiple Infection Influence Basic Virus Dynamics?

This question has been studied with mathematical models in different settings. In the most basic setting, multiple infection can be studied with the following mathematical model. Instead of a single infected cell population, we now assume the existence of several infected cell sub-populations, i.e. cells infected with i copies of a given virus, I_i. The model is given by the following set of ordinary differential equations.

$$\begin{aligned}
\frac{dT}{dt} &= \lambda - dT - \beta TV \\
\frac{dI_1}{dt} &= \beta TV - aI_1 - \beta I_1 V \\
\frac{dI_i}{dt} &= \beta I_{i-1} V - aI_i - \beta I_i V \\
\frac{dI_n}{dt} &= \beta I_{n-1} V - aI_n \\
\frac{dV}{dt} &= k\sum_{i=1}^{n} I_i - uV
\end{aligned} \qquad (2)$$

Cells infected with i viruses die with a rate a, and infection with an additional virus is represented by the term $\beta I_i V$. All Infected cells produce free virus with a rate k. Finally, free virus decays with a rate uV. Note that the end of the infection cascade, I_n, is an artificial feature of the ODE model, and this population should not be concentrated on.

According to this formulation, infected cells produce the same amount of virus regardless of the number of viruses present in these cells. Hence, virus production is completely determined by cellular resources. Adding more virus genomes to the cell reduces the replicative output of the individual viruses in the cells such that the total amount of virus produced remains the same. A more complex variation of this model has been studied by Dixit et al. [74]. They found that in this setting, multiple infection of cells does not influence the basic virus dynamics. This is because an infected cell essentially behaves the same way whether it contains one copy of the virus or whether it contains multiple copies. The opposite assumption has been studied by Cumings et al. [75]. In their model, the rate of virus production increased asymptotically with the number of viruses that are resident within a cell. This leads to a more complicated situation. In particular, the exact properties of the model can depend on the nature of the infection term, for example whether it is a straightforward mass action term or whether the rate of infection saturates with the number of target cells. In such models, the basic dynamics of virus growth can be fundamentally altered (Fig. 3). Whether an infection becomes established or not can depend on the initial conditions, initial virus growth can be super-exponential, and the response to therapy can depend on virus load. These are properties that are not observed in the standard model of virus dynamics. Whether these properties apply to HIV infection is not clear at the moment. Experimental support for such dynamics has been found in adenovirus infections [76],

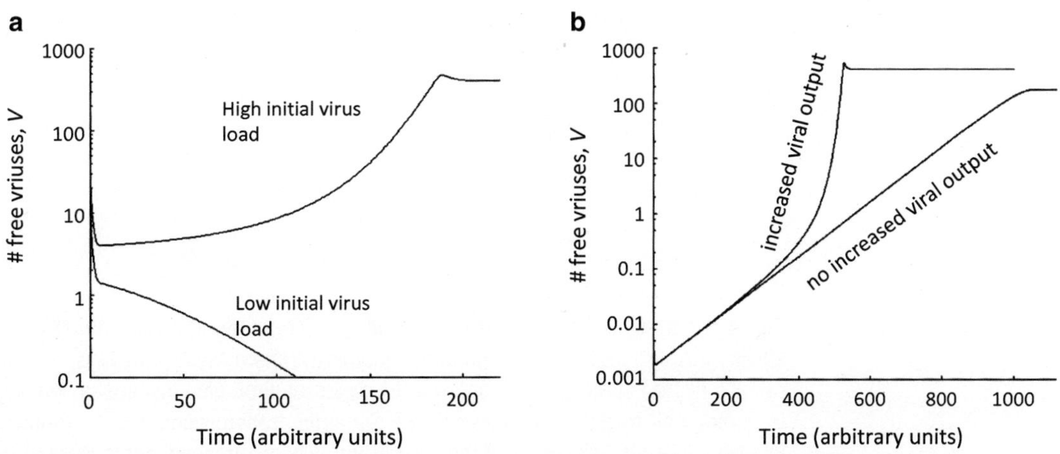

Fig. 3 Basic virus dynamics properties can be altered if an increased viral output from multiply infected cells is assumed in some models. (**a**) Whether the virus establishes a persistent infection can depend on the initial virus load. (**b**) Super-exponential growth can be observed because the rate of viral replication accelerates as virus load rises and more cells become multiply infected. The figure shows computer simulations of a model presented in ref. 75

but these properties have so far not been documented in HIV, although they have not been specifically investigated. With HIV, it will be important to determine whether the rate of virus replication does or does not depend on the number of viruses in the cell, and if it does, what the laws of infection are, i.e. what mathematical infection term describes virus growth best. These issues will be important to investigate. If multiple infection does alter the dynamics of virus growth, then there are obvious implications for the dynamics of the infection and for pathogenesis.

6 Different Virus Transmission Pathways

The rate of viral spread through the target cell population has been shown to influence the level of virus control and the pattern of disease progression [37, 77, 78]. Viral spread through the population of target cells can occur via two basic mechanisms [53, 79–84] (Fig. 4a). (1) In cell-free spread, viruses are released from cells into the extracellular environment and infect susceptible targets that are encountered. (2) In cell–cell spread, viruses can pass directly from one cell to another without entering the extracellular environment, through the formation of virological synapses. On a per cell basis, cell to cell spread has

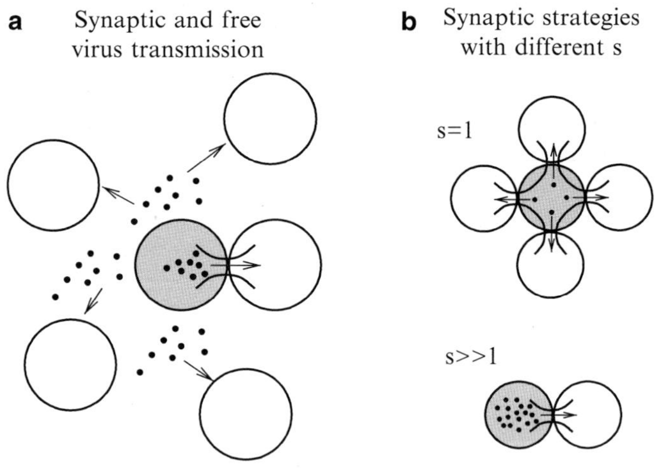

Fig. 4 Synaptic vs. free virus transmission. (**a**) Schematic showing that HIV can spread through its target cell population by two transmission pathways: via the release of free virus, and through the formation of virological synapses, leading to direct cell-to-cell transmission. (**b**) Synaptic transmission can potentially involve different strategies. Either many viruses are passed per synapse, or fewer viruses are transferred per synapse. The higher the number of viruses transferred per synapse, the lower the number of target cells to which a source cell can connect and transfer viruses, because many viruses harbored in the source cell are lost with each synaptic connection

been shown to be very effective [79]. Tens to hundreds of virus particles are transferred through synapses, a certain fraction of which successfully integrates into the genome of the target cell. This has been thought to confer an advantage to the virus population in a variety of settings [79, 85]. Synaptic transmission in HIV infection is considered to be particularly important in tissue sites, such as lymph nodes and the spleen, where cells have a relatively high likelihood to come into contact with each other and to form synapses. This can lead to the frequent multiple infection of target cells. Synapse formation has been shown in vitro to lead to the co-transmission of multiple copies of HIV-1 across a single synapse [86]. This is in contrast to cell-free transmission, which typically leads to the transmission of single viral copies to target cells. Indeed, in the blood where cells mix more readily and synapse formation is less likely to occur, most infected cells have been found to contain a single copy of HIV-1 [87].

The different viral transmission pathways can lead to different rates of virus spread through the target cell population, and the viral spread rate through target cells has been shown to influence the rate of disease progression [77]. Therefore, to better understand the determinants of progression, it is important to quantify the relative contribution of free virus vs. synaptic transmission to virus growth, and this requires mathematical models that take both transmission pathways into account. The following sections will review such mathematical models, demonstrate how application to data can estimate crucial parameters, and investigate synaptic and free virus transmission in an evolutionary light.

7 A Mathematical Model of Synaptic and Free Virus Transmission

We outline a general system of ordinary differential equations (ODEs) that describes the dynamics of target and infected cells in the presence of cell-free and synaptic viral transmission. Let us denote by x_i the number of cells infected by i viruses, $0 \leq i \leq N$. We assume that N is the maximum multiplicity of infection possible in a cell. The variable x_0 stands for uninfected (target) cells. The equations are as follows [88–90]:

$$\dot{x}_0 = \lambda - \mathrm{d}x_0 - \tilde{\beta}\, v x_0 - S\sum_{m=1}^{N} x_m \sum_{j=1}^{N} \gamma_j^{(m)} x_0,$$

$$\dot{x}_i = \tilde{\beta}\, v\left(x_{i-1} - x_i\right) + S\sum_{m=1}^{N} x_m \left(\sum_{j=1}^{i} \gamma_j^{(m)} x_{i-j} - x_i \sum_{j=1}^{N-i} \gamma_j^{(m)} \right) - a^{(i)} x_i, \quad 1 \leq i \leq N-1, \tag{3}$$

$$\dot{x}_N = \tilde{\beta}\, v x_{N-1} + S\sum_{m=1}^{N} x_m \sum_{j=1}^{N} \gamma_j^{(m)} x_{N-j} - a^{(N)} x_N.$$

The various components of this system and notations are explained as follows. The terms describing target cell production and death ($\lambda - dx_0$) are the same as in the standard model of virus dynamics. The terms multiplying β describe infection of cells by free-virus transmission. The variable v stands for the concentration of free virus, and is described by the equation $\dot{v} = \sum_{i=1}^{N} k^{(i)} x_i - uv$ where coefficients $k^{(i)}$ describe the intensity of free virus production by cells of multiplicity i, and u is the viral death rate. Using the standard assumption that the virus population is at a quasi-steady state, adjusting rapidly to the population of infected cells, we have $v = \frac{1}{u} \sum_{i=1}^{N} k^{(i)} x_i$. In system (3), symbol S denotes the rate of synapse formation, and coefficient $\gamma_j^{(m)}$ is the probability for a cell infected with m viruses to transmit j viruses by synapse, with $\sum_{j=0}^{N} \gamma^{(m)}{}_j = 1$. A two-parametric model for these probabilities was used in ref. 89, such that s viruses attempt transfer per synapse, with a probability of successful infection of r per virus. Coefficients $a^{(i)}$ denote death rates of cells with multiplicity of infection i. For simplicity, let us assume that the rate of viral replication and the death rate of infected cells do not depend on the number of viruses in the cell. If kinetics are independent of the number of viruses in cells, a much simpler description is possible. Let us denote $x = x_0$, $y = \sum_{i=1}^{N} x_i$, where x are target cells and y are infected cells. Then we can add the equations in (Eq. 3) with $i = 1, \ldots, N$ and under the quasi-stationarity assumption obtain

$$\dot{x} = \lambda - dx - \left(\beta^{\text{free}} + \beta^{\text{syn}}\right)xy,$$
$$\dot{y} = \left(\beta^{\text{free}} + \beta^{\text{syn}}\right)xy - ay,$$

where $\beta^{\text{free}} = \frac{\beta k}{u}$ and $\beta^{\text{syn}} = S\sum_{j=1}^{N} \gamma_j$.

8 The Relative Contribution of Free Virus and Synaptic Transmission

A variant of the above described simplified system was applied to experimental data in order to estimate the relative contribution of free virus and synaptic transmission to virus growth [88].

In a previous study [84], virus growth was compared in two culture conditions. In a first set of experiments, cultures were kept under gentle shaking conditions, preventing the formation of viral synapses. Thus, only cell-free transmission occurred. A second set of experiments was performed under static conditions, in which both transmission pathways were likely to occur. Data on two types of cells (the transformed CD4+ T cell line Jurkat and primary

CD4+ lymphocytes) were obtained. Time-series of the percentage of infected cells were generated. These data provided a first and interesting picture, but the number of data points in this study was not sufficient for model fitting and parameterization. In order to obtain a larger number of data points a new and analogous set of experiments was performed [88]. We focused on Jurkat cells because the kinetics of cell proliferation and cell death are better defined than in primary CD4 T cells.

The predicted percentage of infected cells, $100y/(x+y)\%$, was fitted to the observed percentages, using standard non-linear least squares procedures. Under static culture conditions, we estimate the replication rate $\beta_{st}=\beta^{syn}+\beta^{free}$. Under shaking conditions, we estimated $\beta_{sh}=\beta^{free}$. From the estimated values of β_{st} and β_{sh}, we calculated the ratio of the viral replication rate for synaptic and free virus transmission, $\beta^{syn}/\beta^{free}=\beta_{st}-\beta_{sh})/(\beta_{sh})$. These calculations indicated that synaptic and free virus transmissions contribute approximately equally to virus spread through the target cells.

This was an interesting result given the observation that on a per cell basis, synaptic transmission has been found to be much more efficient at infecting cells than free virus transmission. It indicates that both pathways can make significant contributions to disease progression, pathogenesis, and responses to treatment. Further work, involving additional experimental methods to dissect the two transmission modes, will be necessary to quantify this further.

9 Synaptic Transmission and Evolutionary Perspectives

The occurrence of synaptic transmission in HIV infection brings up an evolutionary question. What is the optimal number of viruses transferred from a source cell to a target cell such that the rate of viral spread, and hence the potential to cause pathogenesis, is maximized? Along similar lines, how does this optimum depend on the biological assumptions?

Using the full model described above that takes into account both transmission pathways, different synaptic transmission strategies were investigated, defined by the number of viruses transferred per synapse (Fig. 4b) [89, 90]. A number of scenarios were investigated. The most basic scenario gave rise to the prediction that the optimal viral strategy to maximize the rate of virus spread is the transfer of a single virus particle per synapse. Passing a larger number of viruses through synapses leads to the infection of already infected cells. This essentially wastes these viruses because they could be transmitted to uninfected cells instead, thus increasing the rate of viral spread. This result is interesting to consider in the context of a different viral infection. A study examined the in vitro growth and consequent plaque formation with vaccinia virus [91].

It was shown that newly infected cells expressed specific proteins that resulted in the "repulsion" of other viruses that attempted to infect the same cells. Thus, instead of coinfecting the cells, these viruses were "redirected" towards uninfected cells. Hence, vaccinia virus has evolved a mechanism to avoid multiple infection of cells, instead ensuring that more uninfected cells are being targeted. Experiments showed that this mechanism significantly accelerates the rate of virus growth in this system. This observation supports the theoretical notion that transferring many viruses to cells can be ineffective and disadvantageous because virus particles that could in principle enter uninfected cells are wasted by entering already infected cells. Although no viral synapses are formed in the vaccinia system, the spatial arrangement of cells during plaque formation has a similar effect in the sense that viruses released from a source cell are most likely to repeatedly reach the same set of target cells that are in their direct vicinity. The example of vaccinia shows that there is a certain selection pressure against transferring high numbers of virus particles to the same cell.

In the light of this, an explanation is required for the observation that on the order of 10^2 virus particles are transferred through synapses in HIV infection [53, 86]. A number of scenarios were explored that could make an intermediate number of transferred viruses the optimal viral strategy [89], and these scenarios are discussed as follows.

A higher burst size of multiply infected cells could have this effect. While this can indeed elevate the efficiency of passing many viruses per synapse, the increase in burst size must be super-additive for this effect to be observed, e.g. doubly infected cells must produce and transfer more than twice as much virus as singly infected cells. There are currently no data that examine the burst size of infected cells in dependence of the infection multiplicity. A super-additive effect, however, is unlikely to occur unless special cooperative interactions between co-resident viruses occur. Cooperative effects have been observed in the context of unintegrated viral DNA, which could produce offspring virus in the presence of integrated virus rather than becoming a replicative dead end [61], although the contribution of this effect for the overall dynamics is currently unclear. Even if more viruses are produced in multiply infected cells, this could be canceled out by an increased death rate [85]. The effect of multiple infection on the kinetics of virus production and cell death remains to be determined.

Another process that can result in an intermediate optimal number of transferred viruses is the viral saturation of intracellular defense factors. This could make it advantageous to pass many viruses per synapse in the following way. The individual factors bind virus particles with the effect of reducing their probability of successful infection. It is possible that the number of inhibiting factors that can bind the virus particles is limited, and thus a synapse that sends a large number of viruses into a specific target cell has a possibility to "flood" and saturate this defense [92]. Suppose that n_1 immune particles are available in the cell, then the first n_1 viruses will be bound to them, resulting in a low individual probability of infection. If the number of viruses entering the cell by synapse, $s > n_1$, then the remaining $n_2 = s - n_1$ particles will have a higher probability of successfully infecting. The relevance of this mechanism in HIV infection, however, remains unclear. TRIM5α has been identified as an intracellular factor that inhibits HIV replication upon entry into the cell. It has been found to be especially effective at preventing HIV-1 infection in cells derived from Old World monkeys [93–95]. The human version of TRIM5α is less protective against HIV-1. Members of the APOBEC family of restriction factors interfere with reverse transcription, although this effect is countered by viral Vif [96]. It is unlikely, however, that synaptic transmission can lead to the saturation of this factor because it is incorporated into the virion in the source cell before displaying activity during reverse transcription upon infection of the target cell. Similarly, factors such as tetherin probably are not applicable because viral assembly and budding is inhibited [96], and cannot be saturated by multiple infection. Nevertheless, experiments indicate that cells contain saturable targets that inhibit infection of cells [97] and that could be directly relevant to our model scenario, although they remain to be identified [92]. Saturation of antiviral factors in target cells by multiple infection through virological synapses is being investigated increasingly, see ref. 92 for a review.

10 Conclusions

The paper reviews various concepts that are related to HIV pathogenesis, with emphasis on the insights learnt from mathematical models. It starts with a review of some of the classical work which has provided crucial quantitative insights into the processes underlying the infection. The review then concentrates on some of the recent work we have done in the context of multiple infection of cells and synaptic transmission of the virus. These are processes that can determine the rate of viral replication. The rate of virus spread and growth through the target cell population in turn is a major factor influencing the progression and pathogenesis of the disease [77]. Some basic facts about multiple infection and synaptic transmission have been reviewed, as well as some first steps that were taken to describe them mathematically, measuring important parameters and providing some conceptual insights. It will be useful to explore these topics in more detail with a combination of experimental and mathematical approaches, to gain a better understanding of the mechanisms that drive viral replication and pathogenesis.

It is important to point out that the review has covered very specific aspects that are relevant to our understanding of HIV pathogenesis. Obviously, there are many components that are relevant to disease progression and that have also been investigated with mathematical models. The interactions between HIV and HIV-specific immune responses constitute one such topic (for a review, *see* ref. 98). These interactions, and the corresponding mathematical models, are more complex in nature and build upon basic models of virus replication that are discussed here.

References

1. Levy JA (2007) HIV and the pathogenesis of AIDS. ASM press, Washington, DC
2. Moir S, Chun TW, Fauci AS (2011) Pathogenic mechanisms of HIV disease. (Translated from eng). Annu Rev Pathol 6:223–248
3. Lackner A, Lederman MM, Rodriguez B (2012) HIV pathogenesis: the host. Cold Spring Harbor Perspectives in Medicine 2(9)
4. Coffin JM (1995) HIV population dynamics in vivo: implications for genetic variation, pathogenesis, and therapy. Science 267(5197): 483–489
5. Nowak MA, May RM (2000) Virus dynamics. Mathematical principles of immunology and virology. Oxford University Press, Oxford
6. Perelson AS (2002) Modelling viral and immune system dynamics. Nature Rev Immunol 2(1):28–36
7. Perelson AS, Ribeiro RM (2013) Modeling the within-host dynamics of HIV infection. (Translated from Eng). BMC Biol 11(1):96
8. Wodarz D, Nowak MA (2002) Mathematical models of HIV pathogenesis and treatment. Bioessays 24(12):1178–1187
9. Nowak MA (2006) Evolutionary dynamics: exploring the equations of life. Harvard University Press, Cambridge, MA
10. McLean AR (2013) Infectious disease modeling. Infectious diseases. Springer, New York, pp 99–115
11. Nowak MA, Bangham CR (1996) Population dynamics of immune responses to persistent viruses. Science 272(5258):74–79
12. Anderson RM, May RM (1991) Infectious diseases of humans. Oxfors University Press, Oxofrd, England
13. Bonhoeffer S, May RM, Shaw GM, Nowak MA (1997) Virus dynamics and drug therapy. Proc Natl Acad Sci U S A 94(13):6971–6976
14. Ho DD et al (1995) Rapid turnover of plasma virions and Cd4 lymphocytes in HIV-1 infection. Nature 373(6510):123–126
15. Wei XP et al (1995) Viral dynamics in human-immunodeficiency-virus type-1 infection. Nature 373(6510):117–122
16. Perelson AS et al (1997) Decay characteristics of HIV-1-infected compartments during combination therapy. Nature 387(6629):188–191
17. Perelson AS, Neumann AU, Markowitz M, Leonard JM, Ho DD (1996) HIV-1 dynamics in-vivo – virion clearance rate, infected cell life-span, and viral generation time. Science 271(5255):1582–1586
18. Perelson AS, Essunger P, Ho DD (1997) Dynamics of HIV-1 and CD4+ lymphocytes in vivo. AIDS 11(SA):S17–S24
19. De Boer RJ, Ribeiro RM, Perelson AS (2010) Current estimates for HIV-1 production imply rapid viral clearance in lymphoid tissues. PLoS Comput Biol 6(9):e1000906
20. Finzi D et al (1997) Identification of a reservoir for HIV-1 in patients on highly active antiretroviral therapy [see comments]. Science 278(5341):1295–1300
21. Finzi D et al (1999) Latent infection of CD4+ T cells provides a mechanism for lifelong persistence of HIV-1, even in patients on effective combination therapy. (Translated from Eng). Nat Med 5(5):512–517 (in Eng)
22. Eisele E, Siliciano RF (2012) Redefining the viral reservoirs that prevent HIV-1 eradication. (Translated from eng). Immunity 37(3):377–388 (in eng)
23. Nowak MA et al (1997) Viral dynamics of primary viremia and antiretroviral therapy in simian immunodeficiency virus infection. J Virol 71(10):7518–7525
24. Little SJ, McLean AR, Spina CA, Richman DD, Havlir DV (1999) Viral dynamics of acute HIV-1 infection. J Exp Med 190(6):841–850
25. Ribeiro RM et al (2010) Estimation of the initial viral growth rate and basic reproductive number during acute HIV-1 infection. (Translated from eng). J Virol 84(12):6096–6102 (in eng)

26. Asquith B, Edwards CT, Lipsitch M, McLean AR (2006) Inefficient cytotoxic T lymphocyte-mediated killing of HIV-1-infected cells in vivo. PLoS Biol 4(4):e90
27. Wick WD, Yang OO, Corey L, Self SG (2005) How many human immunodeficiency virus type 1-infected target cells can a cytotoxic. T-lymphocyte kill? (Translated from eng). J Virol 79(21):13579–13586 (in eng)
28. Nowak MA (1996) Immune-responses against multiple epitopes – a theory for immunodominance and antigenic variation. Semin Virol 7(1):83–92
29. Nowak MA et al (1991) Antigenic diversity thresholds and the development of AIDS. Science 254(5034):963–969
30. Nowak MA et al (1995) Antigenic oscillations and shifting immunodominance in HIV-1 infections. Nature 375(6532):606–611
31. Wodarz D, Nowak MA (1998) The effect of different immune responses on the evolution of virulent CXCR4 tropic HIV. Proc R Soc Lond B 265(1411):2149–2158
32. Regoes RR, Bonhoeffer S (2005) The HIV coreceptor switch: a population dynamical perspective. Trends Microbiol 13(6):269–277
33. Ball CL, Gilchrist MA, Coombs D (2007) Modeling within-host evolution of HIV: mutation, competition and strain replacement. Bull Math Biol 69(7):2361–2385
34. Stilianakis NI, Schenzle D (2006) On the intra-host dynamics of HIV-1 infections. Math Biosci 199(1):1–25
35. Rouzine IM, Weinberger LS (2013) The quantitative theory of within-host viral evolution. J Stat Mech Theory Exp 2013(01), P01009
36. Lee HY, Perelson AS, Park S-C, Leitner T (2008) Dynamic correlation between intrahost HIV-1 quasispecies evolution and disease progression. PLoS Comput Biol 4(12):e1000240
37. Kimata JT, Kuller L, Anderson DB, Dailey P, Overbaugh J (1999) Emerging cytopathic and antigenic simian immunodeficiency virus variants influence AIDS progression. Nat Med 5(5):535–541
38. Wei X et al (2003) Antibody neutralization and escape by HIV-1. Nature 422(6929):307–312
39. Ganusov VV, De Boer RJ (2006) Estimating costs and benefits of CTL escape mutations in SIV/HIV infection. (Translated from eng). PLoS Comput Biol 2(3):e24
40. Ganusov VV et al (2011) Fitness costs and diversity of the cytotoxic T lymphocyte (CTL) response determine the rate of CTL escape during acute and chronic phases of HIV infection. (Translated from eng). J Virol 85(20): 10518–10528 (in eng)
41. Fryer HR et al (2010) Modelling the evolution and spread of HIV immune escape mutants. PLoS Pathog 6(11):e1001196
42. Kadolsky UD, Asquith B (2010) Quantifying the impact of human immunodeficiency virus-1 escape from cytotoxic T-lymphocytes. PLoS Comput Biol 6(11):e1000981
43. Mostowy R et al (2012) Estimating the fitness cost of escape from HLA presentation in HIV-1 protease and reverse transcriptase. PLoS Comput Biol 8(5):e1002525
44. Ganusov VV, Neher RA, Perelson AS (2013) Mathematical modeling of escape of HIV from cytotoxic T lymphocyte responses. J Stat Mech Theory Exp 2013(01), P01010
45. Lama J (2003) The physiological relevance of CD4 receptor down-modulation during HIV infection. Curr HIV Res 1(2):167–184
46. Levesque K, Finzi A, Binette J, Cohen EA (2004) Role of CD4 receptor down-regulation during HIV-1 infection. Curr HIV Res 2(1): 51–59
47. Michel N, Allespach I, Venzke S, Fackler OT, Keppler OT (2005) The Nef protein of human immunodeficiency virus establishes super infection immunity by a dual strategy to down-regulate cell-surface CCR5 and CD4. Curr Biol 15(8):714–723
48. Chen BK, Gandhi RT, Baltimore D (1996) CD4 down-modulation during infection of human T cells with human immunodeficiency virus type 1 involves independent activities of vpu, env, and nef. J Virol 70(9):6044–6053
49. Wildum S, Schindler M, Munch J, Kirchhoff F (2006) Contribution of Vpu, Env, and Nef to CD4 down-modulation and resistance of human immunodeficiency virus type 1-infected T cells to superinfection. J Virol 80(16): 8047–8059
50. Gratton S, Cheynier R, Dumaurier MJ, Oksenhendler E, Wain-Hobson S (2000) Highly restricted spread of HIV-1 and multiply infected cells within splenic germinal centers. Proc Natl Acad Sci U S A 97(26):14566–14571
51. Jung A et al (2002) Multiply infected spleen cells in HIV patients. Nature 418(6894):144
52. Mattapallil JJ et al (2005) Massive infection and loss of memory CD4+ T cells in multiple tissues during acute SIV infection. Nature 434(7037):1093–1097
53. Hubner W et al (2009) Quantitative 3D video microscopy of HIV transfer across T cell virological synapses. Science 323(5922): 1743–1747
54. Jolly C, Sattentau QJ (2004) Retroviral spread by induction of virological synapses. Traffic 5(9):643–650

55. McDonald D et al (2003) Recruitment of HIV and its receptors to dendritic cell–T cell junctions. Science 300(5623):1295–1297
56. Arganaraz ER, Schindler M, Kirchhoff F, Cortes MJ, Lama J (2003) Enhanced CD4 down-modulation by late stage HIV-1 nef alleles is associated with increased Env incorporation and viral replication. J Biol Chem 278(36):33912–33919
57. Lama J, Mangasarian A, Trono D (1999) Cell-surface expression of CD4 reduces HIV-1 infectivity by blocking Env incorporation in a Nef- and Vpu-inhibitable manner. Curr Biol 9(12):622–631
58. Stoddart CA et al (2003) Human immunodeficiency virus type 1 Nef-mediated downregulation of CD4 correlates with Nef enhancement of viral pathogenesis. J Virol 77(3): 2124–2133
59. Nethe M, Berkhout B, van der Kuyl AC (2005) Retroviral superinfection resistance. Retrovirology 2:52
60. Chen J et al (2005) Mechanisms of nonrandom human immunodeficiency virus type 1 infection and double infection: preference in virus entry is important but is not the sole factor. J Virol 79(7):4140–4149
61. Gelderblom HC et al (2008) Viral complementation allows HIV-1 replication without integration. Retrovirology 5:60
62. Levy DN, Aldrovandi GM, Kutsch O, Shaw GM (2004) Dynamics of HIV-1 recombination in its natural target cells. Proc Natl Acad Sci U S A 101(12):4204–4209
63. Bonhoeffer S, Chappey C, Parkin NT, Whitcomb JM, Petropoulos CJ (2004) Evidence for positive epistasis in HIV-1. Science 306(5701):1547–1550
64. Fraser C (2005) HIV recombination: what is the impact on antiretroviral therapy? J R Soc Interface 2(5):489–503
65. Vijay NNV, Ajmani VR, Perelson AS, Dixit NM (2008) Recombination increases human immunodeficiency virus fitness, but not necessarily diversity. J Gen Virol 89(Pt 6):1467–1477
66. Althaus CL, Bonhoeffer S (2005) Stochastic interplay between mutation and recombination during the acquisition of drug resistance mutations in human immunodeficiency virus type 1. J Virol 79(21):13572–13578
67. Kouyos RD, Althaus CL, Bonhoeffer S (2006) Stochastic or deterministic: what is the effective population size of HIV-1? (Translated from eng). Trends Microbiol 14(12):507–511 (in eng)
68. Kouyos RD, Silander OK, Bonhoeffer S (2007) Epistasis between deleterious mutations and the evolution of recombination. (Translated from eng). Trends Ecol Evol 22(6):308–315 (in eng)
69. Iwabu Y et al (2008) Superinfection of defective human immunodeficiency virus type 1 with different subtypes of wild-type virus efficiently produces infectious variants with the initial viral phenotypes by complementation followed by recombination. Microbes Infect 10(5):504–513
70. Wodarz D, Levy DN (2007) Human immunodeficiency virus evolution towards reduced replicative fitness in vivo and the development of AIDS. Proc Biol Sci 274(1624):2481–2490
71. Wodarz D, Levy DN (2009) Multiple HIV-1 infection of cells and the evolutionary dynamics of cytotoxic T lymphocyte escape mutants. Evolution 63(9):2326–2339
72. Wodarz D, Levy DN (2011) Effect of different modes of viral spread on the dynamics of multiply infected cells in human immunodeficiency virus infection. (Translated from eng). J R Soc Interface 8(55):289–300 (in eng)
73. Wodarz D, Levy DN (2011) Effect of multiple infection of cells on the evolutionary dynamics of HIV in vivo: implications for host adaptation mechanisms. (Translated from eng). Exp Biol Med (Maywood) 236(8):926–937 (in eng)
74. Dixit NM, Perelson AS (2005) HIV dynamics with multiple infections of target cells. Proc Natl Acad Sci U S A 102(23):8198–8203
75. Cummings KW, Levy DN, Wodarz D (2012) Increased burst size in multiply infected cells can alter basic virus dynamics. (Translated from eng). Biol Direct 7:16
76. Hofacre A, Wodarz D, Komarova NL, Fan H (2012) Early infection and spread of a conditionally replicating adenovirus under conditions of plaque formation. (Translated from eng). Virology 423(1):89–96 (in eng)
77. Lifson JD et al (1997) The extent of early viral replication is a critical determinant of the natural history of simian immunodeficiency virus infection. J Virol 71(12):9508–9514
78. Rudensey LM, Kimata JT, Benveniste RE, Overbaugh J (1995) Progression to AIDS in macaques is associated with changes in the replication, tropism, and cytopathic properties of the simian immunodeficiency virus variant population. Virology 207(2):528–542
79. Chen P, Hubner W, Spinelli MA, Chen BK (2007) Predominant mode of human immunodeficiency virus transfer between T cells is mediated by sustained Env-dependent neutralization-resistant virological synapses. (Translated from eng). J Virol 81(22):12582–12595 (in eng)
80. Feldmann J, Schwartz O (2010) HIV-1 virological synapse: live imaging of transmission.

(Translated from eng). Viruses 2(8):1666–1680, www.mdpi.com/journals/viruses (in eng)
81. Martin N, Sattentau Q (2009) Cell-to-cell HIV-1 spread and its implications for immune evasion. (Translated from eng). Curr Opin HIV AIDS 4(2):143–149 (in eng)
82. Sattentau Q (2008) Avoiding the void: cell-to-cell spread of human viruses. Nat Rev Microbiol 6(11):815–826
83. Sattentau QJ (2010) Cell-to-cell spread of retroviruses. (Translated from eng). Viruses 2(6):1306–1321, www.mdpi.com/journals/viruses (in eng)
84. Sourisseau M, Sol-Foulon N, Porrot F, Blanchet F, Schwartz O (2007) Inefficient human immunodeficiency virus replication in mobile lymphocytes. (Translated from eng). J Virol 81(2):1000–1012 (in eng)
85. Sigal A et al (2011) Cell-to-cell spread of HIV permits ongoing replication despite antiretroviral therapy. (Translated from eng). Nature 477(7362):95–98 (in eng)
86. Del Portillo A et al (2011) Multiploid inheritance of HIV-1 during cell-to-cell infection. (Translated from eng). J Virol 85(14):7169–7176 (in eng)
87. Josefsson L et al (2011) Majority of CD4+ T cells from peripheral blood of HIV-1-infected individuals contain only one HIV DNA molecule. (Translated from eng). Proc Natl Acad Sci U S A 108(27):11199–11204 (in eng)
88. Komarova NL et al (2013) Relative contribution of free-virus and synaptic transmission to the spread of HIV-1 through target cell populations. (Translated from eng). Biol Lett 9(1): 20121049
89. Komarova NL, Levy DN, Wodarz D (2012) Effect of synaptic transmission on viral fitness in HIV infection. (Translated from eng). PLoS One 7(11):e48361
90. Komarova NL, Wodarz D (2013) Virus dynamics in the presence of synaptic transmission. (Translated from eng). Math Biosci 242(2): 161–171 (in eng)
91. Doceul V, Hollinshead M, van der Linden L, Smith GL (2010) Repulsion of superinfecting virions: a mechanism for rapid virus spread. (Translated from eng). Science 327(5967):873–876 (in eng)
92. Jolly C (2011) Cell-to-cell transmission of retroviruses: Innate immunity and interferon-induced restriction factors. (Translated from eng). Virology 411(2):251–259 (in eng)
93. Bieniasz PD (2004) Intrinsic immunity: a front-line defense against viral attack. (Translated from eng). Nat Immunol 5(11): 1109–1115 (in eng)
94. Sakuma R, Noser JA, Ohmine S, Ikeda Y (2007) Inhibition of HIV-1 replication by simian restriction factors, TRIM5alpha and APOBEC3G. (Translated from eng). Gene Ther 14(2):185–189 (in eng)
95. Stremlau M et al (2004) The cytoplasmic body component TRIM5alpha restricts HIV-1 infection in old world monkeys. (Translated from eng). Nature 427(6977):848–853 (in eng)
96. Yan N, Chen ZJ (2012) Intrinsic antiviral immunity. (Translated from eng). Nat Immunol 13(3):214–222 (in eng)
97. Sokolskaja E, Luban J (2006) Cyclophilin, TRIM5, and innate immunity to HIV-1. (Translated from eng). Curr Opin Microbiol 9(4):404–408 (in eng)
98. Wodarz D (2007) Killer cell dynamics: mathematical and computational approaches to immunology. Springer, New York

INDEX

Rajat K. De and Namrata Tomar (eds.), *Immunoinformatics*, Methods in Molecular Biology, vol. 1184,
DOI 10.1007/978-1-4939-1115-8, © Springer Science+Business Media New York 2014

Printed by Printforce, the Netherlands